# TechOne: Automotive Suspension and Steering

Jack Erjavec

Professor Emeritus,
Columbus State Community College
Columbus, Ohio

THOMSON
DELMAR LEARNING

Australia Canada Mexico Singapore Spain United Kingdom United States

THOMSON
DELMAR LEARNING™

**TechOne: Automotive Suspension and Steering**

Jack Erjavec

**Vice President, Technology and Trades SBU:**
Alar Elken

**Editorial Director:**
Sandy Clark

**Senior Acquisitions Editor:**
David Boelio

**Developmental Editor:**
Matthew Thouin

**Marketing Director:**
Dave Garza

**Channel Manager:**
William Lawrenson

**Marketing Coordinator:**
Mark Pierro

**Production Director:**
Mary Ellen Black

**Production Editor:**
Barbara L. Diaz

**Art/Design Specialist:**
Cheri Plasse

**Technology Project Manager:**
Kevin Smith

**Technology Project Specialist:**
Linda Verde

**Editorial Assistant:**
Andrea Domkowski



Printed in the United States of America
1 2 3 4 5 XX 06 05 04

For more information contact
Delmar Learning
Executive Woods
5 Maxwell Drive, PO Box 8007,
Clifton Park, NY 12065-8007
Or find us on the World Wide Web at www.delmarlearning.com



Library of Congress Cataloging-in-Publication Data:

Erjavec, Jack.
Tech one. Automotive steering and suspension / Jack Erjavec.
p. cm.
Includes index.
ISBN 0-7668-1166-2
1. Automobiles--Springs and suspension—Maintenance and repair. 2. Automobiles—Steering-gear--Maintenance and repair. I. Title.
TL257.E77 2005
629.2'43'0288--dc22
2005043950

ISBN: 0-7668-1166-2

**NOTICE TO THE READER**

# Contents

# Preface

## THE SERIES

Welcome to Thomson Delmar Learning's *TechOne,* a state-of-the-art series designed to respond to today's automotive instructor and student needs. *TechOne* offers current, concise information on ASE-subject areas, combining classroom theory, diagnosis, and repair into one easy-to-use volume.

You'll notice several differences from a traditional textbook. First, a large number of short chapters divide complex material into chunks. Instructors can give tight, detailed reading assignments that students will find easier to digest. These shorter chapters can be taught in almost any order, allowing teachers to pick and choose the material that best reflects the depth, direction, and pace of their individual classes.

*TechOne* also features an art-intensive approach to suit today's visual learners—images drive the chapters. From drawings to photos, you will find more art to better understand the systems, parts, and procedures under discussion. Look also for helpful graphics that draw attention to key points in features like You Should Know and Interesting Fact.

Just as importantly, each *TechOne* starts off with a section on Safety and Communication, which stresses safe work practices, tool competence, and familiarity with workplace "soft skills," such as customer communication and the roles necessary to succeed as an automotive technician. From there, learners are ready to tackle the technical material in successive sections, ultimately leading them to the real test—an ASE practice exam in the Appendix.

## THE SUPPLEMENTS

*TechOne* comes with an **Instructor's Manual** that includes answers to all chapter-end review questions and a complete correlation of the text to NATEF standards. A **CD-ROM,** included with each Instructor's Manual, consists of **PowerPoint Slides** for classroom presentations and a Examview **Computerized Testbank** with hundreds of questions to aid in creating tests and quizzes. Chapter-end review questions from the text have also been redesigned into adaptable **Electronic Worksheets,** so instructors can modify questions if desired to create in-class assignments or homework.

Flexibility is the key to *TechOne.* For those who would like to purchase jobsheets, Thomson Delmar Learning's NATEF Standards Job Sheets are a good match. Topics cover the eight ASE subjects areas and include:

- Automatic Transmissions and Transaxles
- Automotive Brakes
- Automotive Electrical and Electronic Systems
- Automotive Engine Repair
- Automotive Engine Performance
- Automotive Heating and Air Conditioning
- Automotive Suspension and Steering
- Manual Drive Trains and Axles

Visit **www.autoed.com** for a complete catalog.

## OTHER TITLES IN THIS SERIES

*TechOne* is Thomson Delmar Learning's latest automotive series. We are excited to announce these future titles:

- Advanced Engine Performance
- Automotive Fuels & Emissions

Check with your sales representative for availability.

## A NOTE TO THE STUDENT

There are now more computers on a car than aboard the first spacecraft, and even gifted backyard mechanics long ago turned their cars over to automotive professionals for diagnosis and repair. That's a statement about the nation's need for the knowledge and skills you'll develop as you continue studies. Whether you eventually choose a career as an ASE-certified technician, service writer or manager, or automotive engineer—or even decide to open your own shop—hard work will give you the opportunity to become one of the 840,000 automotive professionals in the United States. As a member of a technically-proficient, cutting-edge workforce, you'll fill a need, and, even better, you'll have a career to feel proud of.

Best of luck in your studies,
The Editors of Thomson Delmar Learning

# About the Author

Jack Erjavec, a master-certified ASE technician and professor emeritus at Columbus State Community College in Columbus, Ohio, has become a fixture in the automotive textbook world with more than 30 works to his credit. In addition to his bestselling comprehensive text, *Automotive Technology,* Jack is also series editor of Thomson Delmar Learning's popular *Today's Technician* series and the new *TechOne* automotive series.

In addition to assuming these editorships, Jack was the Product Development Manager at Thomson Delmar Learning for business, industry, government, and retail automotiove materials, as well as a professional consultant for several automotive manufacturers. He is a long-time affiliate of the North American Council of Automotive Teachers (NACAT), having served on the board of directors and as executive vice-president, and now as Vice-President of Marketing. Jack is also associated with a number of professional organizations, including the Society of Automotive Engineers, and remains active in the industry. In his free time, he most cherishes spending time with his family: Rose, Megan, Craig, Hannah, Grace, Moira, Judah, and Erin.

# Acknowledgments

I'd like to thank the following reviewers, whose technical expertise was invaluable in creating this text:

David Ames
Evergreen Valley College
San Jose, CA

R.J. Ehlers
Linn-Benton Community College
Albany, OR

Randy Jacobsen
Southwestern Michigan College
Dowagiac, MI

Russ Molaski
North East Wisconsin Tech College
Green Bay, WI

Chris Perales
Texas State Technical College
Waco, TX

Dan Perrin
Trident Technical College
Charleston, SC

Christopher VanStavoren
Penn College of Technology
Williamsport, PA

# Features of the Text

*TechOne* includes a variety of learning aids designed to encourage student comprehension of complex automotive concepts, diagnostics, and repair. Look for these helpful features:

**Section Openers** provide students with a **Section Table of Contents** and **Objectives** to focus learners on the section's goals.

**Interesting Facts** spark student attention with industry trivia or history. Interesting facts appear on the section openers and are then scattered throughout the chapters to maintain reader interest.

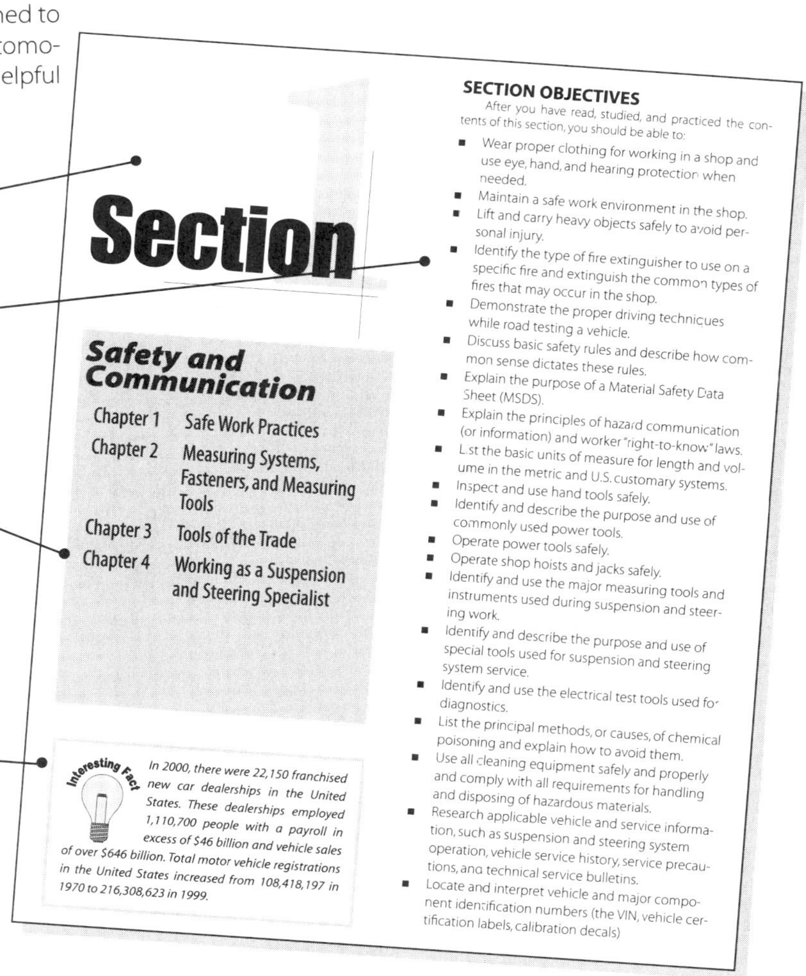
Section 1

Safety and Communication

Chapter 1 Safe Work Practices
Chapter 2 Measuring Systems, Fasteners, and Measuring Tools
Chapter 3 Tools of the Trade
Chapter 4 Working as a Suspension and Steering Specialist

Interesting Fact: In 2000, there were 22,150 franchised new car dealerships in the United States. These dealerships employed 1,110,700 people with a payroll in excess of $46 billion and vehicle sales of over $646 billion. Total motor vehicle registrations in the United States increased from 108,418,197 in 1970 to 216,308,623 in 1999.

SECTION OBJECTIVES

After you have read, studied, and practiced the contents of this section, you should be able to:

- Wear proper clothing for working in a shop and use eye, hand, and hearing protection when needed.
- Maintain a safe work environment in the shop.
- Lift and carry heavy objects safely to avoid personal injury.
- Identify the type of fire extinguisher to use on a specific fire and extinguish the common types of fires that may occur in the shop.
- Demonstrate the proper driving techniques while road testing a vehicle.
- Discuss basic safety rules and describe how common sense dictates these rules.
- Explain the purpose of a Material Safety Data Sheet (MSDS).
- Explain the principles of hazard communication (or information) and worker "right-to-know" laws.
- List the basic units of measure for length and volume in the metric and U.S. customary systems.
- Inspect and use hand tools safely.
- Identify and describe the purpose and use of commonly used power tools.
- Operate power tools safely.
- Operate shop hoists and jacks safely.
- Identify and use the major measuring tools and instruments used during suspension and steering work.
- Identify and describe the purpose and use of special tools used for suspension and steering system service.
- Identify and use the electrical test tools used for diagnostics.
- List the principal methods, or causes, of chemical poisoning and explain how to avoid them.
- Use all cleaning equipment safely and properly and comply with all requirements for handling and disposing of hazardous materials.
- Research applicable vehicle and service information, such as suspension and steering system operation, vehicle service history, service precautions, and technical service bulletins.
- Locate and interpret vehicle and major component identification numbers (the VIN, vehicle certification labels, calibration decals)

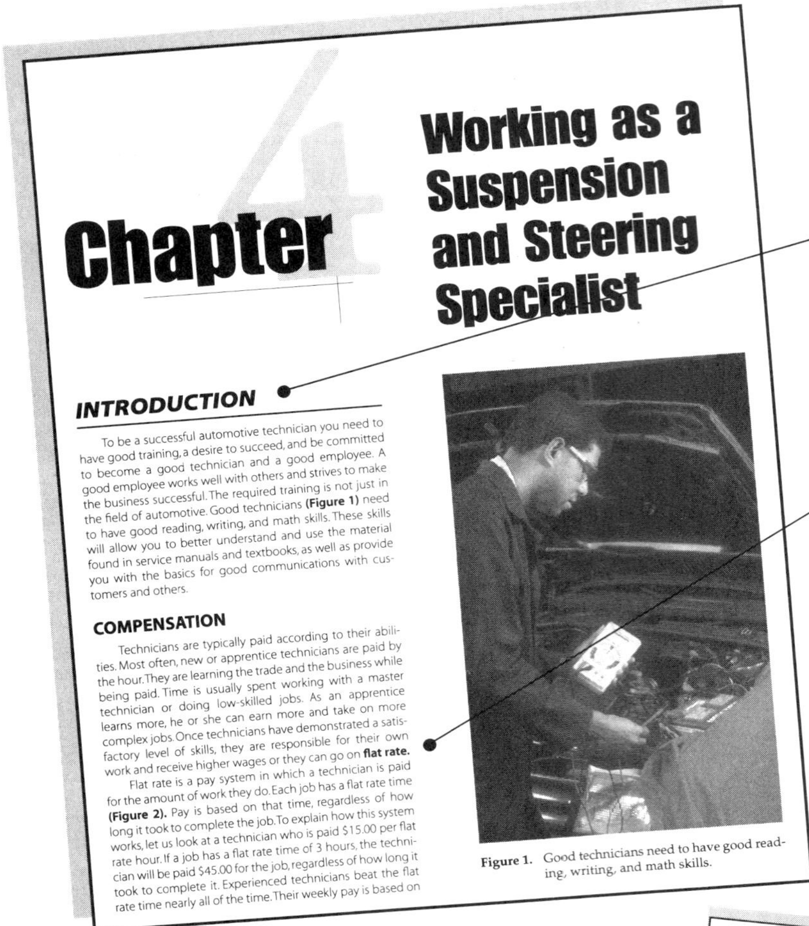

# Chapter 4 Working as a Suspension and Steering Specialist

## INTRODUCTION

To be a successful automotive technician you need to have good training, a desire to succeed, and be committed to become a good technician and a good employee. A good employee works well with others and strives to make the business successful. The required training is not just in the field of automotive. Good technicians **(Figure 1)** need to have good reading, writing, and math skills. These skills will allow you to better understand and use the material found in service manuals and textbooks, as well as provide you with the basics for good communications with customers and others.

### COMPENSATION

Technicians are typically paid according to their abilities. Most often, new or apprentice technicians are paid by the hour. They are learning the trade and the business while being paid. Time is usually spent working with a master technician or doing low-skilled jobs. As an apprentice learns more, he or she can earn more and take on more complex jobs. Once technicians have demonstrated a satisfactory level of skills, they are responsible for their own work and receive higher wages or they can go on **flat rate.**

Flat rate is a pay system in which a technician is paid for the amount of work they do. Each job has a flat rate time **(Figure 2).** Pay is based on that time, regardless of how long it took to complete the job. To explain how this system works, let us look at a technician who is paid $15.00 per flat rate hour. If a job has a flat rate time of 3 hours, the technician will be paid $45.00 for the job, regardless of how long it took to complete it. Experienced technicians beat the flat rate time nearly all of the time. Their weekly pay is based on

**Figure 1.** Good technicians need to have good reading, writing, and math skills.

An **Introduction** orients readers at the beginning of each new chapter. **Technical Terms** are bolded in the text upon first reference and are defined.

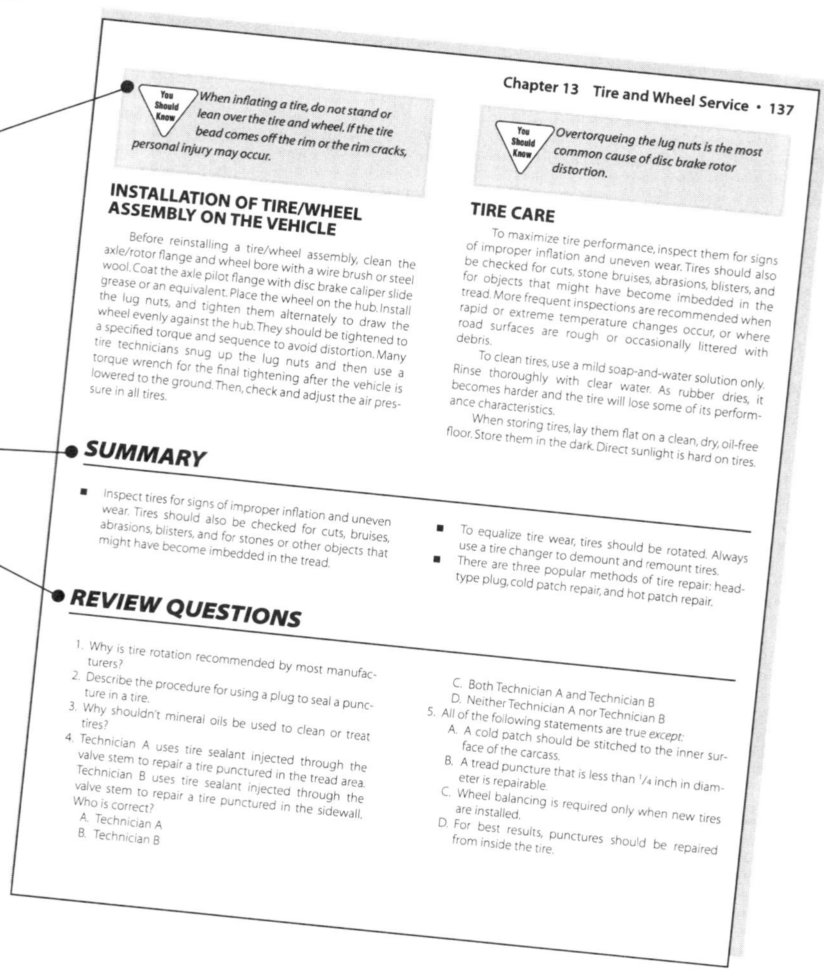

Chapter 13 Tire and Wheel Service • 137

*You Should Know* *When inflating a tire, do not stand or lean over the tire and wheel. If the tire bead comes off the rim or the rim cracks, personal injury may occur.*

### INSTALLATION OF TIRE/WHEEL ASSEMBLY ON THE VEHICLE

Before reinstalling a tire/wheel assembly, clean the axle/rotor flange and wheel bore with a wire brush or steel wool. Coat the axle pilot flange with disc brake caliper slide grease or an equivalent. Place the wheel on the hub. Install the lug nuts, and tighten them alternately to draw the wheel evenly against the hub. They should be tightened to a specified torque and sequence to avoid distortion. Many tire technicians snug up the lug nuts and then use a torque wrench for the final tightening after the vehicle is lowered to the ground. Then, check and adjust the air pressure in all tires.

*You Should Know* *Overtorqueing the lug nuts is the most common cause of disc brake rotor distortion.*

### TIRE CARE

To maximize tire performance, inspect them for signs of improper inflation and uneven wear. Tires should also be checked for cuts, stone bruises, abrasions, blisters, and for objects that might have become imbedded in the tread. More frequent inspections are recommended when rapid or extreme temperature changes occur, or where road surfaces are rough or occasionally littered with debris.

To clean tires, use a mild soap-and-water solution only. Rinse thoroughly with clear water. As rubber dries, it becomes harder and the tire will lose some of its performance characteristics.

When storing tires, lay them flat on a clean, dry, oil-free floor. Store them in the dark. Direct sunlight is hard on tires.

## SUMMARY

- Inspect tires for signs of improper inflation and uneven wear. Tires should also be checked for cuts, bruises, abrasions, blisters, and for stones or other objects that might have become imbedded in the tread.
- To equalize tire wear, tires should be rotated. Always use a tire changer to demount and remount tires.
- There are three popular methods of tire repair: head-type plug, cold patch repair, and hot patch repair.

## REVIEW QUESTIONS

1. Why is tire rotation recommended by most manufacturers?
2. Describe the procedure for using a plug to seal a puncture in a tire.
3. Why shouldn't mineral oils be used to clean or treat tires?
4. Technician A uses tire sealant injected through the valve stem to repair a tire punctured in the tread area. Technician B uses tire sealant injected through the valve stem to repair a tire punctured in the sidewall. Who is correct?
   A. Technician A
   B. Technician B
   C. Both Technician A and Technician B
   D. Neither Technician A nor Technician B
5. All of the following statements are true *except:*
   A. A cold patch should be stitched to the inner surface of the carcass.
   B. A tread puncture that is less than 1/4 inch in diameter is repairable.
   C. Wheel balancing is required only when new tires are installed.
   D. For best results, punctures should be repaired from inside the tire.

**You Should Know** informs the reader whenever special safety cautions, warnings, or other important points deserve emphasis.

A **Summary** concludes each chapter in short, bulleted sentences. **Review Questions** are structured in a variety of formats, including ASE style, challenging students to prove they've mastered the material.

An **ASE Practice Exam** is found in the **Appendix** of every *TechOne* book, followed by a **Bilingual Glossary**, which offers Spanish translations of technical terms alongside their English counterparts.

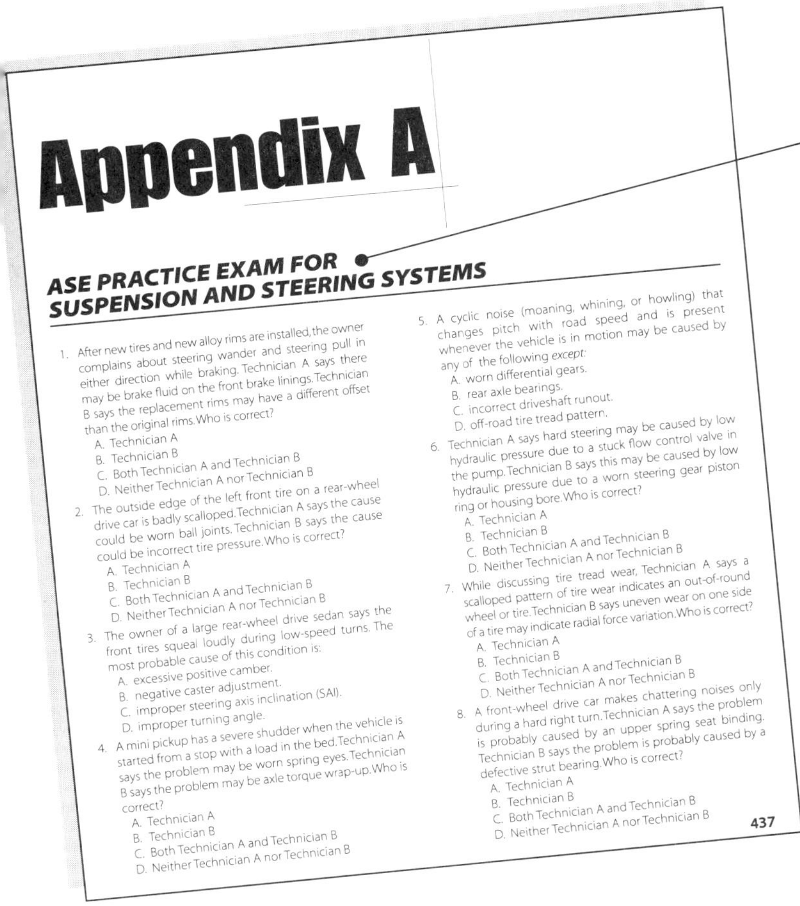

# Appendix A

## ASE PRACTICE EXAM FOR SUSPENSION AND STEERING SYSTEMS

1. After new tires and new alloy rims are installed, the owner complains about steering wander and steering pull in either direction while braking. Technician A says there may be brake fluid on the front brake linings. Technician B says the replacement rims may have a different offset than the original rims. Who is correct?
   A. Technician A
   B. Technician B
   C. Both Technician A and Technician B
   D. Neither Technician A nor Technician B
2. The outside edge of the left front tire on a rear-wheel drive car is badly scalloped. Technician A says the cause could be worn ball joints. Technician B says the cause could be incorrect tire pressure. Who is correct?
   A. Technician A
   B. Technician B
   C. Both Technician A and Technician B
   D. Neither Technician A nor Technician B
3. The owner of a large rear-wheel drive sedan says the front tires squeal loudly during low-speed turns. The most probable cause of this condition is:
   A. excessive positive camber.
   B. negative caster adjustment.
   C. improper steering axis inclination (SAI).
   D. improper turning angle.
4. A mini pickup has a severe shudder when the vehicle is started from a stop with a load in the bed. Technician A says the problem may be worn spring eyes. Technician B says the problem may be axle torque wrap-up. Who is correct?
   A. Technician A
   B. Technician B
   C. Both Technician A and Technician B
   D. Neither Technician A nor Technician B
5. A cyclic noise (moaning, whining, or howling) that changes pitch with road speed and is present whenever the vehicle is in motion may be caused by any of the following *except*:
   A. worn differential gears.
   B. rear axle bearings.
   C. incorrect driveshaft runout.
   D. off-road tire tread pattern.
6. Technician A says hard steering may be caused by low hydraulic pressure due to a stuck flow control valve in the pump. Technician B says this may be caused by low hydraulic pressure due to a worn steering gear piston ring or housing bore. Who is correct?
   A. Technician A
   B. Technician B
   C. Both Technician A and Technician B
   D. Neither Technician A nor Technician B
7. While discussing tire tread wear, Technician A says a scalloped pattern of tire wear indicates an out-of-round wheel or tire. Technician B says uneven wear on one side of a tire may indicate radial force variation. Who is correct?
   A. Technician A
   B. Technician B
   C. Both Technician A and Technician B
   D. Neither Technician A nor Technician B
8. A front-wheel drive car makes chattering noises only during a hard right turn. Technician A says the problem is probably caused by an upper spring seat binding. Technician B says the problem is probably caused by a defective strut bearing. Who is correct?
   A. Technician A
   B. Technician B
   C. Both Technician A and Technician B
   D. Neither Technician A nor Technician B

437

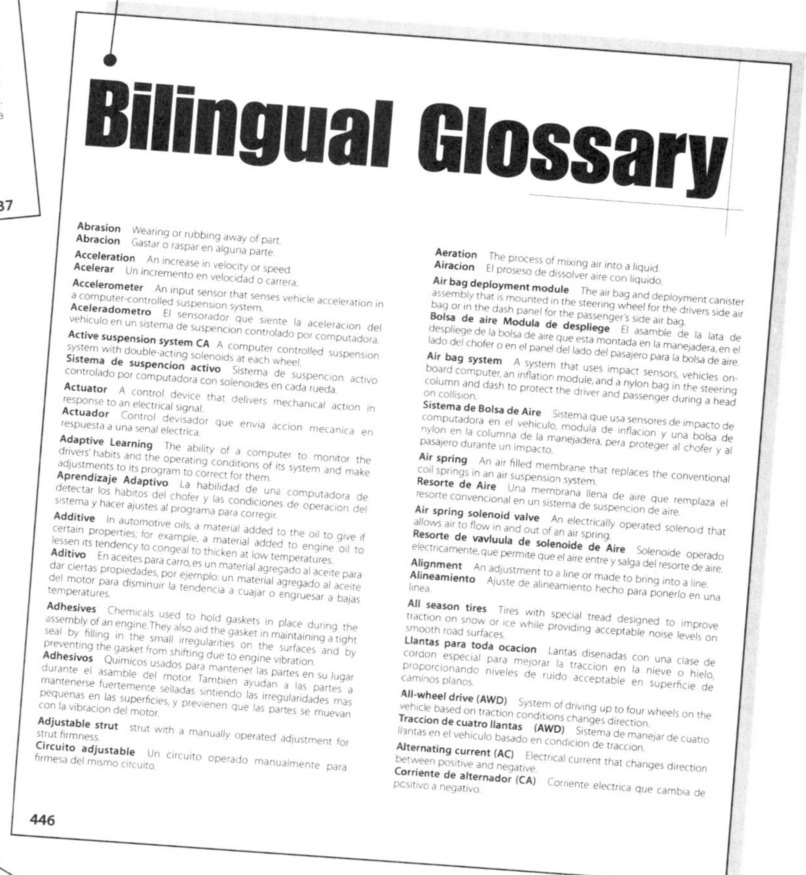

# Bilingual Glossary

**Abrasion** Wearing or rubbing away of part.
**Abracion** Gastar o raspar en alguna parte.
**Acceleration** An increase in velocity or speed.
**Acelerar** Un incremento en velocidad o carrera.
**Accelerometer** An input sensor that senses vehicle acceleration in a computer-controlled suspension system.
**Aceleradometro** El sensorador que siente la aceleracion del vehiculo en un sistema de suspencion controlado por computadora.
**Active suspension system CA** A computer controlled suspension system with double-acting solenoids at each wheel.
**Sistema de suspencion activo** Sistema de suspencion activo controlado por computadora con solenoides en cada rueda.
**Actuator** A control device that delivers mechanical action in response to an electrical signal.
**Actuador** Control devisador que envia accion mecanica en respuesta a una senal electrica.
**Adaptive Learning** The ability of a computer to monitor the drivers' habits and the operating conditions of its system and make adjustments to its program to correct for them.
**Aprendizaje Adaptivo** La habilidad de una computadora de detectar los habitos del chofer y las condiciones de operacion del sistema y hacer ajustes al programa para corregir.
**Additive** In automotive oils, a material added to the oil to give if certain properties; for example, a material added to engine oil to lessen its tendency to congeal to thicken at low temperatures.
**Aditivo** En aceites para carro, es un material agregado al aceite para dar ciertas propiedades, por ejemplo: un material agregado al aceite del motor para disminuir la tendencia a cuajar o engruesar a bajas temperatures.
**Adhesives** Chemicals used to hold gaskets in place during the assembly of an engine. They also aid the gasket in maintaining a tight seal by filling in the small irregularities on the surfaces and by preventing the gasket from shifting due to engine vibration.
**Adhesivos** Quimicos usados para mantener las partes en su lugar durante el asamble del motor. Tambien ayudan a las partes a mantenerse fuertemente selladas sintiendo las irregularidades mas pequenas en las superficies, y previenen que las partes se muevan con la vibracion del motor.
**Adjustable strut** strut with a manually operated adjustment for strut firmness.
**Circuito adjustable** Un circuito operado manualmente para firmesa del mismo circuito.
**Aeration** The process of mixing air into a liquid.
**Airacion** El proseso de dissolver aire con liquido.
**Air bag deployment module** The air bag and deployment canister assembly that is mounted in the steering wheel for the drivers side air bag or in the dash panel for the passenger's side air bag.
**Bolsa de aire Modula de despliege** El asamble de la lata de despliege de la bolsa de aire que esta montada en la manejadera, en el lado del chofer o en el panel del lado del pasajero para la bolsa de aire.
**Air bag system** A system that uses impact sensors, vehicles on-board computer, an inflation module, and a nylon bag in the steering column and dash to protect the driver and passenger during a head on collision.
**Sistema de Bolsa de Aire** Sistema que usa sensores de impacto de computadora en el vehiculo, modula de inflacion y una bolsa de nylon en la columna de la manejadera, pera proteger al chofer y al pasajero durante un impacto.
**Air spring** An air filled membrane that replaces the conventional coil springs in an air suspension system.
**Resorte de Aire** Una membrana llena de aire que remplaza el resorte convencional en un sistema de suspencion de aire.
**Air spring solenoid valve** An electrically operated solenoid that allows air to flow in and out of an air spring.
**Resorte de vavluula de solenoide de Aire** Solenoide operado electricamente, que permite que el aire entre y salga del resorte de aire.
**Alignment** An adjustment to a line or made to bring into a line.
**Alineamiento** Ajuste de alineamiento hecho para ponerlo en una linea.
**All season tires** Tires with special tread designed to improve traction on snow or ice while providing acceptable noise levels on smooth road surfaces.
**Llantas para toda ocacion** Lantas disenadas con una clase de cordon especial para mejorar la traccion en la nieve o hielo, proporcionando niveles de ruido acceptable en superficie de caminos planos.
**All-wheel drive (AWD)** System of driving up to four wheels on the vehicle based on traction conditions changes direction.
**Traccion de cuatro llantas (AWD)** Sistema de manejar de cuatro llantas en el vehiculo basado en condicion de traccion.
**Alternating current (AC)** Electrical current that changes direction between positive and negative.
**Corriente de alternador (CA)** Corriente electrica que cambia de pcsitivo a negativo.

446

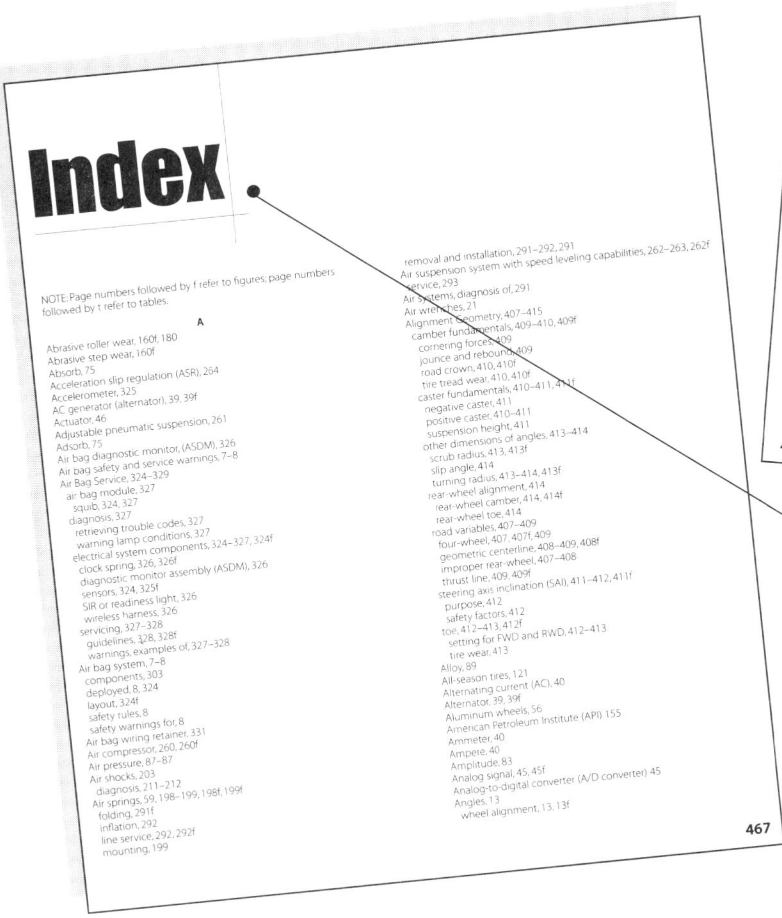

# Index

NOTE: Page numbers followed by f refer to figures; page numbers followed by t refer to tables.



467

A comprehensive **Index** helps instructors and students pinpoint information in the text.

# Section 1

## Safety and Communication



**Interesting Fact**

*In 2000, there were 22,150 franchised new car dealerships in the United States. These dealerships employed 1,110,700 people with a payroll in excess of $46 billion and vehicle sales of over $646 billion. Total motor vehicle registrations in the United States increased from 108,418,197 in 1970 to 216,308,623 in 1999.*

## SECTION OBJECTIVES

After you have read, studied, and practiced the contents of this section, you should be able to:

- Wear proper clothing for working in a shop and use eye, hand, and hearing protection when needed.
- Maintain a safe work environment in the shop.
- Lift and carry heavy objects safely to avoid personal injury.
- Identify the type of fire extinguisher to use on a specific fire and extinguish the common types of fires that may occur in the shop.
- Demonstrate the proper driving techniques while road testing a vehicle.
- Discuss basic safety rules and describe how common sense dictates these rules.
- Explain the purpose of a Material Safety Data Sheet (MSDS).
- Explain the principles of hazard communication (or information) and worker "right-to-know" laws.
- List the basic units of measure for length and volume in the metric and U.S. customary systems.
- Inspect and use hand tools safely.
- Identify and describe the purpose and use of commonly used power tools.
- Operate power tools safely.
- Operate shop hoists and jacks safely.
- Identify and use the major measuring tools and instruments used during suspension and steering work.
- Identify and describe the purpose and use of special tools used for suspension and steering system service.
- Identify and use the electrical test tools used for diagnostics.
- List the principal methods, or causes, of chemical poisoning and explain how to avoid them.
- Use all cleaning equipment safely and properly and comply with all requirements for handling and disposing of hazardous materials.
- Research applicable vehicle and service information, such as suspension and steering system operation, vehicle service history, service precautions, and technical service bulletins.
- Locate and interpret vehicle and major component identification numbers (the VIN, vehicle certification labels, calibration decals)

# Chapter 1 Safe Work Practices

## Introduction

Safety is an important topic, important to you and those around you. Although the safety issues presented in this chapter are categorized, true safe work practices are based on nothing else but common sense and knowledge of the safety equipment. Shop safety is the responsibility of everyone in the shop.

### SAFE WORK AREAS

Your work area should be kept clean and safe. Any oil, coolant, or grease on the floor can make it slippery. Slips can result in serious injuries. To clean up oil, use commercial oil absorbent. Keep all water off the floor. Water is slippery on smooth floors, and electricity flows well through water. Make sure the work areas around machines are large enough to safely operate the machine.

Make sure all drain covers are snugly in place. Open drains or covers that are not flush to the floor can cause toe, ankle, and leg injuries.

Whenever an engine is running in the shop, make sure the vehicle's exhaust is connected to the shop's exhaust ventilation system. Engine exhaust contains carbon monoxide (CO), which can cause death or serious illnesses.

### PERSONAL SAFETY

Personal safety simply involves those precautions you take to protect yourself from injury. Your eyes can become infected or permanently damaged by many items in a shop. Eye protection **(Figure 1)** should be worn whenever you are working in the shop. There are many types of eye protection available. To provide adequate eye protection, safety glasses have lenses made of safety glass. They also offer some sort of side protection.

If chemicals such as battery acid, fuel, or solvents splash your eyes, flush them continuously with clean water. Have someone call a doctor and seek medical help immediately. Many shops have eyewash stations or safety showers that should be used whenever you or someone else has been sprayed or splashed with a chemical.

Your clothing should be well fitted and comfortable but made with strong material. If you have long hair, tie it back or tuck it under a cap. Never wear rings, watches, bracelets, and neck chains. These can easily be caught in moving parts and cause serious injury.

Automotive work involves handling many heavy objects that can be accidentally dropped on your feet or

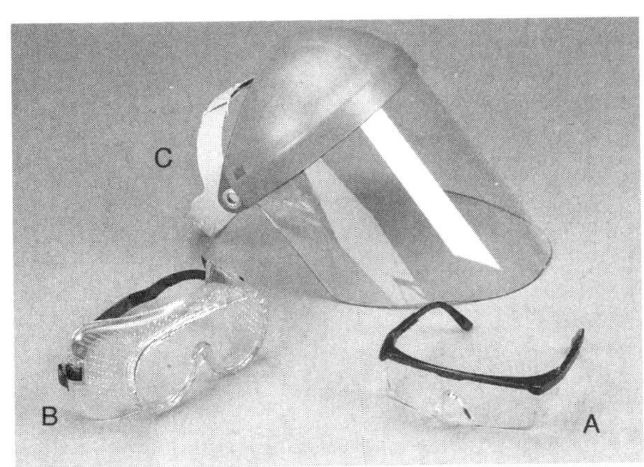


**Figure 1.** Different types of eye protection worn by automotive technicians: (A) safety glasses; (B) goggles; and (C) face shield.

toes. Always wear shoes or boots of leather or similar material with no-slip soles. Steel-tipped safety shoes can give added protection to your feet.

Good hand protection is often overlooked. A scrape, cut, or burn can limit your effectiveness at work for many days. A well-fitted pair of heavy work gloves should be worn during operations such as grinding and welding or when handling high-temperature components. Always wear approved rubber gloves when handling strong and dangerous caustic chemicals. They can easily burn your skin. Many technicians wear thin, surgical-type latex gloves whenever they are working on vehicles. These offer little protection against cuts but do offer protection against disease and grease buildup under and around your fingernails. These gloves are comfortable and are quite inexpensive.

## Lifting and Carrying

When lifting a heavy object like a transmission, use a hoist or have someone else help you. If you must work alone, *always* lift heavy objects with your legs, not your back. Bend down with your legs, securely hold the object you are lifting, and then stand up, keeping the object close to you **(Figure 2).** Trying to "muscle" something with your arms or back can result in severe damage to your back and may end your career and limit what you do the rest of your life!

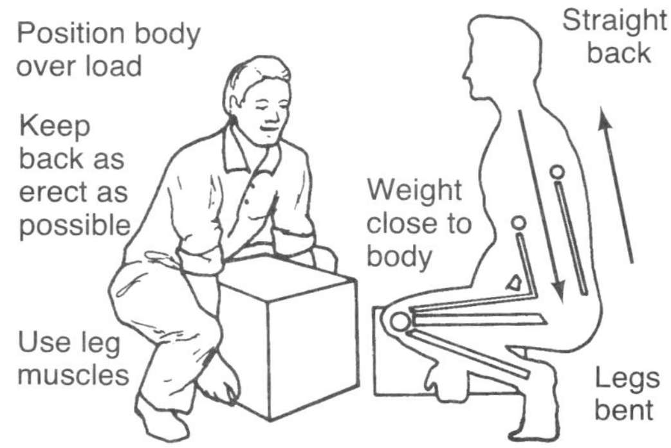


**Figure 2.** Lift with your legs, not your back, and hold the object close to you while lifting or moving it.

## FIRE HAZARDS AND PREVENTION

Many items around a typical shop are potential fire hazards. These include gasoline, diesel fuel, cleaning solvents, and dirty rags. Each of these should be treated as a potential firebomb and must be handled and stored properly **(Figure 3).**

In case of a fire, you should know the location of the fire extinguishers and fire alarms in the shop. You should also know how to use them. You should also be aware of the different types of fires **(Figure 4)** and the fire extinguishers used to put out each type of fire.

Basically, there are four types of fires: Class A fires are those in which wood, paper, and other ordinary materials are burning; Class B fires are those involving flammable liq-

**Figure 3.** To prevent fires and other mishaps, clean up all spills and store flammable materials in approved containers.

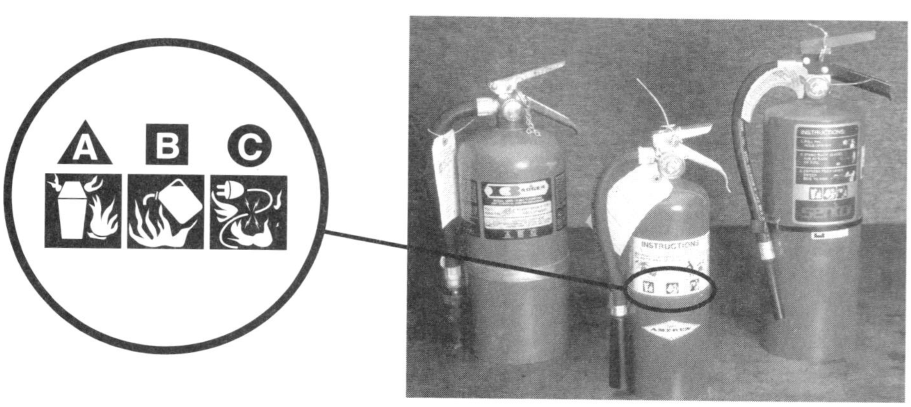


**Figure 4.** Fire extinguishers are labeled to show what type of fire they will safely put out.

uids; such as gasoline, diesel fuel, paint, grease, oil, and other similar liquids; and Class C fires are electrical fires. Class D fires are a unique type of fire because the material burning is a metal. The magnesium used in the construction of some wheels and other parts is a flammable metal and will burn fiercely when subjected to high heat.

## Using A Fire Extinguisher

Remember, never open doors or windows during a fire unless it is absolutely necessary; the extra draft will only make the fire worse. Make sure the fire department is contacted before or during your attempt to extinguish a fire. To extinguish a fire, stand 6 to 10 feet from the fire. Hold the extinguisher firmly in an upright position. Aim the nozzle at the base and use a side-to-side motion, sweeping the entire width of the fire. Stay low to avoid inhaling the smoke. If it becomes too hot or too smoky, get out. Remember: never go back into a burning building for anything. To help remember how to use an extinguisher, remember the word "PASS."

*Pull* the pin from the handle of the extinguisher.

*Aim* the extinguisher's nozzle at the base of the fire.

*Squeeze* the handle.

*Sweep* the entire width of the fire with the contents of the extinguisher.

If there is not a fire extinguisher handy, a blanket or fender cover may be used to smother the flames. Be careful when doing this because the heat may burn you and the blanket. If the fire is too great to smother, move everyone away from the fire and call the local fire department. A simple under-the-hood fire can cause the total destruction of the car and the building and can take lives. You must be able to respond quickly and precisely to avoid a disaster.

## SAFE TOOLS AND EQUIPMENT

Whenever you are using any equipment, make sure it is set up according to the manufacturer's instructions and that you use it properly. All equipment should be properly maintained and periodically inspected for unsafe conditions. Frayed electrical cords or loose mountings can cause serious injuries. All electrical outlets should be equipped to allow for the use of three-pronged electrical cords. All equipment with rotating parts should be equipped with safety guards that reduce the possibility of the parts coming loose and injuring someone.

Never use tools and equipment for purposes other than what they are designed for. Using the proper tool in the correct way will not only be safer but will also allow you to do a better job.

## Lift Safety

Always be careful when raising a vehicle on a lift or a hoist. Adapters and hoist plates must be positioned correctly on twin post-and-rail-type lifts to prevent damage to the underbody of the vehicle. There are specific lift points. These points allow the weight of the vehicle to be evenly supported by the adapters or hoist plates. The correct lift points can be found in the vehicle's service manual **(Figure 5).** Before operating any lift or hoist, carefully read the operating manual and follow the operating instructions.

**You Should Know** ***On many vehicles with computer-controlled or self-leveling suspension systems, this system must be turned off prior to raising the vehicle. These vehicles have a switch in the trunk that allows you to turn the system on or off (Figure 6).***

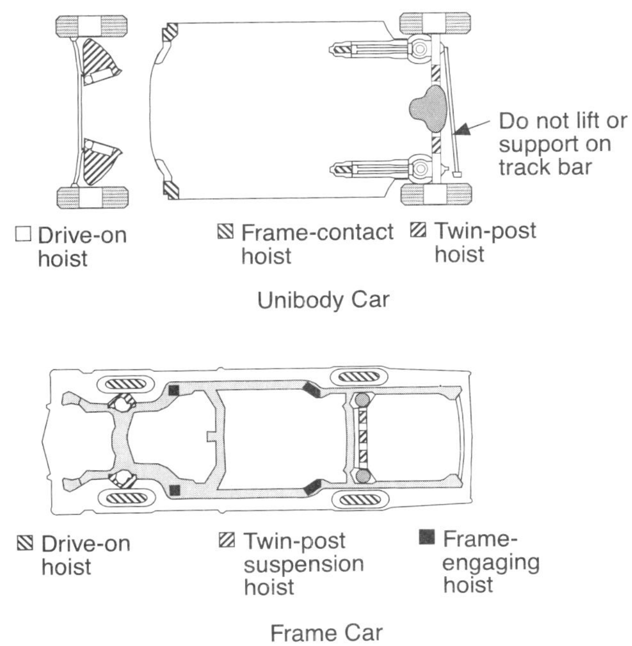


**Figure 5.** Typical lift points for a unibody and a framed vehicle.

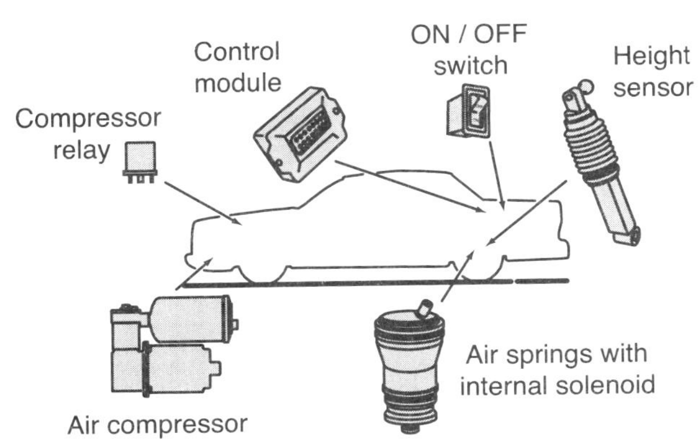


**Figure 6.** Computer-controlled or self-leveling suspension systems often need to be deactivated by depressing a switch in the trunk before the vehicle can be raised.

**Figure 7.** After the vehicle has been raised, make sure the lift pads are seated against the lift points and the safety locks for the hoist are fully engaged before working under the vehicle.

**You Should Know** *When lifting a van or SUV, make sure the overhead clearance is sufficient for the height of the vehicle before it is raised.*

Once the lift supports are properly positioned under the vehicle, raise the lift until the supports contact the vehicle. Check the supports to make sure they are in full contact with the vehicle. Shake the vehicle to make sure it is balanced on the lift, then raise the lift to the desired working height. Also make sure the safety locks for the hoist are fully engaged **(Figure 7).**

## Jack and Jack Stand Safety

A vehicle can also be raised off the ground with a hydraulic jack. The lifting pad of the jack must be positioned under an area of the vehicle's frame or at one of the manufacturer's recommended lift points. Never place the pad under the floor pan or under steering and suspension components because these are easily damaged by the weight of the vehicle. Always position the jack so the wheels of the vehicle can roll as the vehicle is being raised.

**You Should Know** *Never use a lift or jack to move something heavier than it is designed for. If a jack is rated for 2 tons, do not attempt to use it for a job requiring 5 tons. It is dangerous for you and the vehicle.*

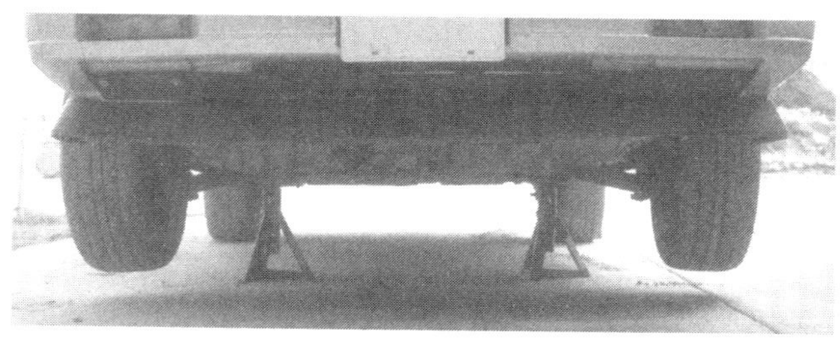

**Figure 8.** Jack stands should be used to support the vehicle after it has been raised by a jack.

Safety (jack) stands are supports of different heights that sit on the floor. They are placed under a sturdy chassis member, such as the frame or axle housing, to support the vehicle **(Figure 8).** Once the safety stands are in position, the hydraulic pressure in the jack should be slowly released until the weight of the vehicle is on the stands. Like jacks, jack stands also have a capacity rating. Always use a jack stand with the correct rating.

Never move under a vehicle when it is only supported by a hydraulic jack; rest the vehicle on the safety stands before moving under the vehicle. The jack should be removed after the jack stands are set in place. This eliminates a hazard, such as a jack handle sticking out into a walkway.

## BATTERIES

When possible, you should disconnect the **battery** of a car before you disconnect any electrical wire or compo-

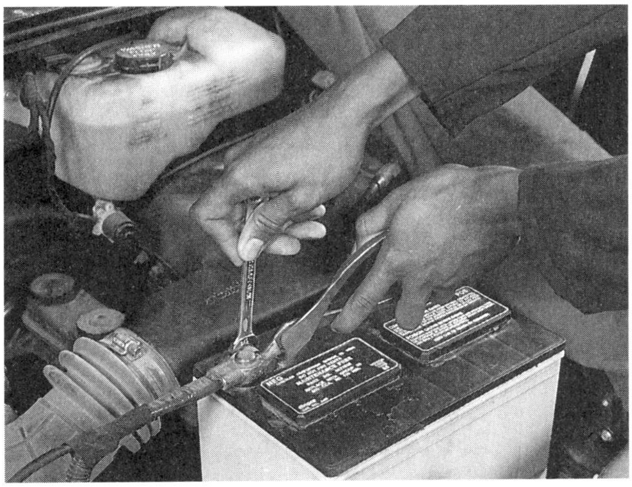

**Figure 9.** When disconnecting a battery, remove the negative cable first.

nent. This prevents the possibility of a fire or electrical shock. It also eliminates the possibility of an accidental short, which can ruin the car's electrical system. This is especially true of newer cars that are equipped with many electronic and computerized controls. Any electrical arcing can cause damage to the components or accidental deployment of the air bag. To properly disconnect the battery, disconnect the negative or ground cable first **(Figure 9)**, then disconnect the positive cable. Since electrical circuits require a ground to be complete, by removing the ground cable you eliminate the possibility of a circuit accidentally becoming completed. When reconnecting the battery, connect the positive cable first, then the negative.

You Should Know

*The active chemical in a battery, the **electrolyte**, is basically sulfuric acid. Sulfuric acid can cause severe skin burns and permanent eye damage, including blindness. If battery acid gets on your skin, wash it off immediately and flush your skin with water for at least 5 minutes. If the electrolyte gets into your eyes, immediately flush them out with water then immediately see a doctor. Never rub your eyes. Working with and around batteries is a common sense time to wear safety glasses or goggles.*

## Electrical System Repairs

Some electronic replacement parts are very sensitive to **static electricity.** These parts will be labeled as such. Whenever you are handling a part that is sensitive to static, you should follow these guidelines to reduce any possible electrostatic charge buildup on your body and the electronic part:

1. Do not open the package until it is time to install the component.
2. Before removing the part from the package, ground the package to a known good ground on the car.
3. Always touch a known good ground before handling the part. This should be repeated while handling the part and more frequently after sliding across the seat, sitting down from a standing position, or walking a distance.
4. Never touch the electrical terminals of the component.
5. Grounding straps are available to prevent damage to components.

## ACCIDENTS

Make sure you are aware of the location and contents of the shop's first aid kit **(Figure 10).** If there are specific first aid rules in your school or shop, make sure you are aware of them and follow them. Some first aid rules apply to all circumstances and are normally included in everyone's rules. If someone is overcome by carbon monoxide, seek fresh air for him or her immediately. Burns should be cooled immediately by rinsing them with water. Whenever there is severe bleeding from a wound, try to stop the bleeding by applying pressure with clean gauze on or around the wound, and seek medical help. Never move someone who may have broken bones unless the person's life is otherwise endangered. Moving that person may cause additional injury. Call for medical assistance.

Your instructor or supervisor should be informed immediately of all accidents. It is a good idea to keep a list of up-to-date emergency telephone numbers posted next to the telephone. The numbers should include a doctor, a hospital, and the fire and police departments.

## AIR BAG SAFETY AND SERVICE WARNINGS

The dash and steering wheel contain the circuits that control the **air bag system**. Whenever working on or

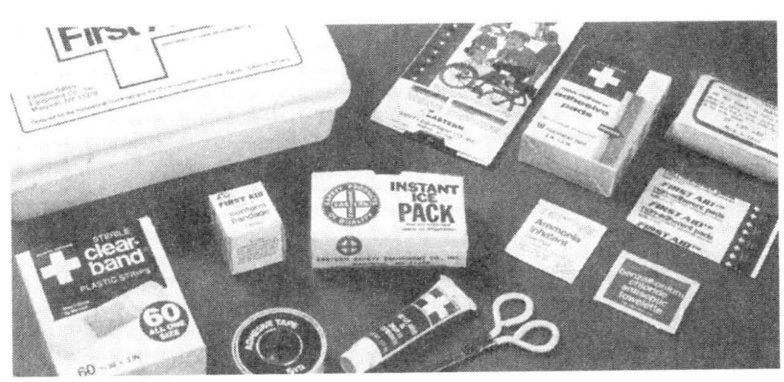


**Figure 10.** Know the location of and the contents of the shop's first aid kit.

around air bag systems, it is important to follow some safety warnings. There are safety concerns with both **deployed** and live (undeployed) air bag modules.

1. Wear safety glasses when servicing the air bag system.
2. Wear safety glasses when handling an air bag module.
3. Wait at least 30 minutes after disconnecting the battery before beginning any service on or around the air bag system. The reserve energy module is capable of storing enough power to deploy the air bag for up to 30 minutes after battery voltage is lost.
4. Handle all air bag sensors with care. Do not strike or jar a sensor in such a manner that deployment may occur.
5. When carrying a live air bag module, face the trim and bag away from your body.
6. Do not carry the module by its wires or connector.
7. When placing a live module on a bench, face the trim and air bag up.
8. Deployed air bags may have a powdery residue on them. Sodium hydroxide is produced by the deployment reaction and is converted to sodium carbonate when it comes into contact with atmospheric moisture. Although it is unlikely that sodium hydroxide will still be present, you should not put yourself at risk. This compound is toxic and can cause serious skin irritations. Wear safety glasses and gloves when handling a deployed air bag, and wash your hands immediately after handling the bag.
9. A live air bag module must be deployed before disposal. Because the deployment of an air bag is through an explosive process, improper disposal may result in injury and in fines. A deployed air bag should be disposed of in a manner consistent with Environmental Protection Agency (EPA) and manufacturer procedures.
10. The air bag electrical system is identified by yellow wires, connectors, and wire conduit. Never use a test light, battery- or AC-powered voltmeter, ohmmeter, or any other type of test equipment that is not specified for use on these circuits.

## HAZARDOUS MATERIALS

Many solvents and other chemicals used in an auto shop have warning and caution labels that should be read and understood by everyone that uses them. These are typically considered hazardous materials. Also, many service procedures generate what are known as **hazardous wastes.** Dirty solvents and liquid cleaners are good examples of these.

Every employee in a shop is protected by **"Right-to-Know" laws** concerning hazardous materials and wastes. The general intent of these laws is for employers to provide a safe working place as it relates to hazardous materials. All employees must be trained about their rights under the legislation, the nature of the hazardous chemicals in their workplace, the labeling of chemicals, and the information about each chemical listed and described on **Material Safety Data Sheets (MSDS).** These sheets are available from the manufacturers and suppliers of the chemicals. They detail the chemical composition and precautionary information for all products that can present health or safety hazards **(Figure 11).**

Employees must be familiar with the intended purposes of the substance, the recommended protective equipment, accident and spill procedures, and any other information regarding the safe handling of the material. This training must be given annually to employees and provided to new employees as part of their job orientation. The Canadian equivalents to the MSDS are called Workplace Hazardous Materials Information Systems (WHMIS).

You Should Know

***When handling any hazardous material, always wear the appropriate safety protection. Always follow the correct procedures while using the material, and be familiar with the information given on the MSDS for that material.***

All hazardous materials should be properly labeled, indicating what health, fire, or reactivity hazard it poses and what protective equipment is necessary when handling each chemical. The manufacturer of the hazardous material must provide all warnings and precautionary information, which must be read and understood by all users before they use it. You should pay great attention to the label information. By doing so, you will use the substance in the proper and safe way, thereby preventing hazardous conditions.

A list of all hazardous materials used in the shop should be posted for the employees to see. Shops must maintain documentation on the hazardous chemicals in the workplace, proof of training programs, records of accidents or spill incidents, satisfaction of employee requests for specific chemical information via the MSDS, and a general right to know compliance procedure manual utilized within the shop.

There are many government agencies charged with ensuring safe work environments for all workers. These include the **Occupational Safety and Health Administration (OSHA),** Mine Safety and Health Administration (MSHA), and National Institute for Occupational Safety and Health (NIOSH). These, in addition to state and local governments, have instituted regulations that must be understood and followed. Everyone in a shop has the responsibility for adhering to these regulations.

```
HEXANE
=========================================================
MSDS Safety Information
=========================================================
Ingredients
=========================================================
Name: HEXANE (N_HEXANE)
% Wt: >97
OSHA PEL: 500 PPM
ACGIH TLV: 50 PPM
EPA Rpt Qty: 1 LB
DOT Rpt Qty: 1 LB
=========================================================
Health Hazards Data
=========================================================
LD50 LC50 Mixture: LD50:(ORAL,RAT) 28.7 KG/MG
Route Of Entry Inds _ Inhalation: YES
Skin: YES
Ingestion: YES
Carcinogenicity Inds _ NTP: NO
IARC: NO
OSHA: NO
Effects of Exposure: ACUTE:INHALATION AND INGESTION ARE HARMFUL AND MAY BE FATAL.
INHALATION AND INGESTION MAY CAUSE HEADACHE, NAUSEA, VOMITING, DIZZINESS, IRRITATION
OF RESPIRATORY TRACT, GASTROINTESTINAL IRRITATION AND UNCONSCIOUSNESS. CONTACT
W/SKIN AND EYES MAY CAUSE IRRITATION. PROLONGED SKIN MAY RESULT IN DERMATITIS (EFTS
OF OVEREXP)
Signs And Symptions Of Overexposure: HLTH HAZ:CHRONIC:MAY INCLUDE CENTRAL
NERVOUS SYSTEM DEPRESSION.
Medical Cond Aggravated By Exposure: NONE IDENTIFIED.
First Aid: CALL A PHYSICIAN. INGEST:DO NOT INDUCE VOMITING. INHAL:REMOVE TO FRESH AIR. IF
NOT BREATHING, GIVE ARTIFICIAL RESPIRATION. IF BREATHING IS DIFFICULT, GIVE OXYGEN.
EYES:IMMED FLUSH W/PLENTY OF WATER FOR AT LEAST 15 MINS. SKIN:IMMED FLUSH W/PLENTY
OF WATER FOR AT LEAST 15 MINS WHILE REMOVING CONTAMD CLTHG & SHOES. WASH CLOTHING
BEFORE REUSE.
=========================================================
Handling and Disposal
=========================================================
Spill Release Procedures: WEAR NIOSH/MSHA SCBA & FULL PROT CLTHG. SHUT OFF
IGNIT SOURCES:NO FLAMES, SMKNG/FLAMES IN AREA. STOP LEAK IF YOU CAN DO SO W/OUT
HARM. USE WATER SPRAY TO REDUCE VAPS. TAKE UP W/SAND OR OTHER NON-COMBUST MATL &
PLACE INTO CNTNR FOR LATER (SU PDAT)
Neutralizing Agent: NONE SPECIFIED BY MANUFACTURER.
Waste Disposal Methods: DISPOSE IN ACCORDANCE WITH ALL APPLICABLE FEDERAL, STATE AND
LOCAL ENVIRONMENTAL REGULATIONS. EPA HAZARDOUS WASTE NUMBER:D001 (IGNITABLE
WASTE).
Handling And Storage Precautions: BOND AND GROUND CONTAINERS WHEN TRANSFERRING LIQUID.
KEEP CONTAINER TIGHTLY CLOSED.
Other Precautions: USE GENERAL OR LOCAL EXHAUST VENTILATION TO MEET
TLV REQUIREMENTS. STORAGE COLOR CODE RED (FLAMMABLE).
=========================================================
Fire and Explosion Hazard Information
=========================================================
Flash Point Method: CC
Flash Point Text: _9F,_23C
Lower Limits: 1.2%
Upper Limits: 77.7%
Extinguishing Media: USE ALCOHOL FOAM, DRY CHEMICAL OR CARBON DIOXIDE. (WATER MAY BE
INEFFECTIVE).
Fire Fighting Procedures: USE NIOSH/MSHA APPROVED SCBA & FULL PROTECTIVE
  EQUIPMENT (FP N).
Unusual Fire/Explosion Hazard: VAP MAY FORM ALONG SURFS TO DIST IGNIT SOURCES & FLASH
BACK. CONT W/STRONG OXIDIZERS MAY CAUSE FIRE. TOX GASES PRDCED MAY INCL:CARBON
MONOXIDE, CARBON DIOXIDE.
=========================================================
```

**Figure 11.** An example of an MSDS.

## OCCUPATIONAL SAFETY AND HEALTH ADMINISTRATION (OSHA)

In 1970, OSHA was formed by the federal government to "assure safe and healthful working conditions for working men and women; by authorizing enforcement of the standards developed under the Act; by assisting and encouraging the states in their efforts to assure safe and healthful working conditions by providing research, information, education, and training in the field of occupational safety and health."

Safety standards have been established that will be consistent across the country. It is the employers' responsibility to provide a place of employment that is free from all recognized hazards and that will be inspected by government agents knowledgeable in the law of working conditions. OSHA controls all safety and health issues of the automotive industry.

## HANDLING SHOP WASTES

Shop wastes are simply those byproducts that are left over when a service procedure is completed. Some of these wastes must be treated as potentially hazardous and should be handled and disposed of accordingly.

- Recycle engine oil. Have a drip or screen table with a used oil collection bucket to collect oils dripping off parts. Place drip pans underneath vehicles that are leaking fluids onto the storage area. Never allow oil to enter into the shop's sewage or drainage system.
- Used oil filters should be drained for at least 24 hours, crushed, and recycled.
- Old batteries should be recycled by sending them to a reclaimer or back to the distributor.
- Collect metal filings when machining metal parts. Keep them separate from other dirt and recycle the metal, if possible. Prevent metal filings from falling into a storm sewer drain.
- Recover and/or recycle refrigerants during the service and disposal of motor vehicle air conditioners and refrigeration equipment. It is illegal to knowingly vent refrigerants to the atmosphere. Recovery and/or recycling during service must be performed by an EPA-certified technician using certified equipment and following specified procedures.
- Replace hazardous chemicals with less toxic alternatives that have equal performance. For example, substitute water-based cleaning solvents for petroleum-based solvent degreasers. Hire a hazardous waste management service to clean and recycle solvents. Store solvents in closed containers to prevent evaporation. Properly label spent solvents.
- Store materials such as scrap metal, old machine parts, and worn tires under a roof or tarpaulin to protect them from the elements and to prevent the potential to create contaminated runoff. Consider recycling tires by retreading them.
- Collect and recycle coolants from radiators. Store transmission and brake fluids and solvents containing chlorinated hydrocarbons separately, and recycle or dispose of them properly.
- Keep dirty shop rags in a closed container marked "Contaminated Shop Towels Only." To reduce costs and liabilities associated with disposal of used towels (which can be classified as hazardous wastes), investigate using a laundry service that is able to treat the wastewater generated from cleaning the towels.

# *SUMMARY*

- Shop safety is the responsibility of everyone in the shop, and everyone must work together to protect everyone in the shop.
- Eye protection should be worn whenever you are working in the shop.
- If chemicals or solvents splash your eyes, flush them continuously with clean water. Have someone call a doctor and seek medical help immediately.
- Dress to protect yourself and to avoid accidents. Wear well-fitted and comfortable clothing as well as foot and hand protection. Never wear anything that may become caught in moving parts or can conduct electricity.
- When lifting heavy objects, use a hoist or have someone help you. Always lift with your legs, not your back.
- Gasoline, diesel fuel, cleaning solvents, and dirty rags should be stored properly.
- Know where the fire extinguishers are located in the shop, and know how to use them.
- Inspect all tools and equipment for unsafe conditions before using them.
- Never use tools and equipment for purposes other than those they are designed for.
- Make sure the contacts of a lift are properly positioned before lifting a vehicle.
- Use jack stands to secure a vehicle after it has been raised by a jack.
- When necessary, disconnect the negative cable of the battery before you work around it, or disconnect any electrical wire or component.
- Follow the proper procedures for eliminating static when handling parts that are static sensitive.
- Whenever working on or around air bag systems, it is important to follow certain procedures.
- Handle and dispose of all hazardous wastes according to local laws and your own common sense.
- Know where the MSDS can be found in the shop, and know how to quickly find the appropriate information on them.

## REVIEW QUESTIONS

1. When should you wear eye protection?
2. What is the correct procedure for putting out a fire with an extinguisher?
3. What do all employees of a shop have the right to know?
4. The general intent of laws is for employers to provide a safe working place as it relates to an awareness of hazardous materials.
5. Technician A says the potential for accidents can be reduced by not having anything dangle near rotating equipment and parts. Technician B says most accidents can be prevented by using common sense. Who is correct?
   A. Technician A only
   B. Technician B only
   C. Both Technician A and Technician B
   D. Neither Technician A nor Technician B
6. Technician A says unsafe equipment should have its power disconnected and marked with a sign to warn others not to use it. Technician B says that all equipment should be inspected for safety hazards before being used. Who is correct?
   A. Technician A only
   B. Technician B only
   C. Both Technician A and Technician B
   D. Neither Technician A nor Technician B
7. While discussing ways to create an accident free work environment, Technician A says everyone in the shop should take full responsibility for ensuring safe work areas. Technician B says the dress and work habits of technicians can help prevent accidents. Who is correct?
   A. Technician A only
   B. Technician B only
   C. Both Technician A and Technician B
   D. Neither Technician A nor Technician B
8. While discussing the car's electrical system, Technician A says you should always disconnect the negative or ground battery cable first, then disconnect the positive cable. Technician B says you should always connect the positive battery cable first, then the negative. Who is correct?
   A. Technician A only
   B. Technician B only
   C. Both Technician A and Technician B
   D. Neither Technician A nor Technician B
9. What should be done to prevent damage to electronic components by static electricity?
10. When working in an automotive shop, hazardous conditions can be avoided by:
    A. storing oily rags in open containers.
    B. storing flammable liquids such as paint in metal cupboards.
    C. leaving sewer covers loose for quick access.
    D. cleaning all spills at the end of the work day.

# Chapter 2 Measuring Systems, Fasteners, and Measuring Tools

## INTRODUCTION

Servicing steering and suspension systems requires the use of various tools. Many of these tools are used to remove and install fasteners. Fasteners of different sizes and shapes are used on today's cars. Basically, the correct tool for the job is the tool that fits the fastener.

In this chapter, the most common measuring systems will be explained. These systems will then be used to describe various fasteners. Also covered in this chapter are the commonly used measuring instruments. The proper use of these instruments ensures that parts are installed properly and according to specifications.

## MEASURING SYSTEMS

Two different systems of weights and measures are currently used in the United States—the United States Customary System (USCS) and the International System (SI), which is commonly referred to as the metric system. The USCS system is sometimes referred to as the English System. The U.S. is slowly changing over to the metric system, and during the changeover, cars produced in the U.S. are being made with both English and metric fasteners and specifications. Most of the world outside the U.S. uses the metric system.

### Measurement of Length

In the USC system, linear measurements are defined by inches, feet, and yards. The inch can be broken down into fractions, such as 1/64, 1/32, 1/16, 1/8, 1/4, and 1/2 of an inch **(Figure 1).** The inch is also commonly broken down by decimals. When an inch is divided into tenths, each part is a tenth of an inch (0.1 in.). Tenths of an inch can be further divided by ten into hundredths of an inch (0.01 in.). The division after hundredths is thousandths (0.001 in.); these are followed by tenthousandths (0.0001 in.).

In the metric system, the basic measurement unit of length is the meter. For exact measurements, the meter is divided into units of ten. The first division is called a decimeter (dm). The second division is the centimeter (cm), and the third and most commonly used division is the millimeter (mm). One decimeter is equal to 0.1 meter, 1 centimeter equals 0.01 meter, and 1 millimeter equals 0.001 meter.

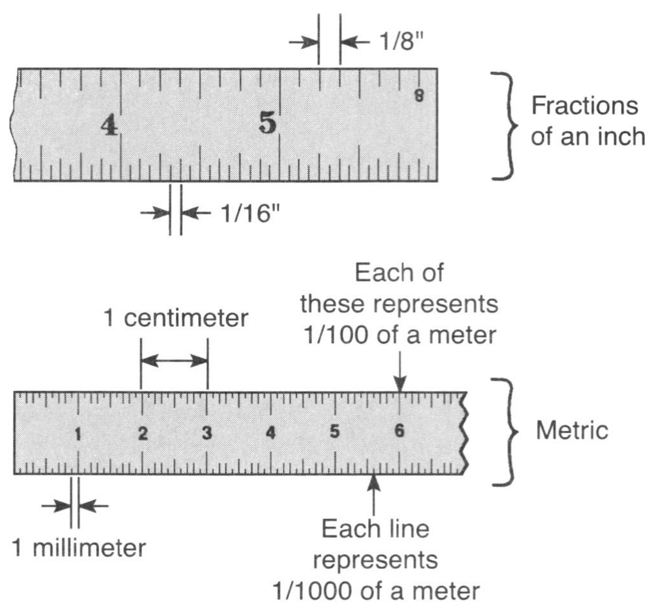


**Figure 1.** (Top) Increments of an inch. (Bottom) Increments of a meter.

## Measurement of Mass

In the metric system, mass is measured in grams, kilograms, or tonnes: 1,000 grams (g) equals 1 kilogram (kg). In the USC system, mass is measured in ounces, pounds, or tons. When converting pounds to kilograms, 1 pound equals 0.453 kilograms.

## Measurement of Volume

In the metric system, volume is measured in milliliters, cubic centimeters, and liters 1 cubic centimeter equals 1 milliliter. If a cube has a length, depth, and height of 10 cm, the volume of the cube is 10 cm x 10 cm x 10 cm equals 1,000 $cm^3$ equals 1 liter. When volume conversions are made between the two systems, 1 cubic inch equals 16.38 cubic centimeters. If an engine has a displacement of 350 cubic inches, 350 x 16.38 equals 5,733 cubic centimeters, and 5,733 ÷ 1,000 equals 5.733 liters.

## Angles

The correct position of the various parts of a suspension and steering system are often set by placing them at the correct angle of another part or a position of the vehicle. Setting these angles is accomplished during a wheel alignment **(Figure 2).** Wheel alignment also includes adjustments to the length of a part or the distance between two parts.

Wheel alignment measures and adjusts the geometric relationship of parts to the vehicle's body or frame. An angle is the shape formed when two straight lines intersect each other. It is also the difference in direction between the lines and is measured in degrees **(Figure 3).**

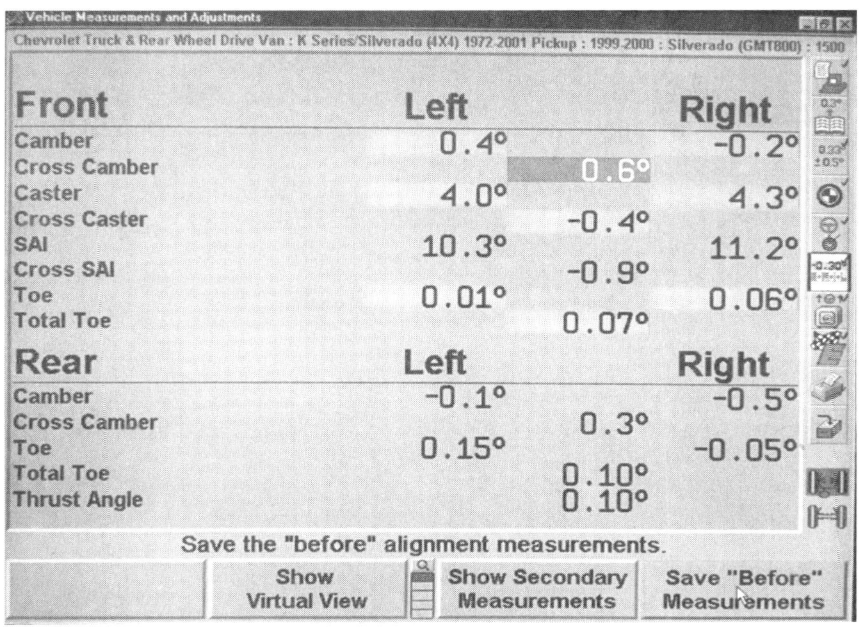


**Figure 2.** Wheel alignment involves measuring and setting various angles of steering and suspension components.

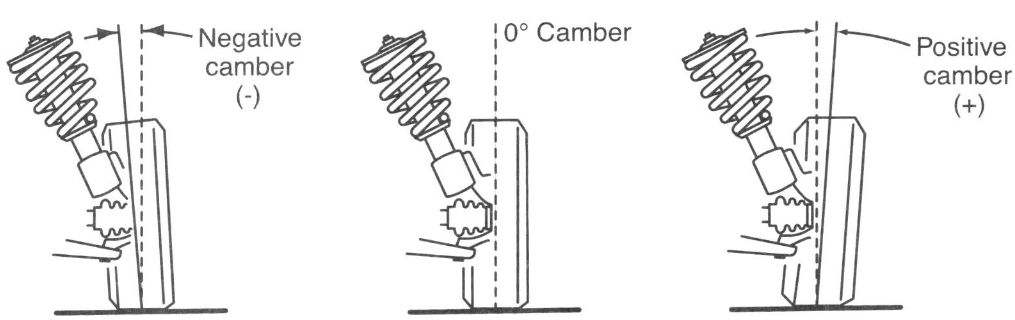


**Figure 3.** Wheel camber is measured and defined in degrees.

## FASTENERS

Fasteners are those items that are used to secure or hold parts of something together. Many types and sizes of fasteners are commonly used. Each fastener is designed for a specific purpose. Threaded fasteners include bolts, nuts, and screws. Because some vehicles have metric fasteners, some have USCS, and others have both, technicians must have both English and metric tools. Vehicle specifications are normally listed in both meters and inches, therefore measuring tools are available in both measuring systems.

Threaded fasteners are available in many sizes, designs, and threads. The threads can be either cut or rolled into the fastener. Rolled threads are 30 percent stronger than cut threads. They also offer better fatigue resistance because there are no sharp notches to create stress points. Fasteners are made to English or metric measurements. There are four classifications of the threads of English fasteners: Unified National Coarse (UNC), Unified National Fine (UNF), Unified National Extra Fine (UNEF), and Unified National Pipe Thread (UNPT or NPT). Metric fasteners are also available in fine and coarse threads.

Coarse threads are used for general-purpose work, especially where rapid assembly and disassembly is required. Fine threaded fasteners are used where greater holding force is necessary. They are also used where greater resistance to vibration is desired.

Bolts have a head on one end and threads on the other. Bolts are identified by their head size, shank diameter, thread pitch, length **(Figure 4),** and grade. Bolts have a shoulder below the head, and the threads do not travel all the way from the head, to the end of the bolt.

Studs are rods with threads on both ends. Most often, the threads on one end are coarse, while the other end is fine thread. One end of the stud is screwed into a threaded bore. A hole in the part to be secured is fitted over the stud and held in place with a nut screwed over the stud. Studs are used when the clamping pressures of a fine thread are needed and a bolt will not work.

Nuts are used with other threaded fasteners when the fastener is not threaded into a piece of work. Many different designs of nuts are found on today's cars. The most common one is the hex nut, which is used with studs and bolts and is tightened with a wrench.

## BOLT IDENTIFICATION

The **bolt head** is used to loosen and tighten the bolt; a socket or wrench fits over the head and is used to screw the bolt in or out. The size of the bolt head varies with the diameter of the bolt and is available in USCS and metric wrench sizes. Many confuse the size of the head with the size of the bolt. The size of a bolt is determined by the diameter of its shank. The size of the bolt head determines the size wrench required to turn it.

Bolt diameter is the measurement across the major diameter of the threaded area or across the bolt shank. The length of a bolt is measured from the bottom surface of the head to the end of the threads.

The **thread pitch** of a bolt in the English system is determined by the number of threads that are in one inch of the threaded bolt length and is expressed in number of threads per inch. A UNF bolt with a 3/8-inch diameter would be a 3/8 x 24 bolt. It would have 24 threads per inch. Likewise, a 3/8-inch UNC bolt would be called a 3/8 x 16.

The distance, in millimeters, between two adjacent threads determines the thread pitch in the metric system. This distance will vary between 1.0 and 2.0 and depends on

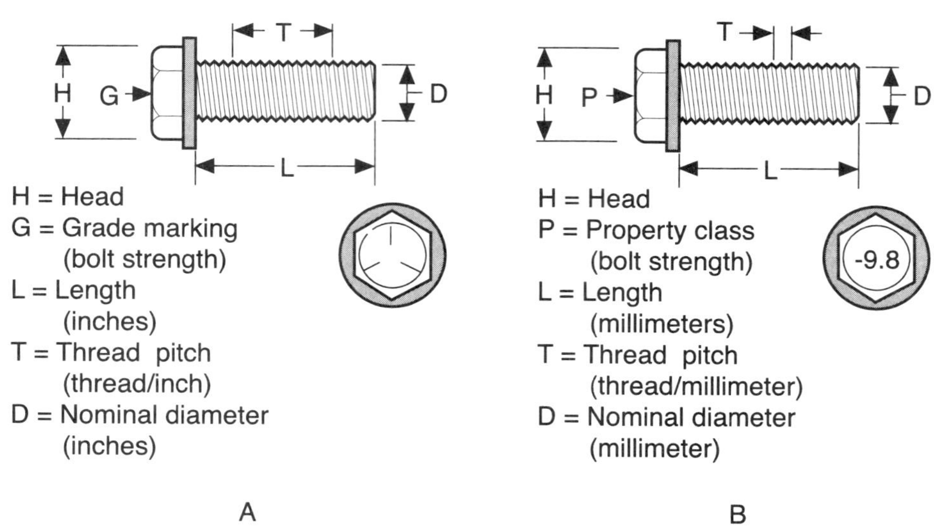


**Figure 4.** Bolt terminology for a USC bolt (A) and a metric bolt (B).

the diameter of the bolt. The lower the number, the closer the threads are placed and the finer the threads are.

The bolt's tensile strength, or grade, is the amount of stress or stretch it is able to withstand before it breaks. The type of material the bolt is made of and the diameter of the bolt determines its grade. In the English system, the tensile strength of a bolt is identified by the number of radial lines **(grade marks)** on the bolt's head. More lines mean higher tensile strength. Count the number of lines and add two to determine the grade of a bolt.

A property class number on the bolt head identifies the grade of metric bolts. This numerical identification is comprised of two numbers. The first number represents the tensile strength of the bolt. The higher the number, the greater the tensile strength. The second number represents the yield strength of the bolt. This number represents how much stress the bolt can take before it is unable to return to its original shape without damage. The second number represents a percentage rating. For example, a 10.9 bolt has a tensile strength of 1000 megapascals (MPa) (145,000 pounds per square inch) and a yield strength of 900 MPa (90 percent of 1000). A 10.9 metric bolt is similar in strength to an SAE grade 8 bolt.

Nuts are graded to match their respective bolts. Grade 8 and critical applications require the use of fully hardened flat washers. These will not dish out when torqued as soft washers will.

Bolt heads can pop off because of **fillet** damage. The fillet is the smooth curve where the shank flows into the bolt head. Scratches in this area introduce stress to the bolt head. Removing any burrs around the edges of holes can protect the bolt head. Also, place flat washers with their rounded, punched side against the bolt head and their sharp side to the work surface.

Fatigue breaks are the most common type of bolt failure. A bolt becomes fatigued from working back and forth when it is too loose. Undertightening the bolt causes this problem. Bolts can also be broken or damaged by overtightening, being forced into a non-matching thread, or bottoming out, which happens when the bolt is too long.

## TIGHTENING BOLTS

Any fastener is near worthless if it is not as tight as it should be. When a bolt is properly tightened, it will be "spring loaded" against the part it is holding. This spring effect is caused by the stretch of the bolt when it is tightened. Normally, a properly tightened bolt is stretched to 70 percent of its elastic limit. The elastic limit of a bolt is that point of stretch where the bolt will not return to its original shape when it is loosened. Not only will an overtightened or stretched bolt not have sufficient clamping force, it will also have distorted threads **(Figure 5).** Always check the service manual to see if there is a torque specification for a bolt before tightening it. If there is, use a torque wrench and tighten the bolt properly.

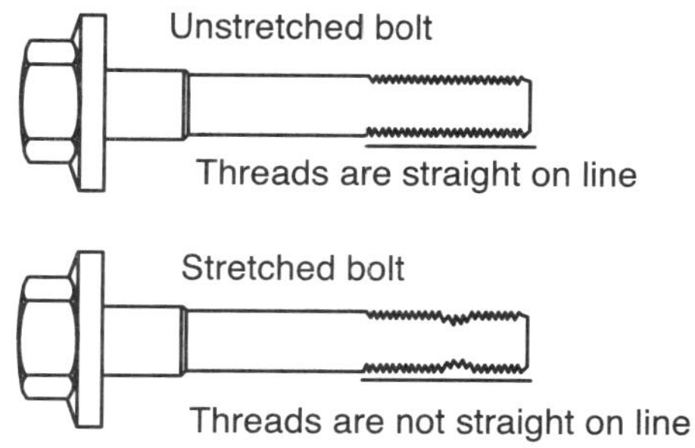


**Figure 5.** A comparison of a normal good bolt and one that has been stretched.

## MEASURING TOOLS

Many of the procedures discussed in this manual require exact measurements of parts and clearances. Accurate measurements require the use of precision measuring devices that are designed to measure in very small increments. Measuring tools are delicate instruments and should be handled with great care. Never strike, pry, drop, or force these tools. Also, make sure you clean them before and after every use.

### Machinist's Rule

The machinist's rule looks very much like an ordinary ruler. Each edge of this basic measuring tool is divided into increments based on a different scale. A typical machinist's rule based on the Imperial system of measurement may have scales based on 1/8-, 1/16-, 1/32-, and 1/64-inch intervals. Of course, metric machinist rules are also available. Metric rules are usually divided into 0.5-mm and 1-mm increments.

Some machinist rules may be based on decimal intervals. These are typically divided into 1/10-, 1/50-, and 1/1,000-inch (0.1, 0.03, and 0.01) increments. Decimal machinist rules are very helpful when measuring dimensions that are specified in decimals.

Tape measures are like machinist rules except they are more compact, less accurate, and are available in many lengths (up to 50 feet). Often, tape measures are used to measure curb riding height **(Figure 6),** which is a common suspension check.

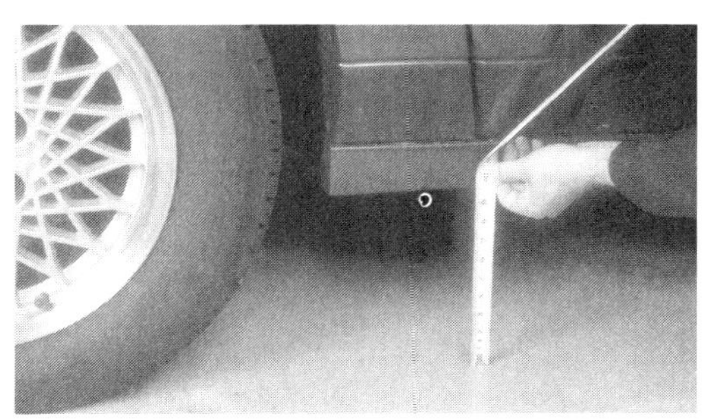

**Figure 6.** Using a tape measure to check curb riding height.

## Feeler Gauge

A **feeler gauge** is a thin strip of metal or plastic of known and closely controlled thickness. Several of these metal strips are often assembled together as a feeler gauge set that looks like a pocketknife. The desired thickness gauge can be pivoted away from others for convenient use. A steel feeler gauge pack usually contains strips or leaves of 0.002- to 0.010-inch thickness (in steps of 0.001 inch) and leaves of 0.012- to 0.024-inch thickness (in steps of 0.002 inch).

A feeler gauge can be used by itself to measure clearances and gaps or it can be used with a precision straightedge to measure alignment and surface warpage.

## Screw Pitch Gauge

The use of a **screw pitch gauge** provides a quick and accurate method of checking the thread pitch of a fastener **(Figure 7).** The leaves of this measuring tool are marked with the various pitches. To check the pitch of threads, simply match the teeth of the gauge with the threads of the fastener. Then, read the pitch from the leaf.

## Calipers and Dividers

Calipers and dividers transfer lengths to another measuring instrument. These are used to measure inside and outside diameters. To take a measurement with these tools, adjust the legs of the tool to fit the inside or outside diameter of the object. Then lay the caliper or divider over a rule and read the length on the rule.

## Vernier Caliper

A **vernier caliper** is a measuring tool. It can make inside, outside, or depth measurements. It is marked in both USCS and metric divisions called a vernier scale. A vernier scale consists of a movable scale in parallel with a stationary scale, in this case the vernier bar to the vernier plate.

Each inch of the stationary scale is divided into ten parts, each equal to 0.100 inch. The area between the 0.100 marks is divided into four, each equaling 0.025 inches. The vernier scale has twenty-five divisions, each one representing 0.001 inch. Measurement readings are taken by combining the main and vernier scales. At all times, only one division line on the main scale will line up with a line on the vernier scale. This is the basis for accurate measurements.

The **dial caliper** is an easier-to-use version of the vernier caliper. A dial caliper has a depth scale, bar scale, dial indicator, inside measurement jaws, and outside measurement jaws.

Both English and metric dial calipers use a thumb-operated roll knob for fine adjustment. When you use a caliper, always move the measuring jaws backward and forward to center the jaws on the object being measured. Make sure the caliper jaws lie flat on or around the object. If the jaws are tilted in any way, you will not obtain an accurate measurement.

Although calipers are precision measuring instruments, they are only accurate to plus or minus two-thousandths (0.002) of an inch. Micrometers are preferred when extremely precise measurements are desired.

## Dial Indicator

The **dial indicator** is calibrated in 0.001-inch (one-thousandth inch) increments. Metric dial indicators are also available. Both types are used to measure movement. Common uses of the dial indicator include measuring axle endplay **(Figure 8),** ball joint movement, brake rotor runout, gear backlash, and shaft endplay.

To use a dial indicator, position the indicator rod against the object to be measured. Then, push the indicator toward

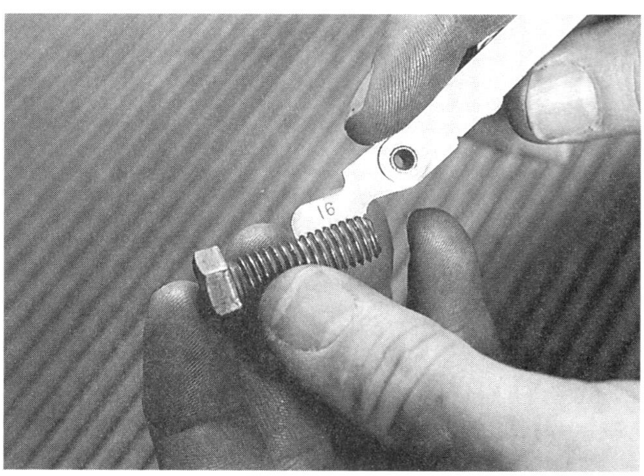


**Figure 7.** A screw pitch gauge.

**Figure 8.** Using a dial indicator to measure the endplay of an axle.

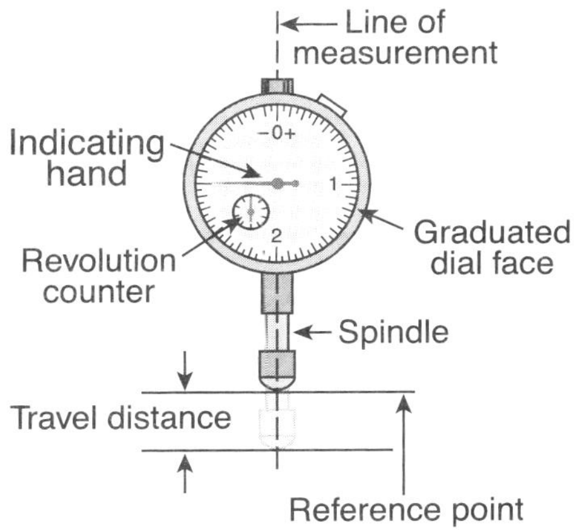


**Figure 9.** Action of a dial indicator when the spindle (plunger) is moved.

the work until the indicator needle travels far enough around the gauge face to permit movement to be read in either direction **(Figure 9).** Zero the indicator needle on the gauge. Always be sure the range of the dial indicator is sufficient to allow the amount of movement required by the measuring procedure. For example, never use a 1-inch indicator on a component that will move 2 inches.

## MICROMETERS

The **micrometer** is used to measure the outside diameter of an object or the inside diameter or a bore or drum. Both outside and inside micrometers are calibrated and read in the same manner. Measurements on both are taken with the measuring points in contact with the surfaces being measured.

The major components and markings of a micrometer include the frame, anvil, spindle, locknut, sleeve, sleeve numbers, sleeve long line, thimble marks, thimble, and ratchet. Micrometers are calibrated in either inch or metric graduations and are available in a range of sizes.

To measure small objects with an outside micrometer, open the jaws of the tool and slip the object between the spindle and the anvil**.** While holding the object against the anvil, turn the thimble using your thumb and forefinger until the spindle contacts the object. Use only enough pressure on the thimble to allow the object to just fit between the tips of the anvil and spindle. The object should slip through with only a very slight resistance. When a satisfactory feel is reached, lock the micrometer. Since each graduation on the sleeve represents 0.025 inches **(Figure 10),** begin reading the measurement by counting the visible lines on the sleeve and multiply that number by 0.025. The graduations on the thimble assembly define the area between the lines on the sleeve; therefore, the number indicated on the thimble should be added to the measurement shown on the sleeve. The sum is the outside diameter of the object.

To measure larger objects, hold the frame of the micrometer and slip it over the object. Turn the thimble while continuing to slip the micrometer over the object until you feel a very slight resistance. Rock the micrometer from side to side while doing this to make sure the spindle cannot be closed any farther. Then, lock the micrometer and take a measurement reading.

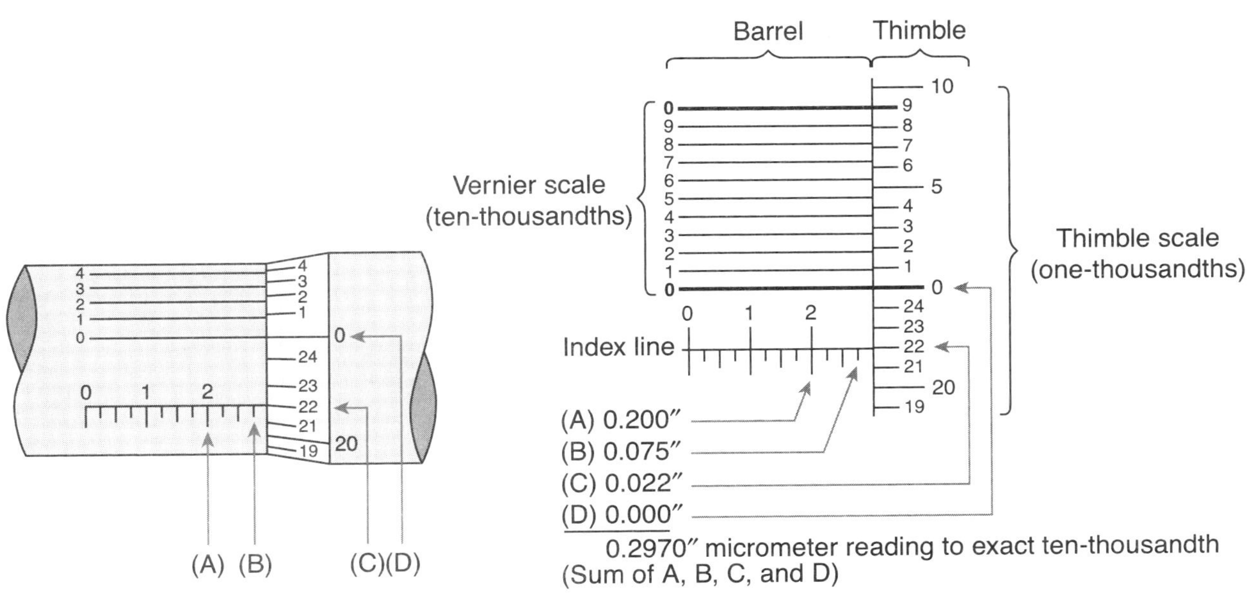


**Figure 10.** How to read a micrometer.

## Reading a Metric Outside Micrometer

The metric micrometer is read in the same manner as the inch-graduated micrometer, except the graduations are expressed in the metric system of measurement **(Figure 11).** Readings are obtained as follows.

1. Each number on the sleeve of the micrometer represents 5 millimeters (mm) or 0.005 meter (m).
2. Each of the ten equal spaces between each number, with index lines alternating above and below the horizontal line, represents 0.5 mm, or five-tenths of a mm. One revolution of the thimble changes the reading one space on the sleeve scale, or 0.5 mm.
3. The beveled edge of the thimble is divided into 50 equal divisions with every fifth line numbered: 0, 5, 10 ... 45. Since one complete revolution of the thimble advances the spindle 0.5 mm, each graduation on the thimble is equal to one-hundredth of a millimeter.
4. As with the inch-graduated micrometer, the three separate readings are added together to obtain the total reading.

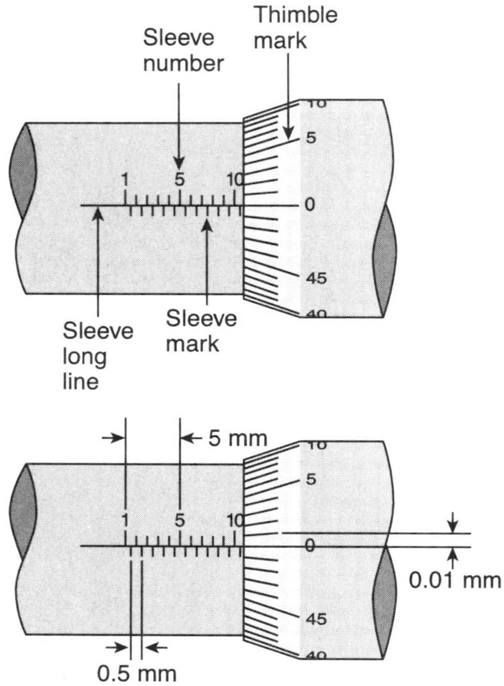


**Figure 11.** The increments on a metric micrometer.

## SUMMARY

- In the United States Customary System, the inch is commonly broken down into fractions or decimals.
- In the metric system, the basic measurement unit of length is the meter, which for exact measurements is divided into units of ten. The most commonly used division is the millimeter (mm).
- The size of a bolt is determined by the diameter of its shank. The size of the bolt head determines what size wrench is required to turn it.
- The length of a bolt is measured from the bottom surface of the head to the end of the threads.
- The thread pitch of a bolt in the English system is expressed by number of threads per inch.
- The bolt's tensile strength, or grade, is the amount of stress or stretch it is able to withstand before it breaks.
- Measuring instruments that are commonly used to service steering and suspension systems include the machinist's rule, screw pitch gauge, dial indicator, and micrometer.

## REVIEW QUESTIONS

1. While discussing automotive fasteners, Technician A says bolt sizes are listed by their appropriate wrench size. Technician B says whenever bolts are replaced, they should be replaced with the same size and type that was originally installed by the manufacturer. Who is correct?
   A. Technician A only
   B. Technician B only
   C. Both Technician A and Technician B
   D. Neither Technician A nor Technician B
2. Technician A says a bolt's tensile strength is the amount of stress or stretch it is able to withstand before it breaks. Technician B says the property class number on a metric bolt head is comprised of two numbers. The first represents the tensile strength of the bolt, and the second represents the yield strength of the bolt. Who is correct?
   A. Technician A only
   B. Technician B only
   C. Both Technician A and Technician B
   D. Neither Technician A nor Technician B
3. List five features of a bolt that are commonly used to identify it.
4. Explain the two primary purposes of a feeler gauge.
5. One millimeter is equal to how many meters?

# Chapter 3 Tools of the Trade

## INTRODUCTION

Although every technician's toolbox contains many different tools, there are certain hand tools that are necessary. These are briefly discussed in this chapter. Also covered are the power and special tools commonly used to work on suspension and steering systems.

## HAND TOOLS

A basic tool set should include a set of box-end and open-end wrenches. Box-end wrenches are available in either 6 or 12 points. Twelve-point box-end wrenches allow you to work in tighter areas than do 6-point wrenches. An open-end wrench is normally the best tool for turning the nut down or holding a bolt head. To be able to work on today's vehicles, you should have complete sets of both USCS and metric wrenches.

**You Should Know** ***Metric and USCS wrenches are not interchangeable. For example, a 9/16-inch wrench is 0.02 inches larger than a 14-millimeter nut. If the 9/16-inch wrench is used to turn or hold a 14 millimeters nut, the wrench will probably slip. This may cause the points of the bolt head or nut to round off and can possibly cause skinned knuckles.***

Your tool set should include both blade and Phillips screwdrivers in a variety of lengths. Some vehicles may require special screwdrivers, such as those with a Torx head design.

In addition, you should have a good variety of pliers and hammers and mallets. Hammers are used with punches and chisels, and mallets are used for tapping parts apart or aligning parts together. Soft-faced mallets will not harm the part it is hitting against, whereas a hammer will.

You should also have an OSHA-approved air blowgun. These are used during cleaning and for drying cleaned parts. Never point a blowgun at yourself or someone else.

### Ratchets and Sockets

A set of USCS and metric sockets, combined with a ratchet handle and a few extensions, should also be included in your tool set. These sockets should be $^{3}/_{8}$-inch drive, although $^{1}/_{4}$- and $^{1}/_{2}$-inch drive sets are also handy. You may also want a long breaker bar to fit your sockets. Breaker bars offer increased leverage when loosening very tight bolts. Extensions allow you to put the handle in the best position while working. They range from 1 inch to 3 feet long. Also available are universal joints that allow a technician to work a bolt or nut at a slight angle.

A six-point socket has stronger walls and improved grip on a bolt compared to a normal 12-point socket. Eight-point sockets are available to use on square nuts or square-headed bolts. Sockets are also available as deep-well sockets that are used to reach a nut when it is on a bolt or stud with long threads.

### Shop Light

Adequate light is necessary when working under and around automobiles. A shop light can be battery powered (like a flashlight) or need to be plugged into a wall socket. Some shops have lights that pull down from a reel suspended from the ceiling. Shop lights use either an incandescent bulb or fluorescent tube. Because incandescent bulbs can pop and burn, it is highly recommended that you only use fluorescent bulbs. Take extra care when using a trouble light. Make sure the cord does not become caught in a rotating object.

## SERVICE INFORMATION AND MANUALS

Service manuals are necessary for all vehicle services. There is no way a technician can remember all of the procedures and specifications needed to repair an automobile correctly. Thus, a good technician relies on service manuals and other information sources for this information **(Figure 1).** Good information plus knowledge allows a technician to fix a problem with the least bit of frustration and at the lowest expense to the customer. Service manuals also provide drawings and photographs that show where and how to perform certain procedures on the particular car you are working on. Special tools or instruments are listed and shown when they are required. Precautions are also given to prevent injury or damage to parts. Perhaps the most important tools you will use are service manuals.

Most automobile manufacturers publish a service manual or set of manuals for each model and year of their cars. These manuals provide the best and most complete information for those cars. The most commonly used specifications and procedures are compiled in comprehensive service manuals. Various editions are available, covering different ranges of model years for both domestic and imported cars. Specialized shop manuals are also produced for special areas, such as transmissions. Although similar to manufacturers' manuals in many ways, they do not provide as much information or detail as do the manufacturers' manuals. Although the manuals from different publishers vary in presentation and arrangement of topics, all service manuals are easy to use after you become familiar with their organization.

Throughout this book, you will be told to refer to the appropriate shop manual to find the correct procedures and specifications. Although the various systems of all automobiles function in much the same way, there are many variations in design. Each design has its own repair and diagnostic procedures. Therefore, it is important that you always follow the recommendations of the manufacturer to identify and repair problems.

Since many technical changes occur on specific vehicles each year, manufacturers' service manuals need to be updated constantly. Updates are published as service bulletins (often referred to as Technical Service Bulletins (TSBs) that show the changes in specifications and repair procedures during the model year. These changes do not appear in the service manual until the next year. The car manufacturer provides these bulletins to dealers and repair facilities on a regular basis.

Automotive manufacturers also publish a series of technician reference books. The publications provide general instructions on the service and repair of their vehicles with their recommended techniques.

A vehicle's service history can also be a valuable diagnostic aid. This information may be found in a shop's records or through online services with the manufacturer or another provider of information. The history records can show the vehicle's maintenance record, recall actions, and previous related services.

### Aftermarket Suppliers' Guides and Catalogs

Many of the larger parts manufacturers have excellent guides on the various parts they manufacture or supply. They also provide updated service bulletins on their products. Other sources for up-to-date technical information are trade magazines and trade associations.

### Computer-Based Information

The same information that is available in service manuals and bulletins is also available electronically on CD-ROMs, DVDs **(Figure 2),** and the Internet. A single

**Figure 1.** Examples of manufacturer's service manuals.

**Figure 2.** DVDs are often used to retrieve service information. These are for a computerized alignment machine.

compact disk can hold 250,000 pages of text. This eliminates the need for a huge library to contain all of the printed manuals. Using electronics to find information is also easier and quicker. The disks are normally updated monthly and not only contain the most recent service bulletins but also engineering and field service fixes. Online data can be updated instantly and requires no space for physical storage.

## POWER TOOLS

Power tools make a technician's job easier. They operate faster and with more torque than hand tools. Power is furnished by air (pneumatic), electricity, or hydraulic fluid. Technicians typically use pneumatic tools because they have more torque, weigh less, and require less maintenance than electric power tools. However, electric power tools tend to cost less than the pneumatics. Electric power tools can be plugged into most electric wall sockets, but to use a pneumatic tool, you must have an air compressor and an air storage tank.

You Should Know *Carelessness or mishandling of power tools can cause serious injury. Make sure you know how to operate a tool before using it.*

Power tools make a technician's job easier. They operate faster and with more torque than hand tools. However, power tools require greater safety measures. Power tools do not stop unless they are turned off.

### Air Wrenches

An impact wrench uses compressed air or electricity to hammer or impact a nut or bolt loose. Light-duty impact wrenches are available in three drive sizes: 1/4, 3/8, and 1/2 inch and two heavy-duty sizes, 3/4 and 1 inch.

You Should Know *Impact wrenches should not be used to tighten critical parts or parts that may be damaged by the hammering force of the wrench.*

To save time, air ratchets are often used during disassembly or reassembly work. Because the ratchet turns the socket without an impact force, these wrenches can be used on most parts and with ordinary sockets. Air ratchets usually have a 3/8-inch drive. Air ratchets are not torque sensitive, therefore, a torque wrench should be used on all fasteners after snugging them up with an air ratchet.

You Should Know *It is possible to injure your hand while tightening bolts with an air ratchet. Once the bolt or nut is tight, the snap of the ratchet may pinch or smash your hand on another object.*

## SHOP TOOLS

Some tools required for working on suspension and steering systems are not part of a technician's tool set. Rather, these are general shop tools. A brief discussion of the common shop tools follows.

### Bench Grinder

A bench grinder is generally bolted to a workbench but may be freestanding. In either case, the grinder should have safety shields and guards.

Bench grinders must be securely bolted to the workbench. When grinding small components on a grinding wheel, wire brush wheel, or buffing wheel, always hold these components with a pair of vise grips to avoid injury to fingers and hands.

You Should Know *Always wear a face shield when using a bench grinder. Also, when grinding small components on a grinding wheel, wire brush wheel, or buffing wheel, always hold these components with a pair of vise grips to avoid injury to fingers and hands.*

### Presses

When two components have a tight precision fit between them, a hydraulic press is used to either separate or press them together. The component being pressed is placed on the press bed with the appropriate supports. A hand-operated hydraulic pump is mounted on the side of the press. When the handle of the press is pumped, hydraulic fluid is forced into the cylinder, and a ram is extended against the component on the press bed to complete the pressing operation. A pressure gauge on the press shows the pressure applied to the object. The press frame is designed for a certain maximum pressure, and this pressure must not be exceeded during operation.

## Floor Jack

The lifting plate (pad) on a hydraulic floor jack is placed under the chassis of the vehicle **(Figure 3),** and the jack handle is operated with a pumping action. To release the hydraulic pressure and lower the vehicle, the handle or release lever must be turned slowly. Another design of portable floor jacks uses compressed air. Pneumatic jacks are operated by controlling air pressure at the jack.

Hydraulic floor jacks are classified by the weights they can lift: 1-1/2, 2, and 2-1/2 tons, etc. The maximum lifting capacity of the floor jack is usually written on the jack decal. Never lift a vehicle that exceeds the jack lifting capacity. This action may cause the jack to break or collapse, resulting in vehicle damage or personal injury.

When a vehicle is raised by a jack, it should be supported by safety stands. Never work under a car with only a jack supporting it—always use safety stands. Seals in the jack can leak and allow the vehicle to drop.

## Lift

A floor lift is the safest lifting tool and is able to raise the vehicle high enough to allow you to walk and work under it. The lift arms must be placed under the car manufacturer's recommended lifting points prior to raising a vehicle. Twin posts are used on some lifts, whereas other lifts have a single post. Some lifts have an electric motor that drives a hydraulic pump to create fluid pressure and force the lift upward. Other lifts use air pressure from the shop air supply to force the lift upward. Always be sure that the safety lock is engaged after the lift is raised. When the safety lock is released, a release lever is operated slowly to lower the vehicle.

## Tire Changer

Tire changers are used to demount and mount tires **(Figure 4).** There is a wide variety of tire changers available, and each one has somewhat different operating procedures. Always follow the procedure in the equipment operator's manual and the directions provided by your instructor.

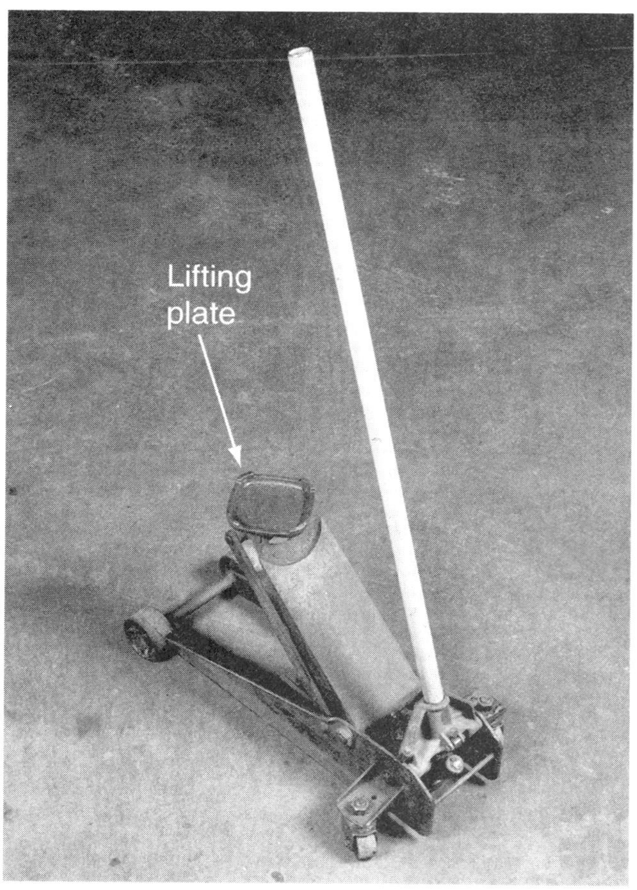


**Figure 3.** A typical hydraulic floor jack.

**Figure 4.** A typical tire changer.

You Should Know *Never operate a tire changer until you are familiar with its operation. This action may lead to serious personal injury.*

## Wheel Balancer—Electronic Type

The most commonly used wheel balancer requires that the tire/wheel assembly be taken off and mounted on the balancer's spindle **(Figure 5).** A switch on the console sets the machine for either static or dynamic balancing. When the wheel balancing assembly is mounted for static balancing, it rotates until the heavy spot falls to the bottom. Weights are added to balance the assembly.

There are several electronic dynamic/static balancer units available that will permit balancing while the wheel and tire are on the car. Often, a strobe light flashes at the heavy point of the tire and wheel assembly. On some machines, two separate tests must be done, one for static and one for dynamic imbalance. One set of weights is placed to correct for static imbalance, and others are placed to correct for dynamic imbalance. Sometimes proper positioning of the static balance weights also corrects dynamic imbalance.

**Figure 5.** A technician operating a tire/wheel balancer.

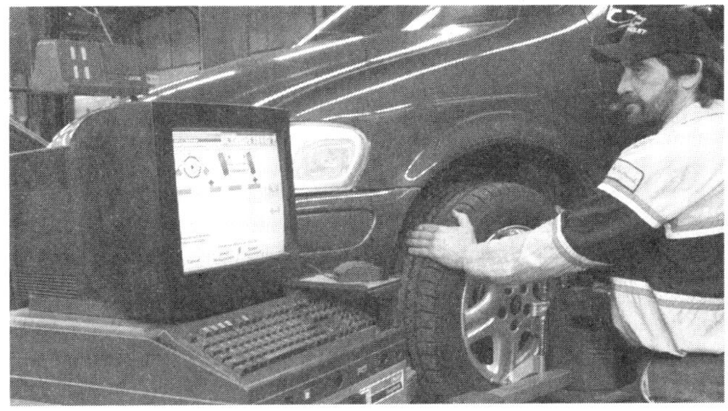

**Figure 6.** A technician checking wheel alignment on a computerized alignment machine.

In the dynamic balance mode, the wheel assembly is rotated at high speed. Observing the balance scale, the operator reads out the amount of weight that has to be added and the location where the weights should be placed.

## Computerized Wheel Aligner

Many automotive shops are equipped with a computerized wheel aligner **(Figure 6)**. These wheel aligners perform all front and rear wheel alignment angles quickly and accurately. A typical computerized system gives information on a screen to guide the technician step by step through the alignment process.

Graphics and text on the screen show the technician where and how to make adjustments. As the adjustments are made on the vehicle, the technician can observe the center block slide toward the target. When the block aligns with the target, the adjustment is within half the specified tolerance.

# SUSPENSION AND STEERING TOOLS

Many different tools and testing and measuring equipment are used to service suspension and steering systems. Often, these tools are for a specific model of vehicle. Most of these tools are listed in the appropriate service manual. Other less model-specific tools are used primarily for servicing suspension and steering systems. These are discussed here.

You Should Know *When using any suspension and steering tools, always follow the vehicle manufacturer's recommended procedure. Improper use of tools may lead to personal injuryImproper use of tools may lead to personal injury.*

## Scan Tools

The introduction of computer-controlled systems brought with it the need for tools capable of troubleshooting electronic control systems. There is a variety of computer scan tools available today that do just that. A **scan tool (Figure 7)** is a microprocessor designed to communicate with the vehicle's computer. Connected to the computer through diagnostic connectors, a scan tool can access trouble codes, run tests to check system operations, and monitor the activity of the system. Trouble codes and test results are displayed on a screen or are printed out on the scanner printer.

The scan tool is connected to specific diagnostic connectors on various vehicles. Most manufacturers have one diagnostic connector. This connects the data wire from each onboard computer to a specific terminal in this connector. Other vehicle manufacturers have several different diagnostic connectors on each vehicle, and each of these connectors may be connected to one or more onboard computers. A set of connectors is supplied with the scan tool to allow tester connection to various diagnostic connectors on different vehicles.

Trouble codes are only set by the vehicle's computer when an input signal is entirely out of its normal range. The codes help technicians identify the cause of the problem when this is the case. If a signal is within its normal range but is still incorrect, the vehicle's computer may not display a trouble code. However, a problem will still exist. As an aid to identify this type of problem, most manufacturers recommend that the signals to and from the computer be carefully looked at. This is done by checking voltage and resistance between specific points within the computer's wiring harness.

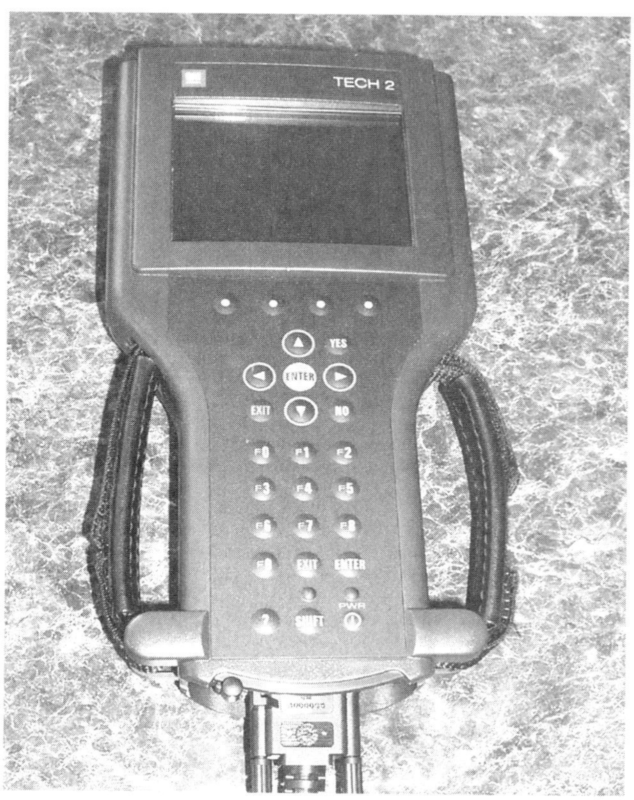


**Figure 7.** A typical scan tool.

## Power Steering Pressure Gauge

A power steering pressure gauge is used to test power steering pump pressure **(Figure 8).** Since the power steering pump delivers extremely high pressure during this test, the recommended procedure in the vehicle manufacturer's service manual must be followed. However, a typical procedure for using a pressure gauge is given here as an example.

To check the pressure of the pump, a pressure gauge with a shut-off valve is needed. With the engine off, disconnect the pressure hose at the pump. Install the pressure gauge between the pump and the steering gear. Use any adapters that may be necessary to make good connections with the vehicle's system. Open the shut-off valve and bleed the system as described in the service manual.

Start the engine and run it for approximately two minutes or until the engine reaches normal operating temperature. Then stop the engine and add fluid to the power steering pump if necessary. Now restart the engine and allow it to idle. Observe the pressure reading. The readings should be about 30 to 80 psi (200 to 550 kPa). If the pressure is lower than what is specified, the pump may be faulty. If the pressure is greater than specifications, the problem may be restricted hoses.

Now close the shut-off valve, observe the pressure reading, and reopen the valve. Do not keep the valve closed

**Figure 8.** A power steering pressure gauge connected to the power steering system.

for more than 5 seconds. With the valve closed, the pressure should have increased to 600 to 1300 psi (4100 to 8950 kPa). Check the pressure reading on the gauge, if the pressure is too high, a faulty pressure relief valve is suggested. If the pressure is too low, the pump may be bad.

## Belt Tension Gauge

A belt tension gauge is used to measure drive belt tension. The belt tension gauge is installed over the belt, and the gauge indicates the amount of belt tension.

## Gear and Bearing Pullers

Many tools are designed for a specific purpose. An example of a special tool is a gear and bearing puller **(Figure 9).** Many gears and bearings have a slight interference fit **(press-fit)** when they are installed on a shaft or in a housing. Something that has a press-fit has an interference fit. For example, the inside diameter of a bore is 0.001 inch smaller than the outside diameter of a shaft. When the shaft is fitted into the bore, it must be pressed in to overcome the 0.001 inch interference fit. This press-fit prevents the parts from moving on each other. The removal of these gears and bearings must be done carefully to prevent damage to the gears, bearings, or shafts. Prying or hammering can break or bind the parts. A puller with the proper jaws and adapters should be used to remove gears and bearings. Using the proper puller, the force required to remove a gear or bearing can be applied with a slight and steady motion.

## Bushing and Seal Pullers and Drivers

Another commonly used group of special tools is the various designs of bushing and seal drivers and pullers. Pullers are either a threaded or slide hammer-type tool. Always make sure you use the correct tool for the job since bushings and seals are easily damaged if the wrong tool or procedure is used. Car manufacturers and specialty tool companies work closely together to design and manufacture special tools required to repair cars. Most of these special tools are listed in the appropriate service manuals.

**You Should Know** ***Never hit the case of a seal with a hammer. This action will damage the seal.***

Seal drivers are designed to fit squarely against the seal case and inside the seal lip. A soft hammer is used to tap the seal driver and drive the seal straight into the housing. Some tool manufacturers market a seal driver kit with drivers to fit many common seals.

## Tie-Rod End and Ball Joint Puller

Ball joint removal and pressing tools are designed to remove and replace pressed-in ball joints on front suspension systems. The size of the removal and pressing tool must match the size of the ball joint. Some car manufacturers recommend a tie-rod end and ball joint puller to remove tie-rod ends and pull ball joint studs from the steering knuckle. A tie-rod end remover **(Figure 10)** is a safer

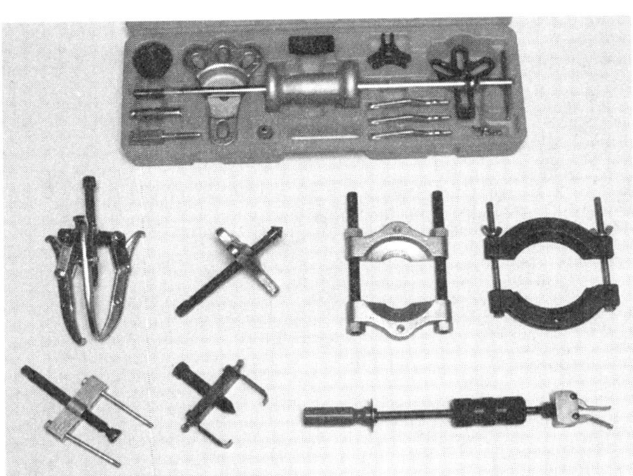

**Figure 9.** A variety of gear and bearing pullers.

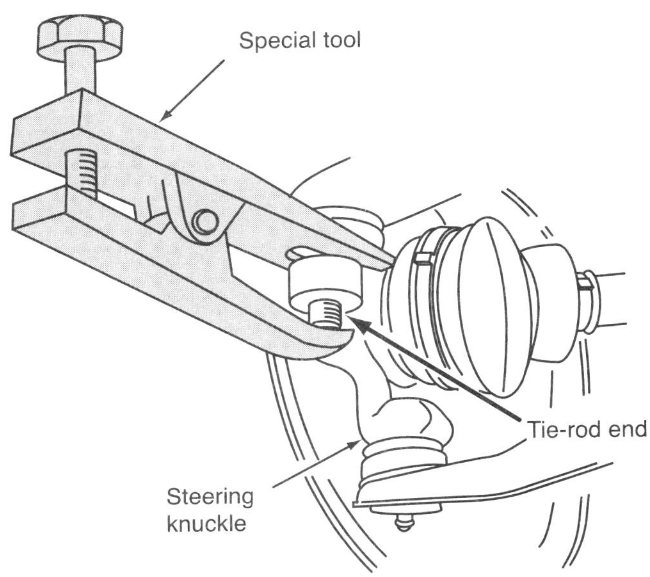


**Figure 10.** A tie-rod end and ball joint puller.

**Figure 11.** A "pickle fork" is often used to separate tie-rod ends and ball joints from the steering knuckle or control arm.

and easier way of separating ball joints than a pickle fork **(Figure 11).**

Some ball joints are riveted to the control arm and the rivets are drilled out for removal.

## Front Bearing Hub Tool

Front bearing hub tools are designed to remove and install front wheel bearings on front-wheel-drive cars. These bearing hub tools are usually designed for a specific make of vehicle, and the correct tools must be used for each application. Failure to do so may result in damage to the steering knuckle or hub. Moreover, the use of the wrong tool will waste quite a bit of your time.

## Special Pliers

Snap- or lock-ring pliers are made with a linkage that allows the movable jaw to stay parallel throughout its range of opening **(Figure 12).** The jaw surface is usually notched or toothed to prevent slippage.

Often, a technician will run into many different styles and sizes of retaining rings that hold subassemblies together or keep them in a fixed location. Using the correct tool to remove and install these rings is the only safe way to work with them. All technicians should have an assortment of retaining ring pliers.

## Stethoscope

Some sounds can be easily heard without using a listening device, but others are impossible to hear unless amplified. A stethoscope is very helpful in locating the cause of a noise by amplifying the sound waves. The procedure for using a stethoscope is simple. Use the metal prod to trace the sound until it reaches its maximum intensity. Once the precise location has been discovered, the sound can be better evaluated. A sounding stick, which is nothing more than a long, hollow tube, works on the same principle, though a stethoscope gives much clearer results.

The best results, however, are obtained with an electronic listening device. With this tool, you can tune into the noise. Doing this allows you to eliminate all other noises that might distract or mislead you.

## Tire Tread Depth Gauge

A tire tread depth gauge measures tire tread depth. This measurement should be taken at three or four locations around the tire's circumference to obtain an average tread depth. This gauge is used to determine the remaining life of a tire, as well as for comparing wear of one tire to the other tires. It is also used when making tire warranty adjustments.

## Wheel Weight Pliers

Wheel weight pliers are actually combination tools designed to install and remove clip-on lead wheel weights. The jaws of the pliers are designed to hook into a hole in the weight's bracket. The pliers are then moved toward the outside of the wheel and the weight is pried off. On one side of the pliers is a plastic hammer head used to tap the weights onto the rim.

## Hand-Held Grease Gun

A hand-operated grease gun forces grease into a grease fitting. These are often preferred because the pressure of the grease can be controlled by the technician. However, many shops use low air pressure to activate a pneumatic grease gun. The suspension and steering system may have several grease or zerk fittings.

## Pitman Arm Puller

A pitman arm puller is a heavy-duty puller designed to remove the pitman arm from the pitman shaft **(Figure 13).**

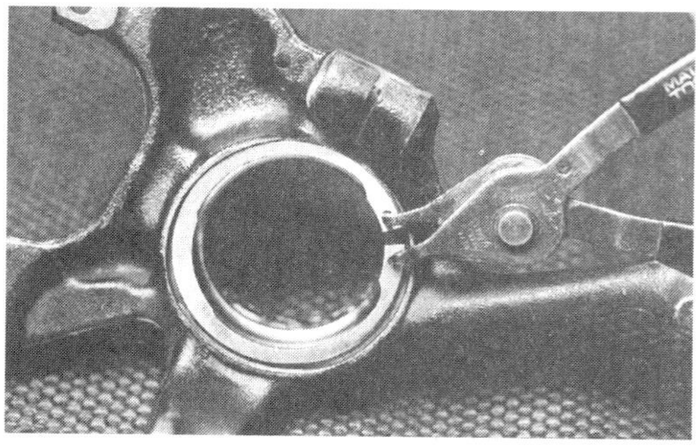

**Figure 12.** Snapring pliers.

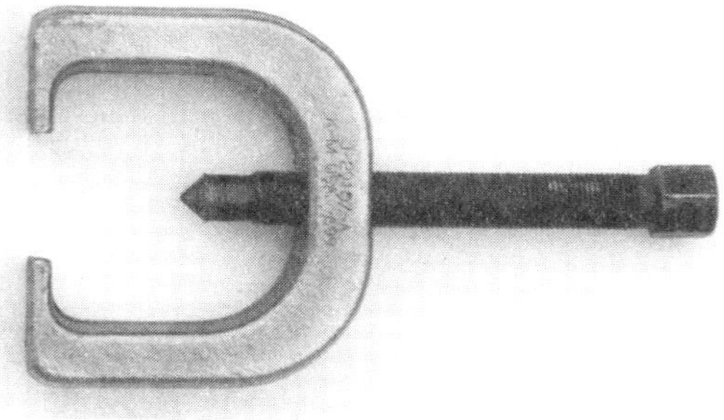

**Figure 13.** A pitman arm puller.

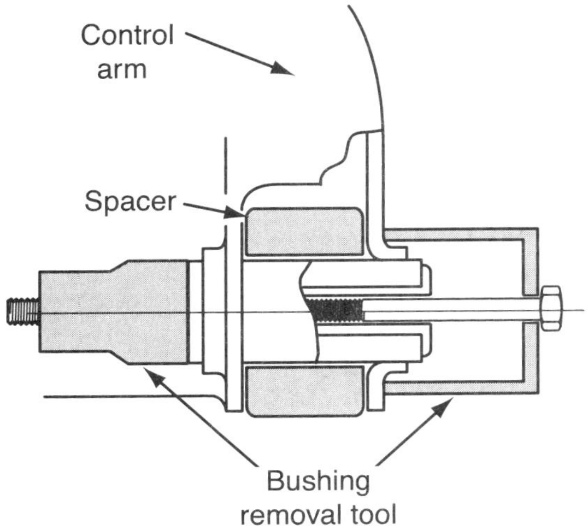


**Figure 14.** A control arm bushing tool.

**Figure 15.** A spring compressor mounted onto a strut assembly.

## Control Arm Bushing Tools

A variety of control arm bushing tools are available to remove and replace control arm bushings. Old bushings are pressed out of the control arm **(Figure 14).** A C-clamp tool can also be used to remove the bushing. The C-clamp is installed over the bushing. An adapter is selected to fit on the bushing and push the bushing through the control arm. Turning the handle on the C-clamp pushes the bushing out of the control arm.

New bushings can be installed by driving or pressing them in place. Adapters are available for the C-clamp tool to install the new bushings. After the correct adapters are selected, position the bushing and tool on the control arm. Turning the C-clamp handle pushes the bushing into the control arm.

## Coil Spring Compressor Tool

There are many types of coil spring compressor tools available to the automotive service industry. These tools are designed to compress the coil spring and hold it in the compressed position while removing the strut from the coil spring **(Figure 15),** removing the spring from a SLA suspension, or performing other suspension work. Various types of spring compressor tools are required on different types of front suspension systems.

One type of spring compressor uses a threaded compression rod that fits through two plates, an upper and lower ball nut, a thrust washer, and a forcing nut. The two plates are positioned at either end of the spring. The compression rod fits through the plates with a ball nut at either end. The upper ball nut is pinned to the rod. The thrust washer and forcing nut are threaded onto the end of the rod. Turning the forcing nut draws the two plates together and compresses the spring.

There is a tremendous amount of energy in a compressed coil spring. Never disconnect any suspension component that will suddenly release this tension. This action may result in serious personal injury and vehicle or property damage.

## Shock Absorber Tools

Often, shock absorbers can be removed with regular hand tools, but there are times when special tools may be necessary. The shocks are under the vehicle and are subject to dirt and moisture, making it difficult to loosen the mounting nut from the stud of the shock. Wrenches are available to hold the stud while attempting to loosen the nut. There are also tools for pneumatic chisels that help to work off the nut.

## Power Steering Pump Pulley Special Tool Set

When a power steering pump pulley must be replaced, it should never be hammered off or on. Doing this will cause internal damage to the pump. Normally, the pulley can be removed with a gear puller, although special pullers are available. To install a pulley, a special tool is used to press the pulley on without a press or the need to drive the pulley in place.

## Steering Column Special Tool Set

A wheel puller is used to remove a steering wheel. Mount the puller over the hub after the horn button and air bag have been removed. Make sure you follow the recommendations for air bag module removal. Attach the puller to the threaded bores in the steering wheel. Then tighten the puller's center bolt against the steering wheel shaft

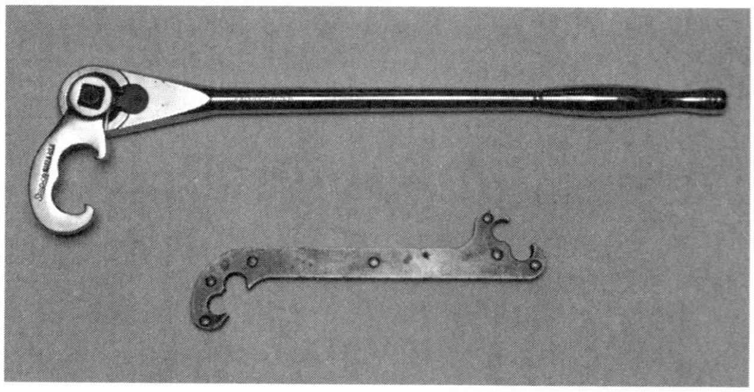

**Figure 16.** Tie-rod sleeve adjusting tools.

until the steering wheel is free. Special tools are also required to service the lock mechanism and ignition switch.

## Tie-Rod Sleeve Adjusting Tool

A tie-rod sleeve adjusting tool is required to rotate the tie-rod sleeves and perform some front wheel adjustments **(Figure 16).** Never use anything except a tie-rod adjusting tool to adjust the tie-rod sleeves. Tools such as a pipe wrench will damage the sleeves.

## Turning Radius Gauge

Turning radius gauge turntables are placed under the front wheels during a wheel alignment. The top plate in the turning radius gauge rotates on the bottom plate to allow the front wheels to be turned during a wheel alignment. A degree scale and a pointer on the gauge indicate the number of degrees the front wheels are turned. If the car has four-wheel steering, the turning radius gauges are also placed under the rear wheels during a wheel alignment.

## Plumb Bob

A plumb bob is a metal weight with a tapered end. This weight is suspended on a string. Plumbers use a plumb bob to locate pipe openings directly below each other at the top and bottom of partitions. Some vehicle manufacturers recommend checking vehicle frame measurements with a plumb bob.

## Tram Gauge

A tram gauge is a long, straight graduated bar with an adjustable pointer at each end. The tram gauge is used for performing frame and body measurements.

## Magnetic Wheel Alignment Gauge

The magnetic wheel alignment gauge is capable of measuring some of the front suspension alignment angles. Each magnetic wheel alignment gauge contains a strong magnet that holds the gauge securely on the front wheel hubs. Magnetic wheel alignment gauge mounting surfaces must be clean with no metal burrs.

## Rim Clamps

When the wheel hub is inaccessible to the magnetic alignment gauge, an adjustable rim clamp may be attached to each front wheel. The magnetic gauges may be attached to the rim clamp. Rim clamps are also used on computer wheel aligners.

## Brake Pedal Depressor

A brake pedal depressor must be installed between the front seat and the brake pedal to apply the brakes while checking some front wheel alignment angles to prevent the vehicle from moving **(Figure 17).**

## Steering Wheel Locking Tool

A steering wheel locking tool is required to lock the steering wheel while performing some front suspension service and during certain steps of the alignment process **(Figure 18).** Its purpose is to prevent the steering wheel and the wheels from turning. Adjusted properly, the steering wheel lock will also help keep the steering wheel centered during alignment.

## Toe Gauge

A toe gauge is a long, straight, graduated bar that may be used to measure front wheel toe.

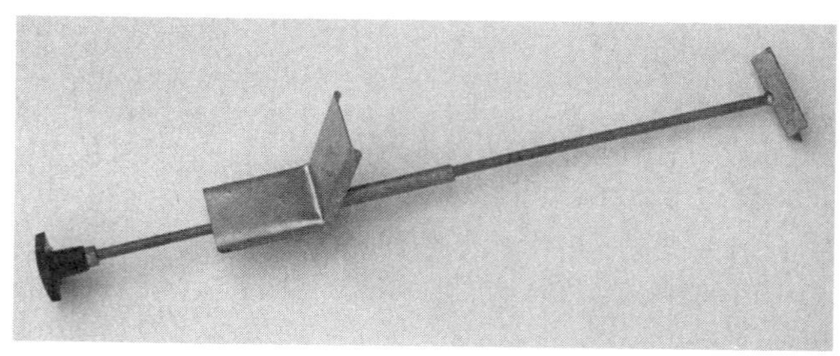

**Figure 17.** A brake pedal depressor (jack).

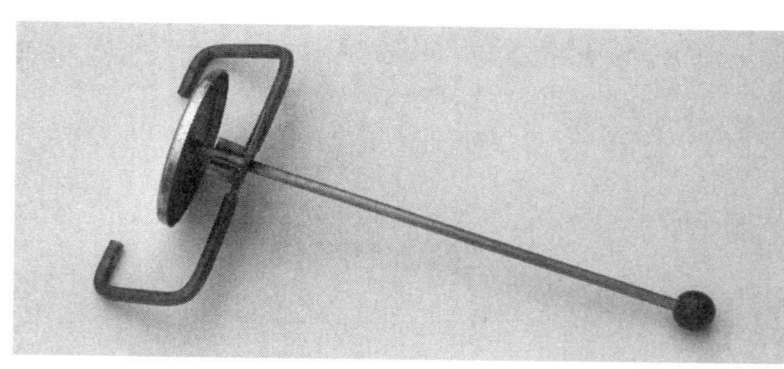

**Figure 18.** A steering wheel locking tool.

## Track Gauge

Some track gauges use a fiber-optic alignment system to measure front wheel toe and to determine if the rear wheels are tracking directly behind the front wheels. The front and rear fiber-optic gauges may be connected to the wheel hubs or to rim clamps attached to the wheel rims. A remote light source in the main control is sent through fiber-optic cables to the wheel gauges. A strong light beam between the front and rear wheel units informs the technician if the rear wheel tracking is correct.

## Axle Pullers

Axle pullers are used to pull rear axles in rear-wheel-drive vehicles. Most rear axle pullers are slide hammer-type.

# SUMMARY

- A technician should have a good assortment of hand tools.
- A bench grinder may have a wire wheel brush, grinding wheel, and/or a buffing wheel.
- Many jobs require the use of a press to assemble or disassemble parts that are press-fit together.
- Service manuals provide the procedures and specifications needed to repair an automobile correctly.
- Flat-rate manuals contain standards for the length of time a specific repair is supposed to require.
- A scan tool is a microprocessor designed to communicate with the vehicle's computer and is used to check the operation of the entire computer system.
- Many different tools and testing and measuring equipment are used to service suspension and steering systems. Often, these tools are for a specific model of vehicle. Most of these tools are listed in the appropriate service manual. Other, less model-specific tools are used primarily for servicing suspension and steering systems.

# REVIEW QUESTIONS

1. What tools should be used to separate tie-rod ends and ball joints from control arms and steering knuckles?
2. Why should you not use an air impact wrench to tighten critical components?
3. Why should you not use regular sockets with an air impact wrench?
4. Describe how most service manuals are divided and how information is organized.
5. While discussing the purpose of torque wrenches, Technician A says they are used to tighten fasteners to a specified torque. Technician B says they are used for added leverage while loosening or tightening a bolt. Who is correct?
   A. Technician A only
   B. Technician B only
   C. Both Technician A and Technician B
   D. Neither Technician A nor Technician B
6. Technician A says one of the most important tools for a technician is a service manual. Technician B says service manuals are not needed if a good selection of TSBs are available. Who is correct?
   A. Technician A only
   B. Technician B only
   C. Both Technician A and Technician B
   D. Neither Technician A nor Technician B
7. Technician A says an open-end wrench should be used to loosen or tighten brake line or tubing fittings. Technician B says flare nut wrenches should only be used on hardened bolts. Who is correct?
   A. Technician A only
   B. Technician B only
   C. Both Technician A and Technician B
   D. Neither Technician A nor Technician B

# Chapter 4

# Working as a Suspension and Steering Specialist

## INTRODUCTION

To be a successful automotive technician you need to have good training, a desire to succeed, and be committed to become a good technician and a good employee. A good employee works well with others and strives to make the business successful. The required training is not just in the field of automotive. Good technicians **(Figure 1)** need to have good reading, writing, and math skills. These skills will allow you to better understand and use the material found in service manuals and textbooks, as well as provide you with the basics for good communications with customers and others.

Figure 1. Good technicians need to have good reading, writing, and math skills.

### COMPENSATION

Technicians are typically paid according to their abilities. Most often, new or apprentice technicians are paid by the hour. They are learning the trade and the business while being paid. Time is usually spent working with a master technician or doing low-skilled jobs. As an apprentice learns more, he or she can earn more and take on more complex jobs. Once technicians have demonstrated a satisfactory level of skills, they are responsible for their own work and receive higher wages or they can go on **flat rate.**

Flat rate is a pay system in which a technician is paid for the amount of work they do. Each job has a flat rate time **(Figure 2).** Pay is based on that time, regardless of how long it took to complete the job. To explain how this system works, let us look at a technician who is paid $15.00 per flat rate hour. If a job has a flat rate time of 3 hours, the technician will be paid $45.00 for the job, regardless of how long it took to complete it. Experienced technicians beat the flat rate time nearly all of the time. Their weekly pay is based on

**Figure 2.** A flat rate manual.

the time "turned in," not on the time spent. If the technician turns in 60 hours of work in a 40-hour workweek, he or she actually earned $22.50 each hour worked. However, if he or she turned in only 30 hours in the 40-hour week, the hourly pay is $11.25.

The flat rate system favors good technicians that work in a shop that has a large volume of work. The use of flat rate times allows for more accurate repair estimates to the customers. It also rewards skilled and productive technicians.

## EMPLOYER-EMPLOYEE RELATIONSHIPS

When you begin employment, you enter into a business agreement with your employer. A business agreement involves an exchange of goods or services that have value. Although the automotive technician may not have a written agreement with his or her employer, the technician exchanges time, skills, and effort for money paid by the employer. Both the employee and the employer have obligations. The automotive technician's obligations include the following:

1. Productivity: As an automotive technician, you have a responsibility to your employer to make the best possible use of time on the job. Each job should be completed in a reasonable length of time. Employees are paid for their skills, effort, and time.
2. Quality: Each repair job should be a quality job! Work should never be done in a careless manner. Nothing improves customer relations like quality workmanship.
3. Teamwork: The shop staff is a team, and everyone including technicians and management personnel are team members. You should cooperate with and care about other team members. Each member of the team should strive for harmonious relations with fellow workers. Cooperative teamwork helps improve shop efficiency, productivity, and customer relations. Customers may be "turned off" by bickering between shop personnel.
4. Ethics: Employers and customers expect and deserve honesty from automotive technicians. Honesty creates a feeling of trust among technicians, employers, and customers.
5. Loyalty: As an employee, you are obliged to act in the best interests of your employer, both on and off the job.
6. Attitude: Employees should maintain a positive attitude at all times. As in other professions, automotive technicians have days when it may be difficult to maintain a positive attitude. For example, there will be days when the technical problems on a certain vehicle are difficult to solve. However, a negative attitude certainly will not help the situation! A positive attitude has a positive effect on the job situation as well as on the customer and employer.
7. Professionalism: You are responsible for your conduct on the job and work-related obligations. These obligations include maintaining good workmanship and customer relations. Pay attention to details, such as always placing fender and seat covers on vehicles before driving or working on them.
8. Following directions: All of us like to do our jobs "our way." Such action, however, may not be in the best interests of the shop, and, as an employee, you have an obligation to follow the supervisor's directions.
9. Punctuality and regular attendance: Employees have an obligation to show up and be on time for work. It is very difficult for a business to operate successfully if it cannot count on its employees to be on the job at the appointed time.
10. Regulations: Automotive technicians should be familiar with all state and federal regulations pertaining to their job situation, such as OSHA and hazardous waste disposal laws. In Canada, employees should be familiar with WHMIS.

### Employer to Employee Obligations

1. Wages: The employer has a responsibility to inform the employee regarding the exact amount of financial remuneration they will receive and when they will be paid.
2. Fringe benefits: A detailed description of all fringe benefits, including holiday pay, sickness and accident insurance, and pension plans.
3. Working conditions: A clean, safe workplace must be provided by the employer. The shop must have adequate safety equipment and first aid supplies. Employers must be certain that all shop personnel maintain the shop area and equipment to provide adequate safety and a healthy workplace atmosphere.
4. Employee instruction: Employers must provide employees with clear job descriptions and be sure that each worker is aware of his or her obligations.

5. Employee supervision: Employers should inform their workers regarding the responsibilities of their immediate supervisors and other management personnel.
6. Employee training: Employers must make sure that each employee is familiar with the safe operation of all the equipment that they are required to use in their job situation. Since automotive technology is changing rapidly, employers should provide regular update training for their technicians. Under the right-to-know laws, employers are required to inform all employees about hazardous materials in the shop. Employees should be familiar with MSDS, which detail the labeling and handling of hazardous waste, and the health problems if exposed to hazardous waste.

## Job Responsibilities

An automotive technician has specific responsibilities regarding each job performed on a customer's vehicle. These job responsibilities include:

1. Do every job to the best of your ability. There is no place in the automotive service industry for careless workmanship! Automotive technicians and students must realize they have a very responsible job. During many repair jobs, you have the customer's life and the safety of his or her vehicle in your hands. Even when careless work does not create a safety hazard, it leads to dissatisfied customers who often take their business to another shop. Nobody benefits when that happens.
2. Treat customers fairly and honestly on every repair job. Do not install parts that are unnecessary to complete the repair job.
3. Use published specifications; do not guess at adjustments.
4. Follow the service procedures in the service manual provided by the vehicle manufacturer or an independent manual publisher.
5. When the repair job is completed, always be sure the customer's complaint has been corrected.
6. Do not be too concerned with work speed when you begin working as an automotive technician. Speed comes with experience.

## Customer Relations

Learn to listen and communicate clearly. Be polite and organized, particularly when dealing with customers. Always be as honest as you possibly can.

Look and present yourself as a professional. Professionals are proud of what they do and they show it. Always dress and act appropriately and watch your language, even when you think no one is near.

Respect the vehicles you work on. They are important to the lives of your customers. Always return the vehicle to the owner in a clean, undamaged condition. Remember, a car is the second largest expense a customer has—treat it that way. It does not matter if you like the car. It belongs to the customer; treat it respectfully.

**Figure 3.** Do your best to explain the problem and services to your customers.

Explain the repair process to the customer in understandable terms **(Figure 3).** Whenever you are explaining something to a customer, make sure you do this in a simple way without making the customer feel stupid. Always show the customers respect and be courteous to them. Not only is this the right approach to take, but it also leads to loyal customers.

## AUTOMOTIVE SERVICE EXCELLENCE (ASE) CERTIFICATION

An obvious sign of your knowledge and abilities, as well as your dedication to the trade, is ASE certification. The National Institute for Automotive Service Excellence (ASE) has established a voluntary certification program for automotive, heavy-duty truck, auto body repair, and engine machine shop technicians. In addition to these programs, ASE also offers individual testing in some specialty areas. This certification system combines voluntary testing with on-the-job experience to confirm that technicians have the skills needed to work on today's vehicles **(Figure 4).** ASE

**Figure 4.** A master automotive technician with his certificate and arm patch.

recognizes two distinct levels of service capability—the automotive technician and the master automotive technician.

To become ASE certified, you must pass one or more tests that stress diagnostic and repair problems. Steering and Suspension Systems is one of the eight basic automotive certification areas.

After passing at least one exam and providing proof of two years of hands-on work experience, you become ASE certified. Retesting is necessary every five years to remain certified. A technician who passes one examination receives an automotive technician shoulder patch. The master automotive technician patch is awarded to technicians who pass all eight of the basic automotive certification exams.

You may receive credit for one of the two years by substituting relevant formal training in one, or a combination, of the following:

- High school training. Three years of training may be substituted for one year of experience.
- Post-high school training. Two years of post-high school training in a trade school, technical institute, or community college may be counted as one year of work experience.
- Short courses. For shorter periods of post-high school training, you may substitute two months of training for one month of work experience.
- Apprenticeship programs. You may receive full credit for the experience requirement by satisfactorily completing a three- or four-year apprenticeship program.

Each certification test consists of forty to eighty multiple-choice questions. The questions are written by a panel of technical service experts, including domestic and import vehicle manufacturers, repair and test equipment and parts manufacturers, working automotive technicians, and automotive instructors. All questions are pre-tested and quality checked on a national sample of technicians before they are included in the actual test. Many test questions force the student to choose between two distinct repair or diagnostic methods. The test questions focus on basic technical knowledge, repair knowledge and skill, and testing and diagnostic knowledge and skill.

## DUTIES OF A SUSPENSION AND STEERING SPECIALIST

As a specialist, you will be responsible for servicing, troubleshooting, and repairing steering and suspension systems and all of the components that affect their operation. Although the most common service is a complete wheel alignment, parts do wear and will need to be replaced. Often these worn parts are discovered during the wheel alignment, but these faulty parts may be discovered during the diagnosis of a customer concern.

## Vehicle Identification

Before any service is done to a vehicle, it is important for you to know exactly what type of vehicle you are working on. The best way to do this is to refer to the vehicle's identification number **(VIN).** The VIN is on a plate behind the lower corner of the driver's side of the windshield, as well as other locations in the vehicle. The VIN is made up of seventeen characters and contains all pertinent information about the vehicle **(Figure 5).** The use of the seventeen number and letter code became mandatory beginning with 1981 vehicles and is used by all manufacturers of both domestic and foreign vehicles.

Each character of a VIN has a particular purpose. The first character identifies the country where the vehicle was manufactured, for example:

- 1 or 4–U.S.A.
- 2–Canada
- 3–Mexico
- J–Japan
- K–Korea
- S–England
- W–Germany

The second character identifies the manufacturer, for example:

- A–Audi
- B–BMW
- C–Chrysler
- D–Mercedes Benz
- F–Ford
- G–General Motors

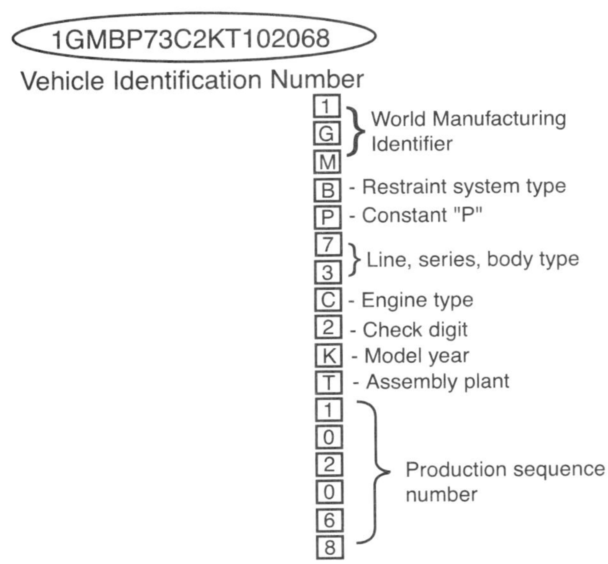


**Figure 5.** A simple explanation of the information contained in a VIN.

- H–Honda
- N–Nissan
- T–Toyota

The third character identifies the vehicle type or manufacturing division (passenger car, truck, bus, etc.). The fourth through eighth characters identify the features of the vehicle, such as the body style, vehicle model, engine type, etc.

The ninth character is used to identify the accuracy of the VIN and is a check digit. The tenth character identifies the model year, for example:

- Y–2000
- 1–2001
- 2–2002
- 3–2003
- 4–2004
- 5–2005

The eleventh character identifies the plant where the vehicle was assembled, and the twelfth to seventeenth characters identify the production sequence of the vehicle as it rolled of the manufacturer's assembly line.

The specifics needed for decoding the characters of the VIN can be found in a service manual for the vehicle.

In addition to the VIN, all vehicles produced since 1972 have an underhood emission control label and/or calibration decal **(Figure 6).** This label gives useful information including engine specifications, a description of the emission control devices, and vacuum hose routing.

## Diagnostics

The true measure of a good technician is an ability to find and correct the cause of problems. Service manuals and other information sources will guide you through the diagnosis and repair of problems. However, those guidelines will not always lead you to the exact cause of the problem. To do this, you must use your knowledge and take a logical approach while troubleshooting. Diagnosis is not guessing and it is more than following a series of interrelated steps in order to find the solution to a specific problem. Diagnosis is a way of looking at systems that are not functioning the way they should and finding out why. It is knowing how the system should work and deciding if it is working correctly. Through an understanding of the purpose and operation of the system, you can accurately diagnose problems.

Most good technicians use the same basic diagnostic approach. Simply because this is a logical approach, it can quickly lead to the cause of a problem. Logical diagnosis follows these steps:

1. Gather information about the problem.
2. Verify that the problem exists.
3. Thoroughly define what the problem is and when it occurs.
4. Research all available information to determine the possible causes of the problem.
5. Isolate the problem by testing.
6. Continue testing to pinpoint the cause of the problem.
7. Locate and repair the problem, then verify the repair.

Vehicle Emission Control Information - XXXXXXX
Engine Family XXXXXXXXX
Evaporative Family XXXX
Displacement XXXXcm$^3$
CATALYST
Tune-up specifications
Tune-up conditions
Engine at normal operating temperature, all accessories turned off, cooling fan off, transmission in neutral
Adjustments to be made in accordance with indications given in shop manual

| | | |
|---|---|---|
| Idle speed | 5 speed transmission | 800 ± 50 rpm |
| | Automatic transmission | 800 ± 50 rpm |
| Ignition timing at idle | | 15 ± 2° BTDC |
| Valve lash | Setting points between camshaft and rocker arm | In. 0.17 ± 0.02 mm cold |
| | | Ex. 0.19 ± 0.02 mm cold |
| No other adjustments needed | | |

**Figure 6.** An example of an underhood emission control label.

# CLEANING WITH CHEMICALS

Parts cleaning is an important part of any repair job. You will often need to clean parts to find problems and measure for wear. If you must disassemble the brakes to service the steering or suspension system, make sure you use the appropriate equipment because of the danger of asbestos exposure.

Face shields and protective gloves must be worn while operating cleaning equipment. The solution in hot and cold cleaning tanks may be caustic, and contact between this solution and skin or eyes must be avoided. When the cleaning solution in hot or cold cleaning tanks is replaced, environmental regulations require that the old solution be handled as hazardous waste. Use caution when placing aluminum or aluminum alloy parts in a cleaning solution. Some cleaning solutions will damage these components. Always follow the cleaning equipment manufacturer's recommendations.

You Should Know

***Vapors from many cleaning solvents may be toxic if inhaled, particularly in large quantities for prolonged periods. Use cleaning solvents only in well-ventilated areas. Wear a respirator and avoid inhaling vapors.***

## Cleaning Equipment Safety

Some parts washers provide electromechanical agitation of the parts to provide improved cleaning action. These parts washers may be heated with gas or electricity. Various water-based hot tank cleaning solutions are available depending on the type of metals being cleaned. Many cleaning solutions contain no ingredients listed as hazardous by the EPA's RCRA Act. This cleaning solution is a blend of sulfur-free hydrocarbons, wetting agents, and detergents. Always use the cleaning solution recommended by the equipment manufacturer.

An aqueous parts cleaning tank uses a water-based environmentally friendly cleaning solution. The immersion tank is heated and agitated for effective parts cleaning. A sparger bar pumps a constant flow of cleaning solution across the surface to push floating oils away, and an integral skimmer removes these oils. This action prevents floating surface oils from redepositing on cleaned parts.

## Chemical Precautions

Some solvents can be absorbed through the skin. This is especially true if you have a cut. Never use compressed air on your hands if they become wet with solvent; this can cause the solvent to penetrate into your skin. Always wear neoprene gloves when washing parts.

Solvents can be ingested by a smoker who lights a cigarette while working with the solvent. Solvents must always be handled carefully and kept in properly labeled containers. When not in use, the containers must be stored away from untrained personnel or children.

Contact with solvents occur most often through inhalation or absorption through your skin. Inhalation is the most serious and has a more immediate effect. Absorption can be just as dangerous, however, and its effects may not be noticeable for several days after exposure.

Exposure to chlorinated hydrocarbons and other solvents can cause nausea, drowsiness, headache, dizziness, and eventually unconsciousness. Prolonged exposure can lead to liver and kidney damage.

Basic common sense will go a long way toward protecting you from the hazards of chemical poisoning.

## Wheel Bearing Grease

Wheel service often includes repacking front or rear wheel bearings. Although some multipurpose greases can be used for both chassis lubrication and for wheel bearings, you must be sure that whatever grease you use is identified as suitable for wheel bearing lubrication.

All greases are made from oils blended with thickening agents so that the grease will stick to the surfaces to be lubricated. Greases are identified by the **National Lubricating Grease Institute (NLGI)** number and by the kinds of thickeners and additives that the grease contains. The higher the NLGI number, the higher viscosity (thicker) the grease is. Almost all wheel bearing and chassis greases are NLGI #2 greases. Most wheel bearing grease uses a lithium-based thickener for temperature resistance. Molybdenum disulfide is another common additive that improves the antiseize properties of wheel bearing grease.

There is no way to tell what kind of grease was used previously on a wheel bearing. Therefore, wipe away all old grease thoroughly when repacking wheel bearings, and use only new grease that meets the vehicle maker's specifications for wheel bearing service.

# SUMMARY

- Technicians are typically paid according to their abilities. New or apprentice technicians are paid by the hour. Once technicians have demonstrated a satisfactory level of skills, they can earn a higher hourly wage or go on flat rate.
- When you begin a job, you enter into a business agreement with your employer. When you become an employee, you sell your time, skills, and efforts. In return, your employer pays you for these resources.
- As part of the employment agreement, your employer also has certain responsibilities: instruction and supervision, clean, safe place to work, wages, fringe benefits, opportunity, and fair treatment.
- Your obligations as an employee to the employer include: regular attendance, following directions, responsibility, productivity, and loyalty.
- When dealing with customers, be polite, respectful, organized, and honest.
- An obvious sign of your knowledge and abilities, as well as your dedication to the trade, is ASE certification.
- The true measure of a good technician is the ability to find and correct the cause of problems.
- Diagnosis is not guessing, and it is more than following a series of interrelated steps in order to find the solution to a specific problem.
- All precautions should be followed before using solvents to clean parts.

## REVIEW QUESTIONS

1. What must a technician do in order to become certified as an ASE Master Automobile Technician?
2. Explain the flat rate pay system.
3. What are your responsibilities as an employee to your employer?
4. List the steps that should be followed while logically diagnosing a problem.
5. Technician A says that after an individual passes a particular ASE certification exam, he or she is certified in that test area. Technician B says all of the questions on an ASE certification exam are written as Technician A and Technician B questions. Who is correct?
   A. Technician A only
   B. Technician B only
   C. Both Technician A and Technician B
   D. Neither Technician A nor Technician B

# Section 2

## Basic Theories and Services



**Interesting Fact**

*Suspension systems evolved slowly. In the 1940s, many front suspensions were changed from I-beam to short-and-long arm designs. MacPherson strut suspensions re-placed the short-and-long arm suspension systems on front-wheel-drive cars introduced in the late 1970s and 1980s. Since then there has been an everexpanding use of electronics in automotive suspension systems. Some computer-controlled suspension systems can react to road or driving conditions in 10 to 12 milliseconds to improve ride and handling quality. Electronics will continue to improve suspension systems so they provide better handling characteristics and safety with longer tire life.*

## SECTION OBJECTIVES

After you have read, studied, and practiced the contents of this section, you should be able to:

- Explain the basic principles of electricity.
- Explain how to use Ohm's law to determine the amount of current, resistance, or voltage in a circuit.
- Name the various electrical components and their uses in electrical circuits.
- Diagnose electrical problems by logic and symptom description.
- Perform troubleshooting procedures using meters, test lights, and jumper wires.
- Inspect, test, and replace basic electrical components.
- Repair electrical wiring and connectors.
- List and describe the operation of the basic parts of suspension and steering systems.
- Define atoms and elements.
- Define compounds and molecules.
- Explain Newton's Laws of Motion.
- Define work and force.
- Define power.
- Define inertia and momentum.
- Explain friction.
- Define mass, weight, and volume.
- Compare static and dynamic unbalance.
- Explain the compressibility of gases and the non-compressibility of liquids.
- Explain how hydraulics can be used to transmit force and motion.
- Explain how vacuum and atmospheric pressure can be used to increase force.
- Identify and interpret suspension and steering concerns.
- Lubricate suspension and steering systems.
- Diagnose vehicle wander, drift, pull, hard steering, bump steer, memory steer, torque steer, and steering return concerns.
- Measure vehicle riding height.

# Chapter 5 Basic Electrical Theory

## INTRODUCTION

A vehicle is fitted with sensors that give vital information to computers that control the systems of the automobile. Some functions of some suspension and steering systems are computer-controlled as well. To understand the operation of these, you must have a good understanding of electricity and electronics. Although the subject is normally covered in a separate course, a quick overview of electricity and its principles is presented here.

### BASIC ELECTRICITY

Electricity is caused by the flow of electrons from one atom to another. The release of energy as one electron leaves the orbit of one atom and jumps into the orbit of another is **electricity.** The key behind creating electricity is to give a reason for the electrons to move.

There is a natural attraction of electrons to protons. Electrons have a negative charge and are attracted to something with a positive charge. An electron moves from one atom to another because the atom next to it appears to be more positive than the one it is orbiting around. An electrical power source provides for a more positive charge and, in order to allow for a continuous flow of electricity, it supplies free electrons. In order to have a continuous flow of electricity, three conditions must be present: an excess of electrons in one place, a lack of electrons in another place, and a path between the two places.

Two power or energy sources are used in an automobile's electrical system—the vehicle's battery **(Figure 1)** and AC generator **(Figure 2).** The chemical reaction in a

**Figure 1.** A typical automotive battery.

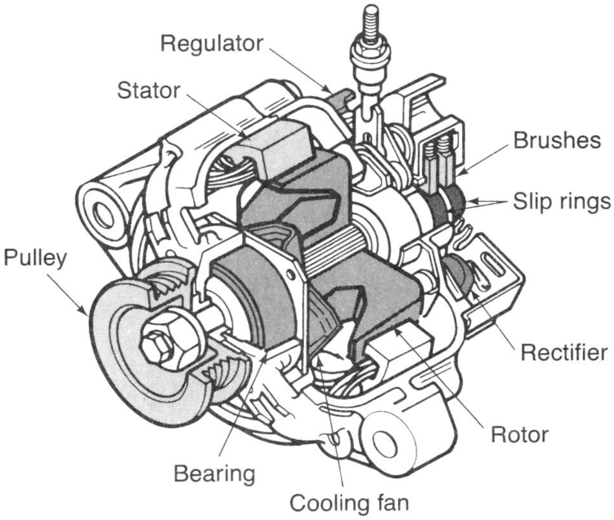


**Figure 2.** An AC generator (alternator).

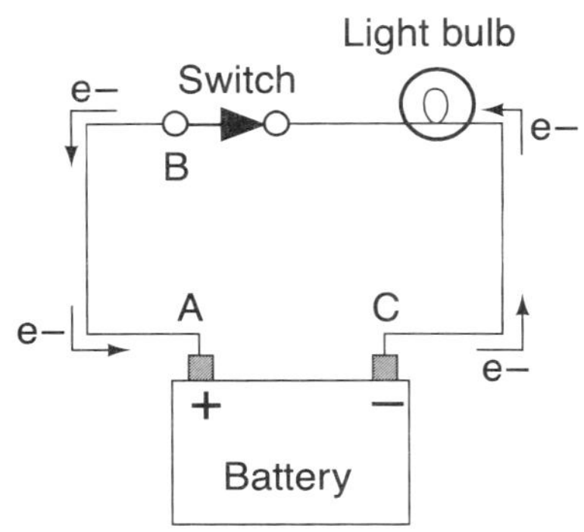


**Figure 3.** A simple light circuit.

battery causes a lack of electrons at the positive (+) terminal and an excess at the negative (B) terminal. This creates an electrical imbalance, causing the electrons to flow through the path provided by a wire. A simple example of this process is shown in the battery and light arrangement in **Figure 3.**

## ELECTRICAL TERMS

Electrical **current** is a term used to describe the movement or flow of electricity. The greater the number of electrons flowing past a given point in a given amount of time, the more current there is in the circuit. **Voltage** is electrical pressure. Voltage is the force developed by the attraction of the electrons to the protons. The more positive one side of the circuit is, the more voltage is available to the circuit. Voltage does not flow rather it is the pressure that causes flow. When any substance flows, it meets resistance. The resistance to electrical flow can be measured.

### Electrical Current

The unit for measuring electrical current is the **ampere,** usually called an amp. The instrument used to measure electrical current flow in a circuit is called an **ammeter.**

When electricity flows, millions of electrons are moving past any given point at the speed of light. The electrical charge from any one moving electron is extremely small. It takes millions of electrons to make a charge that can be measured.

There are two types of current: **direct current** (DC) and **alternating current** (AC). In direct current, the electrons flow in one direction only. In alternating current, the electrons change direction at a fixed rate. Most automobile circuits operate on DC current while the current in homes and buildings is AC.

### Resistance

In every atom, the electrons resist being moved out of their shell. The amount of resistance depends on the type of atom. As explained earlier, in some atoms there is very little resistance to electron flow because the outer electron is loosely held. In other substances, there is more resistance to flow because the outer electrons are tightly held.

The resistance to current flow produces heat. A unit of measured resistance is called an **ohm.** Resistance can be measured by an instrument called an **ohmmeter.**

### Voltage

In electrical flow, some force is needed to move the electrons from one atom to another. This force is the pressure that exists between a positive and negative point within an electrical circuit. This force, also called electromotive force **(EMF),** is measured in units called **volts.** One volt is the amount of pressure (force) required to move 1 ampere of current through a resistance of 1 ohm. Voltage is measured by an instrument called a **voltmeter.**

## ELECTRICAL CIRCUITS

When electrons are able to flow along a path (wire) between two points, an electrical circuit is formed. An electrical circuit is considered a **complete circuit** when there is a path that connects the positive and negative terminals of the electrical power source. In a complete circuit, resistance must be low enough to allow the available voltage to push electrons between the two points. Most automotive circuits contain five basic parts.

- A power source that provides the energy needed to create electron flow.
- Conductors, such as copper wires, that provide a path for the electrons.
- **Loads** or devices that use electricity to perform work, such as lights and motors.
- **Controllers,** such as switches or relays, that direct the flow of electrons.
- A circuit protection device that protects the circuit from damage if there is an overload on the circuit.

A functioning electrical circuit must have a complete path from the power source to the load and back to the source. With the many circuits in an automobile, this would require hundreds of wires connected to both sides of the battery. To avoid this, vehicles have power distribution centers or fuse blocks that distribute battery voltage to the various circuits. The positive side of the battery is connected to the fuse block and power is distributed from there.

As a common return circuit, the vehicle's metal frame **(Figure 4)** is used as part of the return circuit. The loads are often grounded directly to the metal frame. Current passes from the battery, through the load, and into the frame. The

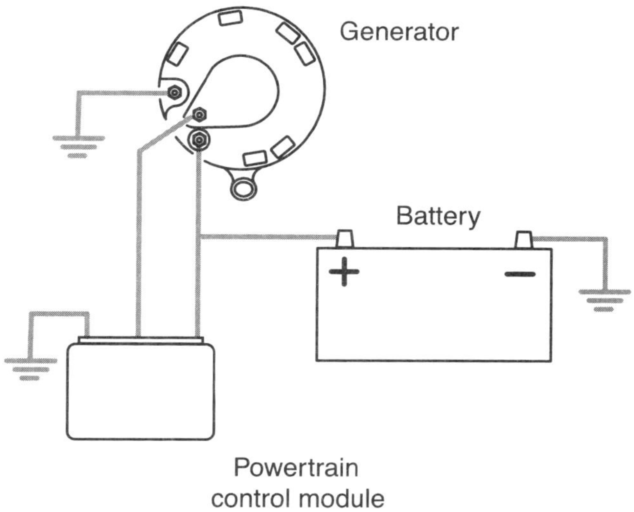


**Figure 4.** To eliminate the need to run separate return wires, many accessories and automotive components are grounded to the vehicle's chassis.

frame is connected to the battery through the battery's ground cable. This completes the circuit.

An electrical component is often mounted directly to the engine block, transmission case, or frame. This grounds the component without the use of a separate ground wire. In other cases, a separate ground wire must be run from the component to the frame or another metal part to ensure a sound return path. The increased use of plastics and other nonmetallic materials in body panels and engine parts has made electrical grounding more difficult. To ensure good grounding, some manufacturers use common grounding terminals and wires.

## CONDUCTORS AND INSULATORS

Controlling and routing electricity requires the use of materials known as conductors and insulators. **Conductors** are materials with a low resistance to the flow of current. Most metals, such as copper, silver, and aluminum are excellent conductors.

Copper wire is by far the most popular conductor used in automotive electrical systems. Wire wound inside of electrical units, such as ignition coils and generators, usually has a very thin baked-on insulating coating. External wiring is often covered with a plastic-type insulating material that is highly resistant to environmental factors like heat, vibration, and moisture. Where flexibility is required, the copper wire will be made of a large number of very small strands of wire woven together.

**Insulators** resist the flow of current. Thermal plastics are the most common electrical insulators used today. They can resist heat, moisture, and corrosion without breaking down.

## OHM'S LAW

To understand the relationship between current, voltage, and resistance, you should understand Ohm's Law. This law states that it takes 1 volt of electrical pressure to push 1 ampere of electrical current through 1 ohm of resistance. The law provides a mathematical formula for determining the amount of current, voltage, or resistance in a circuit when two of these are known.

Although the basic premise of this formula is calculating unknown values in an electrical circuit **(Figure 5),** it also helps to define the behaviors of electrical circuits.

If voltage does not change but there is a change in the resistance of the circuit, the current will change. If resistance increases, current decreases. If resistance decreases, current will increase. If voltage changes, so must the current or resistance. If the resistance stays the same and the current decreases, so will voltage. Likewise, if current increases, so will the voltage.

In a complete circuit, the flow of electricity is controlled and applied to do useful work. Components that use electrical power put a load on the circuit and consume electrical energy. The energy used by a load is measured in volts and is called **voltage drop.**

## CIRCUIT COMPONENTS

Automotive electrical circuits contain a number of different types of electrical devices.

### Resistors

Resistors are used to limit current flow in circuits where full current flow and voltage are not needed. Resistors are

Voltage (E) = Current (I) times Resistance (R), therefore

$$E = I \times R.$$

Current (I) = Voltage (E) divided by Resistance (R), therefore

$$I = E/R.$$

Resistance (R) = Voltage (E) divided by Current (I), therefore

$$R = E/I.$$

**Figure 5.** Ohm's Law.

constructed to introduce a measured amount of electrical resistance into a circuit. Fixed value resistors are designed to have only one rating, which should not change.

Tapped or stepped resistors are designed to have two or more fixed values, available by connecting wires to the several taps on the resistor. Heater motor resistor packs, which provide for different heater fan speeds, are an example of this type of resistor.

Variable resistors are designed to have a range of resistances available through two or more taps and a control. **Rheostats** have two connections **(Figure 6),** one to the fixed end of a resistor and one to a sliding contact with the resistor. Turning the control moves the sliding contact away from or toward the fixed end tap, increasing or decreasing the resistance. **Potentiometers** have three connections **(Figure 7),** one at each end of the resistance and one connected to a sliding contact with the resistor. Turning the control moves the sliding contact away from one end of the resistance but toward the other end.

Another type of variable resistor is the **thermistor.** Thermistors are designed to change in value as their temperature changes. As a temperature sender, a thermistor is connected to a voltmeter calibrated in degrees. As the temperature rises or falls, the resistance also changes. This changes the reading on the meter.

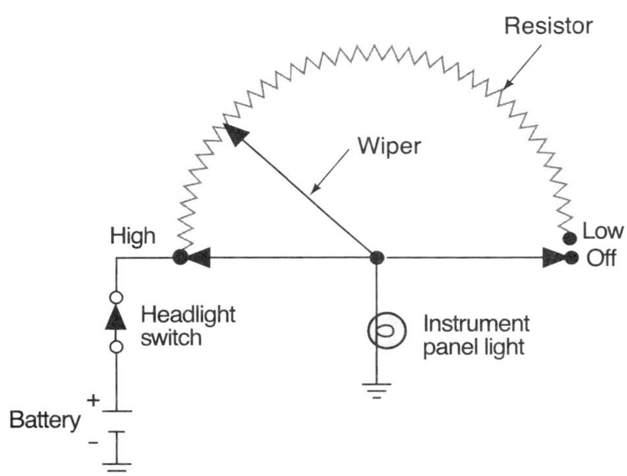


**Figure 6.** A rheostat.

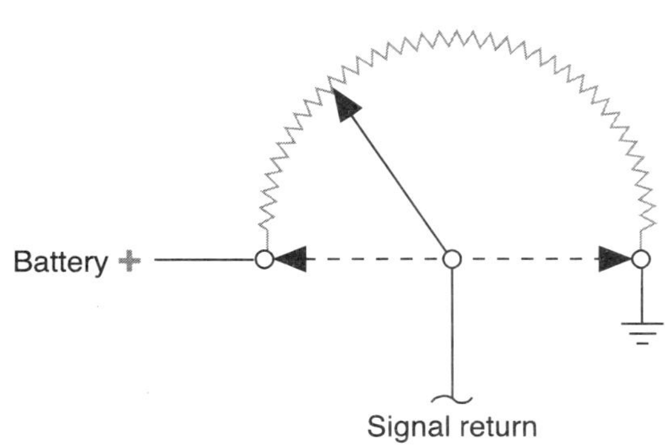


**Figure 7.** A potentiometer.

## Circuit Protective Devices

When overloads or shorts in a circuit cause excessive current flow, the wiring in the circuit heats up, the insulation melts, and a fire can result unless the circuit has some kind of protective device. Fuses, fuse links, maxi-fuses, and circuit breakers are designed to prevent circuit damage caused by high current. Protection devices open the circuit when high current is present and the wiring and components are saved from damage.

## Switches

Electrical circuits are usually controlled by some type of switch. Switches turn the circuit on or off or direct the flow of current in a circuit. Switches can be controlled by the driver or can be self-operating through a condition of the circuit, the vehicle, or the environment. When the switch is before the load, it is controlling voltage to it. When the switch is after the load, the switch is controlling the ground. In both cases, the switch is opening and closing the circuit.

A hinged-pawl switch is the simplest type of switch. It either makes or breaks a single conductor or circuit. It is a single-pole, single-throw **(SPST)** switch. The throw refers to the number of output circuits, and the pole refers to the number of input circuits made by the switch.

Switches can be designed with a great number of poles and throws. Single-pole, double-throw switches have one wire in and two wires out. This type of switch allows the driver to select between two circuits, such as high-beam or low-beam headlights. The transmission neutral start switch may have two poles and six throws and is referred to as a multiple-pole, multiple-throw **(MPMT)** switch **(Figure 8).** It closes the circuit to the starter in either P (park) or N (neutral) and to the back-up lights in R (reverse).

A temperature-sensitive switch, used as a sensor, contains a bimetallic element and is heated electrically or by some component. When engine coolant is at normal or below operating temperature, the engine coolant temperature sensor is in its normally open condition. If the coolant exceeds the temperature limit, the bimetallic element bends the two contacts together and the switch is closed to the indicator or the instrument panel. As a result, the indicator lamp is illuminated.

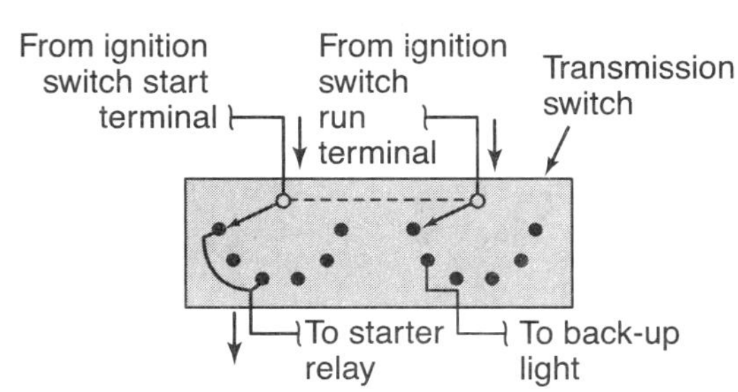


**Figure 8.** A multiple-pole, multiple-throw neutral start safety switch.

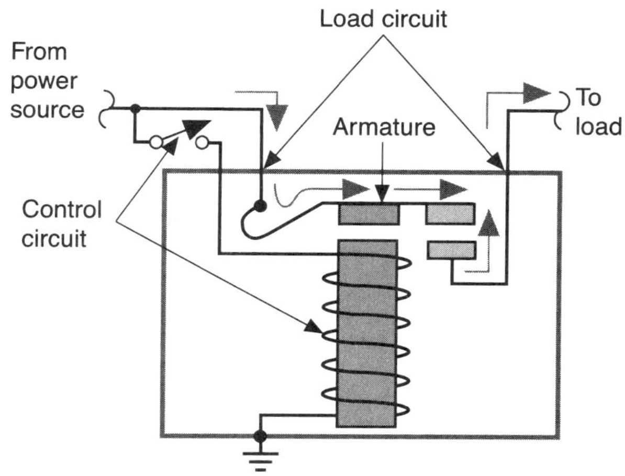


**Figure 9.** A typical relay.

## Relays

A **relay (Figure 9)** is an electric switch that allows a small amount of current to control a circuit with high current. The low-current circuit is called the control circuit. When the control circuit switch is open, no current flows to the coil in the relay, so the coil windings are de-energized. When the switch is closed, the coil is energized, turning the soft iron core into an electromagnet and drawing the armature down. This closes the high current circuit contacts and connects power to the load circuit. When the control switch is opened, the current stops and the electromagnet field disappears. This causes the release of the armature and the circuit is opened.

## Solenoids

**Solenoids** are electromagnetic devices with movable cores and are used to translate electrical current flow into mechanical movement. The movement of the core causes a lever to move or close electrical contacts, acting as a relay at the same time.

# ELECTROMAGNETISM BASICS

Electricity and magnetism are related. One can be used to create the other. Current flowing through a wire creates a magnetic field around the wire. Moving a wire through a magnetic field creates current flow in the wire.

Many automotive components, such as generators, ignition coils, starter solenoids, and magnetic pulse generators operate using principles of electromagnetism.

Something is a magnet if it has the ability to attract substances such as iron, steel, or cobalt. A magnet has two points of maximum attraction, one at each end. These points are called poles, one designated the north pole and the other the south pole. When two magnets are brought together, opposite poles attract, while similar poles repel each other.

A magnetic field, called a **field of flux,** exists around every magnet **(Figure 10).** The field consists of imaginary lines along which a magnetic force acts. These lines emerge from the north pole and enter the south pole, returning to the north pole through the magnet itself.

Magnets can occur naturally in the form of a mineral called magnetite. Artificial magnets can also be made by inserting a bar of magnetic material inside a coil of insulated wire and passing a heavy direct current through the coil. Another way of creating a magnet is by stroking the material with a bar magnet. Both methods force the randomly arranged molecules of the magnetic material to align themselves along north and south poles.

## Induced Voltage

Let's look at how magnetic fields can produce electricity. Consider a straight piece of conducting wire with the terminals of a voltmeter attached to both ends. If the wire is moved across a magnetic field, the voltmeter registers a

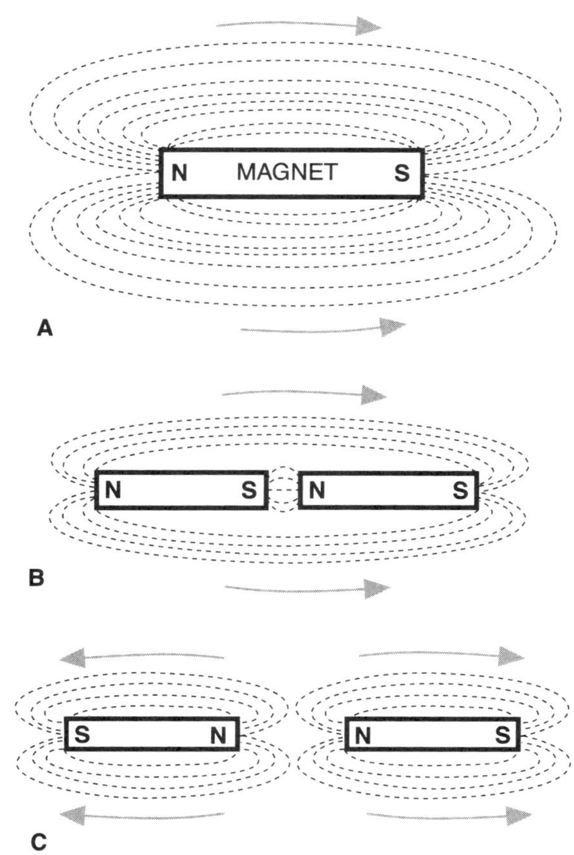


**Figure 10.** Magnetic principles: (A) field of flux around a magnet, (B) unlike poles attract each other, and (C) like poles repel.

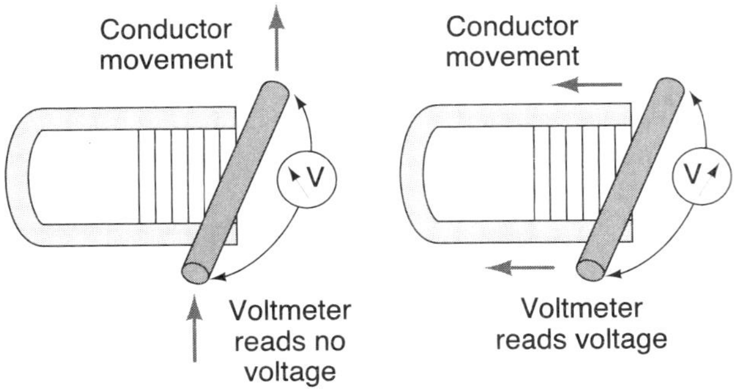


**Figure 11.** Moving a conductor through a magnetic field induces a voltage.

small voltage reading **(Figure 11).** A voltage has been induced in the wire.

It is important to remember that the conducting wire must cut across the flux lines to induce a voltage. Moving the wire parallel to the lines of flux does not induce voltage. Holding the wire still and moving the magnetic field at right angles to it also induces voltage in the wire.

# BASICS OF ELECTRONICS

Computerized engine controls and other features of today's cars would not be possible if it were not for electronics. Electronics is best defined as the technology of controlling electricity.

## Semiconductors

A **semiconductor** is a material or device that can function as either a conductor or an insulator depending on its structure. Semiconductor materials have less resistance than an insulator but more resistance than a conductor. Some common semiconductor materials include silicon (Si) and germanium (Ge).

The **diode** is the simplest semiconductor device. A diode allows current to flow in only one direction. Therefore, it can function as a switch, acting as either a conductor or an insulator depending on the direction of current flow.

A variation of the diode is the zener diode, which acts like a standard diode until a certain voltage is applied to it. When the voltage reaches this point, the zener diode allows current to flow in the reverse direction. Zener diodes are often used in electronic voltage regulators.

A **transistor** is an electronic device produced by joining three sections of semiconductor materials. Like the diode, it is very useful as a switching device, functioning as either a conductor or an insulator.

One transistor or diode is limited in its ability to do complex tasks. However, when many semiconductors are combined into a circuit, they can perform complex functions.

An **integrated circuit** is simply a large number of diodes, transistors, and other electronic components, such as resistors and capacitors, all mounted on a single piece of semiconductor material. This type of circuit is extremely small. Circuitry that used to take up entire rooms can now fit into a pocket. The principles of semiconductor operation remain the same in integrated circuits, only the size has changed.

The computer has taken over many of the tasks in cars and trucks that were formerly performed by vacuum, electromechanical, or mechanical devices. When properly programmed, they can carry out explicit instructions with blinding speed and almost flawless consistency. A typical electronic control system is made up of sensors, actuators, and a computer.

## Sensors

All sensors perform the same basic function. They detect a mechanical condition (movement or position), chemical state, or temperature condition and change it into an electrical signal that can be used by the computer to make decisions. The computer makes decisions based on information it receives from sensors. Each sensor used in a particular system has a specific job to do. Although there are a variety of sensor designs, they all fall under one of two operating categories: reference voltage sensors or voltage generating sensors.

**Reference voltage (Vref) sensors** provide input to the computer by modifying or controlling a constant, predetermined voltage signal **(Figure 12).** This signal, which

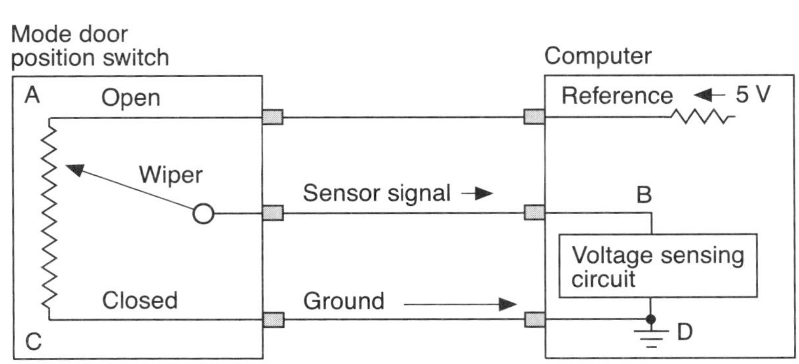


**Figure 12.** A potentiometer alters the reference voltage in response to the movement of the wiper and sends a signal back to the voltage sensing circuit of the computer.

can have a reference value from 5 volts to 9 volts, is generated and sent out to each sensor by a reference voltage regulator located inside the **processor.** The term "processor" is used to describe the actual metal box that houses the computer and its related components. Because the computer knows that a certain voltage value has been sent out, it can indirectly interpret conditions like motion, temperature, and component position based on what comes back.

Most sensors presently in use are variable resistors or potentiometers. They modify a voltage to or from the computer, indicating a constantly changing status that can be calculated, compensated for, and modified. That is, most sensors simply control a voltage signal from the computer. The monitored signal line may be the output signal from the computer to the sensor (one- and two-wire sensors), or the computer may use a separate return line from the sensor to monitor voltage changes (three-wire sensors).

**Voltage-generating sensors (Figure 13)** are commonly Hall-effect switches, permanent magnet (PM) generators, oxygen sensors, and knock sensors, which are capable of producing their own voltage signal. When received by the computer, this varying voltage signal enables the computer to monitor and adjust for changes in the computerized control system.

In addition to variable resistors, two other commonly used reference voltage sensors are switches and thermistors. Regardless of the type of sensors used in electronic control systems, the computer is incapable of functioning properly without input signal voltages from the sensors.

## Communication Signals

Most input sensors are designed to produce a voltage signal that varies within a given range (from high to low, including all points in between). A signal of this type is called an **analog signal.** Unfortunately, the computer does not understand analog signals. It can only read a **digital signal,** a signal that has only two values—on or off **(Figure 14).**

To overcome this communication problem, all analog voltage signals are converted to a digital format by a device known as an analog-to-digital converter (A/D converter) located in a section of the processor. Some sensors, like the Hall-effect switch, produce a digital or square wave signal that can go directly to the computer as input.

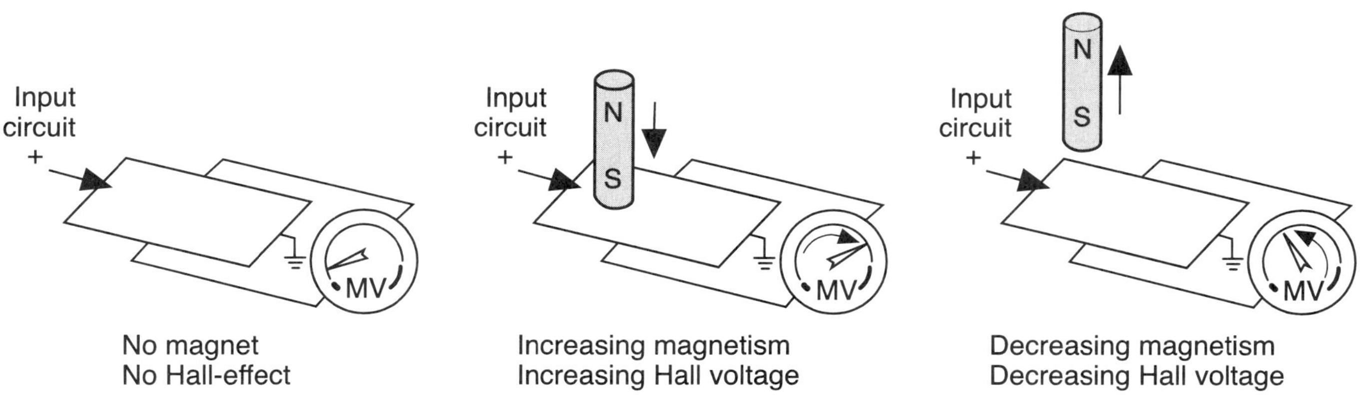


**Figure 13.** Hall-effect principles of voltage induction.

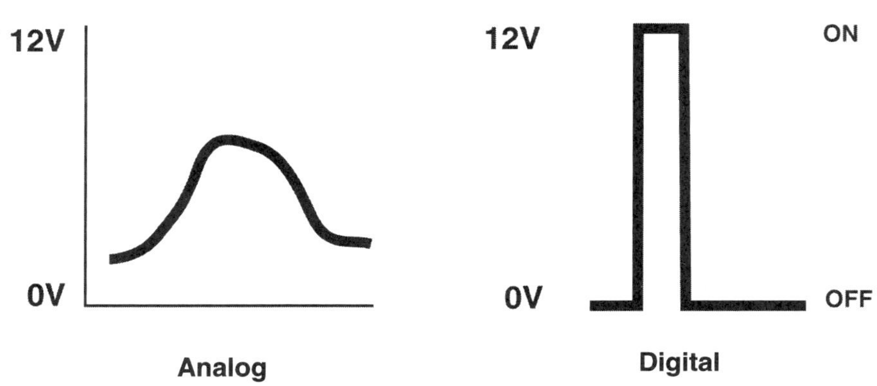


**Figure 14.** Analog signals are constantly variable, whereas digital signals are either on or off or high or low.

A computer's memory holds the programs and other data, such as vehicle calibrations, that the microprocessor refers to in performing calculations. The program is a set of instructions or procedures that the computer must follow. Included in the program is information that tells the microprocessor when to retrieve input (based on temperature, time, etc.), how to process the input, and what to do with it once it has been processed .

## Actuators

After the computer has processed the information, it sends output signals to control devices called actuators. These actuators (solenoids, switches, relays, or motors) physically act or carry out a decision made by the computer.

**Actuators** are electromechanical devices that convert an electrical current into mechanical action. This mechanical action can then be used to open and close valves, control vacuum to other components, or open and close switches. When the computer receives an input signal indicating a change in one or more of the operating conditions, it determines the best strategy for handling the conditions. The computer then controls a set of actuators to achieve the desired effect or strategy goal. In order for the computer to control an actuator, it must rely on a component called an **output driver.**

Output drivers are located in the processor and operate by the digital commands issued by the computer. Basically, the output driver is nothing more than an electronic on/off switch used to control the ground circuit of a specific actuator. At times, a single driver controls four different actuators and is referred to as a quad driver.

# *SUMMARY*

- There is a natural attraction of electrons to protons. Electrons have a negative charge and are attracted to something with a positive charge.
- In order to have a continuous flow of electricity, there must be an excess of electrons in one place, a lack of electrons in another place, and a path between the two places.
- The energy sources used in an automobile's electrical system are the vehicle's battery and its charging system.
- Electrical current describes the movement or flow of electricity. The unit for measuring current is the ampere, usually called an amp. Current is measured with an ammeter.
- Voltage is the force developed by the attraction of the electrons to protons. Voltage is the pressure that causes current flow and is measured with a voltmeter.
- The resistance to electrical flow is measured in ohms, and an ohmmeter is used to measure it.
- An electrical circuit is considered complete when there is a path that connects the positive and negative terminals of the power source.
- Conductors are materials with a low resistance to the flow of current.
- Materials known as insulators resist the flow of current.
- Ohm's Law states that it takes 1 volt of electrical pressure to push 1 ampere of electrical current through 1 ohm of resistance.
- Resistors are used to limit current flow.
- Rheostats have two connections, one to the fixed end of a resistor and one to a sliding contact with the resistor.
- Potentiometers have three connections, one at each end of the resistance and one connected to a sliding contact with the resistor.
- Thermistors change their resistance value as their temperature changes.
- Fuses, fuse links, maxi-fuses, and circuit breakers are designed to prevent circuit damage caused by high current.
- Switches turn the circuit on or off or direct the flow of current in a circuit. Switches can be controlled by the driver or can be self-operating through a condition of the circuit, the vehicle, or the environment.
- A relay is an electric switch that allows a small amount of current to control a circuit with high current.
- Solenoids are electromagnetic devices with movable cores and are used to translate electrical current flow into mechanical movement.
- A semiconductor is a material or device that can function as either a conductor or an insulator depending on its structure.
- A diode is the simplest semiconductor device. It allows current to flow in only one direction and can function as a switch depending on the direction of current flow.
- A transistor is an electronic device produced by joining three sections of semiconductor materials. It can function as either a conductor or an insulator.
- A typical electronic control system is made up of sensors, actuators, and related wiring connected to a computer.
- Sensors detect a mechanical condition (movement or position), chemical state, or temperature condition and change it into an electrical signal that can be used by the computer to make decisions. All sensors fall under one of two operating categories: reference voltage sensors or voltage generating sensors.
- After the computer has processed the information, it sends output signals to control devices called actuators.

## REVIEW QUESTIONS

1. Name the two energy sources used in automobile electrical systems.
2. What is the difference between voltage and current?
3. What is the difference between a fixed resistor and a variable resistor?
4. What types of sensors are typically used in an automotive computer system?
5. Current is measured in __________, electrical voltage is measured in ____________, and electrical resistance is measured in _____________.
6. _______, __________ _________, ___________ ___________, and ________________ are used to protect circuits against current overloads.
7. A computerized circuit depends on two types of signals: ___________ and ________ .
8. While discussing the behavior of electricity, Technician A says that if voltage does not change but there is a change in the resistance of the circuit, the current will change. Technician B says that if resistance increases, current decreases. Who is correct?
   A. Technician A only
   B. Technician B only
   C. Both Technician A and Technician B
   D. Neither Technician A nor Technician B
9. Technician A says that rheostats have three connections, one at each end of the resistance and one connected to a sliding contact with the resistor. Technician B says that potentiometers have two connections, one at the power end of the resistance and one connected to a sliding contact with the resistor. Who is correct?
   A. Technician A only
   B. Technician B only
   C. Both Technician A and Technician B
   D. Neither Technician A nor Technician B
10. Technician A says that electrical resistance is the pressure that causes current to flow in a circuit. Technician B says if there is zero resistance in a circuit, a maximum amount of current will flow in the circuit. Who is correct?
    A. Technician A only
    B. Technician B only
    C. Both Technician A and Technician B
    D. Neither Technician A nor Technician B

# Chapter 6 Basic Electrical Service

## INTRODUCTION

Certain diagnostic basics apply to all electrical systems. Many who have a difficult time diagnosing electrical problems do not understand what electricity is, what electrical meters can and do tell them, and how the different electrical problems affect a circuit. The purpose of this chapter is to help you understand these principles. Also covered in this chapter is the testing and repair of basic electrical systems and the common electrical components.

## QUICK OVERVIEW OF ELECTRICITY

Resistance within an electrical circuit controls the flow of current within that circuit. The amount of voltage and resistance in a circuit determines how much current will flow in the circuit. Voltage pushes current through the circuit's resistors or loads. The amount of voltage within a load decreases as the current passes through the load. By the time the current passes through the last bit of resistance in a circuit, all of the voltage is lost.

In a 12-volt light circuit and the resistance of the bulb being 3 ohms, 4 amps of current will flow through the circuit **(Figure 1).** However, if there is a bad connection at the bulb that creates an additional 1 ohm of resistance, the current will decrease to 3 amps. As a result of the added resistance and decreased current, the bulb will lose about half of its brightness.

This lack of brightness is also caused by the amount of voltage dropped by the bulb. In the normal circuit, the bulb drops 12 volts. However, in the circuit with the added resistance, 3 volts will be dropped by the bad connection and the bulb will drop the remaining 9 volts.

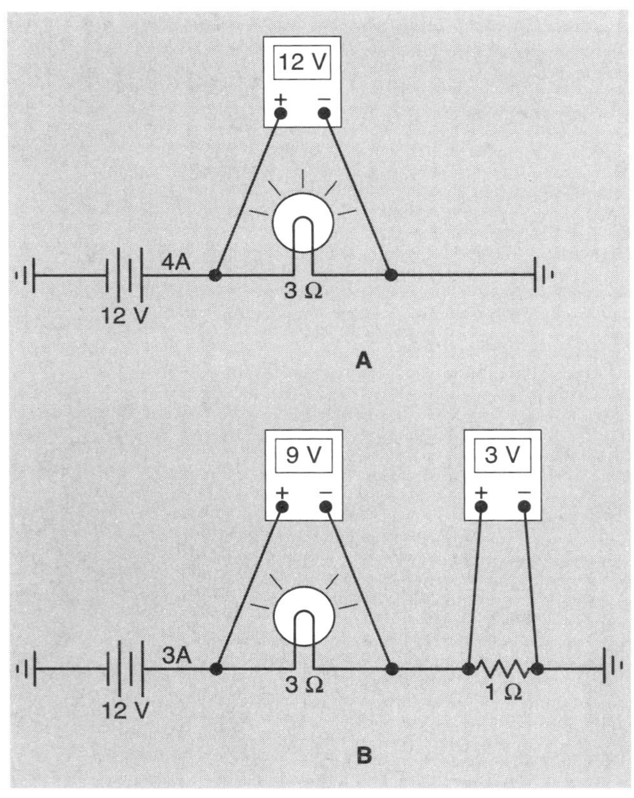


**Figure 1.** (A) A normal 3-ohm light circuit. (B) The same circuit but with a high resistance ground. Notice that the voltage drop across the lamp decreased, as did the circuit current. The bulb will burn dimmer.

The way resistance affects a circuit depends on the placement of the resistance in the circuit. If the resistances (wanted or unwanted) are placed directly into the circuit, the resistances are said to be in series. If a resistor is placed so that it allows an alternative path for current, it is a parallel resistor.

In a **series circuit (Figure 2)** current follows only one path, and the amount of current flow through the circuit depends on the total resistance of the circuit. To calculate the total resistance in a series circuit, all resistance values are added together. At each resistor, voltage is dropped and the total amount of voltage dropped in a series circuit is equal to the voltage of the source (battery). Regardless of the possible differences and resistance values of the loads, current in a series circuit is always constant and is determined by the total resistance in the circuit.

**Parallel circuits (Figure 3)** are designed to allow current to flow in more than one path. This allows one power source to power more than one circuit or load. A car's accessories and other electrical devices can be individually controlled through the use of parallel circuits. Within a parallel circuit, there is a common path to and from the power source. Each branch or leg of a parallel circuit behaves as if it were an individual circuit. Current flows only through the individual circuits when each is closed or completed. Not all legs of the circuit need to be complete in order for current to flow through one of them.

In parallel circuits, the total amperage of the circuit is equal to the sum of the amperages in all of the legs of the circuit. No voltage is dropped when the circuit splits into its branches; therefore, equal amounts of voltage are applied to each branch of the circuit. The total resistance of a parallel circuit is always less than the resistance of the leg with the smallest amount of resistance. The total resistance of two resistors in parallel can be calculated by dividing the product of the two by their sum.

The legs of a parallel circuit may contain a series circuit. To determine the resistance of that leg, the resistance values are added together. The resistance values of each leg are used to calculate the total resistance of a parallel circuit. Total circuit current flows only through the common power and ground paths, therefore a change in a branch's resistance will not only affect the current in the branch but will also affect total circuit current.

## ELECTRICAL PROBLEMS

All electrical problems can be classified into one of three categories: opens, shorts, or high resistance. Identifying the type of problem helps identify the correct tests to conduct when diagnosing an electrical problem.

An **open** occurs when a circuit has a break in the wire **(Figure 4).** Without a completed path, current cannot flow and the load or component cannot work. A disconnected or broken wire or a switch in the OFF position can cause an open circuit. Although voltage will be present up to the open point, there is no current flow. Without current flow, there are no voltage drops across the various loads. If there is an open in one leg of a parallel circuit, the remaining part of that parallel circuit will operate normally.

A **short** results from an unwanted path for current. Shorts cause an increase in current flow, which can burn wires or components. Sometimes two circuits become shorted together resulting in one circuit powering another. This can result in strange happenings, such as the horn

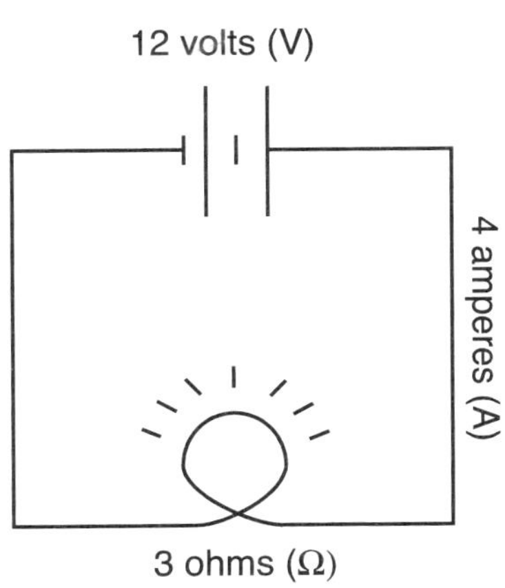


**Figure 2.** In a series circuit, the same amount of current flows through the entire circuit.

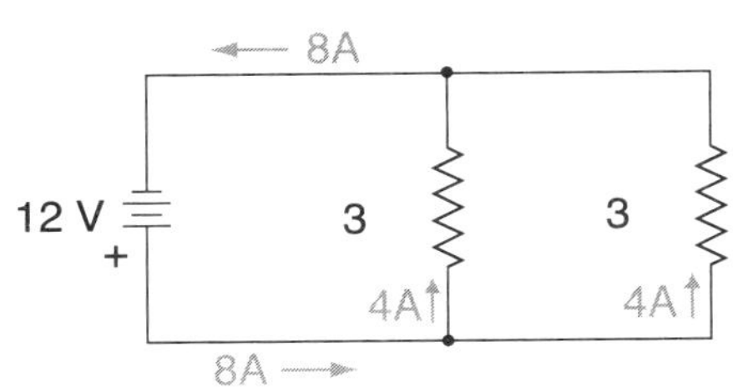


**Figure 3.** A simple parallel circuit.

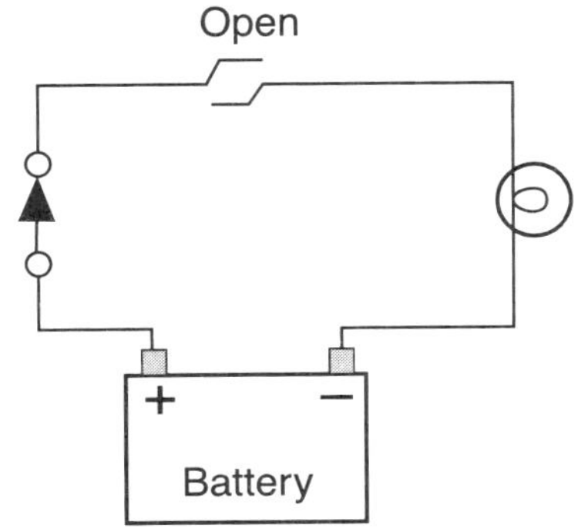


**Figure 4.** An open prevents current flow.

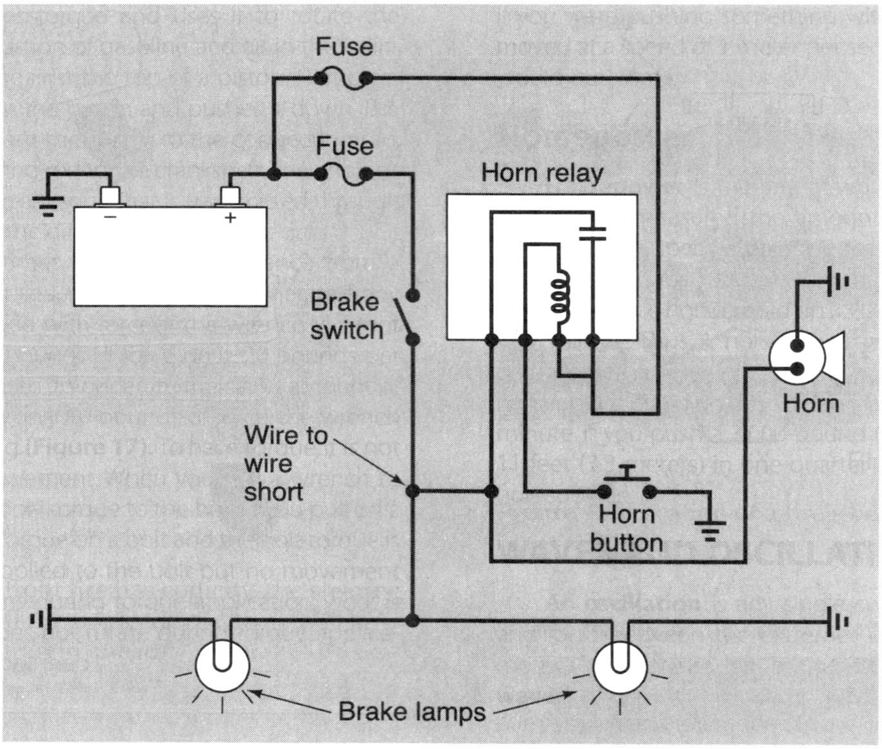


**Figure 5.** The above horn circuit is shorted to the brake light circuit.

sounding every time the brake pedal is depressed. In this case, the brake light circuit is shorted to the horn circuit **(Figure 5).**

A short-to-ground can be present before the load in the circuit or internally within the load. This problem provides a low resistance path to ground **(Figure 6).** Improper wiring and damaged insulation are the two major causes of short circuits.

High resistance problems occur when there is unwanted resistance in the circuit. The higher-than-normal resistance causes current to be lower than normal, and the components in the circuit are unable to operate properly if at all. A common cause for this problem is corrosion at a connector. The corrosion becomes an additional resistance in the circuit. This load not only decreases the circuit's current but also uses some of the circuit's voltage which prevents full voltage to the normal loads in the circuit.

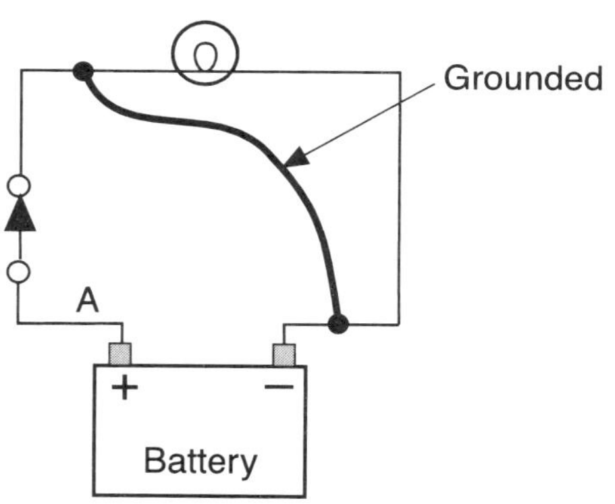


**Figure 6.** A short to ground causes a large incease in circuit current.

## BASIC ELECTRICAL TESTING

To troubleshoot a problem, always begin by verifying the customer's concern. Then operate the system and others to completely understand the problem. There are often other problems that are not as evident or bothersome to the customer that will provide helpful information for diagnostics. Refer to the correct wiring diagram and study the circuit that is affected. From the diagram you should be able to identify testing points and probable trouble areas. Then test and use logic to identify the cause of the problem. A multimeter is used to test and diagnose electrical systems and can be used along with test lights and jumper wires.

### Multimeters

It is unnecessary to own separate voltmeters, ohmmeters, and ammeters. These meters are combined in a single

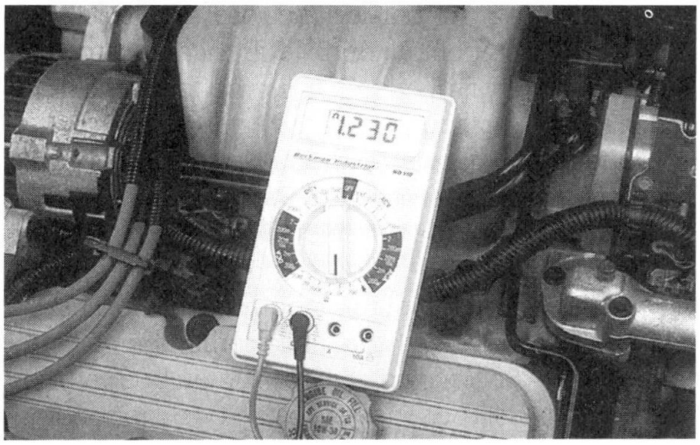

**Figure 7.** A digital multimeter (DMM).

tool called a multimeter **(Figure 7).** The most commonly used multimeter is the digital DVOM or the DMM. These meters display the measurements digitally on the meter.

Most digital meters have high input **impedance,** usually at least 10 megaohms (10 million ohms). The high impedance reduces the risk of damaging sensitive components and delicate computer circuits.

In addition to the basic electrical tests, some multimeters also measure engine revolutions per minute (rpm), ignition dwell, diode condition, distributor conditions, frequency, and even temperature.

DMMs have either an "auto range" feature in which the meter automatically selects the appropriate scale, or they must be set to a particular range. In either case, you should be familiar with the ranges and the different settings available on the meter you are using. To designate particular ranges and readings, meters display a prefix before the reading or range.

## Ammeters

An ammeter must be placed into or in series with the circuit being tested. This normally requires disconnecting a wire or connector from a component and connecting the ammeter between the wire or connector and the component. The red lead of the ammeter should always be connected to the side of the connector closest to the positive side of the battery, and the black lead should be connected to the other side.

**You Should Know**

***Never place the leads of an ammeter across the battery or a load. Ammeters have very low internal resistance! This puts the meter in parallel with the circuit and will blow the fuse in the ammeter or possibly destroy the meter or circuit.***

It is much easier to test with an ammeter if it has an inductive pickup. The pickup clamps around the wire or cable being tested and measures amperage by the strength of the magnetic field created by the current flowing through the wire. This type of pickup eliminates the need to separate the circuit to insert the meter.

Ammeters can be used to diagnose many different circuit problems. For example, assume that a 6-amp fuse protects a circuit with four identical lights wired in parallel. If the circuit constantly blows the fuse, a short exists somewhere in the circuit. To find the short, disconnect all lights by removing them from their sockets. Then, close the switch and read the ammeter. With the load disconnected, the meter should read 0 amperes. If there is any reading, the wire between the fuse block and the socket is shorted to ground.

If zero amps were measured, reconnect each light in sequence; the reading should increase with each bulb. If when making any connection the reading is higher than expected, the problem is in that part of the light circuit.

## Voltmeters

Electrical circuits and components cannot operate properly if the proper amount of voltage is not available. Not only must the battery be able to deliver the proper amount of voltage, the proper amount of voltage must also be available to the intended component. Because of this, voltage is measured at the source, at the electrical loads, and across the loads and circuits.

A voltmeter has two leads, a positive and a negative lead. Connecting the positive lead to any point within a circuit and connecting the negative lead to a ground will measure the available voltage at the point where the positive lead was connected.

To measure the voltage drop of a light bulb in a simple circuit, connect the positive lead to the battery or power lead at the bulb and the other lead to the ground-side of the bulb **(Figure 8).** The meter will read the amount of voltage drop across the bulb. If no other resistances are present in the circuit, the amount of voltage drop will equal source voltage. However, if there is a resist-

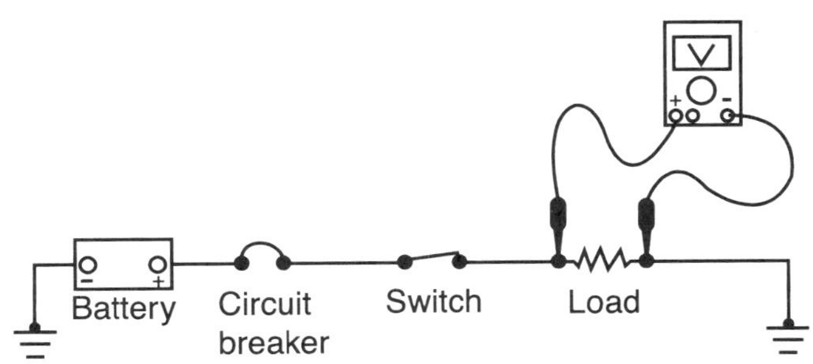


**Figure 8.** To measure voltage drop, the voltmeter is connected across the item being tested.

ance present in the ground connection, the bulb will have a voltage drop that is less than battery voltage. If the voltage drop across the ground is measured, it will equal the source voltage minus the amount of voltage dropped by the light.

The maximum allowable voltage drop across wires, connectors, and other conductors in an automotive circuit is 10 percent of the system voltage. Therefore, in a 12-volt automotive electrical system, this maximum loss is 1.2 volts.

## Ohmmeters

Ohmmeters use their own power to measure the resistance between two points. They should never be connected into an activated circuit.

The amount of resistance is based on the voltage dropped across the component. The scale of an ohmmeter reads from 0 to infinity (∞). A 0 reading means there is no resistance in the circuit and may indicate a short in a component that should show some resistance. An infinite reading indicates a number higher than the meter can measure. Most DMMs display an infinite reading as "O.L." **(Figure 9).** This usually is an indication of an open circuit.

To measure the resistance of something, connect one of the meter's leads to the power side of the component and the other lead to the groundside. If the meter shows a reading of "infinity," it can be assumed that there is no continuity between the two measured points.

Many electrical components have a specified resistance value that is listed in the service manual. This resistance value is important since it controls the amount of voltage dropped and the amount of current that will flow in the circuit. If a component does not have the proper amount of resistance, the circuit will not operate properly.

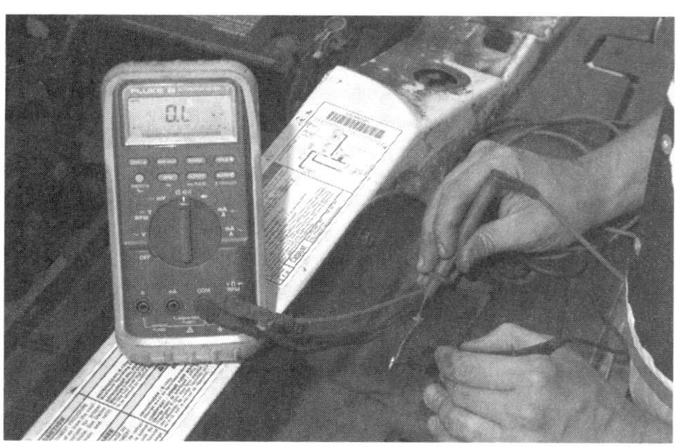

**Figure 9.** Most DMMs display an infinite reading as "O.L."

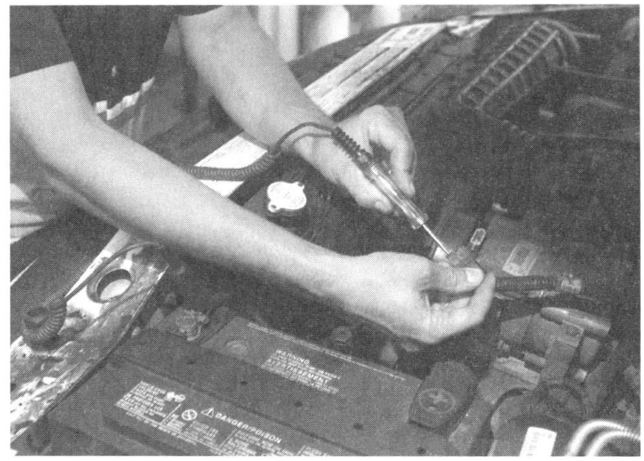

**Figure 10.** A non-powered test light.

## Test Lights

There are two types of test lights commonly used in diagnosing electrical problems: non-powered **(Figure 10)** and self-powered. A **test light** looks like a stubby ice pick. Its handle is transparent and contains a light bulb. A probe extends from one end of the handle and a ground clip and wire from the other end. With the wire lead connected to ground and the tester's probe at a point of voltage, the light turns on with the presence of voltage. The brightness of the bulb is an indication of the amount of voltage present at the test point.

A self-powered test light is often used to test for continuity instead of an ohmmeter. A self-powered test light does not rely on the power of the circuit to light its bulb. Rather, it contains a battery for a power supply and when connected across a completed circuit, the bulb will light. This type test light is connected to the circuit in the same way as an ohmmeter.

**You Should Know** *Test lights should not be used to test electronic circuits, and only use a self-powered test light when the service procedures tell you to do so.*

## Jumper Wires

Jumper wires are used to bypass individual wires, connectors, components, or switches. Bypassing a component or wire helps to determine if that part is faulty. If the problem is no longer evident after the jumper wire is installed, the part bypassed is probably faulty. Technicians typically have jumper wires of various lengths and contain a fuse or circuit breaker in them to protect the circuits being tested.

## ELECTRICAL WIRING DIAGRAMS

Wiring diagrams, sometimes called **schematics,** show how circuits are wired and how the components are connected. A wiring diagram does not show the actual position of the parts on the vehicle or their appearance, nor does it indicate the length of the wire that runs between components. It usually indicates the color of the wire's insulation and sometimes the wire gauge size. Tracing a circuit through a vehicle is basically a matter of following the colored wires.

Wiring diagrams can become quite complex. To avoid this, the vehicle's electrical system may be divided into many diagrams each illustrating only one system, such as the electronic suspension circuit, oil pressure indicator light circuit, or wiper motor circuit.

## TESTING CIRCUIT PROTECTION DEVICES

Circuit protection devices are found in nearly all automotive electrical circuits. They are used to protect the circuit's components and wires. The common types of protection devices are fuses, fusible links, circuit breakers, and maxi-fuses.

### Fuses

There are three basic types of **fuses** found in automobiles: cartridge, blade, and ceramic. To visually check a fuse, look for a break in the internal or external metal strip. Discoloration of the glass or plastic cover around the metal end caps is an indication of overheating.

All fuses are more accurately checked with an ohmmeter or test light. If the fuse is good, there will be continuity through it **(Figure 11).**

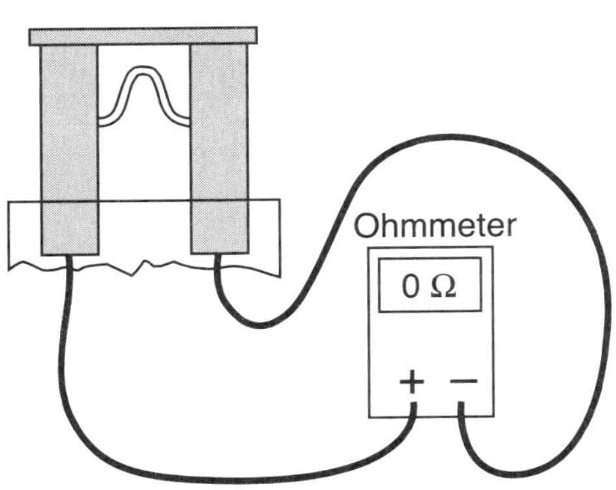


**Figure 11.** An ohmmeter can be used to check the condition of a fuse.

### Fusible Links

**Fusible links** are used in circuits where limiting the maximum current is not extremely critical. They are often installed in the positive battery lead to the ignition switch and other circuits that have power with the key off. Fusible links are also used when it would be awkward to run wiring from the battery, to the fuse panel, and back to the load.

Fusible link wire is covered with a special insulation that bubbles when it overheats, indicating that the link has melted. If the insulation appears good, pull lightly on the wire. If the link stretches, the wire has melted. Of course, when it is hard to determine if the fuse link is burned out, check for continuity through the link with a test light or ohmmeter.

### Maxi-Fuses

Many late-model vehicles use **maxi-fuses** instead of fusible links. Maxi-fuses look and operate like two-prong, blade or spade fuses except they are much larger and can handle more current. Maxi-fuses are located in their own underhood fuse block.

To check a maxi-fuse, look at the fuse element through the transparent plastic housing. If there is a break in the element, the maxi-fuse has blown. To replace it, pull it from its fuse box or panel. Always replace a blown maxi-fuse with a new one having the same ampere rating.

### Circuit Breakers

Some circuits are protected by **circuit breakers.** A circuit breaker will either automatically reset or it must be manually reset by depressing a button.

## TESTING SWITCHES

Switches can be tested with a voltmeter, test light, or ohmmeter. To check the operation of a switch with a voltmeter or a test light, connect the meter's positive lead to the battery side of the switch. With the negative lead attached to a good ground, voltage should be measured at this point. Without closing the switch, move the positive lead to the other side of the switch. If the switch is open, no voltage will be present at that point. The amount of voltage present at this side of the switch should equal the amount on the other side when the switch is closed. If the voltage decreases, the switch is causing a voltage drop due to excessive resistance. If no voltage is present on the groundside of the switch with it closed, the switch is not functioning properly and should be replaced.

If a switch has been removed from the circuit, it can be tested with an ohmmeter or a self-powered test light.

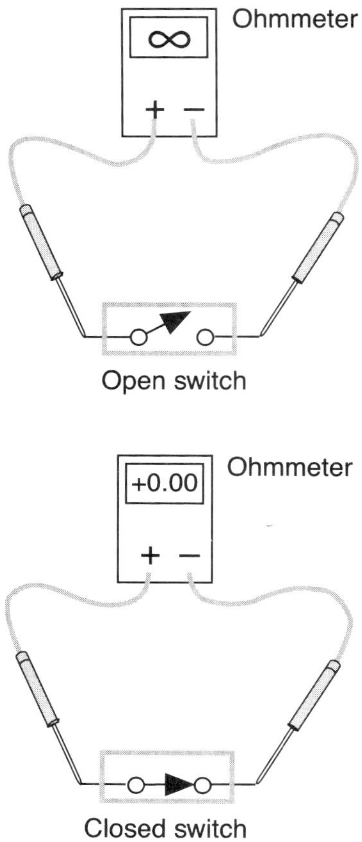


**Figure 12.** An ohmmeter can be used to test the operation of a switch.

By connecting the leads across the switch connections, the action of the switch should open and close the circuit **(Figure 12).**

## BASIC ELECTRICAL REPAIRS

Faulty wiring can cause many electrical problems. Loose or corroded terminals, frayed, broken, or oil-soaked wires, and faulty insulation are the most common causes. Wires, fuses, and connections should be checked carefully during troubleshooting.

Wire end terminals are connecting devices. They are generally made of tin-plated copper and come in many shapes and sizes. They may be either soldered or crimped in place.

When working with wiring and connectors, never pull on the wires to separate the connectors. This can loosen the connector and cause an intermittent problem that may be very difficult to find later. Always follow the correct procedures and use the tools designed for separating connectors.

Never reroute wires. Rerouting wires can result in induced voltages from nearby components. These stray voltages can interfere with the function of electronic circuits.

When installing a terminal, select the appropriate size and type of terminal. Be sure it fits the unit's connecting post and has enough current carrying capacity for the circuit. Also, make sure it is heavy enough to endure normal wire flexing and vibration.

To crimp a connector to a wire, make sure you use the correct size stripping opening on the crimping tool. Remove enough insulation to allow the wire(s) to completely penetrate the connector. Place the wire(s) into the connector and crimp the connector. To achieve a proper crimp, place the open area of the connector facing toward the anvil. Make sure the wire is compressed under the crimp. Use electrical tape or heat shrink tubing to tightly seal the connection. This will provide good protection for the wire and connector.

The preferred way to connect wires or install a connector is by soldering **(Figure 13).** Some car manufacturers used aluminum in their wiring. Aluminum cannot be soldered. Follow the manufacturer's guidelines and use the proper repair kits when repairing aluminum wiring.

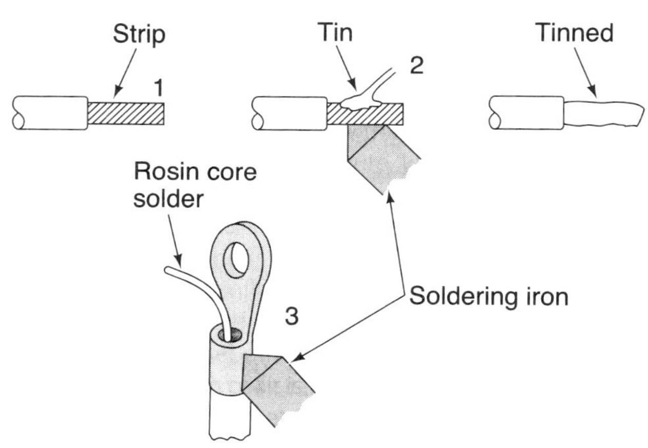


**Figure 13.** The proper procedure for soldering a terminal to a wire.

## SUMMARY

- In a series circuit, current follows one path, and the amount of current through the circuit depends on the total resistance of the circuit. The current is always constant.
- In a parallel circuit, the total amperage of the circuit is equal to the sum of the amperages in all of the legs of the circuit.
- An open occurs when a circuit is not complete and therefore there is no current flow through the circuit.
- A short results from an unwanted path for current and causes an increase in current flow.
- High resistance problems cause lower-than-normal current flow and unwanted voltage drops.
- To measure available voltage, connect the negative lead to the common ground circuit and the positive lead to the point where a measurement is sought.
- To measure the voltage drop of a component, connect the positive lead to the power lead of the component and the other lead to the groundside.
- Ohmmeters should never be connected into an activated circuit.
- On an ohmmeter, a 0 reading means there is no resistance in the circuit, and an infinite reading indicates a number higher than the meter can measure.
- A DMM can measure volts, amps, and ohms, as well as conduct many other tests. A DMM displays the measurements digitally and has high input impedance, which reduces the risk of damaging sensitive components and delicate computer circuits.
- Wiring diagrams show how circuits are wired and how the components are connected.
- Loose or corroded terminals, frayed, broken, or oil soaked wires, and faulty insulation are common electrical problems.
- The preferred way to connect wires or to install a connector is by soldering.

## REVIEW QUESTIONS

1. If you were probing along with a voltmeter on a circuit that has a bad ground, how many volts would you read before and after the load? Why?
2. While discussing how to test a switch, Technician A says the action of the switch can be monitored by a voltmeter. Technician B says continuity across the switch can be checked by measuring the resistance across the switch's terminals when the switch is in its different positions. Who is correct?
   A. Technician A only
   B. Technician B only
   C. Both Technician A and Technician B
   D. Neither Technician A nor Technician B
3. Name the three types of electrical problems and their effect on current flow.
4. Describe what happens to the circuit when a bulb in a parallel circuit burns out.
5. While discussing electricity, Technician A says an open causes unwanted voltage drops. Technician B says high resistance problems cause increased current flow. Who is correct?
   A. Technician A only
   B. Technician B only
   C. Both Technician A and Technician B
   D. Neither Technician A nor Technician B
6. Technician A says a broken or frayed wire can cause an unintentional grounded circuit. Technician B says dirt and grease buildup at terminals and connections can cause the same problem. Who is correct?
   A. Technician A only
   B. Technician B only
   C. Both Technician A and Technician B
   D. Neither Technician A nor Technician B

# Chapter 7 Suspension and Steering Systems

## INTRODUCTION

Like the rest of the systems on a vehicle, the suspension system has greatly changed through the years. Not only has technology brought about these changes, so has the quest for great handling and comfortable vehicles. Suspension systems for the front and rear of a vehicle can be quite complex **(Figure 1).**

Suspension systems must keep the vehicle's wheels lined up the travel of the vehicle, limit the movement of the vehicle's body during cornering and when going over bumps, and provide a smooth and comfortable ride.

The purpose of the steering system is to turn the front wheels. In a few cases, it also turns the rear wheels.

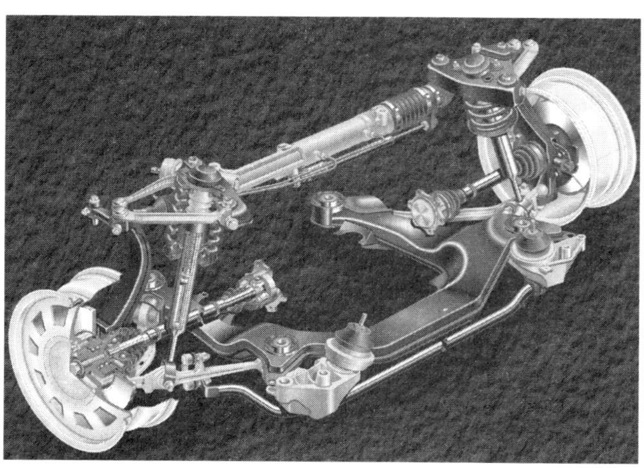

**Figure 1.** A late-model suspension and steering arrangement.

## WHEELS AND TIRES

Wheels are made of either stamped or pressed steel discs riveted or welded together. Die cast and forged aluminum and magnesium wheels are also available (sometimes referred to as mag wheels). Aluminum wheels are lighter in weight when compared with the stamped steel type. This weight savings is important because the wheels and tires on a vehicle are unsprung weight. This means the weight is not supported by the vehicle's springs.

Near the center of the wheel are mounting holes that are tapered to fit tapered mounting nuts **(lug nuts)** that center the wheel over the hub. The rim has a hole for the tire's **valve stem** and a **drop center** area designed to allow for easy tire removal and installation. **Wheel offset** is the vertical distance between the centerline of the rim and the mounting face of the wheel. The offset is considered positive if the centerline of the rim is inboard of the mounting face and negative if outboard of the mounting face. The amount and type of offset is critical.

The wheel is bolted to a **hub,** either by lug bolts that pass through the wheel and thread into the hub, or by studs that protrude from the hub. In the case of studs, special lug nuts are required.

Wheel size is designated by rim width and rim diameter. Rim width is determined by measuring across the rim between the flanges. Rim diameter is measured across the bead seating areas from the top to the bottom of the wheel. Rims have safety ridges that in the event of a blowout, keep the tire from moving into the dropped center and from coming off the wheel.

### Tires

The primary purpose of tires is to provide traction. Tires also help the suspension absorb road shocks. They must perform under a variety of conditions: wet or dry roads, asphalt, concrete, dirt, or gravel surfaces. The vehicle might be traveling fast or slow on a straight road, or may be moving quickly through curves or over hills. All of these conditions call for special requirements that must be present, at least to some degree, in all tires.

Tires are also designed to carry the vehicle's weight, withstand side thrust over varying speeds and conditions, and transfer braking and driving torque to the road **(Figure 2).** As a tire rolls, friction is created between the tire and the road. This friction gives the tire its traction. Although good traction is desirable, it must be limited. Too much traction equals too much friction and rolling resistance. Rolling resistance wastes engine power and fuel, therefore it must be kept to a minimal level. This dilemma is a major concern in the design of today's tires.

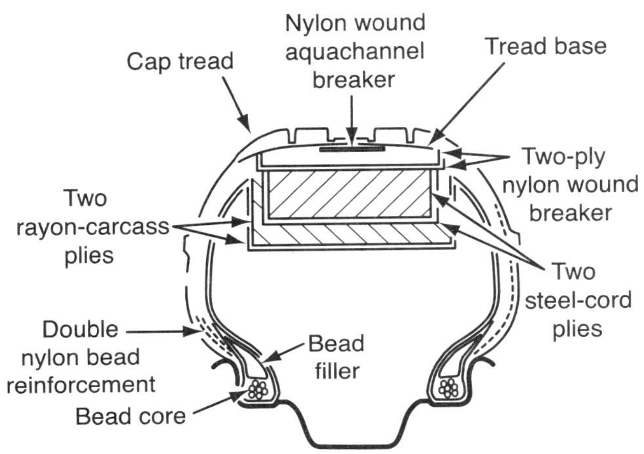


**Figure 2.** The construction of a typical tire.

## Wheel Bearings

All bearings are designed to allow a shaft to rotate smoothly in housing or to allow the housing to rotate smoothly around a shaft. Wheel and axle bearings do this for a vehicle's wheels **(Figure 3).** Typically, the bearings on an axle that drives the wheels are called axle bearings. The wheel is mounted to the hub of an axle shaft and shaft rotates within a housing. Wheel bearings are used on non-driving axles. Axle bearings are usually serviced with the drive axle. Wheel bearings, however, require periodic maintenance and are often serviced with suspension and brake work.

## Axle Shafts

There are two types of axles: dead and live or drive. A **dead axle** does not drive a vehicle; it merely supports the vehicle load and provides a mounting place for the wheels. The rear axle of a front-wheel-drive (FWD) vehicle is a dead axle, as are the axles used on trailers.

A **live axle** is one that drives the vehicle. Drive axles transfer torque from the differential to each driving wheel. Depending on the design, rear axles can also help carry the weight of the vehicle or even act as part of the suspension.

There are three types of driving axles: semifloating, three-quarter floating, and full floating. All three use axle shafts that are splined to the differential side gears. At the wheel ends, the axles can be supported by bearings in

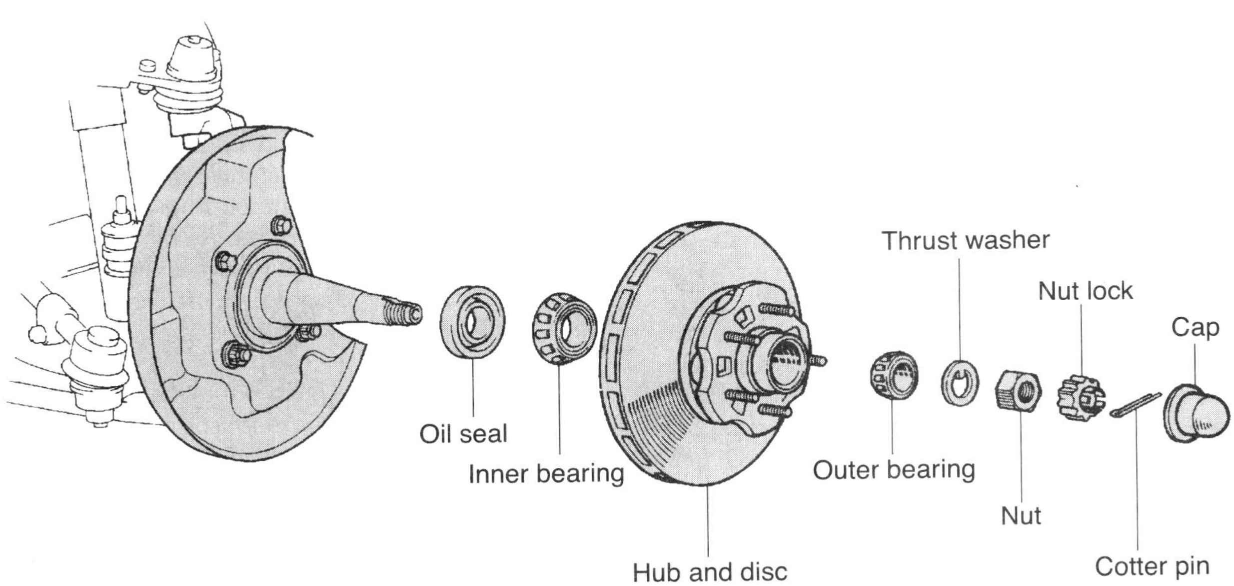


**Figure 3.** A typical wheel bearing assembly for a rear-wheel-drive vehicle.

basically three different ways. Semi-floating axles help to support the weight of the vehicle. The axles are supported by bearings located in the axle housing. The wheel bearing on a three-quarter floating axle is on the outside of the axle housing instead of inside the housing as in the semi-floating axle. The wheel hubs are bolted to the end of the axle shaft and are supported by the bearing. In this arrangement, the axle shaft only supports 25 percent of the vehicle's weight. Most medium- and heavy-duty vehicles use a full-floating axle shaft. This design is similar to the three-quarter floating axle except that two bearings rather than one are used to support the wheel hub.

#### Independently Suspended Axles

In an independently suspended axle system, the axles are usually open instead of being enclosed in a housing.

The **DeDion** axle system resembles a normal driveline. The driving axles look like a drive shaft with U-joints at each end of the axles. A slip joint is attached to the innermost, U-joint. The outboard U-joint is connected to the wheel hub. This allows the driving axle to move up and down as it rotates.

On vehicles that use a **swing axle,** the driving axle shafts can be open or enclosed. An axle fits into the differential by way of a ball-and-socket system. The ball-and-socket system allows the axle to pivot up and down. As the axle pivots, the driving wheel swings up and down. This system best describes the drive axles of a FWD vehicle.

## SUSPENSION SYSTEM COMPONENTS

Most automotive suspension systems have the same basic components and operate similarly. The main difference between the different suspension system designs is the location and arrangement of the basic components.

### Springs

Springs absorb shock forces while maintaining the vehicle's correct riding height. Various types of springs are used: coil, torsion bar, leaf, and air springs. Springs are mounted in rubber-like materials to reduce road shock and noise.

#### Coil Springs

The coil spring is the most commonly used spring. Coil springs are actually a coiled-spring steel bar. When a vehicle wheel strikes a road irregularity, the coil spring compresses to absorb shock and then recoils back to its original height. Coil springs are designed to carry heavy loads, but they must be lightweight. Many coil springs have a vinyl coating which increases corrosion resistance and reduces noise.

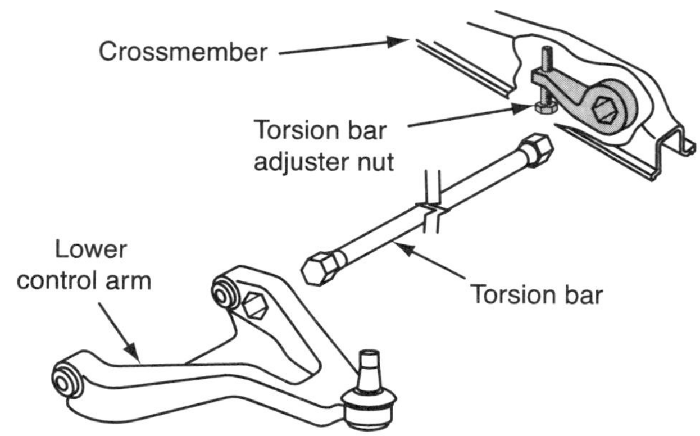


**Figure 4.** A torsion bar suspension system.

#### Torsion Bars

In some suspension systems, torsion bars are used rather than coil springs. As the wheel moves down, the torsion bar twists. As the wheel moves up, the torsion bar unwinds back to its original position. A torsion bar may be thought of as a straight, flattened coil spring. One end of the heat-treated alloy steel torsion bar is attached to the vehicle frame, and the opposite end is connected to the lower control arm. Some light-duty trucks and sport utility vehicles (SUVs) are equipped with longitudinal torsion bars in the front suspension **(Figure 4).** Transversely mounted torsion bars were used in some older front suspensions.

Since the lower control arm moves up and down as the wheel strikes road irregularities, the control arm action twists the torsion bar. The bar's natural resistance to twisting causes it to return to its original position. During the manufacturing process, torsion bars are prestressed to provide fatigue strength. These bars are directional. Torsion bars are marked right or left, and they must be installed on the appropriate side of the vehicle.

Shorter, thicker torsion bars have increased load-carrying capacity compared to longer, thinner bars. Since torsion bars require less space than a coil or leaf spring, they are usually found on front suspensions. However, a few rear suspensions use them. Torsion bars have a riding-height adjustment screw at the end where they are attached to the frame.

#### Leaf Springs

Leaf springs may be multiple leaf or mono leaf. A mono-leaf or single-leaf spring is a tapered plate with a thick center. Multiple-leaf springs have a series of flat steel leaves of varying lengths that are clamped together. A center bolt extends through all the leaves to maintain the leaf position in the spring. The upper leaf is called the main leaf, and this leaf has an eye on each end. An insulating bushing is pressed into each main leaf eye. The front bushing is attached to the frame, and the rear bushing is connected

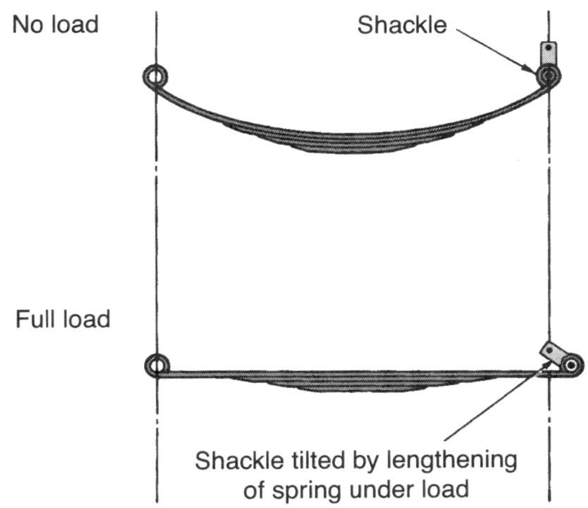


**Figure 5.** The action of a leaf spring and its shackle.

through a shackle to the frame. The shackle provides fore-and-aft movement as the spring compresses **(Figure 5).**

As a leaf spring compresses, it becomes progressively stiffer. When a leaf spring is compressed, its length changes, and the leaves slide on each other. This sliding action can be a source of noise and friction. This noise and friction is reduced by interleaves, or spacers, made from zinc and plastic placed between the steel leaves. The spring center bolt fits into an opening in the axle to position the axle properly and provide proper vehicle tracking. If the center bolt is broken, the axle position may shift and alter vehicle tracking and alignment. Leaf springs are usually mounted at right angles to the axle and provide excellent resistance to lateral movement.

### Air Springs

Air springs are normally part of computer-controlled suspension systems. The air springs provide automatic front and rear load leveling. The springs are located in the same positions as the coil springs they replace. Each spring consists of a reinforced rubber bag pressurized by air. An air compressor controls the air pressure in the springs, thereby controlling the action of the spring.

## Shock Absorbers

**Shock absorbers** dampen or control motion. Without shock absorbers, springs will continue to expand and contract after a bump until all the energy is absorbed. Most of the shock absorbers used on today's vehicles use hydraulics to dampen the movement of the springs. Also common are gas shocks **(Figure 6),** as well as electronically controlled shock absorbers.

## Steering Knuckle

The wheels rotate around a wheel spindle and steering knuckle assembly. The wheel bearings are fit around the outside of the spindle and on the inside of the wheel hub **(Figure 7).** The **steering knuckle** assembly holds the

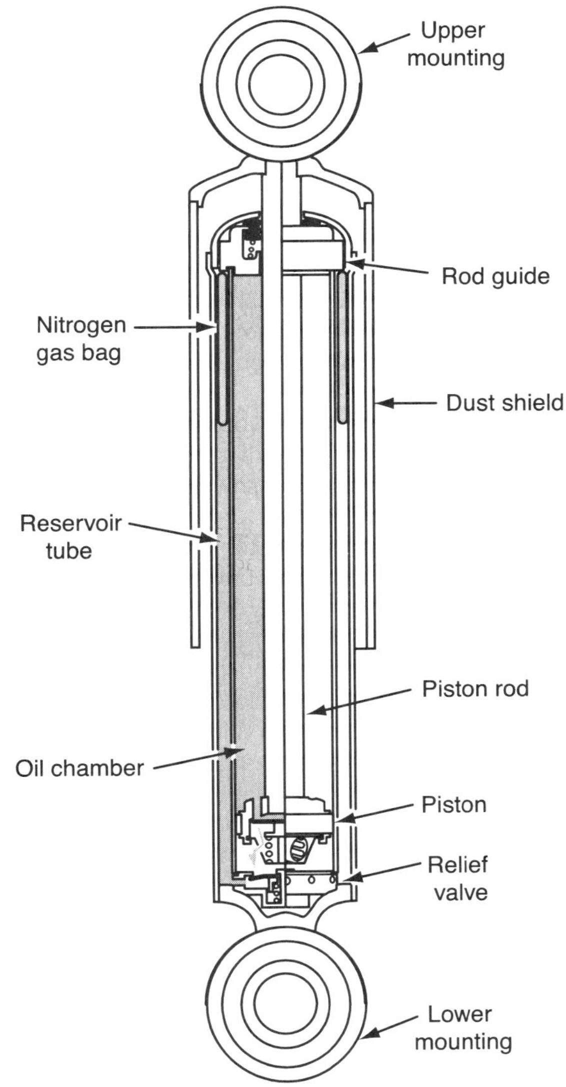


**Figure 6.** A typical gas shock absorber.

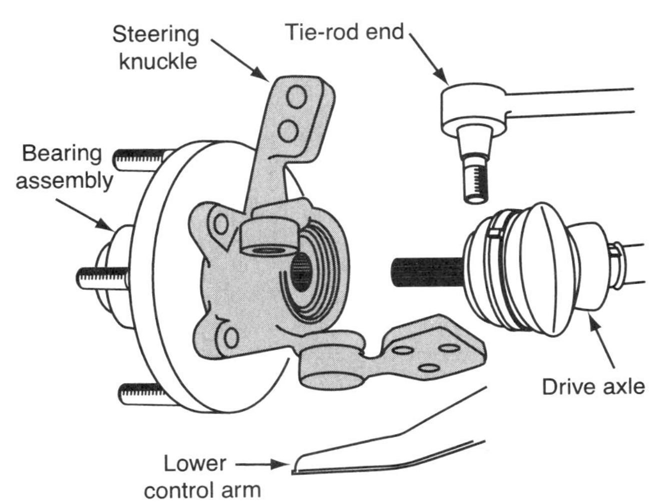


**Figure 7.** A typical steering knuckle assembly.

wheel in place while allowing the steering and suspension system to move in response to road conditions and driver input.

## Control Arms

**Control arms** serve as the locators for the suspension system. Control arms are attached to the vehicle's frame with bushings and allow the wheels to move up and down in response to the road's surface. The outer ends of the control arms are connected to the steering knuckle with ball joints.

## Ball Joints

Ball joints act as pivot points to allow the front wheels and spindles or knuckles to turn between the upper and lower control arms **(Figure 8).** Ball joints may be grouped into two classifications: load carrying and non-load carrying. Ball joints may be manufactured with forged, stamped, cold-formed, or screw-machined housings.

### Load-Carrying Ball Joint

The coil spring is seated on the control arm to which the **load-carrying ball joint** is attached. When the coil spring is mounted between the lower control arm and the chassis, the lower ball joint is a load-carrying joint. In a torsion bar suspension, the load-carrying ball joint is mounted on the control arm to which the torsion bar is attached. A load-carrying ball joint supports the vehicle weight.

**Figure 8.** A lower control arm and ball joint assembly.

In a load-carrying ball joint, the vehicle's weight forces the ball stud into contact with the bearing surface in the joint. Load-carrying ball joints may be compression loaded or tension loaded. If the control arm is mounted above the lower end of the knuckle and rests on the knuckle, the ball joint is compression loaded. In this type of ball joint, the vehicle's weight is pushing downward on the control arm, and the weight is supported on the tire and wheel. Since the ball joint is mounted between the control arm and the steering knuckle, the vehicle's weight squeezes the ball joint together. The ball joint is mounted in the lower control arm, and the ball joint stud faces downward.

When the lower control arm is positioned below the steering knuckle, the vehicle's weight is pulling the ball joint away from the knuckle. This type of ball joint mounting is referred to as tension loaded. The ball joint is mounted in the lower control arm with the ball joint stud facing upward into the knuckle.

Since the load-carrying ball joint supports the vehicle weight, this ball joint wears faster than a non-load-carrying ball joint. Many load-carrying ball joints have built-in wear indicators. These have an indicator on the grease nipple surface that recedes into the housing as the joint wears. If the ball joint is in good condition, the grease nipple shoulder extends a specified distance out of the housing. If the grease nipple shoulder is even with or inside the ball joint housing, the ball joint is worn and replacement is necessary.

### Non-Load-Carrying Ball Joint

A **non-load-carrying ball joint** may be referred to as a stabilizing or follower ball joint. This joint is designed with a preload, which provides damping action. The preload provides improved steering quality and vehicle stability.

### Low-Friction Ball Joint

Low-friction ball joints are standard equipment on many vehicles. Low-friction ball joints provide precise movement of the ball socket in the ball joint. Compared to conventional ball joints, two-thirds of the internal friction is eliminated. Low-friction ball joints have a highly polished ball socket surface surrounded by a high-strength polymer bearing. The smooth ball socket movement provides improved steering performance, better steering wheel return, and longer ball joint life.

## Stabilizer Bar

The **stabilizer bar** is attached to the cross member and connects the lower control arms. Rubber insulating bushings are used at all stabilizer bar attachment points

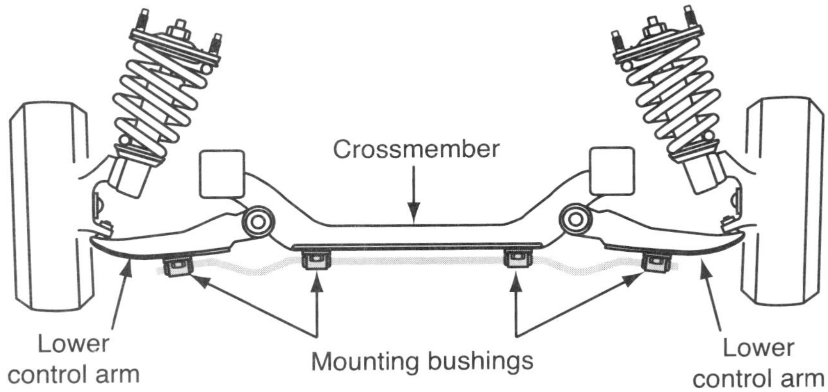


**Figure 9.** The mounting for a stabilizer bar.

**(Figure 9).** When jounce and rebound movements affect one front wheel, the stabilizer bar transmits part of this movement to the opposite lower control arm and wheel, which reduces and stabilizes body roll.

## Strut Rod

On some front suspension systems, a **strut rod** is connected from the lower control arm to the chassis. The strut rod is bolted to the control arm, and a large rubber bushing surrounds the strut rod in the chassis opening. The outer end of the strut rod is threaded, and steel washers are positioned on each side of the strut rod bushing. Two nuts tighten the strut rod into the bushing. The strut rod prevents fore-and-aft movement of the lower control arm. In some suspension systems, the position of the strut rod nuts provides proper front wheel adjustment.

# I-BEAM FRONT SUSPENSION SYSTEMS

Many early automobiles had I-beam front axles and leaf springs. Some I-beam suspension systems had two longitudinally mounted leaf springs, whereas others had one transversely mounted leaf spring. I-beam suspension systems are obsolete on cars today, but they may still be found on some trucks. **Kingpins** are used to connect the I-beam axle to the spindles. Spindle bushings allow the spindle to pivot on the kingpins. When an I-beam front axle is used, if one front wheel strikes a bump, movement is transferred through the axle to the opposite front wheel causing it to move from its normal rotation plane. This action decreases directional control and ride quality and increases tire wear.

## Twin I-Beam Suspension Systems

Some Ford trucks are equipped with twin I-beam front suspension systems. Each front wheel is connected to a separate I-beam. The outer ends of the I-beam are connected to the spindles, and the inner ends of the beams are connected through a rubber pivot bushing to the chassis. Coil springs are positioned between the I-beams and the chassis to support the vehicle's weight. Radius arms are connected rearward from each I-beam to the chassis to prevent longitudinal wheel movement. Since each front wheel can move independently in a twin I-beam suspension system, the problems associated with straight I-beam systems are greatly reduced.

In some I-beam front suspension systems, kingpins are used to attach the I-beams to the spindles. In other systems, ball joints connect the I-beams to the spindles.

# SHORT-AND-LONG ARM (SLA) FRONT SUSPENSION SYSTEMS

As automotive technology evolved, the control arm front suspension system replaced the I-beam front suspension system. This suspension system has coil springs and upper and lower control arms. Since the movement of one front wheel does not directly affect the opposite front wheel, the control arm suspension is called an independent system. Many rear-wheel-drive cars have control arm front suspension systems. Early front suspension systems had equal length upper and lower control arms. On these suspension systems, the bottom of the tire moved in and out as the tire and wheel moved over irregular surfaces. This action caused tire scuffing and wear problems.

In **short-and-long arm (SLA) suspension** systems, the upper control arm is shorter than the lower control arm

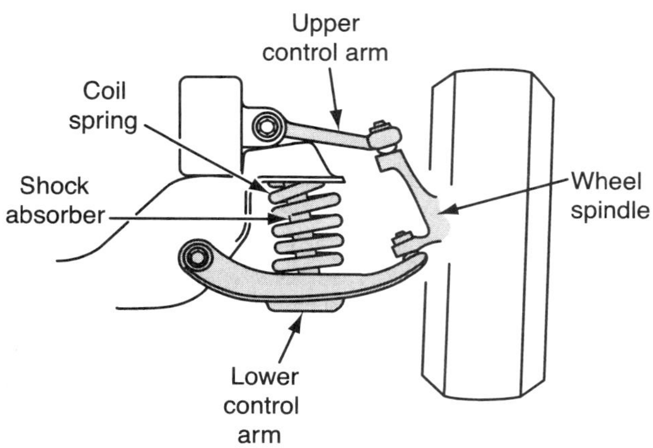


**Figure 10.** An SLA front suspension arrangement.

**(Figure 10).** During wheel travel on irregular surfaces, the upper control arm moves in a shorter arc than the lower control arm. This action moves the top of the tire slightly in and out, but the bottom of the tire remains in a more constant position. This short-and-long arm front suspension system provides reduced tire tread wear, improved ride quality, and better directional stability compared to I-beam suspension systems and suspension systems with equal length upper and lower control arms.

The inner end of the lower control arm contains large rubber insulating bushings, and the ball joint is attached to the outer end of the control arm. The lower control arm is bolted to the front cross member, and the attaching bolts are positioned in the center of the lower control arm bushings. An upper control arm shaft is bolted to the frame, and rubber insulators are located between this shaft and the control arm.

## Steering Knuckle

The upper and lower ball joint studs extend through openings in the steering knuckle. Nuts are threaded onto the ball joint studs to retain the ball joints to the knuckle, and the nuts are secured with cotter pins. When the steering wheel is turned, the steering gear and linkage turn the steering knuckle. During this action, the steering knuckle pivots on the upper and lower ball joints. The upper and lower control arms must be properly positioned to provide correct tracking between the front and rear wheels. The control arm bushings must be in satisfactory condition to position the control arms properly.

The upper control arm is mounted high in the suspension system, and the upper end of the knuckle has a "goose neck" shape. The lower control arm is made from stamped steel to reduce weight. The rear lower control arm bushing carries only fore-and-aft loads and has a softer rear bushing. The horizontal front lower control arm bushing and the lower shock absorber mounting are aligned with the wheel center. This provides a direct path for lateral cornering loads. This design allows the use of a hard front lower control arm bushing.

## Coil Springs and Shock Absorbers

A coil spring is positioned between the lower control arm and the spring seat in the frame. A spring seat is located in the lower control arm, and an insulator is positioned between the top of the coil spring and the spring seat in the frame. The shock absorber is mounted in the center of the coil spring, and the lower shock absorber bushing is bolted to the lower control arm. The top of the shock absorber extends through an opening in the frame above the upper spring seat. Washers, grommets, and a nut retain the top of the shock absorber to the frame. Side roll of the front suspension is controlled by a stabilizer bar mounted to the lower control arms and the frame with rubber bushings. On some short-and-long arm front suspension systems, the coil spring is positioned between the upper control arm and the chassis.

## MacPHERSON STRUT FRONT SUSPENSION SYSTEM DESIGN

When smaller FWD vehicles became popular, most had **MacPherson strut suspension** systems. The **strut** is the shock absorber for the front suspension, and a coil spring is fitted over each strut **(Figure 11).** The lower end of the strut is bolted to the top of the steering knuckle, and

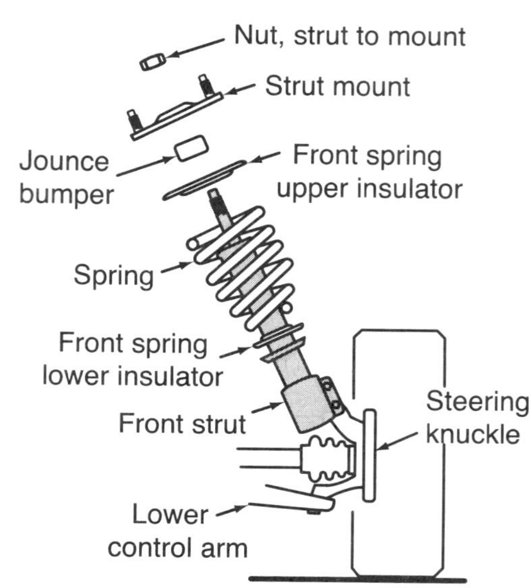


**Figure 11.** A MacPherson strut suspension.

the lower end of the knuckle is attached to the ball joint in the lower control arm. An upper strut mount connects the top of the strut to the chassis. An upper control arm is not required in this type of suspension system because the strut supports the top of the steering knuckle. Since the upper control arm is not required, these suspension systems are more compact.

On some MacPherson strut front suspension systems, a lower control arm is positioned longitudinally on each side of the front suspension and is bolted to the chassis. The inner ends of the lower control arms have large insulating bushings with a bolt opening in the center. The control arm retaining bolts extend through the openings for support.

Road irregularities cause the tire and wheel to move up and down, and the lower control arm bushings pivot on the mounting bolts during this movement. When the vehicle is driven over road irregularities, vibration and noise are applied to the tire and wheel. The control arm bushings help reduce the transfer of this noise and vibration to the passenger compartment. Proper location of the lower control arm is important to provide correct vehicle tracking.

### Steering Knuckle and Bearing Assembly

The front wheel bearing assembly is bolted to the outer end of the steering knuckle, and the brake rotor and wheel rim are retained on the studs in the wheel bearing assembly. The front drive shaft is splined into the center of the wheel bearing hub. Thus, drive axle torque is applied to the front wheel. A tie-rod end connects the steering linkage to the steering knuckle. The top end of the steering knuckle is bolted to the lower end of the strut.

### Strut and Coil Spring Assembly

The strut assembly is normally comprised of a shock absorber (strut) and a coil spring. An insulator is located between the lower spring seat and the bottom of the coil spring. An upper strut mount is retained on top of the strut with a nut threaded onto the upper end of the strut rod. The upper strut mount contains a bearing and upper spring seat, and an insulator is positioned between the top of the coil spring and the seat. The upper and lower insulators help prevent the transfer of noise and vibration from the spring to the strut and body.

A bumper is located on the upper end of the strut rod. This bumper reduces harshness while driving on severe road irregularities. During upward wheel movement, the bumper strikes the upper spring seat before the coils in the spring hit each other. Therefore, this bumper reduces harshness when the wheel and suspension move fully upward. The spring tension is applied against the upper and lower spring seats and insulators. However, the nut on top of the upper mount holds the spring in the compressed position between the upper and lower spring seats. When the steering wheel is turned, the steering linkage turns the steering knuckles to the right or left. During this turning action, the strut and spring assembly pivots on the lower ball joint and the upper strut mount bearing.

### Modified MacPherson Strut Suspension

A modified MacPherson strut front suspension is used on some vehicles. This type of suspension has MacPherson struts with coil springs positioned between the lower control arms and the frame. The struts in these systems may be gas filled or oil filled.

## PERFORMANCE FRONT SUSPENSION SYSTEMS

High-performance suspension systems are usually installed on cars where the driver expects improved steering quality, especially when driving and cornering at higher speeds.

### Multilink Front Suspension System

In a **multilink suspension,** a short upper link is attached to the chassis with a bracket, and the outer end of this upper link is connected to a third link. Large rubber insulating bushings are mounted in each end of the upper link. The lower end of the third link is connected through a heavy pivot bearing to the steering knuckle. The lower link is similar to the conventional lower control arm. A rubber insulating bushing connects the inner end of the lower link to the front cross member, and a ball joint is connected from the outer end of the lower link to the steering knuckle. In the multilink suspension system, the kingpin axis extends through the lower ball joint and upper pivot bearing, but the kingpin axis is independent from the upper and third links. The extra links in a multilink suspension system maintain precise wheel position during cornering to provide excellent directional stability and steering control while minimizing tire wear.

The shock absorbers are connected from the lower end of the third link to the fender reinforcement. A coil spring seat is attached to the lower end of the shock absorber, and the upper spring seat is located on the upper shock absorber mounting insulator. Since the steering knuckle pivots on the lower ball joint and the upper pivot bearing, the coil spring and shock absorber do not rotate with the knuckle as they do in a MacPherson strut suspension. Tension or strut rods are connected from the lower links to tension rod brackets attached to the chassis. A stabilizer bar is mounted on rubber insulating bushings in the tension rod brackets, and the outer ends of this bar are attached to the third link.

## Multilink Suspension with Compression and Lateral Lower Arms

Some multilink front suspensions have compression and lateral lower arms. The lateral arm prevents front wheel movement, and the compression arm prevents fore-and-aft front wheel movement. A rubber insulating bushing in the inner end of the lateral arm is bolted to the chassis. A second rubber insulating bushing near the outer end of the lateral arm is bolted to the damper fork. The upper end of the damper fork is bolted to the front strut. A ball joint in the outer end of the lateral arm is bolted into the steering knuckle. A rubber insulating bushing in the inner end of the compression arm is bolted to the chassis, and a ball joint in the outer end of this arm is bolted into the steering knuckle.

## Double Wishbone Front Suspension System

**Double wishbone suspension** systems provide increased suspension rigidity and maintain precise wheel position under all driving conditions to supply improved directional stability and steering control. In the double wishbone front suspension system, the upper and lower control arms are manufactured from lightweight, high-strength aluminum alloys designed for maximum strength and rigidity. These lighter control arms decrease the unsprung weight of the vehicle, which improves traction and ride quality. Since the upper and lower control arms have a wishbone shape, the term "double wishbone" is used for this type of suspension. Suspension rigidity is also increased by positioning the ball joints and steering knuckle inside the wheel profile. On each side of the car, the front suspension is attached to the chassis by a cast aluminum subframe. This design also reduces vehicle weight. The double wishbones are attached to the chassis at the most efficient locations to maintain precise wheel position and provide improved ride quality.

# TORSION BAR SUSPENSIONS

Some light-duty trucks and sport utility vehicles (SUVs) have torsion bar front suspension systems. These systems have longitudinally mounted torsion bars. The front end of each torsion bar is anchored to the lower control arm, and the rear is anchored to a chassis' cross member. Since the twisting of the torsion bars supports the vehicle's weight, the torsion bars replace the coil springs.

Vehicle ride height is controlled by the torsion bar anchor adjusting bolts in the cross member. Front suspension heights must be within specifications for correct wheel alignment, tire wear, satisfactory ride, and accurate bumper heights. A conventional stabilizer bar is connected between the lower control arms and the cross member. Ball joints are located in the upper and lower control arms. Both ball joints are bolted into the steering knuckle. The shock absorbers are connected between the lower control arms and the cross member support. The inner ends of the lower control arms are bolted to the cross member through an insulating bushing.

# STEERING SYSTEMS

A **steering system** has three major subsystems: the steering linkage, steering gear, and steering column and wheel. As the steering wheel is turned by the driver, the steering gear transfers this motion to the steering linkage. The steering linkage turns the wheels. Although there are many variations to this system, these assemblies are in all steering systems.

# STEERING LINKAGE

The term **steering linkage** is applied to the system of pivots and connecting parts that is placed between the steering gear and the steering arms attached to the front or rear wheels. The steering linkage transfers the motion of the steering gear to the steering arms, turning the wheels to maneuver the vehicle.

The type of suspension (independent wheel suspension as compared to a solid front axle) greatly influences steering geometry. Most passenger cars and many light trucks and recreational vehicles have independent front-wheel suspension systems. Therefore, a steering linkage arrangement that tolerates relatively large wheel movement must be used.

## Parallelogram Steering Linkage

**Parallelogram steering linkage** was the most commonly used type found on passenger cars. Now it is found mostly on larger cars, pickups, and larger SUVs. It is used with the recirculating ball steering gear.

This type of linkage provides good steering and suspension geometry. However, road vibrations and impact forces are transmitted to the linkage from the tires, causing wear and looseness in the system. This looseness causes intermittent changes in the toe setting of the front wheels, allowing further tire wear.

In a parallelogram steering linkage, the tie-rods have ball socket assemblies at each end. One end is attached to the wheel's steering knuckle and the other end to the center link.

The components in a parallelogram steering linkage arrangement are the **pitman arm, idler arm,** links, and **tie-rods (Figure 12).**

### Pitman Arm

The pitman arm connects the linkage to the steering column through the steering gear located at the base of

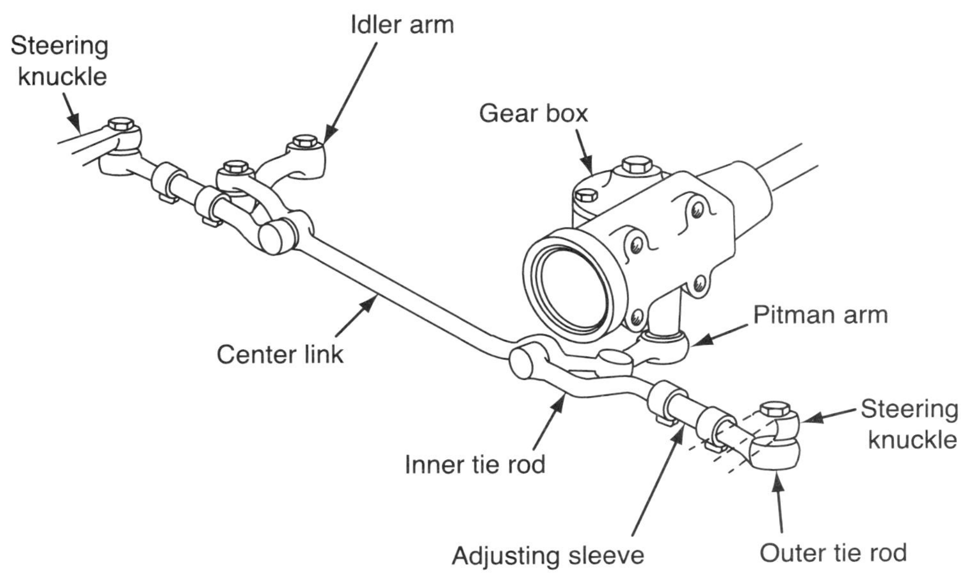


**Figure 12.** Parts of a parallelogram steering linkage setup.

the column. It transmits the motion from the gear to the linkage, causing the linkage to move left or right to turn the wheels. It also maintains the height of the center link. This ensures that the tie-rods are parallel to the control arm movement to prevent unsteady toe settings or **bump steer**. Toe, a critical alignment factor, is a term that defines how well the tires point to the direction of the vehicle.

### Idler Arm

The idler arm or idler arm assembly is normally attached from the pitman arm and to the vehicle's frame, supporting the center link at the correct height. A pivot built into the arm permits sideway movement of the linkage. On some linkages, such as those on a few light-duty trucks, two idler arms are used.

### Links

Links, depending on the design application, can be referred to as **center, drag,** or steering **links.** Their purpose is to control sideway linkage movement which causes changes in the direction of the wheel. Because links are usually mounting locations for tie-rods, they are very important for maintaining correct toe settings. If they are not mounted at the correct height, toe is unstable and toe change or bump steer is produced. Depending on the design of the steering center links and drag links can be used either alone or in conjunction with each other.

### Tie-Rods

Tie-rods make the final connections between the steering linkage and steering knuckles. They consist of inner tie-rod ends (connected to the opposite sides of the center link), outer tie-rod ends (connected to the steering knuckles), and adjusting sleeves or bolts (which join the inner and outer tie-rod ends) permitting the tie-rod length to be adjusted for correct toe settings.

## Rack and Pinion Steering Linkage

**Rack and pinion** is lighter in weight and has fewer components than parallelogram steering **(Figure 13).**

> **You Should Know** *Never apply heat to any part of the steering linkage while servicing it. If any parts must be heated in order to remove them, they must be replaced and not reinstalled.*

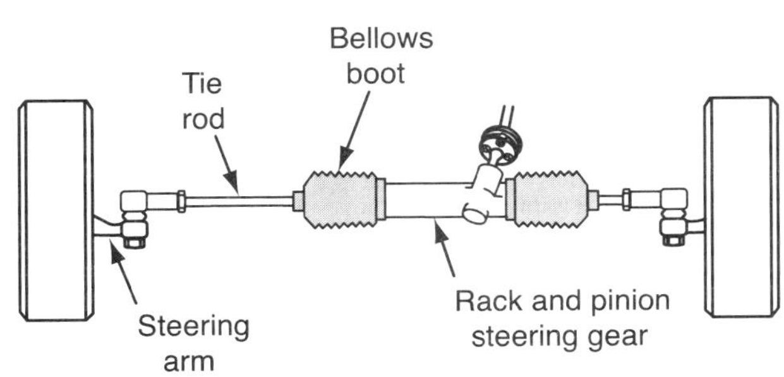


**Figure 13.** A rack and pinion steering assembly.

Steering input is received from a pinion gear attached to the steering column. This gear moves a toothed rack that is attached to the tie-rods.

In the rack and pinion steering arrangement, there is no pitman arm, idler arm assembly, or center link. The movement of the **rack** pushes and pulls the tie-rods to change the wheel's direction. The tie-rods are the only steering linkage parts in a rack and pinion system.

Most rack and pinion systems **(Figure 14)** are comprised of a tube in which the steering rack can slide. The rack is a rod with gear teeth cut along one end-spur and helical. The other end is fitted with two balls to which the ends of the divided track rods are attached. The rack meshes with the teeth of a small pinion at the end of the steering column. The two inner tie-rod ends, which are attached to the rack, are covered by rubber bellows boots that protect the rack from contamination. The inner tie-rods connect to outer tie-rod ends that connect to the steering arms. The rack and pinion housing is fastened to the vehicle at two or three points.

In some cases, the rack and pinion steering gear on unibody cars is bolted directly to a body panel like a cowl. When this is done, the body panel must hold the steering gear in its correct location. The unibody structure must maintain the proper relationship of the steering and suspension parts to each other. Along with other advantages, the rack and pinion steering system, combined with the MacPherson strut suspension system, is found in most front-wheel-drive unibody vehicles because of their weight- and space-saving feature.

### Rack

The rack is a toothed bar contained in a metal housing. The rack maintains the correct height of the steering components so that the tie-rod movement is able to parallel control arm movement. The rack is similar to the parallelogram center link in that its sideway movement in the housing is what pulls or pushes the tie-rods to change wheel directions.

### Pinion

The pinion is a worm gear mounted at the base of the steering column assembly where it is moved by the steering wheel. The pinion gear meshes with the teeth in the rack so that the rack is propelled sideways in response to the turning of the pinion.

### Tie-Rods

Tie-rods are very similar to those used on parallelogram systems. They consist of inner and outer ends and adjusting sleeves or bolts. The inner tie-rod ends on rack and pinion

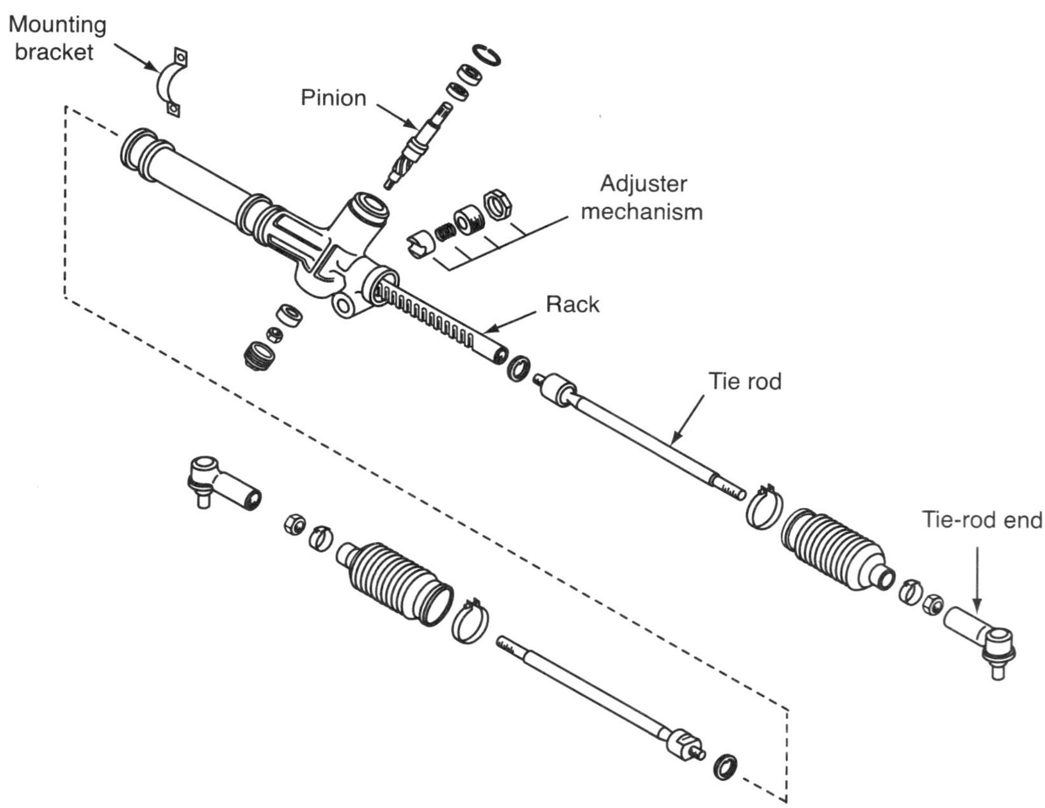


**Figure 14.** Details of a rack and pinion steering assembly.

units are usually spring-loaded ball sockets that screw onto the rack ends. They are preloaded and protected against contaminant entry by a rubber bellows or boot.

## Steering Damper

Some vehicles have a steering damper added to their steering linkage **(Figure 15).** A steering damper reduces the amount of road shock that is transmitted up through the steering column. Steering dampers are mostly found on 4WD, especially those fitted with large tires. The damper serves the same function as a shock absorber but is mounted horizontally to the steering linkage—one end to the center link, the other to the frame.

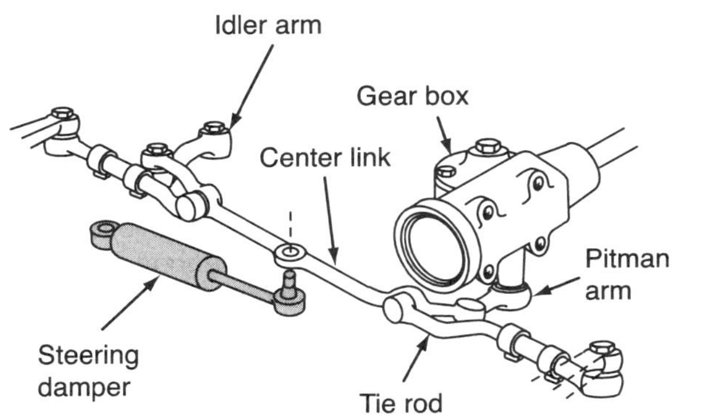


**Figure 15.** The location of a steering damper on a typical steering linkage.

# MANUAL STEERING GEARS

Steering gears change the rotational motion of the steering wheel to a reciprocating motion to move the steering linkage. There are three styles currently in use: the recirculating ball, worm and roller, and the rack and pinion. The latter gear assembly incorporates the already described rack and pinion linkage system and steering gear as a single unit.

The **recirculating ball** is generally found in larger cars, as shown in **Figure 16.** A sector shaft is supported by needle bearings in the housing and a bushing in the sector cover. Ball bearings recirculate through two outside loops, referred to as ball return guide tubes. The ball nut has gear teeth cut on one face that mesh with gear teeth on the sector shaft. As the steering wheel is rotated, the worm shaft rotates, causing the ball nut to move up or down the worm

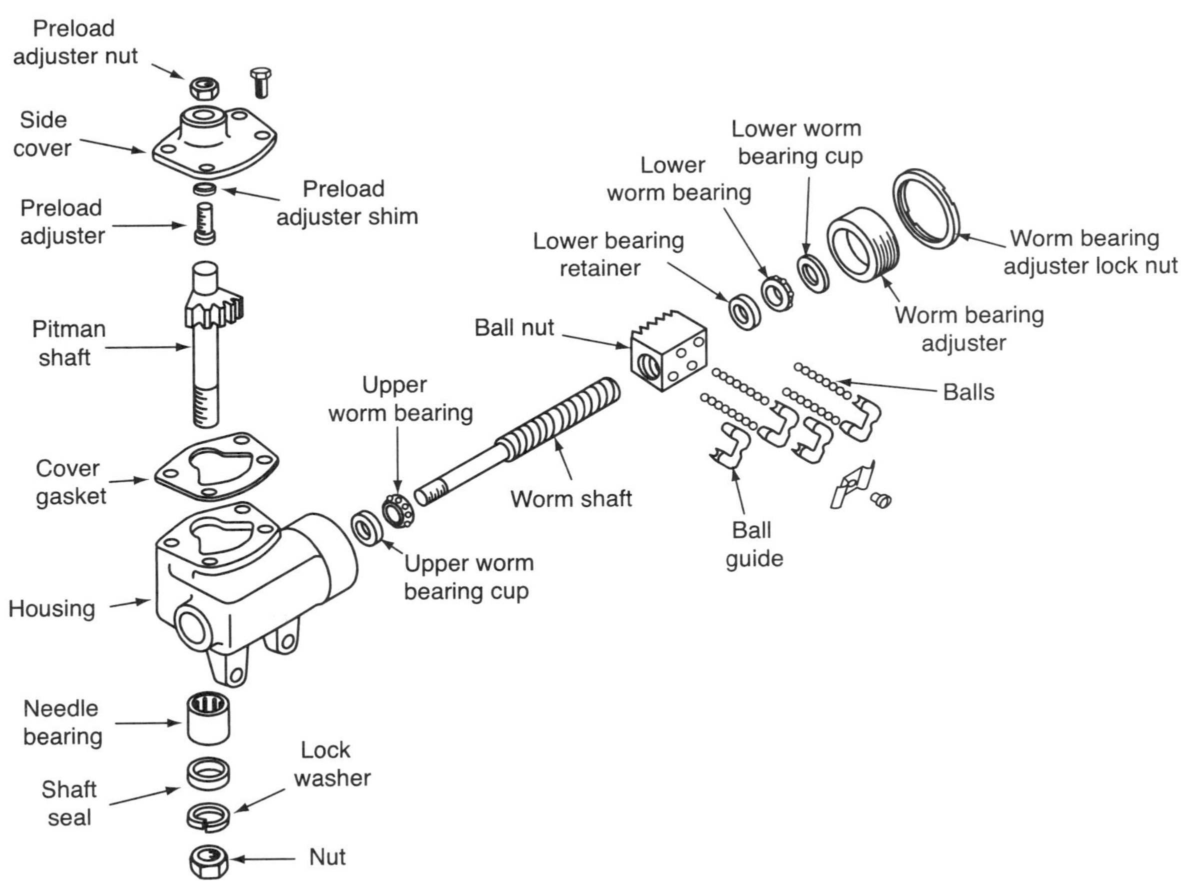


**Figure 16.** A recirculating ball steering gear.

shaft. Since the gear teeth on the ball nut are meshed with the gear teeth on the sector shaft, the movement of the nut causes the sector shaft to rotate and swing the pitman arm.

The design of two separate circuits of balls results in an almost friction-free operation of the ball nut and the worm shaft. When the steering wheel is turned, the ball bearings roll in the ball thread grooves of the worm shaft and ball nut. When the ball bearings reach the end of their respective circuit, they enter the guide tubes and are returned to the other end of the circuits.

The teeth on the sector shaft and the ball nut are designed so that an interference fit exists between the two when the front wheels are straight ahead. This interference fit eliminates gear tooth lash for a positive feel when driving straight ahead. Proper mesh engagement between the sector and ball nut is obtained by an adjusting screw that moves the sector shaft axially.

**You Should Know** *Proper adjustment between the worm gear and the pitman gear is important for adequate steering response. Refer to the vehicle manual for specific adjustment procedures.*

The sector can be either constant or variable ratio. The former's teeth are all identical, while the latter has one long tooth between two shorter teeth that changes the amount of mechanical advantage according to the position of the wheel. This serves to make the steering faster (maybe 13:1) in turns than in a straight direction (15 or 16:1). Variable ratio is ordinarily used only in power-steering units.

The **worm and roller** gearbox is similar to the recirculating ball except a single roller replaces the balls and ball nut. This reduces internal friction, making it ideal for smaller cars. This steering linkage is found on older vehicles, and its components serve the same purpose as a parallelogram steering linkage.

In operation, the steering shaft rotates the worm gear. It, in turn, engages the roller, causing the roller shaft to turn. The shaft moves the pitman arm left or right to steer the vehicle.

## STEERING WHEELS AND COLUMNS

The purpose of the steering wheel and column is to produce the necessary force to turn the steering gear. The exact type of steering wheel and column depends upon the year and the car manufacturer. The steering column, also called a steering shaft, relays the movement of the steering wheel to the steering gear.

The steering wheel is used to produce the turning effort. The lower and upper covers conceal parts. The universal joints rotate at angles. Support brackets are used to hold the steering column in place. Since 1968, all steering columns have a collapsible feature that allows the column to fold into itself upon impact. This feature prevents injury to the driver.

In most vehicles equipped with a driver's side air bag, the air bag assembly is contained in the center portion of the steering wheel. This assembly must be disarmed and removed before the steering wheel can be removed.

**You Should Know** *Always disconnect the negative battery cable and wait 30 minutes before beginning to remove an air bag assembly. Failure to this may result in accidental air bag deployment. Air bag service precautions vary depending on the vehicle. Always follow the manufacturer's service precautions and recommendations when working with an air bag system.*

Different steering wheel and column designs include fixed column, telescoping column, tilt column, manual transmission, floor shift, and automatic transmission column shift. The tilt columns feature at least five driving positions (two up, two down, and a center position). Both fixed and tilt columns may house an emergency warning flasher control, a turn signal switch, ignition key, lights (high/low beams), horn, windshield wipers and washers, and an anti-theft device that locks the steering system. On automatic transmission-equipped vehicles, the transmission linkage locks too.

The shaft is mounted to the tube by a breakaway plastic capsule or a series of inserts, (steel balls held in a plastic retainer that allow the shaft to roll forward inside the tube), a collapsible steel mesh **(Figure 17)** or an accordion-pleated device that gives way under pressure.

The steering wheel is usually held in place on the steering column by either a bolt or nut. When the blocked tooth on the steering gear input shaft is in the 12 o'clock position, the front wheels should be in the straight-ahead

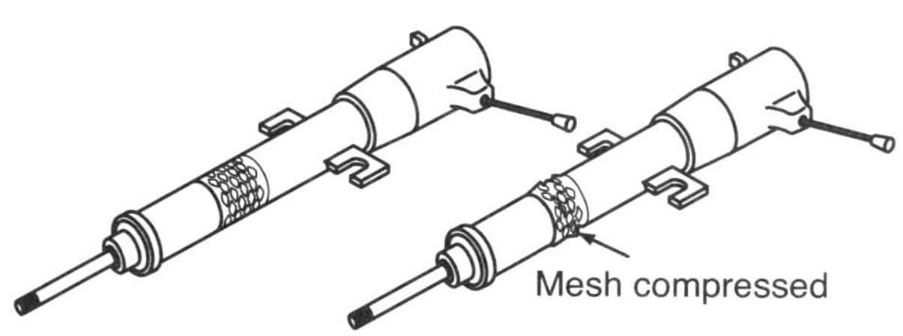


**Figure 17.** The collapsible steel mesh used in many steering columns is designed to give way under pressure in order to protect the driver.

position and the steering wheel spokes in their normal position. If the spokes are not in their normal position, they can be adjusted by changing the toe adjustment. This adjustment can be made only when the steering wheel indexing mark is aligned with the steering column indexing marks. As a rule, indexing teeth or mating flats on the wheel hub and steering shaft prevent misindexing of these components. The alignment of the notches on the steering wheel hub and steering shaft confirm correct orientation.

## POWER STEERING

The power-steering unit is designed to reduce the amount of effort required to turn the steering wheel. It also reduces driver fatigue on long drives and makes it easier to steer the vehicle at slow road speeds, particularly during parking.

Power steering can be broken down into two design arrangements: conventional and non-conventional or electronically controlled. In the conventional arrangement, hydraulic power is used to assist the driver **(Figure 18).** In the non-conventional arrangement, an electric motor and electronic controls provide power assistance in steering.

There are several power steering systems, the most common ones are the external-piston linkage, integral-piston, and power-assisted rack and pinion system.

### External-Piston Linkage System

The **external-piston linkage** system, sometimes called non-integral, is generally found on older cars. It consists of a power-steering pump and reservoir, control valve, power cylinder, and four hoses. The control valve and pressure cylinders are separate components that are not incorporated into the steering gear. The pressure hose connects the pump to the control valve and directs the pressurized fluid into the correct booster hoses to the power cylinder. The power cylinder piston provides the necessary power assistance.

### Integral Piston System

The **integral piston** system is one of the most common conventional power-steering systems in use today. It consists of a power-steering pump and reservoir, power-steering pressure and return hose, and steering gear. The power cylinder and the control valve are in the same housing as the steering gear.

On some recent model cars and light trucks with diesel engines, the hydraulic fluid from the power-steering pump is also used to actuate the brake booster. This brake system is called the hydro-boost system **(Figure 19).**

### Power-Assisted Rack and Pinion System

The power-assisted rack and pinion system is similar to the integral system because the power cylinder and the control valve are in the same housing. The rack housing acts as the cylinder, and the power piston is part of the rack. Control valve location is in the pinion housing. Turning the

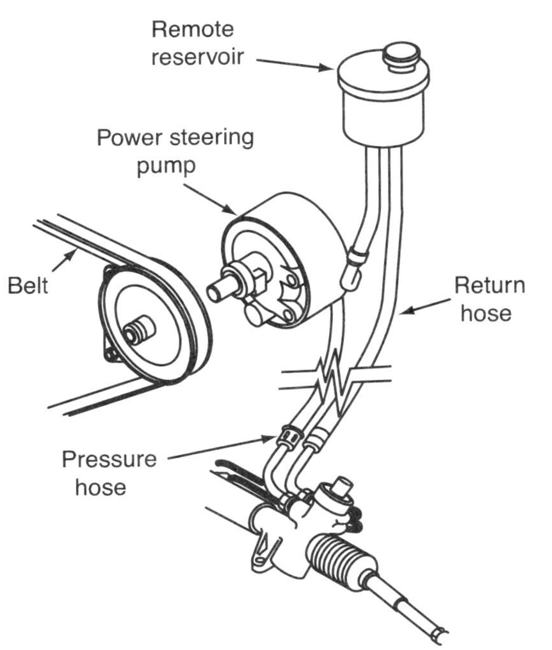


**Figure 18.** Parts of a typical power steering setup.

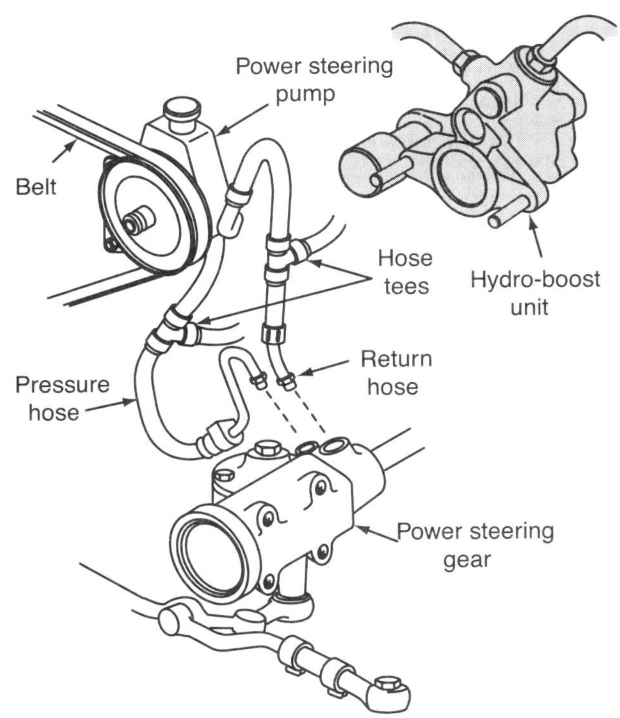


**Figure 19.** A Hydro-boost setup.

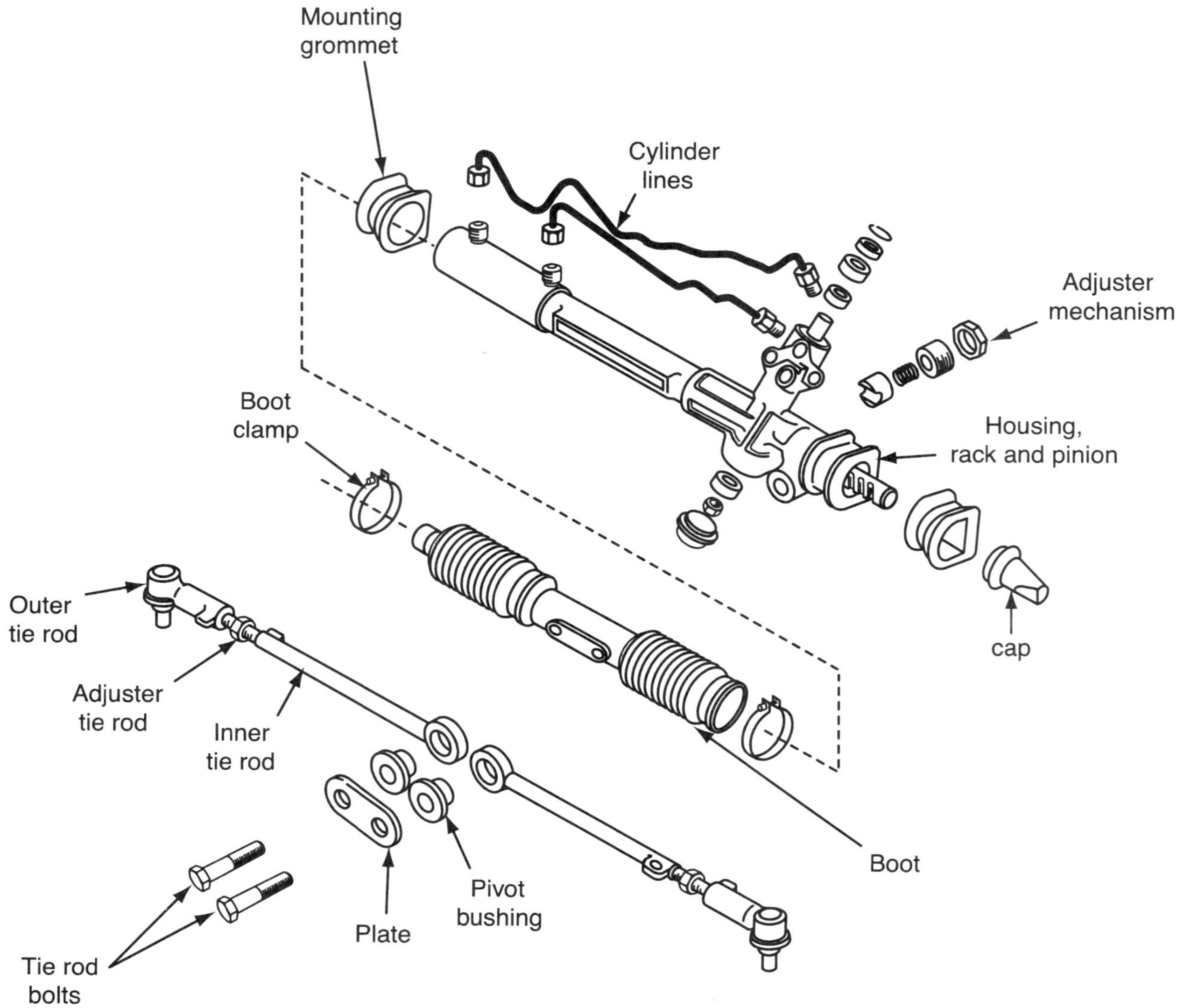


**Figure 20.** A power-assisted rack and pinion steering assembly.

steering wheel moves the valve, directing pressure to either end of the back piston. The system utilizes a pressure hose from the pump to the control valve housing and a return line to the pump reservoir. This type of steering system is common in front-wheel-drive vehicles.

Power-assisted rack and pinion components are basically the same as for manual rack and pinion steering except for the hydraulic control housing **(Figure 20).** In a power rack and pinion gear, the movement of the rack is assisted by hydraulic pressure. When the wheel is turned, the rotary valve changes hydraulic flow to create a pressure differential on either side of the rack. The unequal pressure causes the rack to move toward the lower pressure, thereby reducing the effort required to turn the wheels.

## Components

For the most part, the steering linkage of a power steering system is the same as the linkage for a non-power system. The components that have been added for power steering provide the hydraulic power that drives the system. These are the power-steering pump, flow control and pressure relief valves, reservoir, spool valves and power pistons, hydraulic hose lines, and gearbox or assist assembly on the linkage.

### Power-Steering Pump

The steering pump is used to develop hydraulic flow, which provides the force needed to operate the steering gear. The pump is belt driven from the engine crankshaft, providing flow any time the engine is running. It is usually mounted near the front of the engine. Hydraulic fluid for the pump is stored in a reservoir. Fluid is routed to and from the pump by hoses and lines. Excessive pressure is controlled by a relief valve.

### Flow Control and Pressure Relief Valves

A **pressure relief valve** controls the pressure output from the pump. This valve is necessary because of the variations in engine rpm and the need for consistent steering ability in all ranges from idle to highway speeds. It is positioned in a chamber that is exposed to pump outlet pressure at one end and supply hose pressure at the other. A spring is used at the supply pressure end to help maintain a balance.

When engine speed increases, the pump can deliver more flow than is required to operate the system. Since the outlet orifice restricts the amount of fluid leaving the pump, the difference in pressure at the two ends of the valve becomes greater until pump outlet pressure overcomes the combined force of supply line pressure and spring force. The valve is pushed down against the spring, opening a passage that returns the excess flow back to the inlet side of the pump.

### Power-Steering Gearbox

A power-steering gearbox is basically the same as a manual recirculating ball gearbox with the addition of a hydraulic assist. A power-steering gearbox is filled with hydraulic fluid and uses a control valve.

The integral power steering has the spool valve and a power piston integrated with the gearbox. The spool valve directs the oil pressure to the left or right power chamber to steer the vehicle. The spool valve is actuated by a lever or a small torsion bar.

In linkage systems, the control valve is connected directly to the steering center link and pitman arm. Any movement of the steering wheel compresses the centering spring and moves the valve spool. This opens and closes a series of ports directing fluid under pressure from the pump to one side or the other of the power cylinder piston.

The power cylinder in the linkage system is attached to the steering center link, and the piston shaft is attached to the flame. As fluid under pressure is directed to one side of the piston by the control valve, power assist is provided to aid the driver in moving the steering linkage and road wheels. Two lines connect the cylinder to the control valve. Each one functions as either the return line or the supply line, depending on the direction of the turn.

### Power-Steering Hoses

The primary purpose of power-steering hoses is to transmit power (fluid under pressure) from the pump to the steering gearbox and to return the fluid ultimately to the pump reservoir. Through material and construction, hoses also function as additional reservoirs and act as sound and vibration dampers.

Hoses are generally a reinforced synthetic rubber (neoprene) material coupled to metal tubing at the connecting points. The pressure side must be able to handle pressures up to 1,500 psi. For that reason, wherever there is a metal tubing to a rubber connection, the connection is crimped. Pressure hoses are also subject to surges in pressure and pulsations from the pump. The reinforced construction permits the hose to expand slightly and absorb changes in pressure.

Because of working fluid temperature and adjacent engine temperatures, these hoses must be able to withstand temperatures up to 300°F. Due to various weather conditions, they must also tolerate sub-zero temperatures as well. Hose material is specially formulated to resist breakdown or deterioration due to oil or temperature conditions.

**You Should Know** *Hoses must be carefully routed away from engine manifolds. Power-steering fluid is very flammable. If it comes in contact with hot engine parts, it could start an under-hood fire.*

## Electronically Controlled Power-Steering Systems

The object of power steering is to make steering easier at low speeds, especially while parking. However, higher steering efforts are desirable at higher speeds in order to provide improved down-the-road feel. **Electronically controlled power-steering (EPS) systems** provide both of these benefits. The hydraulic boost of these systems is tapered off by electronic control as road speed increases. Thus, these systems require well under 1 pound of steering effort at low road speeds and 3 pounds plus of steering effort at higher road speeds to enable the driver to maintain control of the steering wheel for improved high-speed handling.

### Active Steering

Active steering improves vehicle stability by turning the wheels more or less than commanded by the turn of the steering wheel during some situations. Through inputs and computer programming, this system can adjust the steering to respond quickly to the threat of skidding. Depending on speed, the system also allows for a variable steering ratio.

## Electric/Electronic Rack and Pinion System

The electric/electronic rack and pinion unit replaces the hydraulic pump, hoses, and fluid associated with conventional power-steering systems with electronic controls and an electric motor located concentric to the rack itself. The design features a DC motor mechanically connected to the ball nut through an internal/external spline arrangement.

This system has the ability to change the rotational direction of the electric motor coupled with the ability to deliver the necessary amount of current to meet torque requirements at the same time. The direction of the turn is controlled by changing the polarity of the signal to the motor.

Unlike conventional power steering, electric/electronic units provide power assistance—even when the engine stalls—since the power source is the battery rather than the engine-driven pump. The feel of the steering can also be adjusted to match the particular driving characteristics of cars and drivers, from high performance to luxury touring cars. It also eliminates hydraulic oil, which means no leaks.

# SUMMARY

- Suspension systems must keep the vehicle's wheels lined up the travel of the vehicle, limit the movement of the vehicle's body during cornering and when going over bumps, and provide a smooth and comfortable ride.
- The steering system turns the front wheels, and in some cases the rear wheels, to change the direction of the vehicle.
- Tires provide traction and help the suspension absorb road shocks.
- Springs are the basis for all suspension systems, and they absorb shock forces while maintaining the vehicle's correct riding height. Various types of springs are used: coil, torsion bar, leaf, and air springs.
- Shock absorbers dampen or control motion in a vehicle.
- The wheels rotate around a wheel spindle and steering knuckle assembly. Wheel bearings are fit around the outside of the spindle and on the inside of the wheel hub. The steering knuckle assembly holds the wheel in place while allowing the steering and suspension system to move in response to road conditions and driver input.
- Control arms serve as the locators for the suspension system.
- Ball joints act as pivot points that allow the front wheels and spindles or knuckles to turn between the upper and lower control arms.
- In SLA suspension systems, the upper control arm is shorter than the lower control arm, which allows the top of the tire to move slightly in and out with the bottom of the tire remaining in a more constant position.
- In MacPherson strut suspension systems, the lower end of the strut is bolted to the top of the steering knuckle, and the lower end of the knuckle is attached to the ball joint in the lower control arm. An upper strut mount connects the top of the strut to the chassis.
- The steering system is comprised of three major subsystems: the steering linkage, steering gear, and steering column and wheel.
- The term steering linkage is applied to the system of pivots and connecting parts that is placed between the steering gear and the steering arms attached to the front or rear wheels, controlling the direction of vehicle travel.
- Parallelogram steering linkage is used with the recirculating ball steering gear and is comprised of a pitman arm, an idler arm, various links, and tie-rods.
- Rack and pinion units have a rack and a small pinion gear located at the end of the steering column. Two inner tie-rod ends attached to the rack are connected to the outer tie-rod ends, which connect to the steering arms.
- The power-steering unit is designed to reduce the amount of effort required to turn the steering wheel. Power steering can rely on hydraulics or electrical and electronic controls.

## REVIEW QUESTIONS

1. How does a stabilizer bar work?
2. What is an integral power-steering system?
3. Explain the difference between sprung and unsprung weight.
4. Describe how a rack and pinion steering, parallelogram steering, and worm and roller system operate.
5. A power-steering hose transmits fluid under pressure from the ________ to the ____________ ____________ __________.
6. The springs of a vehicle
   A. support the weight of the vehicle.
   B. provide axle location.
   C. stabilize the up-and-down motion.
   D. all of the above.
7. What is the primary purpose of power-steering hoses?
   A. to lubricate the pump
   B. to relieve pressure
   C. to transmit power through fluid under pressure
   D. none of the above
8. The main job of the idler arm is to:
   A. support the left side of the center link.
   B. support the right side of the center link.
   C. support the pitman arm.
   D. keep both ends of the steering system level.
9. Rack and pinion steering
   A. is lighter in weight and has fewer components than parallelogram steering.
   B. does not provide as much feel for the road as parallelogram steering.
   C. does not use tie-rods in the same fashion as parallelogram steering.
   D. all of the above.
10. What connects the steering column to the wheels?
    A. steering shaft
    B. steering linkage
    C. gearbox
    D. pitman arm

# Chapter 8 Basic Theories

## INTRODUCTION

This chapter contains many of the principles you have or will learn in other courses. Many of the facts presented in this chapter will be addressed again in greater detail according to the topic covered. These theories explain the reasons why things work the way they do and why engineers have designed things the way they have.

## MATTER

**Matter** is anything that occupies space. All matter exists as a gas, liquid, or solid. Gases and liquids are considered fluids because they move or flow easily and respond easily to pressure. A gas has neither a shape nor volume of its own and tends to expand without limits. A liquid takes a shape and has volume. A solid is matter that does not flow.

### Atoms and Molecules

All matter consists of countless tiny particles called **atoms** and **molecules.** An atom may be defined as the smallest particle of an element in which all the chemical characteristics of the element are present. Atoms are so small they cannot be seen with an electron microscope that magnifies millions of times. A substance with only one type of atom is referred to as an **element.** Over 100 elements are known to exist at present. Of the known elements, ninety two occur naturally, and the remaining elements have been manufactured in laboratories.

Small, positively charged particles called protons are located in the center, or nucleus, of each atom. In most atoms, neutrons are also located in the nucleus. Neutrons have no electrical charge, but they add weight to the atom. The positively charged protons tend to repel each other. This repelling force could destroy the nucleus. The presence of the neutrons with the protons in the nucleus cancels the repelling action of the protons and keeps the nucleus together. Electrons are small, very light particles with a negative electrical charge. Electrons move in orbits around the nucleus of an atom **(Figure 1).**

Since the electrons are orbiting around the nucleus, centrifugal force tends to move the electrons away from the nucleus. However, the attraction between the positively charged protons and the negatively charged electrons holds the electrons in their orbits. The atoms of the different elements have different numbers of protons, electrons, and neutrons.

For some elements, a single atom does not exist because of its structure. An example of this is oxygen, which has a chemical symbol of O. Pure oxygen exists as pairs of

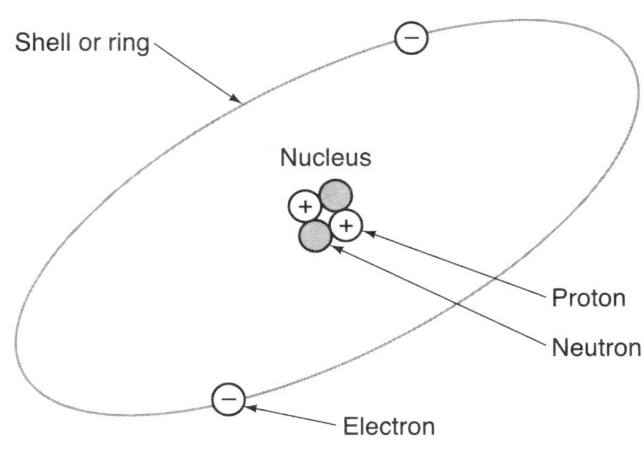


**Figure 1.** The particles of an atom.

oxygen atoms and has a symbol of $O_2$. This is a molecule of oxygen. A molecule is the smallest particle of an element or compound that can exist and still retain the characteristics of that element or compound. Some materials contain only one type of atom, whereas a compound may be described as a liquid, solid, or gas that contains two or more types of atoms. An oxygen atom readily combines with another oxygen atom or atoms of many other elements to form a compound. Many other atoms also have this characteristic.

Water is a compound that contains oxygen and hydrogen atoms. The chemical symbol for water is $H_2O$. This chemical symbol indicates that each molecule of water contains two atoms of hydrogen and one oxygen atom.

## States of Matter

The particles of a solid are held together in a rigid structure. When a solid dissolves into a liquid, its particles break away from this structure and mix evenly in the liquid, forming a **solution.** When they are heated, most liquids **evaporate.** This means that the atoms or molecules of which they are made break free from the body of the liquid to become gas particles. When all of the liquid in a solution has evaporated, the solid is left behind. The particles of the solid normally arrange in a structure called a crystal.

### Absorption and Adsorption

Not all solids will dissolve in a liquid, rather the liquid will be either absorbed or adsorbed. The action of a sponge serves as the best example of absorption. When a dry sponge is put into water, the water is absorbed by the sponge. The sponge does not dissolve, the water merely penetrates into the sponge and the sponge becomes filled with water. There is no change to the atomic structure of the sponge, nor does the structure of the water change. If we take a glass and put it into water, the glass does not absorb the water. The glass, however, still gets wet as a thin layer of water adheres to the glass. This is adsorption. Materials that *absorb* fluids are **permeable** substances. **Impermeable** substances *adsorb* fluids. Some materials are impermeable to most fluids while others are impermeable to just a few.

# ENERGY

Energy may be defined as the ability to do work. Since all matter consists of atoms and molecules that are in constant motion, all matter has energy. Energy is not matter, but it affects the behavior of matter. Everything that happens requires energy.

Each form of energy can change into other forms. However, the total amount of energy never changes; it can only be transferred from one form to another, not created nor destroyed. This is known as the "Principle of the Conservation of Energy."

Energy conversion occurs when one form of energy is changed to another form. Since energy is not always in the desired form, it must be converted to a form we can use. Some of the most common automotive energy conversions are discussed here.

## Kinetic and Potential Energy

When energy is released to do work, it is called **kinetic energy.** This type of energy may also be referred to as energy in motion. Stored energy may be called **potential energy.**

There are many components and systems that have potential energy and, at times, kinetic energy. A vehicle has potential energy before it moves. Once it does move, it does so with kinetic energy. Kinetic energy increases proportionately with the weight of the vehicle **(Figure 2)** and increases exponentially with speed **(Figure 3).**

## Mass and Weight

**Mass** is the amount of matter of an object. **Weight** is a force and is measured in pounds or kilograms. Gravitational force gives the mass its weight. A spacecraft can weigh 500

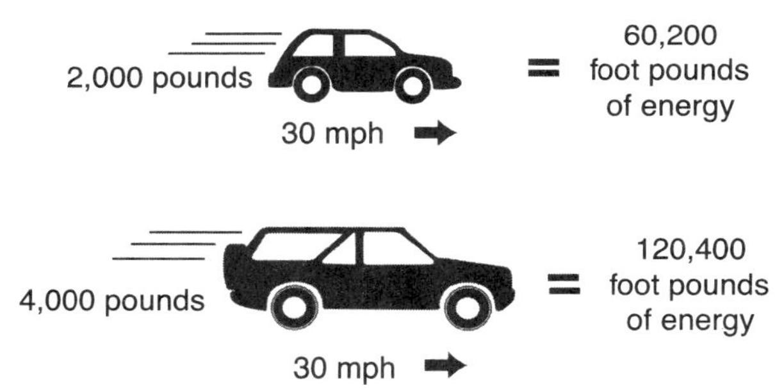


**Figure 2.** Kinetic energy increases proportionately with the weight of the vehicle.

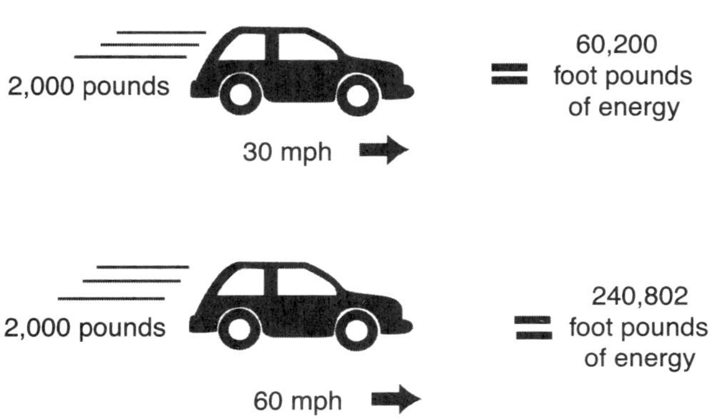


**Figure 3.** Kinetic energy increases exponentially with speed.

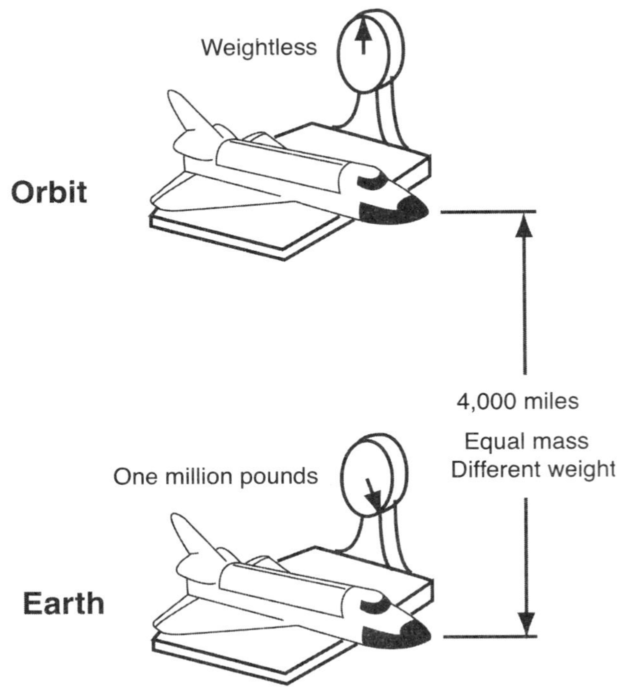


**Figure 4.** An illustration of the effect gravity has on the weight of an object.

tons (1 million pounds) here on earth where it is affected by the earth's gravitational pull. In outer space, beyond the earth's gravity and atmosphere, the spacecraft is nearly weightless **(Figure 4).**

## Size

The size of something is related to its mass. The size of an object defines how much space it occupies. Size dimensions are stated in terms of its length, width, and height. Length is a measurement of how long something is from one end to another. Width is a measurement of how wide something is from one side to another. Obviously, height is a measurement of the distance from something's bottom to its top. All three of these dimensions are measured in inches, feet, yards, and miles in the English system and meters in the metric system.

# VOLUME

**Volume** is a measurement of size and is related to mass and weight. Volume is the amount of space occupied by an object in three dimensions: length, width, and height. For example, a pound of gold and a pound of feathers both have the same weight, but the pound of feathers occupies a much larger volume. In the English system, volume is measured in cubic inches, cubic feet, cubic yards, or gallons. The measurement for volume in the metric system is cubic centimeters or liters.

The volume of a container is calculated by taking the measured length, width, and height of an object and multiplying them. For example, if a box has a length of 2 inches, a width of 3 inches, and is 4 inches in height (2 x 3 x 4 = 24), it has a volume of 24 cubic inches.

## Ratios

Automotive features are often expressed as ratios. A ratio expresses the relationship between two things. If something is twice as large as another, there is a ratio of 2:1. Sometimes ratios are used to compare the movement of an object. For example, if a gear with a 2-inch diameter drives a gear with a 4-inch diameter, the ratio of the gears is 1:2.

## Proportions

Ratios can also be used to express the correct mixture for something. An example of this would be the amount of engine coolant that should be mixed with water before the engine's cooling system is refilled. Typically, specifications call for 50 percent coolant and 50 percent water, or a ratio of 1:1. This mixture allows for maximum hot and cold protection. To apply this ratio, let us look at a cooling system that has a capacity of 9.5 liters. Since most engine coolant is sold in gallon containers, to determine the amount of coolant that should be put in the system we must first convert the liter capacity to gallon capacity. Since 1 gallon equals 3.7854 liters, we need to divide 9.5 liters by 3.7854 (9.5 ÷ 3.7854 = 2.5097). Now we know the total capacity of the cooling system is a little more than 2.5 gallons. To determine how much coolant or antifreeze to put in the system, we divide the total capacity by 2. That will give us the quantity equal to 50 percent of the capacity (2.5 ÷ 2 = 1.25). Therefore, to obtain the correct mixture, $1^1/_4$ gallons of coolant should be mixed with $1^1/_4$ gallons of water.

# FORCE

A **force** is a push or pull, and can be large or small. Force can be applied to objects by direct contact or from a distance. Gravity and electromagnetism are examples of forces that are applied from a distance. Forces can be applied from any direction and with any intensity. For example, if a pulling force on an object is twice that of the pushing force, the object will be pulled at one-half of the pulling force. When two or more forces are applied to an object, the combined force is called the resultant. The resultant is the sum of the size and direction of the forces.

## Automotive Forces

When a vehicle is sitting still, gravity exerts a downward force on the vehicle. The ground exerts an equal and opposite upward force and supports the vehicle. When the engine is running and its power output transferred to the vehicle's drive wheels, the wheels exert a force against the ground in a horizontal direction. This force causes the vehicle to move but is opposed by the mass of the vehicle. To move the vehicle faster, the force supplied by the wheels must increase beyond the opposing forces. As the vehicle does move faster, it pushes against the air as it travels. This becomes a growing opposing force, and the force at the drive wheels must overcome the force in order for the vehicle to increase speed. After the vehicle has achieved the desired speed, no additional force is required at the drive wheels.

### Force—Balanced and Unbalanced

When the applied forces are balanced and there is no overall resultant force, the object is said to be in **equilibrium.** An object sitting on a solid flat surface is in equilibrium because its weight is supported by the surface and there is no resultant force. If the surface is put on an angle, the object will tend to slide down the surface. If the surface is at a slight angle, the force will cause the object to slowly slide down the surface. If the surface is at a severe angle, the downward force will cause the object to quickly slide down the slope. In both cases, the surface is still supplying the force needed to support the object but the pull of gravity is greater, and the resultant force causes the object to slide down the slope.

### Turning Forces

Forces can cause rotation as well as straight-line motion. A force acting on an object that is free to rotate will have a turning effect, or turning force. This force is equal to the size of the force multiplied by the distance of the force from the turning point.

## Forces on Tires and Wheels

If you roll a cone-shaped piece of metal on a smooth surface, the cone does not roll in a straight line. Rather it moves toward the direction of the tilt on the cone. The weight of the cone is applied to the surface, but part of the weight at the large end of the cone is applied at an angle to the small end of the cone. Riding a bicycle is another example of this. When you want to make a left turn, it is easier if you tilt the bicycle to the left. The reason for this action is that a tilted, rolling wheel tends to move in the direction of the tilt. Similarly, if a tire and wheel on an automobile are tilted, the tire and wheel will tend to move in the direction of the tilt **(Figure 5).** This principle is used in front wheel alignment.

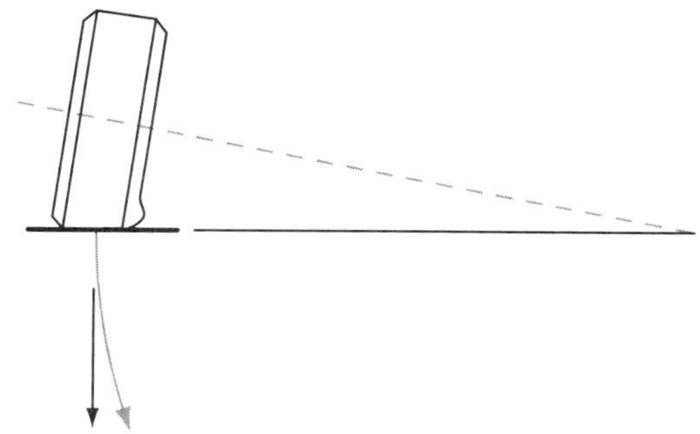

**Figure 5.** A tilted, rolling wheel tends to move in the direction of the tilt.

While riding a bicycle, the force applied to the bicycle by your weight is projected through the bicycle's front fork to the road surface. The centerline of the front fork is tilted rearward in relation to the vertical centerline of the wheel **(Figure 6).** When the handlebars are turned, the tire pivots on the vertical centerline of the wheel. Since the tire's pivot point is behind the front fork centerline where your weight is projected against the surface of the road, the front wheel tends to return to the straight-ahead position after a turn. The wheel also tends to remain in the straight-ahead position as the bicycle is driven. This principle of resultant forces is also the basis for the theories applied during front wheel alignment.

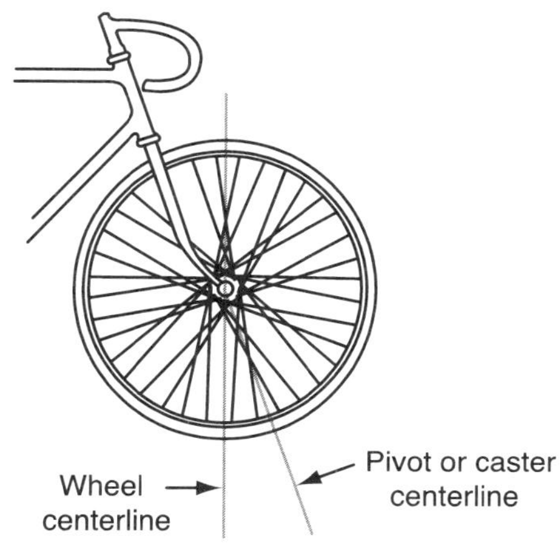


**Figure 6.** If the centerline and pivot point of a wheel is behind the front fork centerline pivot point, the wheel tends to return to the straight-ahead position after a turn. The wheel also tends to remain in the straight-ahead position as the bicycle is driven.

## Centrifugal/Centripetal Forces

When an object moves in a circle, its direction is continuously changing, and all changes in direction require a force. The forces required to maintain circular motion are called **centripetal** and **centrifugal force.** The size of these forces depends on the size of the circle, the mass, and the speed of the object.

Centripetal force tends to pull the object toward the center of the circle, whereas centrifugal force tends to push

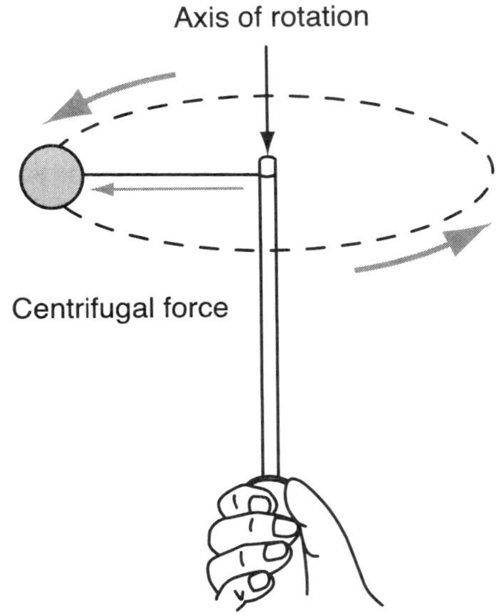


**Figure 7.** A weight tends to rotate at a 90° angle in relation to the axis of rotation due to centrifugal force.

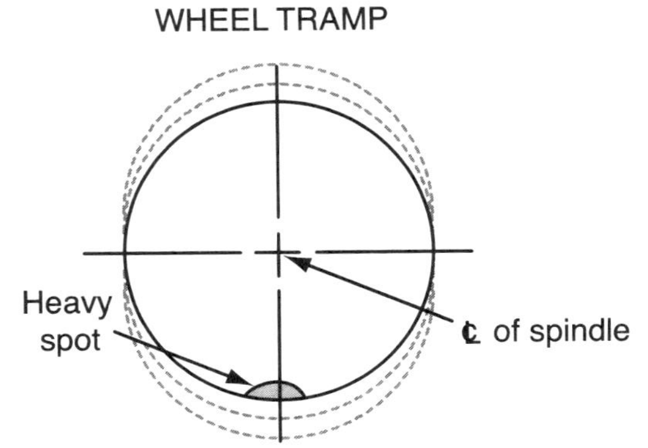


**Figure 8.** Wheels that are statically unbalanced will tend to hop or tramp.

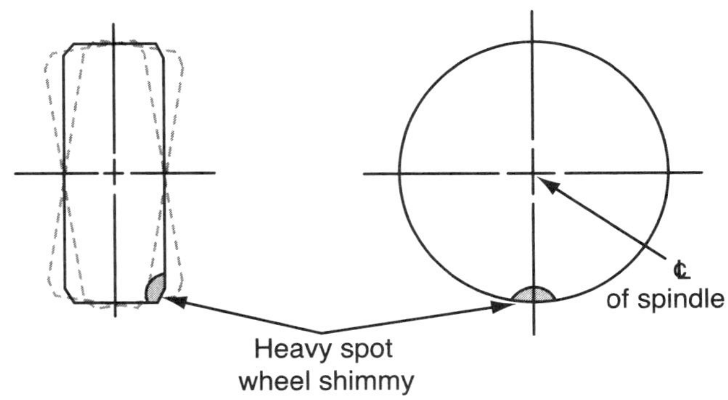


**Figure 9.** Wheels that are dynamically unbalanced will cause a shimmy.

the object away from the center **(Figure 7).** The centripetal force that keeps an object whirling around on the end of a string is caused by **tension** in the string. If the string breaks, there is no longer string tension and centripetal force, and the object will fly off in a straight line because of the centrifugal force on it. Gravity is the centripetal force that keeps the planets orbiting around the sun. Without this centripetal force, the earth would move in a straight line through space.

## Wheel and Tire Balance

When the weight of a tire assembly is distributed equally around the center of wheel rotation, the wheel and tire have proper static balance. Being statically balanced, the wheel and tire assembly will not tend to rotate by itself regardless of the wheel position. If the weight is not distributed equally, the wheel and tire assembly is statically unbalanced. As the wheel and tire rotate, centrifugal force acts on this static unbalance and causes the wheel to "tramp" or "hop" **(Figure 8).**

Dynamic balance exists when the weight thrown to the sides of the tire and wheel assembly are equal when the assembly is rotating **(Figure 9).** To illustrate this balance, assume we have a bar with a ball attached by string to both ends of the bar. If we cause the bar to rotate, the balls will turn with the bar, and centripetal and centrifugal force will keep the balls in an orbit around the rotating bar. If the two balls weigh the same and are at an equal distance from the bar, the bar will rotate smoothly. However, if one of the balls is heavier than the other, the bar will wobble as it rotates. The greater the different in weight of the balls, the greater the wobble. The wobble can eventually destroy the mechanism used to rotate the bar.

Now if we add some weight to the end of the bar that has the lighter ball, the weights and forces can be equalized and the wobble removed. This is basically how we dynamically balance a wheel and tire assembly.

## Pressure

Pressure is a force applied against an object and is measured in units of force per unit of surface area (pounds per square inch or kilograms per square centimeter). Mathematically, pressure is equal to the applied force divided by the area over which the force acts. Consider two 10-pound weights sitting on a table; one occupies an area of 1 square inch and the other an area of 4 square inches. The pressure exerted by the first weight would be 10 pounds per one square inch, or 10 psi. The other weight, although it weighs the same, will exert only 2.5 psi (10 pounds per 4 square inches = 10 ÷ 4 = 2.5). This illustrates an important concept; a force acting over a large area will exert less **pressure** than the same force acting over a small area.

Since pressure is a force, all principles of force apply to pressure. If more than one pressure is applied to an object, the object will respond to the resultant force. Also, all matter (liquids, gases, and solids) tends to move from an area of high pressure to a low-pressure area.

## MOTION

When the forces on an object do not cancel each other out, they will change the motion of the object. The object's speed, direction of motion, or both will change. The greater the mass of an object, the greater the force needed to change its motion. This resistance to change in motion is called **inertia.** Inertia is the tendency of an object at rest to remain at rest, or the tendency of an object in motion to stay in motion. The inertia of an object at rest is called static inertia, whereas dynamic inertia refers to the inertia of an object in motion. Inertia exists in liquids, solids, and gases. When you push and move a parked vehicle, you overcome the static inertia of the vehicle. If you catch a ball in motion, you overcome the dynamic inertia of the ball.

When a force overcomes static inertia and moves an object, the object gains momentum. **Momentum** is the product of an object's weight times its speed and is a type of mechanical energy. An object loses momentum if another force overcomes the dynamic inertia of the moving object.

### Newton's Laws of Motion

How forces change the motion of objects was first explained by Sir Isaac Newton. These explanations are known as Newton's Laws. Newton's first law of motion is called the law of inertia. It states that an object at rest tends to remain at rest, and an object in motion tends to remain in motion unless some force acts on it. When a car is parked on a level street, it remains stationary unless it is driven or pushed.

Newton's second law states when a force acts on an object, the motion of the object will change. This change in motion is equal to the size of the force divided by the mass of the object on which it acts. Trucks have a greater mass than cars. Since a large mass requires a larger force to produce a given acceleration, a truck needs a larger engine than a car.

Newton's third law says that for every action, there is an equal and opposite reaction. A practical application of this law occurs when the wheel on a vehicle strikes a bump in the road surface. This action drives the wheel and suspension upward with a certain force, and a specific amount of energy is stored in the spring. After this action occurs, the spring forces the wheel and suspension downward with a force equal to the initial upward force caused by the bump.

### Friction

**Friction** is a force that slows or prevents motion of two moving objects or surfaces that touch. Friction may occur in solids, liquids, and gases. It is the joining or bonding of the atoms at each of the surfaces that causes the friction. When you attempt to pull an object across a surface, the object will not move until these bonds have been overcome. Smooth surfaces produce little friction, therefore only a small amount of force is needed to break the bonds between the atoms. Rougher surfaces produce a larger friction force because stronger bonds are made between the two surfaces. **Figure 10** shows the different amount of force required to move 100 pounds of iron and ice. The iron has a much rougher surface than the block of ice.

Friction is put to good use in disc brakes **(Figure 11).** The friction force between the disc and brake pad slows down the rotation of the wheel, thereby reducing the vehicle's speed. In doing so, it converts the kinetic energy of the vehicle into heat.

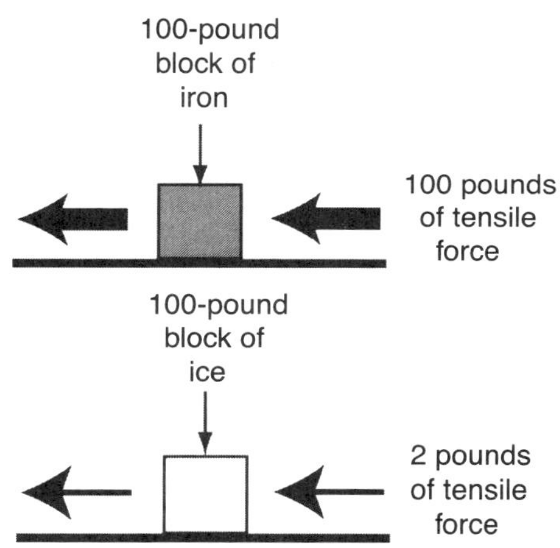


**Figure 10.** The different amounts of force required to move 100 pounds of iron and ice. The iron has a much rougher surface than the block of ice.

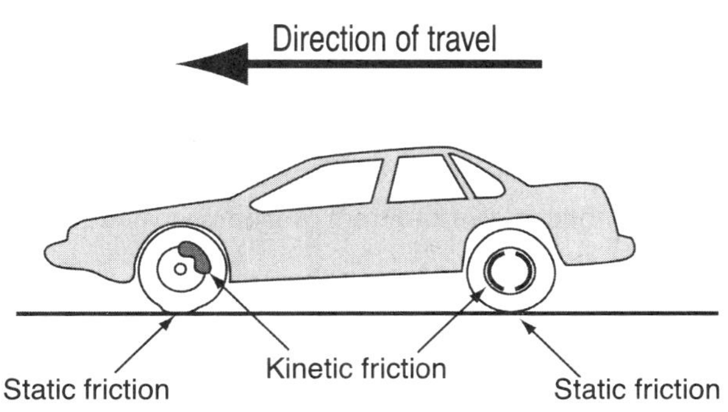


**Figure 11.** Both static and kinetic friction are at work while stopping a vehicle.

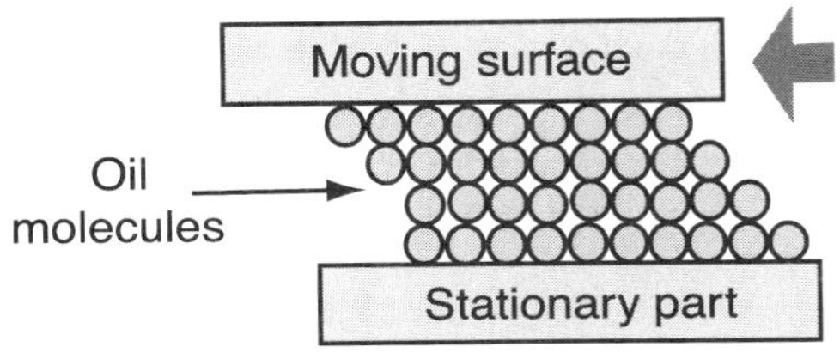


**Figure 12.** Friction can be reduced by the presence of oil between two surfaces keeping the surfaces apart and allowing them to move smoothly past one another.

### Lubrication

Friction can be reduced in two main ways: by lubrication, or by the use of rollers. The presence of oil or another fluid between two surfaces keeps the surfaces apart. Because fluids (liquids and gases) flow, they allow movement between surfaces. The fluid keeps the surfaces apart, allowing them to move smoothly past one another **(Figure 12).**

### Rollers

Rollers placed between two surfaces keep the surfaces apart. An object placed on rollers will move smoothly if pushed or pulled. Rollers actually use friction to grip the surfaces and produce rotation. Instead of sliding against one another, the surfaces produce turning forces, which cause each roller to spin. This leaves very little friction to oppose motion. Bearings are a type of roller used to reduce the friction between moving parts, such as a wheel and its axle **(Figure 13).** As the wheel turns on the axle, the balls in the bearing roll around inside the bearing, drastically reducing the friction between the wheel and axle.

**Figure 13.** Roller bearings are used to reduce the friction between moving parts such as a wheel and its axle or spindle.

## WORK

When a force moves a certain mass a specific distance, **work** is done. When work is accomplished, the mass may be lifted or slid on a surface against a resistance or opposing force **(Figure 14).** Work is equal to the applied force multiplied by the distance the object moved (force x distance = work) and is measured in foot-pounds, watts, or Newton-meters. For example, if a force moves a 3000-pound car 50 feet, 150,000 foot-pounds of work was done.

A force acts on an object to start, stop, or change the direction of the object. It is possible to apply a force to an object and not move the object. For example, you may push with all your strength on a car stuck in a ditch, and not move the car. Under this condition, no work is done. Work is only accomplished when an object is started, stopped, or redirected by a force.

## Simple Machines

A machine is a device that can be used to transmit a force and, in doing so, change the amount of force and/or its direction. The force applied to a machine is called the effort, while the force it overcomes is called the **load.** The effort is often smaller than the load because an effort can overcome a heavy load if the effort is moved a larger distance. The machine is then said to give a mechanical advantage. Although the effort will be smaller when using a machine, the amount of work done (or energy used) will be equal to or greater than that without the machine.

### Inclined Plane

The force required to drag an object up a slope is less than that required to lift it vertically. However, the overall distance moved by the object is greater when pulled up the slope than if it were lifted vertically. A screw is like an inclined plane wrapped around a shaft. The force that turns

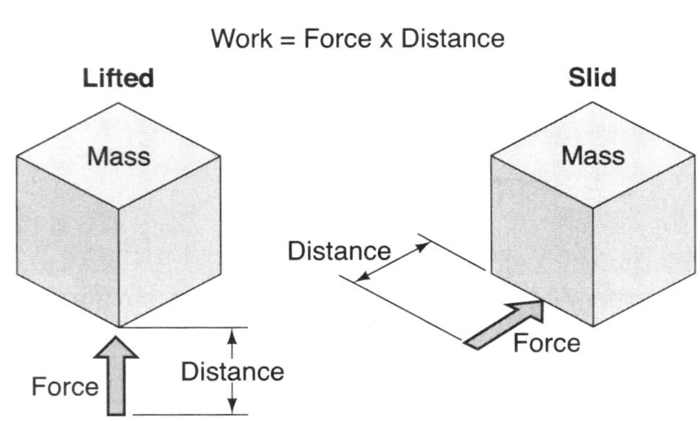


**Figure 14.** Work is performed when a mass is lifted or slid.

the screw is converted to a larger one, which moves a shorter distance and drives the screw in.

### Pulleys

A **pulley** is a wheel with a grooved rim in which a rope, belt, or chain runs to raise something by pulling on the other end of the rope, belt, or chain. A simple pulley changes the direction of a force, but its size and the distance the force moves does not change. By using several pulleys connected together as a block and tackle, the size of the force can be changed, so a heavy load can be lifted using a small force. With a double pulley, the applied force required to move an object can be reduced by one-half, but the distance the force must be moved is doubled. A quadruple pulley can reduce the force by four times, but the distance will be increased by four times. Pulleys of different sizes can change the amount of required applied force, as well as the speed or distance the pulley needs to travel to accomplish work **(Figure 15).**

### Levers

A **lever** is a device made up of a bar turning about a fixed pivot point, called the fulcrum, that uses a force applied at one point to move a mass on the other end of the bar. Types of levers are divided into classes. In a class one lever, the fulcrum is between the effort and the load. The length of the lever determines the amount of effort required to more the load. The longer the less effort is needed **(Figure 16),** but the load will move less. In a class two lever, the load is between the fulcrum and effort. Here again, the load is greater than the effort and moves through a smaller distance. In a class three lever, the effort is between the fulcrum and the load. In this case, the load is less than the effort but it moves through a greater distance.

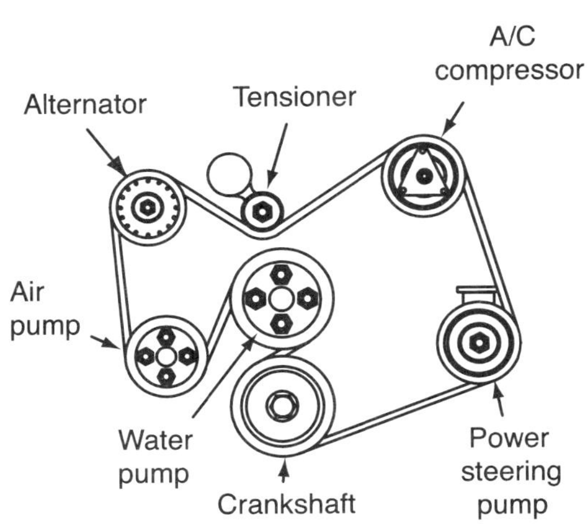


**Figure 15.** Pulleys of different sizes can change the amount of required applied force, as well as the speed or distance the pulley needs to travel to accomplish work.

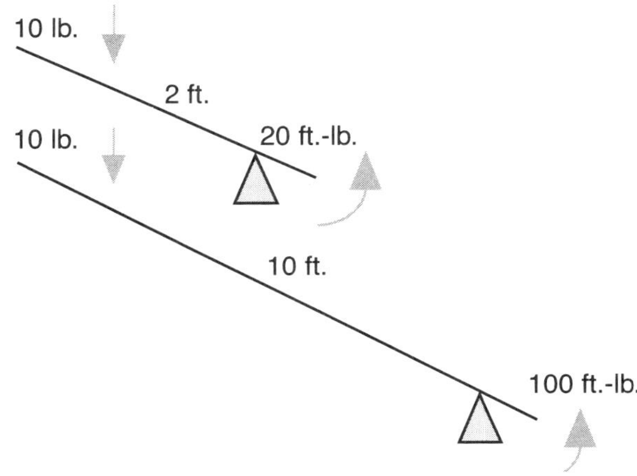


**Figure 16.** Increasing the length of a lever increases input force and makes it easier to do work.

### Gears

A **gear** is a toothed wheel that becomes a machine when it is meshed with another gear. The action of one gear is that of a rotating lever and moves the other gear meshed with it. Based on the size of the gears in mesh, the amount of force applied from one gear to the other can be changed. Keep in mind that this does not change the amount of work performed by the gears because although the force changes, so does the distance of travel. The relationship of force and distance is inverse. Gear ratios express the mathematical relationship (diameter and number of teeth) of one gear to another.

### Wheels and Axles

The most obvious application of a wheel and axle is a vehicle's tires and wheels. These units revolve around an axle and limit the amount of area of a vehicle that contacts the road. Wheels function as rollers to reduce the amount of friction between a vehicle and the road. Basically, the larger the wheel, the less force is required to turn it. However, the wheel must move farther as it gets larger. An example of this is a steering wheel. A steering wheel that is twice the size of another will require one-half the force to turn it but will also require twice the distance to accomplish the same work.

## Torque

**Torque** is a force that tends to rotate or turn things and is measured by the force applied and the distance traveled. The technically correct unit of measurement for torque is pounds per foot (lb-ft). However, it is rather common to see torque stated in terms of foot-pounds (ft-lb). In the metric or SI system, torque is stated in Newton-meters (Nm) or kilogram-meters (kg-m).

An engine creates torque and uses it to rotate the crankshaft. The combustion of gasoline and air in the cylinder creates pressure against the top of a piston. That pressure creates a force on the piston and pushes it down. The force is transmitted from the piston to the connecting rod, and from the connecting rod to the crankshaft. The engine's crankshaft rotates with a torque that is transmitted through the drive train to turn the drive wheels of the vehicle.

Torque is force times leverage, the distance from a pivot point to an applied force. Torque is generated any time a wrench is turned with force. If the wrench is a foot long and you put 20 pounds of force on it, 20 pounds per foot are being generated. To generate the same amount of torque while exerting only 10 pounds of force, the wrench needs to be 2 feet long **(Figure 17).** To have torque, it is not necessary to have movement. When you pull a wrench to tighten a bolt, you supply torque to the bolt. If you pull on a wrench to check the torque on a bolt and the bolt torque is sufficient, torque is applied to the bolt but no movement occurs. If the bolt turns during torque application, work is done. When a bolt does not rotate during torque application, no work is accomplished.

## Power

**Power** is a measurement for the rate, or speed, at which work is done. The metric unit for power is the watt. A watt is equal to one Newton-meter per second. You can multiply the amount of torque in Newton-meters by the rotational speed to determine the power in watts. Power is a unit of speed combined with a unit of force. For example, if you were pushing something with a force of 1 N and it moved at a speed of 1 meter per second, the power output would be 1 watt.

## Horsepower

**Horsepower** is the rate at which torque is produced. James Watt measured the amount of work that a horse could do in a specific time. He found that a horse could move 330 pounds 100 feet in 1 minute. Therefore, he determined that one horse could do 33,000 foot-pounds of work in 1 minute. Thus, 1 horsepower is equal to 33,000 foot-pounds per minute, or 550 foot-pounds per second. Two horsepower could do this same amount of work in one-half minute. If you push a 3,000-pound (1,360-kilogram) car for 11 feet (3.3 meters) in one-quarter minute, you produce 4 horsepower.

# WAVES AND OSCILLATIONS

An **oscillation** is any single swing of an object back and forth between the extremes of its travel. When that motion travels through matter or space, it becomes a **wave.**

## Vibrations

When an object oscillates, it vibrates. To prevent the vibration of one mass from causing a vibration in other masses, the oscillating mass must be isolated from other objects. In an automobile, there are many parts that vibrate as they operate. Through careful engineering,

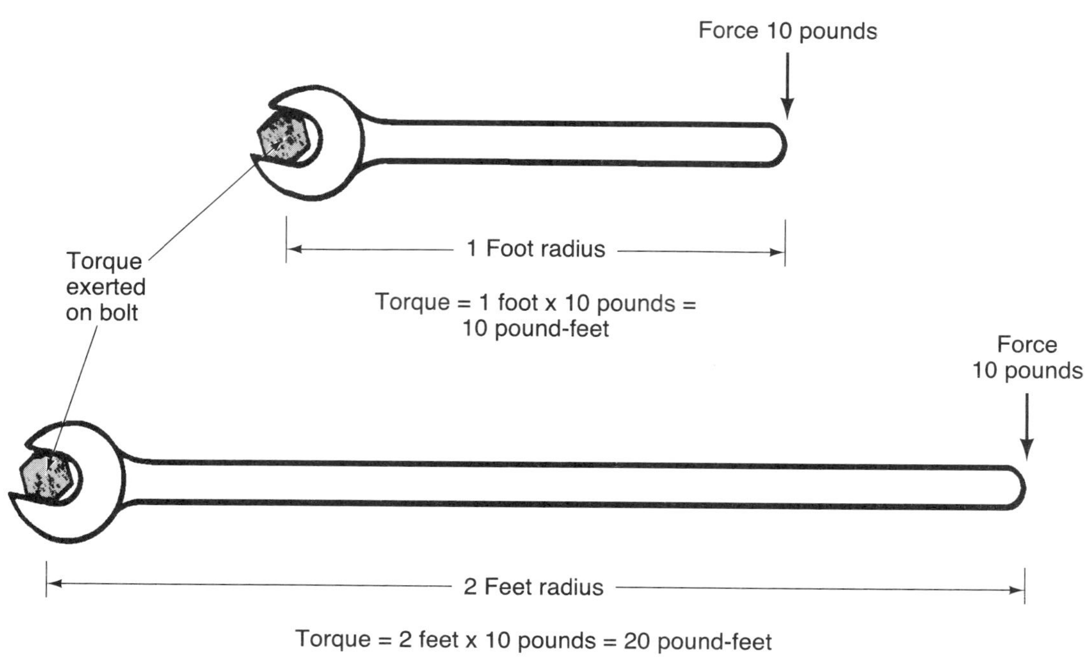


**Figure 17.** When the same force is applied to two wrenches, more torque is available from the longer wrench.

these vibrations can be isolated or insulated, thereby reducing the chances of the vibrations moving through the vehicle. Vibration control is also important for the reliability of components. If the vibrations are not controlled, the object could shake itself to destruction. Vibration control is the best justification for always mounting components in the way they were designed to be mounted.

Unwanted and uncontrolled vibrations are typically the result of one component vibrating at a different frequency than another part. When two waves or vibrations meet, they add up or interfere. This is called the Principle of Superposition and is common to all waves. Making unwanted vibrations tolerable may involve canceling each out with an equal and opposite vibration. This approach to vibration reduction is best illustrated by the use of balance shafts in an engine. These shafts are designed to counter the vibrations that result from the rotation of the engine's crankshaft and pistons. The balance shaft smoothes the engine by operating at an equal but opposite vibration as the crankshaft.

To cancel a vibration, the vibration must be defined. How many times the vibration occurs in one second is called **frequency.** Frequency **(Figure 18)** is most often expressed in **hertz** (Hz). One hertz is equal to one cycle per second. The name is in honor of Heinrich Hertz, an early German investigator of radio wave transmission. The **amplitude** and velocity must also be known. The amplitude of a vibration is its intensity or strength. The velocity of a vibration is the result of its amplitude and its frequency. Since every material has a unique resonant or natural vibration frequency, everything on an automobile can vibrate at its own frequency.

## Sound

Vibration is very common and results in sound. In air, the vibrations that cause sound are transmitted as a wave between air molecules; many other substances transmit sound in a similar way. A vibrating object causes variations in pressure in the surrounding air. Areas of high and low pressure, known as compressions and rarefactions, move through the air as sound waves. Compression makes the sound waves denser, whereas rarefaction makes them less dense. The distance between each compression of a sound wave is called its **wavelength.** Sound waves with a short wavelength have a high frequency and sound high-pitched.

When the rapid variations in pressure occur between about 20 Hz and 20 kHz, sound is potentially audible. Audible sound is the sensation (as detected by the ear) of very small rapid changes in the air pressure above and below atmospheric pressure.

When diagnosing systems, you will often be told to listen to the sound of something. Mostly you will be paying attention to the type of sound and its intensity and frequency. The tone of the sound usually indicates the type of material that is causing the noise. If there is high pitch, you know that the source of the sound is something that can vibrate quickly. This means the source is less rigid than something that vibrates with a low pitch. Although pitch is dependent on the frequency of a sound, the frequency itself can identify the possible sources of the sound. For example, if a sound from an engine increases with an increase in engine speed, you know the source of the sound must be something that is moving faster as a result of the increase in engine speed. If the frequency of the sound appears to be at one-half the speed of the engine, you know the source of the sound is something that is rotating at that speed, such as the camshaft.

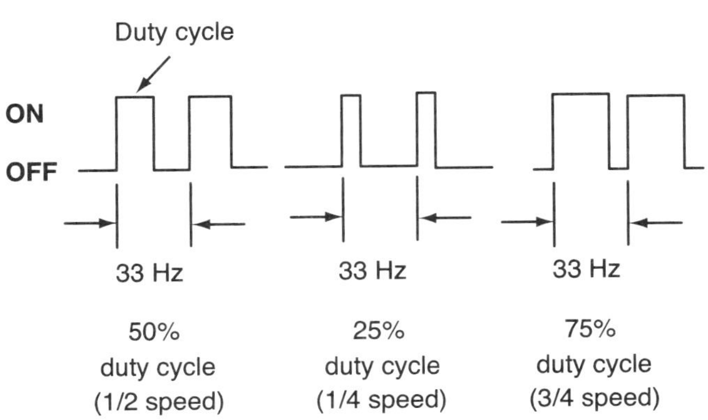


**Figure 18.** A graphic of the same frequency but with different percentages of ON versus OFF time.

## Noise

Noise is any unwanted signal or sound. It can be random or periodic. To identify the source of a noise, it is important to remember that sound and noise are vibrations and the vibration may be traveling through other components. Therefore, the source of the noise is not always where it may appear. There are many potential sources of noise in an automobile, and manufacturers work hard to prevent noise.

Noise can be prevented or reduced by three different approaches. The most effective way is to intervene is at the design stage, to redesign a noisy component so that it produces less noise. A relatively new technique of noise reduction is that of anti-noise or active noise control. This involves producing a sound that is similar to, but out of phase with, the noise. This effectively cancels the original noise. More obvious methods of noise reduction, or passive noise control, involve the use of filters, insulation, and noise barriers.

# LIQUIDS

Fluids—gases and liquids—do not have a definite shape. A characteristic of all fluids is that they will conform to

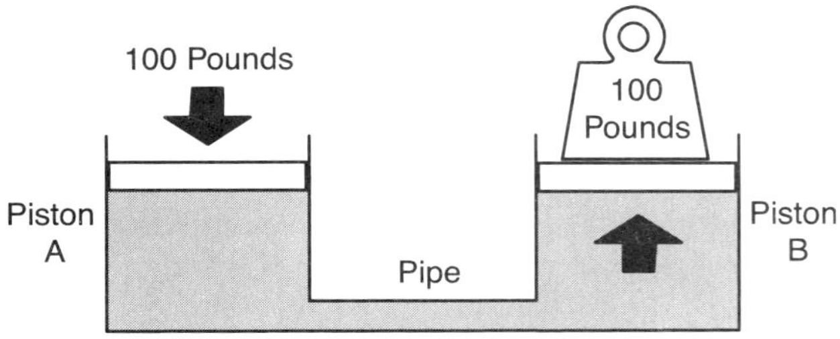


**Figure 19.** Pressure travels through a hydraulic system unchanged if the size of the pistons is equal.

the shape of their container. A major difference between a gas and a liquid is that a gas will always fill a sealed container, whereas a liquid may not. A gas will also readily expand or compress according to the pressure exerted on it. Liquids are basically incompressible, which gives them the ability to transmit force **(Figure 19).** The pressure applied to a liquid in a sealed container is transmitted equally in all directions and to all areas of the system and acts with equal force on all areas. As a result, liquids can provide great increases in the force available to do work. They also always seek a common level. A liquid under pressure may also change from a liquid to a gas in response to temperature changes.

Liquids exert pressure on immersed objects, resulting in an upward resultant force called upthrust. The upthrust is equal to the weight of the liquid displaced by the immersed object. If the upthrust on an object is greater than the weight of the object, then the object will float. Large ships float because they displace huge amounts of water, producing a large upthrust.

## Laws of Hydraulics

Hydraulics is the study of liquids in motion. Liquids will predictably respond to pressures exerted on them. Their reaction to pressure is the basis of all hydraulic applications. This fact allows hydraulics to do work. A simple hydraulic system has liquid, a pump, lines to carry the liquid, control valves, and an output device. The liquid must be available from a continuous source, such as an oil pan or sump. An oil pump is used to move the liquid through the system. The lines that carry the liquid may be pipes, hoses, or a network of internal bores or passages in a single housing. Control valves are used to regulate hydraulic pressure and direct the flow of the liquid. The output device is the unit that uses the pressurized liquid to do work.

Over 300 years ago, a French scientist, Blaise Pascal, determined that if you had a liquid-filled container with only one opening and applied force to the liquid through that opening, the force would be evenly distributed throughout the liquid. This explains how pressurized liquid is used to operate and control systems, such as the brake system and automatic transmissions.

Interesting Fact

*Pascal's work is known as Pascal's Law. Pascal's Law says that pressure at any one point in a confined liquid is the same in every direction and applies equal force on equal areas. One of the most important results of Pascal's work was the discovery that fluids may be used to increase force. Pascal was the person who first demonstrated the relationships of pressure, force, and motion and the inverse relationship of motion and force.*

Pascal constructed the first known hydraulic device, which consisted of two sealed containers connected by a tube. The pistons inside the cylinders seal against the walls of each cylinder and prevent the liquid from leaking out of the cylinder and prevent air from entering into the cylinder. When the piston in the first cylinder has a force applied to it, the pressure moves everywhere within the system. The force is transmitted through the connecting tube to the second cylinder. In the second cylinder, the pressurized fluid exerts force on the bottom of the second piston, moving it upward and lifting the load on the top of it. By using this device, Pascal found he could increase the force available to do work, just as could be done with levers or gears.

Pascal determined that force applied to liquid creates pressure, or the transmission of the force through the liquid. These experiments revealed two important aspects of a liquid when it is confined and put under pressure. The pressure applied to it is transmitted equally in all directions, and this pressure acts with equal force at every point in the container. If a liquid is confined and a force applied, pressure is produced. In order to pressurize a liquid, the liquid must be in a sealed container. Any leak in the container will decrease the pressure.

## Mechanical Advantage with Hydraulics

If a hydraulic pump provides 100 psi, there will be 100 pounds of pressure on every square inch of the system **(Figure 20).** If the system included a piston with an area of 50 square inches, each square inch receives 100 pounds of pressure. This means there will be 5,000 pounds of force applied to that piston, but the output's travel will decrease

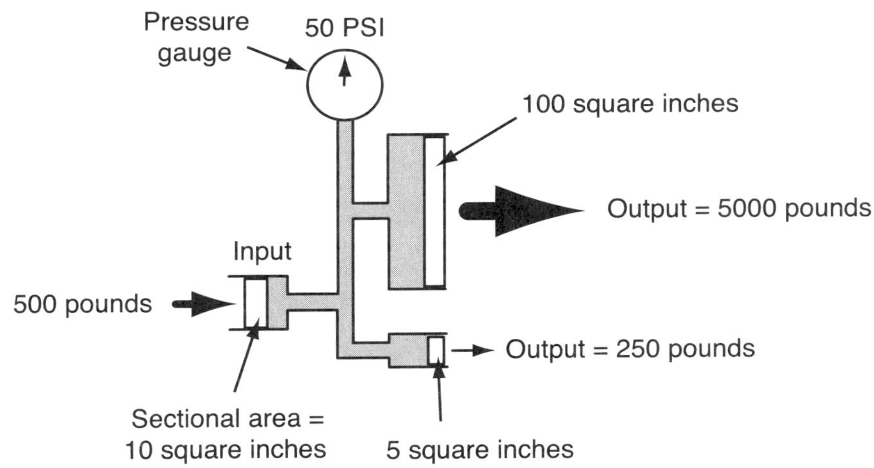


**Figure 20.** This shows the effects of applying hydraulic pressure to different size pistons.

proportionally. The use of the larger piston gives the system a **mechanical advantage** or increase in the force available to do work. The multiplication of force through a hydraulic system is directly proportional to the difference in the piston sizes throughout the system.

By changing the size of the pistons, force is multiplied and, as a result, low amounts of force can be used to move heavy objects. The mechanical advantage of a hydraulic system can be further increased by the use of levers to increase the force applied to a piston.

Although the force available to do work is increased by using a larger piston in one cylinder, the total movement of the larger piston is less than that of the smaller one. A hydraulic system with two cylinders, one with a 1-inch piston and the other with a 2-inch, will double the force at the second piston. However, the total movement of the larger piston will be half the distance of the smaller one.

A hydraulic system responds to the pressure or force applied to it. The mere presence of different sized pistons does not always result in fluid power. Either the pressure applied to the pistons must be different or the size of the pistons must be different in order to cause fluid power. If an equal amount of pressure is exerted onto both pistons in a system and both pistons are the same size, neither piston will move, and the system is balanced or is at equilibrium. The pressure inside the hydraulic system is called **static pressure** because there is no fluid motion.

When an unequal amount of pressure is exerted on the pistons, the piston with the least amount of pressure on it will move in response to the difference between the two pressures. If the size of the two pistons is different and an equal amount of pressure is exerted on the pistons, the fluid will move. The pressure of the fluid while it is in motion is called **dynamic pressure.**

## GASES

A gas will fill any container into which it is placed. The random movement of gas particles ensures that any two gases sharing the same container will totally mix. This is **diffusion.**

The kinetic energy in atoms and molecules increases as the temperature increases. A decrease in temperature reduces the kinetic energy. Molecules in solids move slowly compared to those in liquids or gases. Since gas molecules are in constant motion, they spread out to fill all the space available. At higher temperatures, gas molecules spread out more, whereas at lower temperatures, gas molecules move closer together. The bombardment of particles against the sides of the container produces pressure.

### Behavior of Gases

Three simple laws describe the predictable behavior of gases: Boyle's Law, the Pressure Law, and Charles' Law. Each of these laws describes a relationship between the pressure, volume, and temperature of a gas.

Boyle's Law states that the volume and pressure of a mass of gas at a fixed temperature is inversely proportional. If the pressure on a gas increases, its volume will decrease. Likewise, if the volume is increased, the pressure will decrease.

The Pressure Law states that the volume of a mass of gas depends on its temperature. The higher the temperature, the greater the volume. If the volume cannot change, the pressure of the gas will. Therefore, the pressure and temperature of a gas are also directly related. If you increase one, you also increase the other. This explains why cold air is denser than warm air.

## Air Pressure

Since air is gaseous matter with mass and weight, it exerts pressure on the earth's surface. A 1-square-inch column of air extending from the earth's surface to the outer edge of the atmosphere weighs 14.7 psi at sea level. Therefore, atmospheric pressure is 14.7 psi at sea level. **Atmospheric pressure** may be defined as the total weight of the earth's atmosphere **(Figure 21).** Pressure greater than atmospheric pressure may be measured in psi gauge (psig). Using a standard pressure gauge, air pressure is compared to that of normal atmospheric pressure. When the actual pressure is 19.7 psi, the gauge will read 5 psi showing the pressure differential **(Figure 22).** The actual pressure is referred to as psi absolute (psia).

When air becomes hotter, it expands. Hotter air is lighter compared to an equal volume of cooler air. This hotter, lighter air exerts less pressure on the earth's surface compared to cooler air. This means the weight of the atmosphere changes with weather. This change is rather slight. As the weight changes, so does the atmospheric pressure. The change in atmospheric pressure is measured with a barometer and is called **barometric pressure**. Barometric pressure at normal atmospheric pressure is 29.92 inches of mercury. The increments for measuring barometric pressure are based on the increments of a barometer. A barometer is "J" shaped tube with mercury in it. One end of the tube is exposed to normal atmospheric pressure and the other end to current atmospheric pressure. When the current atmospheric pressure equals normal atmospheric pressure, the level of

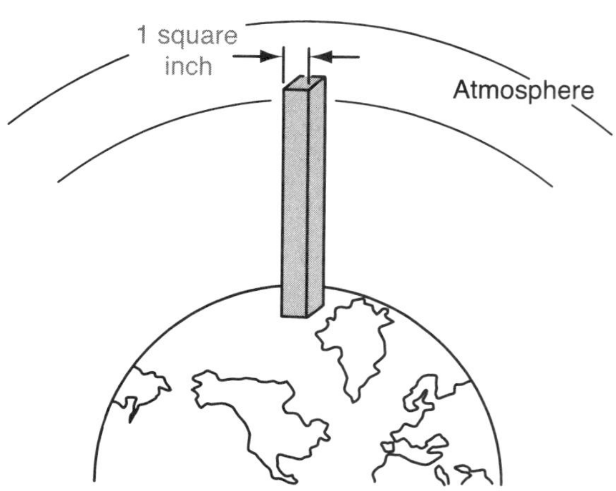


**Figure 21.** Atmospheric pressure is the total weight of one square inch of the earth's atmosphere.

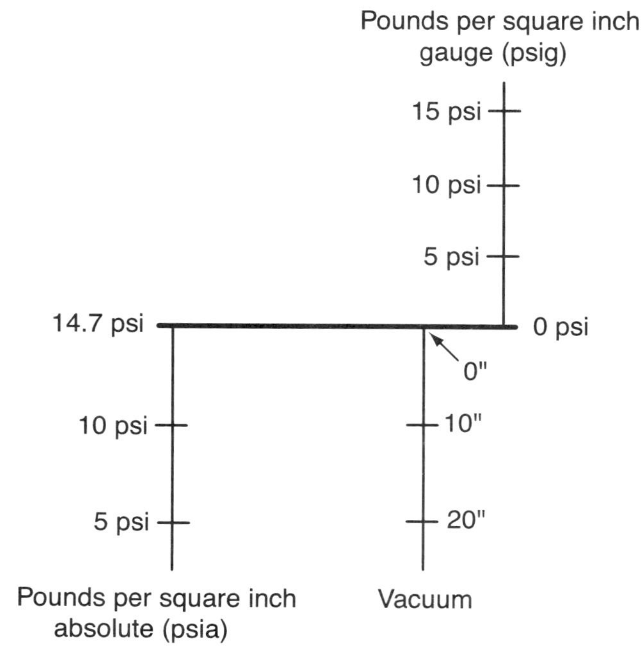


**Figure 22.** The relationship between atmospheric pressure, vacuum, and higher pressures.

the mercury will be 29.92 inches up the tall part of the "J." When the current atmospheric pressure is lower than normal, the normal atmospheric pressure pushes the mercury down. Likewise, when current atmospheric pressure is higher than normal, it will push the mercury up the tube. The amount of mercury movement reflects the difference in the two pressures. This corresponds with a universal law that states a high pressure always moves toward a lower pressure.

To further consider the law that states a high pressure always move to a lower pressure, look at what happens when a nail punctures an automotive tire. The high-pressure air in the tire leaks out until the pressure inside the tire is equal to atmospheric pressure outside the tire. When the tire is repaired and inflated, air with a pressure higher than atmospheric is forced into the tire.

Above sea level, atmospheric pressure decreases. The weight of a column of air is less at an elevation of 5,000 feet (1,524 meters) than it is at sea level. As altitude continues to increase, atmospheric pressure and weight continue to decrease. At an altitude of several hundred miles above sea level, the earth's atmosphere ends, and there is no pressure beyond that point.

### Vacuum

Scientifically, **vacuum** is the absence of atmospheric pressure. However, it is commonly used to refer to any pressure less than atmospheric pressure. Vacuum may also be referred to as low pressure simply because it is a pressure less than atmospheric pressure.

Vacuum can be measured in psig or psia, but inches of mercury (in - Hg) is most commonly used for this measurement. A "U" tube is partially filled with mercury, and atmospheric pressure is allowed to enter one end of the tube. If vacuum is supplied to the other end of the "U" tube, the mercury is forced downward by the atmospheric pressure. When this movement occurs, the mercury also moves upward on the side where the vacuum is supplied. If the mercury moves downward 10 in, or 25.4 cm, where the atmospheric pressure is supplied, and upward 10 in (25.4 cm) where the vacuum is supplied, 20 in-Hg is supplied to the "U" tube. The highest possible, or perfect, vacuum is approximately 29.9 in-Hg.

# HEAT

**Heat** is a form of energy and is used in many ways. The main sources of heat are the sun, the earth, chemical reactions, electricity, friction, and nuclear energy. Heat is the result of the kinetic energy that is present in all matter, therefore everything has heat. Cold objects have low kinetic energy because their atoms and molecules are moving very slowly, whereas hot objects have more kinetic energy because their atoms and molecules are moving fast.

Temperature is an indication of an object's kinetic energy. Temperature is measured with a thermometer, in either Fahrenheit (°F) or Celsius (°C or centigrade). At absolute zero (−273°C, also referred to as 0° Kelvin), particles of matter do not vibrate, but at all other temperatures particles have motion. The temperature of an object states how well the object will transfer heat (or kinetic energy) to or from another object. Heat and temperature are not the same thing; heat is the movement of kinetic energy from one object to another. Temperature is an indication of the amount of kinetic energy something has. Energy from something hot will always move to an object that is colder until both are at the same temperature. The greater the difference in temperature between the two objects, the faster the heat will flow from one to the other.

## Heat Transfer

Heat transfers between two substances at different temperatures through convection, conduction, or radiation. **Convection** is the transfer of heat by the movement of a heated object. Convection can be easily seen by watching a pot of water on a hot stove. The water on the bottom of the pot is the first to be heated by the stove. As the water at the bottom becomes hotter, it expands and becomes lighter than the water at the top of the pan. This causes the heavier water to sink toward the bottom and push the warmer water up. This continues until all of the water in the pan is at the same temperature.

**Conduction** is the movement of heat through a material. Much of the heat generated by a running engine is not used to drive the vehicle. Rather it is "wasted" as it heats the parts of the engine. The immense heat that results from combustion is absorbed by the engine and is used to push the pistons down. The engine's cooling system uses conduction to move the heat off the parts to help cool the engine. Since the energy from something hot moves toward something colder, the heat from the engine moves to the engine's coolant circulating throughout the system. The object that receives conducted heat can be a liquid, gas, or, solid.

**Radiation** does not rely on another material to transfer heat. The moving atoms and molecules within an object create waves of radiant energy. These waves are typically called infrared rays. Hot objects give off more infrared rays than colder objects. Therefore, a hot object will give off infrared rays to anything around it that is colder. No movement is necessary to transfer this heat. You can feel radiation in action by simply putting your hand near something that is hot. This radiation can heat up other objects. The hot object cools as it radiates its heat energy. In an engine's cooling system, the radiator uses radiation to transfer heat from the coolant into the surrounding air.

## The Effects of Temperature Change

Any time the temperature of an object has changed, a transfer of heat has occurred. A transfer of heat may also cause the object to change size or its state of matter. As heat moves in and out of a mass, the movement of atoms and molecules in that mass increases or slows down. With an increase in motion, the size of the mass tends to get bigger or expand. This is commonly called thermal **expansion.** Thermal **contraction** takes place when a mass has heat removed from it and the atoms and molecules slow down. All gases and most liquids and solids expand when heated, with gases expanding the most. Solids, because they are not fluid, expand and contract at a much lower rate. It is important to realize that all materials do not expand and contract at the same rate. For example, an aluminum component will expand at a faster rate than the same component made of iron. This explains why aluminum cylinder heads have unique service requirements and procedures when compared to iron cylinder heads.

Typically, when heat is added to a mass, the temperature of the mass increases. However, in some cases, the additional heat causes no increase in temperature but causes the mass to change its state (solid to liquid or liquid to gas). For example, if we take an ice cube and heat it to

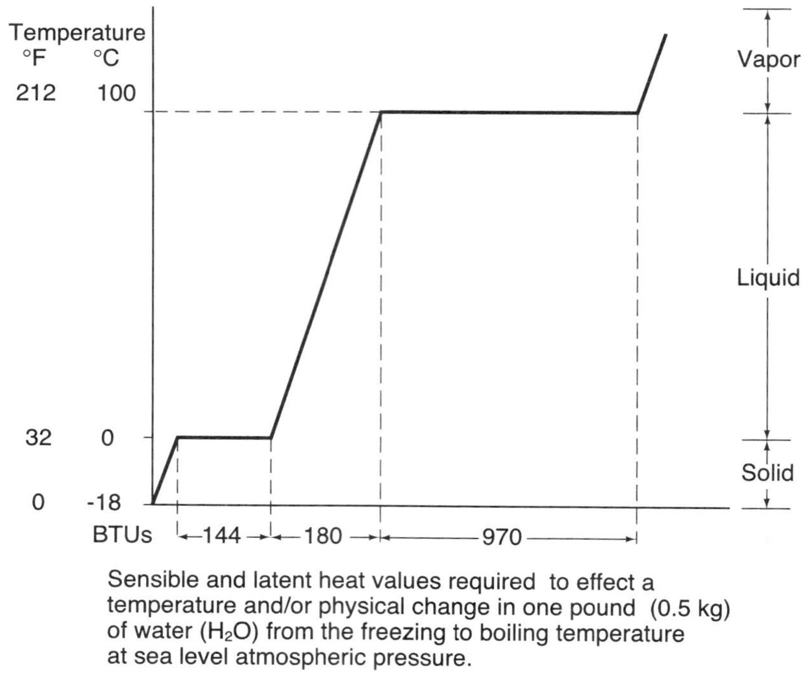


**Figure 23.** A graph showing the amount of heat required to raise the temperature of water and to change it from a solid to a liquid and from a liquid to a vapor.

32°F (0°C), it will begin to melt **(Figure 23).** As heat is added to the ice cube, the temperature of the ice cube will not increase until it becomes a liquid. The heat added to the ice cube that did not raise its temperature but caused it to melt is called **latent heat,** or the heat of fusion. Each gram of ice at 0°C requires 80 calories of heat to melt it to water at 0°C. As more heat is added to the 0°C water, the water's temperature will once again increase. This continues until the temperature of the water reaches 212°F (100°C). This is the boiling temperature of water. At this point, any additional heat applied to the water is latent heat causing the water to change its state to that of a gas. This added heat is called the heat of evaporation.

To change the water gas back to liquid water, the same amount of heat required to change the liquid to a gas must be removed from the gas. At that point, the gas condenses to a liquid. As additional heat is removed, the temperature will drop until enough heat is removed to bring its temperature back down to freezing (melting in reverse) point. At that time, latent heat must be removed from the liquid before the water turns to ice again.

## CHEMICAL PROPERTIES

The properties of something are those characteristics used to identify or describe it. Physical properties are characteristics that are readily observable, such as color, size, luster, and smell. Physical changes are those changes that do not result in the production of a new substance, such as melting an ice cube.

Chemical properties are only observable during a chemical reaction, and they describe how one type of matter reacts with another type of matter to form a new and different substance. Chemical properties are quite different from physical properties. An example of a chemical property is a substance's ability to burn. A chemical property of some metals is the ability to combine with oxygen to form rust (iron and oxygen) or tarnish (silver and sulfur). Another example is hydrogen's ability to combine with oxygen to form water.

A solution is a mixture of two or more substances in varying amounts. Most solutions are liquids, but solutions of gases and solids are possible. An example of a gas solution is the air we breathe; it is composed of mostly oxygen and nitrogen. Brass is a good example of a solid solution since it is composed of copper and zinc. The liquid in a solution is called the **solvent,** and the substance added is the solute. If both are liquids, the one present in the smaller amount is usually considered the solute. Solutions can vary widely in terms of how much of the dissolved substance is actually present. A heavily diluted (much water) acid solution has very little acid and may not be harmful or even noticeably acidic.

## Chemical Reactions

Chemical changes, or chemical reactions, are changes that result in the production of another substance, such as wood turns to carbon after it has burned completely. A chemical reaction is always accompanied by a change in energy. This means energy is given off or taken in during the reaction. Some reactions that release energy need some energy to start the reaction. A reaction takes place when two or more molecules interact and the one of the following happens:

- A chemical change occurs.
- Single reactions occur as part of a large series of reactions.
- Ions, molecules, or pure atoms are formed.

## Reduction and Oxidation

**Oxidation** is a chemical reaction in which a substance combines with oxygen. Rapid oxidation produces heat fast enough to cause a flame. When fuel burns, substances in the fuel combine with oxygen to form other compounds. This chemical reaction is combustion, which produces heat and fire. The rusting of iron is also an example of oxidation. Unlike fire, rusting occurs so slowly that little heat and no flames are produced.

The addition of hydrogen atoms or electrons is called **reduction.** Oxidation and reduction always occur simultaneously; one substance is oxidized by the other, which it reduces. During oxidation, a molecule provides electrons. During reduction, a molecule accepts electrons. Oxidation and reduction reactions are usually called redox reactions. Redox is any chemical reaction in which electrons are transferred. Batteries, also known as voltaic cells, produce an electrical current at a constant voltage through redox reactions.

Iron combines with oxygen to form rust. The rate this occurs depends on several factors: temperature, surface area (more iron exposed for oxygen to get at), and catalysts (speed up a reaction but do not react and change themselves). Rusting is just like the burning of other substances.

Corrosion is the wearing away due to chemical reactions, mainly oxidation. It occurs whenever a gas or liquid chemically attacks an exposed surface and is accelerated by warm temperatures and by acids and salts. Some materials resist corrosion naturally; others can be treated to protect them, such as painting, coatings, galvanizing, or anodizing.

Galvanizing involves the coating of zinc onto iron or steel to protect them against exposure to the atmosphere and rusting. If galvanizing is properly applied, it can protect the metals for 15 to 30 years or more. If the coating is damaged, the iron or steel continues to be protected by sacrificial corrosion, a phenomenon in which atmospheric oxidation spares the iron and affects the zinc.

Anodizing is a method of coating metal for corrosion resistance, electrical insulation, thermal control, abrasion resistance, sealing, improving paint adhesion, and decorative finishing. Anodizing consists of electrically depositing an oxide film from an aqueous solution onto the surface of a metal, often aluminum. During the most common anodizing process, dyes can be added to the oxidation process to give the material a colored surface.

## Metallurgy

Metallurgy is the science of extracting metals from their ores and modifying the metals for a particular use. It also concerns the chemical, physical, and atomic properties and structures of metals and the principles by which metals are combined to form alloys. An **alloy** is a mixture of two or more metals. Steel, for example is an alloy of iron plus carbon and other elements.

Metals can be defined as any substance that has one or more of the following properties:

- Good heat and electric conduction
- Malleability—meaning it can be hammered, pounded, or pressed into a shape without breaking
- Ductility—it can be stretched, drawn, or hammered without breaking
- High light reflectivity—it can make light bounce off its surface
- The capacity to form positive ions in a solution and hydroxides rather than acids when their oxides meet water

About three-quarters of the elements are metals. The most abundant metals are aluminum, iron, calcium, sodium, potassium, and magnesium.

The hardness of something is a statement of its resistance to scratching. **Hardening** is a process that increases the hardness of a metal, deliberately or accidentally, by hammering, rolling, carburizing, heat treating, tempering, or other physical processes. All of these methods deform the metal, but also compact the atoms or molecules to make the material denser. The first few deformations imposed by such treatment weaken the metal, but because of the crystalline structure of metal, its strength increases with continued deformations.

**Carburizing** is a method used to surface-harden steel by heat or mechanical means to increase the hardness of the outer surface while leaving the core relatively soft. This combination of a hard surface and soft interior withstands very high stress and fatigue and also offers low cost and superior flexibility in manufacturing. To carburize, the steel parts are placed in a carbonaceous environment (with charcoal, coke, and carbonates, or carbon dioxide, carbon monoxide, methane, or propane) at a high temperature for several hours. The carbon diffuses into the surface of the steel, thereby altering the crystal structure of the metal. Gears, ball and roller bearings, and piston pins are often carburized.

**Heat treating** is the changing of the properties of a metal (including iron, steel, aluminum, copper, and titanium) by using heat. It is used to harden metals that have different crystal structures at low and high temperatures. The metal is heated and then quenched (cooled rapidly) to retain the high-temperature constituent. Tempering may then be used to attain the desired hardness. Heating followed by slow cooling (annealing) is used to soften metals.

**Tempering** is the heat treating of metal alloys, particularly steel to result in specific properties. For example, raising the temperature of hardened steel from 752°F (4000°C) and holding it for a time before quenching in oil decreases its hardness and brittleness and produces strong steel. Quench and temper heat treating is applied at many different cooling rates, holding times, and temperatures and is a very important means of controlling the properties of steel.

## Solids Under Tension

The atoms of a solid are closely packed, giving it a greater density than most liquids and gases. A solid's rigidity is the result of the strong attraction between its atoms. A force pulling on a solid moves these atoms farther apart, creating an opposing force called tension **(Figure 24).** If a force pushes on a solid, the atoms move closer together, creating compression. These are the principles of how springs function. Springs are used in many automotive systems, the most obvious are those used in suspension systems.

An elastic substance is a solid that becomes larger under tension, becomes smaller under compression, and returns to its original size when no force is acting on it. Most solids show some elastic behavior, but there is usually a limit to the force from which recovery is possible. Stresses beyond its elastic limit cause material to yield, or flow, and the result is permanent distortion or breakage. This limit depends on the material's internal structure. For example, steel, although strong, has a low elastic limit and can be extended only about 1 percent of its length, whereas rubber can be elastically extended to about 1000 percent. Another factor involved in elasticity is the cross-sectional area of the material involved.

**Tensile strength** is the ratio of the maximum load a material can support without breaking when being stretched and is dependent on the cross-sectional area of the material. When stresses less than the tensile strength are removed, the material completely or partially returns to its original size and shape. As the stress approaches that of the tensile strength, the material forms a narrow, constricted region that is easily broken. Tensile strengths are measured in units of force per unit area.

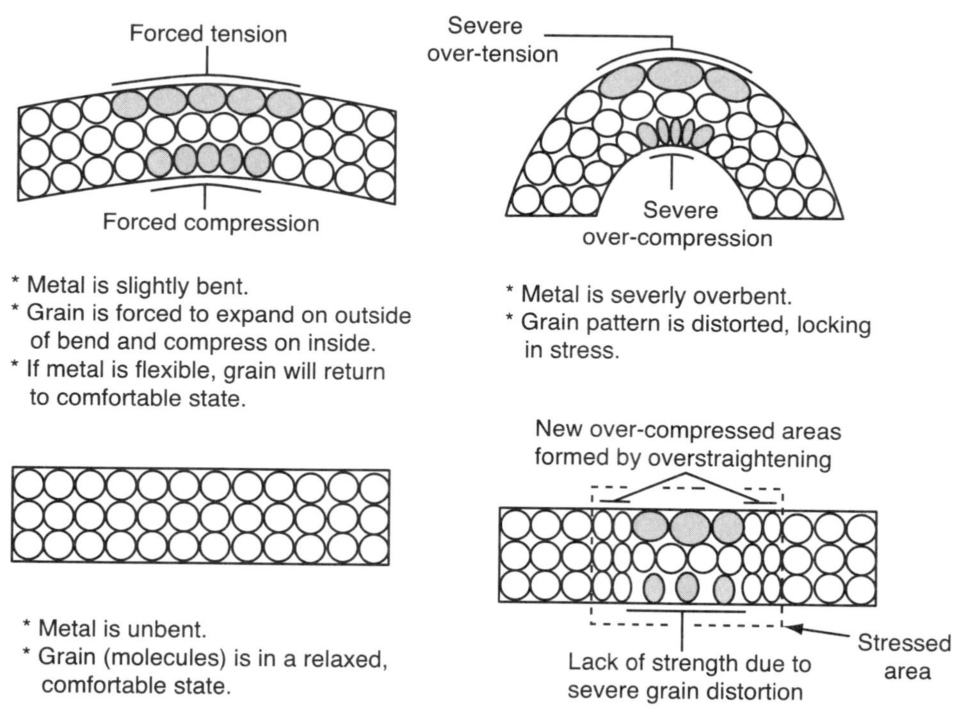


**Figure 24.** The effects of tension or stress on a piece of metal.

## SUMMARY

- All matter consists of countless tiny particles called atoms and molecules, and is anything that occupies space. Matter exists as a gas, liquid, or solid.
- Materials that *absorb* fluids are permeable substances. Impermeable substances *adsorb* fluids.
- Energy is the ability to do work, and all matter has energy.
- The total amount of energy never changes; it can only be transferred from one form to another, not created nor destroyed.
- When energy is released to do work, it is called kinetic energy. Stored energy may be called potential energy.
- Mass is the amount of matter in an object. Weight is a force and is measured in pounds or kilograms. Gravitational force gives the mass its weight.
- Volume is the amount of space occupied by an object.
- A force is a push or pull, can be large or small, and can be applied to objects by direct contact or from a distance.
- When an object moves in a circle, its direction is continuously changing, and all changes in direction require a force. The forces required to maintain circular motion are called centripetal and centrifugal force.
- Pressure is a force applied against an opposing object and is measured in units of force per unit of surface area (pounds per square inch or kilograms per square centimeter).
- The greater the mass of an object, the greater the force needed to change its motion.
- Inertia is the tendency of an object at rest to remain at rest, or the tendency of an object in motion to stay in motion.
- Momentum is the product of an object's weight times its speed.
- Newton's Laws of Motion state that: an object at rest tends to remain at rest, and an object in motion tends to remain in motion.
- Friction is a force that slows or prevents motion of two moving objects that touch and can be reduced by lubrication or by the use of rollers.
- When a force moves a certain mass a specific distance, work is done.
- A machine is any device that can be used to transmit a force and change the amount of force and/or its direction.
- Torque is a force that tends to rotate or turn things and is measured by the force applied and the distance traveled.
- How many times the vibration occurs in one second is called frequency, which is often expressed in hertz (Hz) and is equal to one cycle per second.
- Noise is any unwanted signal or sound and can be random or periodic.
- A gas will always fill a sealed container, whereas a liquid may not. A gas will also readily expand or compress according to the pressure exerted on it. Liquids are basically incompressible, which gives them the ability to transmit force.
- Hydraulics is the study of liquids in motion.
- Pascal constructed the first known hydraulic device and established what is known as Pascal's Law of Hydraulics.
- Boyle's Law states that the volume and pressure of a mass of gas at a fixed temperature is inversely proportional.
- Pressure greater than atmospheric pressure may be measured in psi gauge (psig), and actual pressure is measured in psi absolute (psia).
- Vacuum is commonly used to refer to any pressure less than atmospheric pressure.
- Heat is a form of energy caused by the movement of atoms and molecules.
- Convection is the transfer of heat by the movement of a heated object.
- Conduction is the movement of heat through a material.
- As heat moves in and out of a mass, the size of the mass tends to change.
- Oxidation is a chemical reaction in which a substance combines with oxygen. The addition of hydrogen atoms or electrons is reduction.
- Hardening is a process that increases the hardness of a metal, deliberately or accidentally, by hammering, rolling, carburizing, heat treating, tempering, or other physical processes.
- An elastic substance is a solid that becomes larger under tension, becomes smaller under compression, and returns to its original size when no force is acting on it.
- Tensile strength is the ratio of the maximum load a material can support without breaking when being stretched and is dependent on the cross-sectional area of the material.

# REVIEW QUESTIONS

1. Describe Newton's First Law of Motion, and give an application of this law in automotive theory.
2. Explain Newton's Second Law of Motion, and give an example of how this law is used in automotive theory.
3. Explain why a rotating, tilted wheel moves in the direction of the tilt.
4. Describe the effects of static and dynamic balance.
5. Energy may be defined as the ability to do ________________.
6. When one object is moved over another object, the resistance to motion is called ______________.
7. Torque is a force that does work with a ________________ action.
8. While discussing different types of energy, Technician A says when energy is released to do work, it is called potential energy. Technician B says stored energy is referred to as kinetic energy. Who is correct?
   A. Technician A only
   B. Technician B only
   C. Both Technician A and Technician B
   D. Neither Technician A nor Technician B
9. While discussing mass and weight, Technician A says that mass is the measurement of an object's inertia. Technician B says that mass and weight may be measured in cubic inches. Who is correct?
   A. Technician A only
   B. Technician B only
   C. Both Technician A and Technician B
   D. Neither Technician A nor Technician B
10. When applying the principles of work and force,
    A. work is accomplished when force is applied to an object that does not move.
    B. in the metric system the measurement for work is cubic centimeters.
    C. no work is accomplished when an object is stopped by mechanical force.
    D. if a 50-pound object is moved 10 feet, 500 ft.-lb of work are produced.

Chapter 9

# General Suspension and Steering System Diagnosis

## INTRODUCTION

Drivers may experience many symptoms related to incorrect wheel alignment or defective steering and suspension parts. Not all symptoms are related to wheel alignment, but the customer may request an alignment because of the belief that it will correct the problem. Technicians must diagnose and correct the cause of the concern before aligning the wheels.

## ROAD TEST

As with the diagnosis of any problem, begin with trying to duplicate the customer's concern. Make sure to gather as much information from the customer as possible **(Figure 1).** Knowing what happens and when it happens will greatly help your diagnosis. Verifying the customer's concern is done on a road test. Before taking the road test, make sure the vehicle is safe by checking the tires, brake pedal operation, and steering wheel feel. Note the operation of the ABS and air bag warning lights if the vehicle has these features. During the road test, observe traffic laws and drive in a normal manner so you will not cause any additional wear or damage.

During the road test, drive the vehicle under the conditions when the customer's concern occurred. Road test the vehicle on a surface that has many bumps and a variety of road surfaces. While it may be somewhat inconvenient to seek out a particular type of road surface, this may be the only way to verify the condition. Attempt to determine the problem area, front, back, right, or left side. Determine whether the problem is vehicle speed, torque, gear ratio, or engine speed sensitive. Also, try to determine if it is a tire, suspension, steering alignment, or brake problem. While road testing the vehicle, listen for any unusual noises related to suspension and steering. Note any unusual or unsafe conditions that should be brought to the owner's attention.

**Figure 1.** Get as much information as possible from the customer about the problem.

## NVH

**Noise, vibration,** and **harshness (NVH)** are unwanted sensations that may be caused by suspension and steering problems. Noise occurs from something that moves or rattles, such as a loose or broken part. The type of noise may indicate where to look for it. When diagnosing a noise, pay attention to when the noise gets louder or changes. Often tire noises change with speed or road surfaces. Bearing and axle noises change with load and speed. Suspension noises change with changes in road surface and steering noises are evident during changes in direction.

A vibration is also caused by something that moves and may accompany a noise. Vibrations are felt by the driver as the noise travels through something in the car **(Figure 2).** Tires, wheels, driveshafts are common causes of vibrations. **Harshness** is an unpleasant motion, it is also caused by something that moves or by something that is not able to move properly.

## Causes of Common Symptoms

A good description of the problem can go far to accurately diagnose the cause of the problem. The description should be based on the customer's concern and your observations during the road test. Although customers may describe a concern in different ways, the most common and their typical causes will be discussed next.

### Noises

Determine the cause of the abnormal noises using the below:

- Tire noise varies with road surface conditions, whereas differential noise is not affected when various road surfaces are encountered.
- Uneven tread surfaces may cause tire noises that seem to originate elsewhere in the vehicle. These noises may be confused with differential noise. Differential noise usually varies with acceleration and deceleration. Tire noise is most pronounced on smooth asphalt road surfaces at speeds of 15 to 45 mph.

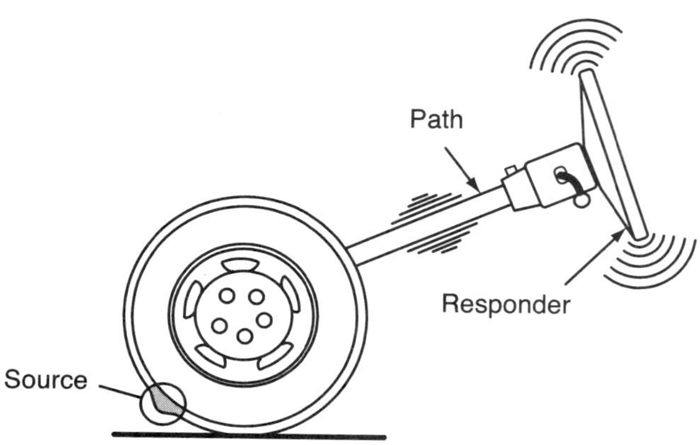


**Figure 2.** A vibration will travel through parts and make it seem as though the problem is at the "responder."

- Rattling on road irregularities can be caused by worn shock absorber bushings or grommets, worn spring insulators, a broken coil spring or spring insulators, worn control arm bushings, worn stabilizer bar bushings, worn strut rod grommets, worn leaf spring shackles and bushings, worn torsion bars, anchors, and bushings.
- Chatter while cornering can be caused by worn upper strut mounts **(Figure 3).**
- Dry or worn control arm bushings may cause a squeaking noise.

There may also be abnormal noises when the steering wheel is turned. The cause is best identified by paying close attention to where the noise is coming from. Some noises may be caused by tires or interference between the steering wheel and the steering column covers. Others can result from a faulty power steering pump or system.

- A hissing noise during low speed turning, such as while parking is normal.
- A chirping or squealing normally means the drive belt is loose or slipping.
- Improperly routed hoses, loose mounting bolts on the power steering pump or the steering gear can cause a rattle.
- A groaning or howling can indicate a restriction in the hydraulic system.
- A sticking control valve will cause a buzzing sound.

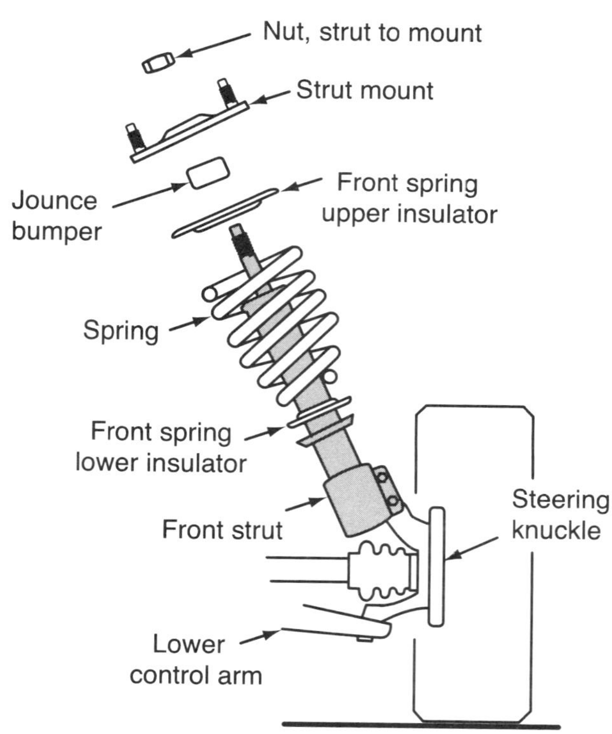


**Figure 3.** Worn upper strut mounts will cause chatter while cornering.

### Harsh Riding

If the ride of the vehicle seems to be harsher or rougher than normal check for low ride height at the front and rear suspension, weak or broken springs, worn torsion bar components such as pivot cushion bushings, and worn or damaged control arm bushings.

### Excessive Sway on Road Irregularities

If the vehicle seems to move or roll to one side when traveling on a rough surface, check for a weakened or disconnected stabilizer bar and worn stabilizer bar bushings **(Figure 4).**

### Reduced Directional Stability

When a vehicle does not seem stable when its direction is changed or when attempting to change direction, check for weak or broken springs or torsion bars and worn bushings or anchors. If the steering seems erratic on irregular road surfaces, inspect the upper and lower control arms for cracks, distortion, and worn bushings.

### Excessive Steering Wheel Play

Excessive play in the steering wheel is apparent when there is too much steering wheel movement before the wheels begin to turn. A small amount of play is normal. This problem can be caused by the following:

- Loose, worn, or damaged steering linkages or tie rod ends
- Loose, worn, or damaged steering column U-joints
- Loose, worn, or damaged steering column bearings
- Damaged or worn steering gear
- Loose steering gear bolts

### Feedback

When the driver feels the surface of the road through the steering wheel, this is called **feedback.** This problem can be caused by the following:

- Loose, worn, or damaged steering linkages or tie rod ends
- Loose, worn, or damaged steering column U-joints
- Loose or damaged steering gear mounting bolts
- Damaged or worn steering column bearings
- Loose suspension bushings, fasteners, or ball joints

### Wandering

When a vehicle wanders, the driver must constantly turn the steering wheel to the left and right to keep the vehicle going straight on a level road. This problem can be caused by the following:

- Loose or worn suspension components
- Poor wheel alignment
- Unevenly loaded or overloaded vehicle
- Loose or damaged steering gear bolts
- Loose steering column U-joint bolts
- Loose, worn, or binding steering linkages or tie rod ends
- Improper steering gear preload adjustment

### Bump Steer

**Bump steer** occurs if the tie rods are not the same height, meaning one of the tie rods is not parallel to the lower control arm. This condition may be caused by a bent or worn idler arm or pitman arm on a parallelogram steering linkage. On a rack and pinion steering system, worn steering gear mounting bushings may cause this unparallel condition between one of the tie rods and the lower control arm.

### Torque Steer

A front-wheel-drive vehicle with unequal drive axles produces some **torque steer** on hard acceleration. On a front-wheel-drive vehicle, torque steer is aggravated by different tire tread designs on the front tires or uneven wear on the front tires.

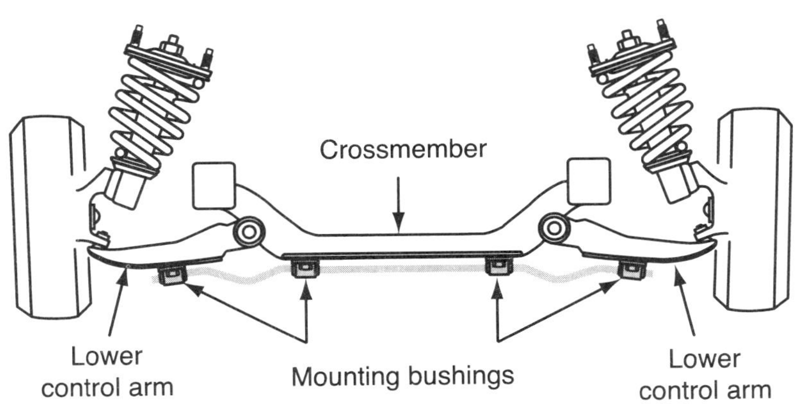


**Figure 4.** The stabilizer bar and its mounting bushings should be checked for damage, looseness, and wear.

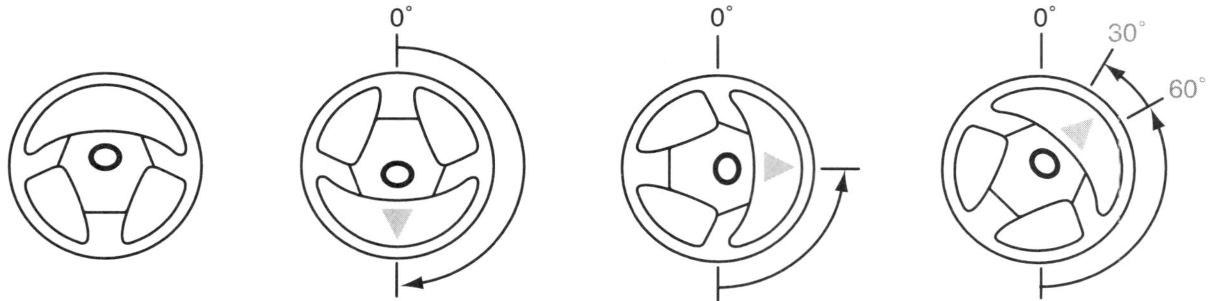


**Figure 5.** Poor steering wheel returnability can be caused by several problems.

### Sticking Steering or Poor Return

Most steering systems are designed so the steering wheel returns to within 30 degrees to 60 degrees of center after a 180 degrees turn **(Figure 5).** If there is a tendency for the steering wheel to return too quickly, suspect low ride height at the rear suspension. If the steering wheel does not properly return and memory steer is evident, check the following:

- Binding steering column U-joints
- Loose, worn, or damaged steering linkages or tie rod ends
- Binding steering gear
- Loose, damaged or worn suspension parts
- Poor wheel alignment
- Binding steering column bearings
- On a front-wheel drive vehicle, the upper strut mounts may be binding.

### Hard Steering

Obviously this complaint results when extra effort is needed to turn the steering wheel. This problem may be simply an absence of power assist. Hard steering problems can occur whenever the steering wheel is turned or just when it has been turned close to its limit. This problem can be caused by the following:

- A faulty power steering pump
- A loose power steering belt
- Low power steering fluid level
- Worn front tires
- Defective steering gear
- Damaged or faulty steering column bearings
- Seized steering column U-joints
- Internal problems in the power rack assembly
- Faulty suspension components
- Inadequately inflated tires
- Low ride height at the rear suspension.

### Steering Pull

A **pull** is a tugging sensation, felt at the steering wheel. The driver must push the steering wheel in the opposite direction as the pull to keep the vehicle going straight. This problem can be caused by the following

- Improper frame alignment
- Brake system problem
- Worn or binding suspension components, especially springs
- Worn upper and lower control arm bushings
- Poor wheel alignment
- Low ride height on one side of the front suspension
- Unevenly loaded or overloaded vehicle
- Loose, worn, or damaged steering linkages or tie rod ends
- Out of balance steering gear valve
- Torque steer
- Tire problems
- Binding strut bearing

**Drifting** is a condition where the vehicle slowly moves to one side of the road when the driver's hands are taken off the steering wheel. If the steering gradually pulls to one side on a smooth, straight road surface, a tire, steering, or suspension defect is present. Check for tires of different types, sizes, designs, or inflation pressures on opposite sides of the vehicle.

> **You Should Know** *If steering pull or drift is evident during the road test, drive the vehicle on the same road in the opposite direction to determine if a crosswind is the cause.*

### Shimmy

When the wheels **shimmy,** the driver will feel large, consistent, rotational oscillations at the steering wheel. These motions are caused by the lateral movement of the tires. This problem can be caused by the following:

- Loose, worn, or damaged steering linkages or tie rod ends
- Loose, worn, or damaged suspension parts
- Out-of-balance tires
- Excessive wheel runout
- A bad tire
- Loose wheel bearings

### Nibble

This feeling is similar to a shimmy. **Nibble** results from the interaction of the tires with the road's surface. The customer's concern may describe the problem as slight rotational oscillations of the steering wheel. This problem can be caused by loose, worn, or damaged steering linkage, suspension parts, or tie rod ends

## BASIC INSPECTION

Once the road test has been completed and it has been determined there is an abnormal condition, a thorough inspection can pinpoint the exact cause.

### Tire Wear

Begin your visual inspection by inspecting the tires. Check for correct pressure, construction, size, wear, and damage, and for defects that include ply separations, sidewall knots, and concentricity problems. As a tire wears out, the wear should occur evenly across the tread. Tire wear is checked by looking at the tread, feeling the tread, and by measuring the depth of the tread. Keep in mind that tire wear patterns are good indicators of steering and suspension problems **(Figure 6).** Tire wear is also a great indicator of wheel alignment problems.

A properly inflated tire gives the best tire life, riding comfort, handling stability, and fuel economy for normal driving conditions. Too little air pressure can result in tire squeal, hard steering, excessive tire heat, abnormal tire

| Conditions | Rapid wear at shoulders | Rapid wear at center | Cracked treads | Wear on one edge | Feathered edge | Bald spots | Scalloped wear |
|---|---|---|---|---|---|---|---|
| Effect | 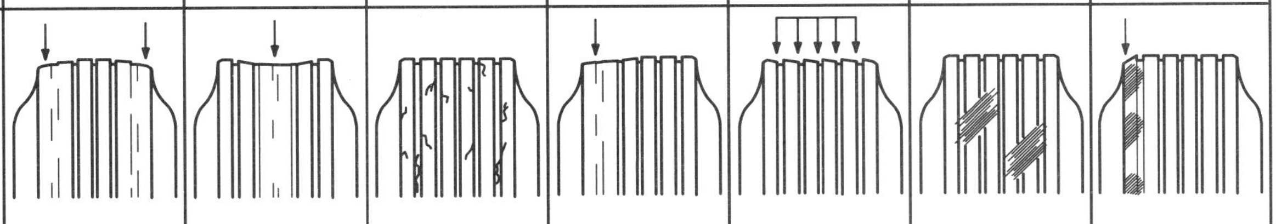 | | | | | | |
| Causes | Underinflation or lack of rotation | Overinflation or lack of rotation | Underinflation or excessive speed | Excessive camber | Incorrect toe | Unbalanced wheel / Or tire defect | Lack of rotation of tires or worn or out-of-alignment suspension |
| | 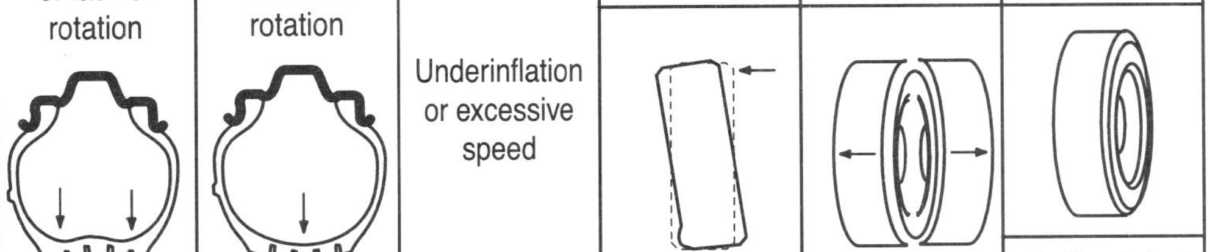 | | | | | | |
| Corrections | Adjust pressure to specifications when tires are cool. Rotate tires. | | | Adjust camber to specs | Adjust toe to specs | Dynamic or static balance wheels | Rotate tires and inspect suspension |

**Figure 6.** Tire wear patterns are good indicators of steering and suspension problems.

| Condition | Rapid wear at center | Rapid wear at shoulders | Cracked treads |
|---|---|---|---|
| Effect | 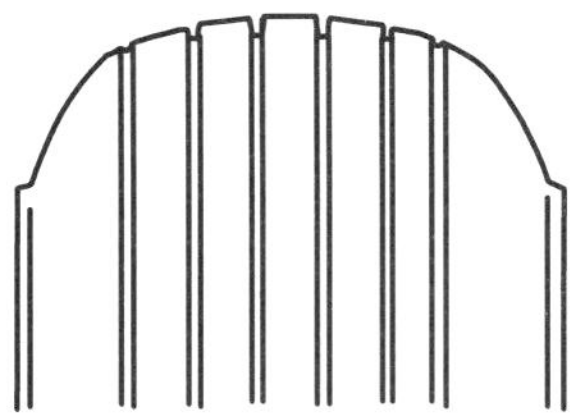 | 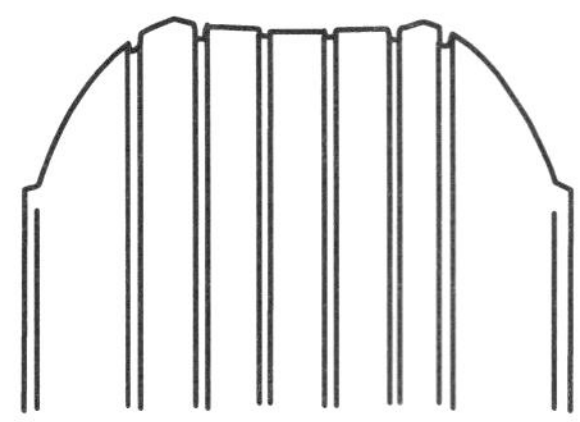 |  |
| Cause | Over inflation or lack of rotation 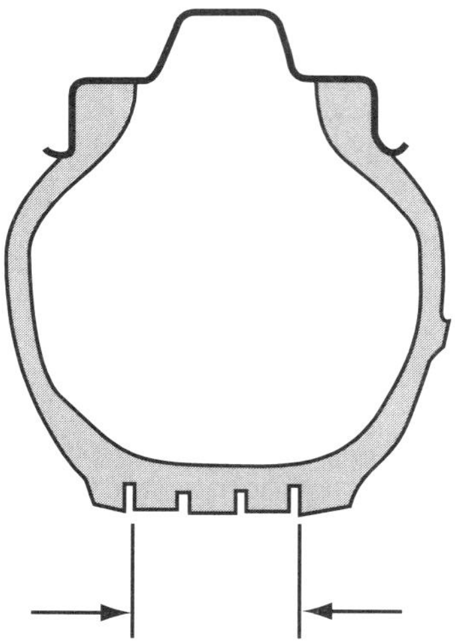 Contact patch area | Under inflation or lack of rotation 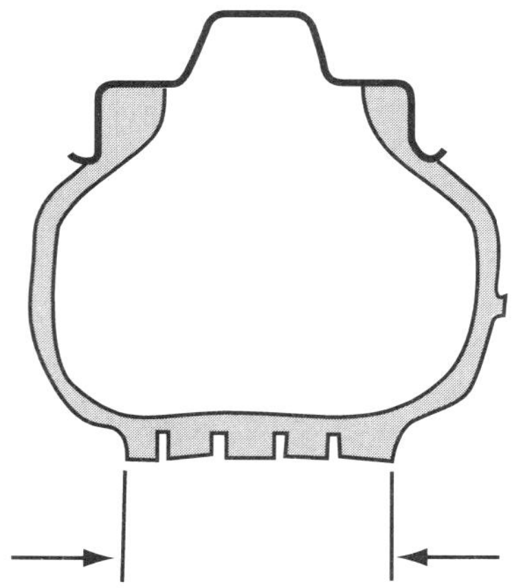 Contact patch area | Under inflation or excessive speed |
| Correction | Adjust pressure to specifications. When tires are cool rotate tires. | | |

**Figure 7.** The wear patterns on a tire that does not have the correct inflation pressure.

wear, and increased fuel consumption by as much as 10 percent. An under-inflated tire shows maximum wear on the outside edges of the tread, there is little or no wear in the center. Conversely, an over-inflated tire **(Figure 7)** shows its wear in the center of the tread and little wear on the outside edges. Higher than recommended tire pressure can cause a hard ride, tire bruising, and rapid wear at the center of the tire.

A few vehicles are fit with tire inflation monitoring systems. These systems have an air pressure sensors strapped around the drop center of each wheel. The sensors monitor the pressure in the tire. When the pressure is below or above a specified range, the vehicle's computer causes a warning light on the dash to illuminate. This alerts the driver of the problem.

## Curb Riding Height

The **curb riding height** is determined mainly by spring condition. Other suspension components such as control arm bushings will affect curb riding height if they are worn. Since incorrect curb riding height affects most of the suspension angles, this measurement is critical. Sagged springs change the normal operating arc of the lower ball joint. This action causes excessive lateral movement of the tire during wheel jounce and rebound with resulting tire wear.

The curb riding height must be measured at the vehicle manufacturer's specified location, which varies depending on the type of suspension system. When the vehicle is on a level floor or an alignment rack, measure the curb riding height from the floor to the manufacturer's specified location on the chassis on the front **(Figure 8)** and rear **(Figure 9)** suspension.

Reduced curb riding height on the front suspension may cause decreased directional stability. If the curb riding height is reduced on one side of the front suspension, the steering may pull to one side. Reduced rear suspension height increases steering effort and causes rapid steering wheel return after turning a corner. Harsh riding occurs when the curb riding height is less than specified. The curb riding height must be measured at the vehicle manufacturer's specified location, which varies depending on the type of suspension system.

Park the vehicle on a level floor or on an alignment rack and measure the curb riding height from the floor to the center of the lower control arm mounting bolt on both sides of the front suspension. Measure the rear suspension from the floor to the center of the strut rod mounting bolt. Compare your measurements to specifications.

If the curb riding height is less than specified, the control arms and bushings should be inspected and replaced as necessary. When the control arms and bushings are in normal condition, the reduced curb riding height may be caused by sagged springs.

**Figure 8.** Measuring the front curb riding height.

**Figure 9.** Measuring the rear curb riding height.

## Shock Absorber and Strut Checks

When performing a shock absorber or strut bounce test, the vehicle's bumper is pushed downward with considerable force. The bumper is then released observed. One free upward bounce should stop the chassis movement if the shock absorber or strut provides proper spring control. Shock absorber replacement is required if more than one free upward bounce occurs.

Some defective shock absorbers or struts may have internal clunking, clicking, and squawking noises, or binding conditions. When these shock absorber noises or conditions are experienced, shock absorber or strut replacement is necessary.

## Dry Park Check

An excellent overall check for worn or loose conventional steering components is the **dry park check.** With the full weight of the vehicle on the wheels, have an assistant rock the steering wheel back and forth. Start your inspection at one side to the other side. Note any looseness in tie-rod, center link, idler arm, or pitman arm sockets. If a second person is not available, reach up under the vehicle and grasp the flexible coupling on the steering shaft. Rock the linkage.

## DETAILED INSPECTION

Raise the vehicle and grasp the upper and lower surfaces of the tire and shake each wheel to check for worn or loose wheel bearings **(Figure 10).** On front-wheel-drive cars, perform this check the bearings at all four wheels.

Clean all excess dirt and grease from the front and rear suspensions. Visually inspect all components for signs of wear or damage. Inspect all rubber boots, bushings, and grommets. Carefully look at the entire system following these guidelines.

**Figure 10.** Grasp the bottom and top of each wheel and rock the tire to check the wheel bearings.

## Shocks, Struts, and Springs

Shock absorbers should be inspected for loose mounting bolts and worn mounting bushings. If these components are loose, they will rattle, and replacement of the bushings and bolts is necessary. The upper mounts of struts should also be carefully checked.

In some shock absorbers, the bushing is permanently mounted in the shock, and the complete unit must be replaced if the bushing is worn. When the mounting bushings are worn, the shock absorber will not provide proper spring control.

Shock absorbers and struts should be inspected for oil leakage. A slight oil film on the lower oil chamber is acceptable. Any indication of oil dripping is not acceptable, and replacement is necessary **(Figure 11).**

Struts and shock absorbers should be inspected for physical damage such as distortion, and severe dents or punctures. When these conditions exist, replace them **(Figure 12).**

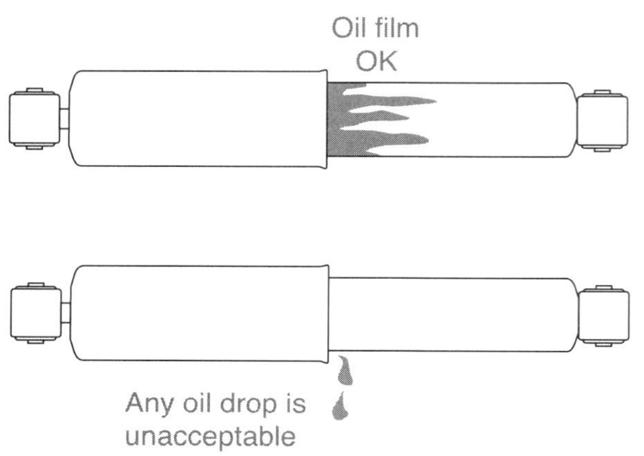


**Figure 11.** When inspecting a shock or strut, look for oil residue.

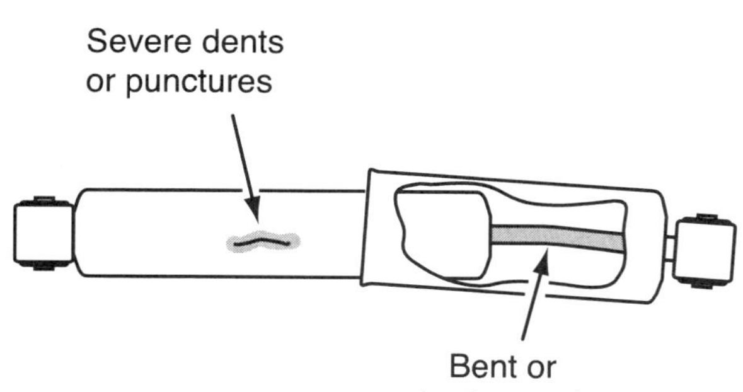


**Figure 12.** Check each shock and strut for physical damage such as a bent condition and severe dents or punctures.

You Should Know *During a visual shock absorber inspection, check the rebound, or strikeout, bumpers on the control arms, or chassis. If the rebound bumpers are severely worn, the shock absorbers may be worn out.*

## Control Arms

Upper and lower control arms shoulc be inspected for cracks, bent conditions, and worn bushings. If the control arm bushings **(Figure 13)** are worn, steering is erratic, especially on irregular road surfaces. Worn control arm bushings may cause a rattling noise on irregular road surfaces. Dry or worn control arm bushings may cause a squeaking noise. Check all bushings for looseness, splits, cracks, misplacement, and noises.

## Ball Joints

Look carefully at the joint's boot. A damaged boot or joint seal will allow lubricant to leak out and allow dirt to enter and contaminate the lubricant. If the boot is damaged, the ball joint must be replaced. If no boot damage is evident, check the ball joint for looseness and wear. Worn ball joints can cause erratic steering and premature tire wear

Wear indicator-type ball joints must remain loaded to check for wear. The joints should be checked with the vehicle at the correct riding height. The most common type ball joint has a small diameter boss, which protrudes from the center of the lower housing. As wear occurs internally, this boss recedes very gradually into the housing. When it is flush with the housing, the ball joint should be replaced.

## Parallelogram Steering

The entire steering system should be carefully checked for looseness, binding, or damage. Use the following guidelines to inspect common steering components.

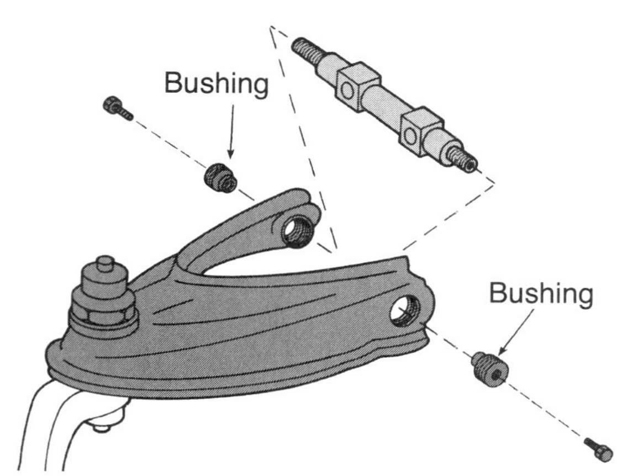


**Figure 13.** Inspect the control arms and bushings.

### Mounts

Check for loose steering gear mounting bolts and worn mounting brackets and bushings.

### Pitman Arm

Because of its function, the pitman arm is the most heavily stressed point in the system. To inspect the pitman arm, grasp it and vigorously shake it to detect any looseness. Check the socket to reveal any damage or looseness. Either condition must be corrected by replacing the worn part.

### Idler Arm

A worn or damaged idler arm can cause instability, uneven tire wear, front-end shimmy, hard steering, excessive play in steering, or poor returnability. Because an idler arm is the weakest link in a parallelogram steering system, they wear before the rest of the system.

To check an idler arm for looseness or wear, make sure the suspension is normally loaded on the ground or an alignment rack. Check the idler arm ends for worn sockets or deteriorated bushings. Grasp the center link firmly with your hand at the idler arm end. Push up with approximately a 25-pound load. Pull down with the same load. The allowable movement of the idler arm and support assembly in one direction is –1/8 inch, for a total acceptable movement of 1/4 inch. The load can be accurately measured by using a dial indicator or pull spring scale located as near the center link end of the idler arm as possible. Keep in mind that the test forces should not exceed 25 pounds, as even a new idler arm might be forced to show movement due to steel flexing when excessive pressure is applied. It is also necessary that a scale or ruler be rested against the frame and used to determine the amount of movement. Observers tend to overestimate the actual movement when a scale is not used. The idler arm should always be replaced if it fails this test.

### Links

Worn or bent center links can cause front-end shimmy, vehicle pull to one side, or change in the toe setting, causing excessive tire wear. When inspecting the center link, look closely to insure it has not been bent or damaged. Grasp the center link firmly and try moving it in all directions. Any movement, or sign of damage, is reason for replacement. Tapered openings seldom wear but should be checked for enlargement caused by a loose connection. Check for worn stabilizer mounting links and bushings

### Tie Rods

Worn tie-rod ends result in incorrect toe-in settings, scalloped and scuffed tires, wheel shimmy, understeering, or front-end noise and tire squeal on turns.

Tie-rod end and center link inspections are similar. Grasp the tie-rod end firmly. Push vertically with the stud, and inspect for movement at the joint with the steering knuckle. Any movement over 1/8 inch or observation of damaged or missing parts, such as seals, is sufficient evidence that replacement is necessary.

Adjusting sleeves should also be carefully inspected. Badly rusted, worn, or damaged adjusting sleeves should be replaced. An additional check of the tie-rods can be made by rotating each tie-rod end to feel for roughness or binding, which could indicate that the socket has probably rusted internally.

### Steering Damper

Steering dampers are generally nonadjustable, nonrefillable, and are not repairable. Inspect the mountings and check the assembly for damage (such as being bent) and fluid leaks. A light film of fluid is permissible on the body of the damper near the shaft seal. A dripping damper should be replaced. A bad steering damper may cause wheel shimmy even though the rest of the suspension and steering systems are fine.

## Rack and Pinion Systems

A rack and pinion system has no idler or pitman arms and no center link: these are replaced by the rack. The number of wear points on rack and pinion systems is reduced to four - the two inner and two outer tie-rod ends. Tie-rod ends are also wear points on the parallelogram steering system. Power rack and pinion assemblies should be carefully checked for leaks. If the pump runs out of fluid, the pump will be damaged.

Check all working components **(Figure 14)** of the systems. Inspect the flexible steering coupling or the universal joints for wear or looseness. If any play is found, they should be replaced. Universal joints can also seize or bind. They should be checked closely.

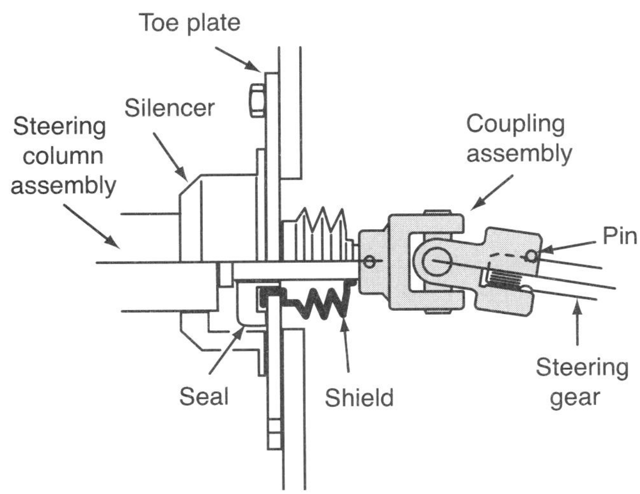


**Figure 14.** Carefully check the connections from the steering column to the steering gear.

Grasp the pinion gear shaft at the flexible steering coupling and try to move it in and out of the gear. If there is movement, the pinion bearing preload might need adjustment. If there is no adjustment, internal components have to be replaced.

Carefully inspect the rack housing. In most cases, the rack and pinion steering assemblies are mounted in rubber bushings. As the vehicle gets older, these mounting bushings deteriorate from heat, age, and oil leakage from the engine. When this happens, the housing moves within its mounting and causes loose and erratic steering.

Stiffness in steering can be caused by a bent rack assembly, tight yoke bearing adjustment, loose power-steering belt, weak pump, internal leaks in power-steering system, and damaged CV joints in front-wheel-drive vehicles.

Check the inner tie-rod socket assemblies located inside the bellows. The most foolproof way of checking these sockets is to loosen the inner bellows clamp and pull the bellows back, giving a clear view of the socket. During the dry park check, observe any looseness. The inner tie-rod socket can also be checked by squeezing the bellows boot until the inner socket can be felt. Push and pull on the tire. If looseness is found in the tie-rod, it should be replaced. On some vehicles, the boot might be made of hard plastic. For this type of boot, lock the steering wheel and push and pull on the tire. Watch for in and out movement of the tie-rod. If movement is observed, replace the inner tie-rod.

One fact to keep in mind is that the condition of the bellows boot determines the life of the inner socket. The bellows boot protects the rack from contamination. It might also contain fluid that helps keep the rack lubricated. If any cracks, splits, or leaks exist, the boot should be replaced. Also, be sure that clamps for the bellows are in their proper place and fastened tightly.

Inspect the outer tie-rod ends. In addition to the dry park check, grab each end and rotate to feel for any roughness that would indicate internal rusting. Be sure to check for bent or damaged forgings and studs; split or deteriorated seals; and, damaged, out-of-round, or loose tapers. If any of these conditions exists, the parts should be replaced.

## POWER STEERING

Continue your inspection by checking the drive belt for the power steering, if so equipped. If the belt shows evidence of cracking or glazing **(Figure 15),** it should be replaced. Adjust belt tension according to service manual instructions and specifications **(Figure 16).**

Also, check the fluid level and condition **(Figure 17).** The fluid is checked at the pump reservoir with a dipstick attached to the reservoir cap. Check the fluid level in the reservoir after the engine has been run at idle for two or three minutes and the wheel has been cycled from lock to lock several times. This warms the fluid to its normal operat-

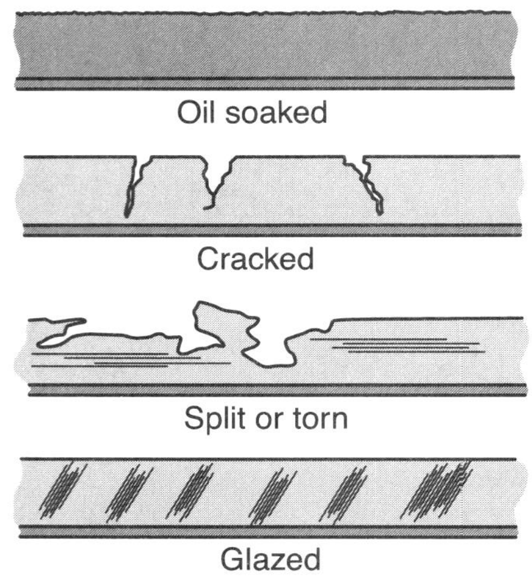


**Figure 15.** Check the condition of the power steering belt.

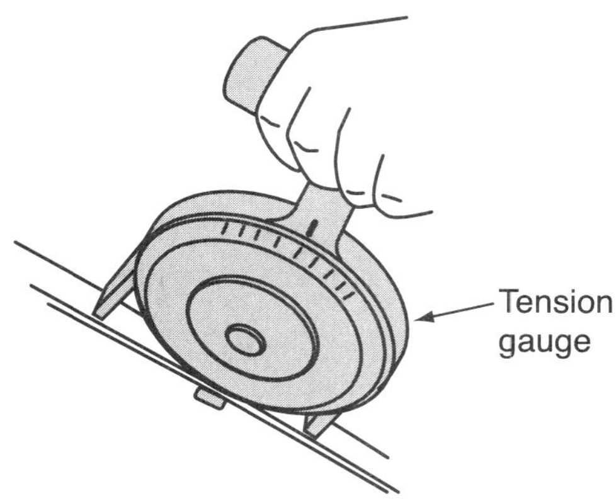


**Figure 16.** Use a belt tension gauge to check the tension of the power steering belt.

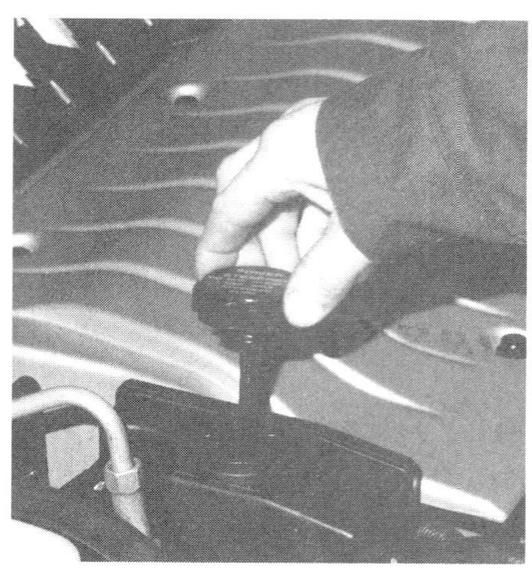

**Figure 17.** Check the level and condition of the power steering fluid.

ing temperature and gives a more accurate reading. Make sure the fluid level is at the full mark.

Examine the condition of the fluid carefully. Check for evidence of contamination such as solid particles or water. If either of these conditions is present or the fluid has a burnt odor, the system should be flushed before returning to service.

Check the fluid for evidence of air. If the fluid looks foamy, it is likely that air is in the system. To verify this, run the engine until it reaches normal operating temperature. Then turn the steering wheel to the left and to the right several times without hitting the stops. If there is air in the system, bubbles will appear in the fluid reservoir. To remove the air the system must be bled. The method of bleeding depends on the type of power-steering system. Follow the procedures given in the service manual. The typical procedure involves connecting a vacuum pump to the cap's opening in the reservoir. While the engine is run, vacuum is applied to the reservoir and maintained for about five minutes. The vacuum is then released and the reservoir refilled with fluid. Vacuum is again applied to the reservoir and the steering wheel is cycled from stop-to-stop every 30 seconds for at least five minutes. After this period of time, the vacuum is release and the reservoir cap installed. Then the system is checked for leaks.

With the ignition OFF, wipe off the outside of the power steering pump, pressure hose, return hose, fluid cooler, and the steering gear. Start the engine and turn the steering wheel several times from stop to stop. Check for leaks. Fluid leakage will not only cause abnormal noises, but may result in unequal and abnormal steering efforts. If no signs of linkage are apparent, repeat the wheel cycling process and inspection several more times. Hoses should also be carefully inspected for swelling and cracks. Always replace power steering hoses with an exact replacement hose. Never attempt to patch or seal a leak in a hose or the hose's fittings.

## LUBRICATE SUSPENSION AND STEERING SYSTEMS

Grease fittings are called **zerk fittings** and are threaded into the part that should be lubricated. Zerk fittings have a one-way spring-loaded check valve that allows grease into the joint but prevents it from leaking out. Make sure to wipe the fitting before applying grease to it. Dirt can plug the check valve allowing grease to leak out and allow water and dirt to leak in.

Some vehicles have threaded plugs instead of zerk fittings. To apply grease at these connections, a special adapter can be threaded into the plug's bore. After grease has been injected into the connection, the plug should be reinstalled or a grease fitting installed.

To properly lubricate the chassis of a vehicle, begin by referring to the service manual for the lubrication points on the steering and suspension system of the vehicle. Then, safely raise the vehicle and lock the lift or set the vehicle on safety stands.

Locate all of the lubrication points and wipe the fittings clean with a shop towel. Carefully look at the joints and determine if the joint boots are sealed or not. If the boots are good, push the grease gun nozzle straight onto a Zerk fitting and pump grease slowly into the joint **(Figure 18).** If the joint has a sealed boot, put just enough grease into the joint to cause the boot to expand slightly. If the boot is not sealed, put in enough grease to push the old grease out. Then wipe off the old grease and any excess grease. Repeat this at all lubrication points.

**Figure 18.** The lube gun nozzle should be placed securely on the grease fitting.

# SUMMARY

- Begin diagnostics with trying to duplicate the customer's concern and gather as much information from the customer as possible.
- Noise, vibration, and harshness (NVH) are unwanted sensations that may be caused by suspension and steering problems.
- Noises are often caused by tires, worn shock absorber bushings or grommets, worn spring insulators, broken springs or spring insulators, worn control arm bushings, worn stabilizer bar bushings, worn strut rod grommets, worn leaf spring shackles and bushings, worn torsion bars, anchors, bushings, worn upper strut mounts, and dry or worn control arm bushings.

- If the vehicle has a harsh ride, check for low ride height at the front and rear suspension, weak or broken springs, worn torsion bar components such as pivot cushion bushings, and worn or damaged control arm bushings.
- If the vehicle moves or rolls to one side when traveling on a rough surface, check for a weakened or disconnected stabilizer bar and worn stabilizer bar bushings.
- If the vehicle is not stable during changes of direction, check for weak or broken springs and worn bushings or anchors.
- When a vehicle wanders, the problem can be caused by loose or worn suspension components, poor wheel alignment, unevenly loaded or overloaded vehicle, loose or damaged steering gear bolts, loose steering column U-joint bolts, loose, worn, or binding steering linkages or tie rod ends, or improper steering gear preload adjustment.
- Most steering systems are designed so the steering wheel returns to within 30°to 60° of center after a 180° turn.
- Hard steering problems can be caused by a faulty power steering pump, loose power steering belt, low power steering fluid level, worn front tires, defective steering gear, damaged or faulty steering column bearings, seized steering column U-joints, internal problems in the power rack assembly, faulty suspension components, inadequately inflated tires, and low ride height at the rear suspension.
- Steering pull can be caused by improper frame alignment, brake system problem, worn or binding suspension components, especially springs, worn upper and lower control arm bushings, poor wheel alignment, low ride height on one side of the front suspension, unevenly loaded or overloaded vehicle, loose, worn, or damaged steering linkages or tie rod ends, out of balance steering gear valve, torque steer, tire problems, or binding strut bearing.
- Shock absorbers should be inspected for loose mounting bolts and worn mounting bushings.
- Inspect the control arms for damage and check the control arm bushings for wear .
- Check all suspension bushings for looseness, splits, cracks, misplacement, and noises.
- The entire steering system should be carefully checked for looseness, binding, or damage.
- Check the drive belt for the power steering.
- Check the fluid level and condition.
- Grease fittings are called zerk fittings and are threaded into the suspension and steering part that should be lubricated.

## REVIEW QUESTIONS

1. What are zerk fittings?
2. What is the difference between a noise and a vibration?
3. Which of the following is not typically caused by worn tie rod ends?
   A. Incorrect caster settings
   B. Scalloped and scuffed tires
   C. Wheel shimmy
   D. Front-end noise and tire squeal on turns.
4. Excessive play in the steering wheel can be caused by the following (except):
   A. Loose, worn, or damaged steering linkages or tie rod ends
   B. Loose, worn, or damaged steering column U-joints
   C. Loose, worn, or damaged steering column bearings
   D. Damaged or worn control arms
5. Technician A says shock absorbers should be inspected for loose mounting bolts and worn mounting bushings. Technician B says shock absorbers and struts should be inspected for oil leakage. Who is correct?
   A. Technician A
   B. Technician B
   C. Both Technician A and Technician B
   D. Neither Technician A nor Technician B

# Section 3

## Tires and Wheels



**Interesting Fact**

*The advances made in tire and suspension system design have led to the use of larger diameter wheels. Forty years ago, 12-, 13-, 14- and 15-inch wheels were the norm. Today, original equipment wheels range from 15 to 20 inches in diameter.*

## SECTION OBJECTIVES

After you have read, studied, and practiced the contents of this section, you should be able to:

- Describe basic wheel and hub design.
- Describe general tire function.
- Explain typical tire construction, and identify the purpose of each component in a tire.
- Explain the differences between the three types of tire construction in use today.
- Explain the tire ratings and designations in use today.
- Describe why certain factors affect tire performance, including inflation pressure, tire rotation, and tread wear.
- Inspect and diagnose tire wear patterns.
- Check and adjust tire air pressure.
- Diagnose wheel/tire vibration, shimmy, and noise.
- Diagnose tire pull (lead) problems.
- Inspect wheel rims.
- Measure tire and wheel radial and lateral runout.
- List the precautions to be observed when selecting replacement tires.
- Inspect and repair a tire.
- Rotate tires according to manufacturers' recommendations.
- Measure wheel, tire, axle, and hub runout.
- Dismount, inspect, repair, and remount a tire on a wheel.
- Reinstall a wheel and torque the lug nuts.
- Describe the differences between static balance and dynamic balance.
- Balance a wheel and tire assembly both on and off a vehicle.
- Define wheel tramp, and explain how static unbalance causes wheel tramp.
- Define wheel shimmy, and describe how dynamic unbalance results in wheel shimmy.

# Chapter 10 Wheels

## INTRODUCTION

The vehicle's tires are mounted to a wheel, which in turn is mounted to a hub **(Figure 1).** The mounting of the wheel to the hub must be secure and allow the wheel to be centered on the hub in all directions at all times. The tires and wheels face many forces. Cornering, irregular road surfaces, acceleration, deceleration, and braking can cause a wheel to move away from the center of the hub. The construction of a wheel stops it from being deflected by these forces.

**Figure 1.** A wheel and tire assembly.

## CONSTRUCTION

A wheel is constructed of two major parts: the rim, which is where the tire is mounted, and the center section, which mounts to the hub. These two parts are typically welded together, but some are bolted or riveted together. There are also aluminum alloy or magnesium alloy rims that are die-cast or forged. These wheels are commonly referred to as mag wheels, although they are commonly made of an aluminum alloy **(Figure 2).** These wheels are lighter than pressed steel wheel rims. This weight savings is important because the wheels and tires on a vehicle are unsprung weight. This means the weight is not supported by the vehicle's springs.

**Figure 2.** An aluminum wheel.

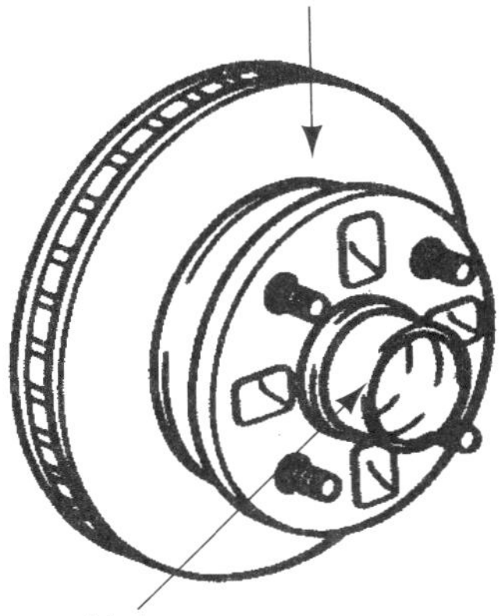


**Figure 3.** A typical wheel hub.

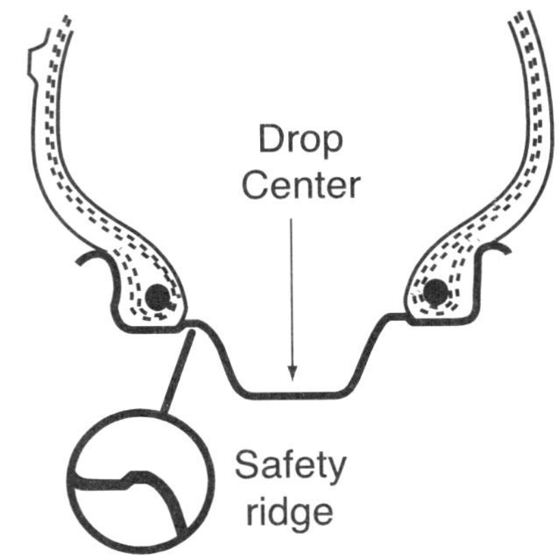


**Figure 4.** Rims have safety ridges behind the tire bead locations that help prevent the beads from moving into the drop center area if the tire blows out.

## Center Section

The center section has a bore in its center that fits over a flange on the hub **(Figure 3)** or over the wheel's spindle. This keeps the wheel centered on the hub. Also in the center section are the bores for the bolts or nuts that attach the wheel to the hub. These nuts and bolts are commonly called **lug nuts** or bolts. The wheel stud mounting holes in the rim are tapered to match the taper on the wheel nuts. The size and number of lug nuts is determined by the anticipated load of the vehicle and driving conditions. For example, most cars have five lug nuts, whereas heavy-duty trucks can have up to eight. The arrangement of the lug nuts is called the bolt pattern and can be different from vehicle to vehicle.

## Rim

The rim has two bead flanges that the tire's beads push against. These flanges must be strong enough to keep the tire beads in place. Just inside the flanges, at a slight angle, are the bead seats. Friction between the tire bead and the wheel bead seats and flanges transmits braking and acceleration forces from the tire to the wheel or vice versa.

Rims have safety ridges **(Figure 4)** behind the tire bead locations that help prevent the beads from moving into the drop center area if the tire blows out. In the event of a tire blowout, these ridges tend to keep the tire from moving deep into the rim center and from coming off the wheel. If a bead drops into the drop center, the tire may come off the wheel.

The rim also has a small hole for the tire's **valve stem (Figure 5).** The **drop center** in the rim makes tire changing easier **(Figure 6).** The drop center is a lowered well section near the center of the rim. The well is deep enough to allow the bead of a tire to pass over the flange on one edge of the rim when the other side of the bead is in the well.

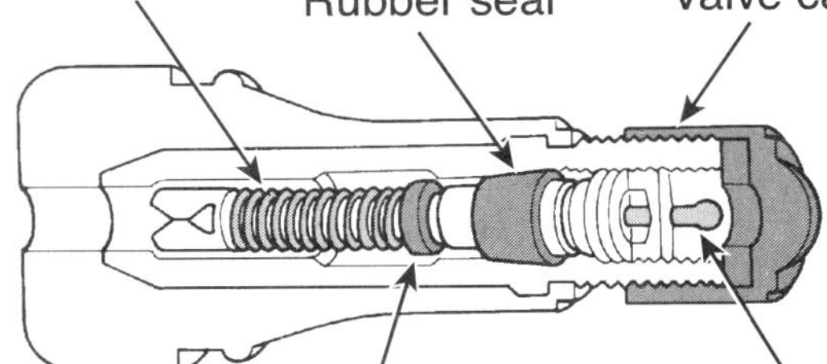


**Figure 5.** Parts of a tire valve stem, which is inserted into the rim of a wheel.

**Figure 6.** One side of the tire's inside diameter is put into the wheel's drop center to make removal from the wheel possible.

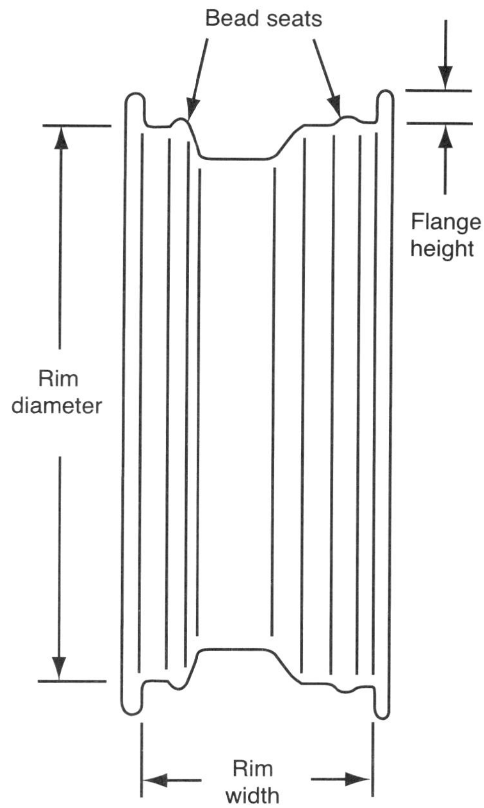


**Figure 7.** The dimensions of a wheel.

## WHEEL SIZES

Wheel size is designated by rim width and rim diameter **(Figure 7)**. These dimensions must match the size of the tire. Rim diameter is determined by measuring across the wheel from the top to the bottom.

Rim width is determined by measuring across the bead seating areas from the top to the bottom of the wheel. A wider rim tends to reduce tire sidewall flexing, which results in increased steering response, a stiffer ride, and increased tread shoulder wear on tires. A narrower rim tends to increase sidewall flexing, sidewall damage, and center tread wear but produces a softer ride.

## WHEEL OFFSET

**Wheel offset** is the distance between the rim centerline and the mounting face of the disc. Offset determines where the wheel mounting face is relative to the rim and determines where the centerline of the tire will be relative to the hub. The rim offset affects front suspension loading and operation. The amount and type of offset is critical because changing the wheel offset changes the loading on the front suspension loading as well as the scrub radius.

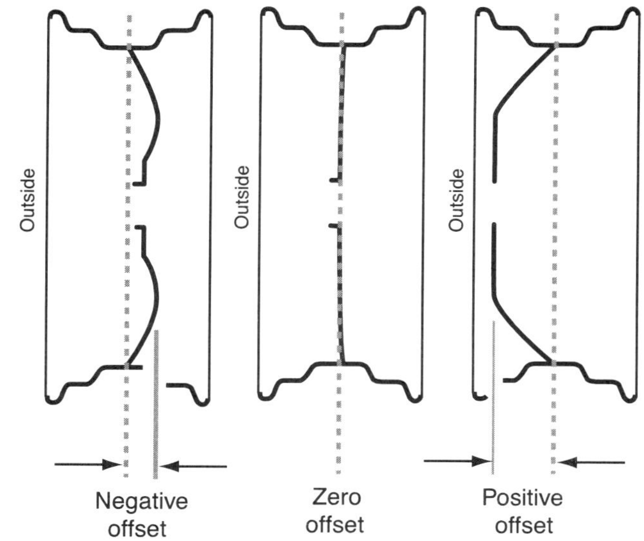


**Figure 8.** The aspects of wheel offset.

If the wheel mounting face is centered and directly in line with the centerline of the wheel, there is zero offset **(Figure 8).** If a rim is designed with positive offset, the rim centerline is inboard of the mounting face. A rim with negative offset has the centerline outboard of the mounting face. Most RWD vehicles position the tire centerline slightly outboard of the steering axis, whereas most FWD cars place the tire centerline inward (positive offset).

To measure offset, measure the width of the wheel between the flanges, and then measure the distance from the mounting flange to the backside of the wheel. If the measurement is greater than one-half the total width of the wheel, the offset is positive by the amount of difference.

Wheel manufacturers will often give a backspacing dimension rather than listing the wheel's offset. The backspacing dimension is the measurement from the mounting flange to the backside of the wheel.

## REPLACEMENT WHEELS

Replacement wheels must be the same as original equipment wheels in load capacity, offset, width, diameter, and mounting configuration. Using a wrong wheel can affect tire life, steering quality, wheel bearing life, vehicle ground clearance, tire clearance, speedometer calibrations, and affect anti-lock brake system operation.

You Should Know *An incorrect wheel may create a safety hazard and cause vehicle damage and/or personal injury.*

If you install wheels with a different than stock width or diameter, make sure you use tires designed that size wheel, and make sure the overall or rolling diameter of the new tires match the original tires. This is important for the correct operation of the vehicle's wheel speed sensors, speedometer, antilock brake system, and many other systems.

## WHEEL ATTACHMENT

The wheel is bolted to a **hub,** either by lug bolts that pass through the wheel and thread into the hub or by studs that protrude from the hub. The face of the lug nuts and bolts are tapered to match the taper of the mounting bores in the wheel. As the lug nuts or bolts are tightened, the tapered face and bore rub against each other and center the connection. If the lug nuts or bolts are tightened to the correct torque and according to the specified sequence, the interference between these two tapered surfaces provides a secure mounting of the wheel onto the hub **(Figure 9).**

A few vehicles have left-hand threads (which turn counterclockwise to tighten) on the driver's side and right-hand threads (which turn clockwise to tighten) on the passenger's side. All other vehicles use right-hand threads on both sides.

## Wheel and Tire Removal

You Should Know *Some wheel covers have fake plastic lug nuts that must be removed to access the real lug nuts. Be careful not to break the fake lug nuts.*

To remove a wheel and tire assembly, follow these steps:

1. Remove the wheel cover. If the vehicle is equipped with anti-theft locking wheel covers, the lock bolt for each wheel cover is located behind the center ornament. A special wrench is supplied to the owner for ornament and lock bolt removal. If the customer's wrench has been lost, a master key is typically available from the vehicle's dealer.
2. Loosen the wheel lug nuts about one-half turn, but do not remove them. Some vehicles are equipped with anti-theft wheel nuts. A special lug nut key is supplied to the vehicle owner. This lug nut key has a hex nut on the outer end and a special internal projection that fits in the wheel nut opening. Install the lug nut key on the

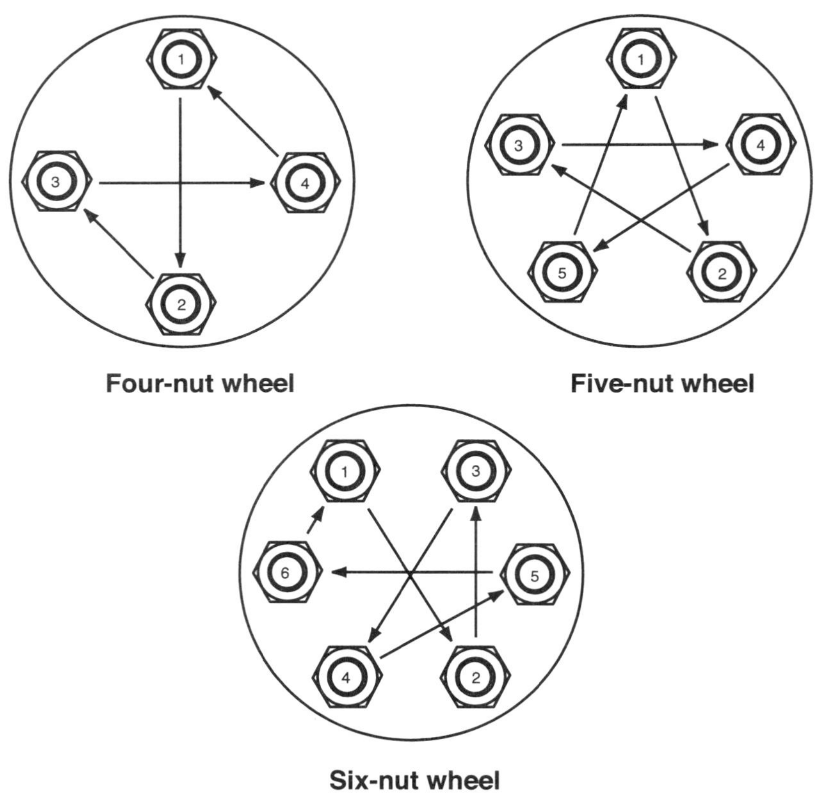


**Figure 9.** Typical tightening sequences for wheels based on the number of lugs.

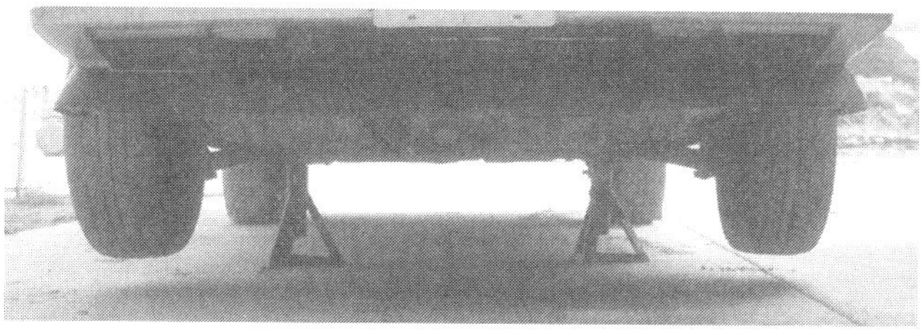

**Figure 10.** If the vehicle is lifted with a floor jack, place safety stands under the suspension or frame and lower the vehicle onto the safety stands.

lug nuts, and connect the lug nut wrench on the key hex nut to loosen the lug nuts.

3. Raise the vehicle on a hoist or with a floor jack to a convenient working level.
4. Chalk mark the tire, wheel, and one of the lug nuts so the wheel and tire can be reinstalled in the same position.

You Should Know *If the vehicle is lifted with a floor jack, place safety stands under the suspension or frame and lower the vehicle onto the safety stands* **(Figure 10-10)**. *Then remove the floor jack from under the vehicle*

5. Remove the lug nuts and the wheel and tire assembly. If the wheel is rusted and will not come off, hit the inside of the wheel with a large rubber mallet. Do not hit the wheel with a steel hammer because this could damage the wheel. Do not heat the wheel.

## SERVICING WHEELS

Steel wheels should be inspected for excessive rust and corrosion, cracks, loose rivets or welds, bent or damaged bead seats, and elongated lug nut holes. Aluminum or magnesium wheels should be inspected for damaged bead seats, elongated lug nut holes, cracks, and evidence of porosity. If any of these conditions exist, replace the wheel.

You Should Know *During many automotive service operations, including tire and wheel service, the technician literally has the customer's life in his or her hands! Always perform tire and wheel service carefully and thoroughly. Always watch for unsafe tire or wheel conditions, and report these problems to the customer. When you prove to the customer that you are concerned about vehicle safety, you will probably have a steady customer.*

**Figure 11.** Steel rims should be spray cleaned witha water hose, and the rim bead seats should be cleaned with a wire brush or coarse steel wool.

Carefully inspect the lug nuts, bolts, and/or studs for signs of damage or wear. If these are defective in any way, they should be replaced. Damaged mounting studs are typically replaced by pressing the old studs out of the hub and carefully pressing new ones in. Make sure you carefully inspect the mounting threads in the hub for damage. If the threads are bad, the lug bolt may give a false holding torque reading as it is tightened. Damaged threads may be able to be repaired by using a tap. If the damage is severe, the hub should be replaced.

Many shops always replace the tire valve assembly when a tire is repaired or replaced. This policy helps prevent future problems with tire valve leaks. The inner end of the valve may be cut off with a pair of diagonal pliers, and then the outer end may be pulled from the rim. Coat the new valve with rubber tire lubricant and pull it into the rim opening with a special puller screwed onto the valve threads.

Steel rims should be spray cleaned with a water hose. The rim bead seats on steel wheel rims should be cleaned with a wire brush or coarse steel wool **(Figure 11).** Aluminum or magnesium wheel rims should be cleaned with a mild soap-and-water solution and rinsed with clean water. Thoroughly clean the rim bead seats with a mild soap-and-water solution.

You Should Know *The use of abrasive cleaners, alkaline-based detergents, or caustic agents on aluminum or magnesium wheel rims may cause discoloration or damage to the protective coating.*

## WHEEL AND TIRE INSTALLATION

The correct procedure for reinstalling a wheel and tire assembly is more than mounting it and tightening the lug nuts. Steps must be taken to ensure the wheel is mounted squarely on the hub and that it is centered. Also, the lugs must be properly tightened to prevent them from coming loose or damaging the wheel, hub, or brake assemblies.

To install a wheel assembly on a vehicle, place the wheel over the studs. Start a lug nut on each of the studs. With your hand, snug down each of lug nuts making sure the tapered face of the nut enters the tapered bore of the wheel. Once the lug nuts are installed, lower the vehicle so that its weight in the wheel.

Now tighten the lug nuts to the correct torque while following the correct tightening sequence. Following this sequence eliminates the chance of distorting the brake rotor or drum and ensures that the wheel is squarely mounted to the hub.

Torque sticks are used by some shops. These special tools are designed to transmit a particular amount of torque and allow the use of an impact wrench to tighten lug nuts **(Figure 12).** If you use torque sticks, make sure you use one that meets the specified torque, and follow the correct tightening sequence after all of the lug nuts have been snugged down.

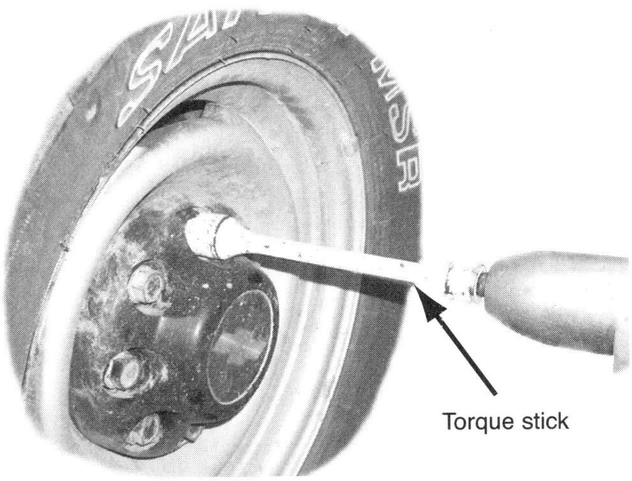


**Figure 12.** A torque stick can be used to properly tighten lug nuts.

## SUMMARY

- A wheel is constructed of two major parts: the rim, which is where the tire is mounted, and the center section, which mounts to the hub.
- Mag wheels are commonly made of an aluminum alloy.
- The rim also has a small hole for the tire's valve stem and a drop center area designed to allow for easy tire removal and installation.
- Wheel size is designated by rim width and rim diameter.
- Wheel offset is the distance between the rim centerline and the mounting face of the disc. The amount and type of offset is critical because changing the wheel offset changes the loading or the front suspension loading as well as the scrub radius.
- Replacement wheel rims must be the same as the original equipment wheels in load capacity, offset, width, diameter, and mounting configuration.
- The wheel is bolted to a hub either by lug bolts that pass through the wheel and thread into the hub or by studs that protrude from the hub.
- Replacement wheel rims must have the same width, diameter, offset, load capacity, and mounting configuration as the original rims to maintain vehicle safety.

## REVIEW QUESTIONS

1. How is a modern tire wheel constructed?
2. Explain the purpose of the wheel rim drop center and safety ridges.
3. The rim offset is the vertical distance between the rim centerline and the ______________________________ ______________________ of the disc.
4. Technician A says steel wheels should be inspected for excessive rust and corrosion, cracks, loose rivets or welds, bent or damaged bead seats, and elongated lug nut holes. Technician B says aluminum or magnesium wheels should be inspected for damaged bead seats, elongated lug nut holes, cracks, and evidence of porosity. Who is correct?
   A. Technician A
   B. Technician B
   C. Both Technician A and Technician B
   D. Neither Technician A nor Technician B
5. Technician A says replacement wheel rims should be the same as the original equipment wheels in load capacity, offset, width, diameter, and mounting configuration. Technician B says if wheels are installed that have a different than stock width or diameter, make sure to use a tire that has the same overall width as the original tires, otherwise the vehicle's wheel speed sensors, speedometer, anti-lock brake system, and many other systems will be affected. Who is correct?
   A. Technician A
   B. Technician B
   C. Both Technician A and Technician B
   D. Neither Technician A nor Technician B

# Chapter 11 Tires

## INTRODUCTION

Tires contribute greatly to the ride and steering quality of a vehicle. Tires also play a significant role in vehicle safety. Improper types of tires, incorrect inflation pressure, and worn-out tires create a safety hazard. When tires and wheels are out of balance, tire wear and driver fatigue are increased, which may create a driving hazard.

The primary purpose of tires is to provide traction. Tires must develop traction to drive and steer the vehicle under a wide variety of road conditions. As a tire rolls on the road, friction is created between the tire and the road. This friction gives the tire its traction. Although good traction is desirable, it must be limited. Too much traction means there is a great deal of friction and a large amount of rolling resistance. Rolling resistance wastes engine power and fuel, therefore it must be kept to a minimal level. This dilemma is a major concern in the design of today's tires.

In addition to providing good traction, tires are also designed to carry the weight of the vehicle, to withstand side thrust over varying speeds and conditions, and to transfer braking and driving torque to the road. Tires also help the suspension absorb road shocks to provide a comfortable ride for the occupant. Tires contribute to directional stability of the vehicle, and they must absorb all the stresses of accelerating, braking, and centrifugal force in turns.

### TIRE CONSTRUCTION

**Figure 1** shows a cutaway view of a typical tire. Tire construction varies depending on the manufacturer and the type of tire. A typical modern tire has a bead wire, sidewall, plies, belts, and tread area. The design and composition of these major parts varies with the anticipated operating conditions and load capacity of the tire.

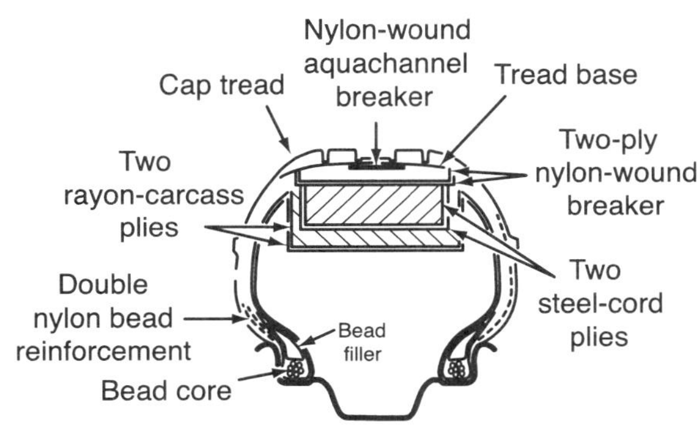


**Figure 1.** A cutaway of a typical tire.

All tires must be uniform in diameter and width. A tire with excessive **radial runout** (variations in tire diameter) causes a tire thumping problem as the car is driven. When a tire has excessive variations in width, this condition is called **lateral runout.** A tire with excessive lateral runout causes the chassis to "waddle" when the car is driven **(Figure 2).**

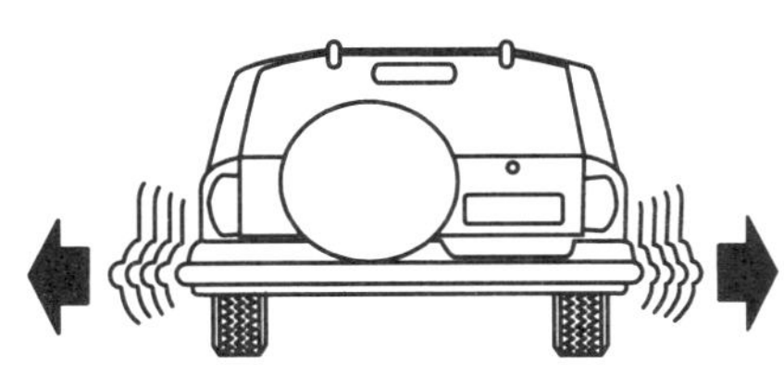


**Figure 2.** A tire with excessive lateral runout causes the chassis to waddle when the car is driven.

Figure 3. When a front tire has conicity, the steering may pull to one side as the vehicle is driven straight ahead.

The tire plies and belts must be level across the tread area. If the plies and/or belts are not level, the tire is cone-shaped **(Figure 3).** This condition is referred to as **tire conicity.** When a front tire has conicity, the steering may pull to one side as the vehicle is driven straight ahead. A rear tire with conicity will not affect the steering as much as a front tire with conicity.

## Commonly Used Materials

The tire body and belt material can be made of rayon, nylon, polyester, fiberglass, steel, amarid, and/or kevlar. Each has its advantages and disadvantages. For instance, rayon tires are low in cost and give a good ride but do not have the inherent strength needed to cope with long high-speed runs or extended periods of abusive use on rough roads.

Nylon tires generally give a slightly harder ride than rayon but offer greater toughness and resistance to road damage. Polyester and fiberglass tires offer many of the best qualities of rayon and nylon but without the disadvantages and provide for a much smoother ride.

Steel is tougher than fiberglass or polyester but it has a rougher ride because steel does not give under impact. Amarid and kevlar belts are lighter than steel ones and stronger than steel.

## Bead

The **bead** is the portion of the tire that helps keep it in contact with the rim of the wheel **(Figure 4).** It also pro-

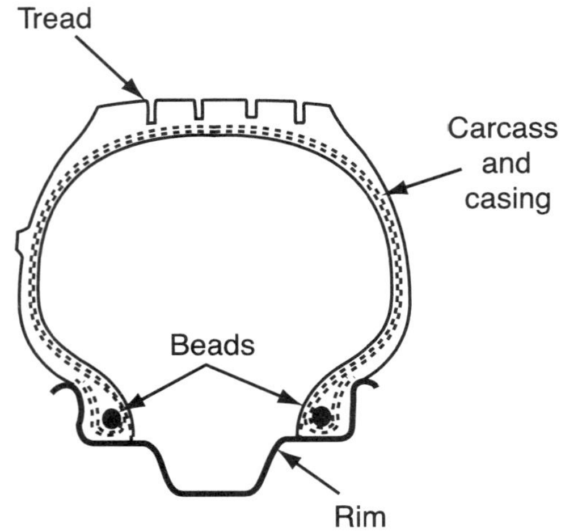


Figure 4. Location of a tire's beads.

vides the air seal on tubeless tires. The bead wire contains several turns of bronze-coated steel wire in a continuous loop. This bead wire is molded into the tire at the inner circumference and is wrapped in the cord plies. The bead wire anchors the tire to the wheel. The bead filler above the bead reinforces the sidewall and acts as a rim extender.

## Sidewall

The **sidewalls** are the sides of the tire's body. Tire sidewalls are made from a blend of rubbers that absorbs shocks and impacts from road irregularities and prevents damage to the plies. They also contain antioxidants and other chemicals that are gradually released to the surface of the sidewall during the life of the tire. These antioxidants help keep the sidewall from cracking and protect it from ultraviolet radiation and ozone attack.

Since the sidewall must be flexible to provide ride quality, they are made as thin as possible. However, they must be strong enough to support the weight of the vehicle. Tire manufacturers have reduced sidewall thickness by 40 percent in recent years to reduce weight and heat buildup and to improve ride quality. A lettering and numbering arrangement for tire identification is molded onto the outside of the sidewall.

## Plies

The cord body (or casing) consists of layers of rubber-impregnated cords called **plies** that are bonded into a solid unit. Typically tires are made of two, four, or eight plies, thus the reference to two-, four-, and eight-ply tires. The plies determine a tire's strength, handling, ride, amount of road noise, traction, and resistance to fatigue, heat, and bruises.

The cord plies surround both beads and extend around the inner surface of the tire to enable the tire to carry its load. The plies are molded into the sidewalls. Each ply is a layer of rubber with parallel cords imbedded in its body. The load capacity of a tire may be increased by adding more cords in each ply or by installing additional plies. Passenger car tires usually have two-cord plies, whereas many trucks and recreation vehicles are equipped with six- or eight-ply tires to carry the heavier loads of these vehicles. In general, tires with more plies have stiffer sidewalls which provide less cushioning and reduced ride quality.

## Belts

The **belt** of a tire restricts ply movement and provides tread stability and resistance to tire deformation. This provides longer tread wear and reduces heat buildup in the tire. Steel is the most common material in tire belts, although other belt materials have been used. Many tires contain two steel belts that expand as wheel speed and tire temperature increases.

Centrifugal force and belt expansion tend to tear the tire apart at high speeds and temperatures. Therefore, high-speed tires usually have a nylon jointless belt cover. This belt cover contracts as it is heated and helps hold the tire together, providing longer tire life, improved stability, and better handling.

## Tread

The **tread** (or crown), is the portion of the tire that comes in contact with the road. It is a pattern of grooves and ribs that provide traction. The grooves are designed to drain off water, while the ribs grip the road. On some tires, small cuts (called **sipes**) are molded into the ribs of the tread **(Figure 5).** These sipes open as the tire flexes on the road offering additional gripping action, especially on wet road surfaces. Tire treads are made from a blend of rubber compounds that are very resistant to abrasion wear. Spaces between the tire treads allow tire distortion on the road without scrubbing, which accelerates wear.

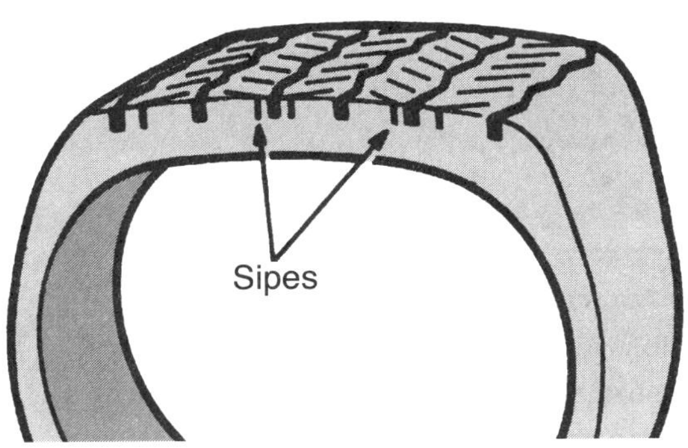


**Figure 5.** Sipes are molded into the ribs of the tread and open as the tire flexes on the road, offering additional gripping action, especially on wet road surfaces.

Most automotive tires contain two layers of tread materials. The first tread layer is designed to provide cool operation, low rolling resistance, and durability. The outer layer (or tread) is designed for long life and maximum traction. Tread rubber is a blend of many different synthetic and natural rubbers. Tire manufacturers may use up to thirty different synthetic rubbers and eight natural rubbers to form the treads. These synthetic and natural rubbers are blended to provide the desired traction and durability.

Traction must be maintained as much as possible on a wide variety of road surfaces. For example, on wet pavement, tire treads must be designed to drain off water between the tire and the road surface. This draining action is extremely important to maintaining adequate acceleration, braking, and directional control. Lines cut across the tread provide a wiping action that helps dry the tire-road contact area.

## Tube and Tubeless Tires

Early vehicle tires were solid rubber. These were replaced with pneumatic tires, which are filled with air. There are two basic types of pneumatic tires: those that use inner tubes and those that do not. On tube-type tires, the air is sealed in the inner tube, and the sealing qualities of the tire and wheel rim are not important. Some heavy-duty truck tires use inner tubes. Nearly all passenger car and light-truck tires are of the tubeless type.

In these tires, the tire bead must provide an airtight seal on the rim, and both the tire and the wheel rim must be completely sealed. A synthetic gum rubber liner is bonded to the inner surface of the tire to seal the tire. On some tires, this inner lining can form a seal around a nail or other object that punctures the tread **(Figure 6).** A self-sealing tire holds in air even after the object is removed. A tubeless tire is mounted on a special rim that retains air between the rim and the tire casing when the tire is inflated.

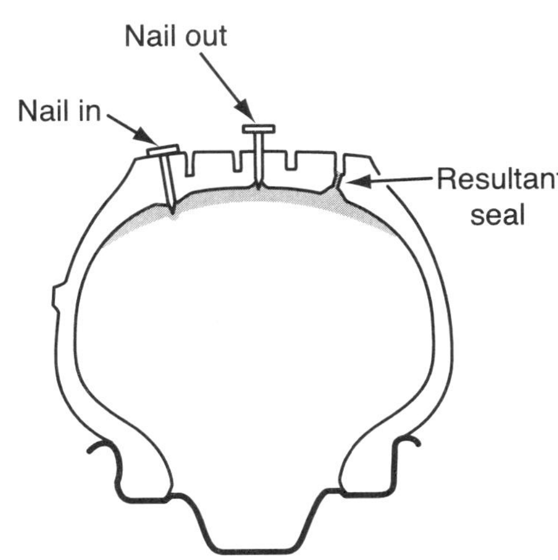


**Figure 6.** A puncture-sealing tire.

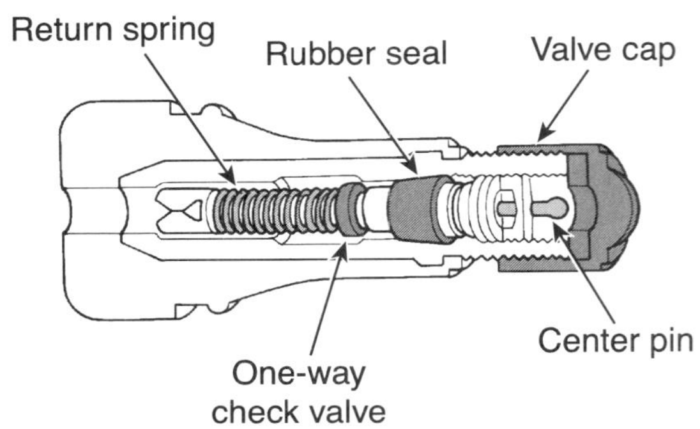


**Figure 7.** A tire air valve.

### Tire Valves

A tubeless tire has a tire valve that allows air to flow into the tire and can also be used to release air from the tire. The core in the center of the valve is spring loaded and allows air to flow inward while the tire is inflated **(Figure 7).** Once the tire is inflated, the valve core seats and prevents air flow out of the tire. The small pin on the outer end of the valve core may be pushed to unseat the valve core and release air from the tire. An airtight cap on the outer end of the valve keeps dirt out of the valve and provides an extra seal against air leakage. A deep groove is cut around the inner end of the tire valve. When the valve assembly is pulled into the wheel's air valve bore, this groove seals the valve in the opening.

## TIRE PLY AND BELT DESIGN

The most commonly used tire designs are bias, belted bias, and belted radial **(Figure 8).** Of these, the oldest is the **bias ply.** This design has fabric plies that run alternately at opposite angles to form a crisscross design. The angle varies from 30 to 38 degrees with the centerline of the tire. Generally speaking, the lower the cord angle, the better the high-speed stability, but also the harsher the ride. Bias ply tires usually are available in two- or four-ply.

**Belted bias** ply tires are similar to bias ply tires but have two or more belts around the tire's circumference. This gives strength to the sidewall and greater stability to the tread. The belts reduce tread motion during road contact, thus improving tread life. Plies and belts of various combinations of rayon, nylon, polyester, fiberglass, and steel are used with in belted bias tires. Two plies and two belts are common, but four plies and four belts may be used in some tires.

Belted bias ply tires generally cost more than bias ply tires but last up to 40 percent longer. Compared to a bias ply tire, a belted bias ply tire has greater tread rigidity.

**Radial ply** tires have body cords that extend from bead to bead at an angle of about 90 degrees—"radial" to the circumferential centerline of the tire—plus two or more layers of relatively inflexible belts under the tread. The belts are made of various combinations of rayon, nylon, fiberglass, and steel (most common) to give strength to the tread area and flexibility to the sidewall. Radial ply tires also offer greater fuel economy, increased skid resistance, more positive braking, less rolling resistance, better steering characteristics, and longer tread life than bias ply tires. Many radial tires have two plies and two belts.

## Tread Designs

The real purpose of a tire is to get a grip on the road. The ideal tire is one that wears little, holds the road well to provide sure handling and braking, and provides a cushion from road shock. The ideal tire should also provide maximum grip on dry roads, wet roads, snow and ice, and operate quietly at any speed. This is a tall requirement, so tires

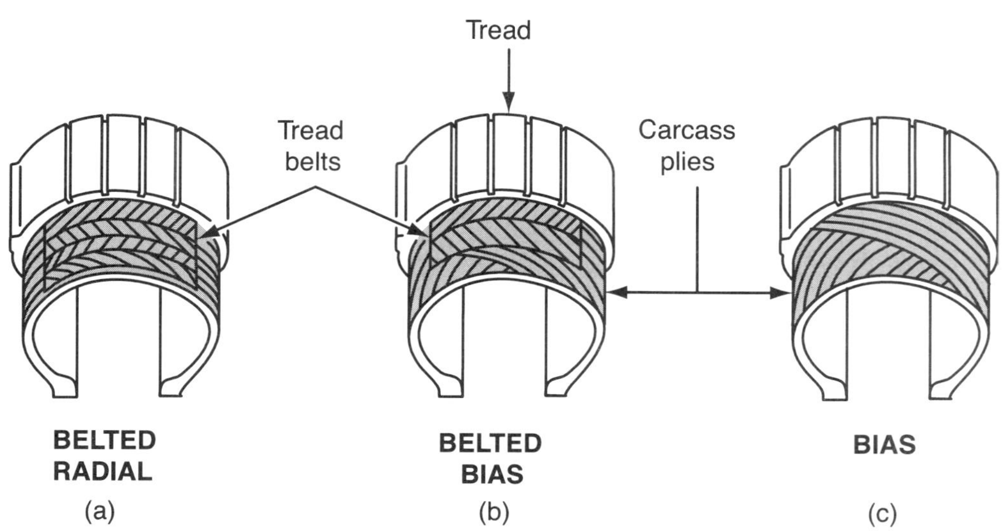


**Figure 8.** Three types of tire construction: (a) belted radial ply. (b) belted bias ply, and (c) bias ply.

are manufactured as compromises. A tire's tread design dictates what the tire will excel at. There are basically three categories of tread patterns: directional, non-directional, and symmetric and asymmetric.

A directional tire is mounted so that it revolves in a particular direction. These tires have an arrow on the sidewalls that show the designed direction of travel. A directional tire performs well only when it is rotating in the direction it was designed to rotate in. A non-directional tire has the same handling qualities in either direction of rotation. A symmetric tire has the same tread pattern on both sides of the tire. An asymmetrical tire has a tread design that is different from one side to the other. Asymmetrical tires are typically designed to provide good grip when traveling straight (the inside half) and good grip in turns (the outside half of the tread). Most asymmetric tires are also directional tires.

The number and size of the blocks, sipes, and grooves on a tire's tread not only determines how much rubber contacts the road and how much water can be displaced, it also determines how quiet the tire will be during travel. The more aggressive the tread, the more noise it will make. This statement is especially true if the tire's tread is made of a hard compound. Softer tires typically make less noise but wear quicker. Soft tires also adhere to the road better.

Channels are cut into a tire's tread to allow water to move away from the direction of travel. The deeper the channel, the more water can move, but road contact is decreased.

**You Should Know** *Knowing what design of tire meets a customer's needs is a science. Always consult with a tire specialist before referring a tire unless you are a tire specialist.*

**Figure 9.** Tread wear indicator bars.

### Tread Wear

Most tires have built-in tread wear indicators to show when they need replacement. These indicators appear as 1/2-inch wide bands when the tire tread depth wears to 1/16 inch **(Figure 9).** When the indicators appear in two or more adjacent grooves at three locations around the tire or when cord or fabric is exposed, tire replacement is recommended.

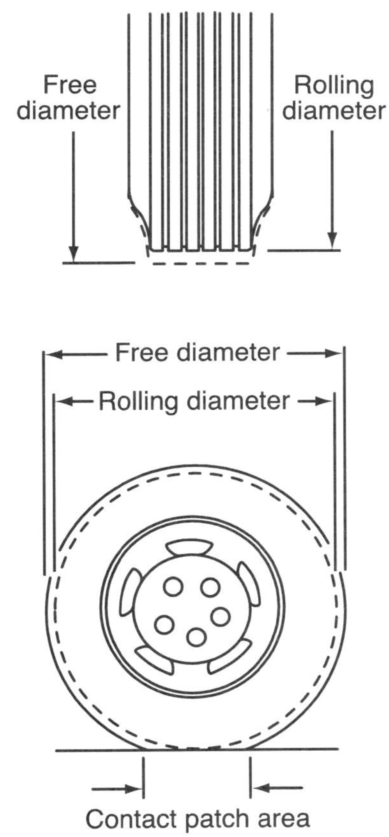


**Figure 10.** Tire rolling diameter, free diameter, and contact area.

## Tire Contact Area

The tire contact area is the area of the tire that contacts the road when the tire is supporting the vehicle's weight. The free diameter is the distance of a horizontal line through the center of the spindle and wheel to the outer edges of the tread. The rolling diameter is the distance of a perpendicular straight line through the center of the spindle to the outer edges of the tread when the tire is supporting the vehicle's weight. The rolling diameter is always less than the free diameter. The difference between the free diameter and the rolling diameter is referred to as deflection. The rolling diameter, free diameter, and contact area are shown in **Figure 10.**

# TIRE RATINGS AND DESIGNATIONS

A tire's rating, as defined by the **Department of Transportation (DOT),** is molded into the sidewall **(Figure 11).** The information contained in a tire's rating include:

1. Size
2. Load range
3. Maximum load
4. Maximum pressure in pounds per square inch (psi)
5. Number of plies under the tread and in the sidewalls
6. Manufacturer's name
7. Tubeless or tube-type construction
8. Type of belt construction
9. DOT approval number.

You Should Know

*It is important to remember that the maximum inflation number on a tire is the maximum inflation **(Figure 12)** and not the recommended inflation. Tires should never be inflated beyond this maximum rating.*

**Figure 12.** The tire's maximum load and maximum inflation air pressure are molded onto the sidewall of the tire.

## Uniform Tire Quality Grading (UTQG) Designations

DOT requirements also specify that all except mud and snow tires must have the **Uniform Tire Quality Grading (UTQG)** System designations molded into the sidewall. The UTQG system designations include the tread wear, traction, and temperature ratings.

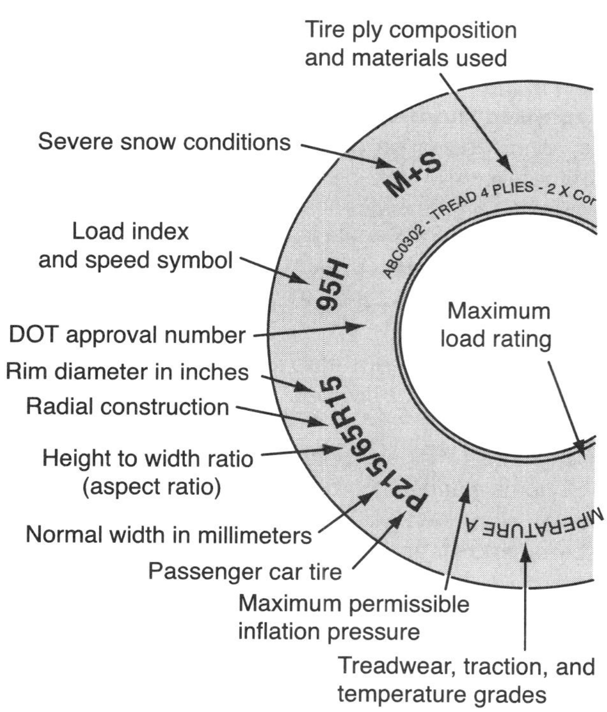


**Figure 11.** Tire sidewall information.

### Tread Wear Rating

Tread wear ratings allow consumers to compare tire life expectancies. When tires are tread wear rated, they are installed on a vehicle and driven on a test course. Tread wear is calculated after this run. This calculation provides a number, or ratio, to compare various tires. For example, a tire with a 150-tread wear rating will provide 50 percent more tire wear mileage than a tire with a 100-tread wear rating.

### Traction Rating

Traction ratings indicate the braking capabilities of the tire. To determine the traction rating, ten skid tests are completed on wet asphalt and concrete surfaces. The results of the skid tests are averaged, and the traction rating is designated A, B, or C. A tire with an A rating has the best traction.

### Temperature Rating

Temperature resistance ratings indicate the tire's ability to withstand heat generated during operation. The National Highway Traffic Safety Administration (NHTSA) has established procedures for temperature resistance testing of tires. Temperature ratings are A, B, or C, and a tire with an A rating has the best temperature resistance. Tires must have a minimum letter C temperature rating to meet NHTSA standards in the United States.

## Tire Size

Nearly all tires today use metrics to define their size. However, the actual size is normally a mix of metric and English system values. The section width (from sidewall to sidewall) is given in millimeters, and the rim diameter is provided in inches. The **aspect ratio** is the section height (from tread to bead) divided by the section width, and the rating is a percentage. In a tire with a 75 aspect ratio, the tire height is 75 percent of its width. The aspect ratio may be

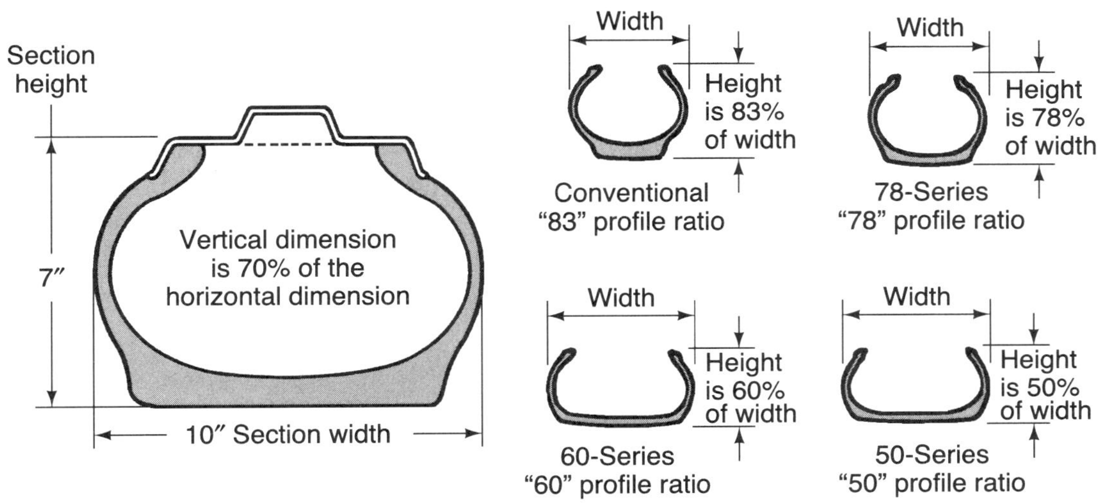


**Figure 13.** Tire aspect ratio is the percentage of the tire's height in relation to its width.

called the **tire profile,** and this ratio number may also be referred to as the series number **(Figure 13).** Belted bias ply and belted radial ply tires usually have lower aspect ratios and profiles than the older bias ply tires.

For many years, the accepted profile ratio for standard bias ply passenger car tires was approximately 83. This meant the tire was 83 percent as high as it was wide. Since the introduction of bias belted and radial ply construction, lower profile tires and ratios of 78, 70, 60, 50, 45, 40, and even 30 have become popular. The lower the number, the wider the tire. As an example, a 50-series tire appears to be low and flat because its height is 50 percent of its width.

In the designation P215/65R15 **(Figure 14),** P identifies passenger car tire. The width in millimeters is 215. The

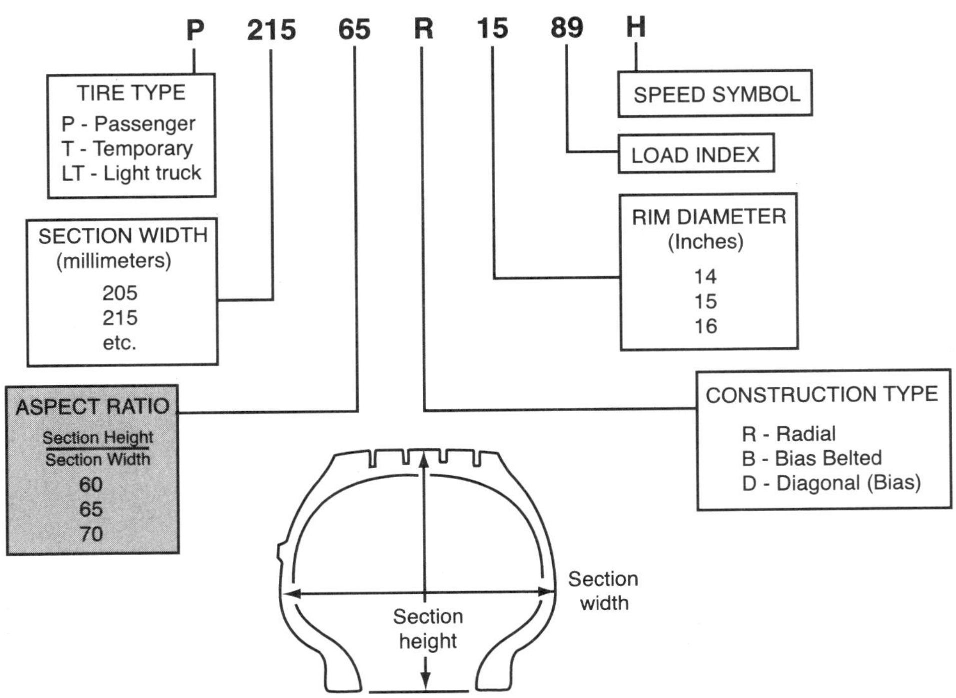


**Figure 14.** The information contained in a tire's rating.

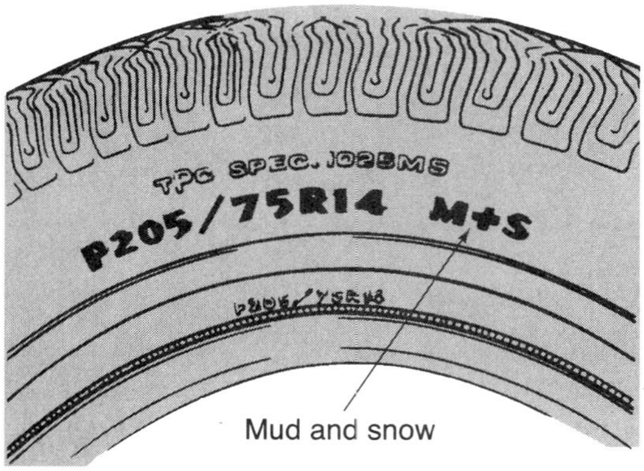


**Figure 15.** If the tire is designed for mud and snow, there will be a symbol denoting it as such. Also, note the TPC rating above the tire size.

height-to-width ratio is 65. R identifies radial construction. The rim diameter is 15. If this designation is followed by M, S, or M+S, the tire's tread is rated for use in mud, snow, or both **(Figure 15).**

For a typical all-metric tire, the rim diameter and width are given in millimeters, the inflation pressure in kilopascals (kPa), and its load capacity in kilograms (kg). One kilopascal equals 6.895 psi. An example is a 190/65R-390 tire that fits a 390-mm diameter wheel.

### Numeric Tire Rating

Prior to using metrics, many tires had a numeric rating. In this rating system, if a tire has an F78R-15 rating, the F indicates the tire is for an intermediate-sized car, and the 78 is the aspect ratio. The R-15 in this rating means that it is a radial tire with a 15-inch rim diameter. Earlier numeric tire ratings just included the cross-sectional width in inches and the rim diameter such as 7.75-14. Metric tire sizes may not have exact corresponding numeric tire size. For example, a P205/75R15 tire may not be exactly the same as a numeric FR78-15 tire. For this reason, replacement tires should have the same TPC number.

### Tire Performance Criteria (TPC) Number

A **tire performance criteria (TPC)** specification number is molded onto the sidewall near the tire rating **(Figure 15).** The TPC assures that the tire meets the manufacturer's performance standards for traction, endurance, dimensions, noise, handling, and rolling resistance.

## Load Rating

Typically, the load rating and the number of tread and sidewall plies are proportional. In most cases, the more plies a tire has, the more weight it can support. A two- or three-digit number is used on some tires as a load index designation. This number precedes the speed-rating letter and indicates the maximum load that a tire can carry at the speed represented by the speed symbol. The maximum tire load is stamped on the tire sidewall with the recommended maximum inflation pressure by some tire manufacturers. If a tire has a sidewall marking of P19575R1490H, the 90 is the load designation and H is the speed rating (see Figure 14).

Some tire manufacturers have used the letters B, C, or D to indicate the tire load capacity. In this load rating system, B indicates the lightest tire design suitable for passenger cars, and a C-rated tire has greater load capacity than a B-rated car. Load range D tires have heavy-duty load capacity for trucks. On many tires, the lettering arrangement for load rating has been replaced by standard load (SL) and extra load (XL) designations. An XL-rated tire has slightly higher load capacity than a D-rated tire in the previous system. The load capacity of an SL-rated tire falls between the previous B and C rating.

## Speed Ratings

Many tires sold in the United States are speed rated to indicate the maximum speed capabilities of a tire. The designation for speed rating is included on the sidewall markings. The following are the most common speed ratings:

S = 112 mph (maximum)
T = 118 mph (maximum)
U = 124 mph (maximum)
H = 130 mph (maximum)
V = 149 mph (maximum)
Z = 149+ mph

Although tires may be speed rated, tire manufacturers do not endorse the operation of a vehicle in an unlawful or unsafe manner. Speed ratings are based on laboratory tests, and these ratings are not valid if tires are worn out, damaged, altered, underinflated, or overloaded. The condition of the vehicle and road may affect high-speed operation.

Driving a vehicle at speeds greater than the speed rating of the tires is risky. The heat generated can cause the tire to come apart. If this happens at high speed, it will be close to impossible for the driver to maintain control of the vehicle.

## SPECIALTY TIRES

Specialty tires reflect the advances made in the conventional tire field. Special snow and mud tires are available in all three construction types. When snow tires are installed on a vehicle, these tires should be the same size and type as the other tires on the vehicle. In areas where snow is encountered, all-season tires have replaced snow tires to a large extent. Studded tires provide improved traction on ice, but these tires are prohibited by law in many states because their use resulted in road surface damage. Because the studs have more contact on the road than the rubber, it is easy for the tire to slide on the road during cornering and stopping. The studs offer much less friction than the rubber tire tread.

Tires designed for wet surfaces have tread designs that move the road's water behind and to the side of the tire. Moving the water is the only way a tire can grip the road. However, when a tire has many directed and open channels for water in its tread, less rubber meets the road.

## All-Season Tires

Many tires are classified as all-season tires. These tires have 37 percent higher average snow traction compared to non-all-season tires. All-season tires may have slightly improved performance in areas such as wet traction, rolling resistance, tread life, and air retention. Improvements in tread design and tread compounds provide the qualities of all-season tires.

These tires are designed to perform well on all types of road conditions, but they will not be excellent performers on any road surface. Tires can be designed to be great on dry surfaces or wet surfaces. However, it is near impossible to have a tire that performs extremely well on both. Tires designed for dry and smooth roads do not really need a tread pattern. They can be "slicks." The smooth surface would give the tires maximum grip on a smooth, dry surface. However when a slick hits a wet spot, there is no traction. The tire simply slides on the water.

### Tire Chains

Tire chains may be used in emergency situations such as driving on snow-covered or ice-covered mountain roads. Most tire manufacturers do not recommend the use of tire chains. Use only SAE class "S" tire chains unless specified by the vehicle and tire manufacturer. These chains must be the proper size and must be installed tightly with the ends secured. Always follow the chain manufacturer's recommended installation procedure. While using chains, driving speed should be reduced. If the chains are heard striking the vehicle, stop immediately and tighten the chains to prevent body damage.

## Run-Flat Tires

A newer tire design has steel reinforced sidewalls that allow the tire to maintain shape with no air pressure. These tires are typically called run-flat tires, but may also be called **extended mobility tires (EMT).** Without air pressure, the sidewalls experience more outward deflection, but these tires support the vehicle sufficiently to allow it to be driven safely for 125 miles (200 km). Since they were designed to be filled with air, their ride and handling characteristics are much better when they have air. Run-flat tires eliminate the need for a spare tire and a jack, and this provides weight and space savings.

A run-flat tire is a self-supporting tire with a specially designed wheel and a flat tire monitor that displays a warning when the tire is flat. The sidewalls feature supplementary reinforcement belts made with a special, heat-resistant rubber compound **(Figure 16).** This allows the deflated tire to rotate without compressing against the wheel's rim flange. These tires use a specially shaped safety rim that ensures the tire remains securely seated. Together these features allow the driver to continue travel without installing the spare tire.

Some run-flat tires have a flexible rubber support ring mounted on a special rim to support the tire if it loses air

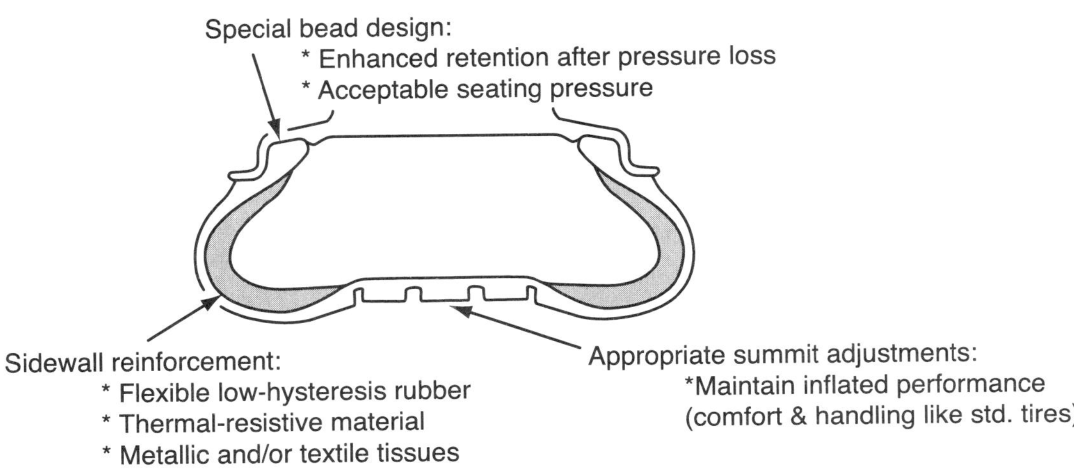


**Figure 16.** Features of a run-flat tire.

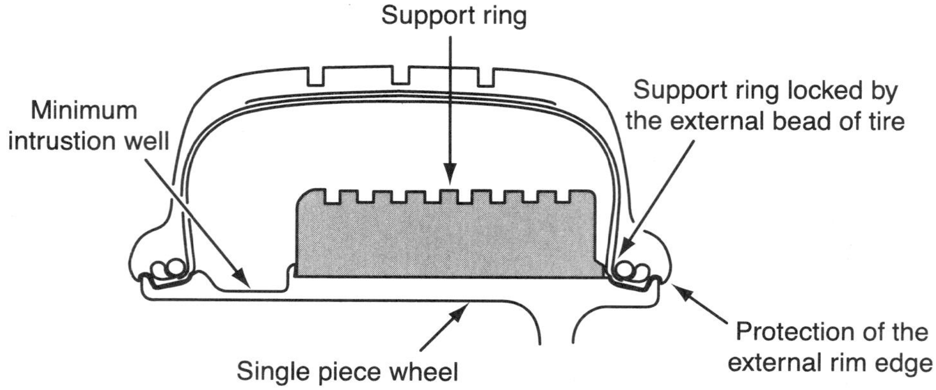


**Figure 17.** A run-flat tire with a support ring.

pressure **(Figure 17).** These tires have a mechanical locking bead system that latches the bead to the rim. The outer part of this bead fits over the outer edge of the rim. Additional load on the tire tends to tighten the latching force on the locking bead.

## Tire Pressure Monitor (TPM)

Run-flat tires are usually installed on vehicles with a tire-pressure monitoring system. The **tire pressure monitor (TPM)** checks the inflation pressures in all four tires at frequent periodic intervals. TPM's sensors keep track of the tire pressures both when the vehicle is moving and when it is stationary. When the TPM detects changes in any tire's inflation pressure, it responds by triggering a warning lamp on the instrument panel. On some vehicles, a graphic will display exactly which tire is flat.

Each wheel is equipped with an electronic sensor assembly located just behind the valve stem. This assembly is made up of a pressure sensor, transmitter, and a battery. The pressure sensor measures the tire's inflation pressure and relays this information via radio waves **(Figure 18).** These signals are picked up by separate sensors for each wheel. A central electronic control unit processes the signals from the wheels and reports any variations to the system.

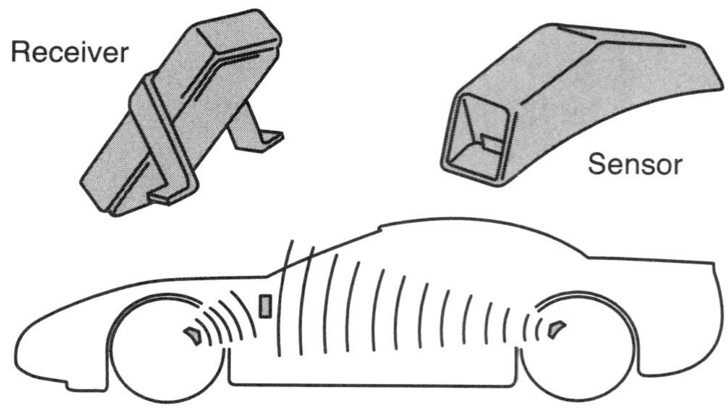


**Figure 18.** A tire-pressure monitoring system. There is a sensor in each tire.

## Spare Tires

Nearly all vehicles are equipped with a spare tire to be used in case one of the vehicle's tires loses air and goes flat. A spare tire can be a tire that matches the tires on the vehicle or can be a compact spare. Compact spares are designed to reduce weight and storage space but still provide the driver with a tire in case of an emergency. Compact tires are typically one of three types: high-pressure mini spare, space-saver spare, and lightweight skin spare.

A high-pressure minispare **(Figure 19)** is a commonly found spare tire, but it is only a temporary tire. It should not be used for extended mileage or for speeds above 50 mph. A space-saver spare must be blown up with a compressor that operates off the cigar lighter or a built-in air compressor. The lightweight skin spare tire is a bias ply tire with a

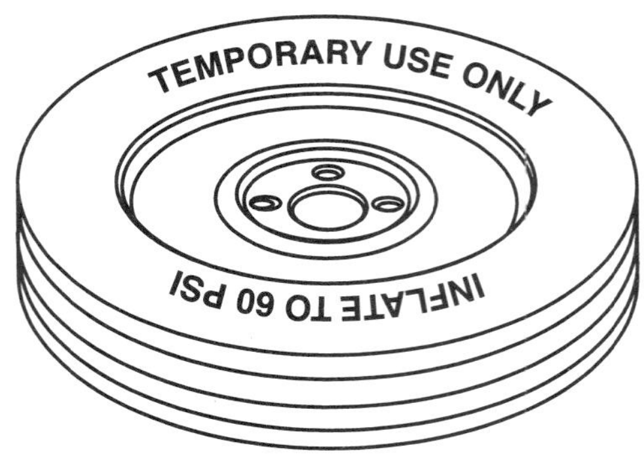


**Figure 19.** A compact, high-pressure mini spare tire.

**You Should Know** *Make sure you warn your customer that a mini space-saver or similar type compact spare should be used only as a temporary tire. It should never be used as a regular tire. Any continuous load use of a temporary spare could result in tire failure, loss of vehicle control, and injury to the vehicle's occupants.*

reduced tread depth to provide an estimated 2,000 miles (3,200 km) of tread life. This type of spare tire is designed for emergency use only, and driving speed must be limited to 50 mph (80 kmph).

## TIRE PLACARD AND INFLATION PRESSURE

The vehicle's weight is supported by the air pressure exerted evenly against all the interior tire surface, which produces tension in the tire carcass. Therefore, tire pressure is extremely important. Underinflation decreases the rolling diameter and increases the contact area, resulting in excessive sidewall flexing and tread wear. Overinflation decreases the contact area, increases the rolling diameter, and stiffens the tire. This results in excessive wear on the center of the tread. Tire pressure should be checked when the tires are cool, since tire pressure normally increases with temperature. Excessive heat buildup in a tire may be caused by underinflation and may lead to severe tire damage.

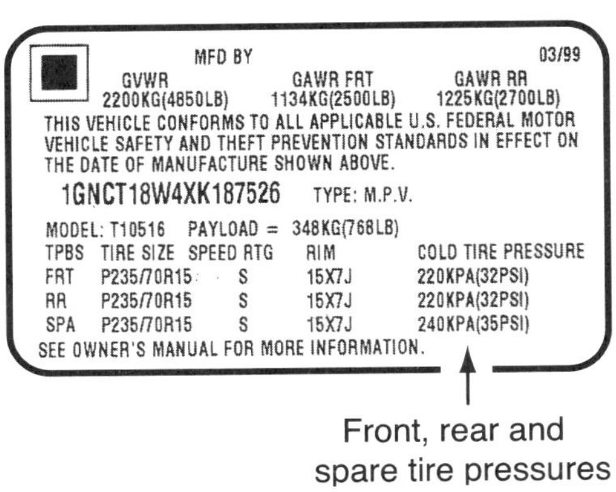


**Figure 20.** A typical tire placard.

### Tire Placard

The **tire placard** (or safety compliance certification label) is generally found on the driver's door jamb. It includes recommended maximum vehicle load, tire size (including spare), and the ***correct*** cold tire inflation for the specific vehicle **(Figure 20).**

Tire pressure is carefully calculated by the manufacturer to provide satisfactory tread life, handling, ride, and load-carrying capacity. Most manufacturers recommend that tire pressures be checked cold once a month or prior to any extended trip. A cold tire is one that has not moved for 3 hours or when the vehicle has been driven less than 1 mile. The tires should be inflated to the pressure indicated on the tire placard. Many inflation pressures listed for imported vehicles are given in kilopascals (kPa) rather than psi.

## REPLACEMENT TIRES

When replacement tires are selected, these tires should have the same size, load capacity, and construction as the original tires. The replacement tires should have the same TPC number to assure that these tires meet the same performance standards as the original tires. If replacement tires have a different size or construction type than the original tires, vehicle handling, ride quality, and speedometer/odometer calibration may be seriously affected. When different-sized tires are installed on these vehicles the ABS operation is abnormal, which may result in serious braking defects. When selecting replacement tires, the following precautions must be observed to maintain vehicle safety.

1. Replacement tires must be installed in pairs on the same axle. Never mix tire sizes or designs on the same axle. If it is necessary to replace only one tire, it should be paired with the tire having the most tread to equalize braking traction.
2. The tire load rating must be adequate for the vehicle on which the tire is installed. Light-duty trucks, station wagons, and trailer-towing vehicles are examples of vehicles that require tires with higher load ratings than passenger car tires.
3. Snow tires should be the same type and size as the other tires on the vehicle.
4. A 4WD vehicle should have the same type and size of tires on all four wheels.
5. Do not install tires with a load rating less than the car manufacturer's recommended rating.
6. Replacement tire ratings should be equivalent to the original tire ratings in all rating designations.
7. When combining different tires front to rear on a vehicle, check the car manufacturer's or tire manufacturer's recommendations.

# SUMMARY

- Tires provide traction and are designed to carry the weight of the vehicle, withstand side thrust over varying speeds and conditions, transfer braking and driving torque to the road, and absorb much of the rock shock from surface irregularities.
- Tires are subjected to severe forces during normal operation.
- There are three types of tread patterns: directional, non-directional, and symmetric and asymmetric.
- The number and size of the blocks, sipes, and grooves on a tire's tread determines how much rubber contacts the road, how much water can be displaced, and how quiet the tire will be during travel.
- Tires may be bias ply, belted bias ply, or radial belted design.
- Tires are rated by their profile, ratio, size, and load range.
- Most tires used today have built-in tread wear indicators to show when tires need replacement.
- The UTQG and DOT designations are standards in the tire industry.
- TPC numbers assure that the tire meets the car manufacturer's performance standards for traction, endurance, dimensions, noise, handling, and rolling resistance.
- Replacement tires must be the same type, ratings, and size as original tires to maintain vehicle safety.
- The tire placard provides valuable information regarding the tires on the vehicle.
- A properly inflated tire gives the best tire life, riding comfort, handling stability, and even gas mileage during normal driving conditions.

# REVIEW QUESTIONS

1. State the basic purposes of a tire.
2. To calculate the tire aspect ratio, the tire section height is divided by the ____________________.
3. Describe the difference between a bias-ply and a radial-ply tire.
4. Car manufacturers recommend that tire inflation pressures should be checked when the tires are ____________________.
5. While discussing tire ply and belt design, Technician A says in a radial tire, the belt core angle in relation to the tire centerline is 10 to 30 degrees. Technician B says in a radial tire, the belt cords are crisscrossed at an angle of 10 to 30 degrees in relation to the tire centerline. Who is correct?
   A. Technician A only
   B. Technician B only
   C. Both Technician A and Technician B
   D. Neither Technician A nor Technician B
6. While discussing tire design and operation, Technician A says a belted radial-ply tire has less rolling resistance than a belted bias-ply tire. Technician B says a bias-ply tire has stronger, but more flexible sidewalls than a belted radial-ply tire. Who is correct?
   A. Technician A only
   B. Technician B only
   C. Both Technician A and Technician B
   D. Neither Technician A nor Technician B
7. When explaining tire ratings to the customer,
   A. a P speed rating indicates the tire has a 130 mph (210 kmph) capability.
   B. tire speed ratings are based on laboratory tests.
   C. tire traction ratings are based on five skid tests.
   D. a tire with a C traction rating has better traction than a tire with an A traction rating.
8. When installing replacement tires,
   A. different tire sizes may be installed on the same axle.
   B. snow tires may be a different type or size than the original tires on the vehicle.
   C. if tire size is different than the original tire size, antilock brake system operation may be adversely affected.
   D. a four-wheel-drive vehicle may have different size tires on the front and rear wheels.
9. Most information about a car's tires can be found on the ____________.
   A. engine block
   B. tire placard
   C. undercarriage of the vehicle
   D. steering wheel
10. A tire that wobbles from side to side is said to have ____________.
    A. radial runout
    B. lateral runout
    C. eccentric runout
    D. none of the above

# Chapter 12 Diagnosis of Tire Problems

## INTRODUCTION

Tire and wheel assemblies provide the only connection between the road and the vehicle. Tire wear that is uneven or premature is usually a good indicator of steering and suspension system problems. Tires, therefore, become not only a good diagnostic aid to a technician **(Figure 1),** but they also can be clear evidence to the customer that there is a need for service.

### TIRE WEAR PATTERNS

Begin a tire inspection by checking each tire's sidewalls for blemishes or damage. Typically, any damage here means the tire(s) should be replaced. If there are cuts or signs of

| Problem Area | Description | Probable Cause |
|---|---|---|
| *Tire Tread Wear* | Wear on center of tread | Overinflation |
| | Wear on edges of tread | Underinflation |
| | Cupped wear on tread | Improper wheel balance |
| | | Radial runout |
| | Premature tread wear | Lack of tire rotation |
| | Shoulder wear | Incorrect camber |
| | Feathered tire wear | Incorrect toe |
| *Noise* | Tire thumping | Tire defects |
| | | Weak spots on tire |
| | | Bulges on tire |
| | | Excessive radial runout |
| | | Cupped tire treads |
| | | Improper wheel balance |
| *Ride Quality* | Vibration above a certain speed | Poor static wheel balance |
| | Shimmy above a certain speed | Poor dynamic wheel balance |
| | | Excessive lateral runout |
| *Vibration* | Chassis waddle | Exessive lateral runout |
| *Directional Control* | Steering pull | Tire conicity |
| | | Unequal tire pressures |
| | | Incorrect camber |
| | | Unequal caster |

**Figure 1.** A basic diagnostic guide for tire and wheel problems.

something rubbing on the tires, look for the source. It is possible that some body or chassis part has moved and is contacting the tire. Make sure to check the inside and outside of the tire.

Continue your inspection by checking the wear pattern on each tire's tread. As a tire wears out, it should show wear evenly across the tread. Abnormal wear may show the effects of poor wheel alignment or tire and wheel problems. It is important to remember that a worn or faulty suspension or steering part can cause poor wheel alignment and that a wheel alignment, by itself, may not correct the problem.

Wear on the edges or shoulders of a tire indicates a camber problem. If the wear is on the inside edge, there is too much negative camber. Excessive positive camber will cause the tire to wear on the outside edge. Wear on the center of the tread indicates an overinflation condition. Underinflation results in wear on both shoulders of the tire and/or cracked tire treads. These problems can also be caused by overloading the vehicle or high-speed cornering.

Toe is a wheel alignment factor that causes abnormal wear. Incorrect toe causes the tread to be feathered or take on a saw-tooth pattern. These are best observed by moving your hand across the width of the tread in both directions. If you feel sharp edges as you move across the tread, the tire is feathered. If you feel these edges as you move toward the center of the tire, there is too little toe-in. If the edges are felt as you move away from the center, there is too much toe-in.

If the tread area has randomly spaced bald or flat spots **(Figure 2),** the wheel and tire assembly are out of balance or there is excessive radial runout. This condition is often called cupping.

If any of these conditions are found, the cause of the problem should be corrected and the tire may need to be replaced if there is much wear.

## Tread Wear Measurement

On most tires, the tread wear indicators appear as wide bands across the tread when tread depth is worn to $^1/_{16}$ in (1.6 mm). Most tire manufacturers recommend tire replacement when the tread wear indicators appear across two or more tread grooves at three locations around the tire. If tires do not have wear indicators, a tread depth gauge may be used to measure the tread depth **(Figure 3).** Tires with $^2/_{32}$ inch of tread depth or less should be replaced.

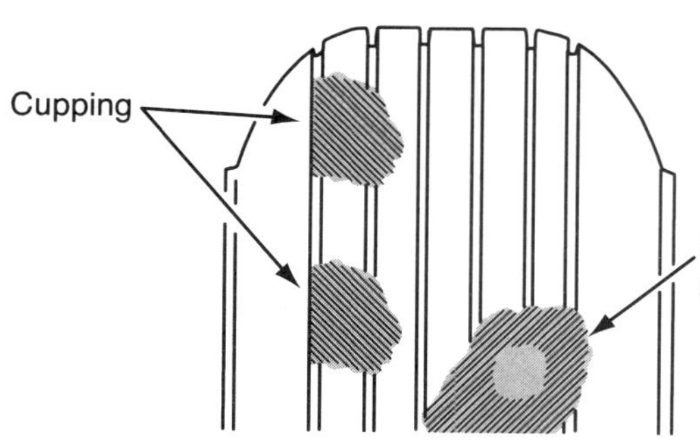


**Figure 2.** Random bald or flat spots on a tire's tread is often called cupping.

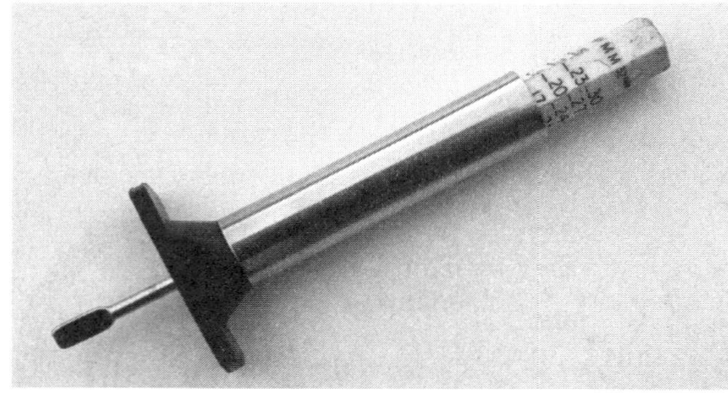

**Figure 3.** A tread depth gauge is used to measure tire wear.

## Air Pressure Checks

Check the pressure of each tire when the tires are cold. Locate the recommended tire pressure of the tires on the tire placard. The tires should be inflated to the manufacturer's recommended pressure. Using an air pressure gauge, check the pressure of each tire. If the pressure is measured as being higher than the recommended amount and the tire is warm or hot, do not reduce the pressure until the tire cools. Since the air inside the tire expands with heat, releasing some air may cause an underinflation condition when the tire is cold. Make sure the tire valves have a valve cap to keep dirt and moisture out of the valve stem.

Excessive inflation pressure causes the following problems:

1. Excessive center tread wear
2. Hard ride
3. Damage to tire carcass

When tires are underinflated, the following problems will be evident:

1. Excessive wear on each side of the tread
2. Hard steering
3. Wheel damage
4. Excessive heat buildup in the tire and possible severe tire damage

## TIRE NOISES AND STEERING PROBLEMS

Road test the vehicle under a variety of speeds and conditions and note all noises and handling problems that occurred during the road test. If any are evident, check the suspension, steering, driveline, and brake system to make sure they are not the cause of the problems before condemning the tires and wheels. If the tires, the suspension, steering, brake systems, and the driveline are in good condition and a noise was evident during the test drive, suspect the tire design as the cause of the problem. Make sure all of the tires have the same tread design.

If the tires felt as if they were bouncing along, the problem is typically called **wheel tramp.** If the tires seemed to vibrate while turning, the problem is called **shimmy.** If wheel tramp **(Figure 4)** is the problem, the tire and wheel assemblies need to be checked for static imbalance. If wheel shimmy is the problem **(Figure 5),** check the tire and wheel assemblies for dynamic imbalance. If the tire and wheel assembly are both statically and dynamically balanced, check the wheel hub or axle for excessive runout. If runout is not the problem, check for loose, worn, or damaged wheel bearings.

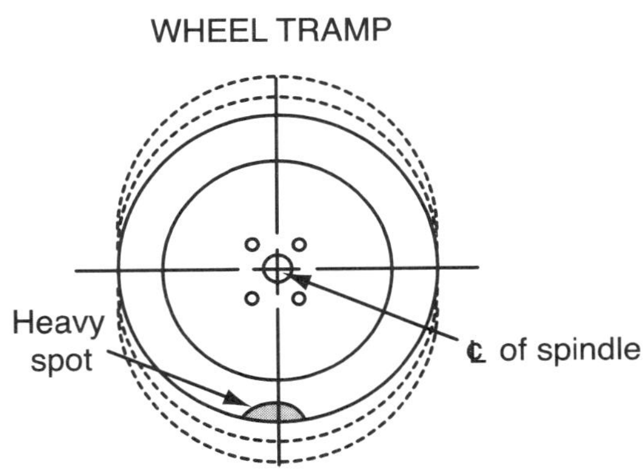


**Figure 4.** A statically unbalanced tire and wheel assembly will cause wheel tramp.

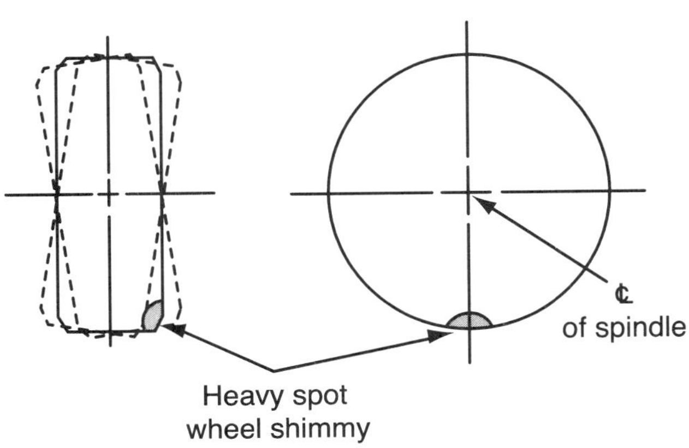


**Figure 5.** Shimmy is cased by a dynamic unbalanced tire and wheel assembly.

## Diagnosis of Tire Noises

One key to discovering the cause of a vibration or noise is its perceived speed compared to the speed of the vehicle. Tires and wheel speed is normally three times greater than driveshaft speed at any vehicle speed. Therefore, if the noise or vibration increases greatly with a small increase in vehicle speed, the probable cause is the tire or wheel. Also, tire noise will vary with road surface conditions, whereas differential and driveshaft noise is not affected when various road surfaces are encountered.

Uneven tread surfaces may cause tire noises that seem to originate elsewhere in the vehicle. These noises may be confused with differential noise. Differential noise usually varies with acceleration and deceleration, whereas tire noise remains more constant in relation to these forces. Tire noise is most pronounced on smooth asphalt road surfaces at speeds of 15 to 45 mph (24 to 72 km/h). Typically, if a vibration is felt at 30 to 70 mph, the cause is normally uneven tire wear. If the vibration is felt at slightly slower speeds (20 to 70 mph), the cause may be radial runout. Lateral runout will cause a vibration at higher speeds (60 to 70 mph). Tire balance problems are heard and felt at speeds of 30 to 60 mph, however wheel bearing problems will be heard at 0 to 70 mph but felt at only 50 to 60 mph.

## TIRE THUMP AND VIBRATIONS

**Tire thump** is a pounding noise that occurs as a tire rotates. Thump normally sets up an accompanying vibration. When tire thump and tire vibration are present, check these items:

1. Cupped tire treads
2. Excessive tire radial runout
3. Tire manufacturing defects such as heavy spots, weak spots, or tread chunking
4. Incorrect wheel balance

## Radial Tire Waddle

If a belt in a radial tire is wrongly positioned during the making of the tire, radial tire waddle will result. **Waddle** feels like someone is rocking the vehicle as it moves. Waddle normally occurs at low vehicle speeds (5 to 30 mph [8 to 50 km/h]). If a tire has a tread or ply separation problem, it will cause a feeling similar to radial wobble but it will become a stronger vibration at higher speeds. Normally these problems are evident by bulges in the tire.

## STEERING PULL

If the steering gradually pulls to one side on a smooth, straight road surface, a tire, steering, or suspension defect is present. Tires of different types, sizes, designs, or inflation pressures on opposite sides of a vehicle cause steering pull.

Sometimes a tire manufacturing defect occurs in which the belts are wound off center. This condition is referred to as **tire conicity**. A cone-shaped object rolls in the direction of its smaller diameter. Similarly, a tire with conicity tends to lead, or pull, to one side, which causes the vehicle to follow the action of the tire **(Figure 6).**

Figure 6. Tire conicity will cause the vehicle to pull to one side.

You Should Know

*Tire conicity is not visible. It can only be diagnosed by changing the tire and wheel position.*

Tire conicity must be diagnosed by switching the two front tires and reversing the front and rear tires. Incorrect front suspension alignment angles also cause steering pull.

## TIRE AND WHEEL RUNOUT

Ideally, a tire-and-wheel assembly should be perfectly round. However, this condition is rarely achieved. A tire-and-wheel assembly that is out of round is said to have **radial runout** or eccentricity. If the radial runout exceeds manufacturer's specifications, a vibration may occur because the radial runout causes the spindle to move up and down. A defective tire with a variation in stiffness may also cause this up-and-down spindle action **(Figure 7).**

A tire that wobbles side to side is said to have **lateral runout.** Excessive lateral runout causes the tire to waddle as it turns, and this waddling sensation may be transmitted to the passenger compartment. Waddling may also be caused by a defective tire in which a belt is not straight. Sometimes tires or wheels can be remounted to lessen or correct runout problems.

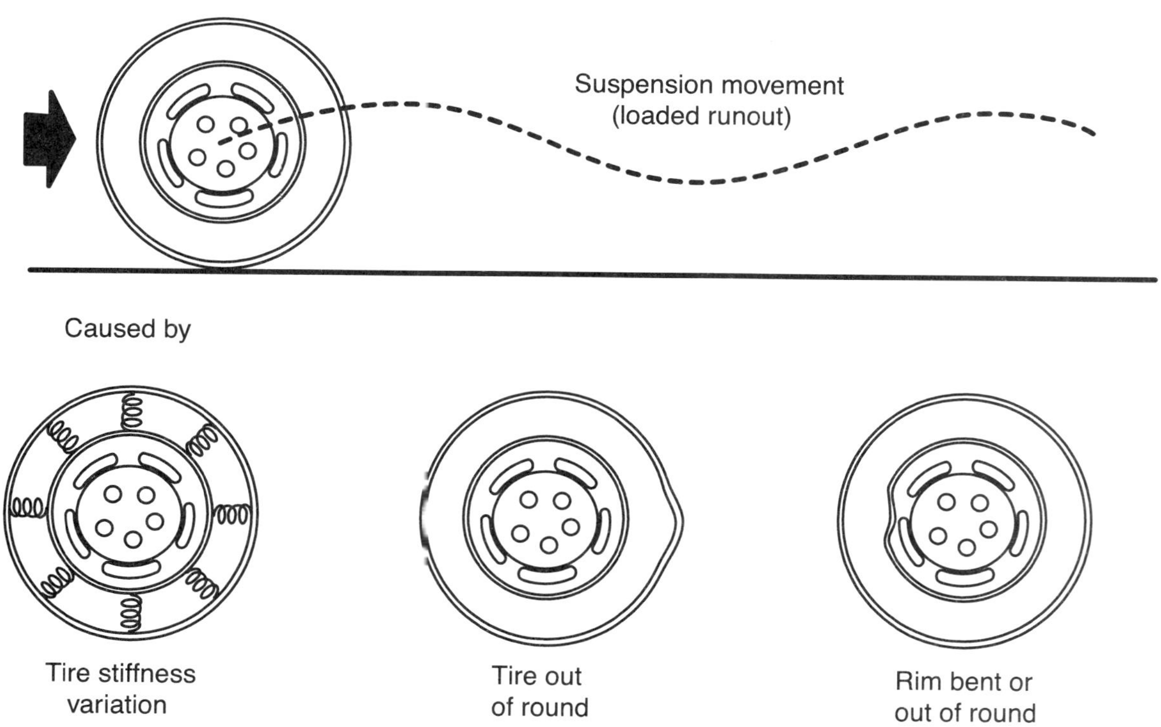


Figure 7. Common causes of wheel vibration.

## Measuring Tire and Wheel Runout

If excessive wheel runout is suspected, visually inspect the tire for abnormal bulges or distortions. All measurements should be made on the vehicle with the tires inflated to the recommended pressures and with the wheel bearing adjusted to specifications. Runout measurements can be taken with the wheel and tire mounted on the vehicle or on a wheel balancer. Measuring runout on a wheel balancer eliminates the possibility of the wheel bearings influencing the runout checks.

Position a dial indicator against the center of the tire tread **(Figure 8)** as the tire is rotated slowly to measure radial runout. Mark the highest point of radial runout on the tire with a crayon, and mark the valve stem position on the tire. If total radial runout of a tire exceeds specified limits, you should check wheel runout to determine whether the wheel or tire is at fault.

Before checking the wheel, measure the lateral runout of the tire by placing the dial indicator just above the buffing rib on the sidewall and rotate the tire slowly. Compare your readings to specifications. If the tire runout was excessive, measure the wheel lateral runout at the wheel rim bead flange just inside the curved lip of the flange as the wheel is rotated.

Wheel radial runout can be measured with the tire on the wheel by placing the dial indicator against the lip of the rim while the rim is rotated **(Figure 9).** Some technicians prefer to demount the tire and measure the wheel on a tire balancer. Mark the highest point of radial runout on the wheel with a crayon. If the total radial runout of a mounted wheel exceeds specified limits, check the runout of the axle flange or hub to determine whether the wheel or hub is at fault. Position the dial indicator on the face near the edge of the wheel hub or axle flange. Rotate the hub or flange slowly to measure its runout.

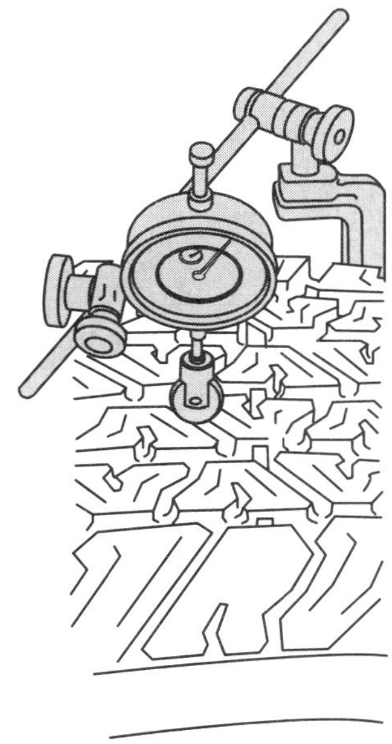

**Figure 8.** Position a dial indicator against the center of the tire tread to measure radial runout.

If either lateral or radial runout exceeds the specified limit but wheel runout is within the specified limit, it might be possible to reduce runout to an acceptable level. This is done by changing the position of the tire on the wheel so that the previously marked high points are 180 degrees apart. Tires or wheels with excessive runout are usually replaced.

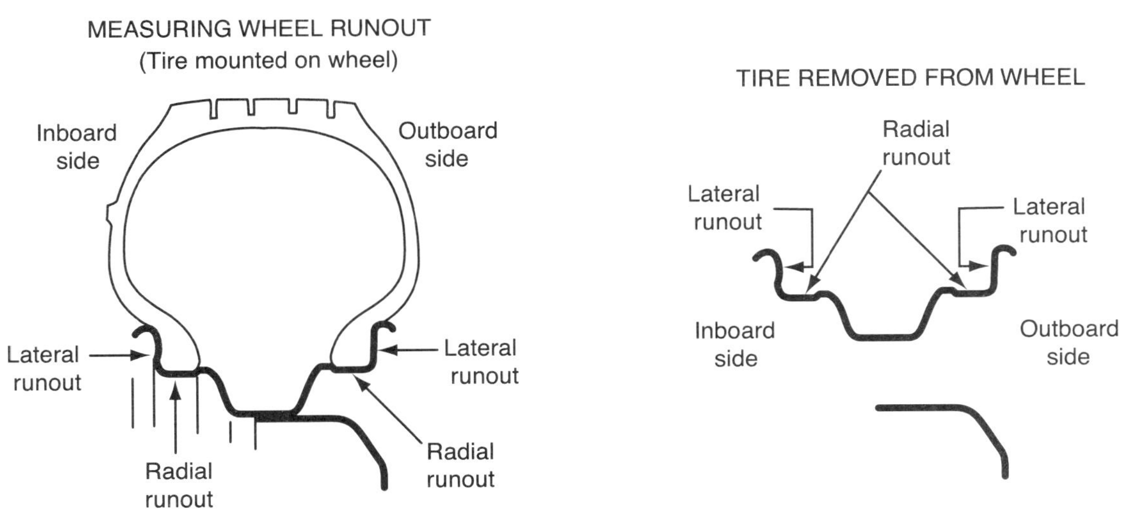


**Figure 9.** The various positions to place the dial indicator to measure wheel runout with the tire on the wheel (left) and with the tire off the wheel (right).

## SUMMARY

- Tire wear that is uneven or premature is usually a good indicator of steering and suspension system problems.
- Carefully check the wear pattern on each tire's tread; they should show wear evenly across the tread.
- Tires with $^2/_{32}$ inch of tread depth or less should be replaced.
- A tire depends on correct inflation pressure to maintain its correct shape and support the vehicle weight.
- Wheel tramp indicates that the tire and wheel assemblies need to be checked for static imbalance.
- Wheel shimmy means the tire and wheel assemblies should be checked for dynamic imbalance.
- Tire thump may be caused by cupped tire treads, excessive tire radial runout, tire manufacturing defects such as heavy spots, weak spots, or tread chunking, and incorrect wheel balance.
- Steering pull may be caused by a defective tire or a suspension and steering system problem.
- A tire with conicity tends to lead, or pull, to one side, which causes the vehicle to follow the action of the tire.
- Excessive radial runout may cause a vibration, and lateral runout may cause waddle.

## REVIEW QUESTIONS

1. What type of tire wear will be caused by incorrect toe? How should you check for this?
2. While diagnosing a tire thumping problem, Technician A says this problem may be caused by cupped tire treads. Technician B says a heavy spot in the tire may cause this complaint. Who is correct?
   A. Technician A only
   B. Technician B only
   C. Both Technician A and Technician B
   D. Neither Technician A nor Technician B
3. While diagnosing a vehicle that pulls to one side, Technician A says that excessive radial runout on the right front tire may cause this problem. Technician B says that tire conicity may be the cause of this complaint. Who is correct?
   A. Technician A only
   B. Technician B only
   C. Both Technician A and Technician B
   D. Neither Technician A nor Technician B
4. Technician A says that static imbalance causes feathered tread wear. Technician B says that dynamic imbalance causes cupped wear and bald spots on the tire tread. Who is correct?
   A. Technician A only
   B. Technician B only
   C. Both Technician A and Technician B
   D. Neither Technician A nor Technician B
5. A front tire has excessive wear on both edges of the tire tread. The most likely cause of this problem is ________________.
   A. overinflation
   B. underinflation
   C. improper static balance
   D. improper dynamic balance

# Chapter 13 Tire and Wheel Service

## INTRODUCTION

For most tire/wheel service, the assembly must be first removed from the vehicle. To repair or replace a tire, the tire and wheel must be separated. These are the most commonly performed tire services. Other common services are covered in this chapter, except the balancing of wheels and tires, which is covered in another chapter.

### WHEEL AND TIRE REMOVAL

When it is necessary to remove a wheel and tire assembly, follow these steps:

1. Remove the wheel cover.
2. Keep the vehicle's wheels on the ground and loosen the wheel lug nuts about one-half turn, but do not remove the wheel nuts.

> **You Should Know** *Before raising the vehicle on a hoist, be sure that the hoist is lifting on the manufacturer's recommended lifting points.*

3. Raise the vehicle on a hoist or with a floor jack) to a convenient working level.

> **You Should Know** *If the vehicle is lifted with a floor jack, position jack stands (**Figure 1**) under the vehicle, remove the jack, place safety stands under the suspension or frame, and lower the vehicle onto the safety stands. then remove the floor jack from under the vehicle.*

**Figure 1.** After the vehicle is at a satisfactory height, place jack stands under the vehicle.

4. Chalk mark the tire, the wheel, and one of the lug nuts so the wheel and tire can be reinstalled in the same position.
5. Remove the lug nuts and the wheel-and-tire assembly.

When tires and wheels are installed on a vehicle, it is very important that the wheel nuts are torqued to manufacturer's specifications in the proper sequence. Do not use an impact wrench when tightening wheel nuts to the specified torque.

### TIRE ROTATION

Driving habits, to a large extent, determine tire life. Severe braking, high-speed driving, turning at high speeds, rapid acceleration and deceleration, and striking curbs are just a few factors that shorten tire life. To equalize tire wear, most manufacturers recommend that the tires be rotated at specified intervals to obtain maximum tire life.

The exact tire rotation procedure depends on the model year, the type of tires, and whether the vehicle has a

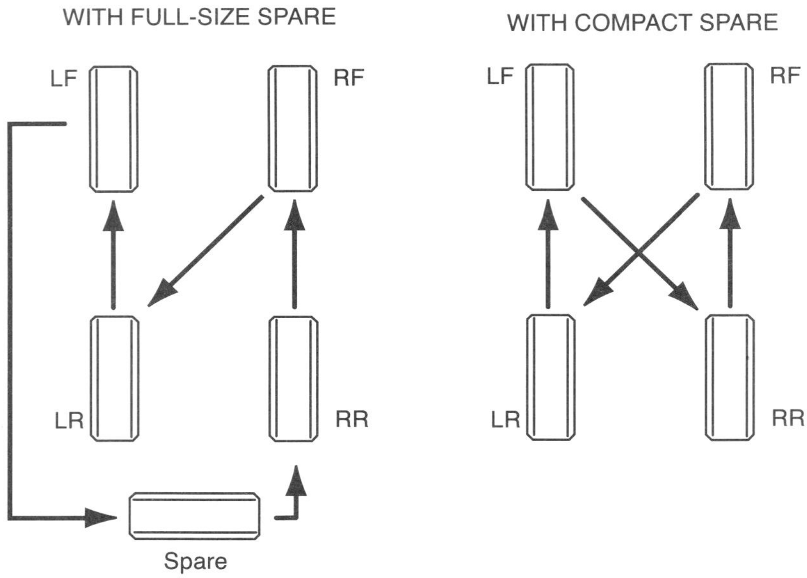


**Figure 2.** The correct procedure for tire rotation depends on a number of factors, including whether the vehicle has a conventional spare or a compact spare.

conventional spare or a compact spare **(Figure 2).** Tire rotation procedures do not include the compact spare. The vehicle manufacturer provides tire rotation information in the owner's manual and service manual. Vehicle manufacturers usually recommend different tire rotation procedures for bias ply tires than for radial tires.

Bias ply and bias-belted tires should be rotated about every 6,000 miles. Radial tires should be initially rotated at 7,500 miles and then at least every 15,000 miles thereafter. It is important that directional tires are kept rotating in the direction they should after the tires have been rotated. This means the tires may need to be dismounted from the wheel, flipped, and reinstalled on the rim before being put on the other side of the car.

## TIRE AND WHEEL SERVICE PRECAUTIONS

There are many different types of tire changing equipment, however, certain precautions apply to all equipment. These include the following:

1. Before you operate any tire changing equipment, always be absolutely certain that you are familiar with the operation of the equipment.
2. When operating tire changing equipment, always follow the manufacturer's recommended procedure.
3. Always completely deflate a tire before attempting to demount the tire.
4. Clean the bead seats on the wheel rim before mounting the tire on the wheel rim.
5. Lubricate the outer surface of the tire beads with rubber lubricant before mounting the tire on the wheel rim.
6. When the tire is mounted on the wheel rim, be sure the tire is positioned evenly on the rim.
7. While inflating a tire, do not stand directly over the tire.
8. Do not overinflate tires.
9. When mounting tires on cast aluminum or magnesium alloy wheels, always use the manufacturer's recommended tools and procedures.

## TIRE DEMOUNTING

Always use a tire changer to demount tires **(Figure 3).** Do not use hand tools or tire irons. Demounting the tire

**Figure 3.** A tire changer.

from the wheel begins with releasing the air and removing the air valve core. Chalk mark the tire at the valve stem opening in the wheel so the tire can be re-installed in the same position **(Figure 4).** If you replacing the tires, this is not needed. When mounting or dismounting tires from aluminum or wire spoke wheels, it is recommended that the appropriate accessories be used to avoid damage to the wheels.

The next step unseats the tire from the wheel. The tire machine does the unseating. The technician merely guides the operating lever **(Figure 5).** Follow the tire changer manufacturer's recommended procedure to attach the wheel to the tire changer.

> **You Should Know** *Do not proceed to demount the tire unless the wheel and tire are securely attached to the tire changer. This action may cause personal injury.*

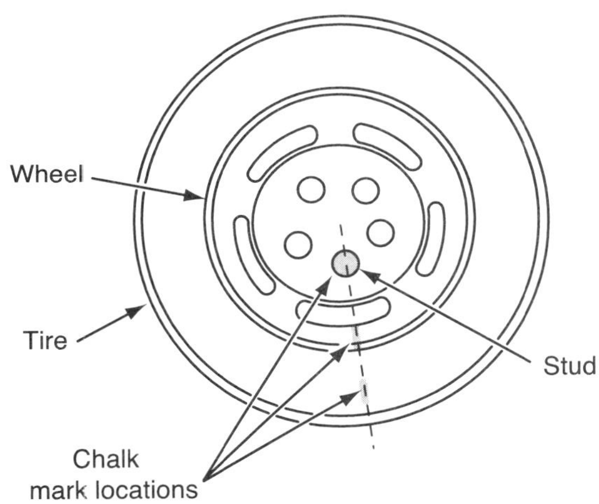


**Figure 4.** Chalk mark the tire at the valve stem opening in the wheel so the tire may be reinstalled in the same position to maintain proper wheel balance.

**Figure 5.** An example of a bead breaker on a tire changer.

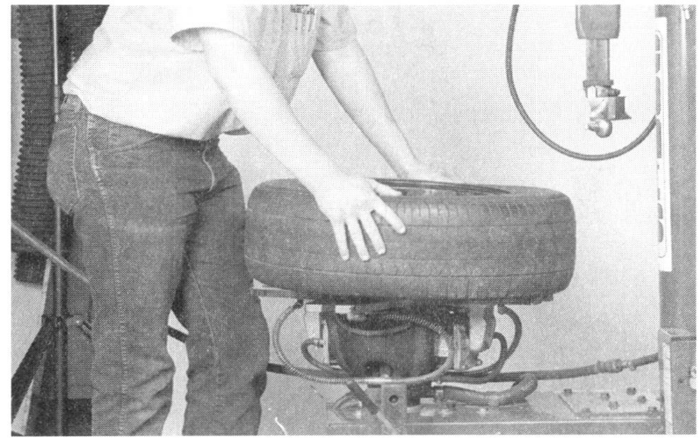

**Figure 6.** The tire and wheel assembly is moved for removing the tire after the tire's beads are unseated.

Follow the procedure recommended by the manufacturer of the tire changer to force the tire bead inward and separate it from the rim on both sides. Typically there is a top and bottom bead breaker arm. The top bead breaker is positioned and operated first. On some machines, the wheel and tire assembly is moved from the bead breaker assembly and mounted to a demounting fixture **(Figure 6).** Again, always follow the procedures of the tire machine's manufacturer.

Once both sides of the tire are unseated, push one edge of the top bead into the drop center of the rim. Position the tire changer bar, lever, or tool under the opposite side of the upper bead from the part of the bead in the drop center of the wheel **(Figure 7).** Operate the tire changer bar, lever, or tool to remove the upper tire bead over the top edge of the wheel. Some tire changers rotate a bar or lever to move the tire beads over the wheel. Other tire changers rotate the tire and wheel to perform this function.

**Figure 7.** Position the tire iron under the opposite side of the upper bead from the part of the bead in the drop center of the wheel.

**Figure 8.** After the tire is totally free from the rim, remove the tire.

Push one part of the lower tire bead into the drop center in the wheel. Position the tire changer bar, lever, or tool under the opposite side of the lower bead from the part of the bead in the drop center of the wheel. Operate the tire changer to rotate the bar or lever and move the bead over the top of the rim. After the tire is totally free from the rim, remove the tire **(Figure 8).** Then remove the wheel from the tire changer. Thoroughly clean the wheel-to-tire bead sealing surfaces with a wire brush. Inspect the wheel for rust or damage.

## TIRE REPAIR

The most common tire problem besides wear is a puncture. When properly repaired, the tire can be put be back in service without the fear of the air leak recurring. Do not attempt to repair punctures over 1/4 inch in diameter. Punctures in the sidewalls or on the tire shoulders should not be repaired. The repairable area in belted bias-ply tires is approximately the width of the belts **(Figure 9).** The belts in radial tires are wider than those in bias-ply tires. The repairable area in radial tires is also the width of the belts. Since compact spare tires have thin treads, do not attempt to repair these tires. Inspect the tire; do not repair a tire with any of the following defects, signs of damage, or excessive wear:

Exposed fabric or belts
Bulges or blisters
Ply separation
Broken or cracked beads
Fabric cracks or cuts
Visible wear indicators

Some car owners attempt to seal punctures with tire sealants. These sealants are injected into the tire through the valve stem. Sometimes the chemicals in the sealant do a great job sealing the hole; other times they fail. The sealants should never be used and will not work on sidewall punctures. Some of the sealants are very flammable

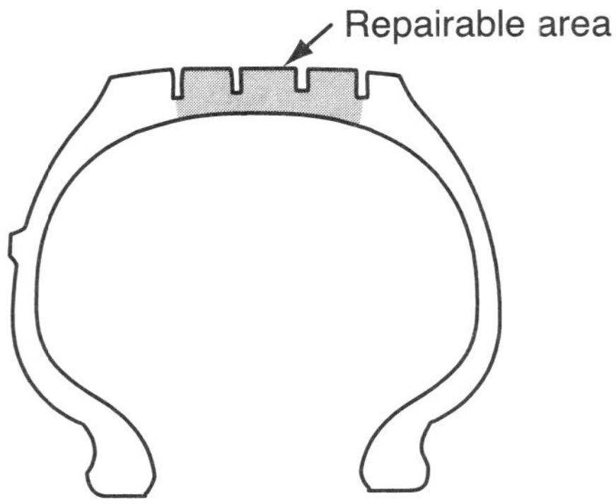


**Figure 9.** The repairable area of a belted bias-ply tire.

and carry a warning that the tire should be marked so that the next technician knows the sealant has been used.

> **You Should Know** *Tire sealants injected through the valve stem can produce wheel rust and tire imbalance.*

To locate a puncture in a tire, inflate it to the maximum inflation pressure indicated on its sidewall. Then submerge the tire/wheel assembly in a tank of water or sponge it with a soapy water solution. Bubbles will appear wherever the leak is located in the tire or wheel. Mark the location of the leak with a crayon so it can be easily found once the tire is removed from the wheel. Also, use the crayon to mark the location of the valve stem so that original tire and wheel balance can be maintained after the tire is put back on the wheel.

Since most vehicles are equipped with tubeless tires, we will discuss this type of tire repair. If the cause of the puncture, such as a nail, is still in the tire, remove it. Most punctures can be repaired from inside the tire with a service plug or vulcanized patch service kit. After the repair has been made and the tire mounted to the wheel, check for air leaks before installing the assembly onto the vehicle. The instructions of the tire service kit manufacturer should be followed, but we will discuss three common tire repair procedures.

### Plug Installation Procedure

The head-type plug is commonly used. A plug that is slightly larger than the size of the puncture is inserted into the hole from the inside of the tire with an insertion tool. Before doing this, buff the area around the puncture with a wire brush or wire buffing wheel. Select a plug slightly larger than the puncture opening, coat the hole, plug, and tool with vulcanizing fluid. Insert the plug in the eye of the insertion tool.

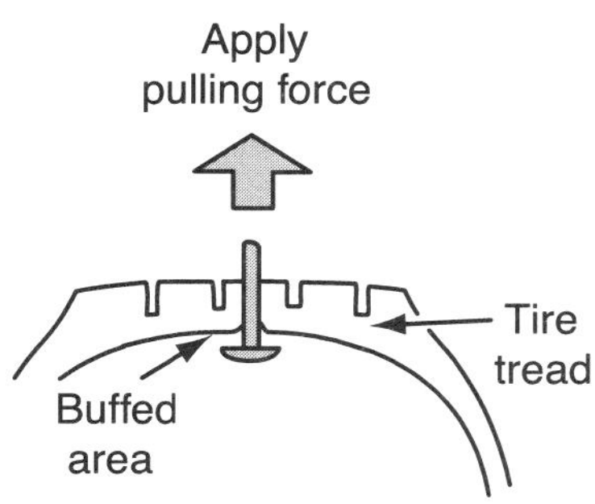


**Figure 10.** Pull the plug through the puncture.

While holding and stretching the long end of the plug, insert it into the hole **(Figure 10).** The plug must extend above both the tread and inner liner surface. If the plug pops through, throw it away and insert a new plug. Once the plug is in place, remove the tool and trim off the plug $^{1}/_{32}$-inch above the inner surface. Be careful not to pull on the plug while cutting it.

## Cold Patch Installation Procedure

A cold patch can seal a puncture by placing it on the inside of the tire. Because this patch does not require heat to seal and adhere, it is called a cold patch. Before adhering the patch to the problem area, buff the area around the puncture with a wire brush or buffing wheel. Then, apply vulcanizing fluid to the buffed area and allow it to dry until it is tacky.

Select a patch that is large enough to cover the problem area and cover a good amount of the buffed area. Carefully remove the backing from the patch. Center the patch base over the punctured area. Run a stitching tool over the patch to help bind it to the tire. When repairing radial tires, use only a patch specially approved for that application. These special patches have arrows that must be lined up parallel to the radial plies. Allow the patch and adhesive to dry before installing the tire onto the wheel.

## Hot Patch Installation Procedure

A hot tire patch application is similar to a cold patch. The difference is that the hot patch is clamped over the puncture, and heat is applied to the patch to make it adhere. The procedure begins by buffing the area around the puncture with a wire brush or buffing wheel. Then apply vulcanizing fluid to the buffed area.

Peel the backing from the patch, and install the patch so it is centered over the puncture on the inside of the tire. Many hot patches are heated with an electric heating element clamped over the patch. This element should be clamped in place for the amount of time recommended by the equipment or patch manufacturer. After the heating element is removed, allow the patch to cool for a few minutes, and be sure the patch is properly bonded to the tire.

# WHEEL RIM SERVICE

> You Should Know *Steel wheel rims must not be welded, heated, or peened with a ball peen hammer. These procedures may weaken the rim and create a safety hazard.*

The wheels should be carefully inspected each time a tire is to be mounted on them. The major causes of wheel failure are improper maintenance, overloading, age, and accidents, including pot hole damage. Wheels must be replaced when they are bent, dented, or heavily rusted, have leaks or elongated bolt holes, and have excessive lateral or radial runout. Wheels with a lateral or radial runout greater than specifications can cause high-speed vibrations. Wobble or shimmy caused by a damaged wheel eventually damages the wheel bearings. Stones wedged between the wheel and disc brake rotor or drum can unbalance the wheel. Also, check the lug nuts to be sure that they are set according to the torque given in the vehicle's service manual. Loose lug nuts can cause shimmy and vibration and can also distort the stud holes in the wheels. Another wheel mounting problem is caused by improperly positioning the wheel on the wheel hub or by improperly tightening the lug nuts.

Make sure you thoroughly inspect the wheel for conditions that would affect its performance or sealing of the tire.

> You Should Know *Installing an inner tube to correct leaks in a tubeless tire or wheel rim is not an approved procedure.*

## Wheel Rim Leak Repair

A wheel rim leak may be repaired if the leak is not caused by excessive rust on a steel rim, and the rim is in satisfactory condition. Follow these steps for wheel rim leak repair:

1. Use #80-grit sandpaper to clean the area on the tire side of the rim.
2. Use a shop towel to remove any grit from the leak area.
3. Be sure the wheel rim is at room temperature, and apply a heavy coating of silicone rubber sealer over the leak area.
4. Spread the sealer over the entire sanded area with a putty knife.
5. Allow the sealer to cure for 6 hours before remounting the tire.

## TIRE REMOUNTING PROCEDURE

To prepare the wheel for the mounting of a new tire, use a wire brush and rags to remove all of the dirt and rust from the sealing surface **(Figure 11).** Coat the tire's beads and the wheel bead seats with rubber tire lubricant. Secure the wheel on the tire changer with the narrow bead ledge facing upward, and place the tire on top of the wheel with the bead of the lower side of the tire on top of the wheel. Push one part of the lower tire bead into the drop center in the wheel. Position the tire changer bar, lever, or tool under the opposite side of the upper bead from the part of the bead in the drop center of the wheel.

Install the lower tire bead over the rim. Push one part of the upper tire bead into the drop center in the wheel. Position the tire changer bar, lever, or tool under the opposite side of the upper bead from the part of the bead in the drop center of the wheel. Install the upper bead over the wheel rim.

Rotate the tire until the crayon mark is aligned with the valve stem. This mark was made during demounting. Install a new valve stem. Most stems are the snap-in type.

**Figure 11.** Use a wire brush to remove all of the dirt and rust from the sealing surface of the rim.

These are installed from inside the wheel with a pulling tool. Make sure that the stem is properly seated. Another style of stem has a retaining nut that must be removed when pulling off the old stem. Be sure to completely tighten the new nut.

Follow the recommended procedure supplied by the manufacturer of the tire changer to inflate the tire. This procedure may involve the use of a bead expander installed around the center of the tire tread to expand the tire beads against the wheel rim.

> **You Should Know** ***When a bead expander is installed around the tire, never exceed 10 psi (69 kPa) pressure in the tire. A higher pressure may cause the expander to break and fly off the tire, causing serious personal injury or property damage.***

Always observe the circular marking around the tire bead and tire locating ring as the tire is inflated. This mark should be centered around the wheel. Also, observe both beads while a tire is inflated. If the circular mark around the tire bead is not centered on the rim, deflate the tire and center it on the wheel rim. If a bead expander is not used, never exceed 40 psi (276 kPa) to move the tire beads out against the rim. A higher pressure may blow the tire bead against the rim with excessive force and could burst the rim or tire, resulting in serious personal injury or property damage. When the tire bead will not move cut against the wheel rim with 40 psi (276 kPa) pressure, deflate the tire and center it on the wheel rim again.

Install and tighten the valve stem core into the valve stem body **(Figure 12).** Connect an air supply hose to the tire valve stem. Inflate the tire to the specified pressure.

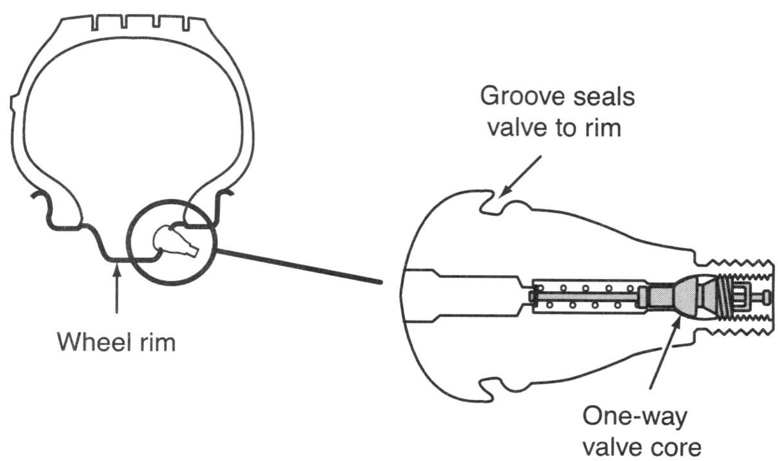


**Figure 12.** After this tire is seated on the wheel, install the air valve core.

**You Should Know** *When inflating a tire, do not stand or lean over the tire and wheel. If the tire bead comes off the rim or the rim cracks, personal injury may occur.*

## INSTALLATION OF TIRE/WHEEL ASSEMBLY ON THE VEHICLE

Before reinstalling a tire/wheel assembly, clean the axle/rotor flange and wheel bore with a wire brush or steel wool. Coat the axle pilot flange with disc brake caliper slide grease or an equivalent. Place the wheel on the hub. Install the lug nuts, and tighten them alternately to draw the wheel evenly against the hub. They should be tightened to a specified torque and sequence to avoid distortion. Many tire technicians snug up the lug nuts and then use a torque wrench for the final tightening after the vehicle is lowered to the ground. Then, check and adjust the air pressure in all tires.

**You Should Know** *Overtorqueing the lug nuts is the most common cause of disc brake rotor distortion.*

## TIRE CARE

To maximize tire performance, inspect them for signs of improper inflation and uneven wear. Tires should also be checked for cuts, stone bruises, abrasions, blisters, and for objects that might have become imbedded in the tread. More frequent inspections are recommended when rapid or extreme temperature changes occur, or where road surfaces are rough or occasionally littered with debris.

To clean tires, use a mild soap-and-water solution only. Rinse thoroughly with clear water. As rubber dries, it becomes harder and the tire will lose some of its performance characteristics.

When storing tires, lay them flat on a clean, dry, oil-free floor. Store them in the dark. Direct sunlight is hard on tires.

# SUMMARY

- Inspect tires for signs of improper inflation and uneven wear. Tires should also be checked for cuts, bruises, abrasions, blisters, and for stones or other objects that might have become imbedded in the tread.
- To equalize tire wear, tires should be rotated. Always use a tire changer to demount and remount tires.
- There are three popular methods of tire repair: head-type plug, cold patch repair, and hot patch repair.

# REVIEW QUESTIONS

1. Why is tire rotation recommended by most manufacturers?
2. Describe the procedure for using a plug to seal a puncture in a tire.
3. Why shouldn't mineral oils be used to clean or treat tires?
4. Technician A uses tire sealant injected through the valve stem to repair a tire punctured in the tread area. Technician B uses tire sealant injected through the valve stem to repair a tire punctured in the sidewall. Who is correct?
   A. Technician A
   B. Technician B
   C. Both Technician A and Technician B
   D. Neither Technician A nor Technician B
5. All of the following statements are true *except:*
   A. A cold patch should be stitched to the inner surface of the carcass.
   B. A tread puncture that is less than $^{1}/_{4}$ inch in diameter is repairable.
   C. Wheel balancing is required only when new tires are installed.
   D. For best results, punctures should be repaired from inside the tire.

# Chapter 14 Balancing Tire and Wheel Assemblies

## *INTRODUCTION*

If an inspection shows uneven or irregular tire wear or when new tires are installed, tire and wheel balance is necessary. Wheel balancing distributes weights along the wheel's rim, which counteracts heavy spots on the wheels and tires and allows them to roll smoothly. An out-of-balance condition can cause increased wear on the tires and ball joints, as well as deterioration of shock absorbers and other suspension components. There are two types of wheel imbalance: static and dynamic.

The main purposes of proper wheel balance may be summarized as follows:

1. Maintains normal tire tread life
2. Provides extended life of suspension and steering components
3. Helps provide directional control of the vehicle
4. Reduces driver fatigue
5. Increases passenger comfort
6. Helps maintain the life of body and chassis components

## TIRE MOTION FORCES

When a vehicle is in motion, wheel rotation subjects the tires to centrifugal force. The tires are also subjected to accelerating and decelerating forces because of their path of travel. If a vehicle is traveling at 55 mph (88.5 km/h), the part of the tire exactly fore and aft of the spindle is also traveling at 55 mph (88.5 km/h). At the very top of the tire, the tire speed is 110 mph (177 km/h). The tire speed actually drops to zero at the exact bottom of the tire where the arc of deceleration ends and the arc of acceleration begins. Since the tires are subjected to strong acceleration and deceleration forces, tire construction must be uniform. This explains why tire condition and balance affects tire wear and the vehicle's handling characteristics.

## STATIC WHEEL BALANCE THEORY

Static balance comes from the equal distribution of weight around the wheel. Statically unbalanced wheels cause a bouncing action called wheel tramp. This eventually causes uneven tire wear. Static balance is balancing a wheel at rest. A statically unbalanced wheel tends to rotate by itself until the heavy portion is down **(Figure 1).**

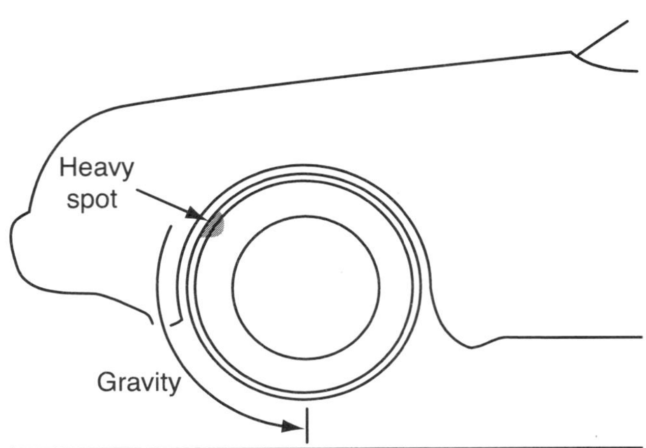


**Figure 1.** A statically unbalanced wheel tends to rotate by itself until the heavy portion is down because of gravity.

Static balance is achieved when the wheel does not rotate by itself regardless of the position in which it is placed on its axis. This is true whether the wheel is mounted vertically, as on a spindle or balancer shaft, or horizontally, as on a bubble-type balancer.

## Results of Static Unbalance

Centrifugal force may be defined as the force that tends to move a rotating mass away from its axis of rotation. The heavy portion of a statically unbalanced wheel is influenced by centrifugal force. This influence attempts to move the heavy spot on a tangent line away from the wheel axis. This action tends to lift the wheel assembly off the road surface.

The wheel-lifting action is wheel tramp. Wheel tramp causes the tire to slip momentarily when it is lifted vertically. When the wheel and tire move downward as the heavy spot decelerates, the tire strikes the road surface with a pounding action. This repeated slipping and pounding action causes severe tire scuffing and cupping.

The vertical wheel motion from static unbalance is transferred to the suspension system and is then absorbed by the chassis and body. This action causes rapid wear on suspension and steering components. The wheel tramp action resulting from static unbalance is also transmitted to the passenger compartment, which causes passenger discomfort and driver fatigue.

When a vehicle is traveling at normal highway cruising speed, the average wheel speed is 850 revolutions per minute (rpm). A statically unbalanced tire-and-wheel assembly is an uncontrolled mass of weight in motion. When a vehicle is traveling at 60 mph (97 km/h) and a tire has 2 ounces (57 grams) of static unbalance, the resultant pounding force is approximately 15 pounds (6.8 kilograms), against the road surface.

# DYNAMIC WHEEL BALANCE THEORY

Dynamic balance is the equal distribution of weight on each side of the centerline. When the tire spins, there is no tendency for the assembly to move from side to side. Dynamic balance, therefore, is balancing a wheel in motion. Once a wheel starts to rotate and is in motion, the static weights try to reach the true plane of rotation of the wheel because of the action of centrifugal force. In an attempt to reach the true plane of rotation when there is an imbalance, the static weights force the spindle to one side. Wheels that are dynamically unbalanced can cause wheel shimmy and a wear pattern.

At 180 degrees of wheel rotation, static weights kick the spindle in the opposite direction. The resultant side thrusts cause the wheel assembly to wobble or wiggle. When severe enough, as already mentioned, it causes vibration and wheel shimmy.

Dynamic wheel balance says the weight of the assembly is distributed equally on both sides of the wheel center viewed from the front. It may be explained by dividing the tire into four sections **(Figure 2).** In Figure 2, if sections A and C have the same weight and sections B and D also have the same weight, the tire has proper dynamic balance. If a tire has dynamic unbalance, section D may have a heavy spot; thus, sections B and D have different **weights (Figure 3).** A tire-and-wheel assembly may be in static balance but have dynamic unbalance. Therefore, wheels must be in balance statically and dynamically.

To correct dynamic unbalance, equal weights are placed 180 degrees opposite each other, one on the inside of the wheel and one on the outside, at the point of unbal-

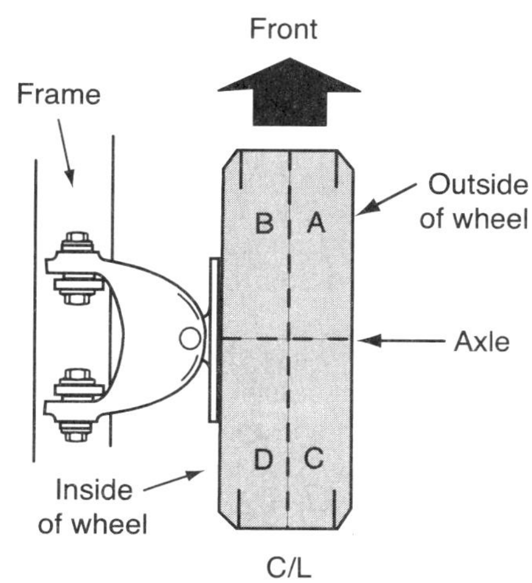


**Figure 2.** An explanation of the effects of dynamic wheel balance.

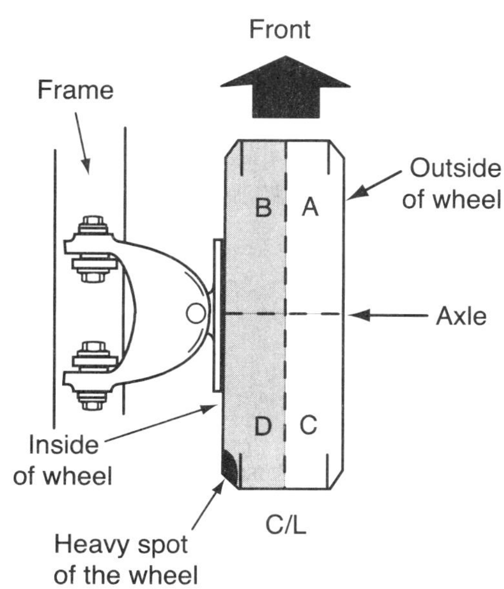


**Figure 3.** The effects of dynamic wheel imbalance.

ance. This corrects the couple action or wiggle of the wheel assembly. Also, note that dynamic balance is obtained while static balance remains unaffected.

## Results of Dynamic Wheel Unbalance

When a dynamically unbalanced wheel is rotating, centrifugal force moves the heavy spot toward the tire centerline. The centerline of the heavy spot arc is at a 90-degree angle to the spindle. This action turns the true centerline of the left front wheel inward when the heavy spot is at the rear of the wheel **(Figure 4).** When the wheel rotates until the heavy spot is at the front of the wheel, the heavy spot movement turns the left front wheel outward **(Figure 5).**

From these explanations, we can understand that dynamic wheel unbalance causes lateral wheel shake or shimmy. This action causes steering wheel oscillations at medium and high speeds with resultant driver fatigue and passenger discomfort. Wheel shimmy and steering wheel oscillations also cause unstable directional control of the vehicle.

A wheel with dynamic unbalance is forced to pivot on the contact area, resulting in excessive tire scuffing and wear. Dynamic wheel unbalance causes premature wear on steering linkage and suspension components. Therefore, dynamic wheel balance is extremely important to provide normal tire life, reduce steering and suspension component wear, increase directional control, and decrease driver fatigue.

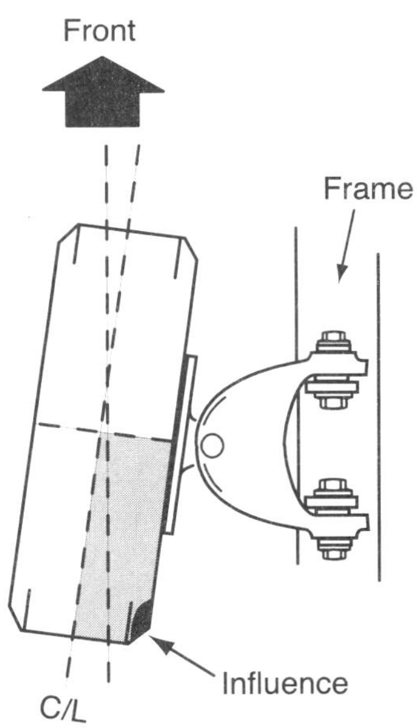


**Figure 4.** A heavy spot at the rear of a left front wheel.

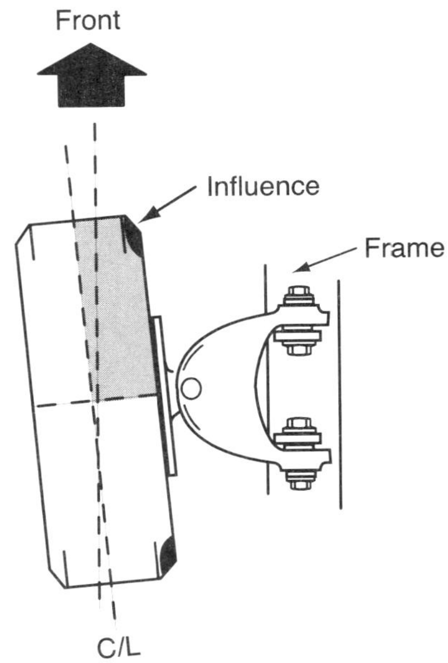


**Figure 5.** The effects of a heavy spot at the rear of a left front wheel.

## PRELIMINARY WHEEL BALANCING CHECKS

Before attempting to statically or dynamically balance a wheel-and-tire assembly, do the following:

1. Check for and remove any objects in tire tread and any mud collected on the inside of the wheel.
2. Inspect the tread and sidewall and replace the tire if necessary.
3. Inflate the tires to the manufacturer's recommended pressure.
4. Measure tire and wheel runout, rotate the tire on the wheel, or replace the wheel or tire as required.
5. If the tires will be balanced on the vehicle, check the adjustment of the wheel bearings.
6. Make sure all wheel weights are attached securely. If not, remove them.
7. Look for tire chunking (a buildup of rubber at one spot in the tread) or tread cupping. If these exist, replace the tire.
8. On the type of wheel balancer that spins the tire and wheel at high speed during the dynamic balance procedure, always attach the tire-and-wheel assembly securely to the balancer. Follow the equipment manufacturer's recommended wheel-mounting procedure. If the tire-and-wheel assembly becomes loose on the balancer at high speed, serious personal injury or property damage may result.
9. Prior to spinning a tire and wheel at high speed on a wheel balancer, make sure the protection shield is over

the tire. This shield provides protection in case anything flies off the tire or wheel.

## Wheel Weights

During the balancing process, weights are added to the wheel. Many **wheel weights** have a spring clamp **(Figure 6)** that holds the weight on the edge of the rim. A special pair of wheel weight pliers is used to tap the wheel weight between the tire bead and rim and to remove a weight from the rim **(Figure 7).** Magnesium or aluminum wheels may require the use of adhesive wheel weights **(Figure 8).** This type of weight must also be used on some steel wheels if the wheel covers will interfere with the conventional weights. Regardless of the type of wheel weight, they must be attached securely to the wheel.

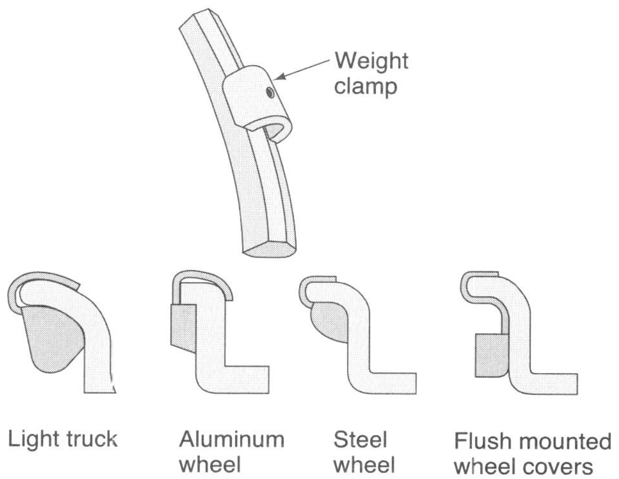


**Figure 6.** The various ways wheel weights are installed on wheels.

**Figure 7.** Wheel weights are installed and removed with special wheel weight pliers.

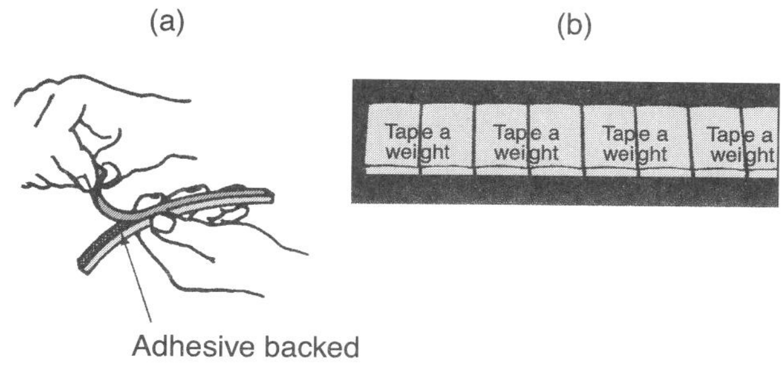


**Figure 8.** Adhesive lead wheel weights.

## STATIC WHEEL BALANCE PROCEDURE

Many different types of wheel balancers are available. The exact wheel balance procedure may vary depending on the type of wheel balancer. One of the simplest types of wheel balancers for static balancing is the bubble balancer. The electronic wheel balancer is the most commonly used **(Figure 9).** With an electronic balancer, the operator must

**Figure 9.** An off-the-vehicle electronic wheel balancer.

enter the wheel diameter, width, and offset. The balancer must have this information to perform its calculations. These balancers are also used to dynamically balance the wheel and tire assembly.

On some electronic wheel balancers during the static balance procedure, the wheel is allowed to rotate by gravity. A heavy spot rotates the tire until this spot is at the bottom. The necessary static balance weights are then added at the top of the wheel 180 degrees from the heavy spot. When the wheel-and-tire assembly is balanced statically, gravity does not rotate the wheel from the at-rest position. Rotate the tire by hand, and check the static balance at 120-degree intervals.

## Bubble Balancer

A **bubble balancer** has a movable flange that rests on a point. A round sight glass filled with liquid containing a bubble rides with the flange **(Figure 10).** The bubble indicates whether the flange is level. When the tire/wheel assembly is placed on the flange; any static imbalance makes it tilt and moves the bubble off center. Wheel balance weights are then placed on the wheel where the bubble points to counteract the imbalance and center the bubble on the sight glass. When balance weights are used in pairs, the total balance-correcting weight is divided in half. One half is installed on the back of the wheel, the other half on the front. This weight-splitting method leaves the wheel's dynamic balance unchanged. Many manufacturers recommend static balancing a wheel at equal distances from the center of the light area.

**Figure 10.** A bubble-type wheel balancer.

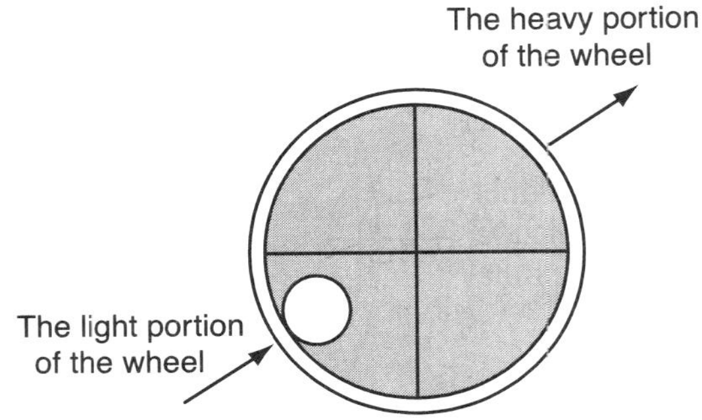


**Figure 11.** The bubble in the balancer's sight glass moves in the opposite direction of the heavy spot on the wheel.

Typically, balancing a tire and wheel on a bubble balancer involves the following steps:

1. Complete the preliminary checks outlined previously.
2. Adjust the legs on the balancer to center the bubble.
3. Position the wheel securely on the balancer, and align the triangular mark with the valve stem on the wheel.
4. Observe the spirit level on the balancer to determine the heavy spot and the light spot on the wheel. The light spot will be directly opposite the heavy spot **(Figure 11).**
5. Place enough weight at the light spot until the spirit level indicates a balanced wheel.
6. Chalk mark from the weight position on the sidewall to the inside bead.
7. Remove the wheel from the balancer and divide the weight by two.
8. Install the required weights on the rim at the chalk-marked positions. The combined weight of the wheel weights must equal the original weight installed in step 5.
9. Reinstall the wheel on the balancer and check for proper balance.

# DYNAMIC WHEEL BALANCE PROCEDURE

The most commonly used dynamic wheel balancer requires that the tire/wheel assembly be taken off and mounted on the balancer's spindle. The machine spins the entire assembly to indicate the heavy spot with a strobe light or other device. Two tests must be done, one for static and one for dynamic imbalance. One set of weights is placed to correct for static imbalance, and others are placed to correct for dynamic imbalance. Sometimes proper positioning of the static balance weights also corrects dynamic imbalance.

When the heavy spot is located on the inside edge of the tread, a correct weight installed on the inside of the

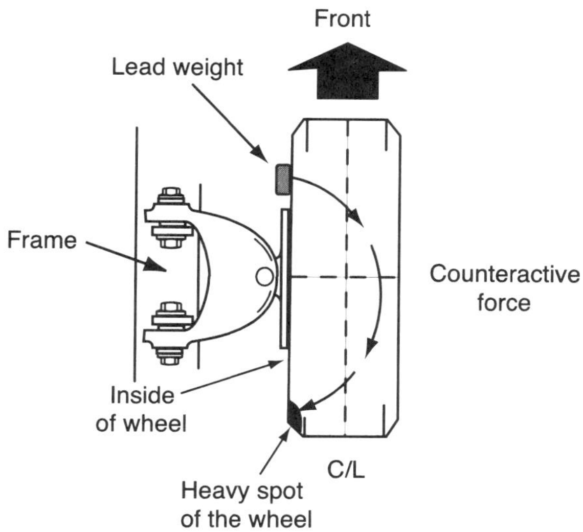


**Figure 12.** Dynamic wheel balance with the heavy spot of the tire on the inside.

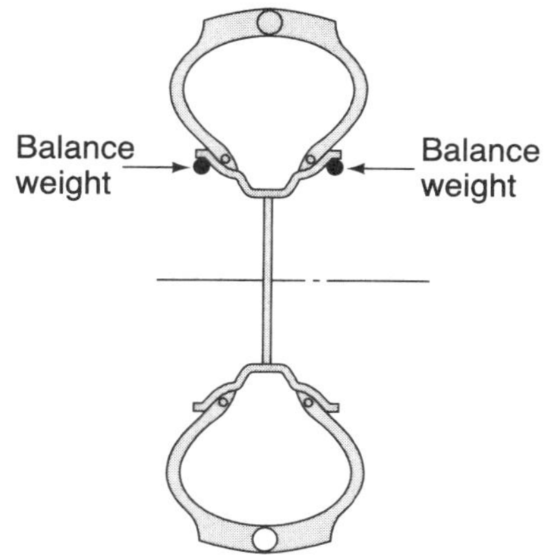


**Figure 13.** The required amount of weight is often divided into two and an equal amount of weight installed on the inside and outside of the wheel.

wheel 180 degrees from the heavy spot provides proper dynamic balance **(Figure 12).** When the heavy spot is on the outside edge of the tread, the correct size wheel weight installed at 180 degrees from the heavy spot on the inside of the wheel provides dynamic balance.

Some dynamic wheel balance procedures require the installation of wheel weights 180 degrees from the heavy spot, and an equal weight on the opposite side of the wheel 180 degrees from the first weight. This second weight maintains static balance **(Figure 13).**

## Off-Car Wheel Balancing

To balance a wheel and tire assembly on a tire balance machine, the assembly must be removed and mounted to the machine. Most electronic wheel balancers have an electric drive motor that spins the tire at high speed. However, with some electronic wheel balancers, the tire is spun by hand during the balance procedure. The electronic-type balancer performs static and dynamic balance calculations simultaneously and indicates the correct weight size and location. The tire-and-wheel assembly must be fastened securely on the balancer with the correct size adapters. Off-car electronic high-speed wheel balancers have a safety hood that must be positioned over the tire prior to dynamic balancing. Wheel balancers must always be operated according to the manufacturer's instructions.

A typical procedure for balancing a wheel and tire assembly includes:

1. Prepare the tire and wheel assembly by performing the preliminary checks.
2. Mount the tire-and-wheel assembly on the wheel balancer with the recommended mounting components **(Figure 14).**

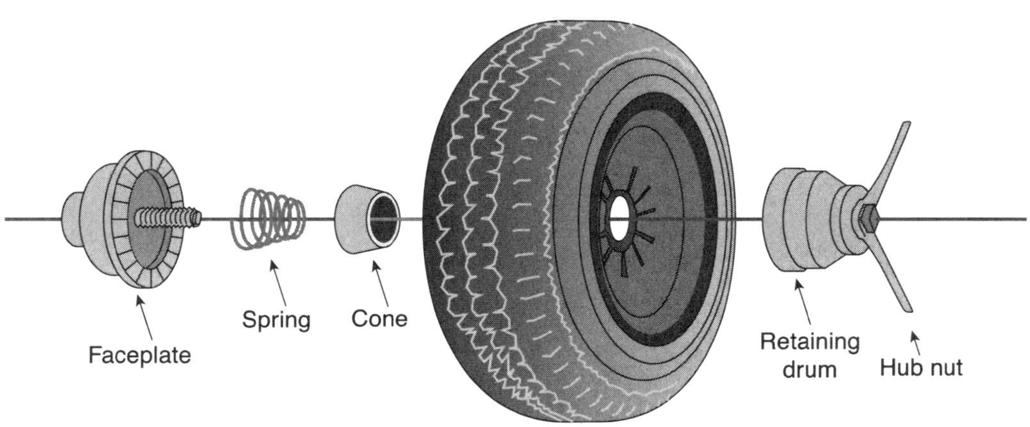


**Figure 14.** A typical wheel mounting assembly for a balancer.

3. Tighten the tire and wheel retaining nut on the wheel balancer to the specified torque.
4. Remove all the old wheel weights from the wheel.
5. Lower the safety hood on the wheel balancer.
6. Enter the required information on the wheel balancer keypad such as wheel diameter, wheel width, and wheel offset or distance.
7. If the wheel balancer has a mode selector, select the desired balance mode.
8. Spin the wheel on the balancer, and then apply the balancer brake to stop the wheel.
9. Observe the balancer display indicating the necessary amount of wheel weights and the required location of these weights. Install the required wheel weights in the locations indicated on the balancer. Be sure the wheel weights are securely attached to the rim.
10. Spin the tire-and-wheel assembly on the balancer again to be sure it is properly balanced.

## On-Car Wheel Balancing

There are several electronic dynamic/static balancer units available that will permit balancing while the wheel and tire are on the vehicle. A switch on the console sets the machine for either static or dynamic balancing. When the wheel balancing assembly is mounted for static balancing, it rotates until the heavy spot falls to the bottom. Weights are added to balance the assembly. In the dynamic balance mode, the wheel assembly is rotated at high speed. Observing the balance scale, the operator reads out the amount of weight that has to be added and the location where the weights should be placed.

On-car wheel balancers **(Figure 15)** contain a drum driven by an electric motor. This drum is positioned against the tire on the vehicle, which allows the electric motor to rotate the wheel. Many on-car balancers have a strobe light with a meter and an electronic vibration sensor. Off-car wheel balancing and on-car wheel balancing are a complementary combination for fine-tuning wheel balance. For example, a wheel vibration problem may still exist after an off-car balance procedure. If this problem occurs, the on-car balancer may be used to correct the problem. The on-car balance procedure corrects imbalance problems in all rotating components, including brake drums or rotors. When balancing a front wheel on a rear-wheel-drive vehicle or a rear wheel on a front-wheel-drive vehicle, follow this procedure:

**Figure 15.** An on-car wheel balancer.

1. Perform the preliminary checks to prepare the wheel and assembly for balancing.
2. Raise the wheel being balanced 5 inches (12 cm) off the floor, and be sure the chassis is supported on safety stands so the wheel drops downward.
3. Install the electronic vibration sensor between the lower control arm and the floor.
4. Chalk mark a reference mark on the outer sidewall of the tire.
5. Spin the wheel just fast enough to produce vibration on the front bumper.
6. When vibration causes strobe light flashing, move the balancer drum away from the tire and allow the tire to spin freely.
7. Shine the strobe light around the tire sidewall and note the chalk-mark position.
8. Note the pointer position on the meter.
9. Use the balancer's brake plate to slow and stop the wheel.
10. Rotate the wheel until the chalk mark is in the exact position as it appeared under the strobe light. The heavy spot is now at the bottom of the wheel, and the balancing weight should be attached 180 degrees from that spot. Install the weight indicated on the meter.
11. Spin the wheel again. If the wheel balance is satisfactory, the meter pointer will read in the balanced position. When the pointer does not indicate a balanced wheel, shine the strobe light on the tire sidewall. If the installed wheel weight is at the 12 o'clock position, additional weight is required, whereas the 6 o'clock weight position indicates excessive weight. A 3 o'clock or 9 o'clock weight position may be corrected by moving the weight 1 inch (2.5 cm) toward the 12 o'clock position.

**You Should Know** *To obtain proper wheel balance, spin the wheel only to the speed where the vibration occurs. Do not spin the wheel at excessive speeds.*

### Rear Wheel Balancing

If an on-car balancer is used on the rear wheels of a rear wheel drive vehicle or the front wheels of a front wheel drive vehicle, determine if the vehicle has a conventional or a limited slip differential. With the transmission in park or in gear with a manual transmission, rotate one rear wheel by hand. If the vehicle has a limited slip differential, the rear wheels will not rotate, whereas a free-turning wheel indicates a conventional differential.

When the vehicle has a conventional differential, raise only the wheel to be balanced, and do not allow the indicated speed to exceed 35 mph (56 km/h). When the other tire on the same axle is resting on the floor, the speed of the wheel being balanced is actually 70 mph (112 km/h).

If the vehicle has a limited slip differential, raise the vehicle and place safety stands under the frame to support the vehicle's weight. Lift the axle housing with a floor jack to reduce the universal and/or CV joint angles, but do not take the vehicle's weight off the safety stands. Remove the wheel that is not being balanced. Install and torque the wheel's lug nuts on the brake disc to hold it in place. Balance the opposite wheel. When that wheel is balanced, leave it on the vehicle and install and balance the other wheel. Be sure to torque the lug nuts properly. Do not allow wheel speed to exceed 55 mph (90 km/h) on the speedometer.

**You Should Know** ***Do not spin the front wheels on a front wheel drive vehicle with the floor jack under the chassis and the suspension dropped downward. Under this condition, severe angles exist in the front drive axle joints, and the joints may be damaged if the wheels are rotated with the balancer. Place the floor jack under the lower control arm to raise the wheel.***

## SUMMARY

- There are two types of wheel balancing: static balance and dynamic balance.
- Static wheel unbalance causes wheel tramp and severe tire cupping.
- Tire tramp may be defined as rapid upward and downward wheel and tire oscillations.
- Dynamic wheel unbalance causes wheel shimmy, increased tire wear, unstable directional control, driver fatigue, and increased wear on suspension and steering components.
- Wheel shimmy may be defined as rapid inward and outward oscillations of a front wheel.
- Many wheel weights have a spring clamp and are installed between the tire bead and rim.
- Magnesium or aluminum wheels may require the use of adhesive wheel weights.
- A bubble balancer may be used to statically balance a tire.
- Electronic tire balancers can statically and dynamically balance a tire.
- The most common wheel balancer requires that the tire/wheel assembly be taken off the vehicle, although some balancers are designed to work with the wheel and tire on the vehicle.

## REVIEW QUESTIONS

1. Define dynamic and static wheel balance.
2. What type of balance is checked on a bubble balancer?
3. All of these statements about improper wheel balance are true *except:*
   A. Dynamic imbalance may cause wheel shimmy.
   B. Dynamic imbalance may cause steering pull in either direction.
   C. Static imbalance causes wheel tramp.
   D. Static imbalance causes rapid wear on suspension components.
4. Dynamic wheel unbalance causes:
   A. cupped tire tread wear.
   B. even wear on one edge of the tire tread.
   C. even wear on the center of the tire tread.
   D. feathered tire tread wear.
5. While balancing tires, Technician A removes all of the old weights before spinning the tire on a balancing machine. Technician B says if mounted correctly, new tires do not need to be balanced. Who is correct?
   A. Technician A only
   B. Technician B only
   C. Both Technician A and Technician B
   D. Neither Technician A nor Technician B

# Section 4

## Wheel and Axle Bearings



**Interesting Fact**

*The first patent application in the United States for a front-wheel-drive vehicle was made by George Selden in 1879. His gasoline buggy had a three-cylinder engine mounted over the front wheels. This made sense since that is where horses were attached to buggies.*

## SECTION OBJECTIVES

After you have read, studied, and practiced the contents of this section, you should be able to:

- Explain the purposes of a bearing.
- Describe three different types of bearing loads.
- Identify the basic parts in ball bearings or roller bearings, and describe the purpose of each part.
- Explain the purposes of bearing snaprings, shields, and seals.
- Describe the components of a wheel bearing assembly.
- Describe three different types of bearing hub units.
- Describe two different types of rear axle bearings in rear-wheel-drive cars, and give the seal location for each bearing type.
- Describe the construction and operation of a rear axle assembly.
- Diagnose wheel and axle bearing problems.
- Name and describe the components of a front-wheel-drive axle.
- Describe the operation of a front-wheel-drive axle.
- Explain the purposes of a FWD car's drive axles and joints.
- Describe the different methods used by manufacturers to offset torque steer.
- Conduct a safe and effective road test to identify axle and joint problems.
- Diagnose rear axle noise, vibration, and fluid leakage problems.
- Diagnose axle shaft and CV joint noise and vibration problems.
- Inspect, service, and replace shaft, boots, and CV joints.
- Perform preventive maintenance on CV joints.
- Remove, inspect, and service or replace front and rear wheel bearings.
- Service wheel bearings, including removal, cleaning, repacking, and installation.
- Remove, inspect, and service or replace front and rear wheel bearings.
- Inspect, service, and replace FWD front wheel bearings and their hubs.
- Measure shaft end play/preload (shim/spacer selection procedure).
- Measure rear axle flange runout and shaft end play.
- Inspect and replace rear axle shaft wheel studs.
- Remove and replace rear axle shafts.
- Inspect and replace rear axle shaft seals, bearings, and retainers.

# Chapter 15 Wheel Bearings

## INTRODUCTION

The purpose of all bearings is to allow a shaft to rotate smoothly in a housing or to allow the housing to rotate smoothly around a shaft. The bearings on an axle that drives the wheels are called *axle bearings*. The wheel is mounted to the hub of an axle shaft, and the shaft rotates within a housing. Wheel bearings are used on non-driving axles. The wheel's hub rotates on a shaft called the spindle. Axle bearings are typically serviced with the drive axle. Wheel bearings, however, require periodic maintenance and are often serviced with suspension and brake work. Although there is a distinction between axle and wheel bearings, the bearings for the front wheels on a FWD and 4WD vehicle are commonly called *wheel bearings* **(Figure 1).** Regardless of what they are called, bad bearings will cause handling and tire wear problems.

Many different types of bearings are used in the automobile. Bearings reduce friction; they also decrease the power required to rotate or move the component. Bearings are machined assemblies that provide smooth operation and long life. When bearings are properly installed and maintained, bearing failure is rare.

### COMMON BEARINGS

The simplest type of **bearing** is a cylindrical hole formed in a piece of material into which the shaft fits freely. The hole is usually lined with a brass or bronze lining, or **bushing,** that not only reduces the friction but also allows

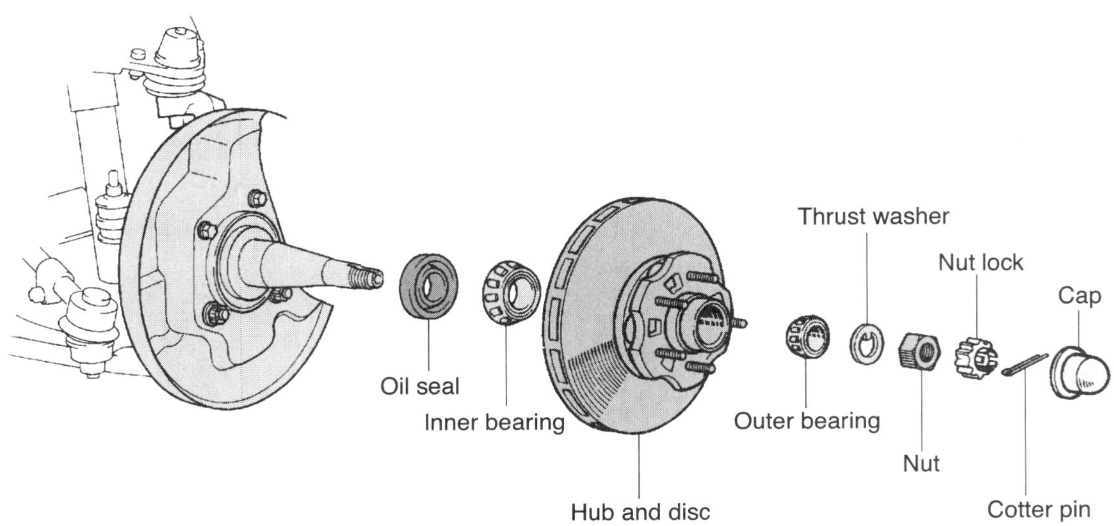


**Figure 1.** A typical wheel bearing assembly.

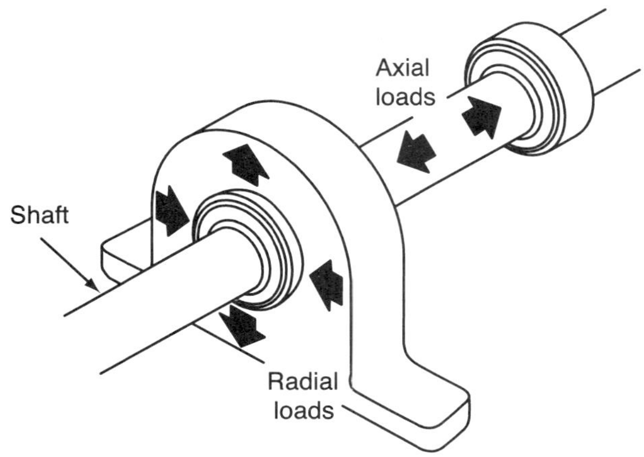


**Figure 2.** The various loads a bearing may need to withstand.

for easy replacement when wear occurs. Bushings usually have a tight fit in the hole in which they fit.

Ball or roller type bearings are used wherever friction must be minimized. Typically, two bearings are used to support a shaft instead of a long bushing. Bearings have three purposes: they support a load **(Figure 2),** maintain alignment of a shaft, and reduce rotating friction.

Most bearings are capable of withstanding only loads, called **radial loads,** that are perpendicular to the axis of the shaft.

To prevent the shaft from moving in the **axial** direction, shoulders or collars may be formed on it or secured to it. If both collars are made integral with the shaft, the bearings must be split or made into halves. The top half or cap is then bolted in place after the shaft has been put in place. Bearings designed this way are termed **thrust bearings.**

A single-row radial **ball bearing** has an inner race made of a ring of case hardened steel with a groove or track formed on its outer circumference for a number of hardened steel balls to run upon. The outer race is another ring that has a track on its inner circumference. The balls fit between the two tracks and roll around in the tracks as either race turns. The balls are kept from rubbing against each other by some form of cage. These bearings can withstand radial loads and can also withstand a considerable amount of axial thrust.

Another type of ball bearing uses two rows of balls. These are designed to withstand considerable amounts of radial and axial loads. Constructed like two single row ball bearings joined together, these bearings are often used in rear axle assemblies.

**Roller bearings** are used wherever there is a large bearing surface and low amount of friction. Large bearing surfaces are needed in areas of extremely heavy loads. The rollers are usually fitted between a journal of a shaft and an outer race. As the shaft rotates, the rollers turn and rotate in the race. Tapered roller bearings are commonly used in drive axle assemblies.

## Ball Bearings

Wheel bearings may be ball or roller bearings with these basic parts **(Figure 3):**

1. Inner race or cone
2. Rolling elements, balls, or rollers
3. Separator, also called a cage or retainer
4. Outer race or cup

The inner race is accurately machined, and its inner surface is mounted on the shaft with a precision fit. The rolling elements are mounted on a very smoothly machined surface on the inner race. Positioned between the inner and outer races, the separator retains the rolling elements and keeps them evenly spaced. The rolling elements are mounted between the inner and outer races. The outer race is the bearing's exterior ring. The outer surface of this race supports the bearing in the housing, and the inner surface is in contact with the rolling elements.

A single-row ball bearing has a crescent-shaped surface in the inner and outer races in which the balls are mounted. When a ball bearing is at rest, the load is distributed equally through the balls and races in the contact area. When one of the races and the balls begin to rotate, the bearing load causes the metal in the race to bulge out in front of the ball and flatten out behind the ball. This action creates a certain amount of friction within the bearing, and the same action is repeated for each ball while the bearing is rotating. If metal-to-metal contact is allowed between the balls and races, very fast wear results. Therefore, bearing lubrication is extremely important to eliminate metal-to-metal contact in the bearing and reduce wear.

Double-row ball bearings contain two rows of balls side by side. As in the single-row ball bearing, the balls are

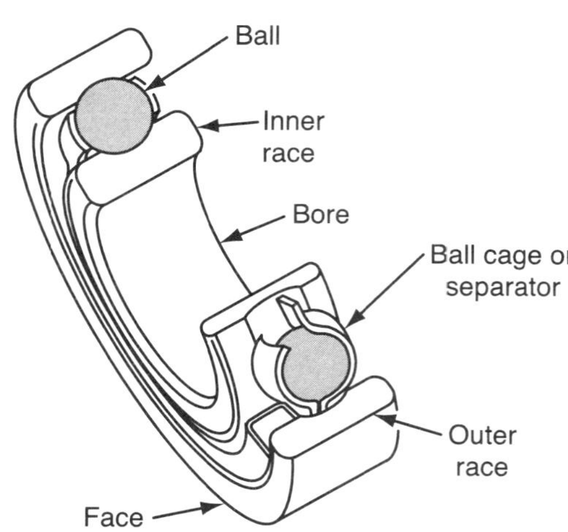


**Figure 3.** Parts of a ball bearing.

mounted in crescent-shaped grooves in the inner and outer races.

## Cylindrical Roller Bearings

> **You Should Know** *Spinning a roller bearing with compressed air may rotate the bearing at extremely high speed and cause the bearing to disintegrate, resulting in serious personal injury.*

A cylindrical roller bearing contains rollers that have the same diameter at both ends. These rollers are mounted in square-cut grooves in the outer and inner races **(Figure 4).** In a cylindrical roller bearing, the races and rollers run parallel to one another. Cylindrical roller bearings are designed primarily to carry radial loads but can withstand some thrust load. Since cylindrical ball bearings do not withstand high thrust loads, they are not used in wheel bearings.

## Tapered Roller Bearings

In a **tapered roller bearing,** the inner and outer races are cone shaped. If imaginary lines extend through the inner and outer races, these lines taper and eventually meet at a point extended through the center of the bearing **(Figure 5).** These bearings have a capability to carry radial, thrust, and angular loads. In the tapered roller bearing, the bearing separator has an open space over each roller. Grooves cut in the side of the separator roller openings match the curvature of the roller. This design allows the rollers to rotate evenly without interference between the rollers and the separator. Lubrication and proper endplay adjustment are critical.

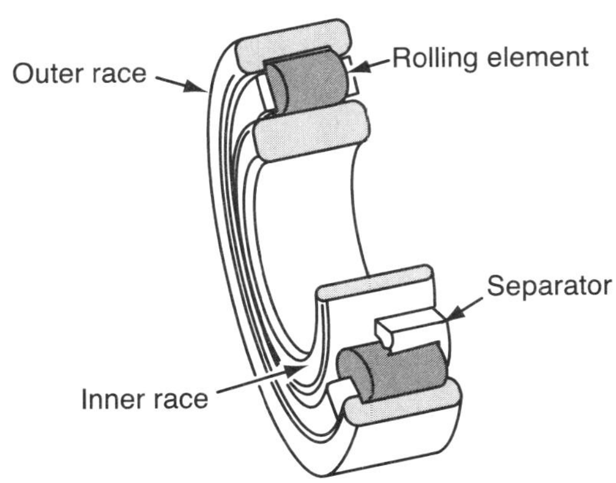


**Figure 4.** Parts of a cylindrical roller bearing.

**Figure 5.** An assortment of tapered roller bearings.

**Figure 6.** A needle roller bearing.

## Needle Roller Bearings

A **needle roller bearing** contains many small-diameter steel rollers in a thin outer race. This type of bearing is very compact, and it is used in steering gears where mounting space is limited. Most needle roller bearings do not have a separator, but the steel rollers push against each other and maintain the roller position. Rather than having an inner race, a machined surface on the mounting shaft contacts the inner surface of the rollers **(Figure 6).** The needle roller bearing is designed to carry radial loads; it does not withstand thrust loads..

# FWD DRIVE AXLES

Because the engine of FWD vehicles is located right above the drive wheels, the drive axles for these vehicles are quite different from those for RWD vehicles. The only exception is when the rear wheels drive the vehicle but the engine is not in the front of the vehicle.

The drive shaft and its joints are often called the driveline. The driveline transmits torque from the transmission to the driving wheels. RWD cars use a long drive shaft that connects the transmission to the rear axle. The engine and driveline of FWD cars are located between the front driving wheels.

## Drive Axle Construction

Two drive shafts or axles are the foundation of a basic FWD driveline. The shafts extend from each side of the differential in the transaxle to supply power to the drive wheels. Typically, the drive axles are solid bars of steel. In all FWD and some 4WD systems, the transaxle is bolted to the

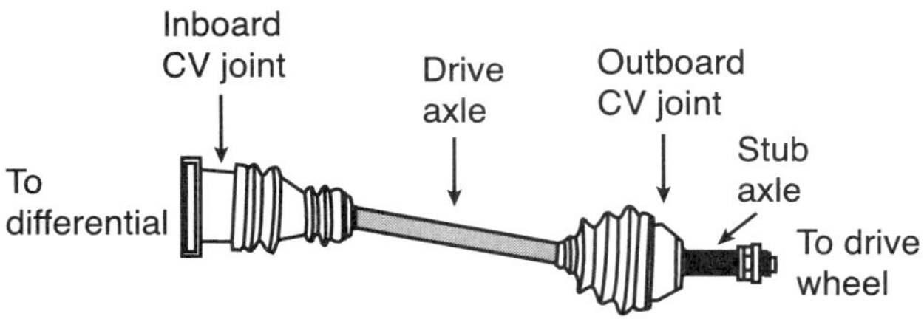


**Figure 7.** The parts of a half-shaft for a FWD vehicle.

chassis, and the axles pivot on **constant velocity (CV) joints.** The outer part of each axle is supported by the steering knuckle, which houses the axle bearings.

The drive axle assembly is comprised of three pieces attached together to allow the wheels to turn and move with the suspension. The axle shaft is connected to the differential by an inboard CV joint. The axle shaft then extends to the outer CV joint. A short spindle shaft, called a stub axle, runs from the outer CV joint to mate with the wheel assembly **(Figure 7).** The drive axle assembly with the inner and outer CV joints is typically called a **half-shaft.**

## RWD DRIVE AXLES

The rear axle housing encloses the complete rear wheel driving axle assembly. In addition to housing the parts, the axle housing also serves as a place to mount the vehicle's rear suspension and braking system. The rear axle assembly changes the direction of the power flow 90 degrees, acts as the final gear reduction unit, and sends torque to the rear drive wheels.

Located within the hollow horizontal tubes of the axle housing are the axle shafts **(Figure 8).** An axle shaft transmits the driving force from the differential to the drive wheels. Axle shafts are heavy steel bars splined at the inner end to mesh with a side gear in the differential. The driving wheel is bolted to the wheel flange at the outer end of the axle shaft.

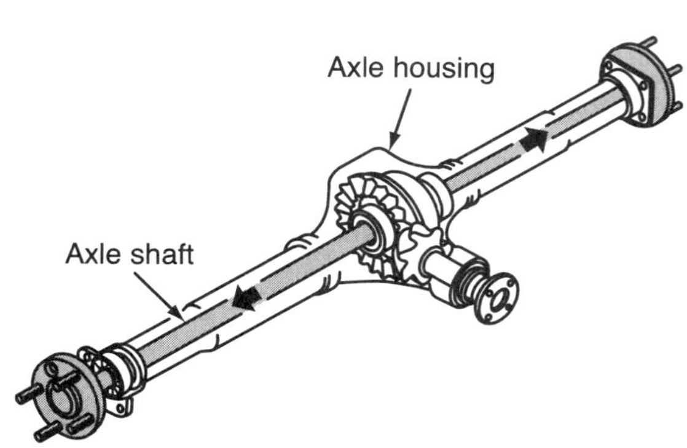


**Figure 8.** The axle shafts move torque from the differential unit to the wheels.

### Types of Axle Shafts

There are basically three designs by which drive axles are supported in a live axle: full floating, three quarter-floating, and semi-floating. These refer to where the axle bearing is placed in relation to the axle and the housing and how much of the vehicle's weight is put on the drive axle. The bearing of a full-floating axle is located on the outside of the housing. This places all of the vehicle's weight on the axle housing with no weight on the axle.

Three-quarter and semi-floating axles are supported by bearings located in the housing and carry some of the vehicle's weight. Most passenger cars have three-quarter or semi-floating axles. Full-floating axles are commonly found on heavy-duty trucks.

The axle shaft bearing supports the vehicle's weight and reduces rotational friction. With semi-floating axles, radial and thrust loads are always present on the axle shaft bearing when the vehicle is moving. Radial bearing loads act at 90 degrees to the axle's center of axis of rotation. Radial loading is always present whether the vehicle is moving or not. Thrust loading acts on the axle bearing parallel with the center of axis of rotation. It is present on the driving wheels, axle shafts, and axle bearings when the vehicle turns corners or curves.

## AXLE BEARING DESIGNS

There are three designs of axle shaft bearings used on semi-floating axles: ball-type, straight-roller, and tapered-roller bearings **(Figure 9).** When a vehicle moves around a corner, centrifugal force acts on the vehicle's body causing it to lean to the outside of the curve. The vehicle's chassis does not lean because of the tire's contact with the road's surface. As the body leans outward, a thrust load is placed on the axle shaft and axle bearing. Each type of axle shaft and bearing assembly handles axle shaft end thrust differently.

The end-to-end movement of the axle is controlled by a C-type retainer on the inner end of the axle shaft or by a bearing retainer and plate at the outer end of the axle shaft.

### Ball-Type Axle Bearings

An axle with ball-type bearings has the axle shaft and bearing assembly held in place in the axle housing by a stamped metal bearing retainer plate bolted to the housing. The bearing is held in place on the axle shaft by a retaining ring that is pressed onto the axle shaft.

A ball-type bearing is designed to absorb radial load as well as axle shaft end thrust. Because both loads are taken at the bearing, there is no axle shaft end thrust absorption or adjustment designed into the rear axle housing.

To seal in the lubricant, an oil seal collar and axle oil seal is used. The oil seal collar is a machined sleeve or finished portion of the axle on which the lips of the seal ride. The axle seal retains the lubricant inside the axle housing and prevents it from leaking into the brakes.

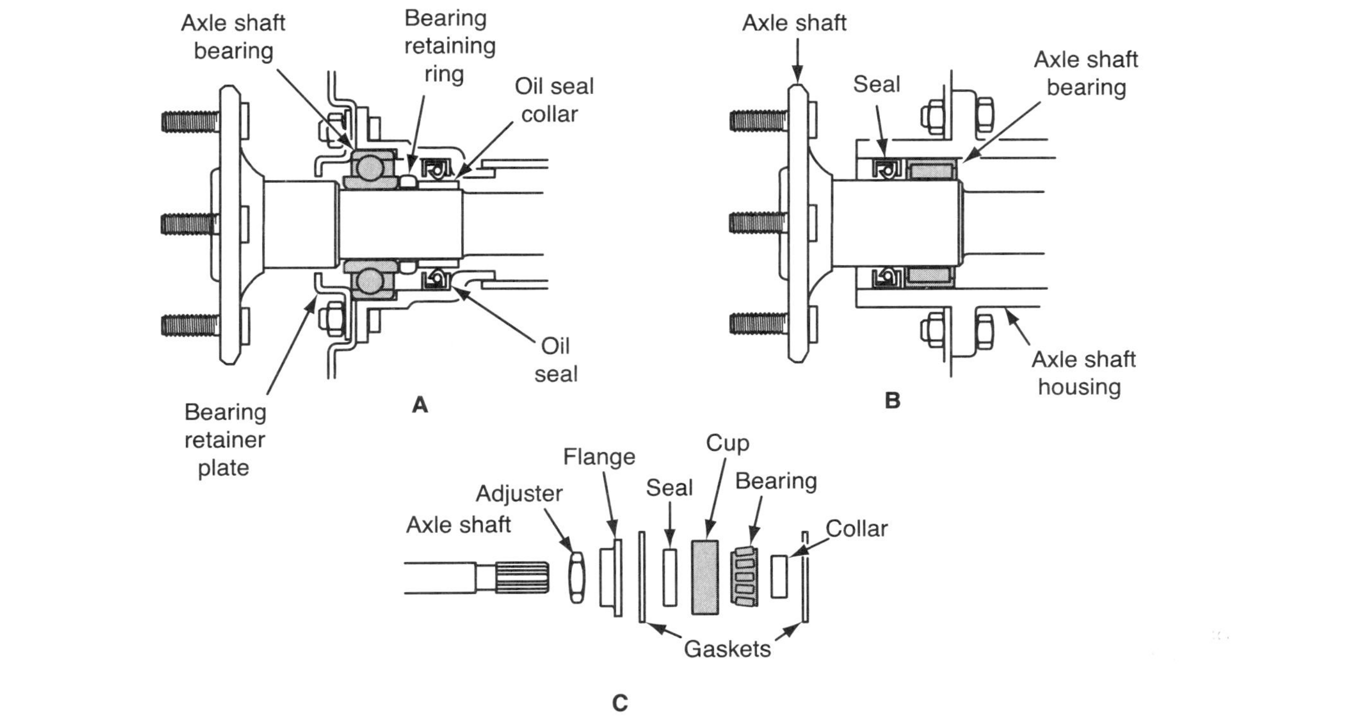


**Figure 9.** There are three designs of axle shaft bearings used on semi-floating axles: (A) ball-type, (B) straight-roller, and (C) tapered-roller bearings.

## Straight Roller Axle Bearings

A straight roller bearing uses the axle shaft as its inner race. The outer bearing race and straight rollers are pressed into the axle tubes of the housing. The inner end of the axle shaft at the differential has a groove machined around its outside diameter where the C-type retainer fits.

The bearing is lubricated by oil in the differential area of the axle housing. A grease seal located outside the axle shaft bearing prevents lubricant from leaking out of the housing.

## Tapered Roller Axle Bearings

The tapered roller bearing and axle shaft assembly are held inside the axle housing by a flange that is bolted to the axle housing. The inside of the flange maybe threaded to receive an adjuster or is machined to accept adjustment shims.

The axle shaft is designed to slightly float with the in-and-out movements of the axles. As the axle shaft moves inward, it contacts a thrust block that separates both axle shafts at the center of the differential. The inward moving axle shaft contacts the thrust block, which passes the thrust force onto the opposite axle shaft. There the axle shaft end thrust becomes an outward moving force that causes the opposite tapered roller axle bearing to seat in its bearing cup. Axle-end thrust adjustments can be made by a threaded adjuster or thin metal shims placed between the brake assembly plate and the axle housing.

The tapered roller bearing is lubricated prior to installation in the axle housing. A seal and two gaskets keep the hypoid lubricant and foreign matter out of the bearing operating area. A collar holds the rear axle bearing in place on the axle shaft.

## Independent Rear Suspension (IRS) Axle Shafts

The drive axles on most new IRS systems use two U- or CV joints per axle to connect the axle between the differential and the wheels **(Figure 10).** They are also equipped

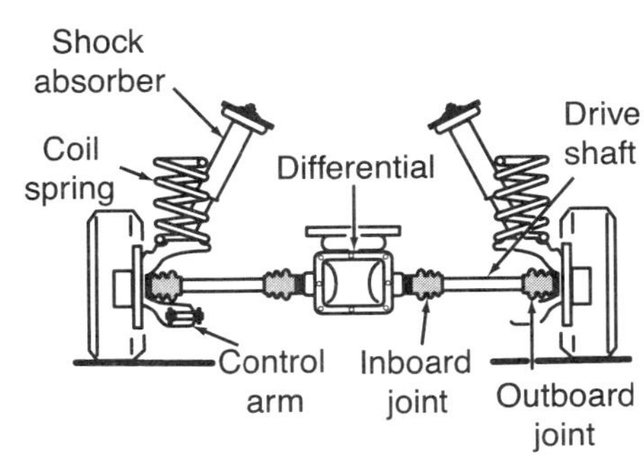


**Figure 10.** The drive axles on most new IRS systems use two CV joints per axle to connect the axle between the differential and the wheels.

with linkages and control arms to limit camber changes. The axles of an IRS system are much like those of a FWD system. An upright or locating member that is also part of the suspension supports the outer portion of the axle.

Swing axles are a form of IRS. In a swing axle car, the differential is bolted to the chassis and the axles are U-jointed to where they meet the differential but not at the wheel end. This makes the wheels move in an arc as they move up and down with the suspension. This causes large camber changes and raises and lowers the rear of the vehicle as it moves over changing road surfaces. These motions make driving difficult which is why swing axles are not found on many vehicles. Swing axles were popular from the late 1940s through the 1960s because they were inexpensive to make and provided a softer ride than a solid live axle.

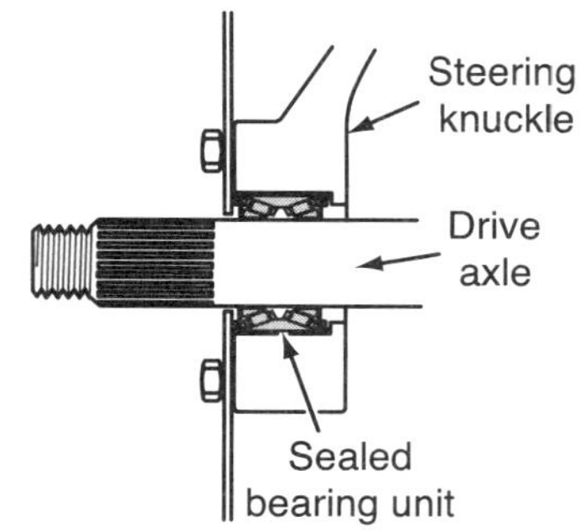


**Figure 12.** Wheel bearings support the drive axles in the steering knuckle on FWD vehicles.

## WHEEL BEARINGS

Adjustable tapered roller bearings **(Figure 11)** used on non-driven wheels must be adjusted correctly. If the bearings are too tight, the hub and brake assembly may overheat with accompanying problems that include steering pull and brake fade. In the worst cases, the bearings may seize on the spindle due to overheating.

If wheel bearings are too loose, wheel runout may be excessive. With excessive runout comes a number of problems such as poor handling, tire wear, and steering and brake pull if the bearings are loose on only one side of the vehicle.

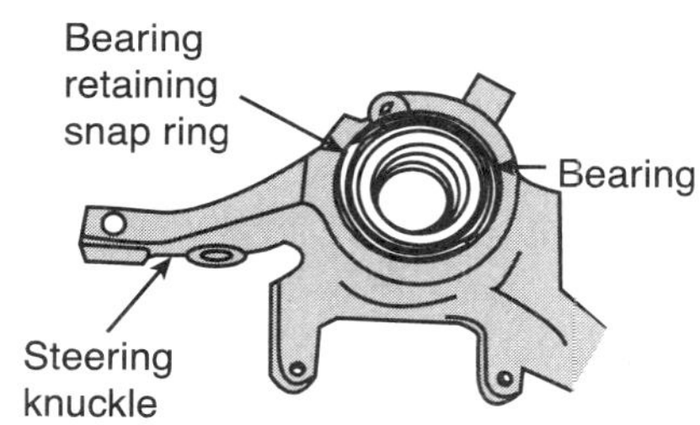


**Figure 13.** Most FWD front wheel bearings are one-piece sealed units that are pressed into the wheel hub assembly or held in the wheel hub by a snapring.

### FWD Wheel Bearings

Drive axles are supported in the steering knuckle by wheel bearings **(Figure 12).** These bearings allow the axle to rotate evenly and smoothly and keep the axle in the center of the steering knuckle's hub. There are two types of FWD wheel bearings. Most GM, Chrysler, and European cars use a double-row, angular-contact ball bearing. These units are simply two rows of ball bearings that are located next to each other. The races for these bearings are slightly offset to control radial loads during cornering. Ford and most Asian cars use opposed tapered roller bearings. Most FWD front wheel bearings are one-piece sealed units. The unit is typically pressed into the wheel hub assembly or is held in the wheel hub by a snapring **(Figure 13).**

On some cars, the bearing is part of a wheel hub assembly, while on others the bearing unit is pressed into the hub. On a few cars, the bearings are retained in the same way as the front wheel bearings on RWD vehicles. Some FWD vehicles have two tapered roller bearings in the rear wheel hubs that are similar to the front wheel bearings in RWD vehicles.

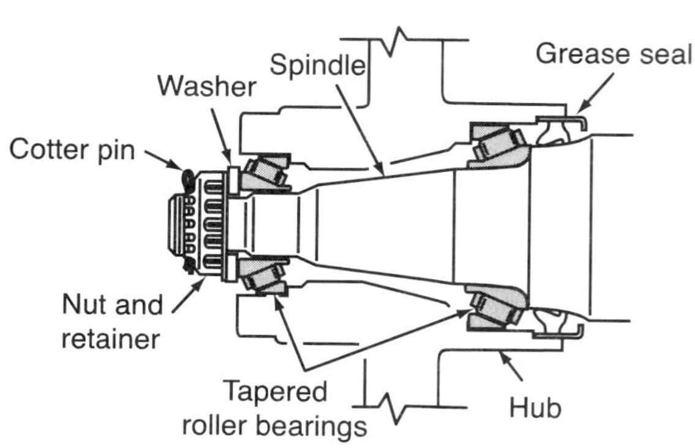


**Figure 11.** The end play of tapered roller bearings must be adjusted correctly in order to minimize tire wear and provide good vehicle handling.

### RWD Wheel Bearings

Many RWD vehicles have two tapered roller bearings in the front hubs to support the hubs and wheels. This type of bearing has its bearing races pressed into the hub. A grease seal is pressed into the inner end of the hub to prevent grease leaks and keep contaminants out of the bearings. The hub and bearing assemblies are retained by a washer, adjusting nut, nut lock, and cotter pin. The adjusting nut must be adjusted to provide the correct bearing end play. A grease cap is pressed into the outer end of the hub to prevent bearing contamination. The rear wheels are attached to axles, which are supported by axle bearings.

## SEALS

Seals are designed to keep lubricant in the bearing and prevent dirt and contaminants from entering the bearing. The metal seal case has a surface coating that resists corrosion and rust and acts as a bonding agent for the seal material. Seals have many different designs including single lip, double lip, and fluted. The seal material is usually a synthetic rubber compound such as nitrile, silicon, polyacrylate, or a fluoroelastomer such as Viton. The actual material depends on the lubricant and contaminants that the seal encounters. All seals may be divided into two groups, springless and spring loaded. **Springless seals (Figure 14)** are used in some front or rear wheel hubs.

In a spring-loaded seal, a **garter spring** provides additional force on the seal lip to compensate for lip wear, shaft movement, and bore eccentricity **(Figure 15).** A fluted lip seal may be used to direct oil back into a housing.

Some seals have a sealer painted on the outside surface of the metal seal housing. When the seal is installed, this sealer prevents leaks between the seal case and the housing.

**Figure 14.** A springless seal.

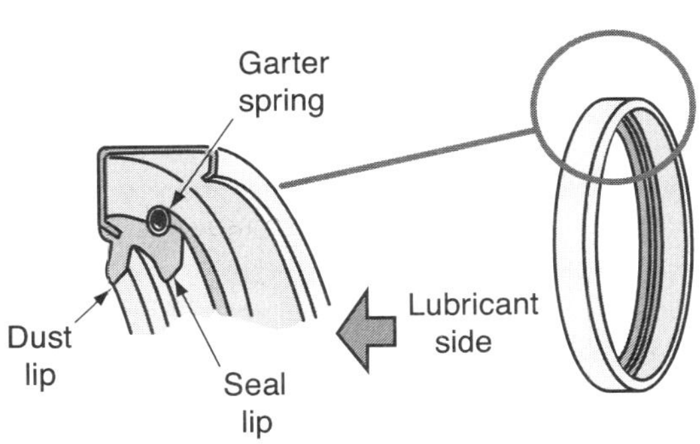


**Figure 15.** In a spring loaded seal, the garter spring behind the seal provides additional force on the seal lip to compensate for lip wear, shaft movement, and bore electricity.

**Bearing shields** cover the space between the two bearing races on one or both sides of the bearing. These shields are usually attached to the outer race, but space is left between the shield and the inner race. Bearing shields prevent dirt from entering the bearing, but excess lubrication can still flow through the bearing.

## BEARING LUBRICATION

Proper bearing lubrication is extremely important to maintain bearing life. Bearing lubricant reduces friction and wear, dissipates heat, and protects surfaces from dirt and corrosion. Sealed or shielded bearings are lubricated during the manufacturing process, and no attempt should be made to wash these bearings or pack them with grease.

Bearings that are not sealed or shielded require cleaning and repacking at intervals specified by the vehicle manufacturer. Always use the bearing grease specified by the vehicle manufacturer. Bearing lubricants may be classified as greases or oils. Many wheel bearings require lithium-based or sodium-based grease.

When rear axle bearings are lubricated from the differential housing, the type and level of oil in the housing are important. Vehicle manufacturers usually recommend SAE No. 90 or SAE No. 140 hypoid gear oil in the differential. In very cold climates, the manufacturer may recommend SAE No. 80 differential gear oil. The **American Petroleum Institute (API)** classifies gear lubricants as GL-1, GL-2, GL-3, GL-4, and GL-5. The GL-4 lubricant is used for hypoid gears under normal conditions. The GL-5 lubricant is used in heavy-duty hypoid gears. Always use the vehicle manufacturer's specified differential gear oil.

If the differential is overfilled, excessive lubricant may be present at the bearings and seals. Under this condition, the lubricant may leak past the seal. When the lubricant level is low in the differential, the lubricant may not be available in the axle housings. When this condition exists, the bearings do not receive enough lubrication, and bearing life is shortened.

### Wheel Bearing Maintenance

Often the front wheel hub bearing assembly for driven and non-driven wheels are two tapered bearings **(Figure 16)**

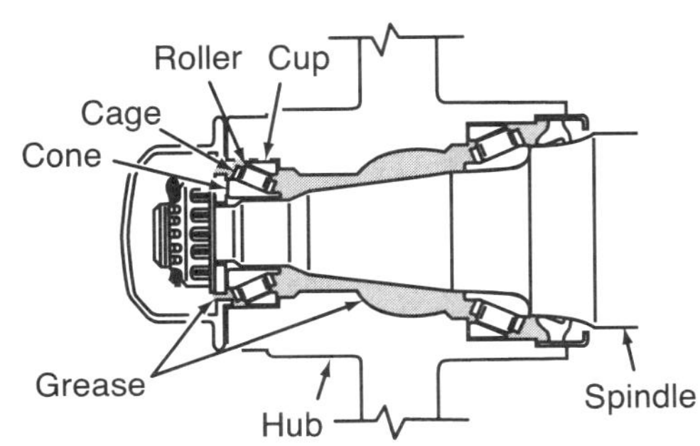


**Figure 16.** Tapered roller bearings must be cleaned and repacked with grease periodically and their end play adjusted.

facing each other. Tapered roller bearings are also used on the rear non-driven wheels of many vehicles. Tapered roller bearings must be cleaned and repacked with grease periodically and their end play adjusted. Straight roller or ball bearings are used on the drive axles of most rear-wheel-drive vehicles. Double-row ball bearings are used on the front hubs of many FWD cars. Straight roller bearings and ball bearings do not require periodic service or adjustment. Also, the front bearings often found on FWD and 4WD vehicles are sealed units and are lubricated for life. They are replaced and serviced as an assembly. Others are serviceable and require periodic lubrication and adjustment. The rear bearings on FWD vehicles are serviced in the same way as the non-driving front wheel bearings.

## SUMMARY

- A bearing reduces friction, carries a load, and guides components such as pivots, shafts, and wheels.
- Radial bearing loads are applied in a vertical direction.
- Thrust bearing loads are applied in a horizontal direction.
- An inner bearing race is positioned at the center and supports the rolling elements.
- The rolling elements in a bearing are positioned between the inner and outer races.
- The outer bearing race forms the outer ring on a bearing.
- Some bearing hub units are bolted to the steering knuckle; others are pressed into the steering knuckle.
- Some steering knuckles contain two separate tapered roller bearings.
- An axle shaft transmits the driving force from the differential to the drive wheels.
- Rear axle bearings are mounted between the drive axles and housing on RWD vehicles.
- Three types of drive axles in a live axle are used: full, three-quarter, and semi-floating.
- The drive axles on IRS systems have two U- or CV joints.
- Bearing seals keep lubricant in the bearing and prevent dirt from entering the bearing.
- The garter spring provides additional force on the seal lip to compensate for lip wear, shaft movement, and bore eccentricity.
- Flutes on seal lips provide a pumping action to direct oil back into a housing.

## REVIEW QUESTIONS

1. List the different ways drive axles are supported in an axle housing, and explain the major characteristics of each one.
2. There are three designs of axle shaft bearings used on semifloating axles: ______________________, ______________________, and ______________________ bearings.
3. What parts make up a typical ball or roller bearing assembly?
4. Define a radial bearing load.
5. Explain the design and purpose of bearing seals.
6. Lubrication and proper ______________________ adjustment are important on tapered roller bearings.
7. While discussing the different designs of rear axles, Technician A says the bearings for full floating shafts are located within the axle tubes of the rear axle housing. Technician B says the names used to classify the different designs actually define the amount of vehicle weight that is supported by the axles. Who is correct?
   A. Technician A only
   B. Technician B only
   C. Both Technician A and Technician B
   D. Neither Technician A nor Technician B
8. Technician A says some axle shafts are retained in the housing by a plate and bolts. Technician B says a C-washer or clip retains some axle shafts in the housing. Who is correct?
   A. Technician A only
   B. Technician B only
   C. Both Technician A and Technician B
   D. Neither Technician A nor Technician B
9. Technician A says if the bearings are too tight, wheel runout may be excessive. Technician B says if the bearings are loose on only one side of the car, steering pull to the opposite side can result. Who is correct?
   A. Technician A only
   B. Technician B only
   C. Both Technician A and Technician B
   D. Neither Technician A nor Technician B
10. Technician A says the wheel bearings on a typical FWD car are pressed in and out of the steering knuckle assembly. Technician B says the rear wheel bearings on a typical FWD car are serviced in the same way as the front bearings for RWD cars. Who is correct?
    A. Technician A
    B. Technician B
    C. Both Technician A and Technician B
    D. Neither Technician A nor Technician B

# Chapter 16 Diagnosis of Bearing Problems

## *INTRODUCTION*

Bearings rarely fail suddenly. Rather, they deteriorate slowly from dirt, lack of lubrication, and improper adjustment. Bearing wear and failure are almost always accompanied by noise and/or vibration.

Wheel bearings are designed for longevity. Their life expectancy, based on metal fatigue, can usually be calculated if general operating conditions are known. Bearing failures not caused by normal material fatigue are called premature failures. The causes can range from improper lubrication or incorrect mounting, to poor condition of the shaft housing or bearing surfaces.

**Figure 1.** During the road test, pay careful attention to when the suspected bearing noise occurs.

## BASIC WHEEL AND AXLE BEARING TROUBLESHOOTING

Wheel bearing problems can create braking, handling, NVH, and tire wear problems. A road test **(Figure 1)** will often determine if a problem is caused by the wheel or axle bearings or by something else. Listen carefully to the customer's complaints, and obtain as much information as possible from him or her. Ask the customer specific questions about abnormal or unusual vehicle noises and operation. Drive the vehicle at a variety of speeds and conditions, and note all noises and handling problems that are evident. Try to verify the customer's complaint by operating the vehicle under the same conditions as he or she says the noise occurs.

Check the suspension, steering, driveline, and brake system to make sure they are not the cause of the problems. Check the wear patterns of the tires to help eliminate the suspension and steering systems as causes of the problem. Check the inflation of all of the tires, and compare the pressure to specifications. If the results from the above all indicate that the tires, the suspension, steering, and brake systems, and the driveline are in good condition and a noise is evident during the test drive, suspect the tire design as the cause of the problem.

If the tires feel as if they are bouncing along, the problem is typically wheel tramp. If the tires seemed to vibrate while turning, the problem is wheel shimmy. If wheel tramp is the problem, the tire and wheel assemblies need to be checked for static imbalance. If wheel shimmy is the problem, check the tire and wheel assemblies for dynamic imbalance. If the tire and wheel assembly are both statically and dynamically balanced, check the wheel hub or axle for excessive runout. If runout is not the problem, check for loose, worn, or damaged wheel bearings.



157

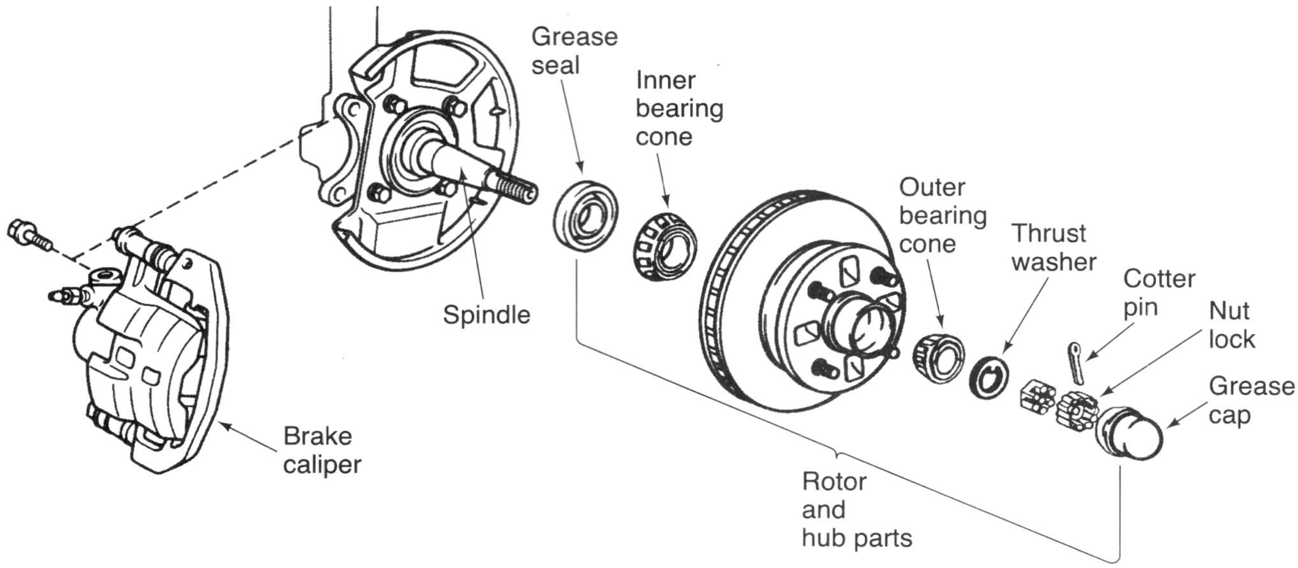


**Figure 2.** There are many probable causes for premature wheel bearing failure.

## Diagnosis of Bearing Defects

Bearings are designed to provide long life, but there are many causes of premature bearing failure. If a bearing fails, the technician must decide if the bearing failure was caused by normal wear or if the bearing failed prematurely. For example, if a front wheel bearing **(Figure 2)** fails on a car that is one year old with an original odometer reading of 15,000 miles (24,000 kilometers), experience tells us the bearing failure is premature because front wheel bearings normally last for a much longer mileage period. If a bearing fails prematurely, there must be some cause for the failure. The causes of premature bearing failure are:

- Lack of lubrication
- Improper type of lubrication
- Incorrect end play adjustment (where applicable)
- Misalignment of related components, such as shafts or housings
- Excessive bearing load
- Improper installation or service procedures
- Excessive heat
- Dirt or contamination

The first indication of bearing failure is usually a howling noise while the bearing is rotating. The howling noise will vary with the load on the bearing. A front wheel bearing usually provides a more noticeable howl when the vehicle is turning a corner because this places additional thrust load on the bearing. A defective rear axle bearing usually provides a howling noise that is more noticeable at lower speeds. A howling noise is more noticeable when driving on a narrow street with buildings on each side because the noise vibrates off the nearby buildings. A rear axle bearing noise is evident during acceleration and deceleration because the vehicle's weight places a load on the bearing. A rear axle bearing noise may be somewhat more noticeable during deceleration because there is less engine noise at that time.

## TAPERED ROLLER BEARING TROUBLESHOOTING

Begin diagnosing possible wheel bearing problems by raising the vehicle by its frame so the wheels can turn freely. Rotate the wheel and listen to the sounds of brake drag and bearing rotation. If the brakes are dragging slightly, they will produce a high-pitched swishing sound. Bearing noise will be a lower-pitched rumble. Low-level bearing noise and brake drag are normal. Normal bearing sounds should be uniform through the wheel revolutions. An uneven rumble or a grinding sound indicates possible bearing problems.

If a bearing is unusually noisy, roughness will be felt as you turn the wheel. Hold the wheel as you rotate it. Listen for a low-frequency rumbling sound, and feel for rough and uneven rotation. Compare left and right wheels to determine whether a problem exists.

While rotating the wheel, try to move it in and out on the spindle, and note the amount of movement. Worn or damaged bearings or bearings that need adjustment will have a noticeable amount of end play. Also, grasp the top

**Figure 3.** Grasp the top and bottom of the tire and try to wobble the wheel and tire back and forth to check for a poorly adjusted or failed wheel bearing.

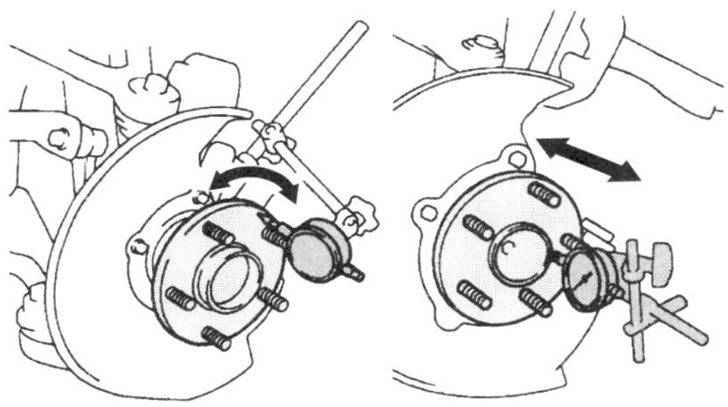

**Figure 4.** Wheel bearing end play can be checked with a dial indicator.

and bottom of the tire and try to wobble it back and forth **(Figure 3).** You should feel little or no wobble from a properly adjusted bearing.

End play can be measured with a dial indicator placed against the wheel hub **(Figure 4).** Set the indicator to zero, move the wheel in and out on the spindle, and note the reading. Tapered roller bearings can have 0.001 to 0.005 inch of end play (0.025 to 0.127 mm). Check the specifications for the exact amount of allowable end play.

Finally, inspect the wheel and the brake drum or rotor for grease that may be leaking past a bad seal. Bearing grease can contaminate brake linings, and a leaking grease seal can let dirt into the bearing. A leaking seal must be replaced, but the bearings must also be cleaned, inspected, and repacked to be sure they have not been damaged. Whether repacking and reinstalling a used bearing or installing a new one, always install a new grease seal.

## WHEEL HUB DIAGNOSIS

When wheel bearings and hubs are an integral assembly, the bearing end play should be measured with a dial indicator stem mounted against the hub. If the end play exceeds 0.005 in. (0.127 mm) as the hub is moved in and

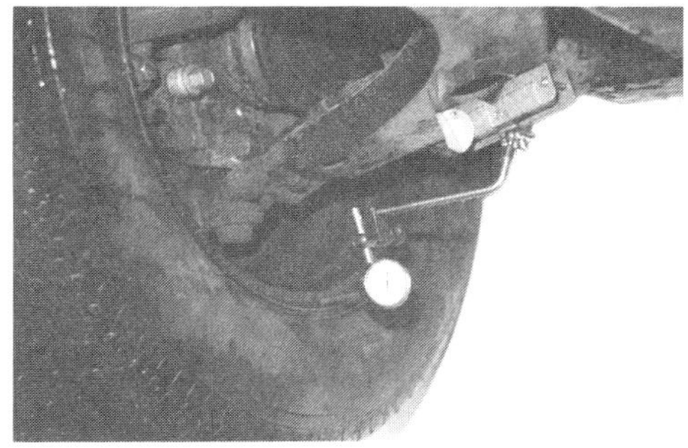

**Figure 5.** When the front wheel bearings are mounted in the steering knuckle the wheel bearings may be checked with the vehicle raised on the hoist and a dial indicator positioned against the wheel's interior rim.

out, the hub and bearing assembly should be replaced. This specification is typical, but the vehicle manufacturer's specifications must be used. Hub and bearing replacement is also necessary if the bearing is rough or noisy. Integral-type bearing and hub assemblies are used on the front and rear wheels on some front-wheel-drive cars.

When the front wheel bearings are mounted in the steering knuckle, the wheel bearings may be checked with the vehicle raised on the hoist and a dial indicator positioned against the outer wheel rim lip as shown in **Figure 5.**

When the wheel is moved in and out, the maximum bearing movement on the dial indicator should be as follows:

0.020 in. (0.508 mm) for 13-in. (33-cm) wheels
0.023 in. (0.584 mm) for 14-in. (35.5-cm) wheels
0.025 in. (0.635 mm) for 15-in. (38-cm) wheels

If the bearing movement is excessive, check the torque of the hub nut before replacing the bearing. When this torque is correct and bearing movement is excessive, the bearing should be replaced.

When a wheel is removed to service the wheel bearings, proper balance must be maintained between the wheel and tire and the hub. Therefore, the tire, wheel, and hub stud should be chalk marked prior to removal.

### Bearing Inspection

When a bearing fails prematurely, the cause of this failure must be corrected to prevent the new bearing from failing. A close inspection of the bearings and races can lead to identifying the cause of the wear. The types of bearing failures and the necessary corrective service procedures are provided in **Figure 6** and **Figure 7**. Bearing fatigue **spalling** appears as flaking of surface metal on bearing rollers and races. Bearing **brinelling** shows up as straight-line indentations on the races and rollers. Bearing smears appear as metal loss in a circular, blotched pattern around the bearing races and rollers. Bearing **frettage** shows up as a fine, corrosive wear pattern around the bearing races and rollers. This wear pattern is circular on the races.

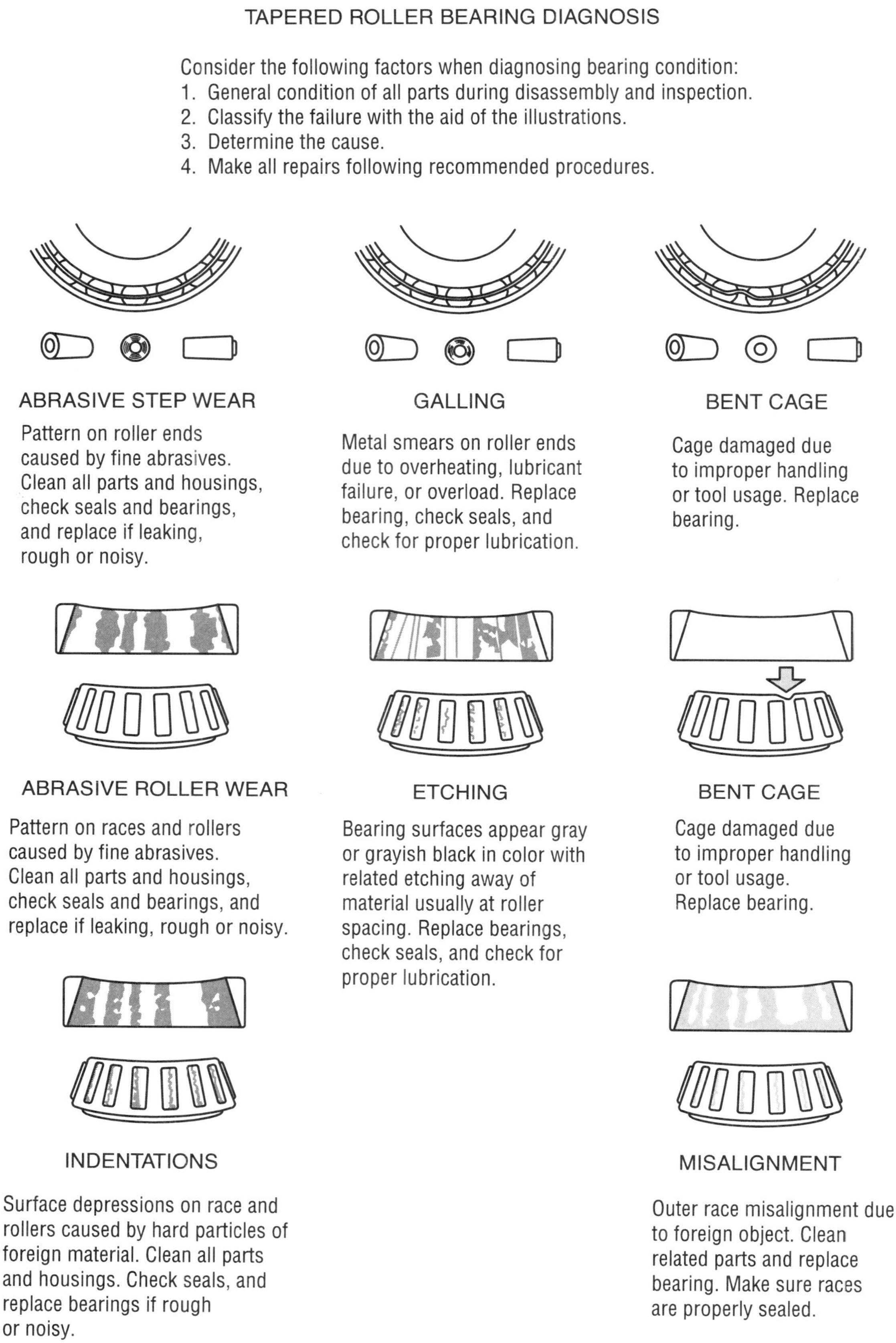


**Figure 6.** Bearing failures and their required corrective procedures.

FATIGUE SPALLING

Flaking of surface metal resulting from fatigue. Replace bearing, clean all related parts.

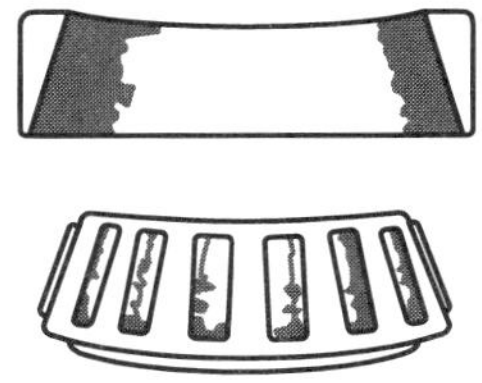

STAIN DISCOLORATION

Discoloration can range from light brown to black caused by incorrect lubricant or moisture. Re-use bearings if stains can be removed by light polishing or if no evidence of overheating is observed. Check seals and related parts for damage.

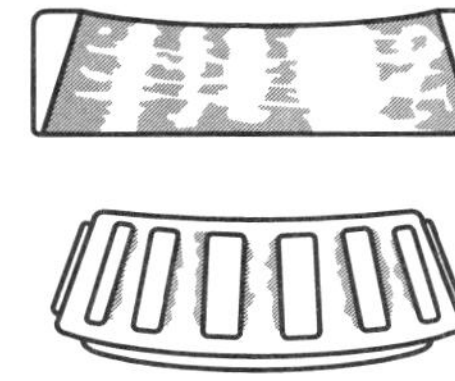

CAGE WEAR

Wear around outside diameter of cage and roller pockets caused by abrasive material and inefficient lubrication. Clean related parts and housings. Check seals and replace bearings.

HEAT DISCOLORATION

Heat discoloration can range from faint yellow to dark blue, resulting from overload or incorrect lubricant. Excessive heat can cause softening of races or rollers. To check for loss of temper on races or rollers, a simple file test may be made. A file drawn over a tempered part will grab and cut metal, whereas, a file drawn over a hard part will glide readily with no metal cutting. Replace bearings if overheating damage is indicated. Check seals and other parts.

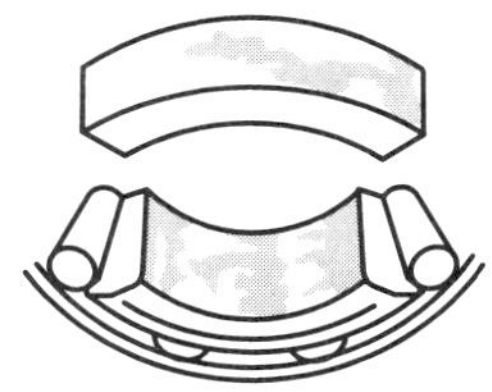

FRETTAGE

Corrosion set up by small relative movement of parts with no lubrication. Replace bearings. Clean related parts. Check seals and check for proper lubrication.

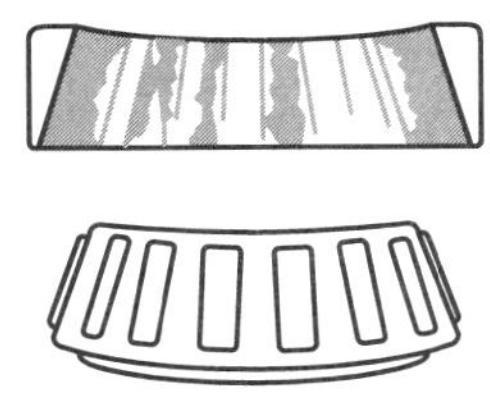

BRINELLING

Surface indentations in raceway caused by rollers either under impact loading or vibration while the bearing is not rotating. Replace bearing if rough or noisy.

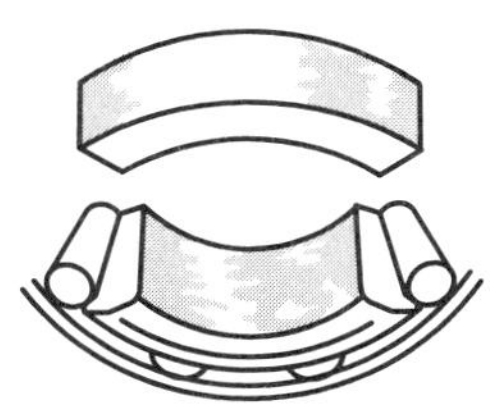

SMEARS

Smearing of metal due to slippage. Slippage can be caused by poor fits, lubrication, overheating, overloads, or handling damage. Replace bearings, clean related parts and check for proper fit and lubrication.

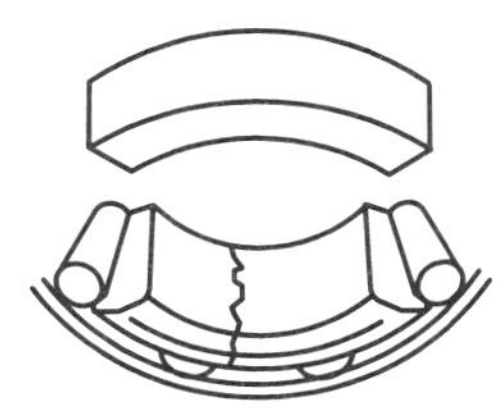

CRACKED INNER RACE

Race cracked due to improper fit, cocking, or poor bearing seats. Replace bearing and correct bearing seats.

**Figure 7.** Bearing failures and their required corrective procedures (continued).

## SUMMARY

- Wheel bearing problems can create braking, handling, NVH, and tire wear problems.
- The cause of bearing failure can range from improper lubrication, incorrect end play adjustment, misalignment of related components, excessive bearing load, improper installation or service procedures, excessive heat, and dirt or contamination.
- The first indication of bearing failure is usually a howling noise while the bearing is rotating.
- Check a wheel bearing by measuring its end play.
- When wheel bearings and hubs are an integral assembly, the hub's end play should be measured.
- When a bearing fails prematurely, the cause of this failure must be corrected to prevent the new bearing from failing.
- A close inspection of the bearings and races can lead to identifying the cause of the wear.
- Bearing fatigue spalling appears as flaking of surface metal on bearing rollers and races.
- Bearing brinelling shows up as straight-line indentations on the races and rollers.
- Bearing smears appear as metal loss in a circular, blotched pattern around the bearing races and rollers.
- Bearing frettage shows up as a fine, corrosive wear pattern around the bearing races and rollers.

## REVIEW QUESTIONS

1. List five problems that could cause premature bearing failure.
2. Most often a rear axle bearing problem can be distinguished from a front wheel bearing problem by the noise that is evident during a road test. Describe the noise from each of these problems.
3. If the grease seal for a bearing assembly is defective and allows lubricant to leak out, what should be done?
4. While inspecting wheel bearings, Technician A says brinelling appears as indentations across the bearing races. Technician B says bearing frettage appears as random patterns of metal loss. Who is correct?
   A. Technician A only
   B. Technician B only
   C. Both Technician A and Technician B
   D. Neither Technician A nor Technician B
5. While discussing front wheel bearing service on a vehicle with the bearing hub unit pressed into the steering knuckle, Technician A says the front wheel bearings may be damaged if this front-wheel-drive vehicle is moved without the hub nut torqued to specifications. Technician B says the hub nut torque supplies the correct wheel bearing adjustment. Who is correct?
   A. Technician A only
   B. Technician B only
   C. Both Technician A and Technician B
   D. Neither Technician A nor Technician B

# Chapter 17 FWD Drive Axles

## INTRODUCTION

Front-wheel-drive axle shafts transfer engine torque from the transaxle's differential to the front wheels. The drive axles must transmit power from the engine to front wheels that must drive, steer, and cope with the severe angles caused by the up-and-down movement of the vehicle's suspension. To accomplish this, these cars must have a compact joint that ensures the driven shaft is rotated at a constant velocity regardless of angle. CV joints also allow the length of the axle assembly to change as the wheel travels up and down.

The half-shafts operate at angles as high as 40 degrees for turning and 20 degrees for suspension travel **(Figure 1).** As a front wheel is turned during steering, the outboard CV joint moves with it around a fixed center. Up-and-down movements of the suspension system force the inboard joint to slide in and out of its race in the transaxle. The inboard joint is often called the plunge joint, and the outboard joint is called the fixed joint. A short spindle shaft, called a stub axle, runs from the outer CV joint to mate with the wheel assembly **(Figure 2).**

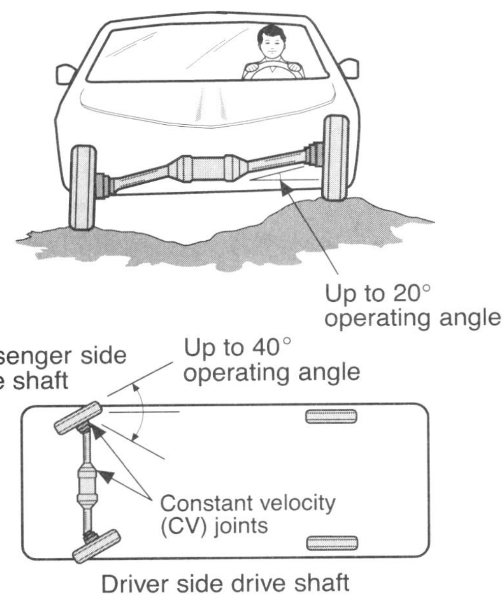


**Figure 1.** FWD drive shaft angles.

**Figure 2.** A short spindle shaft, called a stub axle, runs from the outer CV joint to mate with the wheel assembly.

CV joints are not only used on FWD cars, they are also used on the front axles of many 4WD vehicles. They have also been used on RWD vehicles with the engine mounted in the rear, mid-engine cars, and some RWD cars with IRS.

Axle shafts and CV joints can be the source of many noise and vibration problems and should never be overlooked when noise and vibration are a concern.

### FWD DRIVE AXLES

On most FWD vehicles, the engine is mounted transversely, and the transaxle sits to one side of the engine


163

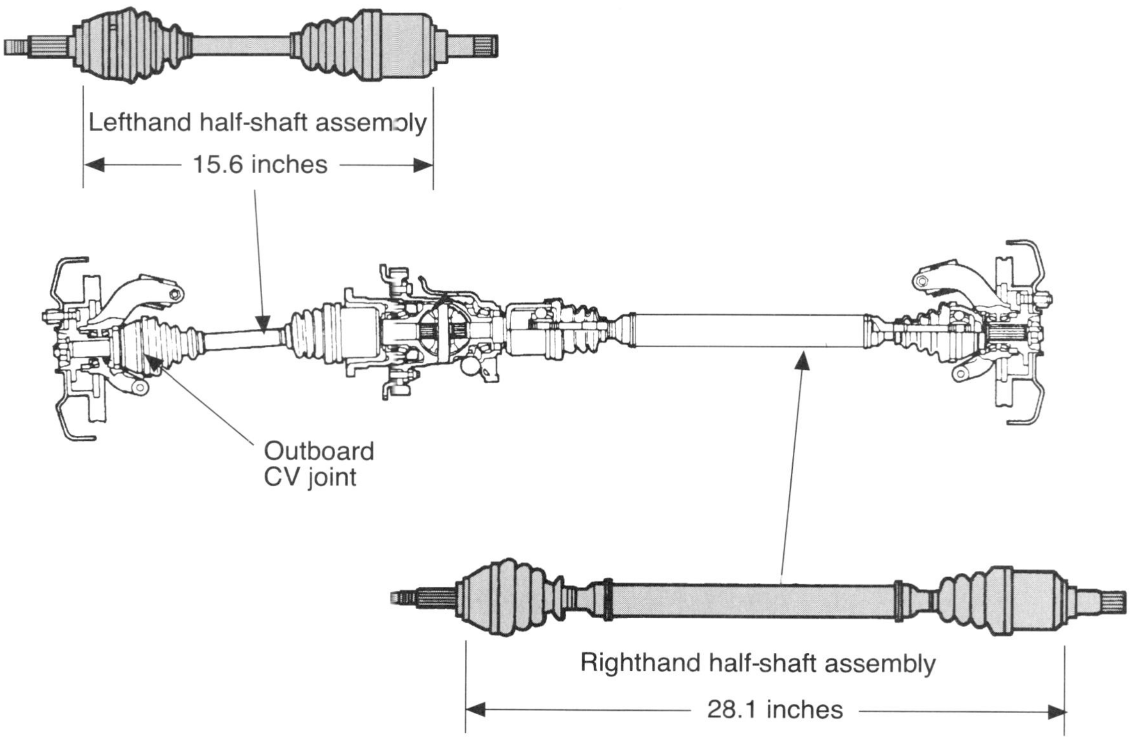


**Figure 3.** The ride side half-shaft is longer than the left side because of the placement of the transaxle.

compartment. Because of this offset, one of the drive axles must be longer than the other **(Figure 3).** The unequal lengths result in torque steer. Torque steer is a term used to describe a condition in which the vehicle tends to steer or pull in one direction as engine power is quickly applied to the drive wheels.

Torque steer occurs when the CV joints of one half-shaft operate at different angles from those on the other shaft. The joints on the longer shaft almost always operate at less of an angle than those on the shorter half-shaft. Therefore, when the differential in a transaxle sends power out to the wheels, the longer shaft will have less resistance, and most of the engine's torque will be sent to that wheel. This is the same thing that happens when one tire has more traction than the other does. The tire with the least traction will receive the most power. This, of course, does not happen if the vehicle has a limited slip differential.

## Equal Length Shafts

Equal length shafts are used in some vehicles to help reduce torque steer. To provide for equal length shafts, manufacturers divide the longer shaft into two pieces **(Figure 4).** An intermediate shaft is used as a link from the transaxle to one of the half-shafts. The other part of the shaft is a half-shaft with the same length as the half-shaft

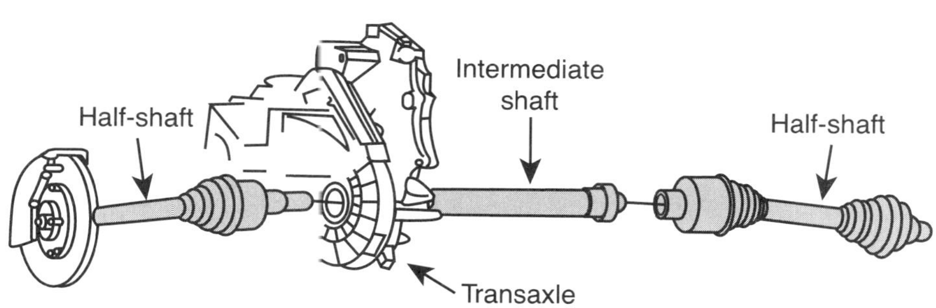


**Figure 4.** Equal length drive axles with an intermediate shaft.

on the other side of the car. When the shafts are effectively the same length, torque steer can still occur but only when the car's ride height on one side is different from the ride height of the other side.

The intermediate shaft is normally linked to the differential by a yoke and ordinary CV joint or is directly splined to the differential. At the outer end of the shaft is a support bracket and bearing assembly that is normally fastened to the engine block. The half-shaft is connected to this bearing assembly with a CV joint. In these applications, looseness in the bearing or bracket can create vibrations. These items should be inspected whenever the customer complains of vibrations and/or noise.

### Vibration Dampers

Instead of using equal length half-shafts, a small damper weight, called a **torsional damper**, may be attached to the long half-shaft to dampen harmonic vibrations in the driveline and to stabilize the shaft as it spins **(Figure 5).** This torsional vibration damper is splined to the axle shaft and is often held in place by a snapring.

The damper is not used to help balance the half-shaft. Unlike the drive shaft of RWD cars, FWD axle shaft balance is not very important because of the axles' relatively slow rotational speeds. In fact, FWD drive shafts operate at about one-third the rotational speed of a RWD drive shaft because the shafts drive the wheels directly, and the final gear reduction at the differential has already occurred. The lower rotational speed has the advantage of eliminating vibrations that result from high rotational shaft speeds.

## DRIVE AXLE SUPPORTS

The drive axles used with transaxles are the full-floating type. This type of axle does not support the weight of the vehicle; rather, all of the vehicle's weight is supported by the suspension. As the car goes over bumps, the axles move up and down, and this changes their length. To allow for this action, the inner CV joints are designed to let the axle slide in and out, thereby changing their length as they move up and down. The outer joints are designed to allow the steering system turn the wheels, as well as allow for the up-and-down movement of the suspension. The weight of the vehicle never rests on the drive axles, so the axles are free to change operating angles and length in response to the steering and suspension systems.

## TYPES OF CV JOINTS

CV joints come in a variety of styles. The different types of joints can be referred to by position (inboard or outboard), by function (fixed or plunge,) or by design (ball-type or tripod).

### Inboard and Outboard Joints

Two CV joints are used on each half-shaft **(Figure 6).** The joint nearer the transaxle or differential is the inner or **inboard joint,** and the one nearer the wheel is the outer or **outboard joint.**

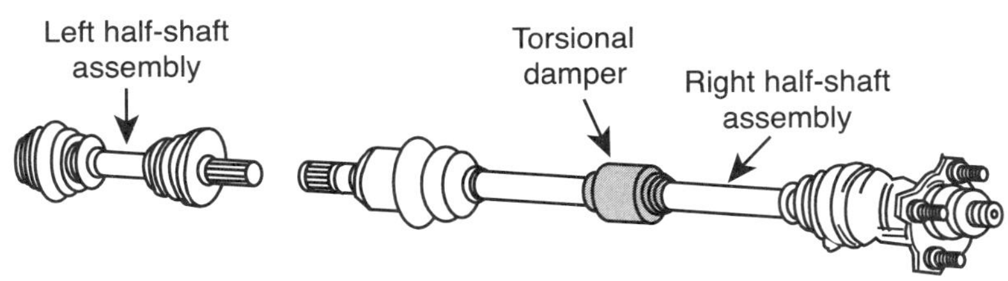


**Figure 5.** The long half-shaft is sometimes fitted with a torsional damper.

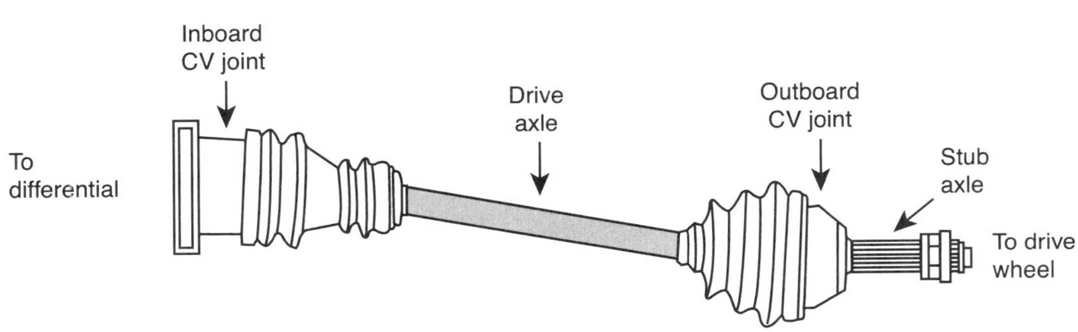


**Figure 6.** Two CV joints are used on each half-shaft.

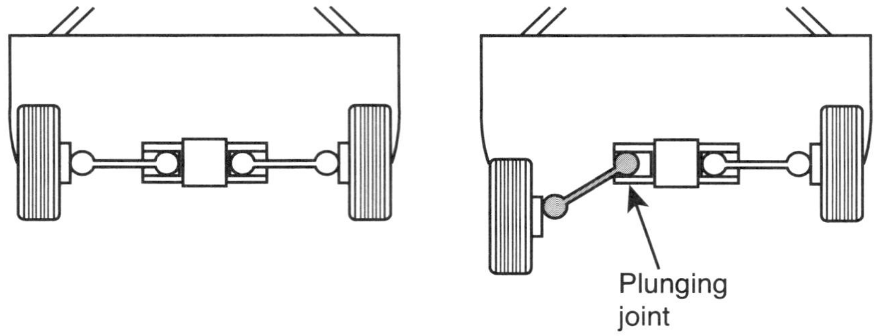


**Figure 7.** Suspension movements and the resulting plunging action of an inboard CV joint.

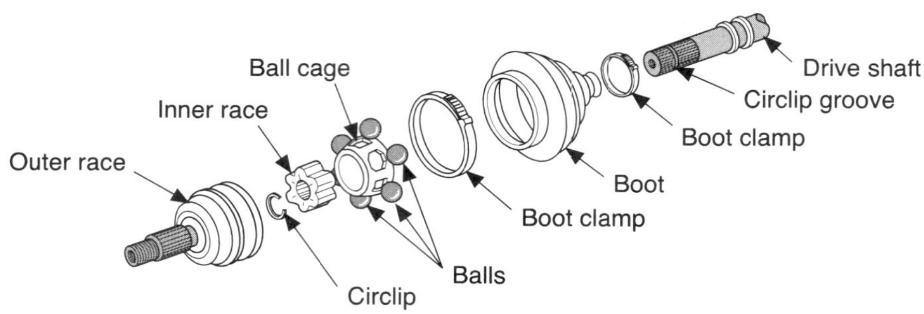


**Figure 8.** An exploded view of a Rzeppa joint.

## Fixed and Plunge Joints

CV joints either are **fixed** (meaning they do not plunge in and out to compensate for changes in length) or are a **plunge joint** (one that is capable of in-and-out movement).

In FWD applications, the inboard joint is also a plunge joint **(Figure 7).** The outboard joint is a fixed joint. Both joints do not need to plunge if one can handle the job. Further, the outdoor joint must also be able to handle much greater operating angles needed for steering.

In RWD applications with IRS, one joint on each axle shaft can be fixed and the other a plunge, or both can be plunge joints. Since the wheels are not used for steering, the operating angles are not as great. Therefore, plunge joints can be used at either or both ends of the axle shafts.

## Ball-Type Joints

There are two basic varieties of CV joints: the **ball-type** and **tripod-type** joints. Both types are used as either inboard or outboard joints, and both are available in fixed or plunge designs.

### Fixed Ball-Type Joints

The fixed ball-type, or **Rzeppa** joint, consists of an inner ball race, six balls, a cage to position the balls, and an outer housing **(Figure 8).** Tracks machined in the inner race and outer housing allow the joint to flex. The inner race and outer housing form a ball-and-socket arrangement. The six balls serve both as bearings between the races and the means of transferring torque from one to the other.

If viewed from the side, the balls within the joint always bisect the angle formed by the shafts on either side of the joint regardless of the operating angle **(Figure 9).** This reduces the effective operating angle of the joint by half

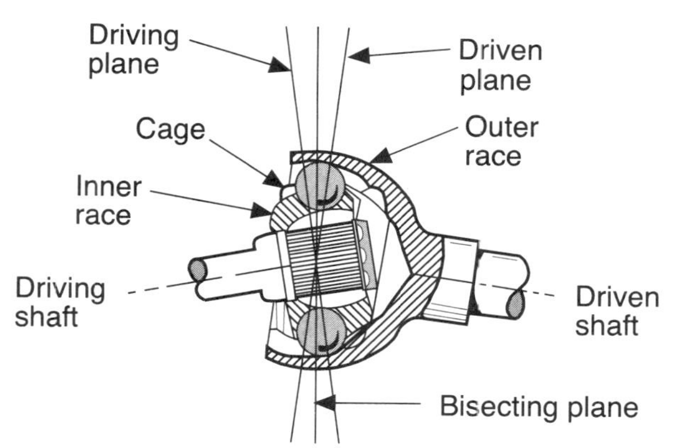


**Figure 9.** The action of a Rzeppa joint.

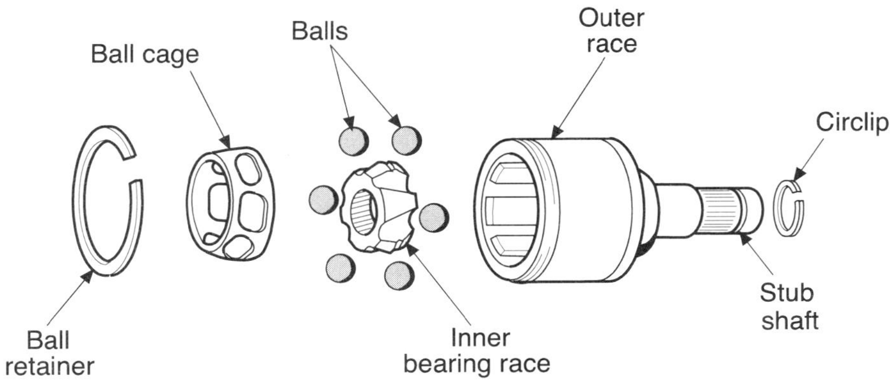


**Figure 10.** A double-offset CV joint.

and virtually eliminates all vibration problems. The input speed to the joint is always equal to the output velocity of the joint—thus the description "constant velocity." The cage helps to maintain this alignment by holding the six balls snugly in its windows. If the cage windows become worn or deformed over time, the resulting play between ball and window typically results in a clicking noise when turning. It is important to note that opposing balls in a Rzeppa CV joint always work together as a pair. Heavy wear in the tracks of one ball almost always results in identical wear in the tracks of the opposing ball.

Another ball-type joint is the dish style, which is used predominantly on Volkswagen as well as on many German rear-wheel-drive models. Its design is very similar to the Rzeppa joint.

### Plunging Ball-Type Joints

There are two basic styles of plunging ball-type joints: the double-offset that uses a cylindrical outer housing with straight grooves, and the cross groove joint. This is a compact design with a flat, doughnut-shaped outer housing and angled grooves.

The **double-offset joint (Figure 10)** is typically used in applications that require higher operating angles (up to 25 degrees) and greater plunge depth (up to 2.4 inches). This type of joint can be found at the inboard position on some front-wheel-drive half-shafts, as well as on the propeller shaft of some four-wheel-drive shafts. The components of a double offset joint are typically held together as an assembly by a retaining ring.

The **cross groove joint (Figure 11)** has a much flatter design than any other plunge joint. It is used as the inboard joint on FWD half-shafts or at either end of RWD independent rear suspension axle shafts. The feature that makes this joint unique is its ability to handle a fair amount of plunge (up to 1.8 inches) in a relatively short distance. The inner and outer races share the plunging motion equally, so less overall depth is needed for a given amount of plunge. The cross groove can handle operating angles up to 22 degrees.

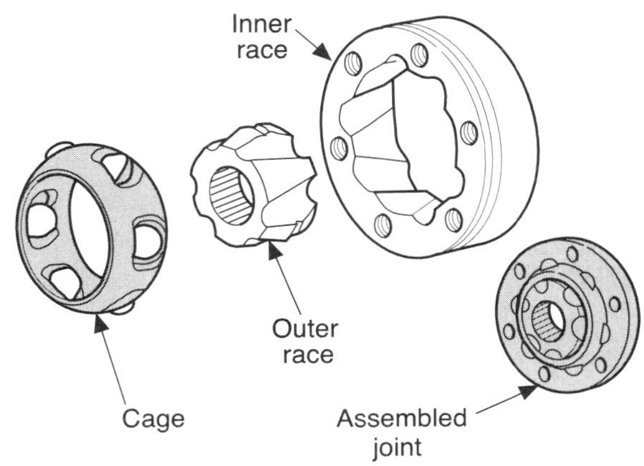


**Figure 11.** A cross-groove CV joint.

## Tripod CV Joints

As with ball-type CV joints, tripod joints come in two varieties: plunge and fixed.

### Tripod Plunging Joints

Tripod plunging joints **(Figure 12)** consist of a central drive part or tripod (also known as spider). This has three trunnions fitted with spherical rollers on needle bearings **(Figure 13)** and an outer housing (sometimes called a tulip because of its three-lobed, flower-like appearance). On some tripod joints, the outer housing is closed, meaning the roller tracks are totally enclosed within it. On others, the tulip is open and the roller tracks are machined out of the housing. Tripod joints are most commonly used as FWD inboard plunge joints.

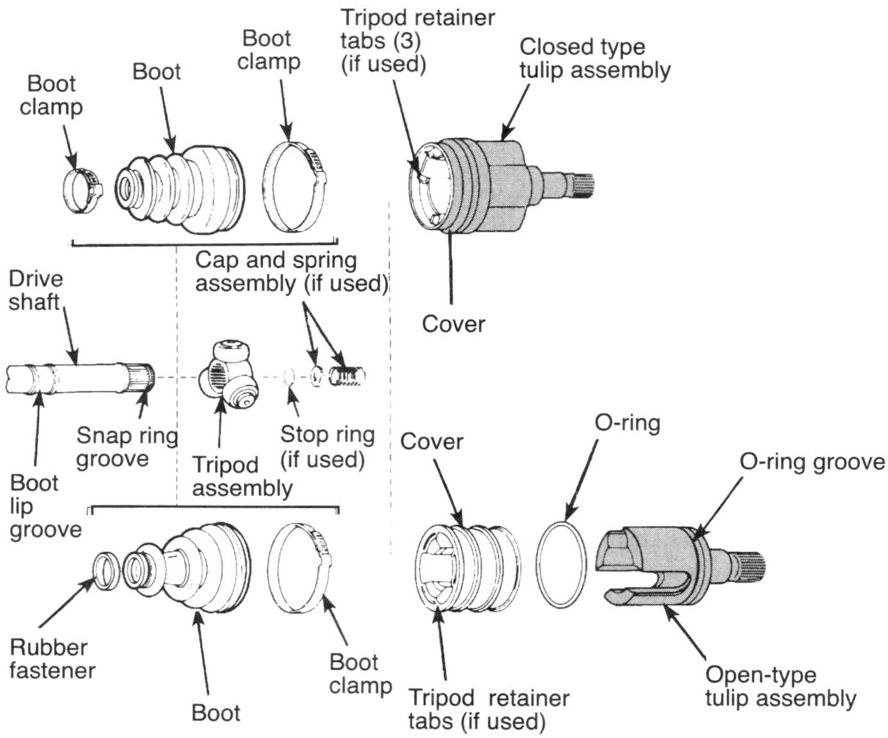


**Figure 12.** Two types of tripod plunging joints: closed and open.

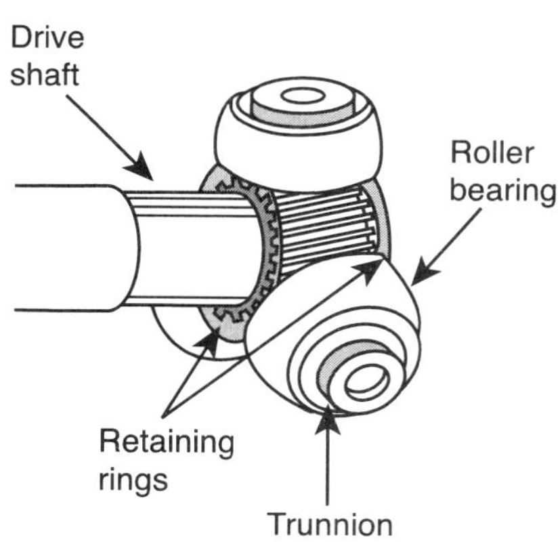


**Figure 13.** A tripod assembly.

### Fixed Tripod Joints

The fixed tripod joint is sometimes used as the outboard joint in FWD applications. In this design, the trunnion is mounted in the outer housing, and the three roller bearings turn against an open tulip on the input shaft. A steel locking spider holds the joint together.

The fixed tripod joint has a much greater angular capability. The only major difference from a service standpoint is that the fixed tripod joint cannot be removed from the halfshaft or disassembled because of the way it is manufactured. The complete joint and shaft assembly must be replaced if the joint goes bad.

## CV JOINT BOOTS

Bellows-type neoprene boots **(Figure 14)** are installed over each CV joint to retain lubricant and to keep moisture and dirt out. These boots must be maintained in good con-

**Figure 14.** Bellows-type neoprene boots are installed over each CV joint to retain lubricant and to keep out moisture and dirt.

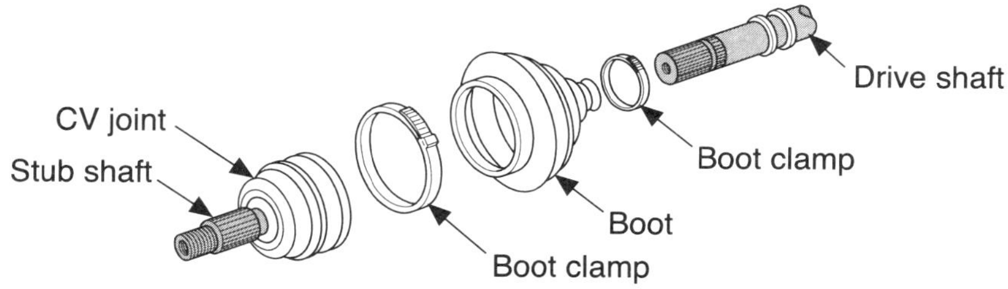


**Figure 15.** Clamps seal the boot against the CV joint and the drive axle.

dition and a CV joint can last more than 100,000 miles if the boot keeps the joint properly sealed. If the boots are torn or damaged, they must be replaced. Often the half-shaft is removed and a direct replacement boot is installed; other times a screw or split boot is installed. These boots are held together by screws or bolts and do not require the disassembly or removal of the half-shaft to install. Each end of a replacement boot is sealed tightly against the shaft or housing by a retaining clamp or strap **(Figure 15).** These straps may be metal or plastic and are available in many sizes and designs.

> **You Should Know** *Because of high failure rates, most of the service industry does not recommend the use of split boots.*

On some vehicles, inboard CV joints operate very close to the engine's exhaust system. Special rubber boots of silicone or thermoplastic materials are required to withstand the heat. In these extremes, the manufacturer may specify a special high-temperature lubricant.

## SUMMARY

- On FWD vehicles, two drive shafts or axles extend out from each side of the differential in the transaxle and pivot on CV joints to supply power to the drive wheels.
- Most problems with front-wheel-drive systems are noted by noise and vibration.
- Torque steer is a term used to describe a condition in which the vehicle tends to steer or pull in one direction as engine power is quickly applied to the drive wheels.
- To reduce torque steer, some vehicles have equal length half-shafts or fit a small weight on the longer half-shaft to dampen harmonic vibrations and stabilize the shaft as it spins.
- The drive axles used with transaxles are the full-floating type that never supports the weight of the vehicle.
- CV joints provide the necessary transfer of uniform torque and a constant speed while operating through a wide range of angles.
- The axle shaft is connected to the differential by the inboard or plunge CV joint. The axle shaft then extends to the outer or fixed CV joint.

## REVIEW QUESTIONS

1. Explain the different methods used by automobile manufacturers to offset torque steer.
2. The type of joint that allows for changes in axle length is the ________________________ joint.
3. Why do some drive axle assemblies include an intermediate shaft?
4. Name the three ways in which CV joints can be classified.
5. In front-wheel drivetrains, the CV joint nearer the transaxle is the
   A. inner joint.
   B. inboard joint.
   C. outboard joint.
   D. both A and B.

# Chapter 18 Diagnosis and Servicing FWD Drive Axles

## INTRODUCTION

Diagnosing FWD drive axle and CV joint problems should follow normal steps for diagnosing any problem. However, make sure those steps include talking to the customer to gather as much information as possible about the problem, road testing the vehicle, and conducting a thorough inspection of all related components. CV joints can have a long life and are relatively trouble-free unless damage to the boot or joint goes unnoticed.

A complete front axle service includes a CV joint inspection and repacking, front hub bearing inspection and possible replacement, and CV joint and/or boot replacement. Boots should be replaced any time any part of the axle assembly is serviced.

### TALK TO THE CUSTOMER

Although your customer probably does not know the cause of the problem, what he or she does know will be of great help while diagnosing the problem. Find out as much as you can about the problem.

Good questions to ask the customer are:

- What is the problem? Is it a shake, vibration, or noise?
- Can you feel and hear the symptom?
- If the problem is a noise, what kind of noise is it?
- If the noise is a vibration or shake, where do you feel it?
- When does the noise or vibration occur?
- When did you first notice it?
- Were you going straight? Turning left? Turning right? Braking the car?
- What work has been done on the vehicle lately?

It is very helpful to write down a summary of the customer's responses. This will give you a quick reference during the road test or other diagnostic work.

### ROAD TEST

A road test can confirm the customer's concern and can also reveal other related problems the customer may not have noticed. Drive the vehicle in an attempt to duplicate the symptoms. You should be listening and feeling for the key indicators of CV joint wear. If possible, take the customer on the road test with you.

A road test on a smooth surface is a good place to begin. The test should include driving at average highway speeds, some sharp turns, acceleration, and coasting. Look and listen for the following signs.

- A popping or clicking noise when turning indicates a possible worn or damaged outer joint. Drive the car through full turns. If only a faint click is heard, put the car in reverse and repeat the test. If the noise is louder, the problem is a damaged or worn joint. The source of this noise can be pinpointed by turning the vehicle to the right and left. Normally the noise is caused by the outboard joint of the wheel that is on the inside of the turn **(Figure 1).**
- A clunk during accelerating, decelerating, or putting an automatic transaxle into *drive* can be caused by excessive play in the inner joint. A clunking noise when putting an automatic transmission into gear or when starting out from a stop usually indicates excessive play in one of the joints. Be careful, this is the kind of noise produced by excessive backlash in the differential gears and transmission. Alternately accelerating and decelerating in reverse while driving straight can reveal worn inner plunge joints. A bad joint clunks or shudders.
- A humming or growling noise may be due to inadequate lubrication of either joint. It is more often due to worn or damaged wheel bearings, a bad intermediate

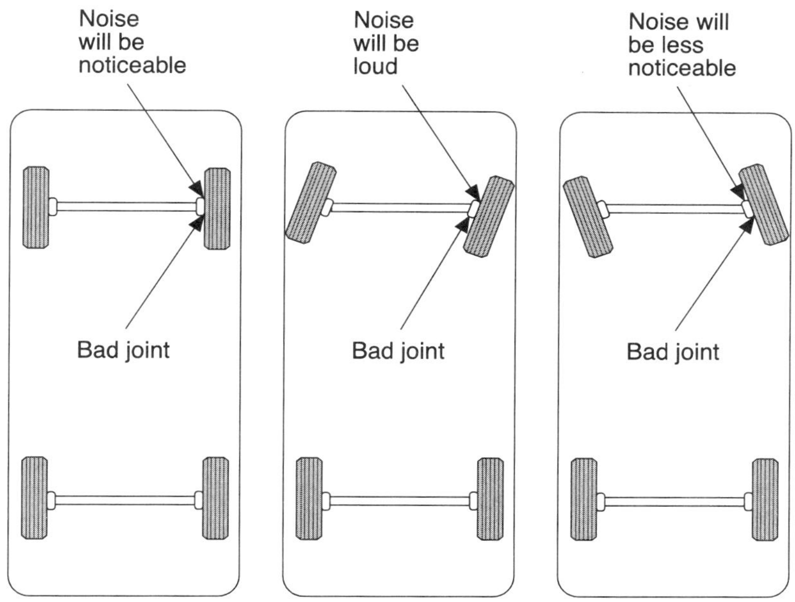


**Figure 1.** A bad outboard joint on the inside of a turn will cause a noticeable noise.

shaft bearing on equal-length half shaft transaxles, or worn shaft bearings within the transmission.

- A **shudder** or vibration when accelerating is often caused by excessive play in the inboard or outboard joint, but more likely it is the inboard plunge joint. These vibrations can also be caused by a bad intermediate shaft bearing on transaxles with equal length half-shafts. On vehicles with transverse-mounted engines, this can also be caused by loose or deteriorated engine/transaxle mounts. Inspect the rubber bushings in the engine's upper torque strap to rule out this possibility. A vibration or shudder that increases with speed or comes and goes at a certain speed may be the result of excessive play in an inner or outer joint or a bent axle shaft.
- A cyclic vibration that comes and goes between 45 mph and 60 mph may seem like an out-of-balance tire; however an out-of-balance wheel produces a continuous vibration. A bad inner tripod CV joint is a likely cause for the vibration.
- If a noise is heard while driving straight but ceases while turning, the problem is usually a bad front wheel bearing.
- A vibration that increases with speed is rarely due to a CV joint or half-shaft. An out-of-balance tire, an out-of-round tire or wheel, or a bent rim are the most likely causes. It is possible that a bent half-shaft, as the result of collision or towing damage, could cause the vibration. A missing damper weight could also be the culprit.

If the road test did not verify the customer's concern, raise the drive wheels. Set jack stands beneath the lower control arms, and carefully lower the suspension down on the stands. Start the engine with the brakes applied. Place the transaxle in gear and release the brakes. Apply power and listen to the drive axles. The maximum speed of the tires during this test should be 25 to 30 mph.

If the source of the noise is found to be in the drive axles, transaxle, or wheel bearings, inspect and test the electrical ground for the vehicle. Drivetrain components can fail because of electrolysis that occurs when they become part of the vehicle's ground circuit.

## VISUAL INSPECTION

Raise the vehicle on a lift to conduct a thorough inspection. The following should be checked and corrected before paying full attention to the drive axles and their joints.

- Check the tires and wheels for damage and proper inflation.
- Check tire runout, and examine each tire's wear pattern for signs of steering and suspension problems.
- Check the control arm, sway bar, ball joints, and shock bushings for damage or wear.
- Check engine and transaxle mounts for wear and deterioration, which can cause clunking during power applications.
- Check the strut, brake hardware, and steering components for wear or damage.

After these components have been inspected and repaired, begin your inspection of the driveline by looking for torn CV joint boots, missing, loose, or incorrect boot clamps, and evidence of grease having been slung or thrown in the area of the CV joint boot. Boots should be

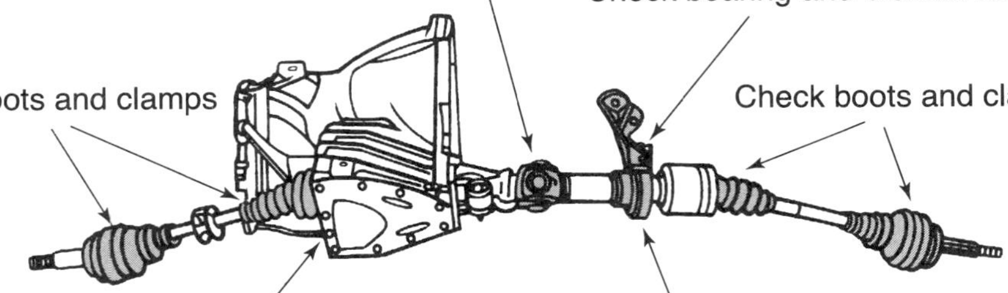


**Figure 2.** Inspection points on FWD vehicles.

airtight and have secure clamps on both ends. Check both the outer and the inner joint on both sides of the vehicle **(Figure 2).** Remember, CV joint boots not only keep the lubricant in but also keep moisture and dirt out of the joint. Defective boots are the most common cause of CV joint failure.

Boots can fail for these reasons:

1. Cuts or tears from foreign objects or sheet metal damage resulting from an accident
2. Improper hookup during towing
3. Ice forming around or near the boot and cutting it when the engine's torque breaks the ice away and turns the axle
4. Deterioration caused by the environment and fluid leaks
5. Boot clamp failure
6. Improper service on a suspension component
7. Improper removal of the half-shaft

It is possible to have a good boot on a defective joint or shaft, therefore continue your inspection by checking the half-shafts for signs of damage. Slowly rotate the wheels and carefully look at the shaft for damage and excessive runout. Some runout is allowed on some cars; always refer to the specifications before coming to a conclusion about the shaft. Also, check the half-shaft mounting flanges and bolts at the transaxle and the wheel spindles for looseness.

If there is any indication of torn boots, failed joints, bad clamps, slung grease from the joints, or worn transaxle carrier assemblies, the shaft should be removed and the problem fixed.

## CV JOINT WEAR

Outer joints wear at a higher rate than inner joints due to their increased range of operating angles. Inner joint angles may change only 10 to 20 degrees, whereas outer joints can undergo changes of up to 40 degrees in addition to **jounce** and **rebound** as the wheels are steered. On an average, nine outer joints are replaced for every inner joint. That does not mean you should overlook the inner joints.

## SERVICING FWD DRIVE AXLES

All CV joints are encased in a protective rubber (neoprene, natural, or silicone) or thermoplastic (Hycrel) boot. The boot retains grease and keeps dirt and water out. The importance of the boot cannot be overemphasized because without its protection the joint does not survive. For all practical purposes, a CV joint is lubed for life. Once packed with grease and installed, it requires no further maintenance. A loose or missing boot clamp or a slit, tear, or small puncture in the boot itself allows grease to leak out and water or dirt to enter. Consequently, the joint is destroyed.

Although outboard joints wear faster than the inboard ones, the decision as to whether to replace both joints when the half-shaft is removed depends on the circumstances. If the vehicle has low mileage and joint failure is the result of a defective boot, there is no reason to replace both joints. On a high-mileage vehicle where the joint is worn, it might be wise to install both CV joints.

### Obtaining Repair Parts

To repair a drive axle, a complete shaft should be installed. Most aftermarket part suppliers offer drive shafts for FWD vehicles. These shafts come fully assembled and ready for installation. This repair method eliminates the need to tear down and rebuild an old shaft.

Most CV joints can be replaced or repaired only as a complete assembly. Joint service kits typically include a CV joint, boot, boot clamps and seals, special grease for

**Figure 3.** The best way to service a joint is to install a complete service kit, which includes a new boot, grease, and the required hardware.

lubrication, retaining rings, and all other attachment parts **(Figure 3).**

Part manufacturers also offer boot sets for each application, including new clamps and the appropriate type and amount of grease for the joint. Use only those clamps supplied with the replacement boot. Follow the directions for positioning and securing them.

Old boots should never be reused when replacing a CV joint. Reusing an old boot on a new joint usually leads to the quick destruction of the joint.

Whenever you replace a CV joint or boot, make sure the replacement parts are correct for that model vehicle. Often manufacturers change the type of joint during production; make sure you have the correct parts.

## DRIVE AXLE REMOVAL

To perform most services on a half-shaft, it must be removed. The exact procedure for removing and installing drive axles will vary with car models. Always refer to your service manual for the special procedures pertaining to the model you are working with. What follows is a general procedure for removing and installing a half-shaft.

1. With the vehicle's tires on the ground, use a large breaker bar or a long-handled ratchet to loosen the axle hub nut. Then remove the hub nut.

> **You Should Know** *Never use an impact wrench to loosen the hub nut or lug nuts. The action of the wrench could cause internal damage to the transaxle.*

2. Raise the vehicle and remove the wheel and tire. Then remove the brake caliper and rotor. Make sure to support the caliper so that it does not hang by its brake hose.
3. Disconnect the suspension from the steering knuckle. In most cases, the easiest way to disconnect the suspension is by separating the lower ball joint from the steering knuckle. After you pull the bolt out of a pinch clamp, the stud should slide out. Remove the **pinch bolt** and use a pry bar to separate the ball joint from the steering knuckle. Disconnect and separate the tie-rod end from the steering knuckle. Inspect the ball joint flange of the steering knuckle. If the ball joint stud was loose or broken, if any out of roundness or damage to the bores is noticeable, or if the flange is deformed, the steering knuckle must be replaced.
4. Remove the axle from the transaxle using the appropriate puller **(Figure 4).**
5. Then remove the drive axle from the hub and bearing assembly using the correct tool. Always use the proper pullers and recommended special tools to remove the shaft from a wheel bearing to prevent damage to the bearings and their races.

> **You Should Know** *On vehicles equipped with ABS, be careful not to damage the sensor, sensor cable, or the trigger wheel mounted on the CV joint housing. The air gap between the sensor and the trigger wheel is critical to proper ABS operation. The gap should be measured with a non-metallic feeler gauge and corrected during reassembly.*

6. Make the necessary repairs to the shaft and joints. Make sure all surfaces of the half-shaft are clean and dry prior to installation.

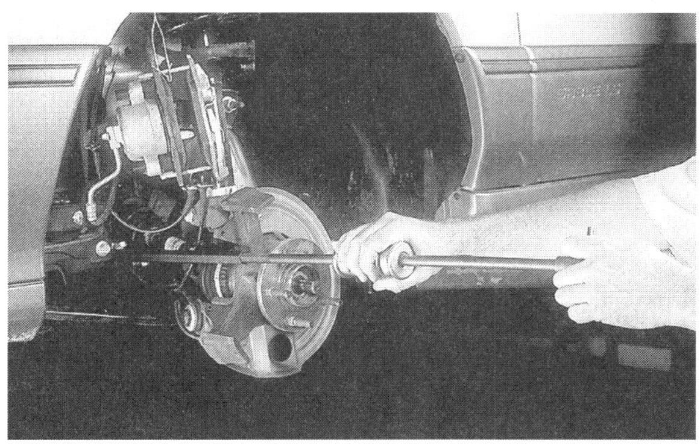

**Figure 4.** Use the appropriate puller to remove the drive axle from the transaxle.

7. Loosely install the half-shaft into the steering knuckle and transaxle. Then tap the shaft into the transaxle or install the flange bolts and tighten them to specifications.

> **You Should Know** *Never let a half-shaft hang on a CV joint during removal or installation. The weight of the shaft can damage the joint.*

8. Then, slide the outer joint into the knuckle **(Figure 5).** Position the bracket for the steering knuckle and strut, then install the attaching bolts.
9. Put the ball joint into the steering knuckle. Install the pinch bolt and tighten it to specifications. Over-torquing the control arms' pinch bolts can lead to bolt failure.
10. Install the hub nut and tighten it by hand. Only use new axle hub nuts.
11. Install the rotor and brake caliper. Torque the caliper bolts to specifications. Then install the wheel and tire assembly.
12. Lower the car and torque the hub nut to the specified amount. Use a staking punch to lock the nut into position **(Figure 6).**

> **You Should Know** *Always torque the hub nuts to specifications. This is necessary to properly preload the wheel bearings. Most axle hub nuts are staked in place after they have been tightened.*

13. Always recheck the alignment after replacing CV joints. Marking the camber bolts is not enough; camber can be off as much as $^3/_4$ degree due to differences between the size of the camber bolts and their bores.

**Figure 5.** Pull the steering knuckle back and slide the splined outer joint into the knuckle.

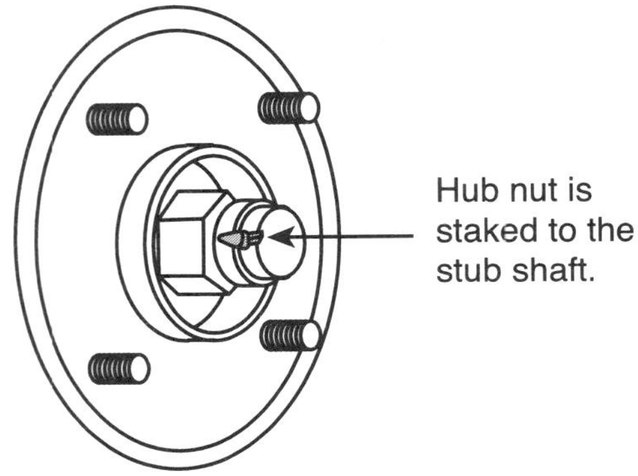


**Figure 6.** The hub nut should be staked after it has been tightened to the proper specifications.

## BENCH INSPECTION

Mount the half-shaft in a soft-jawed vise. Be careful not to overtighten the vise because this could damage the shaft. Mark the shaft where the inner end of the boot sits unless the boot seats in a locating groove **(Figure 7).**

Remove the outer joint boot; wipe any excess grease from around the inner race and cage. Carefully inspect the grease for contamination. Water inside the joint will cause **emulsification,** which gives the grease a milky appearance. If the grease is gritty from sand or dirt contamination and the vehicle has been run for some time with a torn boot or other condition that allowed the contamination, the joint must be replaced or rebuilt. If the grease is not contaminated with water or dirt, only a replacement boot may be required.

Look for cracks, chips, pits, or rust on all parts; including the balls, cage, inner race, or housing. A discolored outer CV joint housing is normal; this is due to the heat treatment it received during manufacturing. A shiny surface on the races does not mean the joint must be rebuilt, however, pits and dents do. Also, check for signs of cracks, wear, or dimpling in the cage windows, and make sure each ball fits snugly in its proper place.

To check for internal wear, tilt the joint to its maximum angle. Attempt to plunge the joint inward and outward. If

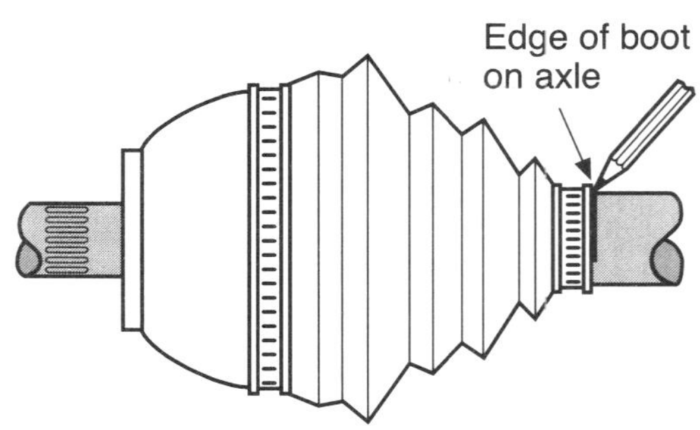


**Figure 7.** Mark the position of the boot on the axle shaft.

the joint catches, stops, or sticks along its path of travel, one or more of the bearings or rollers has a wear spot on it and the joint should be replaced.

If the inner joint has separated, check for a bent or rusted frame member, worn suspension components, and bent, rusted, or worn engine/transaxle mounts.

## GENERAL SERVICE PROCEDURES

The rubber boots need special care when servicing the CV joints, engine, or transaxle. The following tips might save you trouble later.

- Always support the control arm when doing on-the-car balancing the front wheels to avoid high-speed operation at a steep half-shaft angle. Off-the-car balancing might be a wiser choice.
- Do not use half-shafts as lift points for raising a car.
- Use a plastic or metal shield over rubber boots to protect them from accidental tool damage when performing other wheel, brake, suspension, or steering system maintenance.
- Clean the boots only with soap and water.
- Avoid boot contact with gasoline, oil, or degreaser compounds.

### Replacing Joint Boots

The following are some basic guidelines for replacing a boot. The exact procedure varies with the various vehicle models; always refer to the appropriate service manual.

With the axle mounted in a vise, cut and discard both boot clamps. Mark the location of the small end of the old boot before you remove it. Then move the boot. Inspect the grease in the joint for contamination. Clean as much grease as is practical away from the joint. Then, remove the circlip and separate the joint from the shaft.

Now slide the old boot off the shaft. Wipe the axle shaft clean and install the new boot onto the shaft. Place the boot into its proper location on the shaft and install a new clamp. When installing the new clamps, spread them just enough to slip over the axle. The clamp should be positioned so that the end with the holes points in the direction of axle rotation and the holes fit over the locking hooks **(Figure 8).** Using moderate hand pressure, engage the hook at the tightest possible position. Squeeze the ear of the clamp firmly with the clamp pliers **(Figure 9).** If not crimped properly, the clamp will not form a good seal.

Using a new circlip, reinstall the joint onto the shaft. Force the entire packet of grease into the joint and boot **(Figure 10).**

If the boot is dimpled or collapsed after you have installed it, carefully insert a dulled screwdriver between the boot and the shaft to allow air to enter the boot. This should return the boot to its normal shape **(Figure 11).** Then install the new large boot clamp and reinstall the axle into the vehicle.

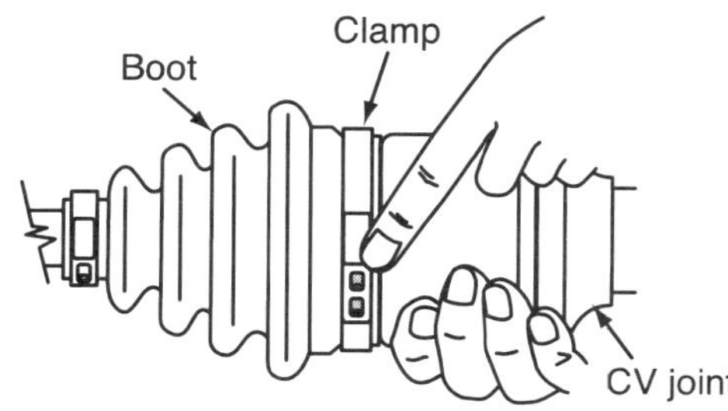


**Figure 8.** Fit the end of the clamp into the clamp's hooks.

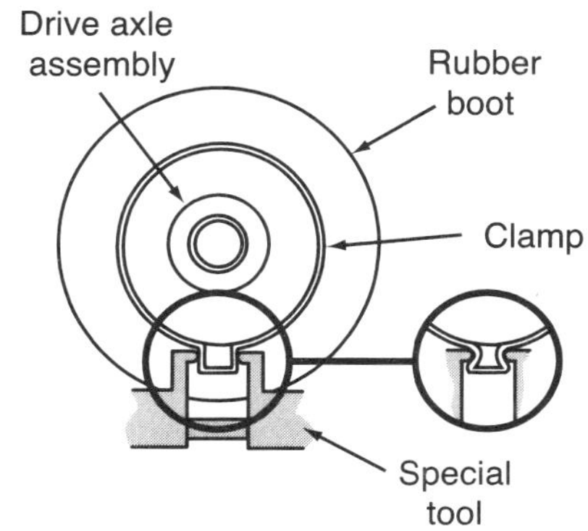


**Figure 9.** Properly installing and tightening a boot clamp.

**Figure 10.** Pack joint grease into the joint and boot. The entire packet of grease that came with the new boot needs to be forced into the joint and boot.

**Figure 11.** If the boot is dimpled or collapsed after you have installed it, carefully insert a dulled screwdriver between the boot and the shaft to allow air to enter the boot. The air should allow the boot to return ot its normal shape.

## Replacing CV Joints

The typical procedure for replacing CV joints begins with the removal of the drive axle. Etch a mark where the rubber boot ends on the half-shaft. Remove the clamps, then the boots. Since the boots will be replaced, you can cut off the old ones to save some time. Clean all excess grease from the shaft and joint. Using your service manual, determine how the joints are retained to the shaft. Replace all snap rings you remove during service.

The Rzeppa joint is most commonly used to fix a CV joint. This joint is held to the half-shaft or intermediate shaft in one of two ways. It may have an external retaining ring that fits into a bearing race. The clip must be expanded to separate the joint from the shaft. An internal retaining clip may be used. This clip fits into a groove on the shaft and expands into the inner race of the joint. The joint can typically be driven off the shaft with a soft-faced hammer.

Three types of plunging joints are most commonly used: the cross groove, double offset, and tripod tulip. The outer race of a cross groove joint is usually a flange bolted to the transaxle. Double offset joints are normally splined to the transaxle. Most often, the shaft and joint are held in position by an internal clip or retaining ring. To remove this type of joint, it must be disassembled. Other double offset joints use a circle clip. This type of joint can be driven off the shaft. The housing of a tripod tulip can be splined to the transaxle or bolted to a flange.

If you are removing a tripod-type joint, index the housing and inner tulip. Remove the retainer or tabs from the outer race, and slide the race from the joint while holding the balls and needle bearings in place. Remove the circlip nearest the end of the half-shaft and discard it. Do not remove the heavy inner spacer ring. It needs to remain on the shaft to help position the joint when you put the assembly back together **(Figure 12).**

To remove the spider from the housing, lift the retaining ring out of its groove and separate the joint. Once the spider is free, wipe away all excess grease and wrap tape or a rubber band around the balls and spider to prevent them from separating while the joint is outside of the housing. Use a brass drift and drive the joint off the shaft.

A Rzeppa plunge joint can be knocked off with a hammer and drift as well. Make sure the parts are indexed prior to disassembly.

**You Should Know** *Never hit the outer surface of the CV joint. This can cause the joint to break.*

Thoroughly clean the shaft and inspect it for wear. Clean the half-shaft with solvent and thoroughly dry it. Slide a new clamp and boot onto the half-shaft. Then, position the new joint on the splines and tap the joint until you hear a click, which indicates that the circlip has seated in the groove inside the joint. Pull on the joint to check that is properly locked on the shaft.

Install a new circlip. Do not overexpand or twist the circlip during installation. Pack all of the supplied grease into the assembly. Line up the boot to the etch mark and tighten the small clamp. Squeeze the remaining grease into the boot over the joint housing. Make sure the boot is not twisted or collapsed. Pull the large end of the boot over the joint housing. Reposition the large end of the boot, install the clamp, and tighten it. The axle can now be reinstalled in the car.

Many FWD vehicles use an intermediate shaft that has support bearings and seals that are serviceable. To provide noise-free operation, the bearings must be well lubricated. The seals retain the lubricant in the bearings as well as seal out contaminants. If the intermediate shaft is wet or covered with grease, carefully check the support bearing and replace the seals.

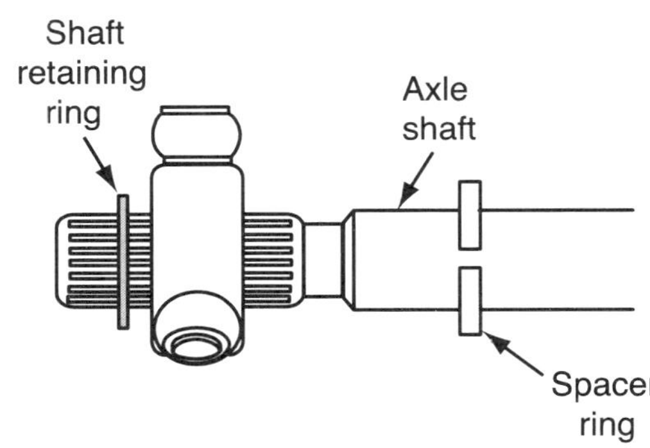


**Figure 12.** A circlip or snap ring is normally used to retain the joint on the shaft.

## SUMMARY

- Diagnosing FWD drive axle and CV joint problems should include gathering as much information as possible about the problem, road testing the vehicle, and conducting a thorough inspection of all related components.
- A clicking or popping noise that occurs only when the vehicle is going around a corner or turning is normally caused by a worn or damaged outer CV joint.
- A clunking noise during torque output changes indicates an inner CV joint wear problem.

- A continuous humming or growling sound may be caused by worn transaxle gears, worn wheel bearings, a faulty intermediate shaft bearing, or poorly lubricated CV joint.
- Defective joint boots are the most common cause of CV joint failure.
- To perform most services on a half-shaft, the unit must be removed from the vehicle.
- Always use the proper tools to remove the axle from the transaxle and its hub.
- Always use new hub nuts. Make sure they are properly tightened and staked in place.
- Check the condition of the grease to get an idea of the condition of the joint.
- A CV joint service kit includes a new boot, grease, and the required hardware.
- When installing the new clamps, they should be positioned so that the end with holes points in the direction of axle rotation and the holes fit over the locking hooks.
- Use the entire packet of grease that came with the new boot into the joint and boot.
- The Rzeppa joint is held to the half-shaft or the intermediate shaft by an external or internal retaining ring.
- The outer race of a cross groove joint is usually a flange bolted to the transaxle. Double offset joints are normally splined to the transaxle and are retained by an internal retaining ring. The housing of a tripod tulip can be splined to the transaxle or bolted to a flange.
- Make sure the bearings of the intermediate shaft are well lubricated and the seals are not damaged.

## REVIEW QUESTIONS

1. What questions should you ask a customer about a problem they are having with their vehicle before you take a road test?
2. Before condemning any part of the driveline as the cause of a vibration or noise problem, what things should be checked first?
3. While discussing the possible causes for a clunking noise during acceleration, Technician A says an outer CV joint could be the cause. Technician B says insufficient runout of the drive axle may be the cause of the problem. Who is correct?
   A. Technician A only
   B. Technician B only
   C. Both Technician A and Technician B
   D. Neither Technician A nor Technician B
4. Technician A says a sticking inboard joint will cause the car to shimmy during acceleration. Technician B says a worn inboard joint will cause a clunking during acceleration. Who is correct?
   A. Technician A only
   B. Technician B only
   C. Both Technician A and Technician B
   D. Neither Technician A nor Technician B
5. While diagnosing the cause of a clicking noise heard only when the car is turning, Technician A says the most probable cause is a worn wheel bearing. Technician B says the most probable cause is a worn outboard CV joint. Who is correct?
   A. Technician A only
   B. Technician B only
   C. Both Technician A and Technician B
   D. Neither Technician A nor Technician B
6. While discussing axle hub nuts, Technician A says they are never to be reused. Technician B says they should be carefully chiseled off the stub axle if they are difficult to remove. Who is correct?
   A. Technician A only
   B. Technician B only
   C. Both Technician A and Technician B
   D. Neither Technician A nor Technician B
7. While installing a new CV joint boot, Technician A disassembles and cleans the joint to remove all old grease. Technician B packs all of the lubricant contained in the service kit into the joint and the boot. Who is correct?
   A. Technician A only
   B. Technician B only
   C. Both Technician A and Technician B
   D. Neither Technician A nor Technician B
8. While reviewing the procedure for replacing an axle boot, Technician A says the position of the old boot should be marked on the shaft prior to removing it. Technician B says that if after the new boot is installed the boot is dimpled or collapsed, a dulled screwdriver should be used to allow air to enter the boot. Who is correct?
   A. Technician A only
   B. Technician B only
   C. Both Technician A and Technician B
   D. Neither Technician A nor Technician B

# Chapter 19 Servicing Non-Drive Axle Bearings

## INTRODUCTION

Wheel bearings are used on non-driving axles. Two separate tapered roller bearings are used in the front wheel hubs of many rear-wheel-drive cars. Each of the bearings rides in its own race. The rear wheel hubs on some front-wheel-drive cars have the same type of bearing setup.

Tapered roller bearings should be serviced at the recommended service intervals. The procedures to remove, clean, lubricate (repack), and install tapered roller bearings are basically the same for servicing a used bearing or installing a new one.

## BEARING REMOVAL

Except when making slight adjustments to the bearings, the wheel's hub and bearing assembly must be removed for all service work. Raise the vehicle to a workable height and take the vehicle's weight off the wheel. Then remove the tire and wheel assembly. Make sure you mark the location of the wheel on the hub so wheel balance is not affected after the wheel is installed on the vehicle. Typically, the procedure for removing the inner and outer tapered roller wheel bearings includes the following steps:

1. Remove the brake disc and hub. Suspend the brake caliper with a wire, and make sure it is out of the way.
2. A dust (grease) cap is in the center of the hub. Using slip-joint pliers or a special removal tool **(Figure 1),** wiggle the cap out of its recess in the hub to expose the bearing nut.
3. Now remove the cotter pin and nut lock from the end of the spindle **(Figure 2).**
4. Loosen and remove the spindle nut.

**Figure 1.** Remove the dust cap with dust cap pliers.

**Figure 2.** Use side cuts to remove the cotter pin from the nut lock and bearing nut.

5. Place a thumb over the thrust washer of the outer bearing and carefully slide the drum or rotor off the spindle. Support the brake drum or rotor with both hands so that it does not drag on the spindle as you remove it. If a brake drum catches on the shoes,

**Figure 3.** Use a large drift punch and a hammer to remove the bearing race.

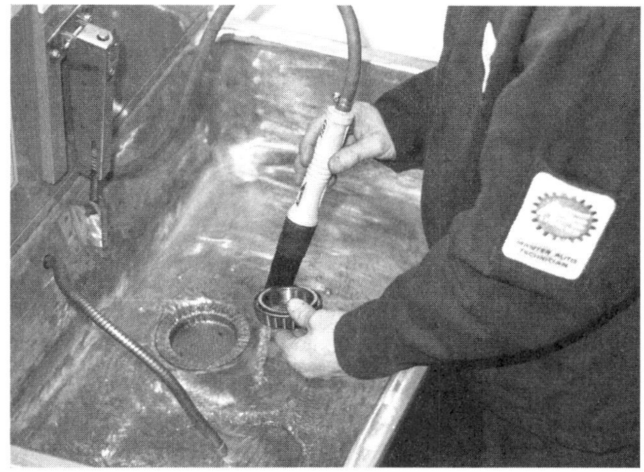

**Figure 4.** Thoroughly clean the bearing with solvent.

reinstall the adjusting nut and back off the brake adjustment. Then remove the drum.

6. Set the drum or rotor on a clean work surface with the inboard side down.
7. Lift the outer bearing and thrust washer from the hub and set them aside.
8. Turn the drum or rotor over so the inboard side is facing up.
9. A grease seal located on the back of the hub normally keeps the inner bearing from falling out when the hub is removed. Remove the inner bearing and grease seal with a seal puller, large screwdriver, or a large, nonmetallic drift punch. Be careful not to damage the hub while doing this, and discard the seal.
10. Keep the outer and inner bearings for each side separated if you plan to reuse them.
11. If the bearings will be replaced, their races must also be replaced. Remove the race by inserting a large drift punch **(Figure 3)** through the opposite side of the hub and placing it against the inner edge of the race. Strike the punch with a hammer at several locations around the circumference of the race until the race is driven from the hub. You can also remove a bearing race from a hub using a press and suitable adapters. Be careful not to damage the bore when removing the races or bearings.
12. If the bearings may be reused, clean and inspect them before proceeding.
13. Check the condition of the spindle. If it is damaged or excessively worn, the steering knuckle assembly should be replaced. Slight burrs can be removed with a fine file or emery cloth.

## CLEANING AND INSPECTION

After the bearings are removed from the hub, wipe the old grease off the spindle with clean, lint-free shop cloths or paper towels. Examine the old grease for metal particles. Any sign of metal particles is a clue to inspect the bearings very closely for wear and damage. Also, inspect the grease for dirt, rust, and signs of moisture that could indicate a leaking grease seal.

Wipe the grease off the bearing races and inspect them closely for wear or damage. Also, wipe as much old grease as possible from the bearings and inspect them for wear or damage. Inspect the grease. Rotate the rollers in the cage, and listen and feel for roughness.

Clean the bearings in a parts washer using clean petroleum-based solvent **(Figure 4).** Wash each bearing separately and keep them sorted to be sure they are reinstalled in their original hub. Rotate the bearings in the cage as you clean them, and remove old grease with a stiff-bristled brush. After cleaning, flush the cleaning solvent from the bearings with a non-petroleum brake-cleaning solvent.

Do not dry bearings or races with paper towels. If you use a shop towel, be sure it is lint free. Lint from shop towels or paper towels may contaminate the bearing.

**You Should Know** *Never wash a sealed bearing or a bearing that is shielded on both sides because solvent may enter the bearing and destroy the lubricant in the bearing resulting in very short bearing life.*

Let the bearings air dry for a few minutes and then blow any remaining solvent from the bearings and cages with low-pressure air. Direct the air through the bearing from side to side along the axis of the rollers. Make sure the shop air supply is free from moisture. If the bearings are not

**Figure 5.** Wrap the cleaned bearing in waterproof papers until it is time to grease it.

going to be installed immediately, cover them with a protective lubricant and wrap them in waterproof paper **(Figure 5).**

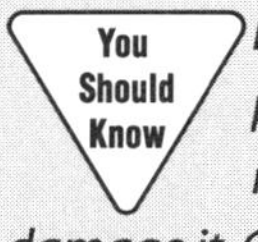


*Do not spin the bearings with compressed air during cleaning. High-speed rotation of an unlubricated bearing will damage it. Compressed air can also dislodge bearings from the cage and cause additional damage or serious personal injury.*

## Inspection

After they are cleaned, inspect the bearings and races. Any damage means they should be replaced. Check the entire surface of the rollers and carefully inspect each race **(Figure 6).**

**Figure 6.** An excessively worn bearing and race is shown on the left, and a good bearng and race is shown on the right.

If a roller or race shows signs of any of these conditions or if you are unsure of its condition, replace the bearing:

- *Galling*—indicated by metal transfer or smears on the ends of the rollers. It is caused by overloading, lubricant failure, or overheating.
- *Etching*—a condition that results in a grayish-black bearing surface. It is caused by insufficient or incorrect lubricant.
- *Brinelling*—occurs when the surface is broken down or indented. It is caused by impact loading, usually because the adjustment is too loose.
- *Abrasive wear*—results in scratched rollers. It is caused by dirty or contaminated bearing lubricant.

Inspect the seal mounting area in the hub for metal burrs. Remove any burrs with a fine round file. Also, inspect the bearing and seal mounting surfaces on the spindle. Again, small metal burrs may be removed from the spindle with a fine-toothed file. If the spindle is severely scored in the bearing or seal mounting areas, the steering knuckle assembly should be replaced.

## Replacement Parts

A replacement wheel bearing consists of the inner and outer races and rollers in their cage. Make sure replacements are the same as the original bearings. Visually compare the old with their replacement. Normally, if the part numbers are same, the new bearings are correct for the application.

Check the fit of the new bearing seal on the spindle and in the hub. The seal lip must fit snugly on the spindle, and the seal case must fit properly in the hub opening. The part number on the old seal and the replacement seal should be the same.

When installing a new bearing, leave the bearing in its protective packaging until you are ready to lubricate it. Be sure your hands are clean before handling a bearing. Some technicians wear latex gloves or apply a light film of motor oil to their fingers before handling a bearing. Doing this prevents the acid on your fingers from damaging the bearing.

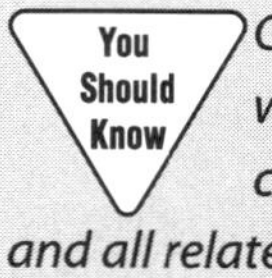


*Cleanliness is very important during wheel bearing service. Always maintain cleanliness of hands, tools, work area, and all related bearing components. One small piece of dirt in a bearing will cause bearing failure.*

A bearing and its outer race must be replaced as a set. Installing a new bearing in a used race will cause an uneven wear pattern that can lead to premature bearing wear or failure.

## BEARING LUBRICATION

After the old bearings and races have been cleaned and inspected or before installing new bearings, they should be packed with grease that is made especially for wheel bearing use. It is important that the grease be forced into and around the outside of the bearing and between the rollers, cone, and cage. Merely coating the outside of the bearing with grease will not do the job.

Tapered roller bearings can be packed with grease with a bearing packer or by hand. Care must be taken not to get grease on the brake disc or on any part that will contact the disc. After packing the bearings, set them aside on clean lint-free paper until you are ready to install them.

### Wheel Bearing Grease Specifications

The grease for wheel bearings should be smooth textured, consist of soaps and oils, and be free of filler and abrasives. Recommended are lithium complex (or equivalent) soaps or solvent-refined petroleum oils. Because wheel bearing grease must operate at higher temperatures than chassis grease, it contains oils and other lubricants that have been made particularly heavy with thickening agents. Additives may have also been added to inhibit corrosion and oxidation. Greases are classified by the National Lubricating Grease Institute (NLGI) to indicate their application **(Figure 7).**

Always use the type of grease specified by the vehicle's manufacturer. Disc brakes cause wheel bearings to run at higher temperatures than drum brakes do, and many wheel bearing greases are labeled for use with disc brakes. If the wrong grease is used, it may not provide the correct protection or it may liquefy from the heat and leak out of the seals. Failure to maintain proper lubrication might result in bearing damage causing a wheel to lock.

**Figure 7.** NLGI labels show the application for the various grades of grease.

**You Should Know** *Keep grease containers covered when not in use. Uncovered grease containers are easily contaminated with dirt and moisture.*

### Packing Tool

Tapered roller bearings are thoroughly packed with grease with a bearing packer **(Figure 8).** Place the bearing in the tool with its taper pointing down. Screw the cone down on the bearing. Apply a hand-operated grease gun to the fitting on the center shaft of the packer and force grease into the bearing and out around the rollers in the cage. Some bearing packers contain a supply of grease and a grease gun is not needed. Pushing on the cone handle forces the bearing down into the packer and lubricant is forced into the bearing.

**Figure 8.** A typical wheel bearing packer.

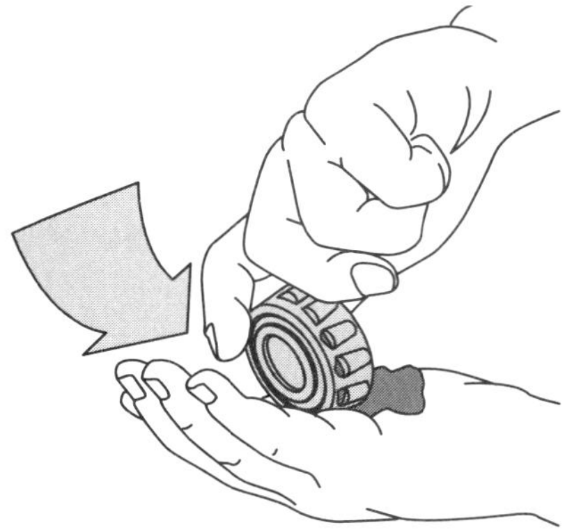

Figure 9. Grease can be forced into the bearing with your hands.

## Hand Packing

A bearing can be packed with your hand if a tool is not available. Place a lump of grease in the palm of one hand and grasp the bearing in the other hand. Force the grease through the large end of the bearing and around the rollers **(Figure 9).** Squeeze the grease into the bearing. Rotate the bearing several times while packing the grease to ensure that grease is forced completely around each roller. Finish by spreading an even film of grease around the outside of all the rollers.

## BEARING REPLACEMENT

To install a tapered roller wheel bearing:

1. With the inboard side of the drum or rotor facing up, apply a thin film of grease to the back of the race and to the hub's bore.

> You Should Know *Always use the recommended grease. If the wrong grease is used, it may not offer the correct protection or it may liquefy from the heat and leak out of the seals.*

2. Position the inner bearing's race squarely at the end of the bore and use a bearing driver **(Figure 10)** or a suitably sized socket and a hammer to drive the race into the hub.
3. Listen for a change in the sound as the race is fully seated in the hub.
4. Apply a light coat of grease to the outer race for the inner bearing.
5. Turn the drum or rotor over and select the correct driver for the outer bearing race.
6. Apply a thin film of grease to the back of the race and to the hub's bore.

Figure 10. A bearing driver that can be used to drive in a bearing race.

7. Position the race squarely at the end of the bore and use a bearing driver or a suitably sized socket and a hammer to drive the race into the hub.
8. Listen for a change in the sound as the race is fully seated in the hub.
9. Apply a light coat of grease to the outer race for the outside bearing.

## Bearing Reassembly

To assemble the wheel bearings in a drum or rotor after new bearing races have been installed or the original ones cleaned, follow these steps:

1. With the inboard side of the drum or rotor facing up, insert the freshly packed bearing into the race.
2. Place a new grease seal into the inner bore of the hub. Make sure the seal is installed in the proper direction with the garter spring and higher part of the lip toward the lubricant in the hub.

> You Should Know *Be very careful not to touch the brake assembly with your greasy hands. Clean your hands before handling the brake parts or use a clean rag to hold the brake assembly.*

3. Drive the seal into place with a seal driver **(Figure 11)** until the outer surface of the seal is flush with the hub.

Figure 11. Drive the new seal into the hub. Make sure it is fully seated.

4. Apply a thin film of bearing grease to the lip of the seal to protect it during installation.
5. Turn the hub over and apply a light coat of grease to the outer race for the outer bearing.
6. Spread a light coating of bearing grease around the entire spindle and inside the hub to a level just below the bearing races.
7. Place a light coating of wheel bearing grease on the spindle and slide the hub assembly onto the spindle. Be careful not to hit the inner bearing on the spindle threads.
8. Support the rotor or drum with one hand and install the freshly packed outer bearing and thrust washer into the hub.
9. Install the bearing and adjust the nut finger tight against the thrust washer.
10. Be sure the bearing, washer, and nut are clean and greased, and then install them.
11. The initial tightening seats the bearings into their races.
12. Place some grease in the wheel hub cavity and tighten the nut until it is finger tight.

## BEARING ADJUSTMENT

Wheel bearing end play is the amount of horizontal wheel bearing hub movement. If a bearing has a preload condition, a slight tension is placed on the bearing. Loose front wheel bearing adjustment results in lateral front wheel movement and reduced directional stability. If the wheel bearing adjusting nut is tightened excessively, the bearings may overheat, resulting in premature bearing failure. The bearing adjustment procedure may vary depending on the make of vehicle. Always follow the procedure in the vehicle manufacturer's service manual.

Normally, the procedure for adjusting wheel bearing end play is the same for new bearings and reused bearings. It is also the same basic procedure for conducting wheel bearing adjustment checks during preventive maintenance services.

If the vehicle is not on jack stands or a hoist, raise it to a comfortable working height. Then remove the wheels and move the disc brake caliper if necessary. Remove the dust cap from the center of the hub and rotate the drum or rotor by hand while lightly tightening the adjusting nut to seat the bearings. Then adjust the bearings by one of the following methods:

- Rotate the drum or rotor and lightly snug up the adjusting nut with a wrench to seat the bearings. Then back off the nut 1/4 to 1/2 turn or until it is just loose while continuing to rotate the drum or rotor. Tighten the nut by hand to a snug fit and lock it.
- Rotate the drum or rotor and tighten the adjusting nut with a torque wrench to the specified torque **(Figure 12),** which is usually 17 to 25 foot-pounds.
- To use a dial indicator, rotate the drum or rotor and tighten the adjusting nut to 12 to 25 foot-pounds. Then back off the nut 1/4 to 1/2 turn or until it is just loose. Mount the base of the indicator as close as possible to the center of the hub. Place the tip of the plunger on the tip of the spindle. Set the indicator to zero. Move the drum or rotor in and out and note the indicator reading. The total movement is the amount of bearing freeplay.

Once the end play is properly adjusted, install the nut lock over the top of the bearing's adjusting nut so its slots align with the cotter pin hole in the spindle **(Figure 13).**

**Figure 12.** Rotate the wheel and tighten the adjusting nut with a torque wrench to the specified torque to seat the bearings.

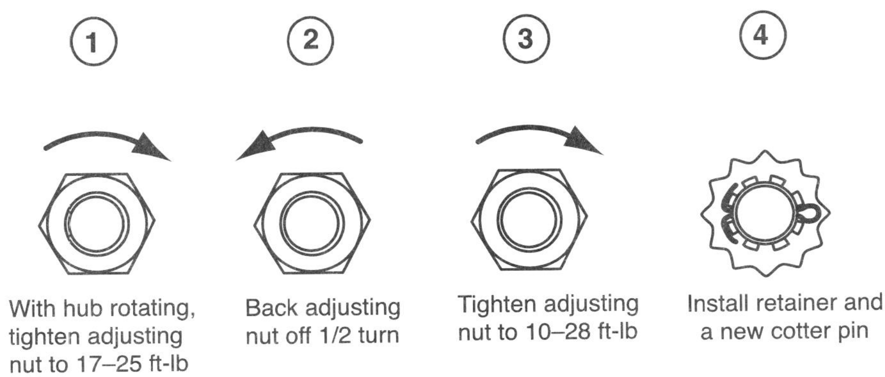


**Figure 13.** Typical spindle nut, nut lock, and cotter pin installation.

**Figure 14.** Install the dust cap by tapping it in place with a rubber mallet.

Then install a new cotter pin through the spindle and nut lock and bend its ends to secure it. Pack grease into the dust cap and install it **(Figure 14).** Make sure the hub rotates freely.

If the axle has drum brakes that were backed off to allow removal of the drum, readjust the brakes. If the axle has disc brakes, reinstall the caliper. Then, reinstall the wheel and tire and lower the vehicle to the ground. Make sure to torque the wheel lugs to specs.

## SUMMARY

- Tapered roller bearings must be periodically cleaned and repacked with grease and then have the end play adjusted.
- The cause of bearing failure can range from improper lubrication or incorrect mounting to poor condition of the shaft housing or bearing surfaces.
- A bearing and its race must be replaced as a set.
- Whether repacking and reinstalling a used bearing or installing a new one, always install a new grease seal.
- After the bearings are removed from the hub, wipe the old grease off the entire assembly and examine the old grease for metal particles from the bearings or their races.
- Clean the bearings in a parts washer using clean petroleum-based solvent.
- After cleaning, flush all of the cleaning solvent from the bearings with a non-petroleum brake-cleaning solvent.
- Bearings can be packed with grease with a bearing packer or by hand.
- Before installing the bearings and the hub, inspect the spindle for wear or damage.
- Spread a light coating of bearing grease around the entire spindle and inside the hub to a level just below the bearing races.

## REVIEW QUESTIONS

1. Why is some wheel bearing grease specifically marked for disc brake use?
2. What can result from incorrectly adjusted wheel bearings?
3. Describe the three ways tapered bearings can be adjusted.
4. While cleaning a used bearing assembly, Technician A wipes the bearings and races with a paper towel after they have been cleaned in solvent. Technician B examines the old grease from the bearings and races before cleaning them. Who is correct?
   A. Technician A only
   B. Technician B only
   C. Both Technician A and Technician B
   D. Neither Technician A nor Technician B
5. While replacing wheel bearings, Technician A says the part number on the old grease seal and the replacement seal should be the same. Technician B says if the original bearing race is not damaged or worn, a new bearing can be used in the race after it is thoroughly lubricated. Who is correct?
   A. Technician A only
   B. Technician B only
   C. Both Technician A and Technician B
   D. Neither Technician A nor Technician B

# Chapter 20 Servicing Drive Axle Bearings

## INTRODUCTION

The bearings on an axle that drives the wheels are called axle bearings. Bad axle and wheel bearings will cause handling and tire wear problems.

Most FWD cars do not use tapered roller bearings. They are usually equipped with one bearing assembly per whee. This assembly has two rows of balls and is permanently sealed so it requires no periodic lubrication. The double-row ball bearings have races that allow the bearing to handle radial and side loads.

Servicing the rear drive axles on most RWD vehicles involves unique procedures when compared to servicing FWD axles. Most RWD axles are enclosed in a single housing, the rear axle assembly, and rotate on bearing. The axles seldom require repair, however axle bearings and seals are commonly replaced.

## SERVICING FWD AXLE BEARINGS

The front bearings on many FWD and 4WD vehicles must be pressed in and out of the hub. To do this, the axle or half-shaft is removed. The bearings may be sealed and require no additional lubrication, or they may need to be packed with grease. In most cases, the bearings are not adjusted. A heavily torqued axle nut is used to retain the assembly on the axle. This nut is typically replaced after it has been removed and is staked in place after it is tightened.

### Axle Bearing Checks

A hub-and-bearing assembly will become noisy when it is excessively worn if spalling or brinelling of the bearings has occurred or if the unit has a cracked bearing race. However, other signs of failure will usually be evident before the unit becomes noisy.

Bearing wear can be checked by grasping a front tire at the 12 o'clock and 6 o'clock positions with the vehicle supported under its lower control arms. Pull outward on the top of the tire while pushing inward at the bottom. Alternate pushing and pulling while you check for play. This is the same procedure used for checking front wheel bearing play on RWD cars that use tapered roller bearings in the front hubs. This check can also be conducted on the rear wheels of GM and other cars that have sealed hubs.

Normally, more than 0.010-inch play indicates a defective axle bearing assembly. Any movement that is felt while conducting the above check without forcing the tire to move will indicate more than the allowable play. Movement of less than 0.010 inch will not be felt.

Replacement is necessary when a front hub-and-bearing assembly has excessive play or is making noise. Not replacing a unit that is failing may cause the assembly to overheat and lock up while the vehicle is driven.

You Should Know *Failure to replace a defective axle bearing can result in the loss of braking and steering control.*

To check for a binding hub and bearing assembly, simply rotate the tire and wheel. If the tire does not turn freely, suspect dragging brakes or a binding hub-and-bearing assembly.

If the axle bearings have failed, check the vehicle's electrical grounds. This is a common cause for bearing failure on some vehicles. With a poor ground circuit, the axle provides the ground, and electrolysis causes the bearings to fail. If poor grounds are the cause of bearing failure, a replacement bearing will fail prematurely unless the grounds are corrected.

## REPLACING FRONT AXLE BEARINGS

The axle bearings on most FWD vehicles are one-piece, unitized assemblies that do not require periodic lubrication but must be replaced if they are worn or damaged. The procedure for removing and replacing front axle bearings is different for different vehicles. Regardless of design, the hub nut and wheel assembly must be removed first.

Have someone apply the brakes with the vehicle's weight on its tires. Loosen the axle hub nut and wheel lug nuts. Then, remove the axle hub nut. If no one is available to apply the brakes for you, use a brake-pedal depressor from an alignment machine to depress and hold the brake pedal down. Then, raise the vehicle and remove the tire and wheel assembly.

There are two basic bearing designs in one design; the bearing is part of the wheel hub assembly. The hub with the bearing assembly is bolted to the steering knuckle. Begin bearing replacement by unbolting the front brake caliper. Once loose, move it out of the way and suspend it with wire **(Figure 1).** Then remove the brake rotor and the hub retaining bolts **(Figure 2).** Remove the brake air deflector or backing plate.

Using the correct type puller, remove the hub assembly from the knuckle **(Figure 3).** Remove the seal from the inside of the knuckle. Inspect and clean the hub's mounting surface on the knuckle. Remove the old bearing and install a new one. Make sure the bearing is fully seated. Put a ight coat of clean transmission fluid on the outside of the bearing.

With the new bearing, place the new seal and the hub assembly onto the knuckle **(Figure 4).** Then tighten the retaining bolts to the specified torque. Install the hub nut and the tire and wheel assembly. Lower the vehicle and tighten the hub nut and lug nuts to specifications.

Some axle bearings are housed in the steering knuckle **(Figure 5).** To service these, the steering knuckle must be

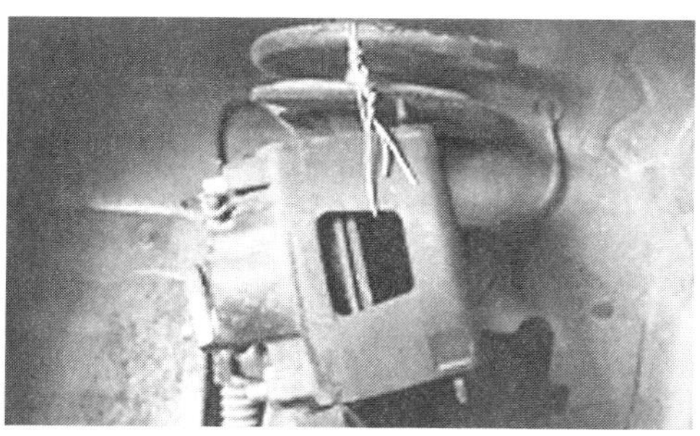

**Figure 1.** After unbolting the front brake caliper, move it out of the way and suspend it with wire.

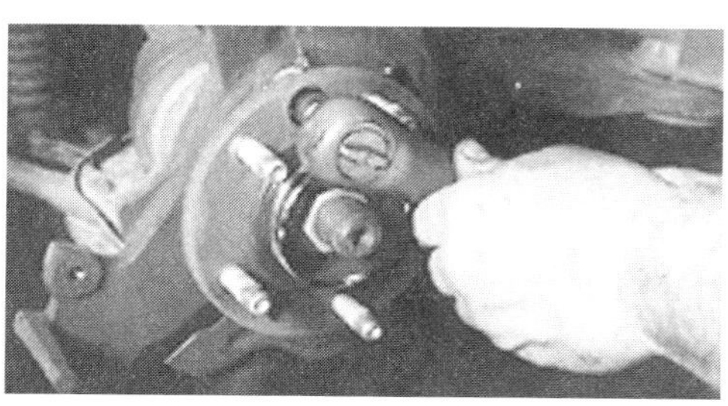

**Figure 2.** Remove the brake rotor and the hub retaining bolts.

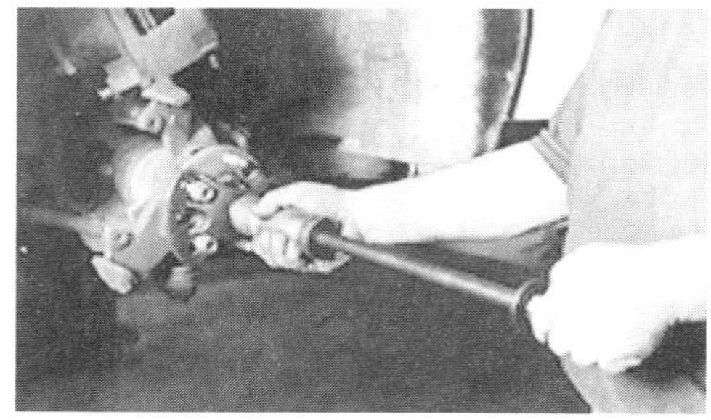

**Figure 3.** Using the correct type puller, remove the hub assembly from the knuckle.

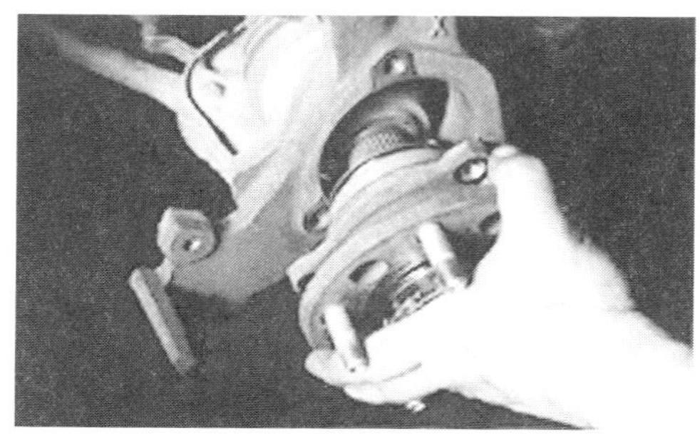

**Figure 4.** Place the new seal and the hub assembly, with the bearing, into the knuckle.

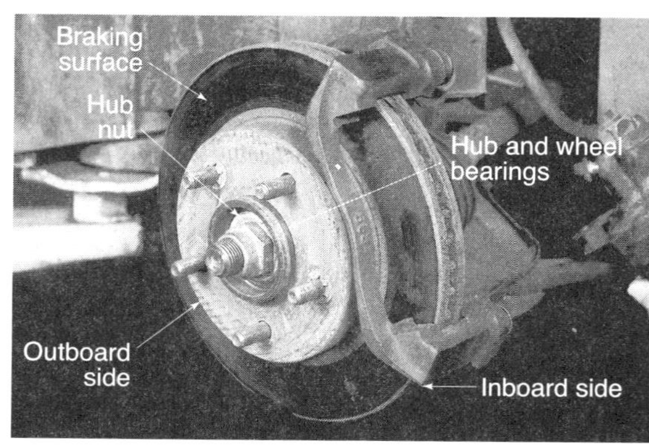


**Figure 5.** Some axle bearings are housed in the steering knuckle, which must be removed and the outer bearing pressed out and the new pressed in. To do this, the hub nut, wheels, brake caliper, and brake rotor must be removed first.

removed and the outer bearing pressed out and the new pressed in. After the hub nut and wheels have been removed, remove the brake caliper. Make sure you support it with a piece of wire so the flexible brake hose does not support its weight. Then remove the brake rotor, which is normally attached with one or two bolts.

Then, separate the lower ball joint. Now remove the cotter pin and castellated nut from the tie-rod end and pull the tie-rod end from the knuckle. The tie-rod end can usually be freed from its bore by using a small gear puller or by tapping on the metal surrounding the bore. On some cars, an **eccentric washer** is located behind one or both of the bolts that connect the spindle to the strut. These eccentric washers are used to adjust camber. Mark the exact location and position of the washer on the strut before you remove it **(Figure 6).** This allows you to reinstall the washer and the assembly without drastically disturbing the car's camber angle. Index the camber eccentric bolt to ensure proper camber adjustment during reassembly.

Now remove the hub-and-bearing to knuckle retaining bolts. Install the appropriate puller and press the hub-and-bearing assembly from the half-shaft. With a soft-faced hammer, tap on the hub assembly to free it from the knuckle. Remove the knuckle to strut bolts **(Figure 7)** and the steering knuckle. Support the drive axle with wire.

With the steering knuckle off, remove the snap ring or collar that retains the bearing. Mount the knuckle on a hydraulic press so that it rests on the base of the press with its hub free. Using a mandrel slightly smaller than the inside diameter of the bearing, press the hub out of the bearing **(Figure 8).** On some cars, this will cause half of the inner race to break out. To remove that half of the inner race, use a small gear puller **(Figure 9).**

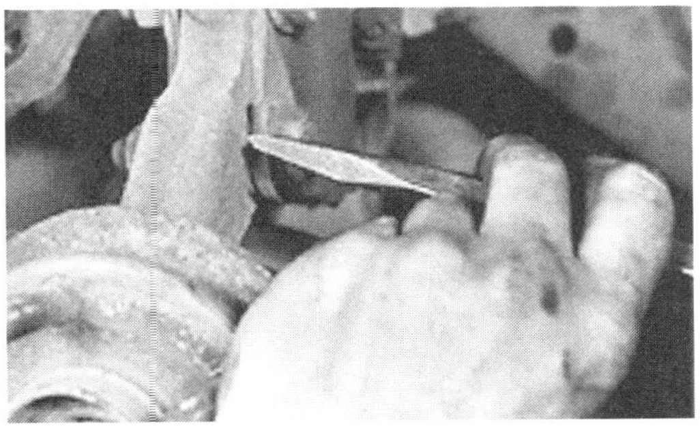

**Figure 6.** On some cars, an eccentric washer is located behind one or both of the bolts that connect the spindle to the strut. Mark the exact location and position of this washer on the strut before you remove it. This allows you to reinstall the washer and the assembly without drastically disturbing the car's camber angle.

**Figure 7.** Remove the knuckle to strut bolts and the steering knuckle. Support the drive axle with wire.

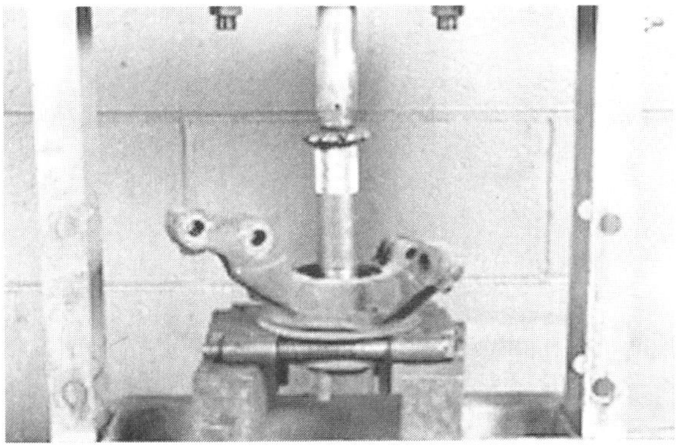

**Figure 8.** Mount the knuckle on a hydraulic press so that it rests on the base of the press with the hub free. Using a mandrel slightly smaller than the inside diameter of the bearing, press the hub out of the bearing.

**Figure 9.** On some cars, pressing the bearing out will cause half of the inner race to break out. To remove the inner race, use a small gear puller.

Remove the snap rings and/or retaining plate that hold the bearing in the knuckle. Then, using a mandrel large enough to touch the outer race of the bearing, press the bearing out of the knuckle. Inspect the bearing's bore in the knuckle for burrs, score marks, cracks, and other damage **(Figure 10).**

If no defects are found, install a new snap ring in the hub side, lightly coat the outer surface of the new bearing and the inner surface of the knuckle bore with moly lube, and then press the bearing into the knuckle **(Figure 11).** Make sure the mandrel applies pressure only to the outer race of the bearing. Applying pressure to any other part of the bearing will ruin it.

Using the proper tool, press a new steering knuckle seal into place. Lubricate the lip seal and coat the inside of the knuckle with a thin coat of grease. With the bearing and seal fully seated against the hub-side snap ring, install the snap ring on the other side of the knuckle. Then, lubricate and install a new O-ring around the rear bearing housing. Install the brake splash shield and hub assembly on the knuckle.

To complete the assembly of the steering knuckle, the hub must be pressed into the inner race. Use a hollow mandrel that contacts only the inner race of the bearing and that has an inner circumference large enough to allow the hub to fully seat in the bearing **(Figure 12).**

To install the steering knuckle slide it over the drive axle while holding the CV joint. Take time to line up the half-shaft splines with those in the bearing. Tighten the bolts to specifications. Then mount the knuckle to the strut and tighten the bolts. Make sure the camber marks to the strut are positioned properly. Reinstall the tie-rod end to the steering knuckle. Then mount the brake caliper and rotor. Torque all bolts to specifications.

Tighten the axle to the hub nut to pre-torque specifications. This ensures that the bearing assembly is properly seated. Then install the tire/wheel assembly and lower the vehicle. Once the car is set on the ground, tighten the new axle hub nut and wheel lug nuts to their specified torque. Road test the car and recheck the torque on the hub nut. Then, stake or use a cotter pin to keep the hub nut in place after it is tightened.

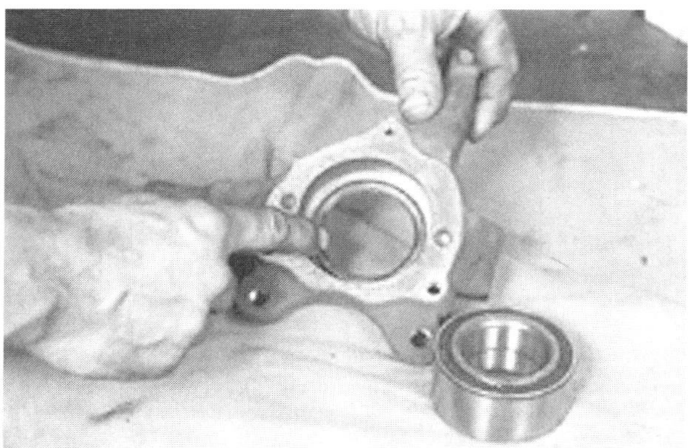

**Figure 10.** Inspect the bearing's bore in the knuckle for burrs, score marks, cracks, and other damage.

**Figure 11.** Lightly coat the outer surface of the new bearing and the inner surface of the knuckle bore with moly lube, and then press the bearing into the knuckle.

**You Should Know** *After you have replaced one or more CV joints or removed a steering knuckle, always check the vehicle's wheel alignment. Although making reference marks during disassembly will get the alignment angles close, they will not be at the desired settings.*

**Figure 12.** Press the hub into the inner race. Use a hollow mandrel that contacts only the inner race of the bearing and that has an inner circumference large enough to allow the hub to fully seat in the bearing.

## SERVICING RWD DRIVE AXLES

RWD axle bearings typically will not cause the types of handling problems caused by faulty FWD axle bearings. The most probable customer concern will be noise. If the rear axle bearings are identified as the source of an abnormal noise, the axle must be removed to replace or service the bearing. There are other rear axle problems that may cause noise and vibration problems, and these should be checked and corrected.

Axle shaft replacement is also required if there is excessive runout of the axle's flange. To check the runout of the **flange,** position a dial indicator against the outer flange surface of the axle **(Figure 13).** Apply slight pressure to the center of the axle to remove the endplay in the axle, and then zero the indicator. Slowly rotate the axle one complete revolution and observe the readings on the indicator. The total amount of indicator movement is the total amount of axle flange lateral runout. Compare the measured runout with specifications.

Inspect the wheel studs in the axle flanges. If they are broken or bent, they should be replaced. Also, check the condition of the threads. If they have minor distortions, run a die over the stud. If the threads are severely damaged, the stud should be replaced. Studs are normally pressed in and out of the flange. Make sure you install the correct size stud.

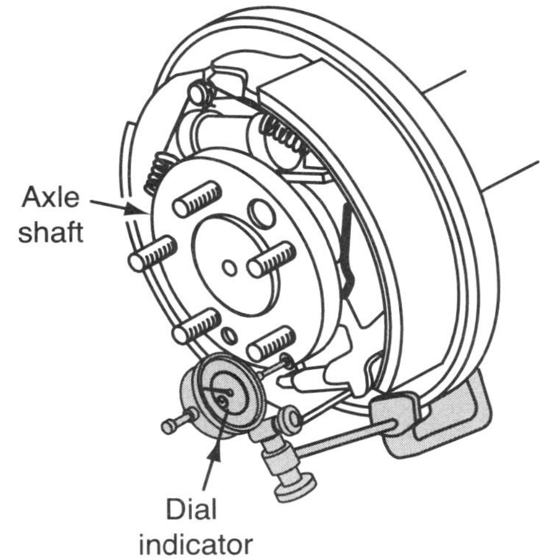


**Figure 13.** Checking axle flange runout.

## REMOVING AXLES

Axles are commonly retained in the housing by a retaining plate. Normally, the way the axles are held in the housing is quite obvious after the rear wheels and brake assemblies have been removed. If a retainer and three or four bolts hold the axle shaft in **(Figure 14),** it is not neces-

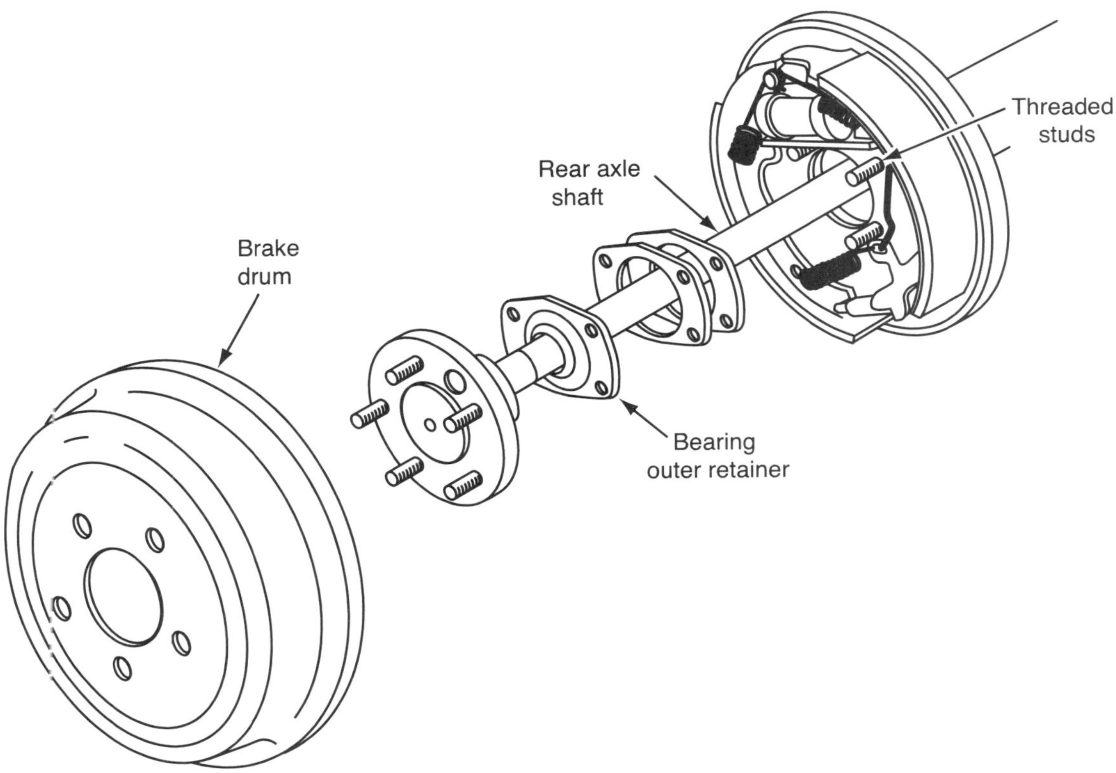


**Figure 14.** Location of an axle bearing retainer.

**Figure 15.** Sometimes an axle puller is needed to free the axle from the housing.

sary to remove the differential cover to remove the axle. Most ball and tapered roller bearing-supported axle shafts are retained in this manner. To remove the axle, remove the four bolts that hold the retainer to the backing plate, and then pull the axle out. Normally the axle shaft will slide out without the aid of a puller, but sometimes a puller is required **(Figure 15).**

A straight roller bearing-supported axle shaft does not rely on a retainer to secure it. Rather, a **C-lock** (washer) is used to retain the axle shaft **(Figure 16).** This C-lock is located inside the differential, and the differential cover must be removed to gain access to it. To remove this type of axle, first remove the wheel, brake drum, and differential cover. Then, remove the differential pinion shaft retaining bolt and differential pinion shaft. Now push the axle shaft in and remove the C-lock. The axle can now be pulled out of the housing.

You Should Know *You can also identify if the axles are held by C-locks by pushing the vehicle from side to side. You will hear a slight clunking noise if there are C-locks.*

## AXLE BEARINGS

There are three types of rear axle bearings used in today's vehicles: ball, roller, and tapered roller bearings. Ball bearings are lubricated with grease packed in the bearing. Rear axle gear oil does not lubricate this type of bearing. An inner seal, designed to keep the gear oil from the bearing, rides on the axle shaft just in front of the retaining ring **(Figure 17).**

This type bearing also has an outer seal to prevent grease from spraying onto the rear brakes. This type bearing is pressed on and off the axle shaft. A retainer ring is made of soft metal and is pressed onto the shaft against the wheel bearing. Never use a torch to remove the ring. Rather, drill or notch it in several places with a cold chisel to break the seal. The ring can then be slid off the shaft easily.

The gear oil in the axle housing lubricates straight roller axle bearings. Therefore, only a seal to protect the brakes is necessary with these bearings. These bearings are pressed into the axle housing. To remove them, the axle must be removed and the bearing pulled out of the housing. With the axle out, inspect the area where it rides on the bearing for pits or scores. If pits or score marks are present, replace the axle.

Tapered roller axle bearings are not lubricated by gear oil; rather, they are sealed and lubricated with wheel grease. This type of bearing uses two seals like the ball bearing type. Tapered roller bearings must be pressed on and off the axle

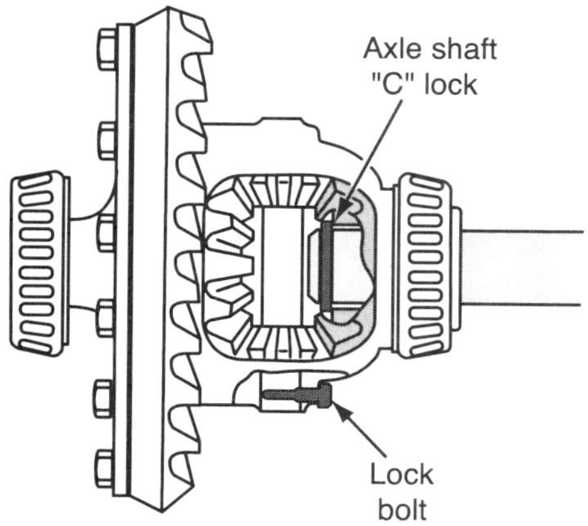


**Figure 16.** A rear axle "C" lock and lock bolt, which is the pinion shaft retaining bolt.

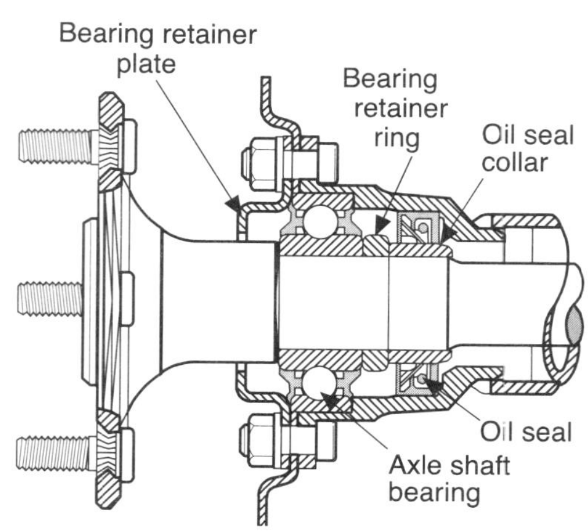


**Figure 17.** Location of oil seals on a ball bearing-type axle shaft.

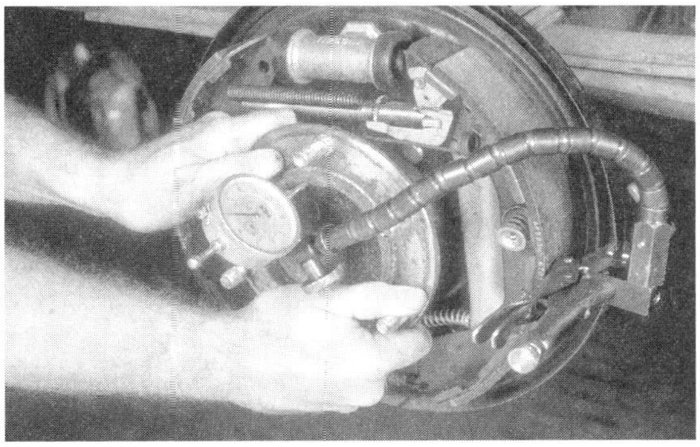

Figure 18. Check axle shaft end play.

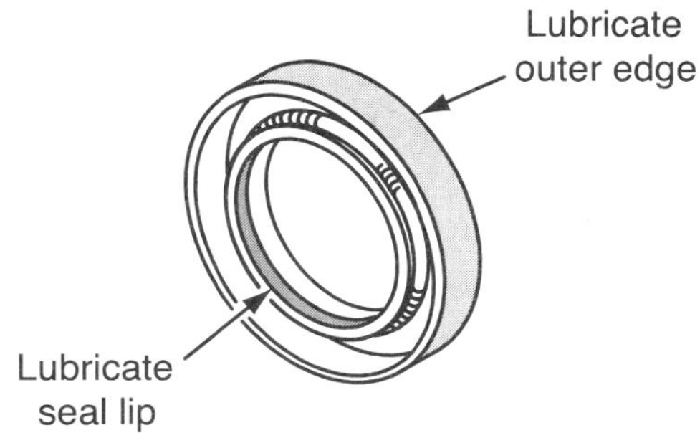


Figure 19. Lubricate the outer edge and inner seal lip prior to installing a new oil seal.

shaft using a press. After the bearing is pressed onto the shaft, it must be packed with wheel bearing grease.

After packing the bearing, install the axle in the housing and check the shaft's end play **(Figure 18).** Adjust the end play to specifications. If the end play is not within specifications, change the size of the bearing shim or correct it with the adjusting nut.

## Servicing the Bearing Bore

When an axle is removed from a housing, the bearing's bore in the housing should be inspected and cleaned. If the bore is corroded, sand it lightly with fine emery cloth and coat it with non-water-soluble grease. This will help prevent corrosion of the bearing's outer race.

## AXLE SHAFT SEALS

The installation of new axle shaft seals is recommended whenever the axle shafts have been removed. Some axle seals are identified as being either right or left side. When installing new seals, make sure to install the correct seal in each side. Check the seals for markings of "right" or "left" or for color coding.

Coat the outer edge of the new seal with gasket sealer and apply a coat of lubricant the seal's inner lip **(Figure 19).** Make sure the lip seal is facing the inside of the axle housing, then use the correct sized seal driver to install the seal squarely in the axle tube. Some General Motors products have axle seals on the inside the bearing and an O-ring on the outside of the bearing. Both of these seals should be lubricated prior to installation.

**You Should Know** *Be careful not to damage a new seal when installing the axle shaft. Support the entire length of the axle, and keep it flat while inserting it into the axle tube.*

## INSTALLING AXLES

In most cases, the installation of axle shafts is a simple procedure. Prior to installing the shafts, make sure you have installed all of the bearings, seals, and retaining plates on the shaft.

Some axles require an end play adjustment after installation. This adjustment is made by an adjusting nut or by selective shims. The shims are normally positioned between the retainer plate and the axle housing. The adjusting nut threads into the axle retainer plate. End play adjustments are made by adding or subtracting shims or by turning the adjusting nut. This adjustment is done on one side of the housing but sets the end play for both sides.

To check the end play, position the dial indicator so it is able to measure the end-to-end movement of the axle. Push the axle into the housing and set the indicator to zero. Then, pull the axle out and note the reading on the indicator. This reading is the amount of end play in the shaft. Compare the reading against specifications and correct the end play as necessary.

# SUMMARY

- Replace a front hub-and-bearing assembly when it has excessive play or is making noise.
- Rotate the tire. If it does not turn freely, suspect dragging brakes or a binding hubandbearing assembly.
- Basically there are two FWD front bearing designs; the bearing may be part of the wheel hub assembly, or the bearing is pressed into the backside of the knuckle.
- Replace a drive axle if it is damaged or bent, or if its flange has excessive runout.
- The axle and its flange can be checked for runout with a dial indicator.
- Wheel studs should be carefully checked and replaced if they are broken or bent.
- A straight roller bearing-supported axle shaft uses a C-lock to retain the axle shaft.
- Some axles require end play adjustments that are made with an adjusting nut or selective shims.
- Ball bearings are lubricated with grease packed in the bearing at the factory and have an inner seal to keep the gear oil from the bearing and an outer seal to prevent grease from spraying onto the rear brakes.
- Ball bearings are pressed on and off the axle shaft after the retainer ring is removed.
- The gear oil in the axle housing lubricates straight roller axle bearings, and this type axle has an outer seal to protect the brakes.
- Tapered roller axle bearings are sealed and lubricated with wheel grease and have an inner and outer seal.
- Axle bearings should be inspected for wear or other damage. If they show any evidence of damage, they should be replaced.
- When an axle is removed from a housing, the bearing's bore in the axle housing should be inspected and cleaned.
- New axle seals should be installed whenever the shafts have been removed.

# REVIEW QUESTIONS

1. What type of scheduled maintenance is normally required for the front wheel bearings of a FWD vehicle? How often should this be performed?
2. What is the easiest way to check for a binding hub and bearing assembly?
3. How many basic types of FWD front wheel bearings are there, and how do the procedures for replacing the bearings differ between the designs?
4. What is the procedure for checking bearing wear on the front wheels of FWD vehicles?
5. How should you check the runout of a drive axle flange?
6. How do you check and correct axle end play, and do you need to do this on all axles?
7. While servicing FWD front wheel bearings, Technician A says the axle hub nut sets the amount of bearing play at the wheel. Technician B says some FWD front bearings are removed from the steering knuckle after the knuckle has been removed from the car. Who is correct?
   A. Technician A only
   B. Technician B only
   C. Both Technician A and Technician B
   D. Neither Technician A nor Technician B
8. Technician A says some axle shafts are retained in the housing by a plate and bolts. Technician B says a C-lock retains some axle shafts in the housing. Who is correct?
   A. Technician A only
   B. Technician B only
   C. Both Technician A and Technician B
   D. Neither Technician A nor Technician B

# Section 5

## Springs, Shocks, and Struts



**Interesting Fact**

*The transverse leaf spring front and rear suspension is one of the oldest suspension systems in the automotive industry. It was used on Ford products from the early 1900s until 1948.*

## SECTION OBJECTIVES

After you have read, studied, and practiced the contents of this section, you should be able to:

- Explain the causes of coil spring failure.
- Describe the design and spring rate of a linear-rate coil spring.
- Describe the functions of three different types of coils in a variable-rate coil spring.
- Explain basic torsion bar action as a front wheel strikes a road irregularity.
- Explain how friction and noise problems are reduced in a multiple-leaf spring.
- State the three purposes of shock absorbers.
- Explain spring operation during wheel jounce and rebound.
- Describe shock absorber operation.
- Describe the advantages of nitrogen-gas-filled shock absorbers and struts.
- Explain shock absorber ratios.
- Describe the operation of an adjustable shock absorber.
- Explain the operation of load-leveling struts and shock absorbers.
- Explain the effect of sagged front springs on caster angle and directional stability.
- Explain the effect of sagged front springs on camber angle.
- Remove and replace coil springs, and check spring and insulator condition.
- Diagnose, remove, and replace leaf springs.
- Replace torsion bars, and check torsion bar condition.
- Remove, inspect, and install stabilizer bar bushings, brackets, and links.
- Perform a visual shock absorber inspection.
- Perform a shock absorber bounce test, and determine shock absorber condition.
- Determine shock absorber condition from a manual shock absorber test.
- Remove and replace shock absorbers.
- Diagnose shock absorber and strut noise complaints.
- Remove and replace front and rear struts.
- Remove struts from coil springs.
- Install coil springs on struts.
- Follow the vehicle manufacturer's recommended strut disposal procedure.
- Perform off-car strut cartridge replacement procedures.
- Perform on-car strut cartridge replacement procedures.

# Chapter 21 Springs

## INTRODUCTION

The spring is the core of all suspension systems. It absorbs shock forces while maintaining correct riding height. If the spring is worn or damaged, the other suspension parts shift out of their proper positions and are subject to increased wear.

Various types of springs are used in suspension systems—coil, leaf, air springs, and torsion bar **(Figure 1).** Springs are mounted in rubber to reduce road shock and noise.

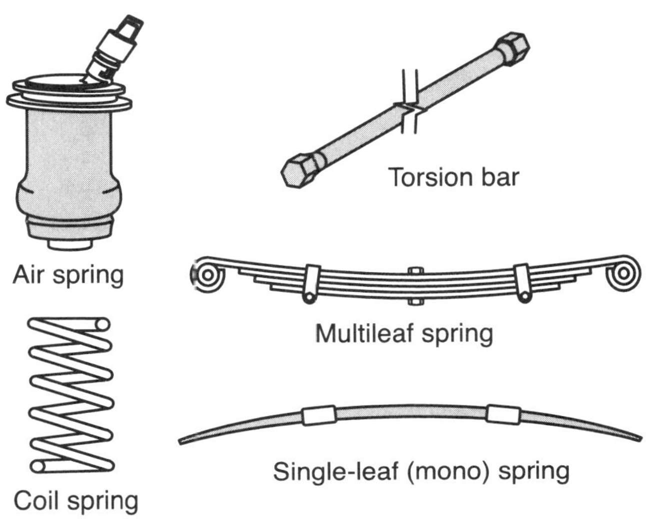


**Figure 1.** The various types of automotive springs.

## SPRING TERMINOLOGY

Automotive springs are generally classified by the amount of deflection exhibited under a specific load. This is referred to as the **spring rate.** According to the law of physics, a force (weight) applied to a spring causes it to compress in direct proportion to the force applied. When that force is removed, the spring returns to its original position if not overloaded. Remember that a heavy vehicle requires stiffer springs than a lightweight car.

The springs take care of two fundamental vertical actions: jounce and rebound **(Figure 2).** Jounce occurs

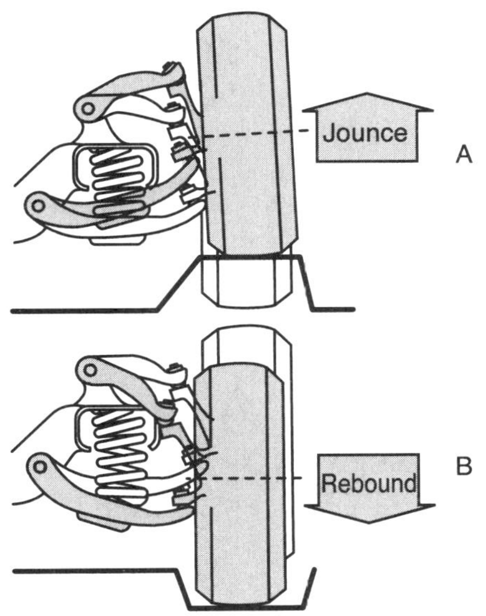


**Figure 2.** (A) Upward and (B) downward suspension movement.

when a wheel hits a bump and moves up. When this happens, the suspension system acts to pull in the top of the wheel, thereby maintaining an equal distance between the two wheels and preventing a sideways scrubbing action as the wheel moves up and down. Rebound occurs when the wheel hits a dip or hole and moves downward. In this case, the suspension system acts to move the wheel in at both the top and bottom equally while maintaining an equal distance between the wheels.

The spring moves back and forth from jounce to rebound. Each time, the movement becomes smaller. This is caused by friction of the spring's molecular structure and its pivot joints. A shock absorber is added to the suspension to dampen the motion of the spring.

All of the vehicle's weight supported by the suspension system is known as **sprung weight.** The weight of those components not supported by the springs is known as **unsprung weight.** The vehicle's body, frame, engine, transmission, and all of its components are considered sprung weight. Undercar parts classified as unsprung weight include the steering knuckle and rear axle assemblies (but not always the differentials). Keep in mind that, in general, the lower the ratio of unsprung weight to sprung weight, the better the vehicle's ride will be.

## COIL SPRINGS

The coil spring is the most commonly used spring for suspension systems. A coil spring **(Figure 3)** is actually a coiled-spring steel bar. When a wheel strikes a road irregularity, the coil spring compresses to absorb shock and then recoils back to its original installed height. Many coil springs are made of a steel alloy that contains other elements such as silicon or chromium. Coil springs may be manufactured by a cold or hot coiling process. The hot coiling process includes procedures for tempering and hardening the steel alloy. Coil springs are designed to carry heavy loads, but they must be light in weight. Many coil springs have a vinyl coating that increases corrosion resistance and reduces noise.

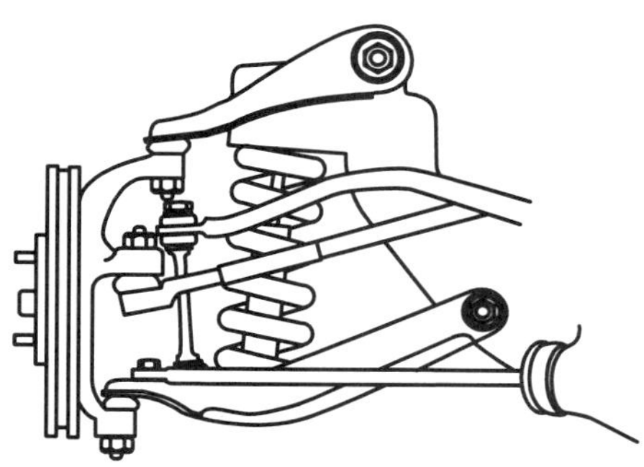

**Figure 3.** A front suspension with a coil spring.

You Should Know *Never disconnect a suspension component that quickly releases the tension on a compressed coil spring. This action may cause serious personal injury.*

Coil spring failures may be caused by these conditions:

1. Constant overloading
2. Continual jounce and rebound action
3. Metal fatigue
4. A crack or nick on the surface layer or coating

Coil springs do not resist lateral movement. However, when coil springs are used on the drive wheels, the suspension usually has special bars to prevent lateral movement.

## Linear-Rate Coil Springs

**Linear-rate** springs characteristically have equal spacing between the coils and one basic shape with a consistent wire diameter. All linear springs are wound from a steel rod into a cylindrical shape. When the load is increased on a linear-rate spring, the spring compresses and the coils twist or deflect. As the load is removed from the spring, the coils unwind—or flex—back to their original position. The amount of load necessary to deflect the spring one inch (25.4 mm) is the spring rate. Linear-rate coil springs have a constant spring rate regardless of the load or the amount the spring is compressed. For example, if 200 pounds deflect the spring 1 inch, 400 pounds will deflect the spring 2 inches. The spring rate on linear springs is usually calculated between 20 percent and 60 percent of the total spring deflection.

### Heavy-Duty Coil Springs

Since heavy-duty springs are designed to carry 3 to 5 percent greater loads than regular duty springs **(Figure 4),** they are somewhat different. The first difference that can

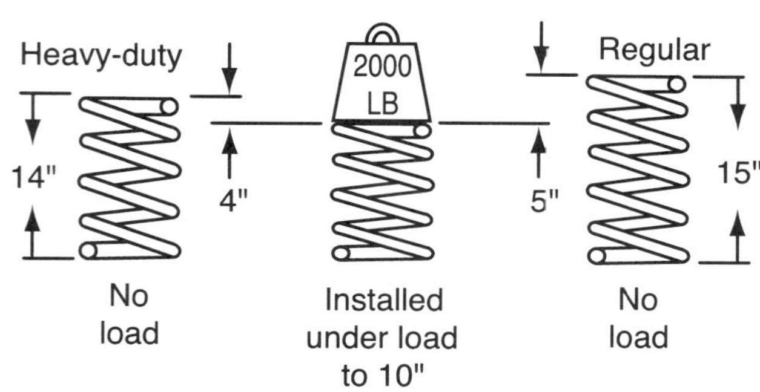


**Figure 4.** Heavy-duty springs are designed to carry 3 to 5 percent greater loads than regular-duty springs.

be noticed is wire diameter. On heavy-duty springs, wire diameter can be up to 0.100 inch (2.5mm) greater than the regular-duty spring for the same application. This larger diameter wire increases the load-carrying capacity of the spring.

The other difference is free length. A heavy-duty spring is up to 2-1/2 inches (64 mm) shorter than a regular-duty spring for the same application.

## Variable-Rate Coil Springs

**Variable-rate** spring designs are characterized by a combination of wire sizes and shapes. The most commonly used variable-rate springs have a consistent wire diameter, are wound in a cylindrical shape, and have unequally spaced coils **(Figure 5).** This type of spring is called a progressive rate coil spring.

The design of the coil spacing gives the spring three functional ranges of coils: inactive, transitional, and active. The inactive coils at the end of the spring introduce force into the spring when the wheel strikes a road irregularity. When the transitional coils are compressed to their point of maximum load-carrying capacity, these coils become inactive. Active coils work throughout the entire range of spring loading.

Theoretically, when a stationary load is applied to a variable-rate coil spring, the inactive coils support the vehicle's weight. If the load is increased, the transitional coils support the load until they reach their maximum load-carrying capacity, and the active coils carry the remaining overload. This action provides automatic load adjustment while maintaining vehicle height.

Some variable-rate coil springs have a tapered wire in which the active coils have the larger diameter, and the inactive coils have the smaller diameter.

The latest variable-rate spring designs deviate from the old cylindrical shape. These include the truncated cone, the double cone, and the barrel spring. The major advantage of these designs is the ability of the coils to nest or bottom out within each other without touching, which lessens the amount of space needed to store the springs in the vehicle.

Unlike a linear-rate spring, a variable-rate spring has no predictable standard spring rate. Instead, it has an average spring rate based on the load of a predetermined spring deflection. This makes it impossible to compare a linear-rate spring to a variable-rate spring. Variable-rate coil springs usually have more load-carrying capacity than linear-rate springs in the same application.

# LEAF SPRINGS

Although leaf springs were the first type of suspension spring used on automobiles, today they are generally found only on trucks, vans, SUVs, and a few passenger cars. There are two basic types of leaf springs: multiple leaf and monoleaf.

## Multiple-Leaf Springs

Multiple-leaf springs consist of a series of flat steel leaves of varying lengths that are bundled together and held with clips or by a bolt placed slightly ahead of the center of the bundle. One leaf, called the **main leaf,** runs the entire length of the spring and has an eye on each end. The next leaf is a little shorter and attaches to the main leaf. The next leaf is shorter yet and attaches to the second leaf, and so on. The more leaves and the thicker and shorter the leaves, the stronger the spring. This arrangement of leaves means that nearly any number of leaves can be used to support the vehicle's weight. It also gives a progressively stiffer spring.

The front eye of the main leaf is attached to a bracket on the vehicle's frame with a bolt and bushing **(Figure 6).** The

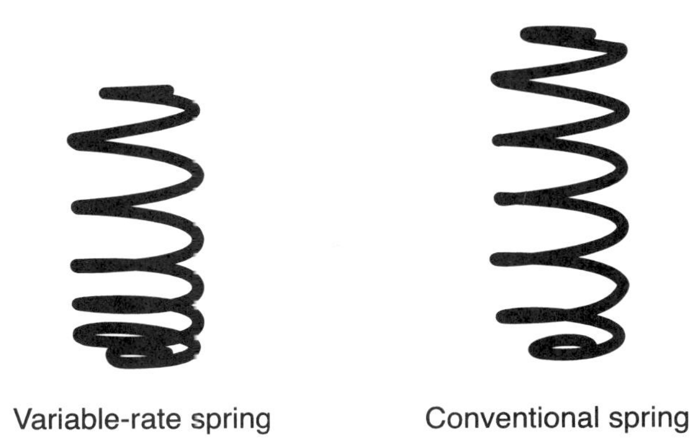


**Figure 5.** Variable-rate coil springs have consistent wire diameter and unequally spaced coils.

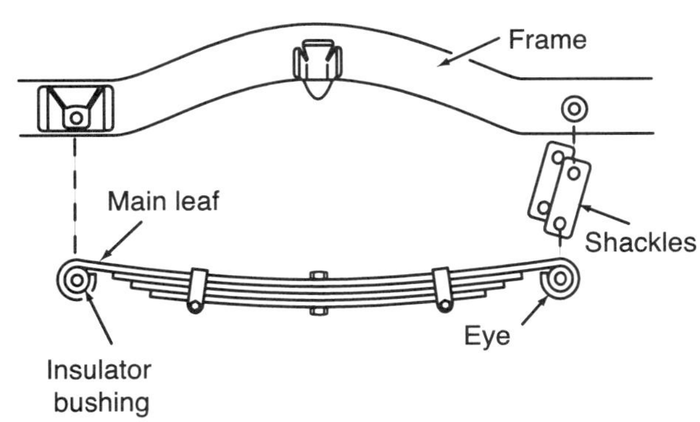


**Figure 6.** The front eye of the main leaf is attached to a bracket on the frame of the vehicle, and the rear eye is secured to the frame with a shackle.

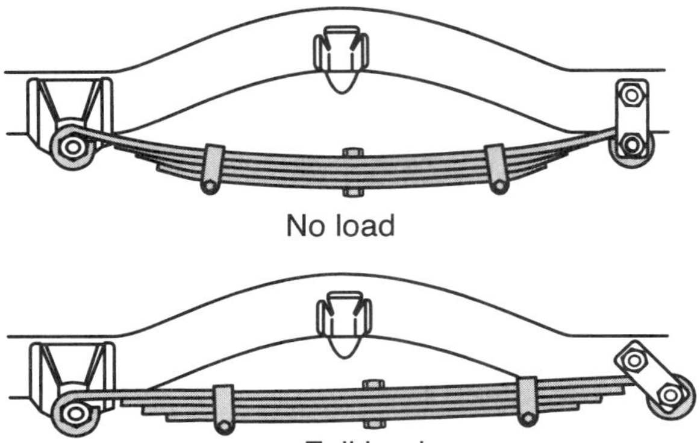


**Figure 7.** When a leaf spring is compressed, the length of the leaves changes and the leaves slide on each other.

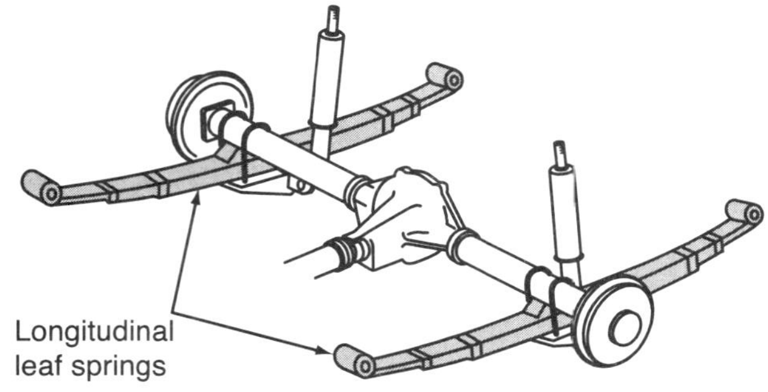


**Figure 8.** Leaf springs are usually mounted at right angles, or longitudinally, to the axle. This position provides excellent resistance to lateral wheel movement.

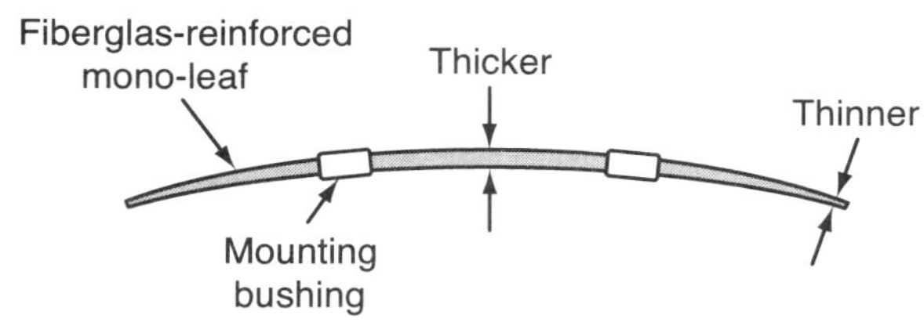


**Figure 9.** A mono-leaf spring constructed of fiberglass-reinforced plastic to reduce weight.

rear eye is secured to the frame with a shackle, which permits fore and aft movement in response to spring compression and other forces resulting from acceleration, deceleration, and braking. An insulating bushing is pressed into each main leaf eye.

When a leaf spring is compressed, the length of the leaves changes and the leaves slide on each other **(Figure 7).** This sliding action could be a source of noise and can also produce friction. The friction can produce a harsh ride as the spring flexes and can also dampen the motion of the spring. These noise and friction problems are reduced by interleaves, or spacers, made from zinc and plastic that are placed between the steel leaves.

Each spring leaf is curved in the manufacturing process. If this curve were doubled, it would form an ellipse. Leaf springs are referred to as semi-elliptical or quarter elliptical. The ellipse designation refers to how much of the ellipse the spring actually describes. The vast majority of leaf springs are semi-elliptical.

A center bolt extends through all the leaves to maintain the leaf position in the spring. In addition to absorbing blows from road forces, they also serve as suspension locators by fixing the position of the suspension with respect to the front and rear of the vehicle. Leaf springs are usually mounted at right angles to the axle **(Figure 8).** Because of this location, they provide excellent resistance to lateral wheel movement.

The head on the spring center bolt fits into an opening in the axle to position the axle. If the center bolt is broken, the axle position may shift and alter vehicle tracking and alignment.

## Mono-Leaf Springs

Some leaf springs are made of a single steel leaf, and these springs are referred to as mono-leaf springs. **Mono-leaf** springs are usually tapered with a heavy or thick center section tapering off at both ends. This provides a variable spring rate for a smooth ride and adequate load-carrying capacity. In addition, single-leaf springs do not have the noise and static friction characteristic of multiple-leaf springs. Mono-leaf springs may be mounted longitudinally or transversely and may be used in front or rear suspensions. Some cars, such as the Corvette, use a fiberglass-reinforced plastic mono-leaf spring **(Figure 9)** to reduce weight.

## Air Springs

An air spring **(Figure 10)** is used in electronically controlled systems that provide automatic front and rear level-

**Figure 10.** An air spring.

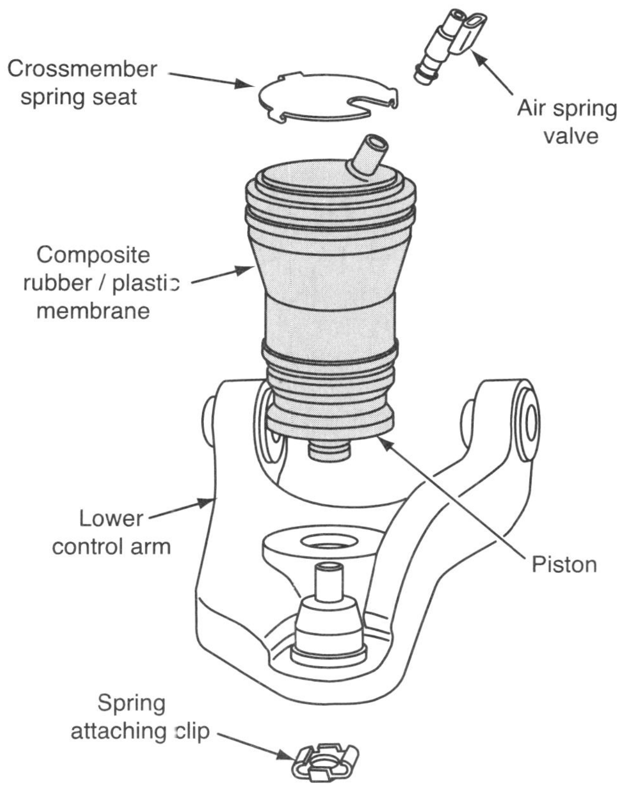


**Figure 11.** An air spring may be added to the existing leaf-spring suspensions and serve as an adjustable and additional spring at each end of the axle.

ing. The air springs are located in the same positions where coil springs are found. Each spring consists of a reinforced rubber bag pressurized with air. The bottom of each bag is attached to an inverted piston-like mount that reduces the interior volume of the air bag during jounce. This has the effect of increasing air pressure inside the spring as it is compressed, making it progressively stiffer. A vehicle equipped with an electronic air suspension system is able to provide a ride that is about one-third softer than conventional coil springs. At the same time, its variable spring rate helps absorb bumps and protect against bottoming.

Some late-model pickups and SUVs offer air suspension systems. These systems are added to existing leaf-spring suspensions. The air spring is positioned between the center of the leaf spring and the frame of the truck **(Figure 11).** The air spring serves as an adjustable and additional spring at each end of the axle.

## TORSION BARS

In some front suspension systems, torsion bars **(Figure 12)** replace the coil springs. However, instead of compressing like coil springs, a torsion bar twists during wheel jounce and straightens out to its original position during wheel rebound **(Figure 13).** A torsion bar is capable of storing a higher maximum energy compared to a loaded coil or leaf spring.

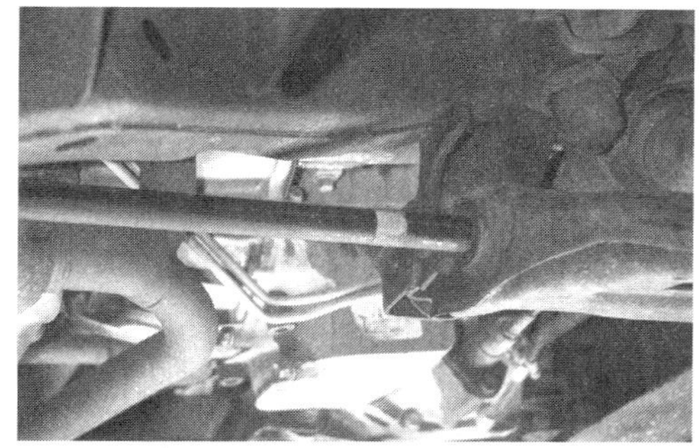

**Figure 12.** A longitudinally mounted torsion bar.

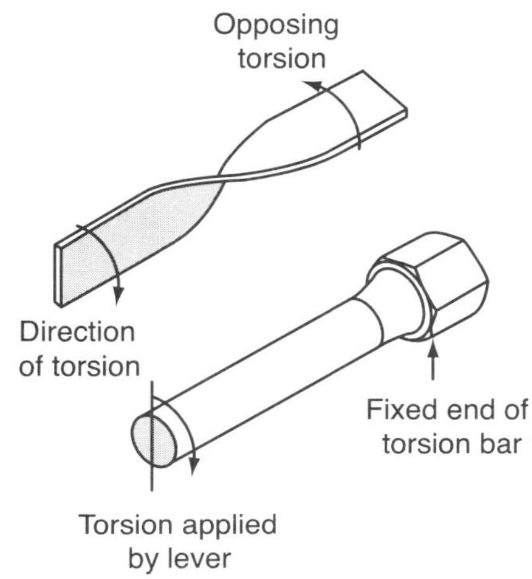


**Figure 13.** The action of a torsion bar.

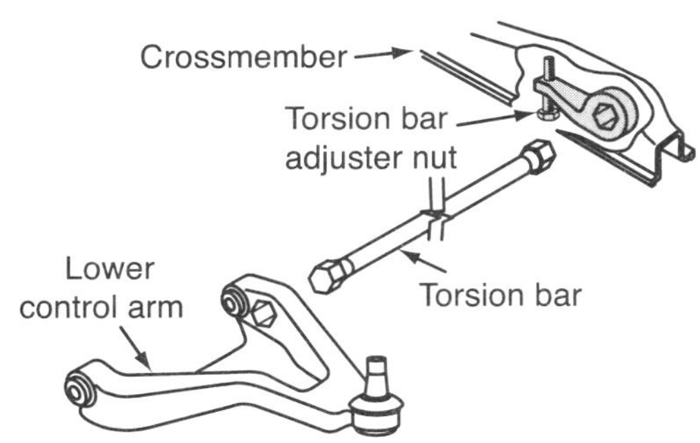


**Figure 14.** One end of a torsion bar is attached to the vehicle frame; the other end is attached to the lower control arm.

One end of a **torsion bar**—made of heat-treated alloy spring steel—is attached to the vehicle frame. The other end is attached to the lower control arm **(Figure 14).** When the wheel moves up and down, the lower control arm is raised and lowered. This twists the torsion bar, which causes it to absorb road shocks. The bar's natural resistance to twisting quickly restores it to its original position, returning the wheel to the road.

During the manufacturing process, torsion bars are prestressed to provide fatigue strength. Because of directional prestressing, torsion bars are directional. Torsion bars are marked right or left, and they must be installed on the appropriate side of the vehicle. Left and right on a vehicle is always viewed from the driver's seat.

Torsion bars have a riding height adjustment screw at the end where they are attached to the frame.

Many late-model pickups and SUVs use longitudinal torsion bars in their front suspensions. They are used in this type vehicle because they require less space compared to coil or leaf springs. They can also be mounted low and out of the way of the driveline components. Transversely mounted torsion bars were used in some front suspensions on older cars.

## STABILIZER BARS

A variety of devices are added to basic suspension components to provide additional stability. One of the most common is the **sway bar,** which is also known as the anti-sway or stabilizer bar. This is a metal rod running between the opposite lower control arms. As the suspension at one wheel responds to the road surface, the sway bar transfers a similar movement to the suspension at the other wheel **(Figure 15).** For example, if the right wheel is drawn down by a dip in the road, the sway bar creates a downward draw on the left wheel as well. This produces a more level ride, and sway or lean during cornering is also reduced.

If both wheels go into a jounce, the sway bar simply rotates in its insulator bushings. It is a different matter when only one wheel goes into jounce. The stabilizer bar twists, just like a torsion bar, to lift the frame and the opposite suspension arm. This action reduces body roll.

The sway bar can be a one-piece, U-shaped rod fastened directly into the control arms with rubber bushings **(Figure 16),** or it can be attached to each control arm by a separate sway bar link. The arm is held to the links with nuts and rubber bushings and is also mounted to the frame in the center with rubber bushings **(Figure 17).** If the sway bar is too large, it causes the vehicle to wander. If it is too small, it has little effect on stability.

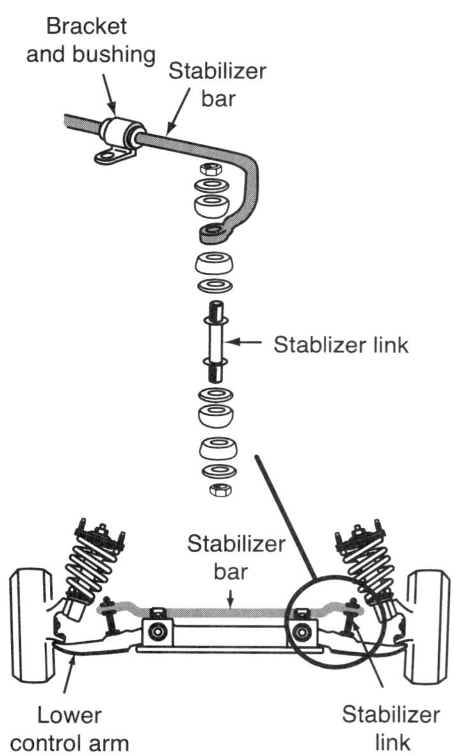


**Figure 15.** As the suspension at one wheel responds to the road surface, the sway bar transfers a similar movement to the suspension at the other wheel.

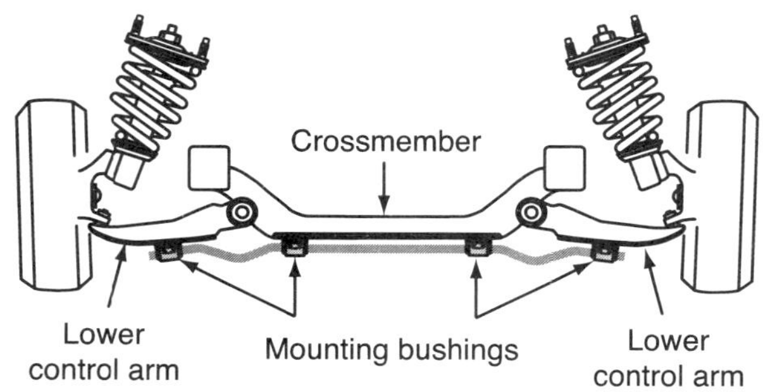


**Figure 16.** The sway bar can be a one-piece, U-shaped rod fastened directly into the control arms with rubber bushings.

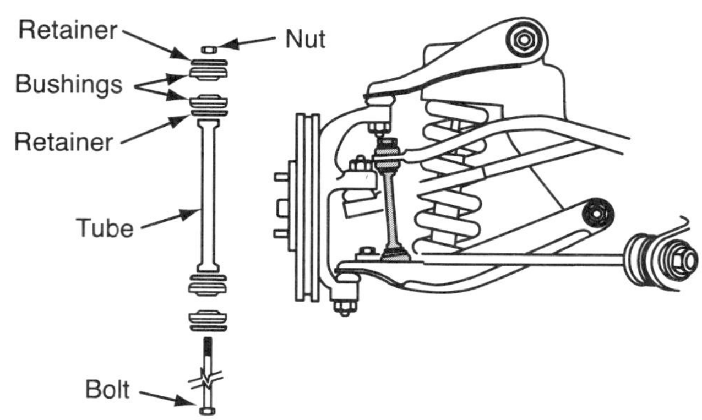


**Figure 17.** The sway bar can be attached to each control arm by links with nuts and rubber bushings.

On suspensions that use single-housing lower control arms instead of wishbone types, the sway bar can also be used to add lateral stability to the control arm. Strut rods **(Figure 18)** are used on models that do not use the sway bar for this purpose. Strut rods are attached to the control arm and frame with bushings, allowing the arm a limited amount of forward and backward movement. Strut rods are directly affected by braking forces and road shocks, and their failure can quickly lead to failure of the entire suspension system.

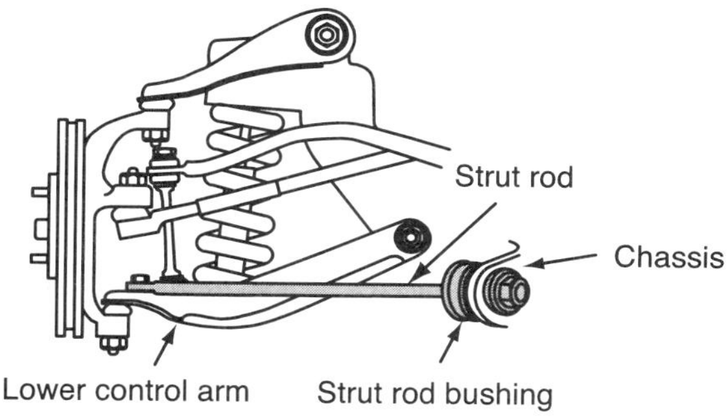


**Figure 18.** Strut rods are attached to the control arm and frame with bushings that allow the arm a limited amount of forward and backward movement.

## SUMMARY

- Coil springs may be classified as linear rate or variable rate.
- A linear-rate coil spring has a constant spring rate regardless of spring load.
- Heavy-duty coil springs have larger diameter wire than regular-duty coil springs.
- Variable-rate coil springs have unequally spaced coils.
- In a leaf spring, zinc and plastic spacers between the leaves reduce noise and friction.
- Leaf springs may be mono-leaf or multiple-leaf design.
- Spring leaves may be manufactured from steel or fiberglass.
- Most leaf springs are mounted longitudinally, but may be mounted transversely.
- In a torsion bar suspension system, the torsion bars replace the coil springs.
- Torsion bars are directional, and they are marked right or left.
- To reduce sway and body lean during cornering, many vehicles have a stabilizer bar.

## REVIEW QUESTIONS

1. Explain the meaning of constant spring rate.
2. Describe the main difference in the design of heavy-duty coil springs compared to regular-duty coil springs, and compare the free length of each type of spring.
3. Explain torsion bar action during wheel jounce and rebound.
4. Describe the action of a leaf spring as it compresses.
5. A variable-rate coil spring provides automatic __________________ adjustment.
6. In a variable-rate spring, the __________________ coils operate during the complete range of spring loading.
7. While discussing types of coil springs, Technician A says linear-rate coil springs have equal spacing between the coils. Technician B says a variable-rate spring may have a cylindrical shape with unequally spaced coils. Who is correct?
   A. Technician A only
   B. Technician B only
   C. Both Technician A and Technician B
   D. Neither Technician A nor Technician B
8. Which of the following statements about leaf springs is *true*?
   A. A fiberglass mono-leaf spring is heavier than a steel mono-leaf spring.
   B. Some mono-leaf springs are mounted transversely.
   C. In a multiple-leaf spring, the lower leaf is called the main leaf.
   D. A multiple-leaf spring has the same stiffness regardless of how much it is compressed.

# Chapter 22

# Shocks and Struts

## INTRODUCTION

A shock absorber is used to control, or dampen, spring oscillations. A strut serves the same purpose as a shock absorber, but it also supports the steering knuckle. Typically, there is one shock absorber or strut at each wheel.

Shock absorbers have three main purposes:

- To control spring action and oscillations to provide the desired ride quality
- To prevent body sway and lean while cornering
- To reduce the tendency of a tire to lift off the road, which improves tire life, traction, and directional stability

Since shock absorbers control spring action, spring oscillations, and chassis oscillations, they contribute to vehicle safety and passenger comfort. If the shock absorbers are worn out, excessive chassis oscillations may occur, particularly on rough road surfaces. These excessive chassis oscillations may result in loss of steering control. Worn-out shock absorbers also cause excessive body lean and sway while cornering, which may cause the driver to lose control of the vehicle. Shock absorbers are extremely important to providing longer tire life, and improving vehicle handling, steering quality, and ride quality.

Shock absorber design is matched to the deflection rate of the spring to control the spring's action.

## SHOCK ABSORBER DESIGN

The lower half of a shock absorber is a twin tube steel unit filled with hydraulic oil and nitrogen gas **(Figure 1).** In some shock absorbers, the nitrogen gas is omitted. A relief valve is located in the bottom of the unit, and a circular lower mounting is attached to the lower tube. This mounting contains a rubber isolating bushing, or grommets. A pis-

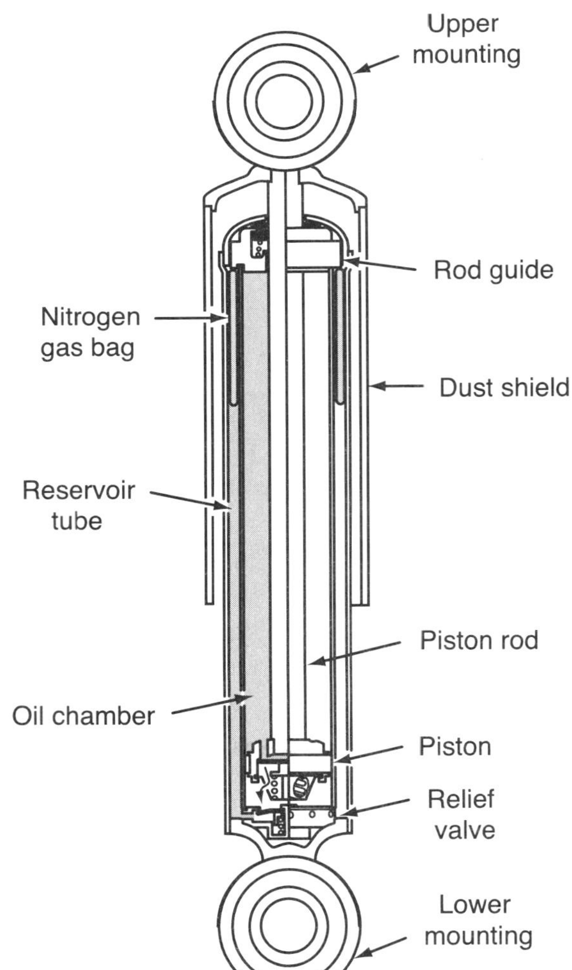


**Figure 1.** A shock absorber filled with hydraulic oil and nitrogen gas.

ton and rod assembly is connected to the upper half of the shock absorber. The upper portion has a dust shield that surrounds the lower tube unit. The piston is precision fit in the inner cylinder of the lower unit. A piston rod guide and seal are located in the top of the lower unit. A circular upper mounting with a rubber bushing is attached to the top of the shock absorber. A nitrogen gas charge is located in the oil reservoir to prevent oil cavitation and foaming, which provides more positive action but does not provide for spring support.

## Heavy-Duty Shock Absorbers

Some heavy-duty shock absorbers have a dividing piston in the lower oil chamber. The area below this piston is pressurized with nitrogen gas to 360 psi or 2,482 kPa. Hydraulic oil is contained in the oil chamber above the dividing piston. The operation of the heavy-duty shock absorber is similar to that of the conventional type **(Figure 2).** The other main features of a heavy-duty shock absorber include a higher quality seal for longer life, single tube design to prevent excessive heat buildup, and a rising rate valve to provide precise spring control under all conditions.

Aftermarket spring-assisted shock absorbers are also available. These are conventional shock absorbers with a small coil spring mounted over them. The coil springs on the shock absorbers help the springs support the vehicle's weight.

## Load-Leveling Shock Absorbers

Load-leveling rear struts or shock absorbers, often called air shocks, are used with an electronic height control system. An onboard air compressor pumps air into the rear shocks to raise the rear of the vehicle, and an electric solenoid releases air from the shocks to lower the rear chassis. An electromagnetic height sensor may be contained in the shock absorber, or an external sensor may be used **(Figure 3).** This sensor sends a signal to an electronic control module in relation to the rear suspension height. The module controls the air compressor and the exhaust solenoid to control air pressure in the shock absorbers. This action maintains a specific rear suspension trim height regardless of the load on the rear suspension. If a heavy package is

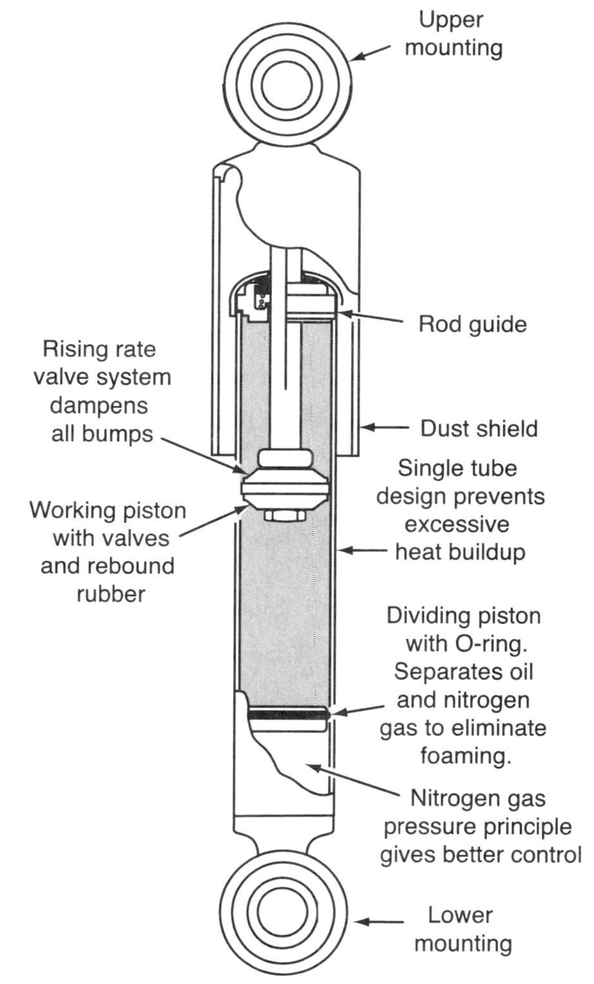


**Figure 2.** A heavy-duty shock absorber.

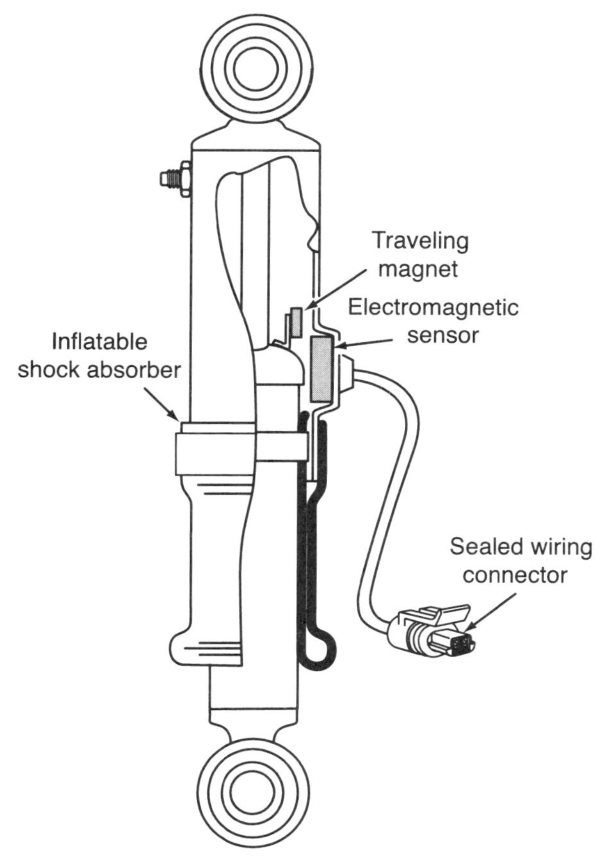


**Figure 3.** A load-leveling shock.

placed in the trunk, the vehicle chassis is forced downward. However, the load-leveling shock absorbers extend to restore the original rear suspension height.

Aftermarket air shock absorbers are available. These contain an air valve connection, and a shop air hose is used to supply the desired pressure in these air shock absorbers.

## Shock Absorber Ratios

Most automotive shock absorbers are a double acting and control spring action during jounce and rebound. The piston and valves in many shock absorbers are designed to provide more extension control than compression control. An average shock absorber may have 70 percent of the total control on the extension cycle, and thus 30 percent of the total control is on the compression cycle. Shock absorbers are so designed because they must control the heavier sprung body weight on the extension cycle. The lighter unsprung axle, wheel, and tire weight are controlled by the shock absorber on the compression cycle. A shock absorber with this type of design is referred to as a 70/30 type. Shock absorber ratios vary from 50/50 to 80/20.

# SHOCK ABSORBER OPERATION

A shock absorber is bolted between the axle and chassis **(Figure 4).** The mounting bolts pass through isolating bushings on each end of the shock absorber. The isolating bushings prevent vibration and noise.

Front shock absorbers are usually mounted between the lower control arms and the chassis. When a vehicle wheel strikes a bump, the wheel and suspension move upward in relation to the chassis. Upward wheel movement is referred to as **jounce travel.** This jounce action causes the spring to deflect or compress. Under this condition, the spring stores energy and springs back downward with all the energy absorbed when it deflected upward. This downward spring and wheel action is called **rebound travel.** If this spring action were not controlled, the wheel would strike the road with a strong downward force, and the wheel jounce would occur again. Therefore, some device must be installed to control the spring action, or the wheel would bounce up and down many times after it hit a bump, causing passenger discomfort, directional instability, and suspension component wear.

Shock absorbers are installed on suspension systems to control spring action. When a wheel strikes a bump and jounce travel occurs, the shock absorber lower tube unit is forced upward. This action forces the piston downward in the lower tube unit. Since oil cannot leak past the piston, the oil in the lower unit is forced through the piston valves to the upper oil chamber. These valves provide precise oil flow control and control the upward action of the wheel and suspension, which is referred to as a shock absorber compression stroke **(Figure 5).**

When the spring expands downward in rebound travel, the lower shock absorber unit is also forced downward. When this occurs, the piston moves upward in the lower tube unit, and hydraulic oil is forced through the piston valves from the upper oil chamber to the lower oil chamber. Since the valves restrict oil flow with precise control, the downward suspension and wheel movement is controlled.

When the shock absorber's piston moves, oil is forced through it. Since the piston's valves and orifices resist the

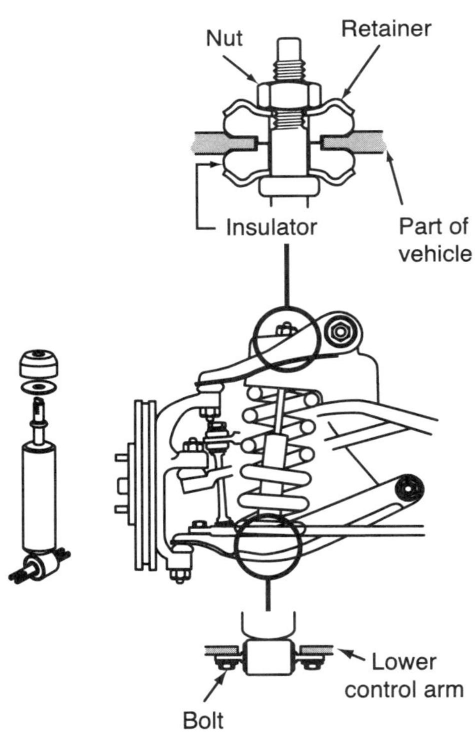


**Figure 4.** A shock absorber is bolted between a control arm or an axle and the chassis.

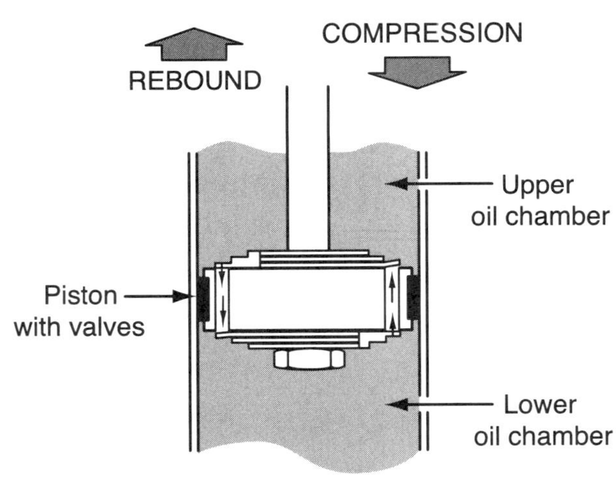


**Figure 5.** Shock absorber action.

flow of oil, friction and heat are created. The resistance of the oil moving through the piston must be calibrated as close as possible to the spring's deflection rate. Suspension systems deflect at different speeds depending on the type and size of bump and the vehicle's speed. The resistance of a shock absorber increases with the square of its speed. If a wheel strikes a large bump at high speed, the wheel deflection and rebound can be effectively locked by the shock absorber. Shock absorber engineers prevent this action by designing shock absorber valves and orifices to provide enough friction to prevent the spring from overextending on the rebound stroke. The valves and orifices must not create excessive friction, which would slow the wheel from returning to its original position.

Shock absorber pistons have many different valves and orifices. In some, small orifices control the oil flow during slow wheel and suspension movements. Stacked steel valves control the oil flow during medium speed wheel and suspension movements. During maximum wheel and suspension movements, larger orifices provide oil flow control. In other pistons, the stacked steel valves provide oil flow control. Regardless of the design of the orifice and valve design, a shock absorber must be precisely matched to absorb the spring's energy.

During fast upward wheel movement on the compression stroke, excessive pressure in the lower oil chamber forces the base valve open and allows oil to flow through the valve to the reservoir **(Figure 6).** Nitrogen gas provides a compensating space for the oil. Since the gas exerts pressure on the oil, cavitation—or foaming—of the oil is eliminated. When oil bubbles are eliminated, the shock absorber can provide continuous damping for wheel deflections as small as 0.078 in. (2.0 mm). A rebound rubber is located on top of the piston. If the wheel drops far, the shock absorber

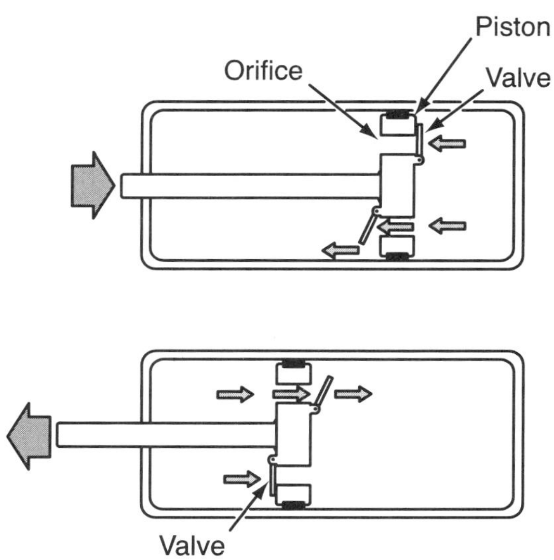


**Figure 6.** Fluid is forced through an orifice to dampen spring action.

can become fully extended, and the rebound rubber cushions the action.

## STRUT DESIGN

> **Interesting Fact** *For many years, rear-wheel-drive cars were equipped with front and rear shock absorbers. Most front-wheel-drive cars are equipped with front struts, and some of these cars also have rear struts. Front and rear struts are now very common.*

A strut-type front suspension is used on most FWD vehicles and some RWD cars. Internally, the design of a **strut** is very similar to a shock absorber and performs the same duties. In many strut-type suspension systems, the coil spring is mounted on the strut. The coil spring is largely responsible for proper curb riding height. A weak or broken coil spring reduces curb riding height and results in a harsh ride. The lower end of the front strut is bolted to the steering knuckle **(Figure 7).** An upper strut mount is bolted to

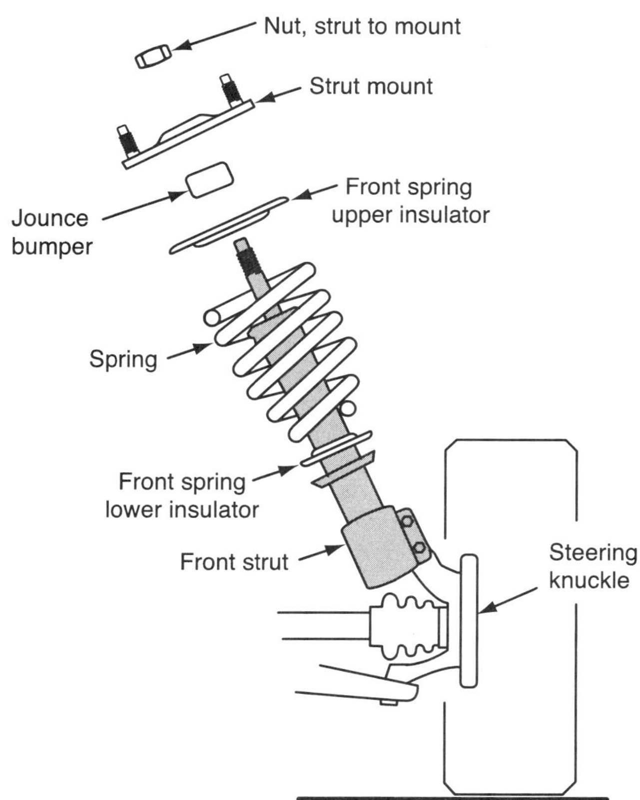


**Figure 7.** A front strut assembly.

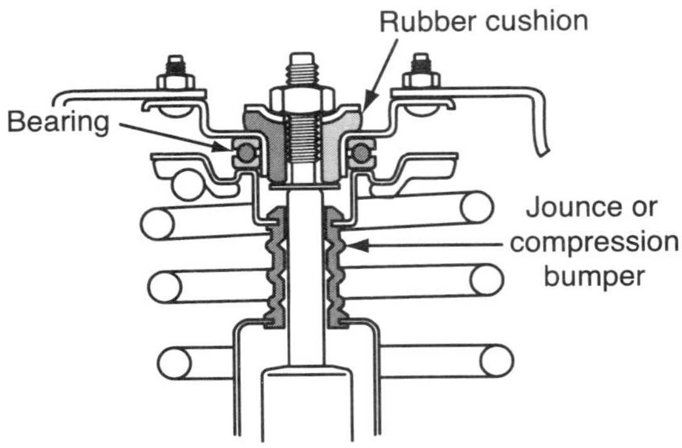


**Figure 8.** An upper strut mount.

the chassis strut tower. A lower spring seat is part of the strut assembly, and a lower insulator is positioned between the coil spring and the spring seat on the strut. Another spring insulator is located between the coil spring and the upper strut mount. The two insulators prevent metal-to-metal contact between the spring and the strut, or mount. These insulators reduce the transmission of noise and harshness from the suspension to the chassis. A rubber spring bumper is positioned around the strut piston rod. When the strut is fully compressed, the spring bumper provides a cushioning action between the top of the strut and the upper support. The upper strut mount contains a bearing, upper spring seat, and jounce bumper **(Figure 8).**

When the front wheels are turned, the front strut and coil spring rotate with the steering knuckle. The strut and spring assembly rotates on the upper strut mount bearing.

Some cars have a multi-link front suspension with an upper link connected from the chassis to the steering knuckle. The strut is connected from the upper link to the strut tower **(Figure 9).** A bearing is mounted between the upper link and the steering knuckle, and the wheel and knuckle turn on this bearing and the lower ball joint. Therefore, the coil spring and strut do not turn when the front wheels are turned, and a bearing in the upper strut mount is not required **(Figure 10).**

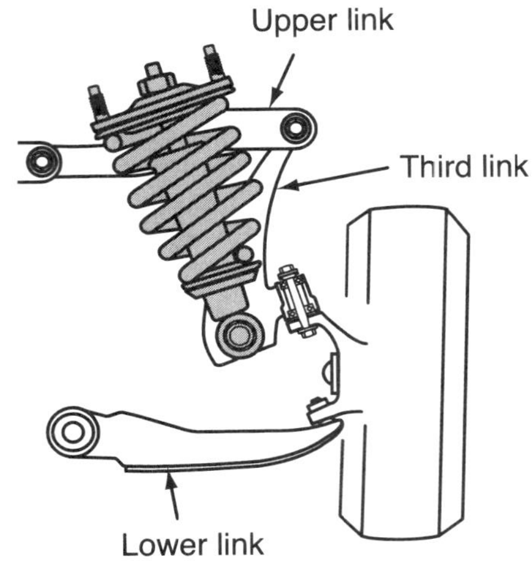


**Figure 10.** Multi-link front suspension with knuckle and wheel pivots at the upper link bearing and lower ball joint, and a non-rotating upper strut mount.

## Strut Design, Rear Suspension

In some rear suspension systems, the lower end of the strut is bolted to the spindle, and the top is connected through a strut mounted to the chassis. The rear coil springs are mounted between the lower control arms and the chassis separate from the struts **(Figure 11).**

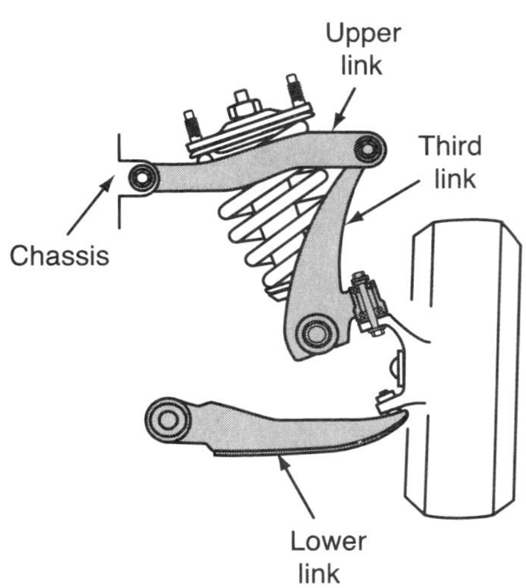


**Figure 9.** A multi-link front suspension with the strut connected between the upper link and the strut tower.

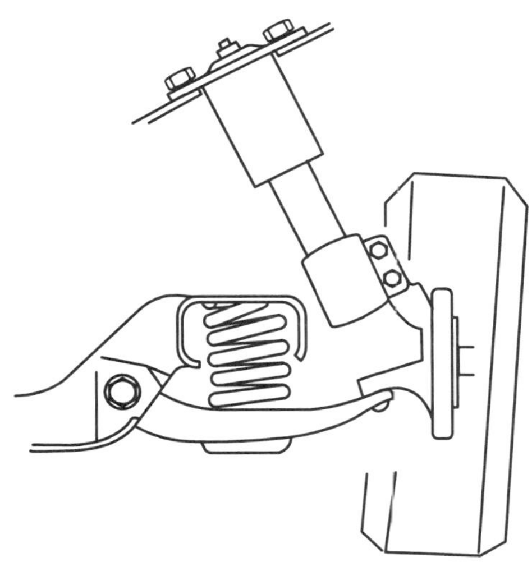

**Figure 11.** A rear suspension system with coil springs mounted separately from the struts.

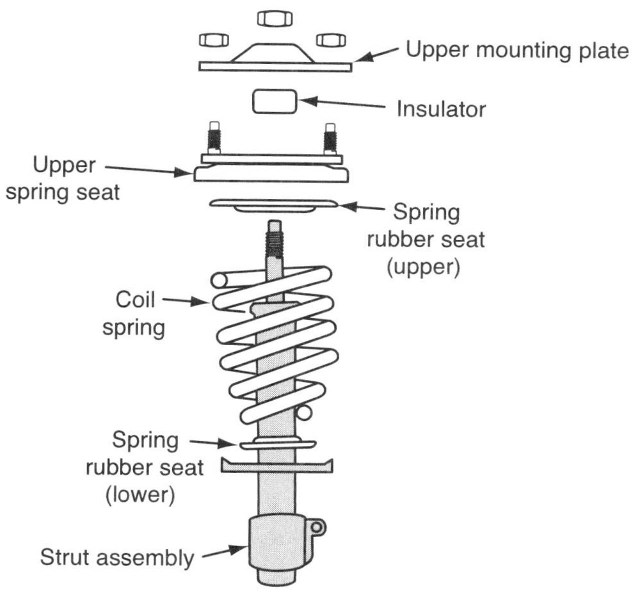


**Figure 12.** A rear suspension system with coil springs mounted on the struts.

In other rear suspension systems, the coil springs are mounted on the struts **(Figure 12).** An upper insulator is positioned between the top of the spring and the upper spring support, and a lower insulator is located between the bottom of the spring and the spring mount on the strut.

## Travel-Sensitive Strut

Some travel-sensitive struts contain narrow longitudinal grooves in the lower oil chamber **(Figure 13).** These grooves are parallel to the piston's orifices, and some oil flows through the grooves as well as the orifices. Under normal driving and road conditions, the orifices and grooves are calibrated to provide normal spring damping and control. If the front wheel drops suddenly, such as when it strikes a large hole, the piston moves into the narrow portion of the oil chamber. Under this condition, all the oil flow must flow through the piston orifices, which greatly increases the strut's resistance to movement and the suspension's damping action. This strut action prevents harsh impacts against the internal strut rebound rubber.

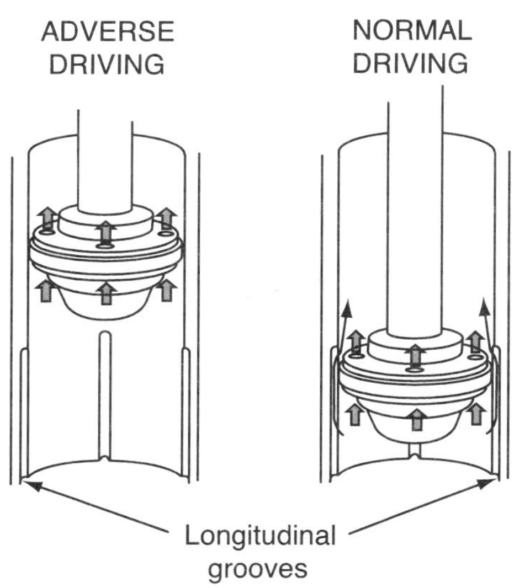


**Figure 13.** A travel-sensitive strut.

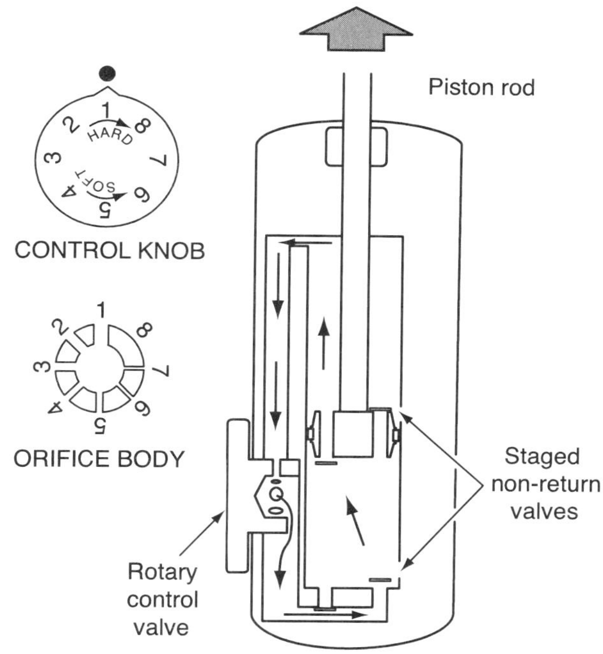


**Figure 14.** An adjustable strut.

## Adjustable Struts

Some adjustable struts have a manual adjustment that allows the vehicle owner or technician to adjust the struts to suit driving conditions **(Figure 14)**. The strut adjusting knob varies the strut orifice opening. This knob has eight possible settings. The factory setting is #3, which provides average suspension control. The #1 setting provides reduced spring control and the softest ride, whereas a #8 adjustment gives increased spring control and the hardest ride. The adjustment knob is usually accessible without raising the vehicle.

# ELECTRONICALLY CONTROLLED SHOCK ABSORBERS AND STRUTS

Many cars are now equipped with computer-controlled **active suspension** systems. In these systems, a computer-controlled actuator is positioned at the top of

Figure 15. A computer-controlled strut actuator.

each shock absorber or strut **(Figure 15).** The actuators rotate a shaft connected to the shock's valve inside the piston rod. Many of these systems have two modes, soft and firm. In the soft mode, the actuators position the valves so there is less restriction to the movement of oil. When the computer changes the actuators to the firm mode, the actuators position the valves so they provide more restriction to oil movement, which results in a firmer ride.

## MagneRide

MagneRide is a semi-active suspension system that features shocks or struts without electromechanical valves or small moving parts. Instead of valve-controlled orifices, MagneRide regulates the flow of fluid by a variable magnetic field produced by a small electric coil mounted in the shock. The shocks are filled with magneto-rheological (MR) fluid. MR fluid consists of magnetically soft particles, such as iron, suspended in a synthetic fluid.

The action of the shock forces the fluid through an opening. When the shock is in its off state, the fluid is not magnetized and flows freely through the orifice. When current is sent to the coil, the fluid becomes magnetized and its viscosity changes instantly **(Figure 16).**

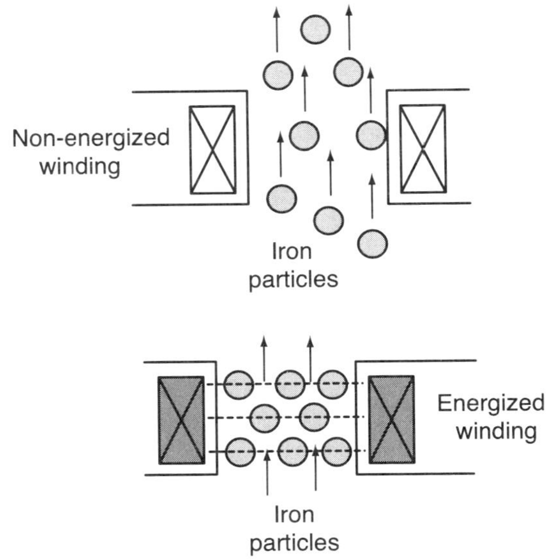


Figure 16. Action of the MR fluid in a MagneRide shock.

The material changes from a fluid state to a semi-solid state that is directly proportional to the magnetic field applied to it. With little or no electrical current, the iron particles are randomly distributed, and the fluid passes freely through the piston orifice. When a strong electrical current is applied to the coil, the resulting magnetic field aligns the iron particles so that the fluid stiffens and the flow is resisted. This causes heavy damping. The resulting damping force is proportional to the viscosity, which is proportional to the strength of the magnetic field.

Sensors monitoring wheel position, lateral acceleration, vehicle speed, steering wheel angle, and brake pedal angle are inputs to the control module that sends current to the coil in the shocks. This system provides extremely quick response time, typically about 5 ms, and the fluid is capable of reacting 30,000 times per second.

# SUMMARY


- Shock absorbers or struts are important to ride quality, steering control, and tire life.
- The shock absorbers control spring action and prevent excessive tire oscillations.
- During jounce travel, the piston moves downward in the lower shock absorber chamber; during rebound travel, this piston moves upward.
- Valves and openings in the shock absorber's piston provide control of the oil flow past the piston to control spring action. Shock absorber valves are matched to the amount of energy that may be stored in the spring.
- A nitrogen gas charge is located in the oil reservoir to prevent oil cavitation and foaming.
- Load-leveling shock absorbers have air pumped into them from an onboard compressor to maintain a specific rear suspension height regardless of the rear suspension load.

- Shock absorber ratio refers to the difference between the shock absorber control on the compression and extension cycle.
- Most front struts rotate on the upper strut mount bearing as the front wheels are turned.
- A travel-sensitive shock absorber provides increased resistance to piston movement as the shock absorber is extended.
- Adjustable struts have a manual adjustment that allows the technician or owner to adjust the strut orifice opening.

## REVIEW QUESTIONS

1. Describe uncontrolled spring action without a shock absorber.
2. Explain shock absorber operation.
3. Explain shock absorber ratio.
4. Explain the purpose of the nitrogen gas charge in shock absorbers and struts.
5. Identify the purposes of an upper strut mount.
6. Shock absorber design is matched to the ______________________ rate of the spring.
7. While discussing shock absorber operation, Technician A says that the oil flow through the shock absorber valves and orifices must be calibrated in relation to the spring's strength and deflection rate. Technician B says an average shock absorber has 70 percent of the total control on the extension cycle. Who is correct?
   A. Technician A only
   B. Technician B only
   C. Both Technician A and Technician B
   D. Neither Technician A nor Technician B
8. While discussing heavy-duty shock absorbers, Technician A says that heavy-duty shock absorbers have a high-quality seal for longer life. Technician B says that heavy-duty shock absorbers have a single tube design to prevent excessive heat buildup. Who is correct?
   A. Technician A only
   B. Technician B only
   C. Both Technician A and Technician B
   D. Neither Technician A nor Technician B

# Chapter 23

# Shock and Strut Diagnosis

## INTRODUCTION

A shock absorber or strut that is functioning properly ensures vehicle stability, handling, and rideability. Most motorists fail to notice the gradual changes caused by worn shock absorbers. Some common indications of shock absorber failure are: that steering and handling are more difficult, braking is not smooth, bouncing is excessive after stops, tire wear patterns are unusual, especially cupping, and springs are bottoming out.

Vibrations set up by a worn shock absorber can cause premature wear in many of the undercar systems. They can cause wear in the front and rear parts of the suspension system, the linkage of the steering system, and the U-joints and motor or transmission mounts of the driveline. Also, vibrations can cause unnatural wear patterns on the tires.

A strut suspension system a triangular design. The strut shaft is a structural member that eliminates the upper control arm bushings and upper ball joint. Since the shaft is also the shock absorber shaft, it receives a tremendous amount of force vertically and horizontally. Therefore, this assembly should be inspected very closely for leakage, damage, and poor damping.

## DIAGNOSING FRONT SUSPENSIONS

When diagnosing possible shock- and strut-related problems, check the entire suspension system before condemning the shocks or struts. This should involve a road test, measuring the vehicle's riding height **(Figure 1),** and inspecting basic suspension components. On the road test, note any abnormal or questionable noises and handling characteristics.

**Figure 1.** Checking the vehicle's riding height is part of conducting an inspection of the suspension system.

### Noises

Abnormal noises from the suspension system can be caused by a number of problems. Tire noise varies with road surface conditions, whereas differential noise is not affected when various road surfaces are encountered. Uneven tread surfaces may cause tire noises that seem to originate elsewhere in the vehicle. These noises may be confused with differential noise. Differential noise usually varies with acceleration and deceleration, whereas tire noise remains more constant in relation tc these forces. Tire noise is most pronounced on smooth asphalt road surfaces at speeds of 15 to 45 mph.

Rattling on road irregularities can be caused by worn shock absorber bushings or grommets, worn spring insulators, a broken coil spring or spring insulators, worn control

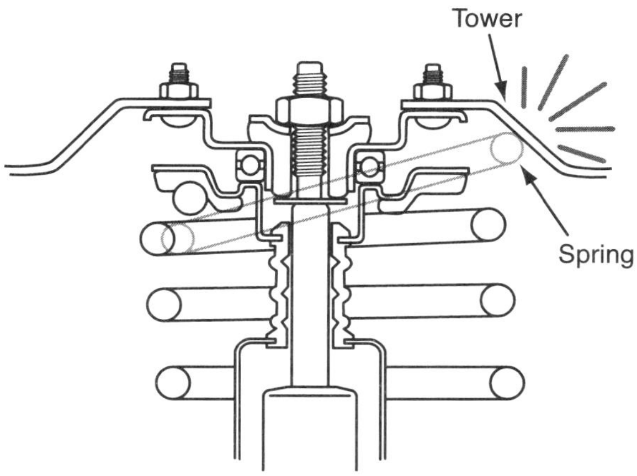


**Figure 2.** Noise can be caused by coil spring to upper strut tower interference.

arm bushings, worn stabilizer bar bushings, worn strut rod grommets, worn leaf spring shackles and bushings, or worn torsion bars, anchors, and bushings. Dry or worn control arm bushings may cause a squeaking noise on irregular road surfaces.

Chatter while cornering can be caused by worn upper strut mounts. Front strut noise on sharp turns or during suspension jounce may be caused by interference between the coil spring and the strut tower **(Figure 2),** or between the coil spring and the upper mount.

## Curb Riding Height

The curb riding height is determined mainly by spring condition. Other suspension components, such as control arm bushings, affect curb riding height if they are worn. Since incorrect curb riding height affects most of the other suspension angles, this measurement is critical. Reduced curb riding height on the front suspension may cause decreased directional stability. If the curb riding height is reduced on one side of the front suspension, the steering may pull to one side. Reduced rear suspension height increases steering effort and causes rapid steering wheel return after turning a corner. Harsh riding occurs when the curb riding height is less than specified. The curb riding height must be measured at the vehicle manufacturer's specified location, which varies depending on the type of suspension system.

If curb riding height is less than specified, the control arms and bushings should be checked. If the control arms and bushings are good, the reduced curb riding height may be caused by sagged or weak springs, worn-out air shocks, or bad air shock lines and fittings.

## Component Inspection

During the road test, pay attention to all abnormal or questionable handling behaviors of the vehicle. Use the following to diagnose the cause of the handling concerns:

### Steering Pull

A vehicle should maintain a forward direction on smooth, straight road surfaces without excessive steering wheel correction by the driver. If the steering gradually pulls to one side on a smooth, straight road surface, a tire, steering, or suspension defect is present.

- Check for tires of different types, sizes, designs, or inflation pressures.
- Check for incorrect front suspension alignment angles that also cause steering pull.
- Check the condition of the upper and lower control arm bushings. If worn, these can affect the caster and camber angles on the front suspension.
- Check for low ride height on one side of the front suspension.
- Check for a broken center bolt in leaf springs. This problem allows one side of the front suspension to move rearward in relation to the other side.

### Excessive Chassis Oscillations

Check for worn or leaking struts or shock absorbers.

### Excessive Sway on Road Irregularities

Check for a weakened or disconnected stabilizer bar and for worn stabilizer bar bushings.

### Reduced Directional Stability

Check for weak or broken coil springs that affect front suspension alignment angles. This problem may cause reduced directional stability. Also, check for weak or broken leaf springs and for a weak torsion bar or worn bushings or anchors.

### Erratic Steering on Irregular Road Surfaces

Inspect the upper and lower control arms for cracks, distortion, and worn bushings.

### Increased Steering Effort

Check for low ride height at the rear suspension.

### Rapid Steering Wheel Return

Check for low ride height at the rear suspension.

### Harsh Riding

Check for low ride height, weak or broken springs, worn torsion bar components such as pivot cushion bushings, and worn or damaged control arm bushings.

## AIR SHOCK ABSORBER DIAGNOSIS

Air shock absorbers have air pressure applied to them to control chassis curb height. Some air shock absorbers must be pressurized with the shop air hose and have a valve for inflation purposes. Other air shock absorbers are inflated by an onboard compressor with interconnecting plastic lines between the compressor and the shocks. Shock absorber lines must be secured to the chassis and must not rub against other components.

When air shock absorbers slowly lose their air pressure and reduce curb riding height, shock replacement is required. Before removing an air shock, relieve the air pressure.

## ELECTRONIC AIR SUSPENSION SYSTEM DIAGNOSIS

On vehicles with electronic suspensions **(Figure 3),** shocks and struts need to be checked along with the entire system. Every electronic air suspension system has a different diagnostic procedure. What follows is a typical procedure for a typical system:

1. Be sure the air suspension system switch is turned on.
2. Turn the ignition switch on for 5 seconds and then turn it off. Leave the driver's door open and the other doors closed.
3. Ground the diagnostic lead located near the control module and close the driver's door, leaving the window down.
4. Turn the ignition switch on. The warning lamp should blink continuously at 1.8 times per second to indicate that the system is in the diagnostic mode. Open and close the driver's door. These things should happen:

- The rear suspension should be raised for 30 seconds, lowered for 30 seconds, and raised for 30 seconds.
- The left front suspension should be raised for 30 seconds, lowered for 30 seconds, and raised for 30 seconds.
- The right front rear suspension should be raised for 30 seconds, lowered for 30 seconds, and raised for 30 seconds.
- The compressor is cycled on and off at 0.25 cycles per second.
- The vent solenoid opens and closes at 1 cycle per second.
- Open and close the driver's door. The left front air valve opens and closes at 1 cycle per second, and the vent solenoid is opened. When this occurs, the left front corner of the vehicle should drop slowly.
- The right front air valve opens and closes at 1 cycle per second, and the vent solenoid is opened. This action causes the right front corner of the vehicle to drop slowly.
- During this test, the right rear air valve opens and closes at 1 cycle per second and the vent valve opens. This should cause the vehicle's right rear corner to drop slowly.

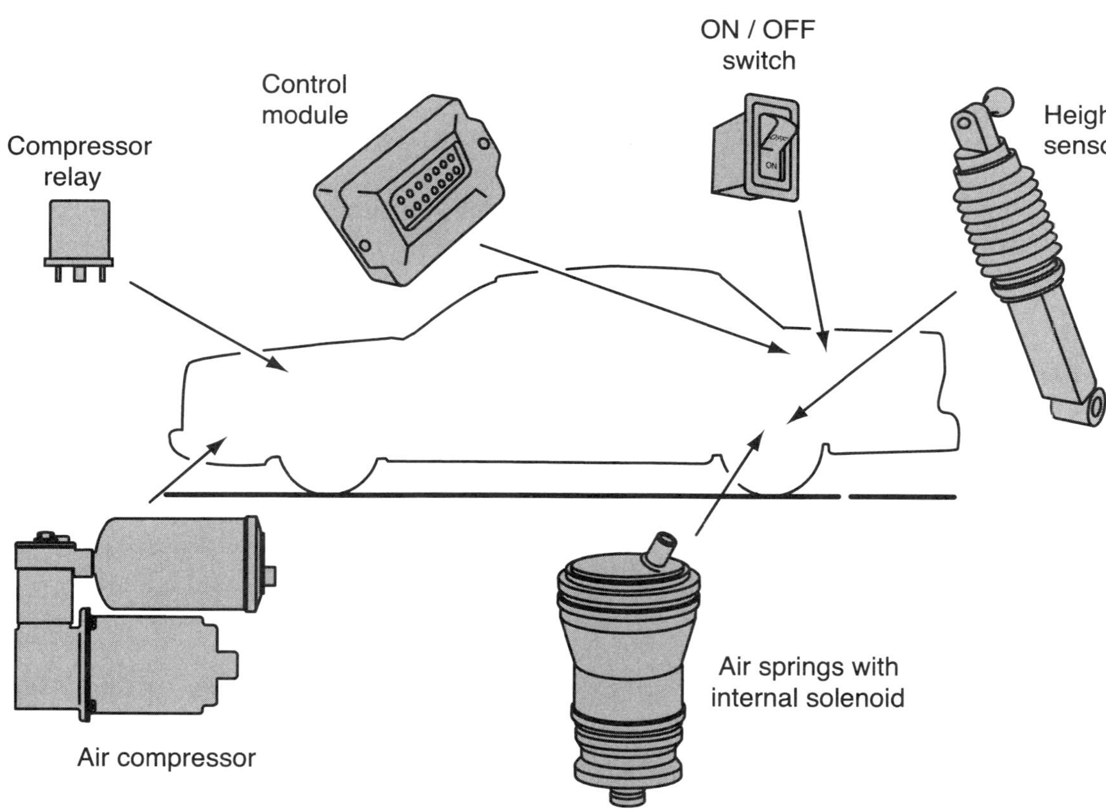


**Figure 3.** A typical electronic air suspension system.

- The left rear solenoid opens and closes at 1 cycle per second and the vent valve is opened, which should cause the left rear corner of the vehicle to drop slowly.

5. Return the module from the diagnostic mode to normal operation by disconnecting the diagnostic lead from ground, turning the ignition switch off, and depressing the brake pedal.

## Diagnosis of Electronically Controlled Shock Absorbers

To test the actuators on electronically controlled shock absorbers, remove them by pushing inward simultaneously on the two actuator retaining tabs and lifting the actuator off the top of the strut **(Figure 4).**

With the actuator removed from the strut and its wiring harness connected, turn the ignition switch on. Move the ride control switch to the auto position and wait 5 seconds. Move the ride control switch to the firm position and wait 5 seconds. If the actuator control tube on the bottom of the actuator rotated, the actuator is operating. If the actuator control tube did not rotate, proceed with testing the actuator.

Wiring colors vary with system design. The wiring colors in the following steps are based on Ford vehicles. If you are working on a different make of vehicle, refer to the wire colors in the vehicle's service information. To test the actuator:

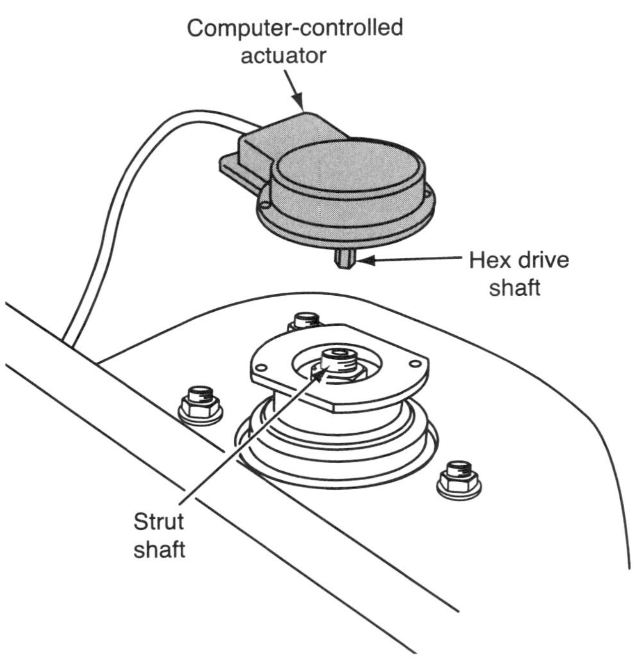


**Figure 4.** To test the actuators on electronically controlled shock absorbers, remove them by pushing inward simultaneously on the two actuator retaining tabs and lifting the actuator off the top of the strut.

> **You Should Know** *If the strut piston rod nut and actuator mounting bracket are removed, do not move or raise the vehicle. This releases the coil spring tension and may result in personal injury and vehicle damage.*

1. Connect the ohmmeter leads to the position sensor wire and the signal return wire. The position sensor wire is white or white with colored tracer, and the signal return wire is yellow with a black tracer. With the actuator in the S position, the position sensor should be closed and the ohmmeter should indicate less than 10 ohms.
2. Rotate the actuator to the H position. This should open the position sensor switch, and the ohmmeter should read over 1,000 ohms. If the ohmmeter readings are not within specifications, replace the actuator.
3. Connect the ohmmeter leads to the position sensor and power terminals in the actuator's electrical connector. The power wire is tan with a red tracer. If the ohmmeter reads over 1,000 ohms, the actuator is good. If the reading is below 10 ohms, replace the actuator.
4. Connect the ohmmeter leads from the signal return terminal in the wiring harness side of the actuator connector to a chassis ground. The ohmmeter should read less than 10 ohms. If the reading is higher than specified, check the signal return wire and the programmed ride control (PRC) module.

## SHOCK ABSORBER OR STRUT BOUNCE TEST

With the vehicle on the ground, a quick check, called the **bounce test,** can be performed to check the operation of the shocks and struts. When the bounce test is performed, the bumper is pushed downward with considerable weight applied two or three times on each corner of the vehicle. The bumper is released after each push, and the vehicle should oscillate about one and one-half cycles and then settle. One free upward bounce should stop the vertical chassis movement if the shock absorber or strut provides proper spring control. If the vehicle's bumper does more than one and one-half free upward bounces, the shock absorber or strut is defective.

## SHOCK ABSORBER VISUAL INSPECTION

Carefully inspect the shocks and/or struts. Also, check the rebound, or strikeout, bumpers on the control arms, or

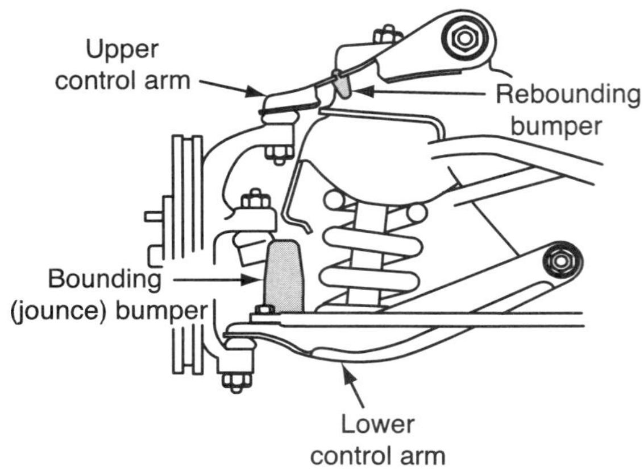


**Figure 5.** Carefully innspect the shock and its mountings, and look carefully at the suspension's bumpers.

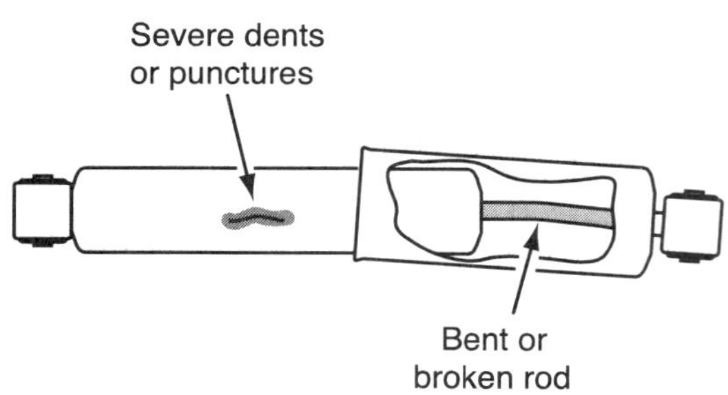


**Figure 6.** A shock that has dents, punctures, or is bent must be replaced.

chassis **(Figure 5).** If the rebound bumpers are severely worn, the shock absorbers may be worn out. If the shock absorbers or struts are bent, dented, or punctured **(Figure 6),** they should be replaced.

## Bolt Mounting

The mounting of a shock absorber should be checked for loose bolts and worn bushings. A strut assembly should also be inspected; carefully look at the condition of the insulators, spring bearing, and mounting bolts **(Figure 7).** If these components are loose or worn, a rattling noise will result. It may be necessary to replace the bushings and bolts or strut or shock.

## Bushing Condition

In some shock absorbers, the bushing is permanently mounted in the shock, and the complete unit must be replaced if the bushing is worn. When the mounting bushings are worn, the shock absorber will not provide proper spring control.

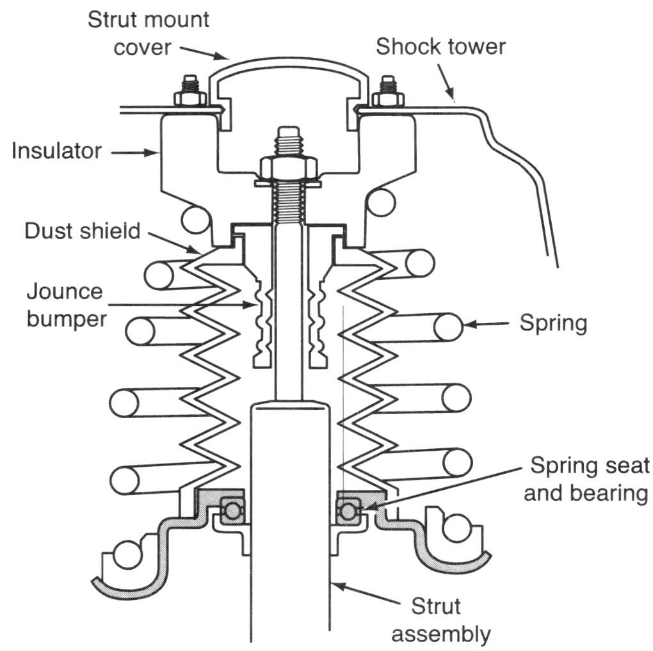


**Figure 7.** The mounting points and insulators in a strut assembly should be carefully checked.

## Shock Condition

Shock absorbers and struts should be inspected for oil leakage. A slight oil film on the lower oil chamber is acceptable. Any indication of oil dripping is unaccceptable, and replacement is necessary **(Figure 8).** Struts and shock absorbers should also be inspected for physical damage.

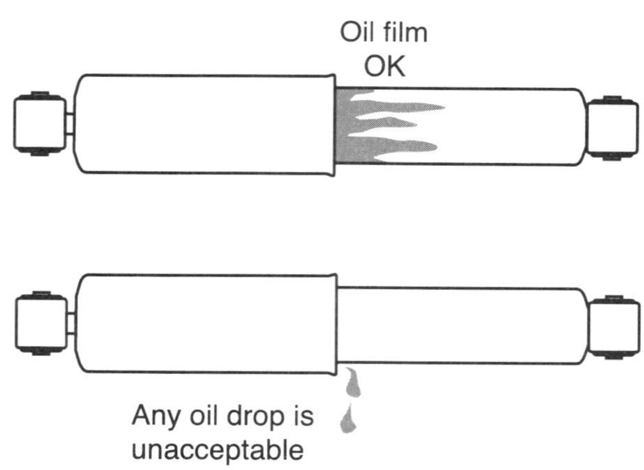


**Figure 8.** Shock absorbers and struts should be inspected for oil leakage. A slight oil film on the lower oil chamber is acceptable; however, any indication of oil dripping indicates the shock should be replaced.

## SHOCK ABSORBER MANUAL TEST

A shock absorber can be tested with a **manual test** of the shock. After removing the shock from the vehicle, turn it up in the same direction it occupies in the vehicle. Then, extend it fully. Next, turn it upside down and fully compress it. Repeat this operation several times.

During a manual test, the shock absorber should offer a strong, steady resistance to movement on the entire compression and rebound strokes. The amount of resistance may be different on the compression stroke compared to the rebound stroke. If a loss of resistance is experienced during either stroke, shock replacement is essential. The amount of resistance may be greater on the rebound stroke than on the compression stroke depending on shock absorber ratio. If there is a reduction or loss of resistance on either stroke, shock absorber replacement is necessary. Install a new shock absorber if a lag or skip occurs near mid-stroke of the shaft's change in travel direction, or if the shaft seizes at any point in its travel except at the ends.

Also, install a new shock absorber if noise (other than a switch or click) is encountered when the stroke is reversed rapidly, if there are any leaks, or if the action remains erratic after purging air. Some defective shock absorbers or struts may have internal clunking, clicking, and squawking noises, or binding conditions. When these shock absorber noises or conditions are experienced, shock absorber or strut replacement is necessary.

# SUMMARY

- A good shock absorber or strut ensures vehicle stability, handling, and rideability.
- Common indications of shock absorber failure include more difficult steering and handling, unsmooth braking, excessive bouncing after stops, unusual tire wear patterns, and springs bottoming out.
- Abnormal noises can be caused by a number of suspension and non-suspension problems.
- The curb riding height is determined mainly by spring condition, but other components should be checked if the height is incorrect.
- A vehicle should maintain the straight-ahead forward direction on smooth, straight road surfaces without excessive steering wheel correction by the driver.
- A bounce test can be preformed to check the operation of the shocks and struts.
- Carefully inspect the shocks and/or struts and their rebound bumpers.
- A shock absorber can be bench tested by conducting a manual test of the shock.

# REVIEW QUESTIONS

1. While discussing a shock absorber and strut bounce test, Technician A says the shock absorber is satisfactory if the bumper makes two free upward bounces. Technician B says the bumper must be pushed downward with considerable force. Who is correct?
   A. Technician A only
   B. Technician B only
   C. Both Technician A and Technician B
   D. Neither Technician A nor Technician B
2. When the front wheels are turned on a vehicle equipped with front struts, the left front coil spring provides a chattering action and noise. Technician A says the strut has internal defects and strut replacement is necessary. Technician B says the upper strut bearing and mount is defective. Who is correct?
   A. Technician A only
   B. Technician B only
   C. Both Technician A and Technician B
   D. Neither Technician A nor Technician B
3. During wheel jounce, a thumping noise is heard in the right front suspension, but the strut bounce test is satisfactory. All of these defects could cause the problem *except:*
   A. a worn upper strut mount.
   B. worn spring insulators.
   C. a broken coil spring.
   D. a fluid leak in the strut.
4. A customer says his MacPherson strut front suspension is making a chattering noise when the steering wheel is turned hard to the left. He says he also feels a "kind of vibration" in the steering wheel when the chatter is heard. To correct this problem, you should:
   A. replace the front struts.
   B. check the lower strut spring seating.
   C. check the stabilizer links.
   D. check the upper bearing and strut mounting.
5. Technician A says any visible presence of oil on a strut or shock requires replacement. Technician B says a slight film of oil on a strut or shock is okay if it performs satisfactorily in a bounce test. Who is correct?
   A. Technician A only
   B. Technician B only
   C. Both Technician A and Technician B
   D. Neither Technician A nor Technician B

# Chapter 24 Spring Service

## INTRODUCTION

Weak or broken springs can cause a number of problems, whether they are coil, air, torsion bar, or leaf springs. This chapter covers the typical replacement of the various types of springs, as well as the adjustment of torsion bars.

## EFFECTS OF WEAK SPRINGS

Sagged springs cause insufficient curb riding height. Therefore, the distance is reduced between the rebound bumper and its stop. This distance reduction causes the bumper to hit the stop frequently, resulting in harsh ride quality. If both front springs are sagged, the caster angle tilts excessively toward the front of the vehicle. When the front springs are sagged and caster is negative, directional stability is decreased and the front wheels tend to wander.

The caster angle is the number of degrees between the true vertical centerline of the wheel and tire and an imaginary line through the center of the upper strut mount and lower ball joint. Positive caster angle occurs when the caster line is tilted toward the rear and negative caster angle exists when the caster line is tilted toward the front of the vehicle **(Figure 1).**

When both rear springs are sagged, the caster angle tilts excessively toward the rear of the vehicle **(Figure 2).** This caster angle, called positive caster, results in increased steering effort and rapid steering wheel return after a turn. Rear spring sag and excessive positive caster increase steering effort and cause rapid steering wheel return after a turn is completed.

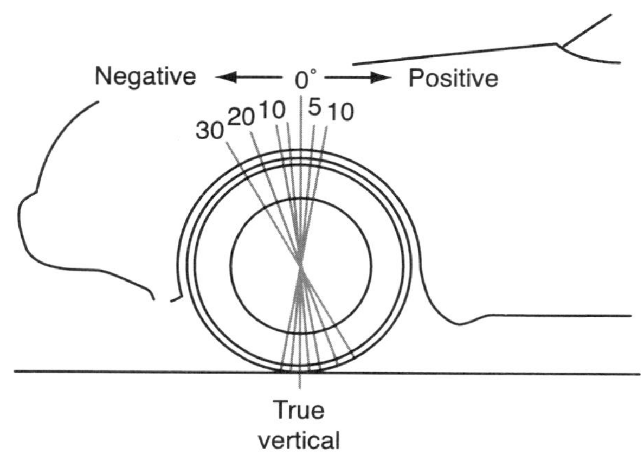

**Figure 1.** An illustration of caster angles.

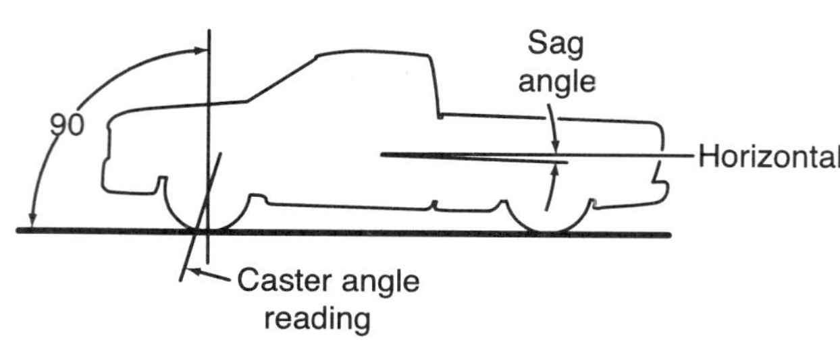

**Figure 2.** The effects of rear spring sag on wheel caster angle.

If a spring is sagged on one side of the front suspension, the camber angle is negative on that side. Excessive negative camber results in rapid wear on the inside edge of the tire tread and a decrease in directional stability. There-

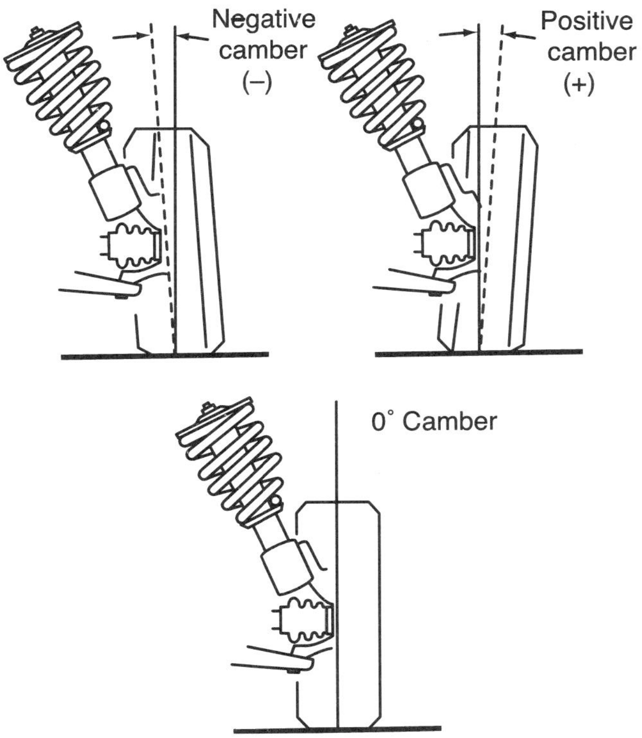


**Figure 3.** An illustration of camber angles.

fore, curb riding height is extremely important to maintaining correct wheel alignment angles, normal tire wear, and satisfactory directional stability. When the curb riding height does not meet the vehicle manufacturer's specifications, check for bent control arms and worn control arm bushings or mounting bolts. If these components are satisfactory, spring replacement is required to correct the curb riding height.

Positive camber occurs when the camber line is tilted outward from the vertical centerline of the wheel. If the camber line is tilted inward from the vertical centerline of the wheel, the camber is said to be negative **(Figure 3).**

## SELECTING REPLACEMENT SPRINGS

When springs must be replaced, the correct replacement spring must be selected. The original part number is usually on a tag wrapped around one of the coils. However, this tag may have fallen off. Some aftermarket suppliers stamp the part number on the end of the coil spring. If the original part number is available, the replacement springs should be ordered with the same number. Vehicles are manufactured with springs that match the vehicle's tires and wheels, suspension, and the installed optional

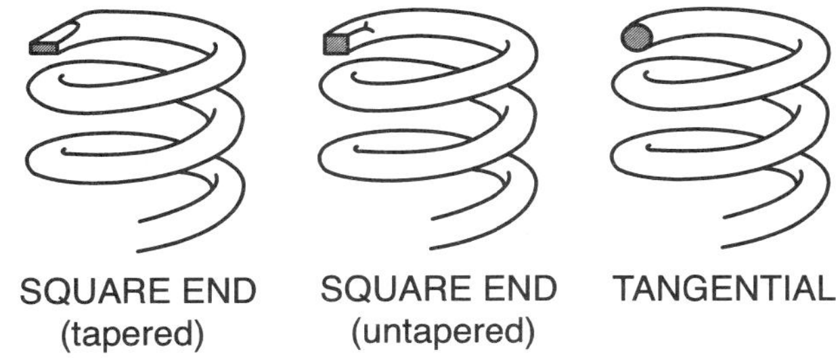


**Figure 4.** Types of coil spring ends.

equipment. Refer to the vehicle's VIN when selecting a replacement spring. If the customer desires a heavy-duty replacement for the springs, make sure the springs are an exact fit for the application.

Most manufacturers recommend that front and rear springs be replaced at the same time. Replacement springs must have the same type of ends as the springs in the vehicle **(Figure 4).**

Full-wire open-end springs have the ends cut straight off, and sometimes these ends are flattened, squared, or ground to a D-shape. Taper-wire closed-end springs are ground to a taper and wound to ensure squareness. Pigtail spring ends are wound to a smaller diameter.

Linear-rate coil springs are usually found in regular-duty, heavy-duty, and sport suspension packages. Heavy-duty coil springs are required when the vehicle is carrying a continuous heavy load, such as trailer towing.

Variable-rate coil springs are generally used when automatic load leveling is required under increased loads. Variable-rate coil springs maintain the correct vehicle height under various loads and provide increased load-carrying capacity compared to heavy-duty coil springs.

## COIL SPRINGS

Broken coil springs may cause a rattling noise while driving on irregular road surfaces. Weak or broken coil springs also reduce curb riding height. A rattling noise while driving on road irregularities may also be caused by worn or broken spring insulators. Since weak or broken coil springs affect front suspension alignment angles, this problem may cause reduced directional stability, excessive tire wear, and harsh riding.

### Front Coil Spring Replacement

The exact procedure for replacing a coil spring in front suspensions varies with the model of vehicle and the type of suspension. The following procedure describes what is usually required to replace a lower control arm and/or spring on

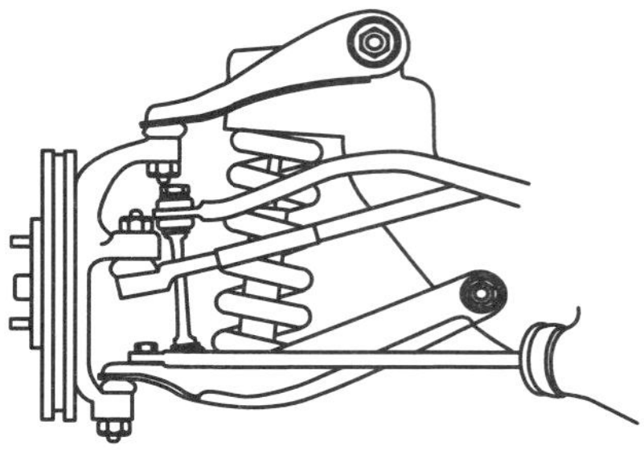

Figure 5. A suspension system with the coil spring positioned between the lower control arm and the frame.

a vehicle with the coil springs positioned between the lower control arm and the frame **(Figure 5)**:

1. Lift the vehicle on a hoist until the tires are a short distance off the floor, and allow the front suspension to drop downward. An alternate method is to lift the vehicle with a floor jack, and then support the chassis securely on safety stands so the front suspension drops downward.
2. Disconnect the lower end of the shock absorber. On some applications, the shock absorber must be removed.
3. Disconnect the stabilizer bar from the lower control arm.
4. Install a spring compressor and tighten the compressor until the spring is compressed **(Figure 6).** Make sure all the spring tension is supported by the compressing tool.

> **You Should Know** *During control arm and spring replacement, coil spring tension is supported by a compressing tool or floor jack. Serious personal injury or property damage may occur if the spring tension is released suddenly.*

Figure 6. A coil spring compressor is used to compress the spring to remove it or before disconnecting other suspension parts.

5. Place a floor jack under the lower control arm and raise the jack until the control arm is raisec and the rebound bumper is not making contact.
6. Remove the lower ball joint cotter pin and nut, and use a threaded expansion tool to loosen the lower ball joint stud.
7. Lower the floor jack very slowly to ower the control arm and coil spring.
8. Disconnect the lower control arm inner mounting bolts and remove the lower control arm.
9. Rotate the compressing tool bolt to release the spring tension, and remove the spring from the control arm.
10. Inspect the lower control arm for a bent condition or cracks. If either of these conditions is present, replace the control arm. Visually inspect all control arm bushings. Loose or worn bushings must be replaced. Visually inspect upper and lower spring insulators for cracks and wear, and inspect the spring seat areas in the chassis and lower control arm. Worn or cracked spring insulators must be replaced.
11. Reverse steps 1 through 9 to install the lower control arm. Be sure the coil spring and insulators are properly seated in the lower control arm and in the upper spring seat.

## Servicing Rear Coil Springs

Raise the vehicle on a frame contact hoist or position jack stands under the frame forward of the rear-axle assembly. This allows the shock absorbers to fully extend. Place a floor jack under the center of the rear-axle housing and support the weight of the rear axle, but do not lift the vehicle off the jack stands. Disconnect the lower end of the shock absorber. Then, lower the floor jack until the tension on the spring is relieved. If a coil spring positioner is used, remove it from the center of the coil spring. If the suspension has strut bars **(Figure 7),** disconnect them at one end. It is best to disconnect them at the point that allows for easy access. At this time, the coil spring can usually be removed from the vehicle by lifting it from its spring seat. If the springs are to be used again, mark or tag each one so it

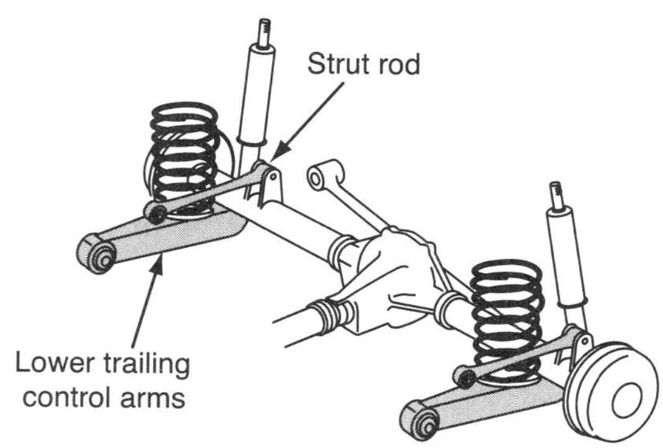


Figure 7. Typical location of the strut bars in a coil spring rear suspension.

can be returned to its original location. When a replacement is needed, always replace coil springs in pairs. This ensures equal height.

To install a spring, place the insulator on top of the coil spring and position the spring on the spring seat. The end of the top coil must be positioned to line up with the recess in the spring seat. Jack up the rear axle housing so the spring is properly seated at the lower end and the shock absorbers line up. Reconnect the shock absorbers.

There are some definite advantages to working on one spring at a time. First of all, the assembled side of the vehicle helps support the disassembled side. It also keeps the parts aligned and eliminates the possibility of putting the parts on the wrong side of the vehicle.

## TORSION BARS

Worn torsion bar components such as pivot cushion bushings and control arm bushings result in harsh riding and suspension noise when driving on road irregularities. A worn torsion bar hex and anchor may cause a rattling noise while driving on irregular road surfaces. Weak torsion bars or those with worn bushings or anchors cause reduced curb riding height, which may result in reduced directional stability and excessive tire tread wear.

When torsion bar **(Figure 8)** replacement is necessary, make sure to follow the manufacturer's recommended procedure for torsion bar removal and replacement. The following is a typical torsion bar removal and replacement procedure:

> **You Should Know** *Some torsion bar front suspension systems are combined with an air suspension system. When servicing these systems, the air suspension switch in the rear jack storage area must be shut off before servicing the suspension or hoisting, jacking, or towing the vehicle. If this procedure is not followed, personal injury and vehicle damage may occur.*

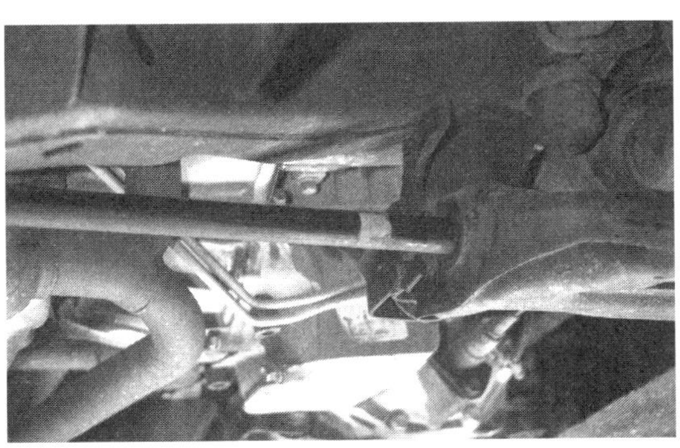

**Figure 8.** A longitudinally mounted torsion bar.

1. Raise the vehicle on a lift with the tires supported on the lift.
2. Remove the torsion bar cover plate bolts and the cover plate **(Figure 9).**
3. Measure and record the distance from the lip on the head of the torsion bar adjusting bolt to the casting surface this bolt is threaded into **(Figure 10).** When reinstalled, this bolt must be adjusted to this same measurement.
4. Install the torsion bar tool and adapters **(Figure 11)**. Tighten this tool until the torsion bar adjuster lifts off the adjustment bolt.

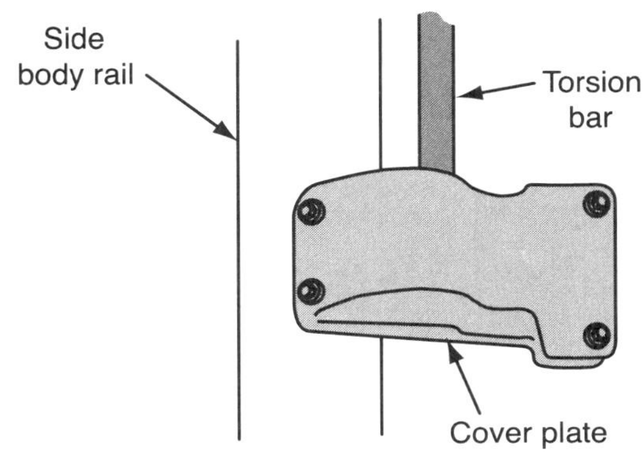


**Figure 9.** The cover plate for a typical torsion bar.

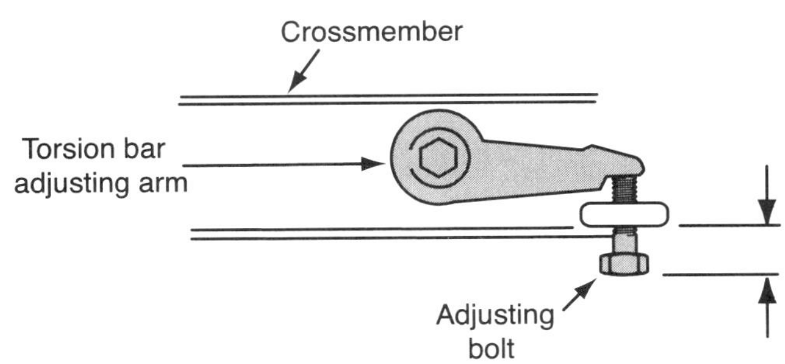


**Figure 10.** Measuring the position of the torsion bar adjusting bolt.

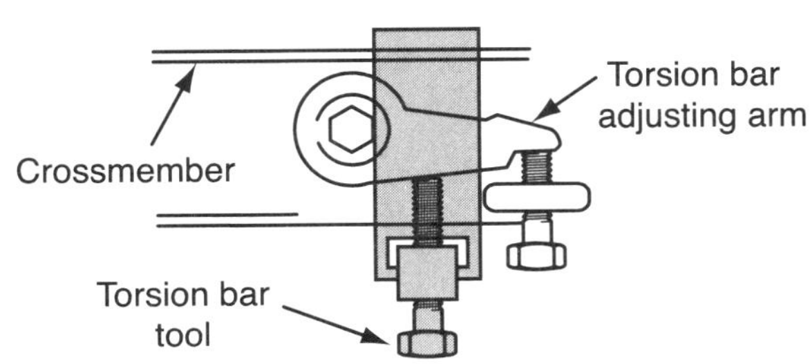


**Figure 11.** A torsion bar tool and adapter.

**You Should Know** *The torsion bar adjusting bolt is coated with a dry adhesive. This bolt must be replaced if it is backed off or removed. Failure to follow this procedure may cause the bolt to loosen, resulting in improper wheel alignment, reduced steering control, increased tire wear, and reduced ride quality.*

5. Remove the torsion bar adjustment nut and bolt.
6. Loosen the torsion bar tool to remove all the tension from the torsion bar.
7. Use the end of a screwdriver to place matching alignment marks on the torsion bar and the adjuster so these components may be reassembled in the same position. Remove the torsion bar insulator. Pull the torsion bar to the rear to remove this bar from the lower control arm.
8. Position the torsion bar in the lower control arm.
9. Install the torsion bar adjuster on the torsion bar with the alignment marks properly aligned.
10. Install the torsion bar tool and adapters. Tighten the torsion bar tool until a new adjustment bolt can be installed. Tighten this bolt until the measurement obtained in step 3 is obtained between the head of the bolt and the casting the bolt is threaded into.
11. Install the torsion bar cover plate, and tighten the plate retaining bolts to the specified torque.
12. Measure the suspension riding height. If the suspension riding height is not within specifications, adjust the torsion bar adjusting bolt to obtain the specified riding height.

## Torsion Bar Adjustment

On torsion bar front suspension systems, the torsion bars is adjustable. On some torsion bar front suspension systems, the curb riding height is measured with the vehicle on a lift and the tires supported on the lift. Measure the distance from the center of the front lower control arm bushing to the lift. Then measure the distance from the lower end of the front spindle to the lift. The difference between these two readings is the curb riding height.

If the curb riding height is not correct, the torsion bar anchor adjusting bolts must be rotated until the curb riding height equals the manufacturer's specifications **(Figure 12).**

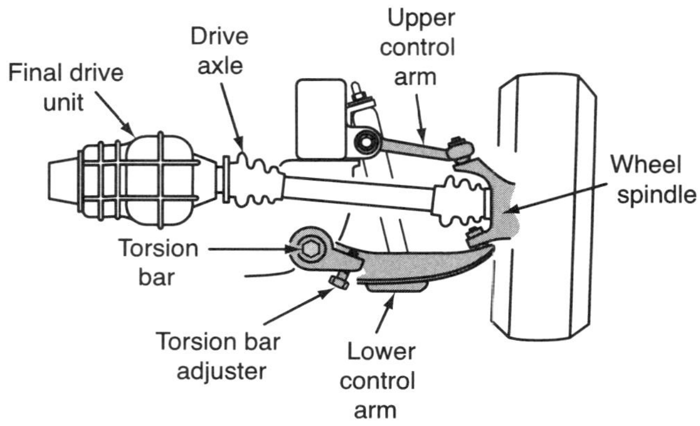


**Figure 12.** Curb riding height adjuster for a torsion bar suspension.

## LEAF SPRINGS

Typical leaf spring service includes the replacement of shock absorbers or springs. Bushings, shackles, or control arms do not need replacement frequently. However, they are commonly replaced with the spring. Always follow the procedures outlined in the service manual whenever servicing a leaf spring suspension.

Leaf springs are mounted longitudinally with a spring on each side of the suspension **(Figure 13).** Many leaf springs have plastic silencers between the spring leafs. If these silencers are worn out, creaking and squawking noises will be heard when the vehicle is driven over road irregularities at low speeds.

When the silencers require checking or replacing, lift the vehicle with a floor jack and support the frame on safety stands so the suspension hangs downward. With the vehicle weight no longer applied to the springs, the spring leafs may be pried apart with a pry bar to remove and replace the silencers. On some leaf springs, the clamps must be removed to replace the silencers.

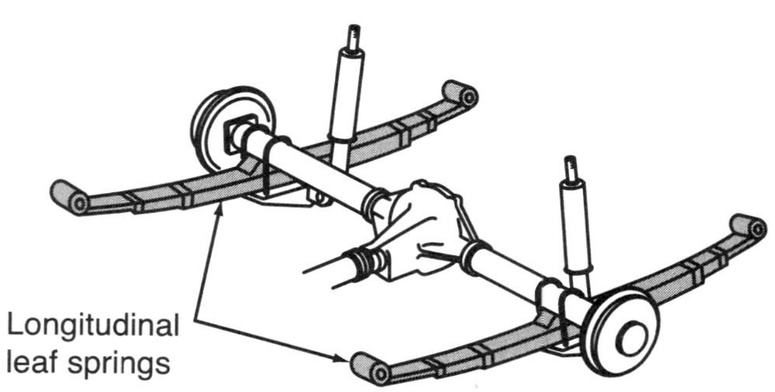


**Figure 13.** Leaf springs are mounted longitudinally with a spring on each side of the suspension.

Leaf springs should be inspected for a sagged condition, which causes the curb riding height to be less than specified. Leaf springs should also be visually inspected for broken leafs, broken center bolts, and worn shackles or bushings. Weak or broken leaf springs affect suspension alignment angles and cause excessive tire tread wear, reduced directional stability, and harsh riding. A rattling noise while driving on irregular road surfaces may be caused by worn shackles or bushings. Worn shackles and bushings can also cause a lower curb riding height. A broken center bolt usually allows one side of the suspension to move rearward in relation to the other side. This action may cause steering pull to one side.

## Front Leaf Springs

Front leaf springs are used in some truck suspension systems. The following is a typical front leaf spring replacement procedure:

1. Lift the front end of the vehicle with a floor jack under the front axle and place safety stands under the frame. Lower the vehicle onto the safety stands, but leave the floor jack under the front axle to support some of the weight of the front suspension.
2. Remove the nuts from the spring U-bolts, and remove the U-bolts and lower spring plate. If the spring plate is attached to the shock absorber, it may be left on the shock absorber and moved out of the way. If shims are positioned between the spring plate and the front axle, note their position and the number of shims. These shims must be reinstalled in their original position because they set the front wheel caster angle.
3. Lower the floor jack enough to relieve the vehicle's weight from the springs.
4. Remove the front shackle assembly.
5. Remove the rear spring mounting bolt, and then the spring **(Figure 14).**

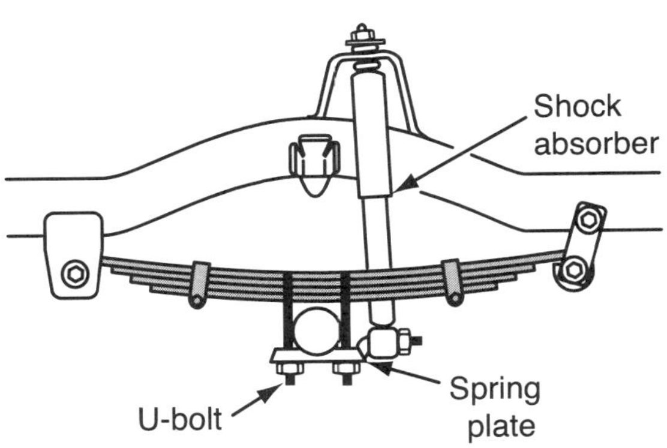


**Figure 14.** To remove a leaf spring, remove the shackle and the mounting bolt after disconnecting the shock absorber.

6. Check all the spring hangers, bolts, bushings, and shackle plates for wear, and replace them as required. Springs with broken or sagged leafs are usually sent to a spring rebuilding shop for repair.
7. Check the spring center bolt to be sure it is not broken.
8. Reverse steps 1 through 5 to install the front springs. Tighten all bolts to the manufacturer's specifications.

## Rear Leaf Springs

Servicing the leaf springs on a rear suspension is very similar to servicing leaf springs on a front suspension. If the springs squeak as the vehicle goes over a bump, the leaf spring insulators are undoubtedly bad. To replace the insulators, lift the vehicle with a floor jack and support the frame on safety stands so the rear suspension can move downward. With the vehicle's weight no longer on the springs, the spring leafs may be pried apart with a pry bar to remove and replace the silencers.

If there is a rattle coming from the area around the springs, the shackle bushings, brackets, and mounts may be worn. To check for worn shackles, shackle bushings, and brackets or mounts, allow the normal weight of the vehicle to rest on the springs. Then, insert a pry bar between the rear outer end of the spring and the frame. Apply downward pressure on the bar and observe the rear shackle for movement. Shackle bushings, brackets, or mounts must be replaced if there is movement in the shackle. The same procedure may be followed to check the front bushing in the main leaf.

Also, make sure to check the condition of the center bolt in both spring assemblies. A broken spring center bolt may allow the rear axle assembly to move rearward on one side. This will cause vehicle handling, braking, and tracking problems.

Check the curb riding height and compare your measurements to the specifications. If the springs are sagging and need to be replaced, raise the vehicle with a floor jack and place safety stands under the frame. Lower the vehicle's weight onto the safety stands, and leave the floor jack under the differential housing to support the rear suspension weight. Remove the nuts from the spring U-bolts, and remove the U-bolts and lower spring plate. The spring plate may be left on the rear shock absorber and moved out of the way.

Make sure the floor jack is positioned safely and that it is supporting the weight of the vehicle. Once all of the vehicle's weight is off the springs, remove the rear shackle nuts, plate, shackle, and bushings. Then, remove the front spring mounting bolt and remove the spring from the chassis. Check the spring center bolt to be sure it is not broken. Also, check the condition of the front hanger, bushing, and bolt, and the rear shackle, bushings, plate, and mount. Replace all worn or damaged parts. To install a spring, reverse the steps for removal. Make sure all parts are positioned properly, and all bolts and nuts should be tightened to the specified torque.

## STABILIZER BARS

Worn stabilizer bar mounting bushings, grommets, or mounting bolts cause a rattling noise as the vehicle is driven on irregular road surfaces. A weak stabilizer bar or worn bushings and grommets cause harsh riding and excessive body sway while driving on irregular road surfaces. Worn or very dry stabilizer bar bushings may cause a squeaking noise on irregular road surfaces. All stabilizer bar components should be visually inspected for wear.

Stabilizer bar removal and replacement procedures vary depending on the vehicle. Always follow the manufacturer's recommended procedure in the service manual. The following is a typical stabilizer bar removal and replacement procedure:

1. Lift the vehicle on a hoist and allow both sides of the front suspension to drop downward as the vehicle's chassis is supported on the hoist.
2. Remove the mounting bolts at the outer ends of the stabilizer bar and remove the bushings, grommets, brackets, or spacers **(Figure 15).**
3. Remove the mounting bolts in the center area of the stabilizer bar.
4. Remove the stabilizer bar from the chassis.
5. Visually inspect all stabilizer bar components such as bushings, bolts, and spacer sleeves. Replace the stabilizer bar, grommets, bushings, brackets, or spacers as required. Split bushings may be removed by slipping them over the stabilizer bar. Bushings that are not split must be pulled from the bar.

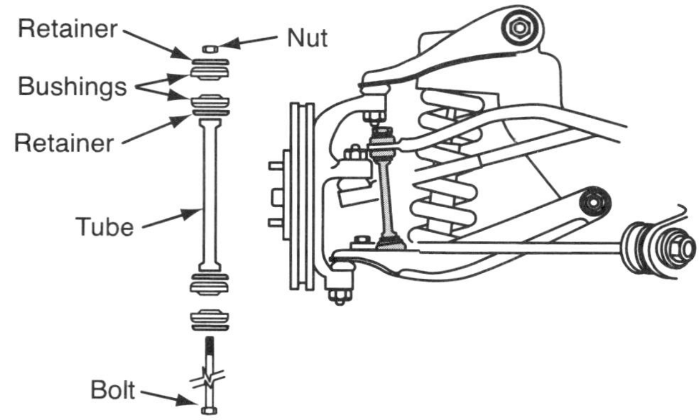


**Figure 15.** The mounting bolts, bushings, grommets, brackets, or spacers for a typical stabilizer bar.

6. Reverse steps 2 through 4 to install the stabilizer bar. Make sure all components are installed in the original position, and tighten all fasteners to the specified torque.

Some manufacturers specify that stabilizer bars must have equal distances between the outer bar ends and the lower control arms. Always refer to the manufacturer's recommended measurement procedure. If this measurement is required, adjust the nut on the outer stabilizer bar mounting bolt until equal distances are obtained between the outer bar ends and the lower control arms. Worn grommets cause these distances to be unequal.

# SUMMARY

- Sagged springs cause insufficient curb riding height, which results in handling problems.
- When replacing springs, the correct spring for the vehicle's tire and wheel package, suspension package, and the optional equipment installed on the vehicle must be used.
- Broken coil springs and worn or broken spring insulators may cause a rattling noise while driving on irregular road surfaces.
- Weak or broken coil springs reduce curb riding height, which affects front suspension alignment angles. This, in turn, can cause reduced directional stability, excessive tire wear, and harsh riding.
- Worn torsion bar components such as pivot cushion bushings and control arm bushings result in harsh riding and suspension noise when driving on road irregularities.
- A worn torsion bar hex and anchor may cause a rattling noise on irregular road surfaces.
- Weak torsion bars and worn bushings or anchors cause reduced curb riding height.
- Torsion bars are often adjusted to correct the curb riding height.
- Many leaf springs have plastic silencers between the spring leafs that can serviced without removing the spring.
- Weak or broken leaf springs affect suspension alignment angles and cause excessive tire tread wear, reduced directional stability, and harsh riding.

- A rattling noise on irregular road surfaces may be caused by worn shackles or bushings.
- A broken center bolt usually allows one side of the suspension to move rearward in relation to the other side. This action may cause steering pull to one side.
- Worn stabilizer bar mounting bushings, grommets, or mounting bolts cause a rattling noise on irregular road surfaces.
- A weak stabilizer bar or worn bushings and grommets cause harsh riding and excessive body sway on irregular road surfaces.
- Worn or dry stabilizer bar bushings may cause a squeaking noise.

## REVIEW QUESTIONS

1. A car experiences excessive body sway when cornering, but there is no abnormal noise in the suspension. The most likely cause of this problem is:
   A. the stabilizer bar bushing is missing.
   B. a weak stabilizer bar.
   C. the stabilizer bar grommets are worn out.
   D. a broken stabilizer bar.
2. Technician A says worn stabilizer bushings may cause squeaks as the vehicle moves down a bumpy road. Technician B says worn or dry control arm bushings may squeak as the vehicle moves down a bumpy road. Who is correct?
   A. Technician A
   B. Technician B
   C. Both Technician A and Technician B
   D. Neither Technician A nor Technician B
3. Technician A says a weak stabilizer bar can cause excessive body sway when driving on an irregular road surface. Technician B says a weak stabilizer bar can cause wear on the inside edges of the front tires. Who is correct?
   A. Technician A
   B. Technician B
   C. Both Technician A and Technician B
   D. Neither Technician A nor Technician B
4. While diagnosing a rattling noise that is very apparent when the vehicle is driven on bumpy roads, Technician A says the problem can be worn spring insulators. Technician B says the problem can be worn silencers between the leaves of the spring. Who is correct?
   A. Technician A
   B. Technician B
   C. Both Technician A and Technician B
   D. Neither Technician A nor Technician B
5. Which of the following is *not* likely to be caused by worn or broken coil springs?
   A. Excessive tire wear
   B. Increased riding height
   C. Directional instability
   D. Abnormal noises

# Chapter 25 Shock and Strut Service

## INTRODUCTION

Shock absorbers and struts should be replaced in pairs to ensure good handling and control of the vehicle. Replacement shock absorbers and struts should be exact replacements unless the customer desires to upgrade the shocks for improved handling or load carrying ability. In either case, the shocks or struts must be of the appropriate size for the model of vehicle.

Shocks are replaced as units, whereas struts may be replaced as units or have some of their main components replaced. The two most common items replaced on a strut are the cartridge and coil spring. The cartridge is the internal shock absorber in a strut. In some struts, cartridge replacement is not possible, and the struts must be replaced if they are defective.

### ON-CAR INSTALLATION OF A STRUT CARTRIDGE

On some vehicles, the front strut cartridge can be removed and replaced with the strut still installed in the vehicle. Always refer to a service manual for the proper procedure. A typical procedure for installing a cartridge while the strut is still in the vehicle follows:

1. Remove the nuts from the upper strut mount and remove the strut mount cover **(Figure 1).**
2. Use the special tool designed for this vehicle and the appropriate wrench to remove the nut from the strut piston rod **(Figure 2).**

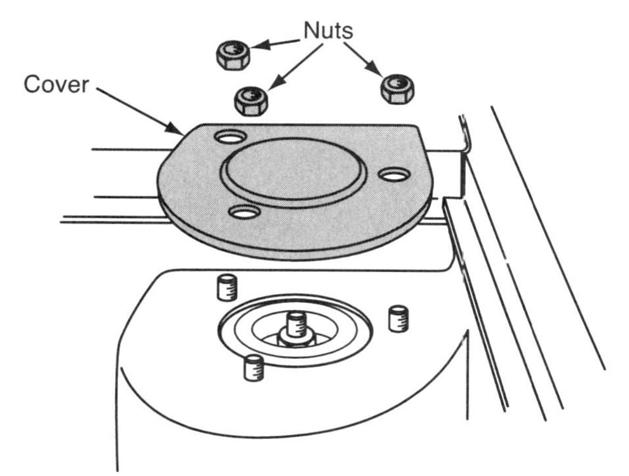


**Figure 1.** An upper strut mount cover and retaining nuts.

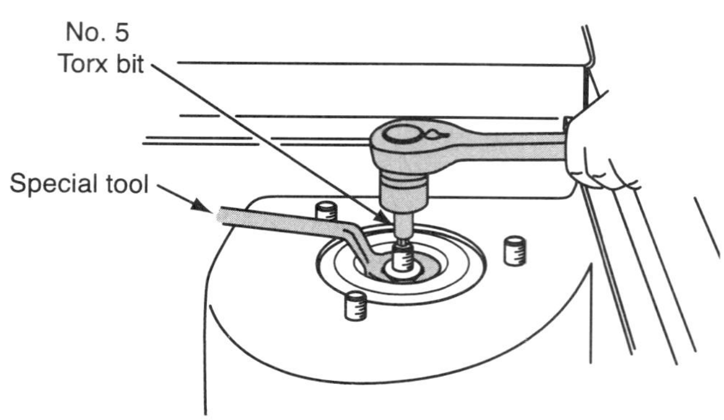


**Figure 2.** Use a special tool and a Torx bit to remove the nut from the strut piston.

> **You Should Know** *If a vehicle is hoisted or lifted in any way while performing this procedure, the coil spring may fly off the strut, causing vehicle damage and personal injury.*

3. Remove the bushing from the upper strut mount.
4. Thread the appropriate special tool onto the top of the strut piston rod and push the rod downward into the strut. Remove this tool and then remove the jounce bumper retainer.
5. Remove the jounce bumper **(Figure 3).**
6. Thread the special tool onto the strut piston rod and extend the rod. To remove the closure nut on top of the strut **(Figure 4),** another special tool must be used.
7. Grasp the top of the strut piston rod and remove the cartridge.
8. Remove the oil from the strut tube with a hand-operated suction pump.
9. Install the new cartridge in the strut **(Figure 5).**
10. Add the correct type of oil in the correct amount to the strut tube.
11. Install the closure nut and with the correct tool. Tighten it to specifications.
12. Install the jounce bumper.
13. Install the jounce bumper retainer.

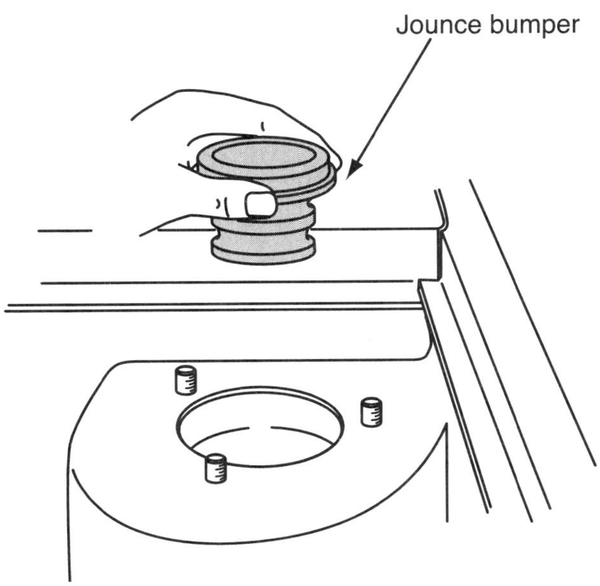


**Figure 3.** The jounce bumper is pulled off the strut rod.

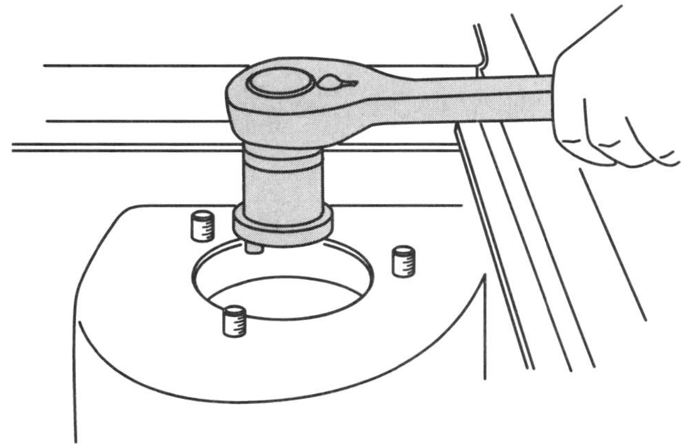

**Figure 4.** A special tool and ratchet are used to loosen and remove the strut closure nut.

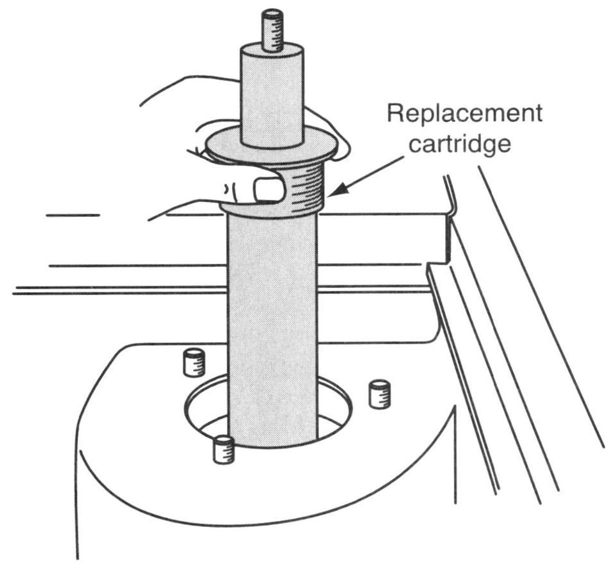


**Figure 5.** Install the new cartridge into the strut.

14. Place a light coating of the recommended oil on the upper strut mount bushing.
15. Install the bushing into the upper strut mount.
16. Use the special tool and the appropriate wrench to tighten the nut on the strut piston rod. Tighten this nut to specifications.
17. Install the strut mount cover and nuts.

## FRONT STRUT REMOVAL

The first step in removing a strut assembly is to scribe or make alignment marks from the upper strut mounting

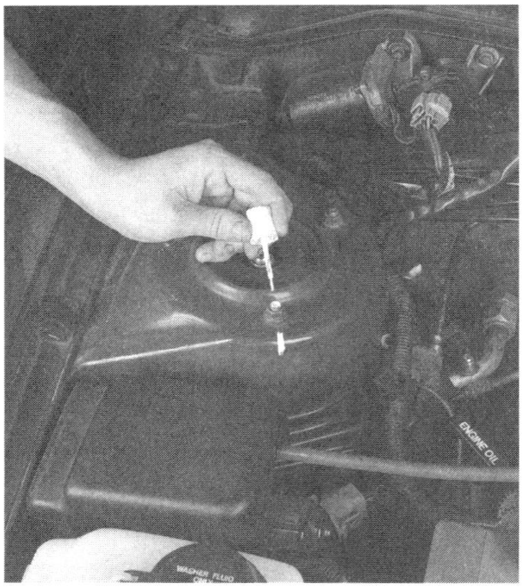

Figure 6. Before loosening the strut-to-chassis bolts, make an alignment mark from a bolt to the strut tower.

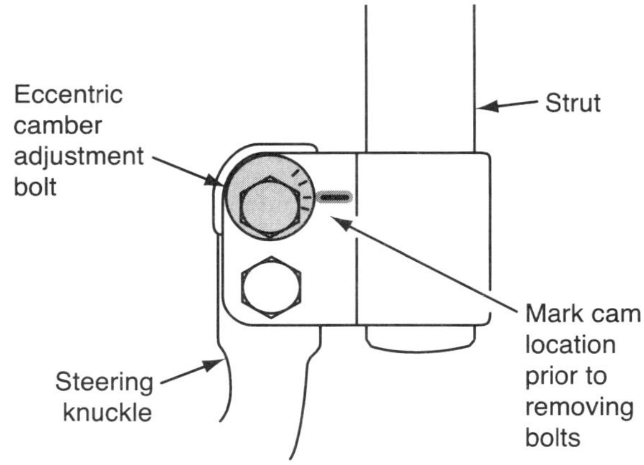


Figure 7. If one of the mounting bolts has an eccentric cam, the cam position should be marked on the strut prior to loosening the bolts.

bolts and the chassis of the vehicle **(Figure 6).** Many vehicles have oblong bolt holes and, to minimize the work required to align the wheels, the strut should be placed in its original position after service. After doing this, raise the vehicle on a hoist or with a floor jack. If a floor jack is used, lower the vehicle onto safety stands placed under the chassis. Allow the lower control arms and front wheels to drop downward with no vehicle weight on them. Then, remove the floor jack from under the vehicle.

Remove the front wheels. The strut is accessible from the wheel well after the wheel is removed. Now, disconnect and remove the brake line and antilock brake system (ABS) wheel-speed sensor wire from the clamps at the bottom of the strut. In some cases, these clamps may need to be removed from the strut to remove the brake line and wires.

If the strut-to-steering knuckle mounting bolts have an eccentric cam, the cam position should be marked on the strut prior to loosening the bolts **(Figure 7).** Returning the bolts to the same location will help maintain the correct camber angle. Then, remove the strut-to-steering knuckle retaining bolts. Separate the lower end of the strut from the knuckle. Support the steering knuckle with wire to prevent a strain on the brake line. Then, remove the upper strut mounting bolts at the top of the strut tower. Now the strut assembly can be removed.

## Separating the Spring from the Strut

Once the strut assembly is removed, install a bolt and two nuts into the upper strut-to-knuckle mounting bolt-hole. Place one of the nuts on the inside of the strut flange and the other nut on the same flange, but on the outside of the flange. Holding the bolt with a wrench, tighten the two nuts into each other. Put the bolt into the jaws of a bench vise, and clamp it securely in the vise to hold the strut and coil assembly with the compressing tool **(Figure 8).**

**You Should Know** *Never clamp the lower strut or shock absorber chamber in a vise with excessive force. This will distort the chamber and affect piston movement.*

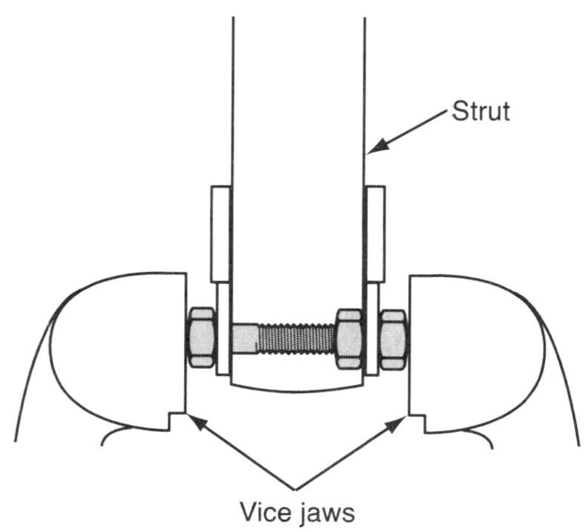


Figure 8. A bolt and two nuts installed in the upper strut-to-knuckle bolt hole can be used to hold the strut in a bench vice.

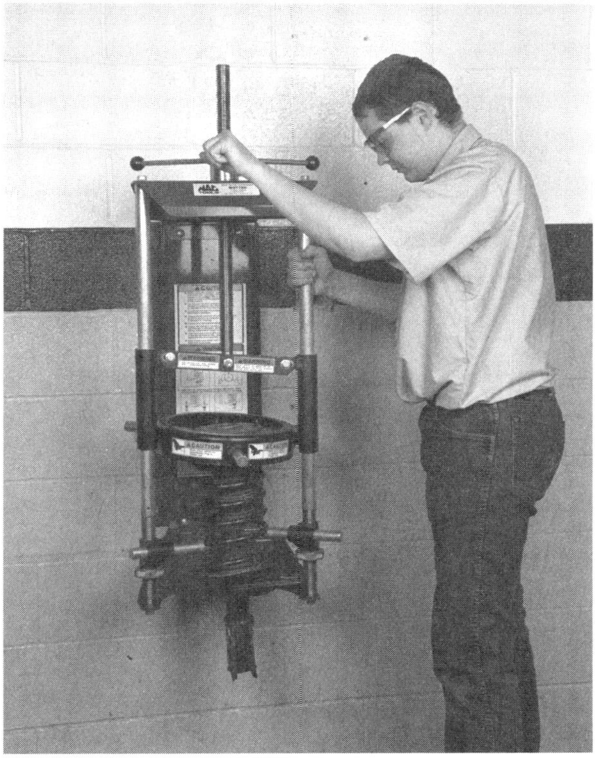

Figure 9. A wall-mounted strut fixture.

Special fixtures are designed to allow strut service without securing the strut in a vise. These fixtures are mounted to a wall and make strut service safer and easier **(Figure 9).**

The coil spring must be compressed with a spring compressor before the strut can be disassembled. All of the spring's tension must be removed from the upper spring seat before loosening the upper strut piston rod nut. Many different spring-compressing tools are available, and they must always be used according to the recommended procedure. If the coil spring has an enamel-type coating, tape the spring where the spring compressor will contact the spring. If this coating is chipped by the tool or during service, the spring may break at that point. Make sure the compressor is properly installed on the spring. If the spring slips off the tool while the spring is compressed, serious personal injury or property damage can result.

> **You Should Know** *If the upper strut nut is loosened with spring tension still on the upper mount, the mount can become a dangerous projectile and cause serious injury.*

Figure 10. Compress the spring until all tensioin is off the upper strut mount.

Tighten the spring compressor **(Figure 10)** to compress the spring until all of the tension is removed from the upper strut mount. Use the bar on the spring compressor to prevent the strut and spring assembly from turning while loosening the nut on the upper strut mount.

Remove the nut, upper strut mount, upper insulator, coil spring, spring bumper, and lower insulator. Then pull the strut from the bottom of the spring and remove it. Make sure the compressed spring is secure in the compressor.

During the disassembly of the strut, make sure you check the strut pivot bearing. Move the bearing with your hand. If the bearing is hard to move or seems to bind, it must be replaced. When replacing the bearing, make sure the correct side is up. Manufacturers normally mark the up side with paint or some other marking. Also, inspect the upper strut mount, coil spring, spring insulators, and spring bumper. Make sure you check each part for deterioration and other damage and replace them if necessary.

## STRUT CARTRIDGE REPLACEMENT

Many strut assemblies are sealed units, therefore a new cartridge cannot be simply installed. However, some manufacturers supply a replacement cartridge along with the required special tools that may be installed in the strut housing after the strut has been removed from the vehicle. Always follow the strut cartridge manufacturer's recommended installation procedure. The procedure for installing a cartridge in a serviceable strut removed from the vehicle is the same as replacing the cartridge while the strut is in the vehicle. The following is a typical procedure for replacing a cartridge in a sealed strut **(Figure 11):**

1. With the strut mounted in a vise, locate the groove near the top of the strut body.
2. Position a pipe cutter into the groove and cut the top of the strut body off.
3. After the cutting, remove the strut piston assembly from the strut.
4. Remove the strut from the vise and dump out the oil from the strut.
5. Place the special tool supplied by the vehicle's or cartridge's manufacturer on top of the strut body and strike it with a plastic hammer until its shoulder contacts the top of the strut body. This removes any burrs from cutting and places a slight flare on the body.
6. Remove the tool from the strut body.
7. Place the new cartridge in the strut body and rotate the cartridge until it settles into the indentations in the bottom of the strut body.
8. Place the retaining nut that came with the cartridge over the cartridge.
9. Using the correct tool, tighten the nut to the specified torque.
10. Move the strut piston rod in and out several times to check for proper operation.

**You Should Know** ***Gas-filled struts have an internal catch that holds them in the compressed position. To release this catch, hold the piston rod and rotate the strut counterclockwise 45 degrees.***

## STRUT REASSEMBLY

A typical procedure for installing a coi spring on a strut follows:

1. With the strut securely mounted in a vise or fixture, install the lower insulator on the lcwer strut spring

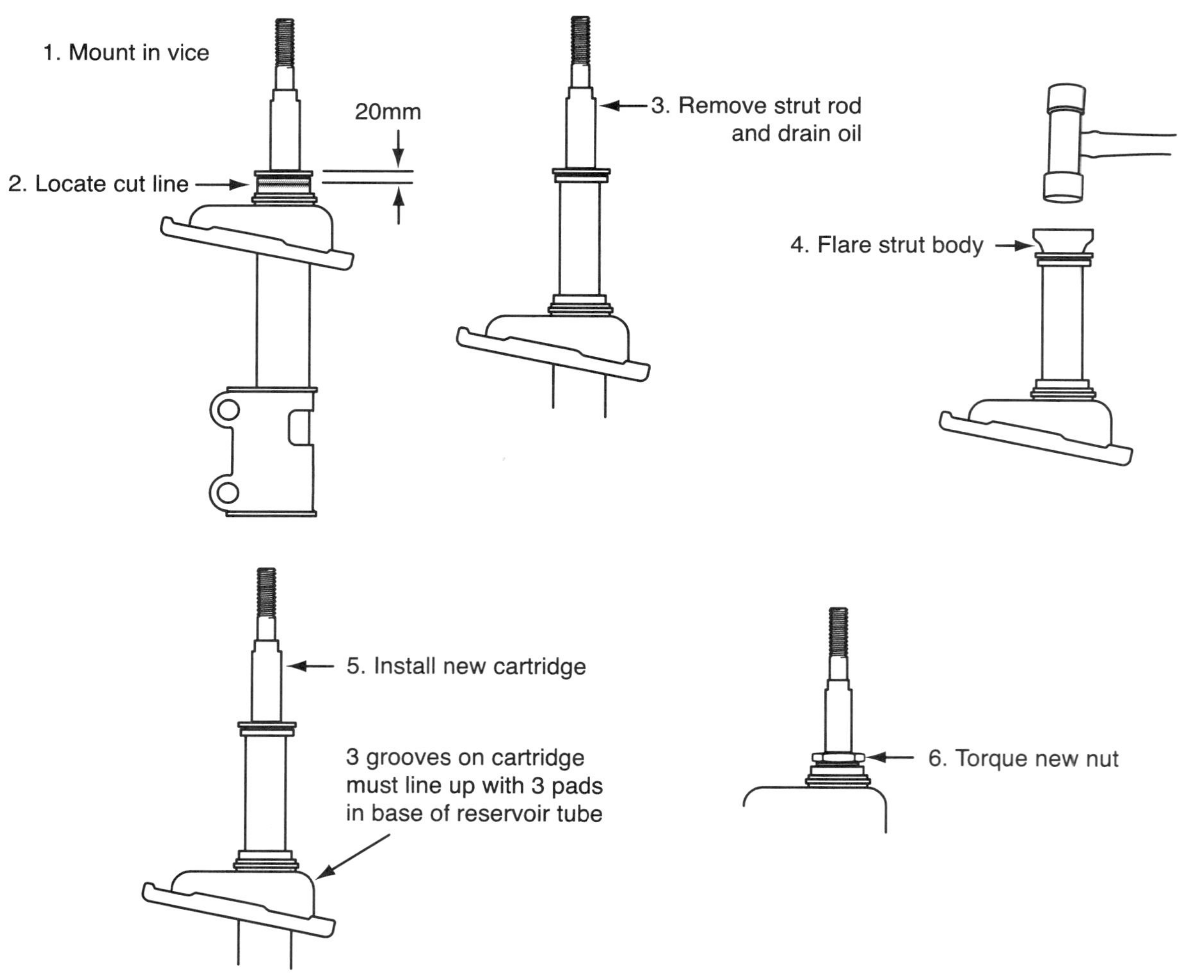


**Figure 11.** A typical procedure for installing a strut cartridge.

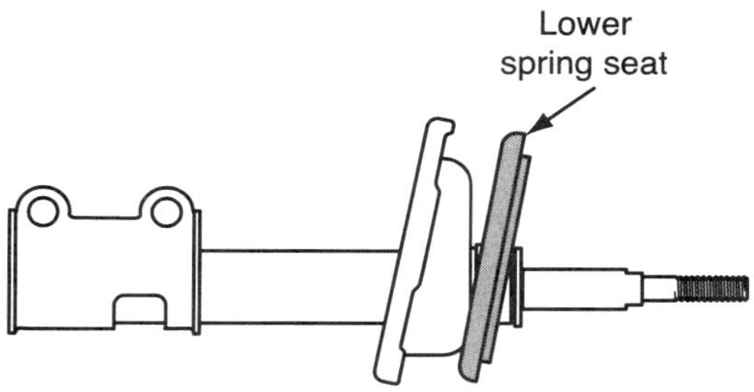


**Figure 12.** Before installing the spring onto the strut, make sure the lower spring seat is in its proper position.

seat, and be sure the insulator is properly seated **(Figure 12).**

2. Install the spring bumper onto the strut rod.
3. With the coil spring compressed in the spring compressor, install the spring on the strut. Make sure the spring is properly seated on the lower insulator spring seat.
4. Make sure the strut piston rod is fully extended, then install the upper insulator on top of the coil spring.
5. Install the upper strut mount on the upper insulator.
6. Make sure the spring, upper insulator, and upper strut mount are properly positioned and seated on the coil spring and strut piston rod.
7. Using the bar on the spring compressor, prevent the strut and spring from turning as the strut piston rod nut is tightened to the specified torque.
8. Rotate the upper strut mount until the lowest bolt in the mount is aligned with the tab on the lower spring seat **(Figure 13).**
9. Gradually loosen the spring compressor until all of the spring tension is released from the tool, then remove the tool from the spring.
10. When a strut and coil spring are reassembled, always make sure the spring and insulators are properly seated on the spring seats.

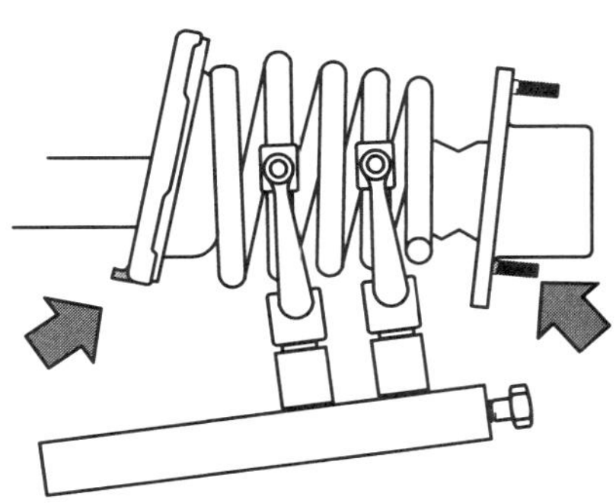

**Figure 13.** Align the lowest bolt on the upper strut mount with the tab on the lower spring seat.

## STRUT ASSEMBLY INSTALLATION

Typically, installing a front strut is a straightforward procedure. The procedure is almost the reverse of the removal procedure. Begin by installing the strut and spring assembly with the upper strut mounting bolts extending through the bolt holes in the strut tower. Tighten the nuts on the upper strut mounting bolts to the specified torque.

Insert the strut and spring assembly into its basic position with the upper strut mounting bolts extending through the holes in the strut tower. Align the strut bolt to chassis index marks, then tighten the nuts on the upper strut mounting bolts to the specified torque.

Then, install the lower end of the strut in the steering knuckle to the proper depth, and tighten the strut-to-knuckle retaining bolts to the specified torque. If one of the strut-to-knuckle bolts has an eccentric washer, make sure the eccentric is aligned with the mark placed on the strut during the removal procedure.

Install the brake hose in its clamp on the strut. Also, place the ABS wheel-speed sensor wire in the clamp. Make sure the clamp will not crimp the wire or brake line, then tighten the clamp. Make sure all parts that were removed or disconnected during strut removal are installed. Then, install the wheels and tighten the lug nuts to specifications. Now, raise the vehicle, remove the safety stands, and lower the vehicle onto the shop floor. The wheels should be aligned prior to returning the vehicle to the customer.

## STRUT DISPOSAL PROCEDURE

Always follow the vehicle manufacturer's recommended procedure for strut disposal. Do not throw gas-filled struts or shock absorbers in a fire of any kind nor in a dumpster. If the vehicle manufacturer recommends drilling the strut to release the pressure, drill the strut at the manufacturer's recommended location. Never drill a hole in a gas-filled strut or shock absorber unless instructed to do by the manufacturer.

The following is a typical strut drilling procedure that is performed prior to strut disposal:

1. Fully extend the strut rod.
2. Center punch the strut at the recommended drilling location **(Figure 14).**
3. Drill a small hole at the center-punched position.

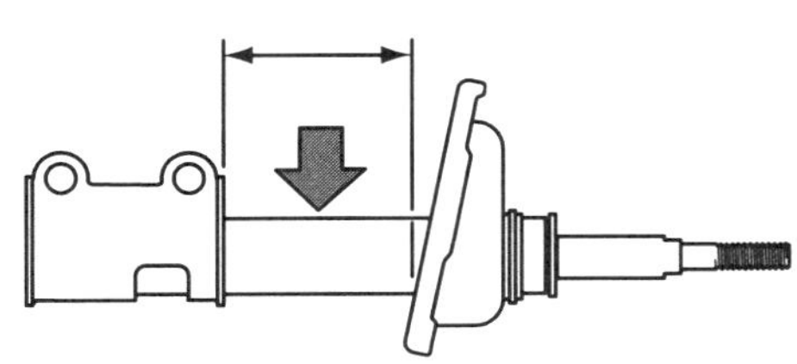

**Figure 14.** The typical location for drilling a hole in a strut to relieve pressure before disposing of the strut.

## REAR STRUT REPLACEMENT

The correct procedure for replacing a rear strut varies with the different types of rear suspensions. Always follow the recommended procedure when doing this. On some vehicles, it may be necessary to remove the rear seat and the package tray behind the seat **(Figure 15).** Doing this will give access to the upper mounting bolts and nuts for the strut.

Lift the vehicle with a floor jack, and lower the chassis onto safety stands so the rear suspension is allowed to drop. Place a wooden block between the floor jack and the rear spindle on the side where the strut will be removed. Raise the floor jack to support some of the weight on the suspension.

Remove the rear wheel, disconnect the nut or clamp on the lower part of the strut, and remove the brake hose and ABS sensor wire from the strut. Remove the stabilizer bar link from the strut, and loosen and remove the strut-to-spindle mounting bolts. Pull upward on the strut to remove the strut from the spindle. If necessary, lower the floor jack slightly to remove the strut from the spindle.

It may be wise to make alignment marks from the upper mounting bolts to the chassis. Doing this will bring the strut and wheel close to proper alignment after the strut has been installed. After doing this, remove the upper support nuts and remove the strut.

Visually inspect the coil spring, upper and lower mounts, insulators, and spring bumper. If any of these components are damaged, worn, or distorted, replacement is necessary. The rear coil springs may be removed from the rear struts using the same basic procedure for spring removal from the front struts.

To install the strut, place it in the vehicle, then position the upper mounting bolts at the alignment marks made during removal. Then tighten the upper support nuts to specifications.

Push the bottom of the strut onto the spindle. If this is difficult, raise or lower the floor jack under the spindle to align the strut-to-spindle mounting bores. Connect the stabilizer bar link to the strut, and install and tighten the strut-to-spindle mounting bolts.

Place the brake line and ABS sensor wire in the clamp and then tighten the clamp. Then, install the wheel and tighten the lug nuts to specifications. Now install the rear package tray and rear seat, if necessary. Then, check the wheel alignment of the vehicle.

## SHOCK ABSORBER REPLACEMENT

Front shock absorbers are usually mounted to the lower control arm by nuts or bolts and are bolted to the spring tower through rubber insulator bushings or grom-

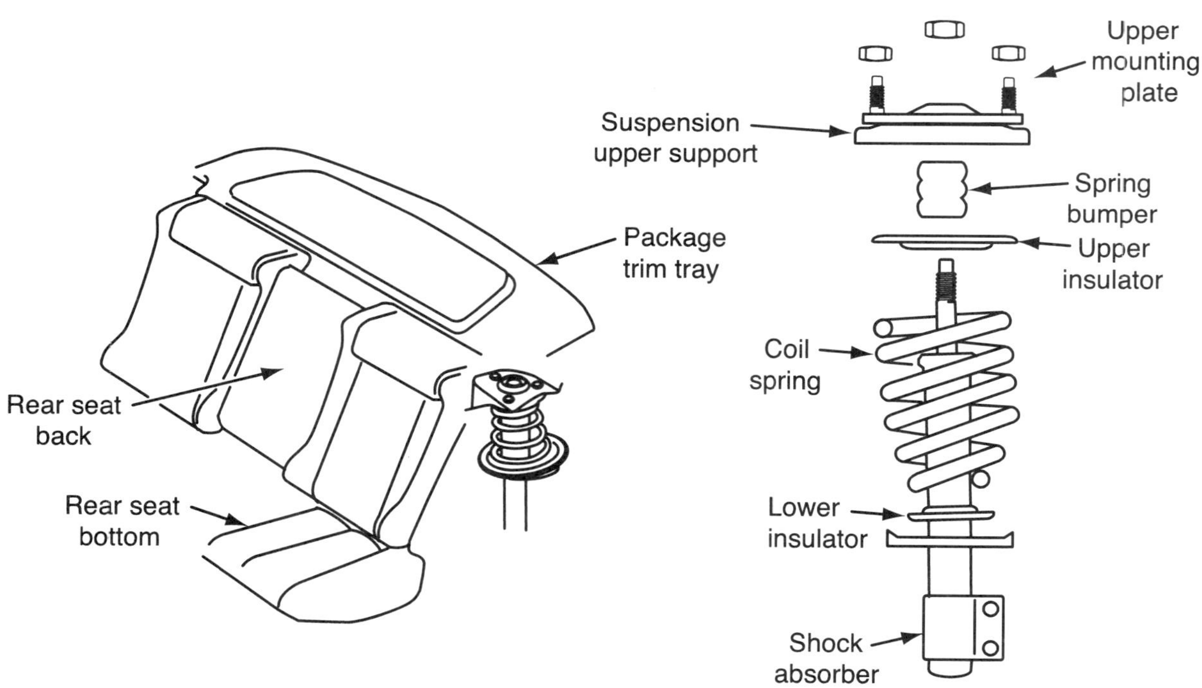


**Figure 15.** On some vehicles, it may be necessary to remove the rear seat and the package tray behind the seat to gain access to the upper mounting bolts and nuts for the strut.

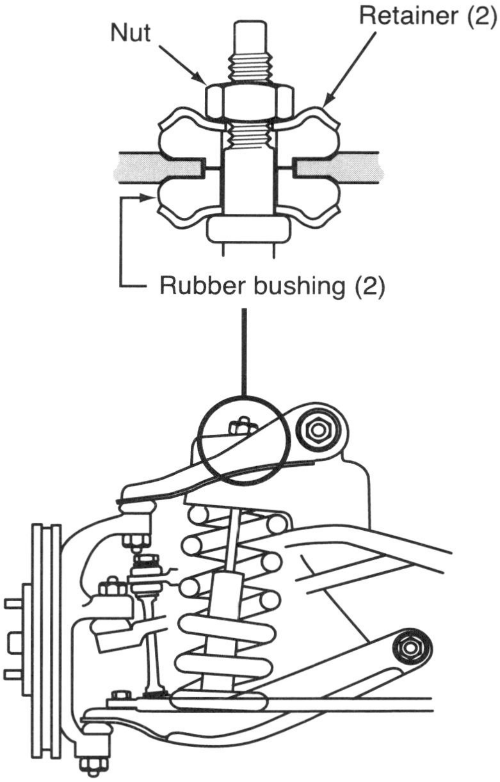


**Figure 16.** Front shock absorbers are usually mounted to the lower control arm by nuts or bolts and are bolted to the spring tower through rubber insulator bushings or grommets.

mets **(Figure 16).** When a front shock absorber must be changed, lift the front end on the vehicle with a floor jack, then place safety stands under the lower control arms. Lower the vehicle onto the safety stands and remove the floor jack. If the nut or top mounting bolts are easily accessible from under the hood, loosen and remove them before raising the vehicle.

The nut on single-stud shock absorbers tends to rust to the stud. This makes removal very difficult because the stud, which is the end of the piston rod inside the shock, will rotate as the nut is turned. When this is the case, begin removal by soaking the nut and stud with penetrating oil. After a period of time, try to loosen the nut. If this did not loosen it, use a special tool that is designed to fit around the flats on the stud. Hold the stud with the tool and attempt to loosen the nut. Sometimes, the use of an air impact wrench will free the nut. A special tool, called a nut breaker, may also work. This tool fits over the nut. When is tightened, it can split the nut. This, of course, will allow it to be removed. If none of these methods works, the stud can be broken off by bending the stud back and forth..

> **You Should Know** *Never apply heat to the lower shock camber of a shock or strut with an acetylene torch. Excessive heat may cause a shock or strut to explode, which could result in a serious personal injury.*

Disconnect the upper shock mounting nut and grommet. Then, remove the lower shock mounting nut or bolts and remove the shock absorber.

Reverse the removal procedure to install new shocks with new grommets. Before installing new shock absorbers, purge the air from them by compressing the shock while it is upside down. Then put the shock in the upright position and extend it. Repeat this until the shock has a constant resistance to movement in both directions.

With the full weight supported on the suspension, tighten the shock mounting nuts to the specified torque.

## Rear Shock Absorbers

The top of rear shocks are mounted to the frame or body of the vehicle while the bottom is bolted to a bracket on the rear axle housing **(Figure 17).** When rear shock absorber replacement is necessary, follow this procedure:

1. Raise the vehicle on a lift and support the rear axle with safety stands to prevent the shock absorbers from extending fully.

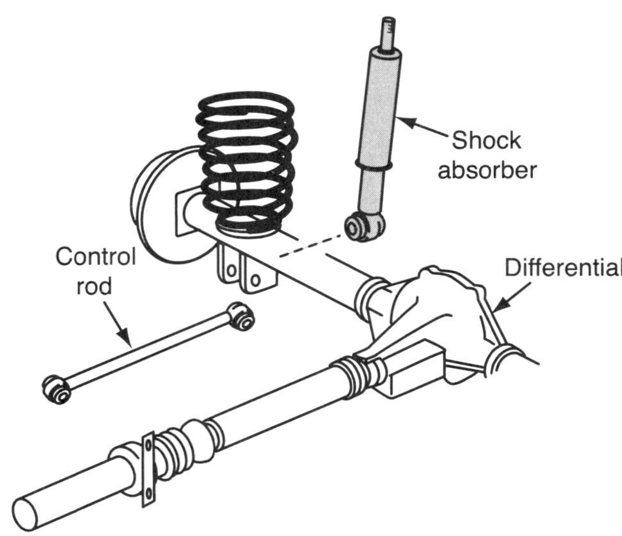


**Figure 17.** The top of rear shocks are mounted to the frame or body of the vehicle while the bottom is bolted to a bracket on the rear axle housing.

2. Place a floor jack under the rear axle to support it and prevent a strain on the brake lines.
3. Remove the wheel and tire assembly.
4. Check the mounting bolts and bushings for wear, looseness, or damage. If these are loose and a rattling noise is evident, replacement of the bushings and bolts is necessary.
5. Remove the upper shock mounting nut(s) or bolt(s) and the grommet.
6. Remove the lower shock mounting nut(s) or bolt(s).
7. Remove the shock absorber.
8. Purge air from the shock by compressing the shock while it is upside down. Then put the shock in the upright position and extend it. Repeat this until the shock has a constant resistance to movement in both directions.
9. Install the new shock absorber and grommet.
10. Install and tighten the lower shock mounting nut(s) or bolt(s) to specifications.
11. Install and tighten the upper shock mounting nut(s) or bolt(s) to specifications.
12. Remove the transmission jack from under the axle.
13. Install the wheel-and-tire assembly.
14. Lower the vehicle so that the vehicle's weight is on the tires.
15. Tighten the wheel's lug nuts to the specified torque.

## SUMMARY

- On some vehicles, the front strut cartridge can be removed and replaced with the strut installed on the vehicle.
- To service the strut on many vehicles, the strut must be removed from the vehicle.
- A strut is normally serviced by replacing the entire strut, replacing the coil spring, or replacing the strut cartridge and its related rubber parts.
- If a coil spring has an enamel-type coating, it may break if the coating is chipped.
- The strut pivot bearing should always be checked when servicing a strut.
- Do not throw gas-filled struts or shock absorbers in a fire of any kind nor in a dumpster. Dispose of them in the way recommended by the manufacturer.
- Front shock absorbers are usually mounted to the lower control arm by nuts or bolts and are bolted to the spring tower through rubber insulator bushings or grommets.
- The top of rear shocks are mounted to the frame or body of the vehicle while the bottom is bolted to a bracket on the rear axle housing.

## REVIEW QUESTIONS

1. Describe the typical procedure for disposing of a strut cartridge or shock absorber.
2. What should you do to safely mount a strut in a bench vise before servicing it?
3. Which of the following is *not* a correct statement about the removal of a front strut in a MacPherson strut suspension?
   A. The eccentric washer on some strut to knuckle bolts should be marked in relation to the strut.
   B. The strut rod nut may be loosened with spring tension applied to the upper strut mount.
   C. The strut's lower chamber should not be clamped in a vise to support the strut.
   D. The spring compressor must be installed before the strut is removed from the vehicle.
4. Technician A supports the rear axle with a floor jack before replacing the rear shocks. Technician B removes all weight from the axle and wheel before removing the shocks. Who is correct?
   A. Technician A only
   B. Technician B only
   C. Both Technician A and Technician B
   D. Neither Technician A nor Technician B
5. While replacing a strut cartridge while the strut is still in the vehicle, Technician A uses a pipe cutter to cut the top of the strut body off. Technician B removes the oil from the strut tube with a hand-operated suction pump after the old cartridge is removed. Who is correct?
   A. Technician A only
   B. Technician B only
   C. Both Technician A and Technician B
   D. Neither Technician A nor Technician B

# Section 6

## Suspension Systems



**Interesting Fact**

*RWD vehicles usually have short/long arm front suspension systems. Since space and weight are important on smaller FWD cars, the lighter, more compact MacPherson strut front suspension is used on most of these vehicles.*

## SECTION OBJECTIVES

After you have read, studied, and practiced the contents of this section, you should be able to:

- Identify different types of frame construction.
- Define vehicle tracking.
- Define setback and the effect setback has on vehicle steering.
- Define rear axle offset, and describe the effect of this problem on steering control.
- Visually inspect frames.
- Explain how frames are measured.
- Identify two types of load-carrying ball joints and the location of the control arm for each type.
- Describe a live-axle rear suspension system.
- Describe the movement of the rear axle housing during vehicle acceleration.
- Compare a semi-independent with an independent rear suspension system.
- Describe the major components in an electronic air suspension system.
- Diagnose short/long arm (SLA) suspension system noises, body sway, and uneven riding height concerns.
- Diagnose strut suspension system noises, body sway, and uneven riding height concerns.
- Remove, inspect, and install upper and lower control arms, bushings, shafts, and rebound bumpers.
- Inspect, remove, and replace ball joints.
- Remove, inspect, install, and adjust strut rods (compression/tension) and bushings.
- Remove, inspect, and install steering knuckle assemblies.
- Remove, inspect, and install transverse links, control arms, bushings, and mounts.
- Diagnose rear suspension noises.
- Diagnose rear suspension sway and lateral movement.
- Remove, replace, and inflate air springs.
- Adjust front and rear trim height on electronic air suspension systems.
- Test and diagnose components of electronically controlled suspension systems using a scan tool.

# Chapter 26 Frames

## INTRODUCTION

A vehicle without a frame could not support its own weight nor the weight of its passenger or cargo load. A vehicle's frame also allows it to absorb stress from road irregularities, absorb torque from the engine and drive wheels, and provides a main member for the attachment of the body and other components.

The frame, together with the front and rear suspension systems, must position the wheels properly to minimize tire tread wear and provide accurate steering control.

## FRAME CONSTRUCTION

Frames may contain channel (partial box), complete box, or tubular steel members **(Figure 1).** Most frames on today's vehicles include all three types of members.

### Ladder-Type Frames

The **ladder frame design** is used on trucks and full-size vans. In this type of frame, the side rails have very little offset and are in a straight line between the front and rear wheels. The ladder-type frame has more crossmembers than other designs. This provides for increased load-carrying capacity and rigidity **(Figure 2).**

The body and its related components are bolted to the frame, but rubber insulating bushings are positioned between the body and frame mounting locations. These bushings help prevent the transfer of road noise and vibration from the suspension and frame to the body. Rubber insulating mounts are also positioned between the engine and transmission. These mounts reduce the transfer of engine vibration to the frame, body, and vehicle interior. Many suspension components are also connected to the frame through rubber insulating bushings to reduce the transfer of harshness and vibration from the suspension to the frame.

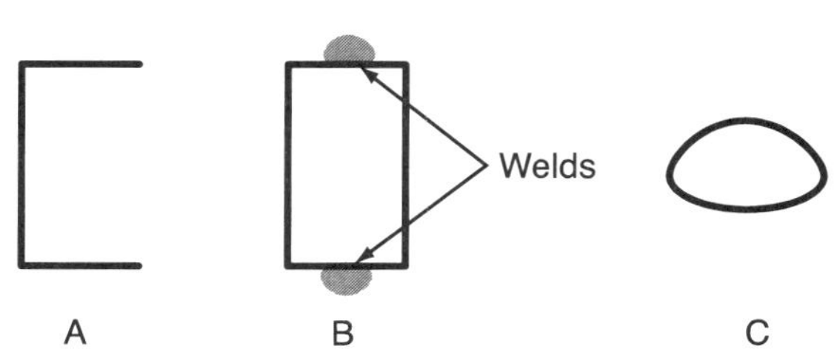


**Figure 1.** Types of frame construction: (A) channel or partial box, (B) complete box, (C) tubular crossmember.

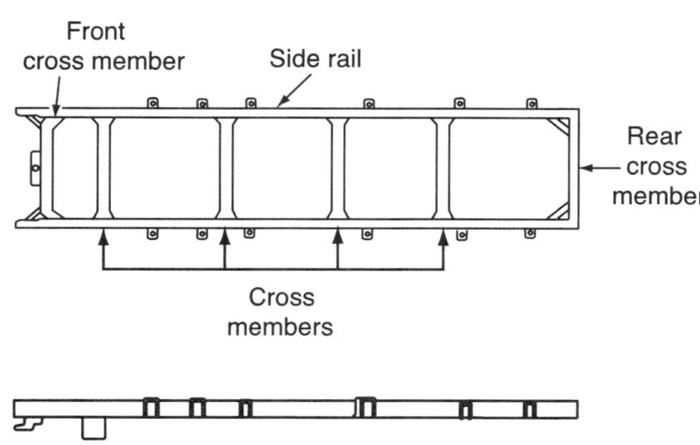


**Figure 2.** A ladder-type frame.

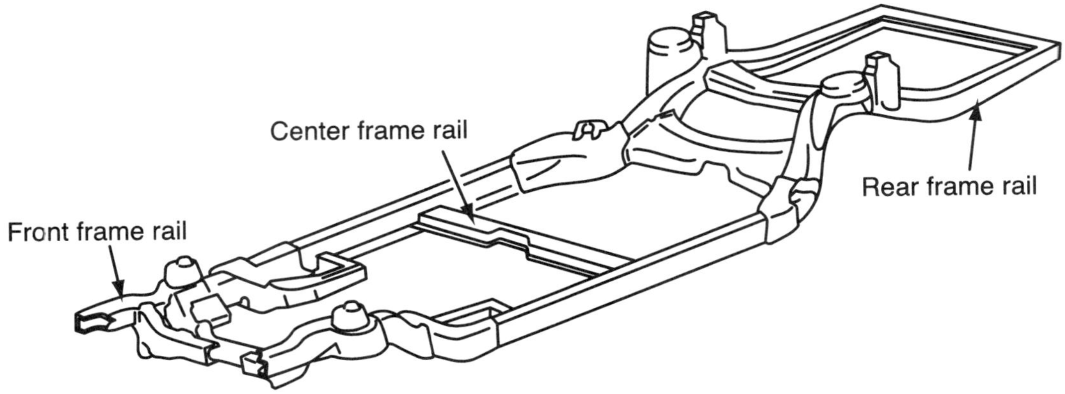


**Figure 3.** Perimeter-type frame components.

## Perimeter-Type Frames

A **perimeter-type frame** is used on some RWD vehicles. This frame forms a border around the passenger compartment. The frame rails are stepped inward at the cowl area to provide increased strength to support the engine and front suspension. This inward step also provides room for the movement of the front wheels. Lateral support is provided by crossmembers welded between the frame rails near the front and rear of the frame. A transmission support member is welded or bolted between the frame rails at the back of the transmission. The rear frame kick-up side rails support the rear suspension and the rear portion of the body. A front torque box is positioned ahead of the transmission support member, and a rear torque box is at the front of the rear kick-up side rails **(Figure 3).** The torque boxes are designed to absorb the impact during a side collision and reduce damage to other body components. They also provide some protection for the occupants during a side collision.

The body and other components are bolted to the frame. Like a ladder-framed vehicle, rubber insulating bushings are positioned between the body and frame mounting locations. Vibration and noise insulating mounts are also used to isolate the engine, transmission, and suspension parts.

## Unitized Body Design

Most front-wheel-drive vehicles and some small vans, SUVs, and pick-up cars have a **unitized body**. In this design, the frame and body are constructed as one unit. The strength and rigidity of the body is achieved by body design rather than by a heavy steel frame. All members of a unitized body are load-carrying components. The floor pan, roof, inner aprons, quarter panels, pillars, and rocker panels are integrally joined to form a unitized body **(Figure 4).**

**Interesting Fact**

*During the late 1940s, American Motors Corporation introduced the unitized body design in the United States. Some of these first-generation unitized bodies did not have partial frames. However, unitized bodies did not become popular until 1980 and 1981. These FWD cars had partial frames and may be referred to as second-generation unitized bodies.*

Most unitized bodies have bolt-on partial frames at the front and rear of the vehicle. Some of the components in a unitized body are made from **high-strength steels (HSS)** to provide additional protection in a collision. Ultra-HSS may be used for some body components such as door beams and bumper reinforcements. The unitized body has a complex design that spreads collision forces throughout the body to help protect the vehicle's occupants.

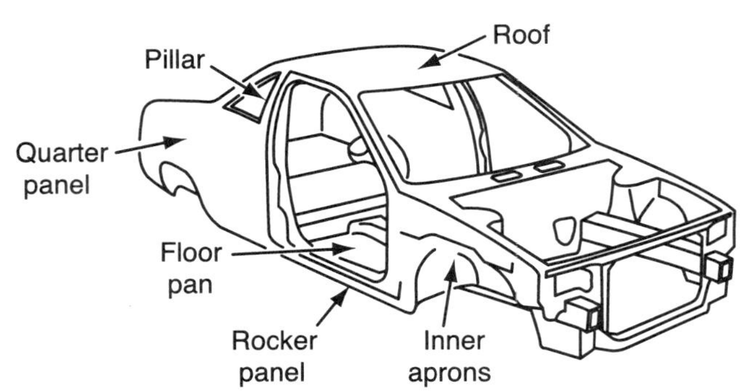


**Figure 4.** Unitized body design.

## VEHICLE DIRECTIONAL STABILITY

The front and rear suspension systems are attached to the frame, partial frame, or unitized body. Therefore, the frame or unitized body are indirectly responsible for good **directional stability**. Vehicle **tracking** is the parallel relationship between the front and rear wheels during forward motion. To provide proper tracking, each front wheel must be the same distance from the vehicle's centerline, and each rear wheel must be an equal distance from the same centerline. The distance between the front wheels and rear wheels does not necessarily need to be the same, but all four wheels must have a parallel relationship to provide proper tracking. The vehicle's **wheelbase** is the distance between the centers of the front and rear wheels. Equal wheelbase measurements on each side of the vehicle and proper tracking are absolutely essential to providing directional stability.

> **You Should Know** *Improper vehicle tracking may cause steering pull that may be worse during hard braking. Therefore, improper tracking creates a safety hazard that must be corrected..*

### Setback

If one of the front wheels has **setback,** the spindle on this wheel is rearward in relation to the other front wheel. The centerline of each front wheel is still at a 90-degree angle to the vehicle centerline, but the front wheels no longer share the same centerline **(Figure 5).**

When the left front wheel has setback, the vehicle will tend to steer to the left as it is driven straight ahead. Under this condition, the driver must continually turn the steering wheel to the right to keep the vehicle moving straight.

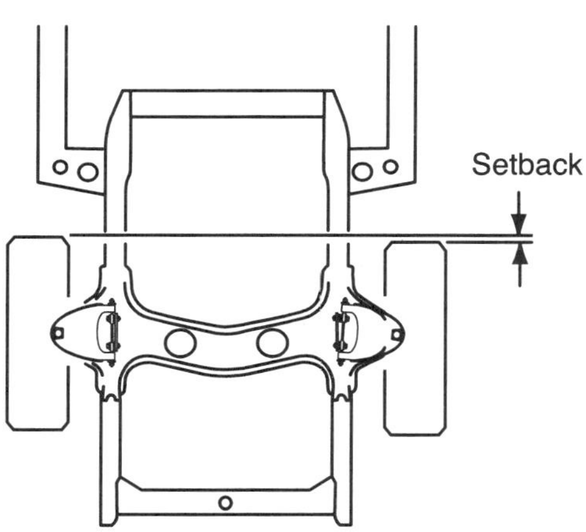


**Figure 5.** When one front wheel is to the rear of the other wheel, the vehicle has setback.

### Axle Offset

If the rear axle has shifted and its centerline is no longer at a 90 degree angle to the vehicle's centerline **(Figure 6)** a condition called **axle offset** results. When the left side of the rear axle has shifted rearward, the steering will pull continuously to the right. Under this condition, the driver must turn the steering wheel to the left to keep the vehicle moving straight.

### Axle Sideset

When **axle sideset** occurs, the rear axle moves inward or outward, but the axle and vehicle's centerlines remain at a 90-degree angle in relation to each other **(Figure 7).** When the front-to-rear axle centerline is no longer at the vehicle's centerline, this causes steering pull.

### Indications of Frame Damage

The most common cause of frame damage on cars and light-duty trucks is collision. In some cases, collision damage may be repaired to make the appearance of the vehicle satisfactory, but the frame damage may not always be correctable. Some indications of frame damage are:

1. Excessive tire wear when the front suspension alignment angles are correct
2. Steering pull when the front suspension alignment angles are correct
3. Steering wheel not centered when driving straight ahead, but the steering wheel was centered in the shop

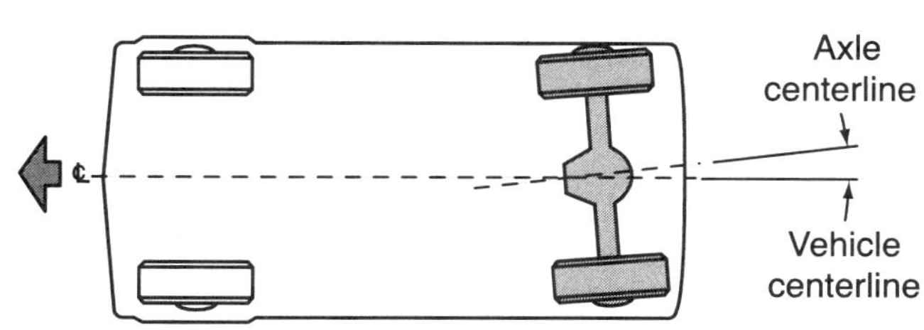


**Figure 6.** An example of rear axle offset.

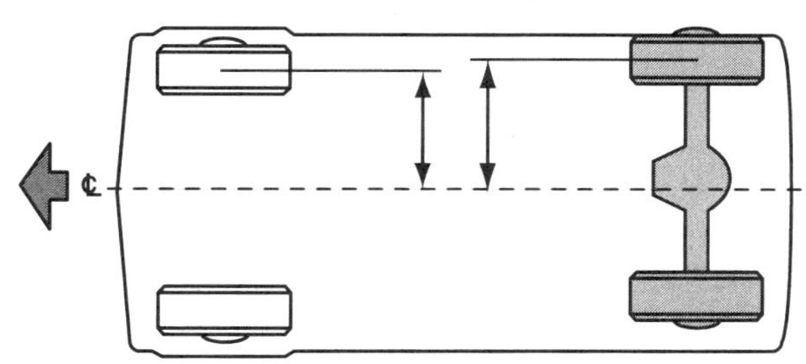

**Figure 7.** An example of axle sideset.

# FRAME CHECKS

Special equipment is used to check the squareness of a frame. These measurements are not normally taken by a steering and suspension technician. Rather, they are done by body repair technicians. However, you should be able to conduct a good visual inspection of a frame and be familiar with the different techniques used to measure and check a frame.

## Visual Inspection

When frame damage is suspected, it should be checked and measured. However, before having this done the frame and suspension should be visually inspected. Check for wrinkles on the upper flange of the frame, which indicate a sag problem. Visually inspect the lower flange for wrinkles, which are definite evidence of frame buckle. Since suspension or axle problems may appear as frame problems, the suspension components should be inspected for wear and damage. Check all suspension mounting bushings and inspect leaf-spring shackles and center bolts.

The frame should also be inspected for cracks, bends, and severe corrosion. Minor frame bends are not visible. Straight cracks may occur at the edge of the frame flange, and sunburst cracks may radiate from a hole in the frame web or crossmember **(Figure 8).**

## Frame Measurement

The two most commonly used tools to check and measure a frame or front cradle are the plumb bob and tram gauge. A plumb bob is a weight with a sharp, tapered point suspended by a string. The plumb bob will always indicate a true vertical line; therefore, measuring horizontally from the plumb bob provides a true horizontal measurement. Frame measurements are taken between specified locations. When measuring a frame, horizontal measurements are made straight across the frame and diagonally forward and rearward on the frame. Vertical measurements from specific locations on the frame to an imaginary standard line are also necessary to detect such conditions as frame sag and twist.

Horizontal frame measurements may also be done with a tram gauge rather than a plumb bob. A tram gauge is a long, straight bar with two adjustable pointers. The distance between the pointers is adjustable, as is the height of the pointers. A tram gauge may be called a track gauge.

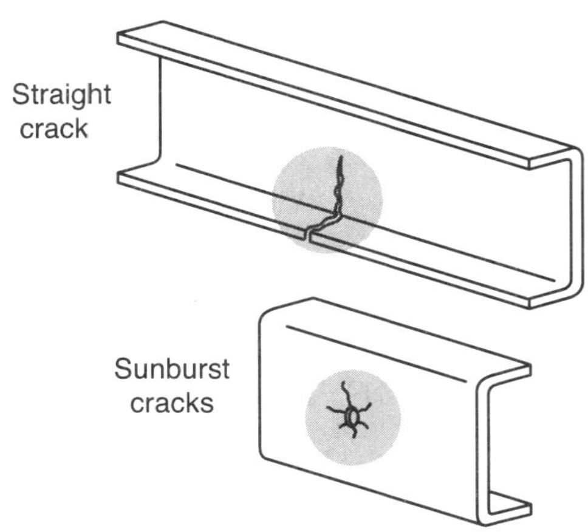


**Figure 8.** Straight cracks may occur in the frame flange, and sunburst cracks may radiate from a hole in the frame or crossmember.

## Measuring Unitized Body Alignment

Since the lower control arms are attached to the front crossmember or cradle, front wheel alignment angles are affected by a bent crossmember. The engine and transaxle mounts are also connected to the front crossmember. Therefore, a bent crossmember may cause improper engine and transaxle position, which results in front drive axle vibration. Loose or worn front crossmember mounts may cause improper front wheel alignment and engine position.

### Tram Gauge

A tram gauge may be used to make measurements on unitized bodies such as the upper strut towers **(Figure 9)** and front crossmember. The tram gauge may also be used to perform vertical measurements on the front and rear subframes on a unitized body vehicle.

### Bench

A dedicated bench system is necessary to check many unitized bodies after medium or heavy collision damage. The dedicated bench system has three main parts: the bench, transverse beams, and the dedicated fixtures. When used together, this equipment will perform many undercar unitized body measurements, including length, width, and height at the same time. The bench contains strong steel beams mounted on heavy casters.

The transverse beams are mounted perpendicular to the bench. There is a variety of holes in the bench for proper transverse beam attachment. Various holes in the transverse beams provide the correct dedicated fixture positions.

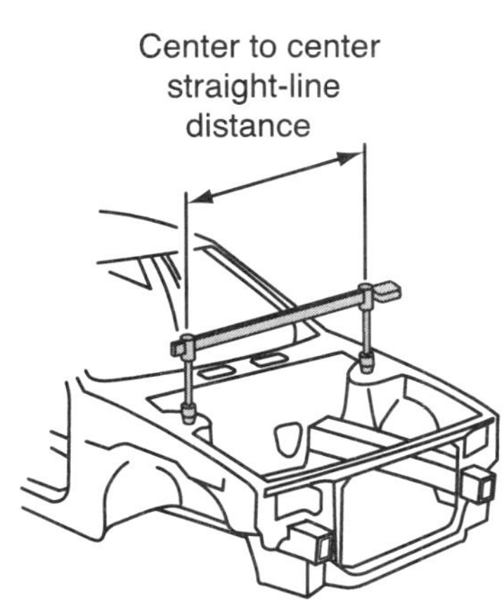


**Figure 9.** Use a tram gauge to measure the upper strut towers on a unitized body.

The dedicated fixtures are specific to the body style. These fixtures are bolted to the transverse beams, and the holes in the upper end of the fixtures are aligned with specific body openings.

The unitized body must be straightened until the holes in the dedicated fixtures and the body openings are aligned. If the holes in the unitized body are behind and above the dedicated fixture openings, the unitized body must be pulled forward and downward in that area.

Universal benches are now available with computer measuring systems. Measuring a unitized body with a bench system and straightening these bodies with special body pulling equipment is usually done in an auto body shop rather than an automotive repair shop.

## FRAME STRAIGHTENING

Frame straightening is usually done with special hydraulically operated bending equipment and by experienced body technicians. During the straightening process on a unitized body, the body is securely bolted to the dedicated bench at all the possible locations except those where misalignment is present. The dedicated bench is securely chained to special holding fixtures. Hydraulically operated pulling equipment is then attached securely to the unitized body to pull the body in the intended direction. Pulling equipment may be single pull in one location, or multi-pull in several locations at the same time **(Figure 10).** The unitized body is pulled until the holes in the body are aligned with the holes in the dedicated fixtures.

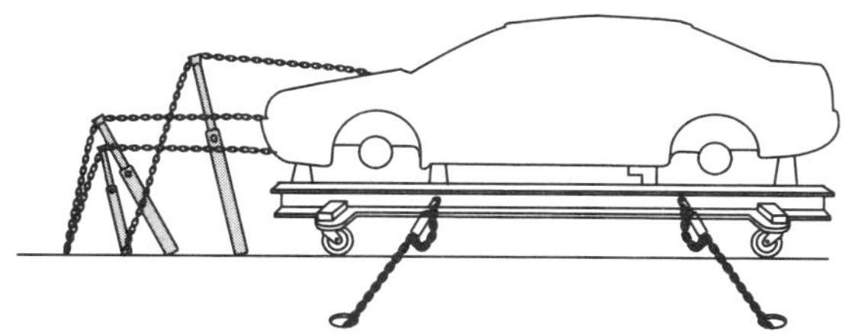

**Figure 10.** A multi-pull unitized body straightening system.

# SUMMARY

- The frame enables a vehicle to support its weight and absorb stress and torque. It also provides a main member for attachment of other components.
- In a unitized body design, all body members are load-carrying components that are welded together.
- Proper tracking and wheelbase are essential to providing directional stability.
- Each front wheel and each rear wheel must be an equal distance from the vehicle centerline to provide correct tracking.
- To provide proper wheelbase, the centers of the front and rear suspensions must be at a 90-degree angle to the vehicle's centerline.
- Front suspension setback, rear axle offset, and rear axle sideset cause the steering to pull.
- Frames should be inspected for wrinkles on the upper and lower flanges and for cracks or bends.
- A plumb bob or tram gauge can be used to check the squareness of a frame.
- A dedicated bench system is necessary to check many unitized body measurements after medium or heavy collision damage.

# REVIEW QUESTIONS

1. Explain the necessary wheel position to provide proper tracking.
2. Explain the effects of front wheel setback.
3. While discussing unitized body design, Technician A says the strength in a unitized body is obtained by shape and design. Technician B says the pillars and quarter panels are the only load-carrying components in a unitized body. Who is correct?
   A. Technician A only
   B. Technician B only
   C. Both Technician A and Technician B
   D. Neither Technician A nor Technician B
4. *True or False:* Steering pull can be caused by axle sideset.
5. While discussing frame damage, Technician A says a possible indication of frame damage is an uncentered steering wheel when driving straight ahead. Technician B says suspension alignment may be impossible if the frame is bent. Who is correct?
   A. Technician A only
   B. Technician B only
   C. Both Technician A and Technician B
   D. Neither Technician A nor Technician B

# Chapter 27 Front Suspensions

## INTRODUCTION

The front and rear suspension systems have many important purposes, one of which is to provide safe handling. The suspension system must allow for steering control under all road conditions. The suspension system provides a comfortable ride, must maintain proper vehicle tracking and directional stability, provide proper wheel alignment, and minimize tire wear.

## BASIC OPERATION

The impact of the front tires striking road irregularities must be absorbed and dissipated by the front suspension system. These impacts are distributed throughout the suspension system and are prevented from moving into the passenger compartment. The vehicle's ride characteristics are determined by the amount of impact energy that can be absorbed by the suspension and by the rate at which the suspension dissipates these impacts.

The front suspension may be designed to provide a very soft ride in which all tire impacts are absorbed and dissipated quickly. This type of suspension also allows excessive body lean during cornering, which reduces high-speed cornering and handling capabilities. Sports cars and SUVs are normally designed with a suspension system that provides a firm ride that absorbs and dissipates tire impacts more slowly. Although high-speed cornering and handling capabilities are improved, this type of suspension may transfer some road shock to the passenger compartment.

When the vehicle is driven over road irregularities, the front suspension must allow the front wheels to move vertically **(Figure 1)** while maintaining the tire's proper hori-

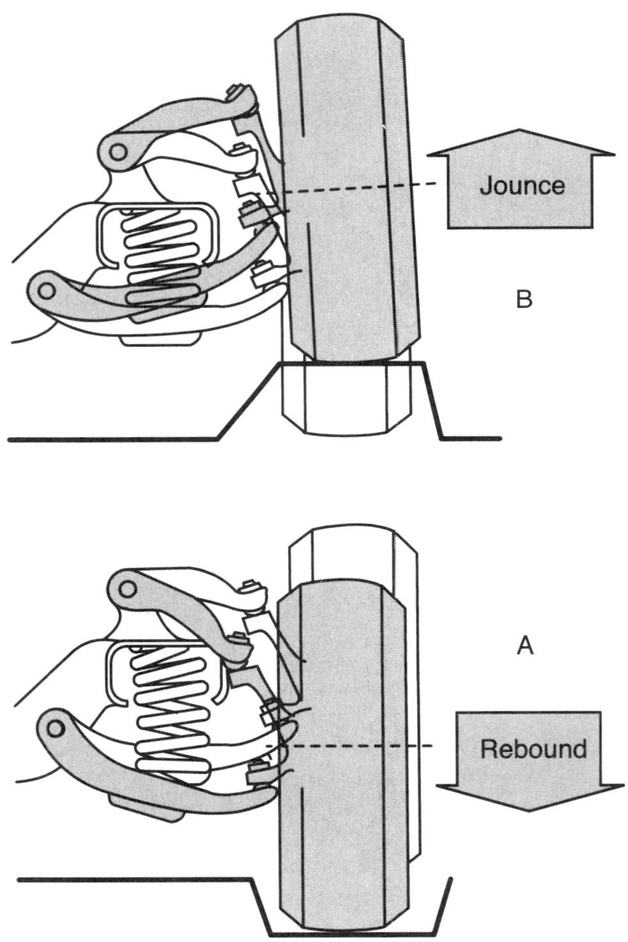


Figure 1. (A) Upward and (B) downward suspension movement.

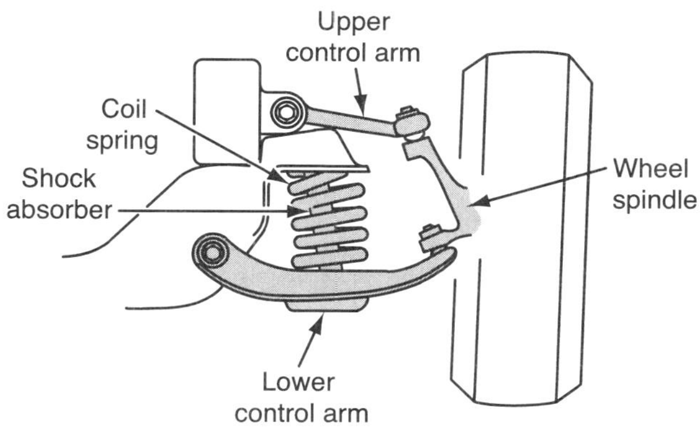


**Figure 2.** The wheel spindle (steering knuckle) is located between the upper and lower control arms.

**Figure 3.** A lower ball joint.

zontal position in relation to the road. To provide for this, the steering knuckle must be mounted between the upper and lower control arms on short/long arm (SLA) suspension systems **(Figure 2)** or between the lower control arm and the strut on a MacPherson strut suspension system. The upper and lower control arms must be able to pivot on their inner ends.

Coil springs absorb much of the shock from the road. When a front wheel strikes a bump and moves upward, the spring compresses and absorbs the energy resulting from the bump. The spring immediately dissipates this energy as the spring moves back to its original state. Shock absorbers are installed in the front suspension to dampen the oscillations of the springs.

A stabilizer bar is mounted on bushings and is bolted to the chassis. The outer ends of the stabilizer bar are attached through links to the lower control arms. When one front wheel moves upward, the stabilizer bar transfers the lower control arm movement to the opposite lower control arm which reduces and stabilizes body roll. Therefore, the stabilizer bar helps to define the suspension characteristics related to body roll.

## BALL JOINTS

A **ball joint (Figure 3)** connects the steering knuckle to the control arm allowing it to pivot on the control arm during steering. Ball joints also permit up-and-down movement of the control arm as the suspension reacts to road conditions. The ball joint stud protrudes from its socket through a rubber seal that keeps lubricating grease in the housing and keeps dirt out. Some ball joints require periodic lubrication, while most do not. These maintenance-free ball joints move in a prelubricated nylon bearing. Ball joints may be manufactured with forged, stamped, cold-formed, or screw machined housings.

### Load-Carrying and Non-Load-Carrying Ball Joints

Ball joints may be grouped into two classifications, load carrying and non-load carrying. A **load-carrying ball joint** supports the car's weight. It is generally the control arm that holds or seats the spring **(Figure 4).** In a torsion bar suspension, the load-carrying ball joint is mounted on the control arm to which the torsion bar is attached.

In a load-carrying ball joint, the vehicle's weight forces the ball stud into contact with the bearing surface in the joint. Load-carrying ball joints may be called compression loaded or tension loaded. The correct term depends on whether the force of the load tends to push the ball into the socket (compression) or pull it out of the socket (tension). If the control arm is mounted above the lower end of the knuckle and rests on the knuckle, the ball joint is com-

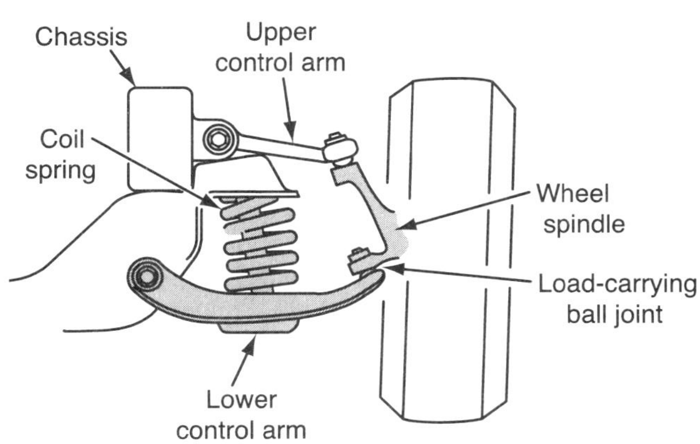


**Figure 4.** A load-carrying ball joint mounted in the control arm on which the spring is seated.

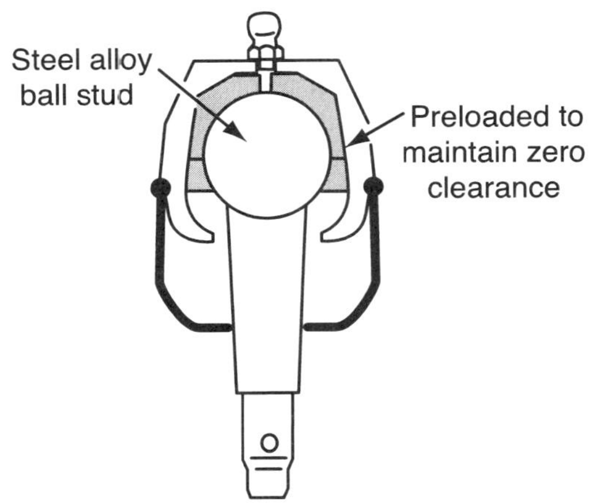


**Figure 5.** A compression-loaded ball joint.

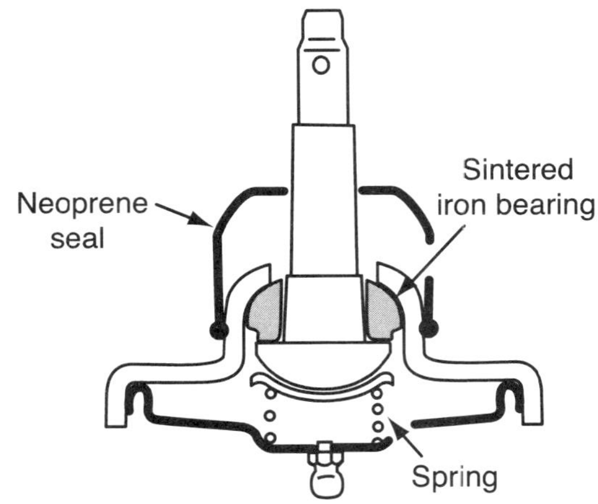


**Figure 7.** A non-load-carrying ball joint.

pression loaded. In this type of ball joint, the vehicle's weight is pushing downward on the control arm. This weight is supported by the tire and wheel, which are attached to the steering knuckle. Since the ball joint is mounted between the control arm and the steering knuckle, the vehicle's weight squeezes the ball joint together **(Figure 5).** This type of ball joint is mounted in the lower control arm, and the ball joint stud faces downward.

When the lower control arm is positioned below the steering knuckle, the vehicle's weight is pulling the ball joint, away from the knuckle **(Figure 6).** This is a tension-loaded joint and it is mounted in the lower control arm with its stud facing upward into the knuckle.

**Non-load-carrying ball joints** are often called follower or friction-loaded ball joints. A follower ball joint mounts on the control arm that does not provide a seat for the spring. The follower does not support vehicle weight and does not receive the same stress as the load carrier. A non-load-carrying ball joint is designed with a preload which provides damping action **(Figure 7).** The preload provides improved steering quality and vehicle stability.

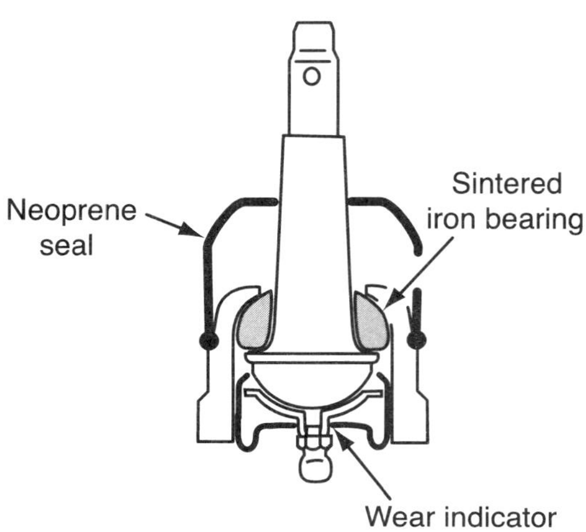


**Figure 6.** A tension-loaded ball joint.

Depending on the location of the spring, either the upper or lower ball joint will be the load-carrying joint. In a MacPherson strut suspension, there is usually only one ball joint on each side, and it is typically a follower. In modified strut suspensions, the ball joint is a load-carrying joint because the spring is positioned between the frame crossmember and the lower control arm.

Low-friction ball joints are standard equipment on many vehicles. Low-friction ball joints provide precise low-friction movement of the ball socket in the ball joint. Compared to conventional ball joints, two-thirds of the internal friction is eliminated in a low-friction ball joint. The smooth ball socket movement in a low-friction ball joint provides improved steering performance, better steering wheel return, and longer ball joint life. Low-friction ball joints have a highly polished ball socket surface surrounded by a high-strength polymer bearing.

A ball joint is nothing more than a ball-in-socket joint. As long as the ball is firmly in the socket and the ball and/or socket is not worn, the joint provides a solid connection. Once the ball or socket is worn, the connection becomes sloppy. Load-carrying ball joints rely on the vehicle's weight to keep the ball in the socket. As weight is removed from the joint, the ball relaxes and will feel loose. Follower ball joints are held in place by friction inside the joint. A spring inside the joint keeps the ball tight. This type joint should never have any play.

Since a load-carrying ball joint supports the vehicle's weight, it wears faster compared to a non-load-carrying ball joint. Many load-carrying ball joints have built-in wear indicators. These ball joints have an indicator on the grease nipple surface that recedes into the housing as the joint wears. If the ball joint is in good condition, the grease nipple shoulder extends a specified distance out of the

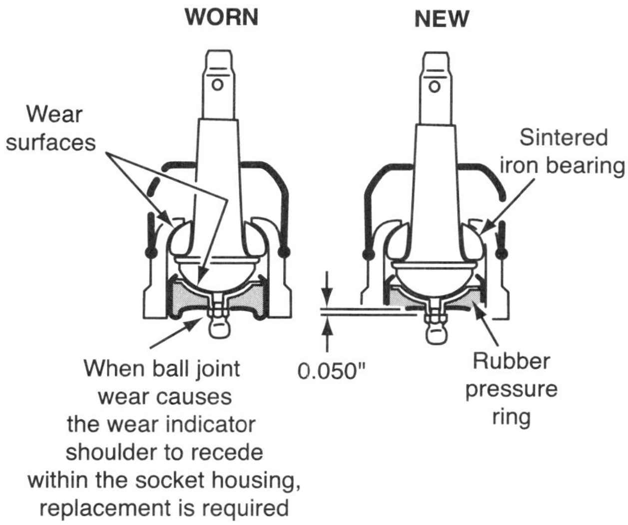


**Figure 8.** A ball joint wear indicator.

housing. If the grease nipple shoulder is even with or inside the ball joint housing, the ball joint is worn, and replacement is necessary **(Figure 8).**

## SHORT/LONG ARM FRONT SUSPENSION SYSTEMS

The unequal length control arm or **short/long arm (SLA)** suspension system has been common on domestic-made vehicles for many years **(Figure 9).** Each wheel is independently connected to the frame by a steering knuckle, ball joint assemblies, and short upper and longer lower control arms. During wheel jounce and rebound travel in this suspension system, the upper control arm moves in a shorter arc than the lower control arm. This action moves the top of the tire in and out slightly, but the tire's road contact remains constant **(Figure 10).** Since wheel jounce or rebound movement of one front wheel does not directly affect the opposite front wheel, this suspension is an independent system.

Early front suspension systems had equal-length upper and lower control arms. On these suspension systems, the bottom of the tire moved in and out with wheel jounce and rebound travel. This action causes tire scuffing and wear problems. SLA systems provide reduced tire tread wear, improved ride quality, and better directional stability compared to suspension systems with equal-length upper and lower control arms.

The inner end of the lower control arm contains large rubber insulating bushings, and the ball joint is attached to the outer end. The lower control arm is bolted to the front crossmember, and the attaching bolts fit through the center of the bushings. The ball joint may be riveted, bolted, pressed, or threaded into the control arm **(Figure 11).** A spring seat is located in the lower control arm. An upper

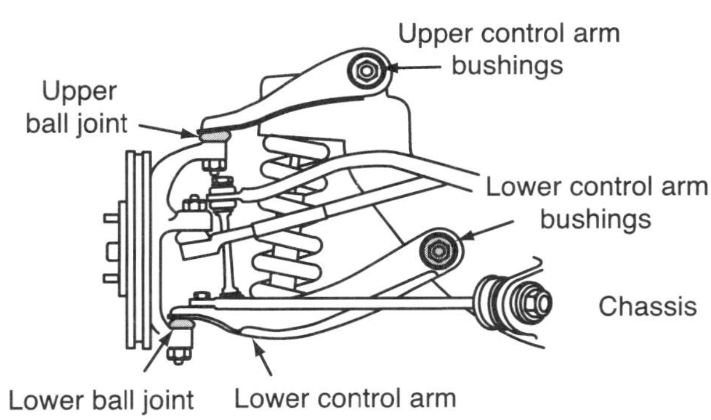


**Figure 9.** A short/long arm (SLA) suspension system.

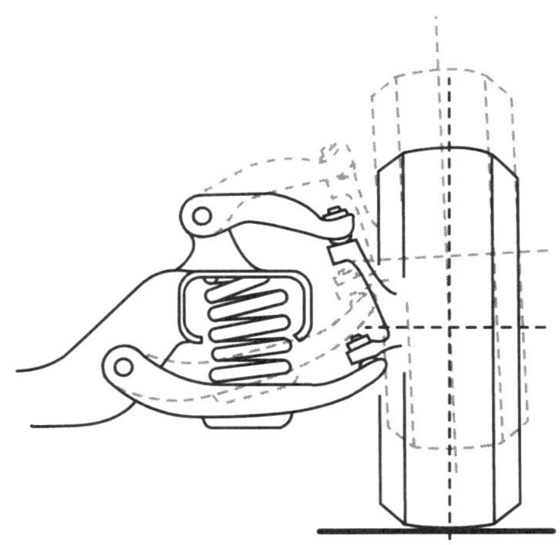

**Figure 10.** The action of a wheel on an SLA suspension system.

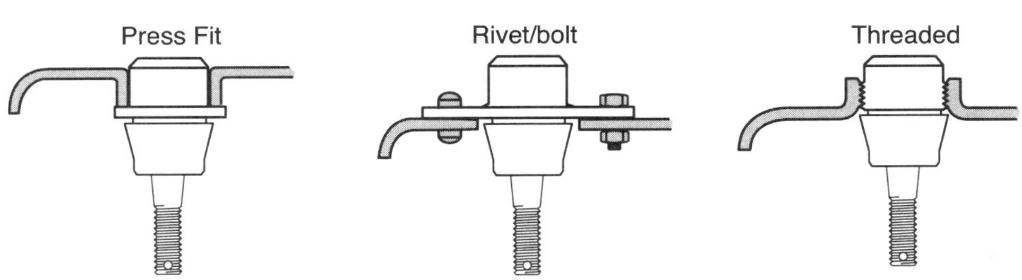


**Figure 11.** The ball joint may be riveted, bolted, pressed, or threaded into the control arm.

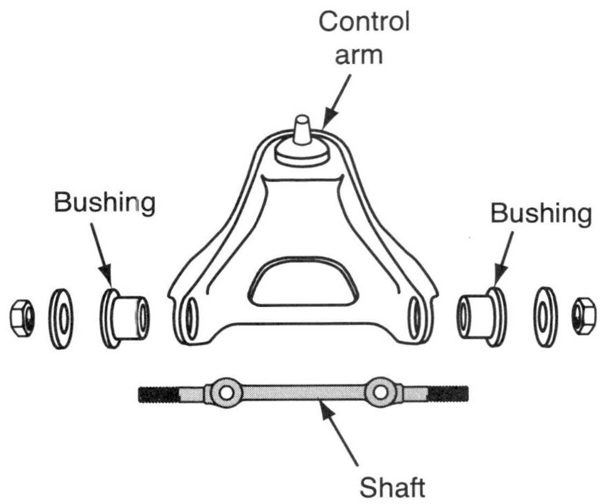


**Figure 12.** An upper control arm shaft is bolted to the frame, and bushings are located between this shaft and the control arm.

control arm shaft is bolted to the frame, and rubber insulators are located between this shaft and the control arm **(Figure 12).** On some short/long arm front suspension systems, the coil spring is positioned between the upper control arm and the chassis. In these suspension systems, the upper ball joint is compression loaded.

One design of SLA uses a narrow lower control arm, shaped like an "I," and a strut rod to hold the control arm in place **(Figure 13).** The strut rod is attached to the control arm close to the steering knuckle and to the frame in front of the wheel assembly. Rubber bushings at the frame mounting allow the strut rod to move a little when the tire hits a bump. The bushing dampens the shock and prevents it from transmitting through the vehicle's frame.

## Steering Knuckle

The upper and lower ball joint studs extend through openings in the steering knuckle. Nuts retain the ball joints

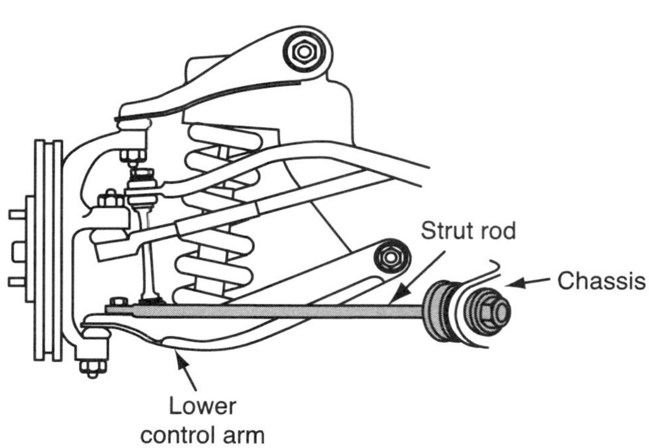


**Figure 13.** A strut, or radius, rod and bushings.

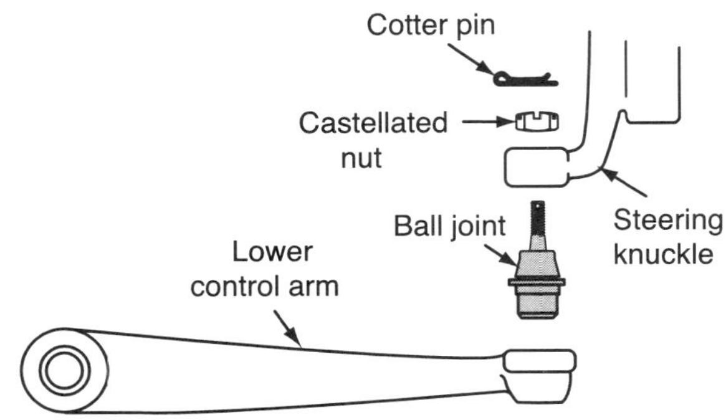


**Figure 14.** Ball joint studs extend through openings in the steering knuckle. Nuts on the ball joint studs may be secured with cotter pins.

in the knuckle, and the nuts are secured with cotter pins **(Figure 14).** The wheel hub and bearings are positioned on the steering knuckle extension, and the wheel assembly is bolted to the wheel hub. When the steering wheel is turned, the steering gear and linkage turn the steering knuckle. During turning, the steering knuckle pivots on the upper and lower ball joints. The upper and lower control arms must be positioned properly to provide correct tracking and wheelbase between the front and rear wheels. To keep the control arms in their proper positions, the control arm bushings must be in good condition.

## Coil Spring and Shock Absorber

The coil spring is positioned between the lower control arm, and the spring seat in the frame. A spring seat is located in the lower control arm, and an insulator is positioned between the top of the coil spring and the spring seat in the frame. The shock absorber is mounted in the center of the coil spring, and the lower shock absorber bushing is bolted to the lower control arm. The top of the shock absorber extends through an opening in the frame above the upper spring seat. Washers, grommets, and a nut retain the top of the shock absorber to the frame.

On some later model short/long arm systems, the lower end of the coil spring is mounted on a seat attached to the shock absorber. A rubber bushing is installed in the lower end of the shock absorber and is bolted to the lower control arm. The upper end of the coil spring is seated on an upper shock absorber mount. The rod in the center of the shock absorber extends through this mount, and a nut retains the mount on the shock absorber. Bolts in the top of the upper shock absorber mount extend through the upper control arm.

The upper control arm is mounted high in the suspension system, and the upper end of the knuckle has a "goose neck" shape. The lower control arm is made from stamped steel to reduce weight. The rear lower control arm bushing

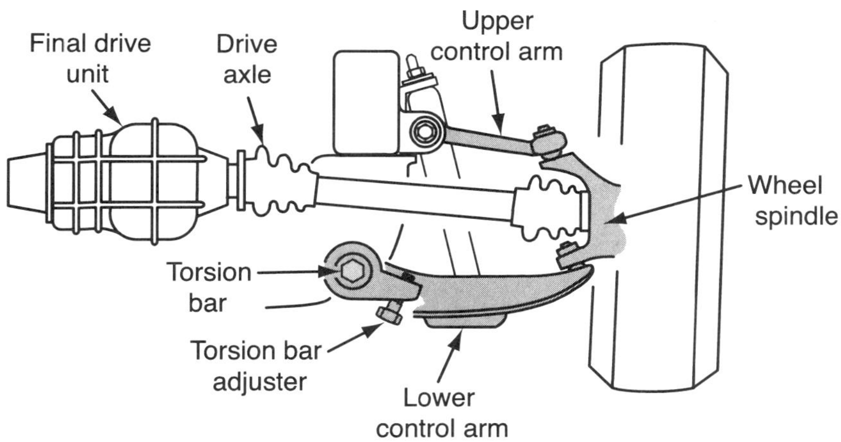


**Figure 15.** A front suspension system with longitudinal torsion bars.

is mounted vertically and carries only fore-and-aft loads. This mounting allows the use of a softer rear bushing in the lower control arm. The horizontal front lower control arm bushing and the lower shock absorber mounting are aligned with the wheel center. This provides a direct path for lateral cornering loads. This design allows the use of a hard front lower control arm bushing.

## Torsion Bar Suspension

Some light-duty trucks and SUVs have torsion bar front suspension systems. These suspension systems have longitudinally mounted torsion bars. The front and rear ends of the torsion bars have a hex shape. The front of each torsion bar is anchored in the lower control arm, and the rear is anchored to a chassis crossmember **(Figure 15).**

A conventional stabilizer bar is connected between the lower control arms and the crossmember. Ball joints are located in the upper and lower control arms, and both ball joints are bolted into the steering knuckle. The shock absorbers are connected between the lower control arms and the crossmember support, and the inner ends of the lower control arms are bolted to the crossmember through an insulating bushing.

# MacPHERSON STRUT FRONT SUSPENSION SYSTEMS

In **MacPherson strut suspension** systems **(Figure 16),** the lower end of the strut is bolted to the top of the steering knuckle, and the lower end of the knuckle is attached to the ball joint in the lower control arm. An upper strut mount connects the top of the strut to the chassis. Since the upper control arm is not required, these systems are compact and therefore very suitable for smaller cars.

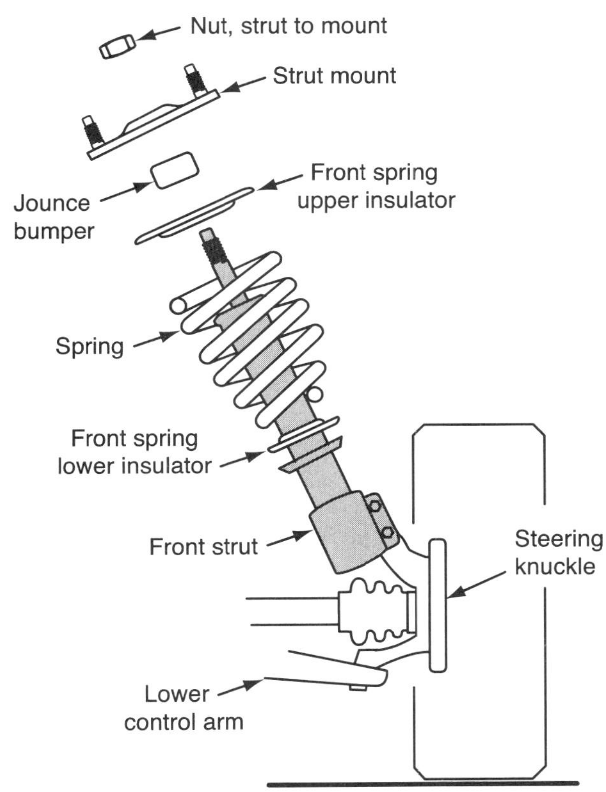


**Figure 16.** A MacPherson strut front suspension system.

On some MacPherson strut front suspension systems, a steel support is positioned longitudinally on each side of the front suspension. These supports are bolted to the body. The inner ends of the lower control arms contain large insulating bushings. The control arm retaining bolts extend through the center of these bushings and openings in the support.

Road irregularities cause the tire and wheel to move up and down, and the lower control arm bushings pivot on the mounting bolts during this movement. When driven over road irregularities, vibration and noise are applied to the tire and wheel. The control arm bushings help prevent the transfer of this noise and vibration to the body. Proper location of the support and lower control arm is important to provide correct vehicle tracking.

### Lower Ball Joint

The lower ball joint is attached to the outer end of the lower control arm. The ball joints are attached to the control arms by bolting, riveting, pressing, and threading. A threaded stud extends from the top of the lower ball joint. This stud fits snugly into a hole in the bottom of the steering knuckle. When the ball joint stud is installed in the steering knuckle opening, a nut and cotter pin retain the ball joint.

### Steering Knuckle and Bearing Assembly

The front wheel bearing assembly is bolted to the outer end of the steering knuckle, and the brake rotor and wheel rim are retained on the studs in the wheel bearing assembly. This front wheel bearing assembly is a complete, nonserviceable, sealed unit. The front axle shaft is splined into the center of the wheel bearing hub. Thus, drive axle torque is applied to the front wheel. A tie-rod end connects the steering linkage from the steering gear to the steering knuckle. The top end of the steering knuckle is bolted to the lower end of the strut.

### Strut and Coil Spring Assembly

The strut is the shock absorber, and the lower spring seat is attached near the center area of the strut. An insulator is located between the lower spring seat and the bottom of the coil spring. An upper strut mount is retained on top of the strut with a nut threaded onto the upper end of the strut rod. The upper strut mount contains a bearing and upper spring seat, and an insulator is positioned between the top of the coil spring and the seat. The upper and lower insulators help prevent the transfer of noise and vibration from the spring to the strut and body.

A bumper is located on the upper end of the strut rod. This bumper reduces harshness while driving on severe road irregularities. During upward wheel movement, the bumper strikes the upper spring seat before the coils in the spring hit each other. Therefore, this bumper reduces harshness when the suspension moves fully upward. The spring tension is applied against the upper and lower spring seats and insulators. However, the nut on top of the upper mount holds the spring in the compressed position. When the steering wheel is turned, the steering linkage turns the steering knuckles. During this turning action, the strut and spring assembly pivots on the lower ball joint and the upper strut mount bearing.

### Modified MacPherson Strut Suspension

A modified MacPherson strut front suspension is used on some vehicles. This type of suspension has MacPherson struts with coil springs positioned between the lower control arms and the frame. The struts in these systems may be gas-filled or oil-filled.

## PERFORMANCE-ORIENTED SUSPENSIONS

Performance suspension systems are usually installed on performance or sports cars. The drivers of this type of car expect good handling and steering quality when driving and cornering at higher speeds. Many of the performance-type suspension systems found in today's cars are based on a strut suspension. The two most common performance suspension designs are the multi-link and double wishbone systems.

### Multi-link Front Suspension System

In a **multi-link front suspension,** a short upper link is attached to the chassis with a bracket, and the outer end of this upper link is connected to a third link. Large rubber insulating bushings are mounted in each end of the upper link. The lower end of the third link is connected through a heavy pivot bearing to the steering knuckle **(Figure 17).** A

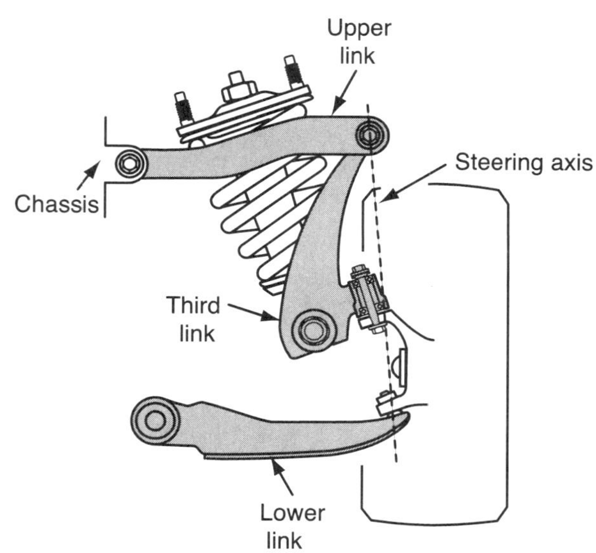


**Figure 17.** A multi-link front suspension system.

rubber insulating bushing connects the inner end of the lower link to the front crossmember, and a ball joint is connected from the outer end of the lower link to the steering knuckle. The extra links in a multi-link suspension system maintain precise wheel position during cornering to provide directional stability and steering control while minimizing tire wear.

The shock absorbers are connected from the lower end of the third link to the body. A spring seat is attached to the lower end of the shock absorber, and the upper spring seat is located on the upper shock absorber mounting insulator. Since the steering knuckle pivots on the lower ball joint and upper pivot bearing, the coil spring and shock absorber do not rotate with the knuckle as they do in a strut suspension. Tension or strut rods are connected from the lower links to tension rod brackets attached to the chassis. A stabilizer bar is mounted on rubber insulating bushings in the tension rod brackets, and the outer ends of this bar are attached to the third link.

### Four-Link Front Suspension

A four-link front suspension fixes the wheel with four rod-type control arms and a tie rod. The strut supports the vehicle's weight against the body via the load-bearing link. By separating wheel attachment and suspension elements, this suspension optimizes ride quality and movement.

### Multi-link Suspension with Compression and Lateral Lower Arms

Some multi-link front suspensions have compression and lateral lower arms **(Figure 18).** The lateral arm prevents front wheel movement, and the compression arm prevents fore-and-aft front wheel movement. A rubber insulating bushing in the inner end of the lateral arm is bolted to the chassis and a second bushing is in the damper fork. The upper end of the damper fork is bolted to the front strut. A ball joint in the outer end of the lateral arm is bolted into the steering knuckle, as is the ball joint for the compression arm.

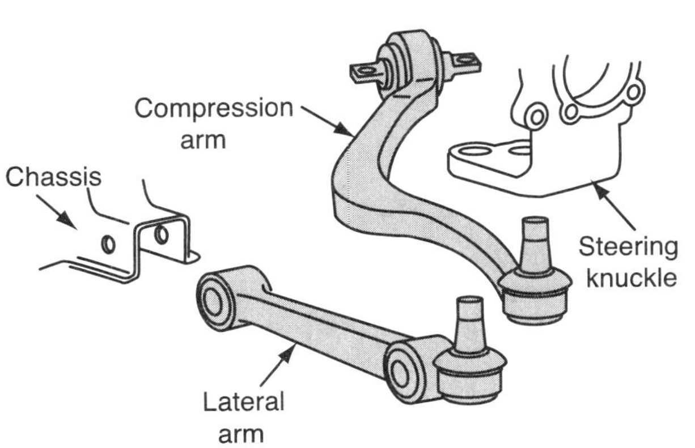


**Figure 18.** A multi-link suspension system with lateral and compression arms.

The upper control arm is mounted above the front tire. This, along with the lateral and compression lower arms, provides excellent suspension stability and steering control, especially during high speed cornering or when driving on irregular road surfaces. Two shafts at the inner end of the upper control arm are bolted into the strut tower. There are no provisions for camber or caster adjustments on this multi-link front suspension.

## Double Wishbone Front Suspension System

**Double wishbone suspension systems** have increased suspension rigidity and can maintain precise wheel position during all driving conditions. In double wishbone suspension systems, the upper and lower control arms are lightweight, high-strength aluminum alloys. These lighter control arms decrease the vehicle's unsprung weight, which improves traction and ride quality. Since the upper and lower control arms have a wishbone shape, the design is called a double wishbone suspension. Suspension rigidity is also increased by positioning the ball joints and steering knuckle inside the wheel's profile **(Figure 19).** On each side of the car, the front suspension is attached to the chassis by a cast aluminum subframe.

The front of the upper and lower control arms are attached to a compliance pivot assembly. When one of the front wheels is subjected to rearward force by hard braking or a road irregularity, the coil spring is compressed and the ride height is lowered. This rearward force on the front wheel twists the compliance pivot, allowing both control arms to pivot slightly. This movement of the upper and lower control arms allows the front wheel to move rear-

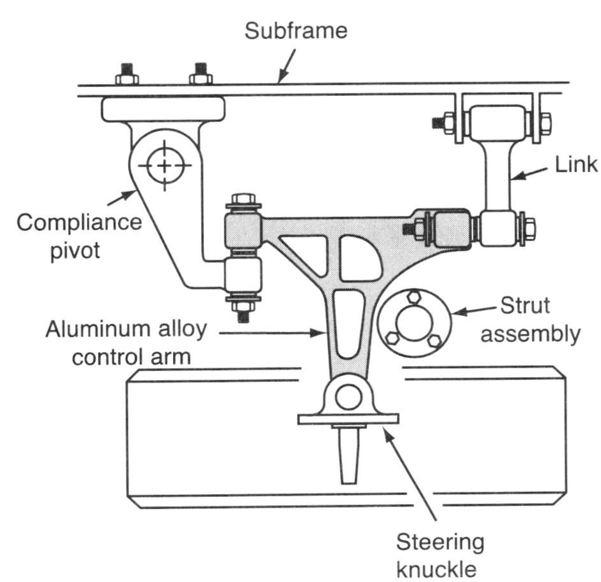


**Figure 19.** A double wishbone front suspension.

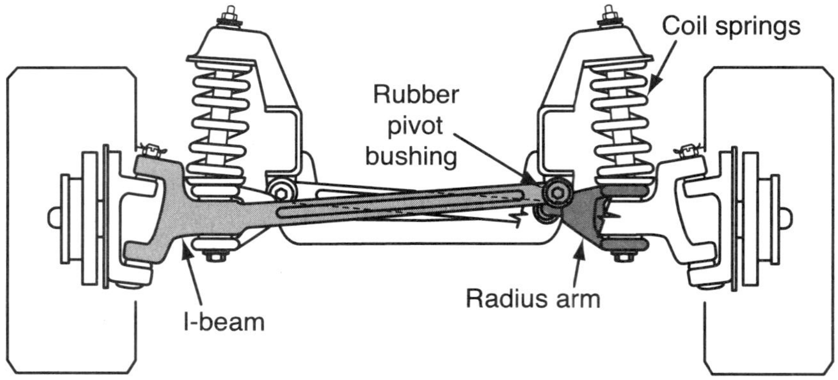


**Figure 20.** A twin I-beam front suspension system.

ward a small amount and absorbs energy to significantly improve ride quality. During this movement, track width and wheel geometry changes are minimal and do not affect steering control. While cornering, the compliance pivot does not move, and lateral suspension stiffness is maintained to supply excellent steering control.

## OTHER FRONT SUSPENSION DESIGNS

When the automobile first became popular, most cars had I-beam front axles and leaf springs. Some I-beam suspension systems had two longitudinally mounted leaf springs, whereas other applications of this suspension system had one transversely mounted leaf spring. These I-beam suspension systems are obsolete on cars today, but they may still be found on some trucks.

In a typical I-beam front suspension system, **kingpins** are used to connect the axle to the spindles. Spindle bushings allow the spindle to pivot on the kingpins. On an I-beam front axle, if one wheel strikes a bump, movement is transferred through the axle to the opposite front wheel, causing it to move from its normal position. This decreases directional control and ride quality and increases tire wear. These are non-independent suspension systems. Some I-beam suspension systems use ball joints to connect the I-beams to the spindles rather than kingpins.

I-beam front suspension systems are very strong and have excellent weight-carrying capabilities. In an I-beam front suspension system, the springs and axle must be properly positioned to maintain correct wheel tracking and equal wheelbase on each side of the vehicle.

### Twin I-Beam Suspension Systems

Some Ford trucks are equipped with twin I-beam front suspension systems where each front wheel is connected to a separate I-beam. The outer ends of the I-beam are connected to the spindles, and the inner ends of the beams are connected through a rubber pivot bushing to the chassis **(Figure 20).** Coil springs are positioned between the I-beams and the chassis to support the vehicle's weight. Radius arms are connected from each I-beam to the chassis to prevent longitudinal wheel movement. Upper and lower ball joints are connected to the steering knuckle. Since each front wheel can move independently in a twin I-beam suspension system, the problems associated with straight I-beam systems are greatly reduced.

### Truck Four-Wheel-Drive Front Suspension Systems

Ford 4WD trucks use twin I-beam front suspension systems. F-150 models have a coil spring twin I-beam front suspension. Universal joints in the outer ends of the drive axles allow simultaneous wheel rotation and wheel turning. The right-side drive axle also has an inner universal joint near the differential.

Ford F-250 and F-350 4WD trucks have a leaf spring twin I-beam front suspension system. The leaf spring eye is bolted into a frame bracket. The rear spring eye is connected to the frame through a conventional spring shackle. Since the leaf springs maintain the axle position, radius arms are not required.

Some 4WD light-duty trucks have a straight front drive axle housing and coil springs. Upper and lower radius rods control fore-and-aft drive axle movement, and a track bar controls lateral axle movement. Ball joints connect the steering knuckles to the drive axle housing.

Other 4WD light-duty trucks use torsion bars with upper and lower control arms. The front of the torsion bars have hex-shaped ends mounted to a matching hex in the lower control arm. The hex-shaped rear ends of the bars are mounted in adjusting arms retained in the torsion bar crossmember. An adjusting bolt contacts the outer end of each adjusting arm. Rotation of the adjusting bolt changes the tension on the torsion bar to adjust the curb riding height.

## BUSHINGS

Suspension and steering parts are typically mounted in bushings. Bushings isolate the metal parts from each other and from the frame. Typically, two materials are used to make bushings, each with their own characteristics. Neoprene bushings provide for a softer ride by offering better isolation of the parts fitted together. Polyurethane bushings wear less than neoprene bushings and offer more vehicle control but also provide for a harsher ride.

# SUMMARY

- A suspension system must provide safe handling, allow for steering control under all road conditions, provide a comfortable ride, maintain proper vehicle tracking and directional stability, provide proper wheel alignment, and minimize tire wear.
- Ball joints permit a control arm to move up and down as the suspension reacts to the road.
- Ball joints may be classified as load carrying or non-load carrying.
- A load-carrying ball joint may be compression loaded or tension loaded.
- The load-carrying ball joint wears faster than the non-load-carrying ball joint.
- Many ball joints have wear indicators that provide visual ball joint wear inspection.
- A stabilizer bar reduces body roll or sway.
- A strut rod prevents fore-and-aft lower control arm movement.
- In a short/long arm suspension system, the coil springs may be mounted between the lower control arm and the frame or between the upper control arm and the chassis.
- In many MacPherson strut front suspension systems, the lower end of the strut is bolted to the steering knuckle, and the upper end of the strut is connected through an upper mount to the fender reinforcement.
- Torsion bars may be transversely or longitudinally mounted in a front suspension system.
- In a twin I-beam suspension system, the spindles may be connected to the I-beams with kingpins or ball joints.

# REVIEW QUESTIONS

1. Explain the position of the lower control arm when the lower ball joint is compression loaded.
2. Describe the position of the wear indicator when a ball joint is worn.
3. If the lower control arm is mounted above the steering knuckle and rests on the knuckle, the lower ball joint is ____________________ ____________________.
4. All these statements about ball joints are true *except:*
   A. A load-carrying ball joint wears faster than a non-load-carrying ball joint.
   B. A non-load-carrying ball joint is designed with a preload.
   C. Wear indicators may be positioned in the side of the ball joint housing.
   D. Low-friction ball joints have a highly polished ball socket surrounded by a high-strength polymer bearing.
5. Which of the following statements about MacPherson strut suspension systems is true?
   A. During a turn, the strut, spring, and knuckle rotate on the upper strut mount.
   B. The lower end of the coil spring is seated on the lower control arm.
   C. The strut is welded into the knuckle, and these components are replaced as an assembly.
   D. the inner ends of the lower control arms are bolted to the vehicle frame.
6. Which of the following statements about short/long arm front suspension systems is true?
   A. The coil springs are positioned between upper and lower control arms.
   B. Compared to a suspension system with equal length control arms, the short/long arm design reduces track width change and tire wear.
   C. During wheel jounce and rebound travel in this suspension system, the upper control arm moves in a shorter arc than the lower control arm.
   D. The coil spring is retained in the lower control arm with a clamp and bolt.

# Chapter 28 Rear Suspensions

## INTRODUCTION

The rear suspension system is important to a vehicle's overall ride and handling. It also helps control suspension and differential noise, vibration, and shock. The rear suspension is also vital to steering and must provide adequate tire life and maintain tire traction on the road's surface. Rear suspension systems include live-axle, semi-independent, and independent. Live-axle suspension systems are found on RWD trucks and vans, a few RWD cars, and some 4WD vehicles. Most FWD vehicles have semi-independent or independent rear suspensions. Independent rear suspensions are also found on some RWD cars and 4WD vehicles.

The vehicle's sprung weight is supported by the springs through the rear axle housing and wheels. When the vehicle accelerates, the rear wheels turn counterclockwise when viewed from the left vehicle side. One of Newton's Laws of Motion states that for every action there is an equal and opposite reaction. Therefore, when the wheels turn counterclockwise (when viewed from the left), the rear axle housing tries to rotate clockwise. This rear axle torque action is absorbed by the rear springs or suspension parts and the chassis moves downward **(Figure 1).** Engine torque supplied through the driveshaft to the differential tends to twist the differential housing and the springs. This twisting action may be referred to as **axle windup.**

When braking and decelerating, the rear axle housing tries to turn counterclockwise. As this torque is applied to the springs, it tends to lift the chassis and is called braking and deceleration torque.

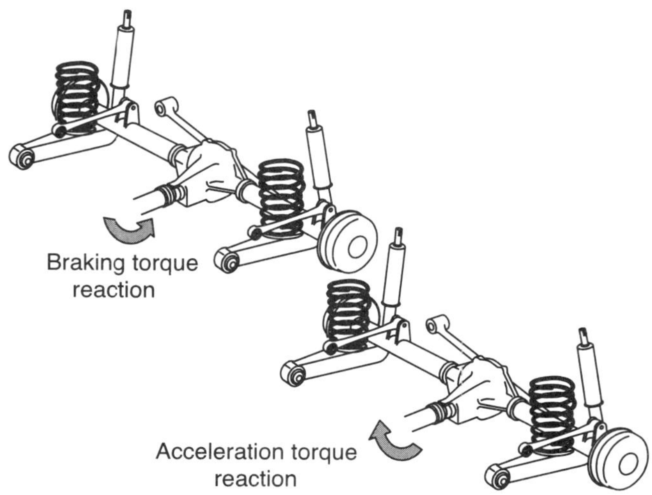


**Figure 1.** Rear axle torque action during acceleration and braking.

During hard acceleration, the entire power train twists in the opposite direction of driveshaft rotation. The engine and transmission mounts absorb this torque. However, the twisting action of the driveshaft tends to lift the rear wheel on the passenger's side of the vehicle. Extremely hard acceleration may cause this wheel to lift off the road. Once this wheel slips on the road's surface, engine torque is reduced, and the leaf spring forces the wheel downward. When the tire contacts the road, engine torque increases and the cycle repeats. The repeated lifting of the rear axle is called **axle tramp**, and occurs on live-axle rear suspension systems.

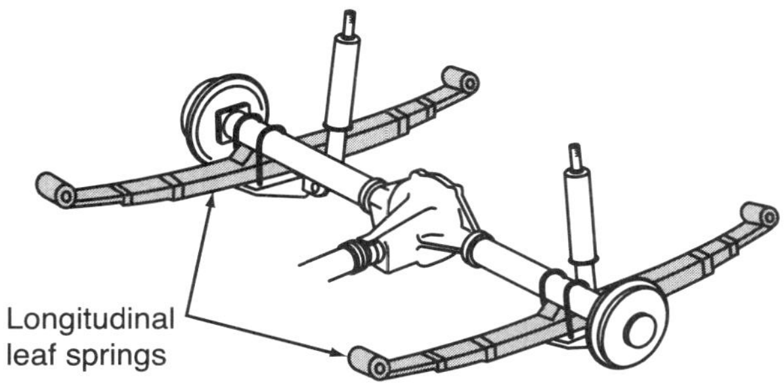


**Figure 2.** A leaf-spring rear suspension system.

## LEAF-SPRING REAR SUSPENSION

A leaf spring is mounted longitudinally on each side of the rear suspension **(Figure 2).** These relatively flat springs provide excellent lateral stability and reduce side sway, which contributes to a well-controlled ride with very good handling characteristics. However, leaf-spring systems have much unsprung weight and require a considerable amount of space.

The semi-elliptical springs have steel leaves and zinc or plastic interleaves to reduce corrosion, friction, and noise. A large rubber bushing is installed in the front eye of the main spring leaf, and a bolt retains this bushing to the front spring hanger. The rear spring shackle is bolted to a rubber bushing in the rear main leaf eye, and the upper shackle bolt extends through a similar rubber bushing in the rear spring hanger. Some rear spring shackles contain threaded steel bushings or a slipper mount in which the end of the spring slides through the shackle. Shackle insulating bushings help prevent the transfer of noise and road shock from the suspension to the chassis. When a rear wheel strikes a road irregularity, the spring compresses and changes length. The rear shackle provides movement for the variations in spring length.

Since the rear axle housing is a one-piece unit, jounce and rebound of one rear wheel affects the other rear wheel. This increases tire wear and decreases ride quality and traction.

The axle housing is mounted above or below the springs, and a spring plate with an insulating clamp and U-bolts retains the springs to the rear axle housing **(Figure 3).** The shock absorbers are mounted between the spring plates and the frame.

Many leaf springs have a shorter distance from the center bolt to the front of the spring compared to the distance from the center bolt to the rear of the spring. This type of leaf spring is referred to as asymmetrical. The shorter front part of the spring resists axle windup. A symmetrical leaf spring has the center bolt in the center of the spring.

Axle tramp is more noticeable on live-axle suspension systems when the springs must absorb all of the driving

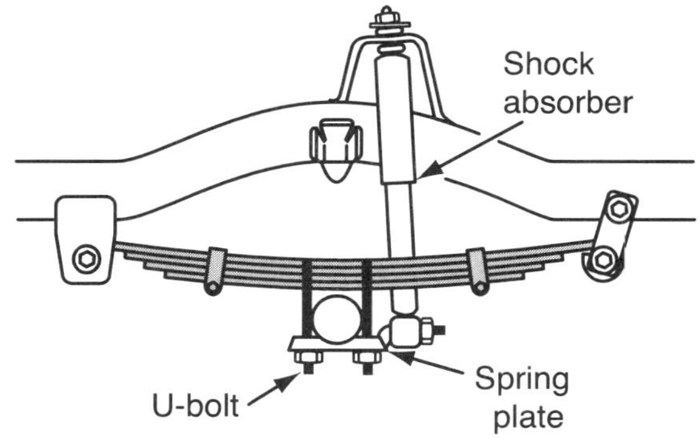


**Figure 3.** A spring plate with an insulating clamp and U-bolts retain the springs to the rear axle housing.

torque. Suspension and axle parts, such as spring mounts, shock absorbers, and wheel bearings may be damaged by axle tramp. Mounting one rear shock absorber in front of the rear axle and the other rear shock behind it helps reduce axle tramp.

## COIL-SPRING REAR SUSPENSION

**You Should Know** *Compressed coil springs contain a large amount of energy. Never disconnect any suspension component that suddenly releases coil-spring tension. This may result in personal injury and vehicle damage.*

Some RWD vehicles have a coil-spring rear suspension. Upper and lower suspension arms with insulating bushings are connected between the rear axle housing and the frame **(Figure 4).** The upper arms control lateral

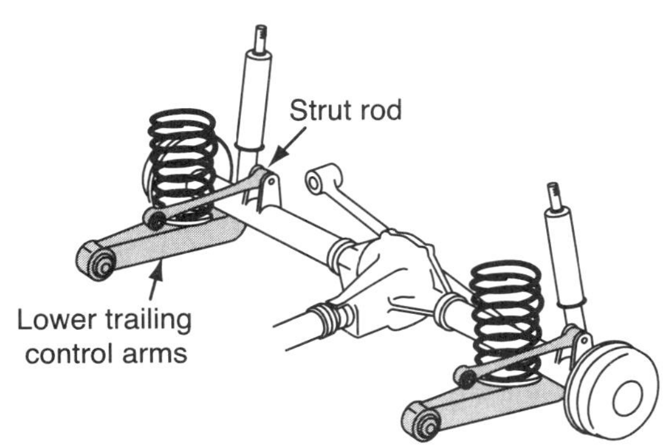


**Figure 4.** A live-axle coil-spring rear suspension.

movement, while the lower trailing arms absorb differential torque. In some rear suspension systems, the upper arms are replaced with strut rods. The coil springs are usually mounted between the lower suspension arms and the frame, whereas the shock absorbers are mounted between the back of the suspension arms and the frame.

Some rear suspension systems have a **track bar** connected from one side of the axle housing to the chassis to prevent lateral chassis movement. A tracking bar may be referred to as a Panhard rod or Watts rod.

In some cars with higher torque engines, a long torque arm is bolted to the axle housing **(Figure 5).** This arm helps prevent differential rotation during hard acceleration and braking. The front of the arm is mounted in a rubber insulator and bracket bolted to the back of the transmission housing. This long torque arm helps prevent differential rotation when high torque is delivered from the engine to the differential. The center bearing assembly on the driveshaft is bolted to the long torque arm. This system has a track bar connected from the left side of the axle housing to the chassis to help prevent lateral axle movement. A track bar brace is connected from chassis end of the track bar to the other side of the chassis to provide extra rigidity.

Some SUVs have a coil spring rear suspension with shock absorbers mounted separately from the springs **(Figure 6).** Control arms are mounted in bushings between the axle housing and the chassis. This design has an upper and lower control arm on each side of the suspension. These control arms (rods) prevent rear axle windup during hard acceleration.

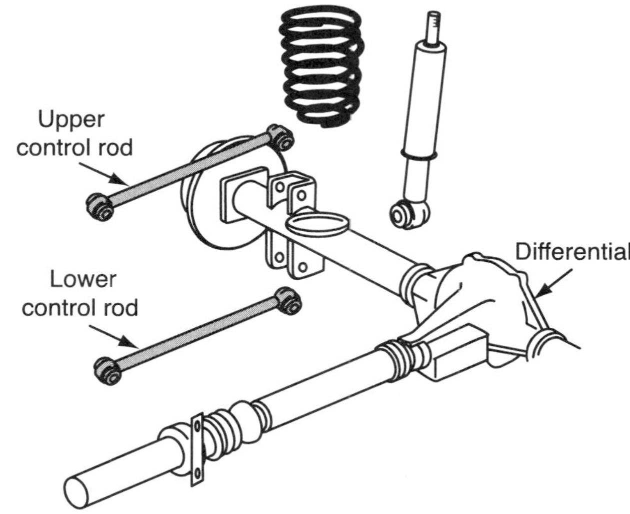


**Figure 6.** A coil-spring rear suspension with two control rods on each side.

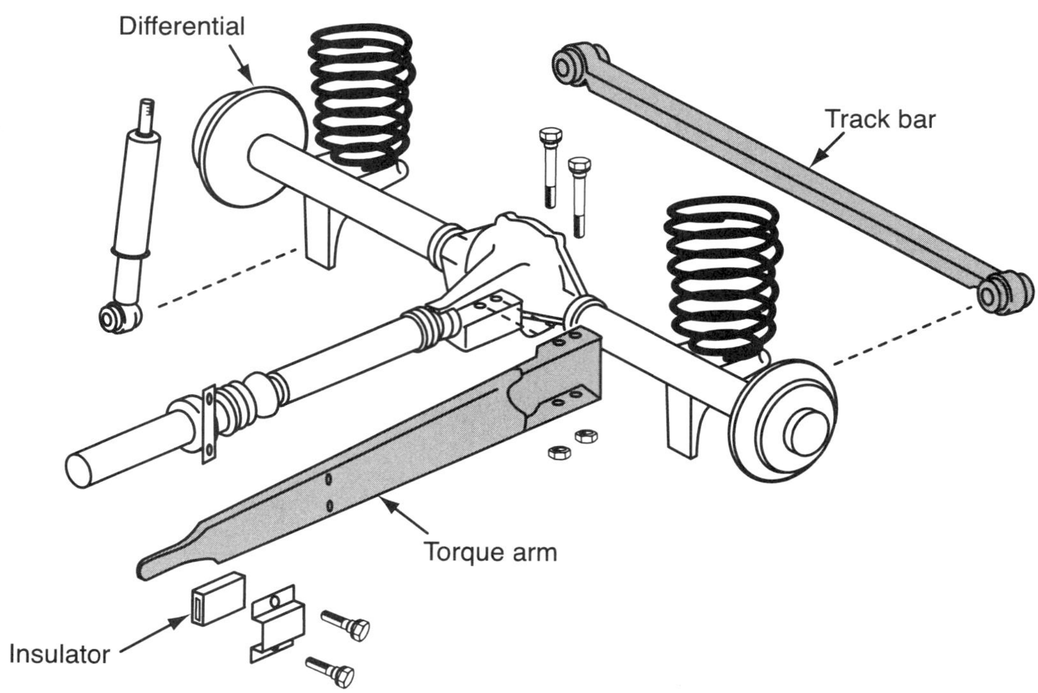


**Figure 5.** A rear suspension system with a long torque arm and track bar.

## SEMI-INDEPENDENT REAR SUSPENSION SYSTEMS

Many front-wheel-drive vehicles have a semi-independent rear suspension that has a solid axle beam connected between the rear trailing arms **(Figure 7).** Some of these rear axle beams are fabricated from a transverse inverted U-section channel.

In some rear suspension systems, the inverted U-section channel contains an integral tubular stabilizer bar. When one rear wheel strikes a road irregularity and the wheel moves upward, the inverted channel twists, which allows some independent rear wheel movement. The trailing arms are connected to the chassis through insulating bushings. In some semi-independent rear suspension systems, the coil springs are mounted on the rear struts and the lower spring seat is located on the strut with the upper spring seat positioned on the upper strut mount.

In other semi-independent suspension systems, coil springs are mounted separately from the shock absorbers.

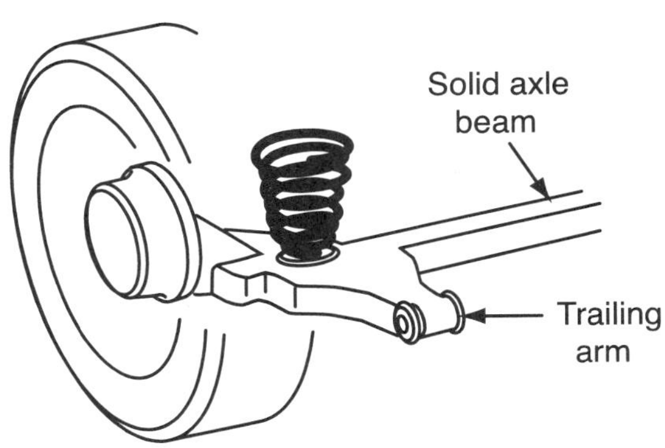


**Figure 7.** A semi-independent rear suspension system with a solid axle beam.

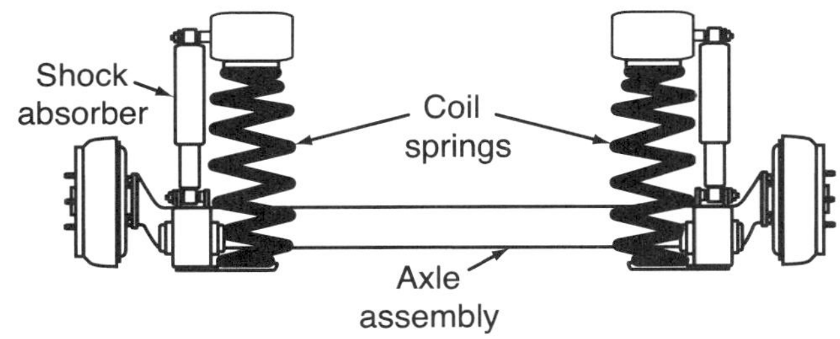


**Figure 8.** A semi-independent rear suspension system with coil springs and shock absorbers mounted separately.

Coil-spring seats are located on the trailing arms, and the shock absorbers are connected from the trailing arms to the chassis. A crossmember connected between the trailing arms provides a twisting action and some independent wheel movement **(Figure 8).**

Some semi-independent suspension systems have a track bar connected from a rear axle bracket to the chassis. Some also have an extra brace connected from the chassis to the rear upper crossmember **(Figure 9).** The track bar and brace prevent lateral rear axle movement.

## INDEPENDENT REAR SUSPENSIONS

Some RWD vehicles have independent rear suspensions. Most of these have a centrally mounted final drive with axles extending from it. In an independent rear suspension system, each rear wheel can move independently of the opposite rear wheel. Independent rear suspension systems may be found on FWD and RWD vehicles.

A multi-link rear suspension uses several control arms to guide the wheel. There are different types of

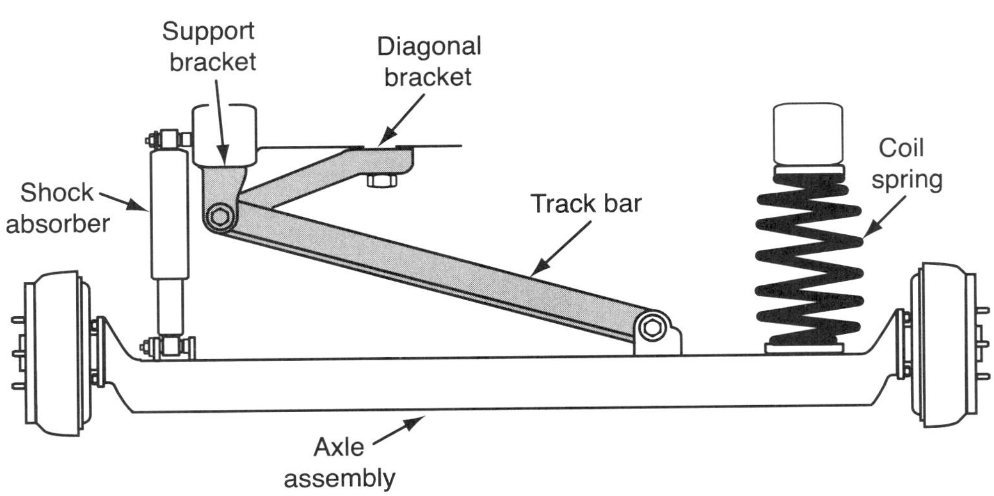


**Figure 9.** A semi-independent rear suspension with a track bar and brace.

multi-link rear suspensions used to meet the varying demands of vehicle dynamics, ride comfort, and space requirements. These include double wishbone suspensions, trailing link double wishbone suspensions, and trapezoidal-link suspensions.

In a double wishbone suspension, the wheel is guided by two triangulated lateral control arms (the wishbones) and a tie rod. A strut is attached to the lower wishbone to provide vertical support.

Trailing-link double wishbone suspensions have a trailing link that also carries the wheel and upper and lower wishbones. The spring is located on the trailing link ahead of the center of the wheel; the shock absorber is behind it.

Trapezoidal-link suspensions permit excellent performance, handling, and comfort. The rear wheel is fixed by an upper lateral control arm and a trapezoidal lower link with a tie rod. For reduced weight, the trapezoidal link and upper control are hollow aluminum castings.

## Multi-link Independent Rear Suspension Systems

The multi-link independent rear suspension system is used on some imported rear-wheel-drive cars. In this suspension system, the differential housing is bolted to a large rear suspension member. This member is attached to the chassis through four hydraulic mounts **(Figure 10).** These hydraulic mounts contain a silicone oil that helps prevent noise, vibration, and shock transfer from the differential and suspension to the chassis and vehicle interior.

The rear suspension is connected to the outer shell of the hydraulic mounts, and the inner shell is attached to the chassis. Silicone oil fills the area between the inner and outer shells in each mount. Noise, vibration, and shock are transferred from the differential and suspension to the rear suspension, but the oil in the mounts prevents the transfer of these undesirable forces to the chassis. These hydraulic mounts have superior noise and vibration dampening characteristics compared to rubber bushings.

A rear upper link is connected from each side of the rear suspension to the top of the knuckles. Both ends of this upper link contain rubber-insulating bushings. The lower end of the shock absorber extends into the rear upper link. A front upper link is also connected from the rear suspension to the knuckle. The top of the knuckle is supported by the front and rear upper links rather than being supported by the shock absorber. The coil springs are mounted on the shock absorbers. An upper spring seat is attached to the top of the shock absorber strut and is bolted to the inner fender reinforcement. The lower end of the shock absorber is connected to the back of the lower control arm, which is connected to the lower end of the knuckle.

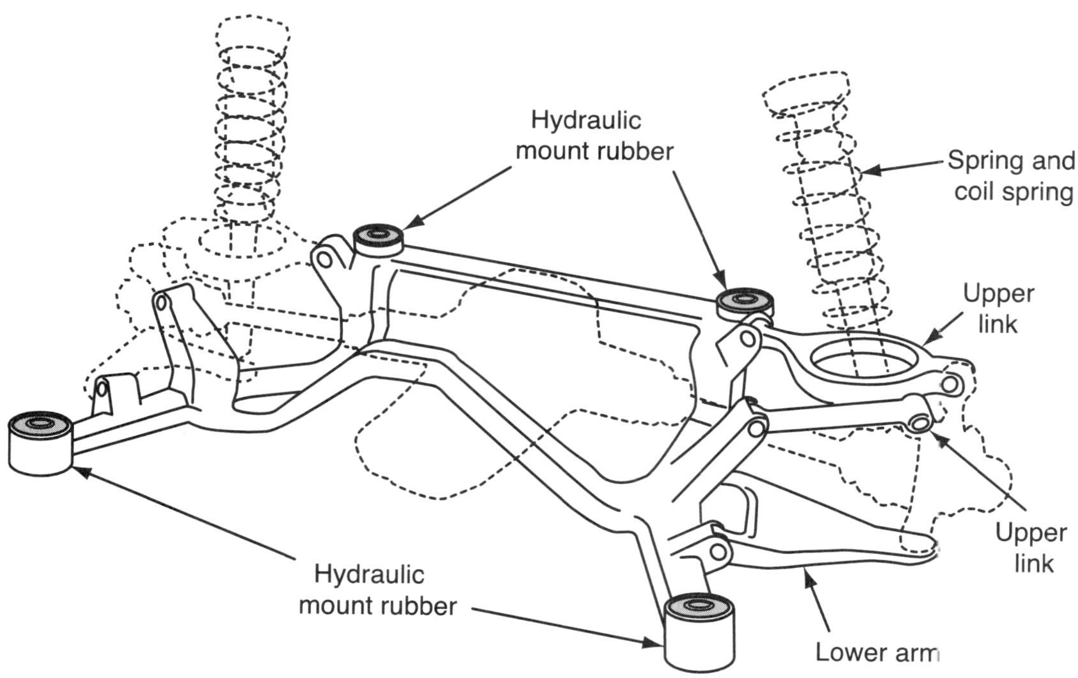


**Figure 10.** A multi-link independent rear suspension system.

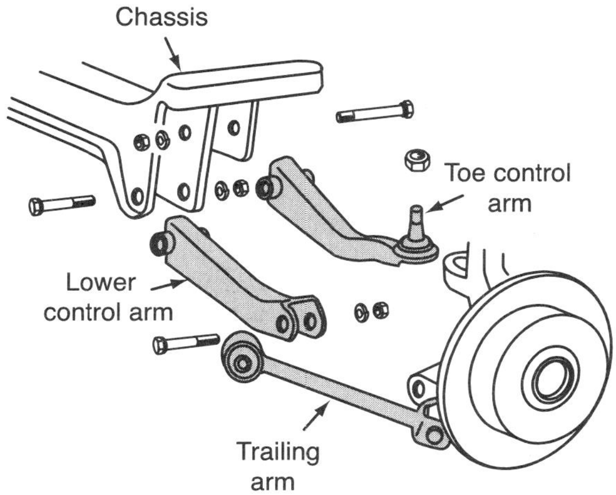


**Figure 11.** A multi-link rear suspension with a parallel lower control arm and toe control arm.

Some multi-link rear suspension systems have a lower control arm and a parallel toe control arm **(Figure 11).** The lower control arm and toe control arm are connected to the lower end of the steering knuckle. A ball joint is used to connect the toe control arm to the knuckle. Rubber insulating bushings are pressed into the inner ends of the toe control arm and lower control arm. The trailing arm, lower control arm, and toe control arm prevent lateral and fore-and-aft wheel movement and provide excellent suspension rigidity when cornering or driving on irregular road surfaces. An eccentric cam bolt on the inner end of the toe control arm provides for toe adjustment.

The upper end of the knuckle is bolted to the outer end of the upper control arm **(Figure 12).** Front and rear bushings in the inner ends of the upper control arm are bolted to brackets that are in turn bolted to the chassis.

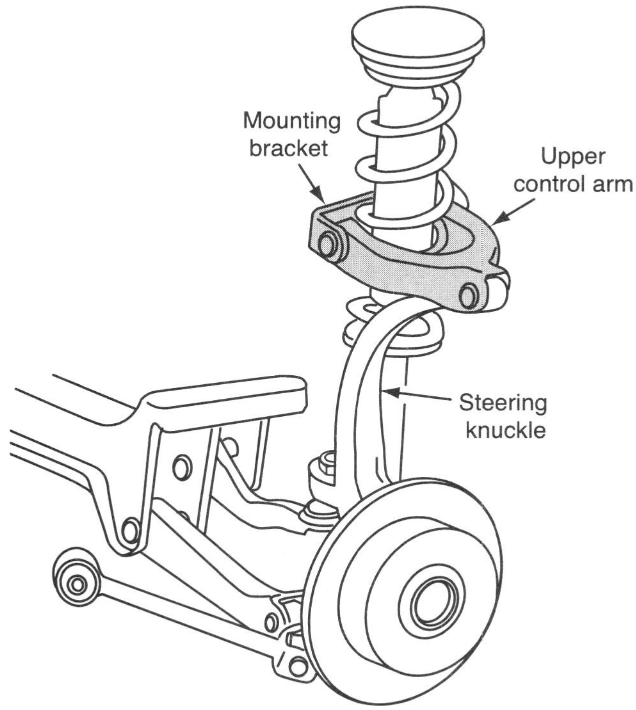


**Figure 12.** Upper control arms and struts in a multi-link rear suspension.

## Double Wishbone Rear Suspension

In the double wishbone rear suspension system, the upper and lower control arms are made from lightweight, high-strength aluminum alloys designed for maximum strength and rigidity. These lighter control arms decrease the unsprung weight. Suspension rigidity is also increased by positioning the ball joints inside the wheel's profile. Since the upper and lower control arms have a wishbone shape, the suspension system is called a double wishbone system. On each side of the car the suspension is attached to the chassis by a cast aluminum subframe. This transmits suspension loads to the chassis at the most efficient locations.

Some independent rear suspension systems experience undesirable toe changes during wheel jounce and rebound. These toe changes cause vehicle instability during cornering and acceleration. In the double wishbone rear suspension system, the control arm design and pivot locations on the rear toe control arm provide minimal change in toe-in during wheel jounce. The toe control arm may be lengthened or shortened to adjust rear wheel toe. An eccentric cam on the rear upper control arm bolt provides a rear wheel camber adjustment **(Figure 13).**

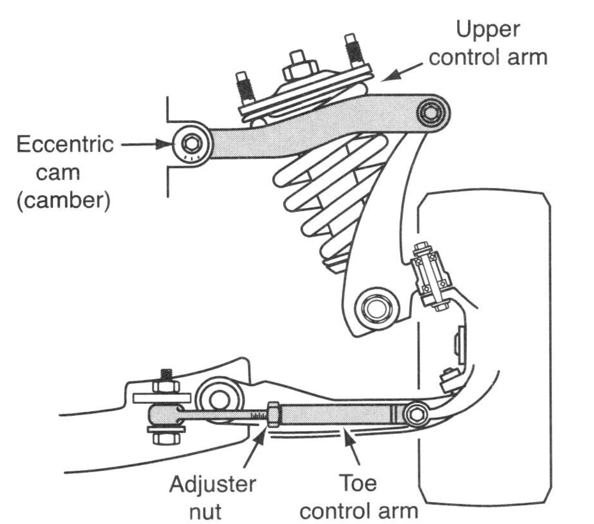


**Figure 13.** Rear suspension camber and toe adjustments.

## MacPherson Strut Independent Rear Suspension Systems

In a MacPherson strut rear suspension system, the coil springs are mounted on the rear struts. A lower spring seat is located on the strut, and the upper spring seat is positioned on the upper strut mount. This upper strut mount is bolted into the inner fender reinforcement. Dual lower control arms on each side of the suspension are connected from the chassis to the lower end of the spindle **(Figure 14).**

The lower end of each strut is bolted to the spindle. Two strut rods are connected from the spindles to the chassis. Bushings are used at both ends of the strut rods. A stabilizer bar mounted in bushings is connected to the chassis and its ends are linked to the struts.

Insulators are mounted between the lower end of the coil spring and the lower spring seat, and the top of the coil spring and the upper spring support. These insulators help prevent the transfer of spring noise and vibration to the chassis and passenger compartment.

### Modified MacPherson Strut Rear Suspension System

Some FWD vehicles have a modified MacPherson strut rear suspension. Each side of the rear suspension has a shock strut, lower control arm, tie rod, forged spindle, and a coil spring mounted between the lower control arm and crossmember side rail.

The strut has a rubber isolated top mount with a one-piece jounce bouncer dust shield. This top mount is attached to the body side panel with a nut, and the lower end of the strut is bolted to the spindle. The stamped lower control arms are bolted to the crossmember and the spindle. The control arm controls the lateral (side-to-side) wheel movement and contains the lower spring seat. A tie rod is connected from the spindle to the underbody and controls fore-and-aft wheel movement and positions the spindle properly.

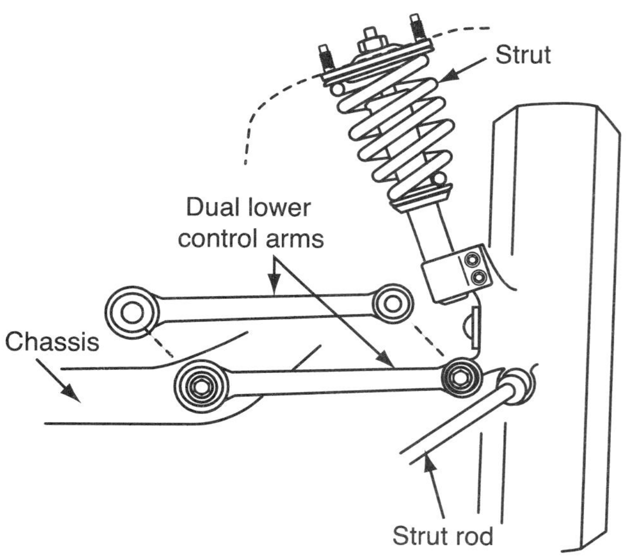


**Figure 14.** A MacPherson strut independent rear suspension system.

### Independent Rear Suspension with Lower Control Arm and Ball Joint

Some FWD cars have an independent rear suspension with a ball joint pressed into the outer end of the lower control arm. The ball joint is bolted into the knuckle. The inner end of the lower control arm is connected to the chassis through two bushings. The lower end of the strut is bolted to the knuckle, and the upper strut mount is bolted to the inner fender reinforcement. A stabilizer bar is attached to the chassis with its outer ends linked to brackets connected to the strut.

A suspension adjustment link is connected from the lower control arm to the knuckle **(Figure 15).** This link provides a rear wheel toe adjustment. A coil-spring seat is located in the lower control arm, and the coil spring is mounted between this seat and an upper seat in the chassis. Upper and lower insulators are mounted between the coil spring and the seats.

### Independent Rear Suspension System with Rear Axle Carrier

Some RWD vehicles have an independent rear suspension with a final drive unit attached to the chassis. This reduces unsprung weight. Bushings connect the large trailing arms to the rear axles. Lower coil-spring seats are located on the trailing arms and upper coil-spring seats are positioned in the chassis. Shock

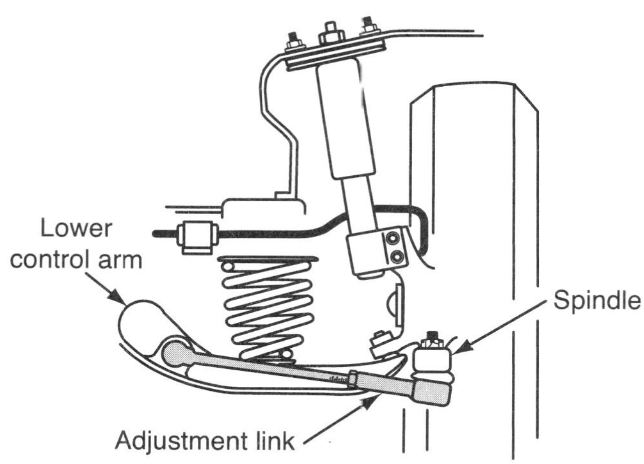


**Figure 15.** A suspension adjustment link.

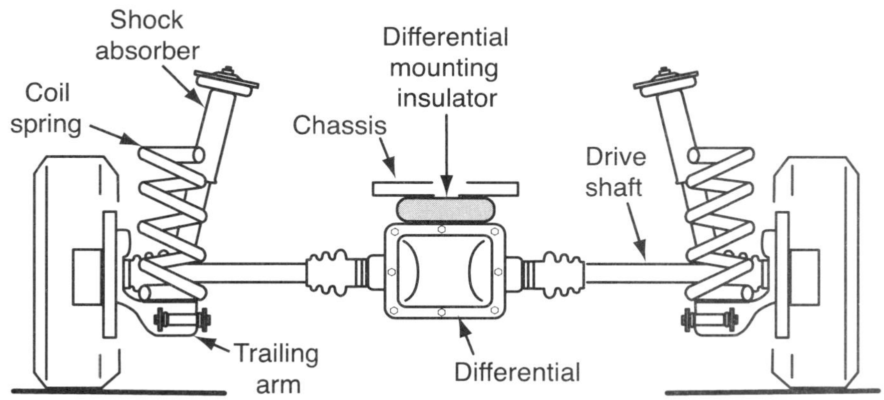


**Figure 16.** An independent rear suspension with rear axle carrier.

absorbers are connected from the back of the trailing arms to the chassis **(Figure 16).**

The rear wheel bearings are in the outer ends of the trailing arms. Drive axles with inner and outer drive joints are connected from the differential to the wheels. With this design, rear wheel camber change is minimized during wheel jounce and rebound. On some early independent rear suspension systems in RWD cars, the drive axles only had inner joints and camber change was excessive causing wear on the edges of the tire.

### Independent Rear Suspension with Transverse Leaf Spring

Some cars have an independent rear suspension system with a transverse mono-leaf fiberglass spring. This spring is compact, lightweight, and corrosion free. Dual trailing arms are connected rearward from the chassis to the knuckle on each side of the suspension, and spindle support rods are attached from the center of the suspension to the bottom of the knuckle **(Figure 17).** Tie rods are connected from the rear of the knuckle to the center of the suspension.

Other independent rear suspension systems have a multiple-leaf transversely mounted rear spring. In these systems, control arms extend rearward from the chassis to the knuckles, and strut rods are connected from the bottom of the knuckles to the center of the suspension. The shock absorbers are connected from the lower end of the knuckles to the chassis. A suspension member connects the differential housing to the chassis.

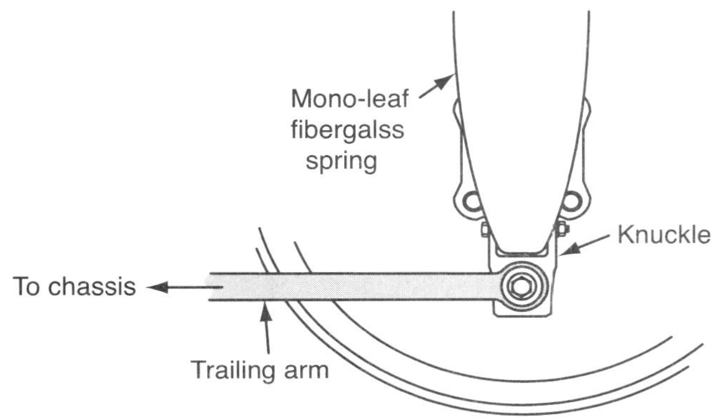


**Figure 17.** An independent rear suspension with a mono-leaf transverse fiberglass spring.

## SUMMARY

- A live-axle rear suspension is one in which the rear axle housing, wheel bearings, and brakes act as a unit.
- A live-axle rear suspension system provides good lateral stability, sway control, and steering characteristics, but increases the chance for axle tramp, causes increased tire wear, and decreased ride quality and traction compared to other types of suspensions.
- In a live-axle coil-spring rear suspension system, the lower control arms absorb differential torque and the upper arms control lateral movement.
- Live-axle coil-spring rear suspension systems may have a tracking bar to control lateral movement.
- A semi-independent rear suspension has a limited amount of individual rear wheel movement provided by a steel U-section channel or crossmember.

- In an independent rear suspension, each rear wheel can move individually without affecting the opposite rear wheel.
- Compared to a live-axle rear suspension system, an independent rear suspension provides improved ride quality, steering control, tire life, and traction.
- In a MacPherson strut rear suspension, the coil springs are mounted on the struts.
- In a modified MacPherson strut rear suspension, the coil springs are mounted separately from the struts.
- In some independent rear suspension systems, the knuckle is positioned by a ball joint on the lower end and the strut on the upper end, and the coil spring is positioned between the lower control arm and the chassis.
- In a multi-link independent rear suspension, the top of the knuckle is positioned by front and rear upper links, and the lower end of the knuckle is connected to a lower arm.
- Some RWD vehicles have an independent rear suspension with the differential connected through insulating mounts to the chassis.
- Some FWD and RWD vehicles have an independent rear suspension with a mono-leaf or multiple-leaf transversely mounted leaf spring.

## REVIEW QUESTIONS

1. A live-axle leaf-spring rear suspension provides excellent ____________________ stability.
2. In many live-axle coil-spring rear suspensions, the differential torque is absorbed by the trailing lower ____________________ ____________________.
3. Explain the differential torque action during acceleration, and describe how this torque is absorbed in a live-axle coil-spring rear suspension.
4. Define axle tramp.
5. In a MacPherson strut independent rear suspension system,
   A. the spring seats are positioned on the struts.
   B. the upper spring seat is positioned in the upper strut mount.
   C. the lower end of each strut is attached to the wheel spindle or knuckle.
   D. all of the above
6. In an independent rear suspension with a lower control arm and ball joint,
   A. the ball joint is pressed into the tie rod.
   B. the top of the strut is bolted to the top of the knuckle.
   C. the lower coil-spring seat is mounted on the strut.
   D. the suspension adjustment link is connected from the lower control arm to the steering knuckle.
7. While discussing multi-link independent rear suspensions, Technician A says the rear suspension member mounts contain silicone oil. Technician B says two upper links and one lower control arm are connected from the rear suspension member to each rear knuckle. Who is correct?
   A. Technician A only
   B. Technician B only
   C. Both Technician A and Technician B
   D. Neither Technician A nor Technician B
8. While discussing live-axle leaf-spring rear suspensions, Technician A says this type of rear suspension has a great deal of unsprung weight. Technician B says this type of rear suspension often has semi-elliptical springs. Who is correct?
   A. Technician A only
   B. Technician B only
   C. Both Technician A and Technician B
   D. Neither Technician A nor Technician B

# Chapter 29

# Computer-Controlled Suspension Systems

## INTRODUCTION

**Interesting Fact**

*Suspension systems have evolved slowly. In the 1940s, many front suspensions were changed from I-beam to the short/long arm. MacPherson strut suspensions replaced the short/long arm suspension systems on front-wheel-drive cars introduced in the late 1970s and 1980s. Now, we are experiencing the use of electronics in suspension systems. Some suspension systems can now react to road or driving conditions in less than 10 milliseconds (0.010 second) to improve ride and handling quality.*

## AIR SYSTEMS

In an air suspension system, air springs replace the coil springs used in conventional systems. The air springs have a composite rubber and plastic membrane with an air spring valve. The air springs are inflated or deflated to provide a constant vehicle trim height. Front air springs are mounted between the control arms and the crossmember **(Figure 1).** The lower end of the air springs is retained in the control arm with a clip, and the upper end sits in a spring seat.

The lower ends of the rear springs are bolted to the rear suspension arms, and the upper ends of these springs are attached to the frame.

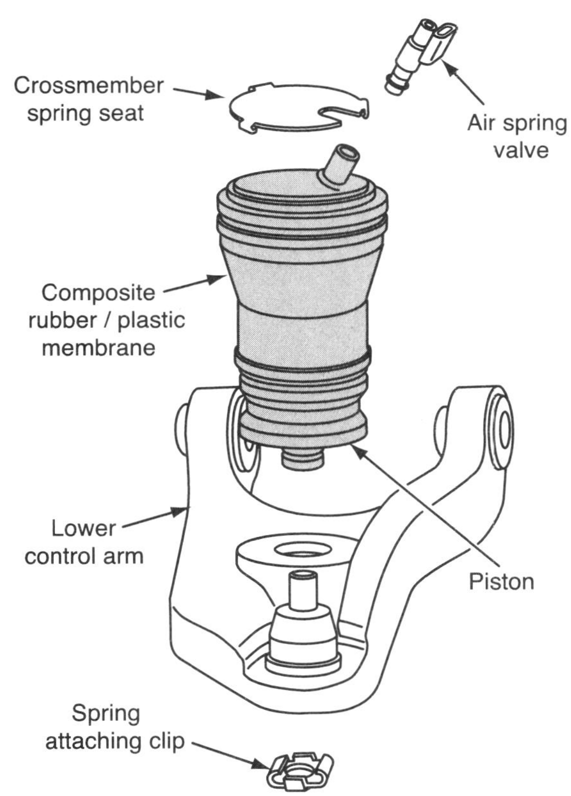


**Figure 1.** A front air spring.

### Air Spring Valves

A solenoid valve is mounted at the top of each air spring. These valves are normally closed. When the solenoid is energized, plunger movement opens the air passage to

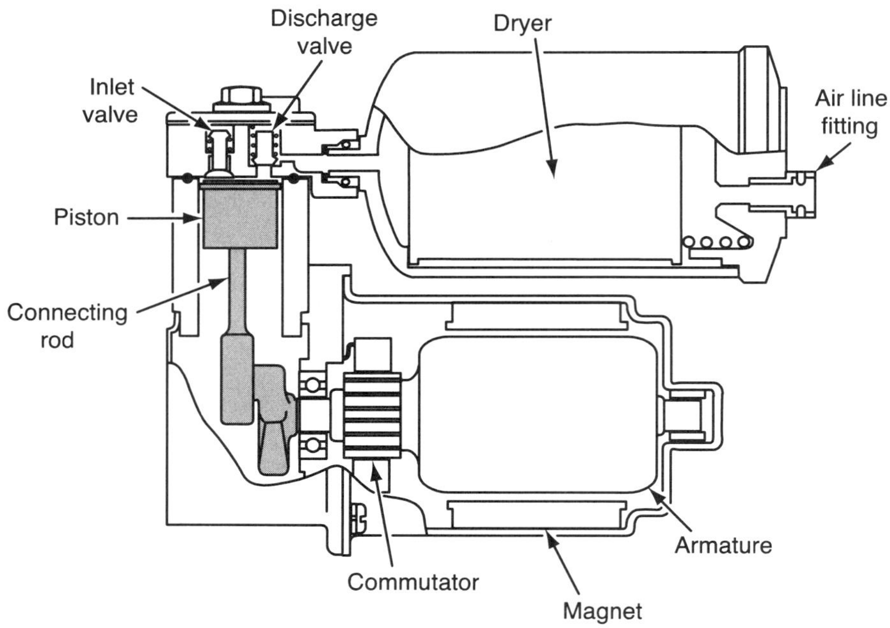


**Figure 2.** A typical air compressor.

the air spring and air enters or exhausts from the spring. O-ring seals are at the end of the valves to seal them.

## Air Compressor

The air compressor **(Figure 2)** in most systems is activated by a computer or control module. An air dryer that contains a silica gel is mounted on the compressor to remove moisture from the air as it enters the system. Nylon air lines connect the compressor outlets to the air spring valves. The compressor operates when it is necessary to force air into one or more air springs to restore the vehicle's trim height.

An air vent valve allows air to vent from the system. When it is necessary to exhaust air from an air spring, the air spring valve and vent valve are energized at the same time and the compressor is shut off. Air exhausts when the vehicle's trim height is too high.

The control module operates the compressor, vent valve, and air spring valves and is typically located in the trunk. The module also turns on the service indicator light when a system defect is present. If the air suspension system is working normally, the warning lamp comes on for 1 second when the ignition switch is turned from the OFF to the RUN position. After this time, the lamp should remain off.

An on/off switch opens the power supply circuit to the control module. This switch is located in the trunk near the control module **(Figure 3).** When the air suspension system is serviced or some other vehicle service is performed, the switch must be in the OFF position.

**You Should Know** *The on/off switch must be in the OFF position before the car is hoisted, jacked, towed, raised off the ground, or jump started. If this precaution is not observed, it may result in personal injury, component damage, or vehicle damage.*

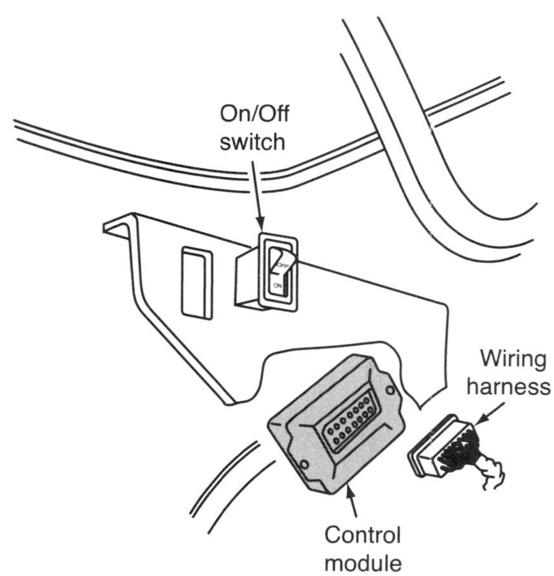


**Figure 3.** An on/off switch located in the trunk close to the control module.

## Height Sensors

There are two front height sensors between the two lower control arms and crossmember. A single rear height sensor is positioned between the suspension arm and frame **(Figure 4).** Each height sensor contains a magnet slide attached to the upper end of the sensor. The magnet slide moves up and down in the lower sensor housing as changes in vehicle trim height occur **(Figure 5).** The lower sensor housing contains two switches connected to the control module.

When the vehicle is at trim height, the switches remain closed and the control module receives a trim height signal. If the magnet slide moves upward, the above trim switch opens and a "lower the vehicle" command is sent to the module. This signal causes the module to open the appropriate air spring and vent valve. This exhausts air from the spring and corrects the trim height. Downward magnet slide movement closes the above trim switch and opens the below trim switch. This sends a command to raise the vehicle, and the module energizes the compressor relay to start the compressor and open the appropriate air spring valve to allow air to enter the spring.

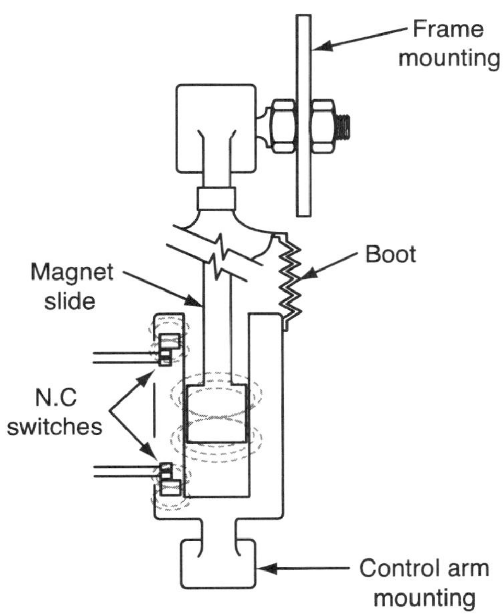


**Figure 4.** Rear height sensor mounting.

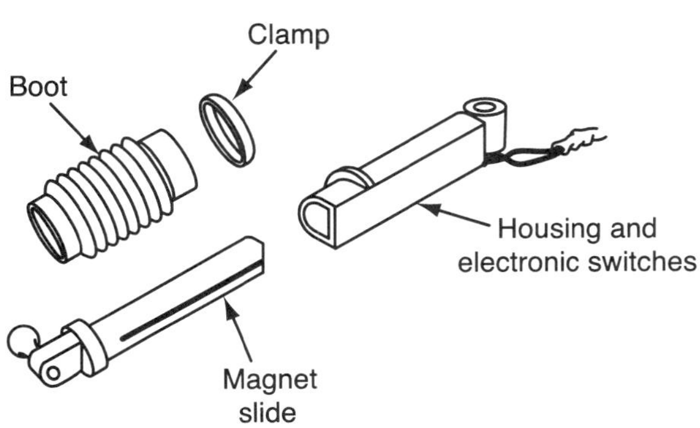


**Figure 5.** A height sensor.

## Adjustable Pneumatic Suspension

An automatically adjustable pneumatic suspension can vary the vehicle's ground clearance so it can be used off-road but still perform and handle well on the highway. There are four ride-height positions that can be selected either manually or automatically, with a total range of over 8 inches of ground clearance. At highway speeds, the vehicle has a normal ground clearance. Urban mode raises it a full inch. For moderate off-road driving, ground clearance increases another 1 to 2 inches. For severe off-road conditions at speeds under 25 mph, maximum clearance is achieved. The vehicle will adjust to the desired height based on vehicle speed, or the driver can temporarily override the setting by depressing a button.

## Automatic Air Suspension System

An automatic air suspension system maintains the vehicle at the proper level under various suspension loads. The programmed ride control system switches the shock absorbers from soft to firm in response to the input sensor signals. This system is completely automatic with no driver-controlled inputs. Variable rate air springs at each corner of the vehicle are integral with the struts. Two front and one rear electronic rotary height sensors send inputs to the air suspension control module.

The height sensors contain a permanent magnetic rotor and a Hall element **(Figure 6).** Suspension movement rotates the permanent magnet rotor and changes the voltage signal in the Hall element. The voltage signal is

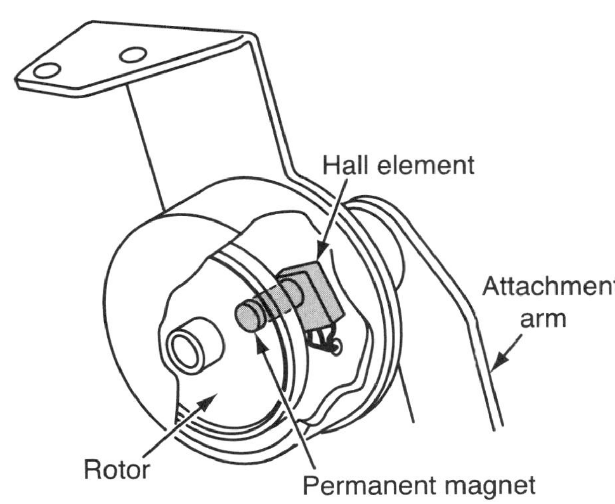


**Figure 6.** An electronic rotary height sensor.

proportional to trim height, above trim height, and below trim height.

The air springs are attached to the struts, and each spring has a fill/vent solenoid valve. Since the front air springs support more weight than the rear air springs, the front air springs are larger. The upper end of each front strut is mounted to the strut tower through a ball bearing and rubber mount system. These ball bearings provide a smooth, durable, pivot point for the strut, air spring, and wheel assembly **(Figure 7).**

Each rear strut has a dual path mount that separates the strut and air spring mounting surfaces to the chassis to provide maximum isolation. In the soft mode, there are two orifices in the strut piston that allow oil movement. When the actuator moves to the Firm position, there is only one path for oil movement, which provides firmer strut action.

## Air Suspension System with Speed Leveling Capabilities

In an air suspension system with speed leveling capabilities **(Figure 8),** the front springs are combined with the struts, but the struts do not have actuators. The rear springs are separated from the shock absorbers. A **vehicle speed sensor (VSS)** is mounted in the transmission and sends a voltage signal to the suspension control module. When the

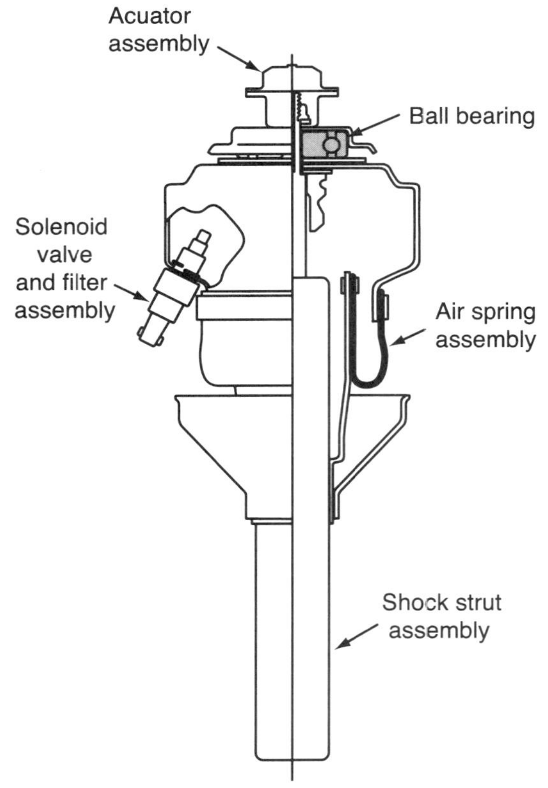


**Figure 7.** A front air spring and strut assembly.

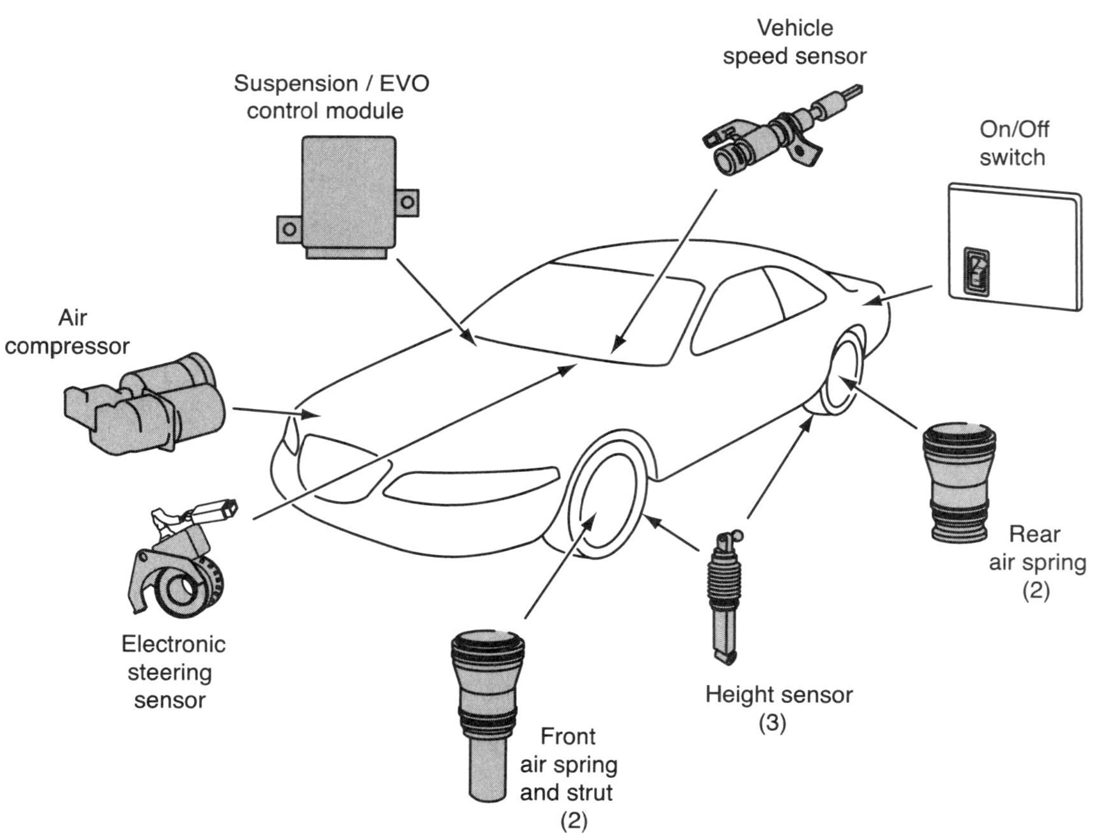


**Figure 8.** An air suspension system with speed leveling capabilities.

VSS signal indicates a speed of 65 mph (105 kmph), the control module energizes the spring fill/vent solenoids and the vent solenoid in the compressor. This vents air from the air springs and lowers the ride height a specific amount. If the speed is below 65 mph (105 kmph), the control module operates the spring fill/vent solenoids to supply the specified trim height.

If weight is added to the vehicle, the rear suspension height sensor sends a voltage signal to the control module indicating the rear trim height is below the specified ride height. The control module then starts the compressor and energizes the fill/vent solenoids in the rear air springs. The compressor runs until the rear height sensor signal indicates the specified rear suspension trim height. Once this signal is received, the control module shuts the compressor off and the fill/vent solenoids on the rear air springs.

When weight is removed, the rear height sensor signal indicates the rear trim height is above normal. The control module now energizes the rear spring solenoids and vent solenoid in the compressor to vent air from the rear springs until the rear trim height returns to normal.

## Automatic Ride Control (ARC), Four-Wheel-Drive Vehicles

Unlike other air suspension systems in which the air springs replace the conventional coil springs, the **automatic ride control (ARC)** system has a front torsion bar suspension and a rear leaf-spring suspension. The empty vehicle is fully supported by these springs, and ride height is maintained at the specified height. The air suspension system operates only when additional weight is added to the vehicle or if the driver selects four-wheel-drive high or low.

The ARC system has one front and one rear ride height sensor. The height sensors are connected from the suspension to the chassis **(Figure 9).** The height sensors send normal, below normal, or above normal trim height voltage signals to the ARC module. The ARC system can control the suspension height from the specified ride height to 2 in. (50 mm) above it.

**Figure 9.** Height sensors.

The ARC system has air shock absorbers in place of conventional shock absorbers. Air pressure is supplied from the compressor through nylon air lines to each air shock absorber to increase trim height. Each air shock absorber contains a DC electric motor and a hollow piston rod.

In the Soft ride mode, the ARC module allows the DC motor and bypass valve to remain open. The fluid in the shock absorber flows through the hollow piston rod and the fixed base valve in the shock absorber during wheel jounce and rebound to provide a softer ride.

If the Firm ride mode is required, the ARC module operates the DC motors to close the bypass valves in each shock absorber. Fluid flows only through the fixed base valve in the shock absorber during wheel jounce and rebound, and this provides a firm ride.

If the transfer case is in two-wheel drive, the suspension remains at the base height at all vehicle speeds. When the driver selects the four-wheel-drive high mode, the ARC module operates the compressor and fill solenoids to supply more air to the air shock absorbers to increase suspension height. If the vehicle speed exceeds 58 mph (93 kmph) in the four-wheel-drive high mode, the ARC module returns the suspension to base height.

When the driver selects four-wheel-drive low, the ARC module operates the compressor and fill solenoids to supply more air to the shock absorbers and increase suspension height by 2 in. (50.8 mm). With the higher pressure in the air shock absorbers, the suspension is much stiffer.

The ARC system is also load sensitive. If extra weight is added to the vehicle, the ARC module reacts to the suspension height sensor signals and returns the suspension to the proper height for the selected transfer case mode.

## Rear Electronic Level Control

The **electronic level control (ELC)** system maintains the rear trim height regardless of rear suspension load. If a heavy object is placed in the trunk, the rear wheel position sensors send below-trim-height signals to the control module. When this signal is received, the module sends a voltage signal to the air compressor to turn it on.

The compressor supplies air through the nylon lines to the rear air shocks and raises the rear suspension height. When trim height signals are received from the rear wheel position sensors, the module stops the compressor.

When the heavy object is removed from the trunk, the rear wheel position sensors send above trim height signals to the module. The module then energizes the exhaust solenoid in the compressor to release air from the rear air shocks.

## ELECTRONICALLY CONTROLLED SHOCK ABSORBERS AND STRUTS

Many cars are now equipped with computer-controlled active suspension systems. In these systems, a computer-

**Figure 10.** A strut actuator.

controlled actuator is positioned in the top of each shock absorber or strut **(Figure 10).** The actuators rotate a shaft inside the piston rod. The shaft is connected to the shock valve. Many of these systems have two modes, Soft and Firm. In the Soft mode, the actuators position the shock absorber valves so there is less restriction to the movement of oil. When the computer changes the actuators to the firm mode, the actuators position the shock valves so they provide more restriction to oil movement, which provides a firmer ride.

## Integrated Electronic Systems

With the rapid advances in electronic technology, there is a trend toward integrating the computer-controlled systems. Rather than having a separate computer for each electronic system, several of these systems may be controlled by one computer. This integration solves computer mounting problems and reduces the length and number of wires in a vehicle. Some vehicles have an integrated antilock brake system (ABS), acceleration slip regulation (ASR), and vehicle dynamics control (VDC) system.

Yaw force occurs when the rear end of a vehicle begins to swerve sideways from the intended vehicle course. This yaw force may occur because of a slippery road surface or excessive vehicle speed when cornering. The yaw velocity sensor operates on a vibrating gyro principle. It contains a metal cylinder whose rim vibrates in elliptical shapes. The vibration and rotation of the metal cylinder is proportional to the rotational speed of the vehicle around the center of the cylinder.

### Yaw Control

In response to hazards, the driver may oversteer the vehicle. This can place the vehicle in an uncontrolled swerve in which the driver has lost steering control. When cornering at high speeds or driving on a slippery road surface, the rear of the vehicle tends to swing around as the tires lose traction on the road. If vehicle directional stability is automatically maintained in these situations, the possibility of driver oversteer and uncontrolled vehicle swerve are greatly reduced. Yaw may be defined as deviating erratically from side to side.

On a car equipped with VDC, if the rear of the vehicle begins to swing around, the yaw velocity sensor and lateral acceleration sensor send voltage signals to the computer. When these signals indicate a slight amount of rear vehicle swerve, the computer applies the brake to the appropriate front wheel to prevent more swerving. For example, if the rear of the vehicle is swerving toward the right, the system applies the brake on the left front wheel.

## Programmed Ride Control System

The **programmed ride control (PRC)** system **(Figure 11)** is available on some vehicles. The damping action of the front and rear struts and shocks is automatically controlled by the system to provide improved ride and han-

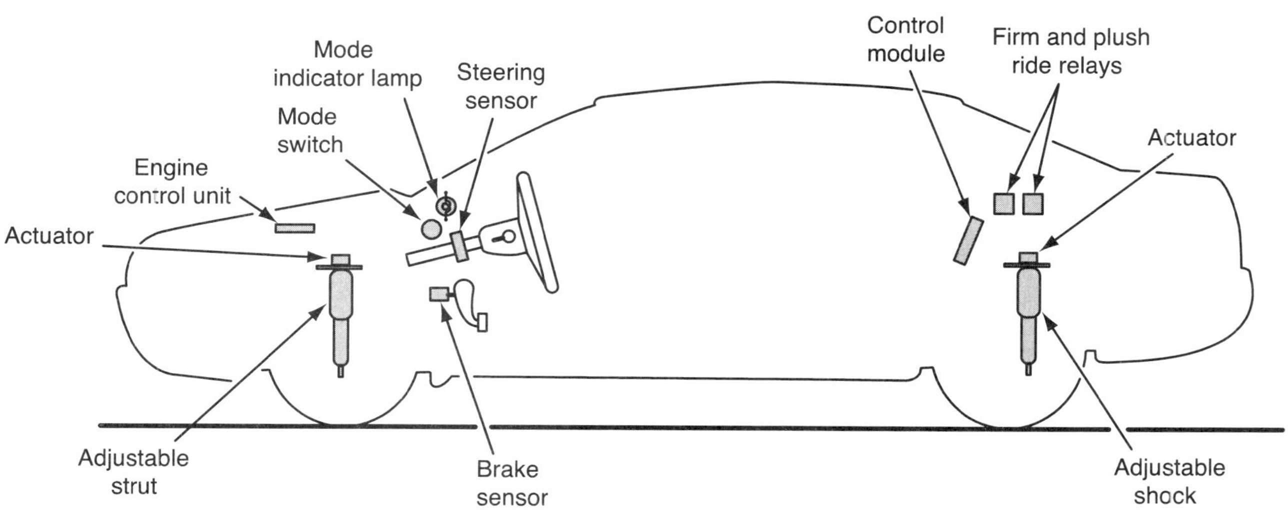


**Figure 11.** Programmed ride control (PRC) system components.

dling characteristics under various driving conditions. The main components in the PRC system are several sensors, struts and shocks with electric actuators, control module, and mode relays and switches.

### Sensors

A steering sensor sends signals to the control module in response to the amount and speed of steering wheel rotation. The brake sensor sends a signal to indicate when the brakes have been applied. A speed sensor is also used to inform the module of the vehicle's speed.

### Strut and Shock Actuators

An actuator is positioned in the top of each strut and shock. Each actuator contains a single pole armature, a pair of permanent magnets, and a position switch. In the plush mode, current is applied to the armature and the magnetic fields of the armature and the permanent magnets repel each other. This action causes clockwise armature rotation until the armature hits the internal stop. Under this condition, the leaf-spring switch is open in the position sensor circuit, and no signal is returned to the PRC control module.

In the Firm mode, current is applied to the armature, but there is an attraction between the magnetic fields of the armature and permanent magnets. This attraction causes counterclockwise armature rotation until the armature contacts the internal stop. The armature rotates the strut or shock valve to restrict oil movement and provide increased damping. In the Firm position, the leaf-spring switch closes and sends a feedback signal to the control module. In this mode, the shocks/struts provide approximately three times the damping action on the extension stroke compared to the normal mode.

When the mode select switch is in the "Auto" position, the system operates in the plush or soft mode until vehicle speed reaches a predetermined level, the engine is heavily accelerated, large amounts of brake system pressure has been developed, or when the vehicle's lateral acceleration reaches a predetermined level. Any of these conditions will cause the system to switch from the Normal mode to the Firm mode.

## Computer Command Ride Control System

The **computer command ride (CCR) system** is designed to prevent vehicle lift during acceleration, dive during deceleration and braking, and roll while cornering. If the mode switch is in the Normal position, the control module will change the struts to the Firm position under certain conditions. The module can also cause the strut actuator motors to rotate to the Comfort position under other conditions.

This system or a variation is used on various vehicles manufactured by General Motors. The vehicle speeds at which the control module operates the strut actuators to change strut modes varies depending on the vehicle make and model year.

### Struts

The control module operates an actuator motor in each strut, and these motors rotate variable strut orifices. In the Normal mode, the motor positions the variable orifice in the Comfort position and provides less restriction to fluid movement in the strut. This reduces damping action in each strut for a softer ride. When the module drives the actuators and variable orifices to the firm position, fluid movement in the strut is restricted by the variable orifice **(Figure 12).** Under this condition, strut damping action is increased to provide a firm ride.

## Road Sensing Suspension System

The **road sensing suspension (RSS)** system is referred to as a *real time damping* system. The RSS controls damping forces in the front struts and rear shock absorbers in response to various road and driving conditions. The RSS system changes shock and strut damping forces in 10 to 12 milliseconds, whereas other damping systems require a much longer time interval to change damping forces. It requires about 200 milliseconds to blink your eye.

The RSS module receives inputs regarding vertical acceleration, wheel-to-body position, and speed of wheel movement, vehicle speed, and lift/div. The module controls a solenoid in each shock or strut to provide damping control.

### Sensors

A wheel position sensor is mounted at each corner of the vehicle between a control arm and the chassis. These

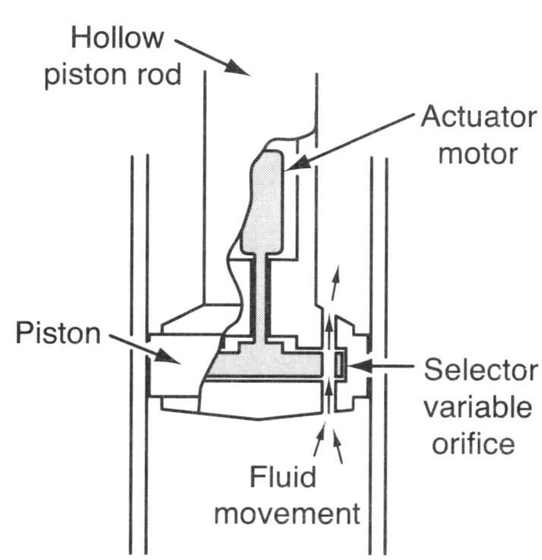


**Figure 12.** Strut actuator and variable orifice positions.

sensors provide voltage signals to the module regarding relative wheel-to-body movement and the velocity of wheel movement. The rear position sensor also provides rear suspension height information to the module.

An accelerometer is mounted on each corner of the vehicle. These send information to the module in relation to vertical acceleration of the body. A VSS sends a voltage signal to the PCM, which in turn sends vehicle speed information to the RSS module.

The lift/dive input is also sent from the PCM to the RSS module. Lift information is based on throttle position, vehicle speed, and transaxle gear input signals. The PCM calculates suspension dive information based on the rate of vehicle speed change when decelerating.

### Shocks and Struts

Each strut or shock contains a solenoid that is controlled by the RSS module. Each damper solenoid valve provides two levels of damping, firm or soft. In the Soft mode, the solenoid is switched on, causing the oil in the shock or strut to bypass the main damper valving. When the system switches to the Firm mode, the damper solenoid switches off and causes the oil to flow through the main damper valving to provide a firm ride.

## Electronic Damping Control (EDC)

**Electronic Damping Control (EDC)** adjusts the shock absorbers through an infinitely variable range to adapt to variations in vehicle load, driving style, and road conditions. It reduces pitch and dive during braking and body movement in curves, during acceleration, and on irregular road surfaces.

Sensors continuously monitor all vehicular motion. The computer evaluates the monitored data and relays the required commands to the shock absorbers. Solenoid valves steplessly adjust compliance rates to the desired levels within a matter of milliseconds. When the vehicle is driven on smooth road surfaces, EDC adjusts the shock absorbers to provide the best possible ride comfort. EDC immediately reacts to rapid cornering by reducing the damping action of the shock absorbers.

## MagneRide

**MagneRide** is a semi-active suspension system that features shocks or struts with no electromechanical valves or small moving parts. Instead, MagneRide regulates the flow of fluid by a variable magnetic field produced by a small electric coil mounted in the shock. The shocks are filled with MR fluid. MR fluid consists of magnetically soft particles, such as iron, suspended in synthetic hydrocarbon fluid.

The action of the shock forces the MR fluid through a magnetized opening in each shock. When the shock is in its off state, the fluid is not magnetized and flows freely through the orifice. When current is sent to the coil, the fluid becomes magnetized and its viscosity changes instantly **(Figure 13).**

The material changes from a fluid state to a semi-solid state that is directly proportional to the magnetic field applied to it. With little or no electrical current, the iron particles are randomly distributed, and the fluid passes freely through the piston orifice. When a strong electrical current is applied to the coil, the resulting magnetic field aligns the iron particles so that the fluid stiffens and the flow is resisted. This causes heavy damping. The resulting damping force is proportional to the viscosity, which is proportional to the strength of the magnetic field.

Sensors monitoring wheel position, lateral acceleration, vehicle speed, steering wheel angle, and brake pedal angle are inputs to the control module that sends current to the coil in the shocks. This system provides an extremely quick response time, typically about 5 ms, and the fluid is capable of reacting 30,000 times per second.

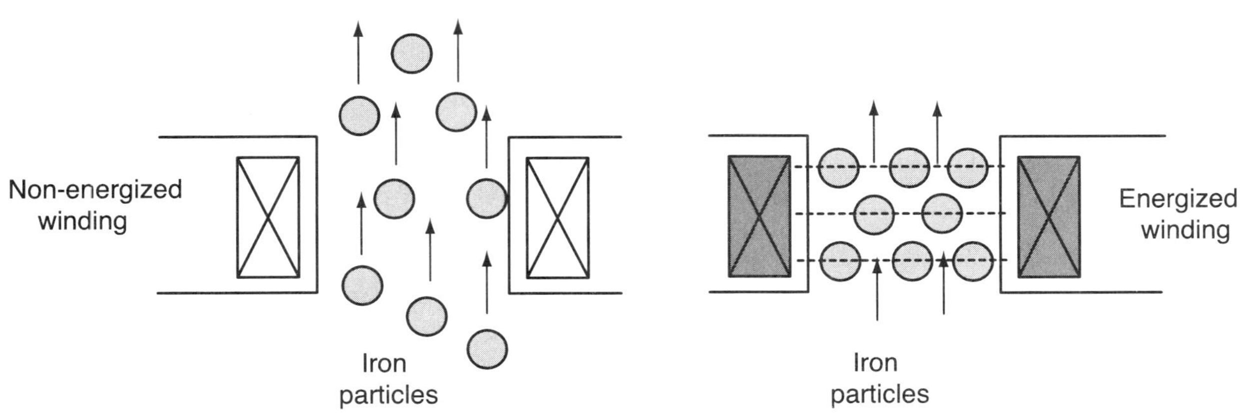


**Figure 13.** Magneto-rheological fluid action in a chock absorber or strut.

## SUMMARY

- An electronic air suspension system maintains a constant trim height regardless of passenger or cargo load.
- To exhaust air from an air spring, its solenoid valve and the vent valve must be energized.
- To force air into an air spring, the air spring solenoid valve must be energized.
- The on/off switch in an electronic air suspension system must be off before the car is hoisted, jacked, towed, or raised off the ground.
- Rotary height sensors send voltage signals to the control module in relation to the amount and speed of wheel jounce and rebound.
- Some air suspension systems reduce trim height at speeds above 65 mph (105 kmph) to improve handling and fuel economy.
- Automatic ride control (ARC) systems on 4WD vehicles can provide firmer shock absorber valving in relation to transfer case mode selection, vehicle speed, and operating conditions.
- In a PRC system, the steering sensor sends signals to the control module regarding the amount and speed of steering wheel rotation.
- The struts and shock absorbers in some PRC systems provide three times as much damping action in the Firm mode as in the Normal mode.
- The accelerometer in a CCR system contains a mercury switch and sends a vehicle acceleration signal to the control module.
- In a CCR system, the accelerometer signal or the vehicle speed signal may inform the control module to switch from the Normal to the Firm mode.
- The road sensing suspension system changes shock and strut damping forces in 10 to 12 milliseconds.
- A vehicle dynamics control (VDC) system applies one of the front brakes if the rear of the car begins to swerve out of control. This action maintains vehicle direction control.

## REVIEW QUESTIONS

1. Describe the operation of the steering sensor in a PRC system.
2. Describe the purpose of the accelerometer sensor.
3. Explain how air is forced into an air spring in a rear load-leveling air suspension system.
4. List the conditions when the on/off switch in an electronic air suspension system must be turned off.
5. In a PRC system, if the car is accelerating with the throttle wide open, the PRC system is in the ____________ mode.
6. An electronic rotary height sensor contains a ____________ ____________.
7. While discussing a road sensing suspension system, Technician A says this system has three wheel position sensors. Technician B says the wheel position sensors send a signal to the module in relation to the amount and velocity of wheel-to-body movement. Who is correct?
   A. Technician A only
   B. Technician B only
   C. Both Technician A and Technician B
   D. Neither Technician A nor Technician B
8. All of these statements about a rear load-leveling suspension system are true *except:*
   A. The on/off switch is mounted in the vehicle trunk.
   B. The control module operates the compressor relay.
   C. If a door is open, the control module completes lower suspension height commands.
   D. The rear suspension has one, nonserviceable suspension height sensor.

# Chapter 30 Front Suspension Diagnosis

## INTRODUCTION

There are number of factors that can cause a customer concern of poor handling or poor ride quality. Some of these may be due to poor wheel alignment or bad tires; others are caused by suspension problems. Several checks can be made on individual suspension parts. However, diagnosis of the suspension system should begin with a road test and inspection of the vehicle.

## ROAD TEST

Before road testing the vehicle, gather as much information from the customer as possible. Knowing what happens and when it happens will greatly help your diagnosis. Before taking the road test, make sure the vehicle is safe to operate by quickly checking the tires, brake pedal operation, and steering wheel feel. Note the operation of the ABS and air bag warning lights. During the road test, make sure you observe all traffic laws and drive the vehicle in a normal manner so you will not cause any additional wear or damage.

During the road test, try to confirm the customer's complaint by driving under the conditions the customer is concerned about. While it may be somewhat inconvenient to duplicate a particular type of road surface, this may be the only way to verify the condition. Attempt to determine the problem area, front or back, or right or left side. Determine whether the problem is vehicle speed sensitive and at what speed it is most noticeable, whether it is torque sensitive, and whether it is engine speed sensitive or gear ratio sensitive. While road testing the vehicle, listen for any unusual noises. Also note any unusual or unsafe condition.

### Causes of Common Symptoms

The most obvious problems experienced on a road test are abnormal noises, harsh riding, excessive sway, and wheel shimmy. The latter is often related to steering problems but may be the source of a customer's concern.

#### Noises

Often, the source of an abnormal noise can be identified by the driving conditions. Tire noise varies with road surface conditions and is most pronounced on smooth asphalt road surfaces at speeds of 15 to 45 mph. Rattling on road irregularities can be caused by worn shock absorber bushings or grommets, worn spring insulators, a broken coil spring or spring insulators, worn control arm bushings, worn stabilizer bar bushings, worn strut rod grommets, worn leaf spring shackles and bushings, worn torsion bars, anchors, or bushings. While chattering during cornering is often caused by worn upper strut bearings, squeaking on road irregularities can be caused by worn leaf spring silencers, dry or worn control arm bushings, dry or worn stabilizer bushings, or worn-out silencers between the spring leafs.

#### Harsh Riding

If the ride of the vehicle seems to be harsher or rougher than normal, check for low ride height at the front and rear suspension, weak or broken springs, worn torsion bar components such as pivot cushion bushings, and worn or damaged control arm bushings.

#### Excessive Sway

If the vehicle seems to roll to one side when traveling on a rough surface, check for a weak or broken torsion bar, weakened or disconnected stabilizer bar, and worn stabilizer bar bushings. When a vehicle is not stable when its direction is changed or when attempting to change direction, check for weak or broken springs that affect front suspension alignment angles, worn control arm bushings, worn ball joint stud openings in the knuckle, a bent steering knuckle, or a broken spring center bolt. If the steering seems erratic on irregular surfaces, inspect the upper and lower control arms for cracks, distortion, and worn bushings.

### Shimmy

When the wheels shimmy, the driver feels large, consistent, rotational oscillations at the steering wheel. These motions are caused by the lateral movement of the tires and can be caused by loose, worn, or damaged steering linkages or suspension parts, out-of-balance tires, excessive wheel runout, an unevenly worn or damaged tire, or loose wheel bearings.

## GENERAL FRONT-SUSPENSION INSPECTION

Once the road test has been completed and it has been determined there is an abnormal condition, a thorough inspection can pinpoint the exact cause. Inspect the tires and the vehicle's ride height and make a quick check of the shocks and/or struts.

> **You Should Know** *If the vehicle has a harsher-than-normal ride, check it for optional suspension equipment, including trailer towing packages. They have a firmer ride quality.*

### Tire Wear

Check the tires for correct pressure, construction, and size, and for wear, damage, and defects. The wear on a tire should be even across the tread. Tires are checked by looking at the tread, feeling the tread, and by measuring the depth of the tread. Wear patterns are good indicators of problems **(Figure 1).** Abnormal tire wear may be caused by worn control arm bushings, worn ball joints, a bent steering knuckle, and weak or broken springs.

### Curb Riding Height

Curb riding height **(Figure 2)** is determined mainly by spring condition. Other suspension components, such as control arm bushings, will affect curb riding height if they are worn. The curb riding height must be measured at the vehicle manufacturer's specified location, which varies depending on the type of suspension system.

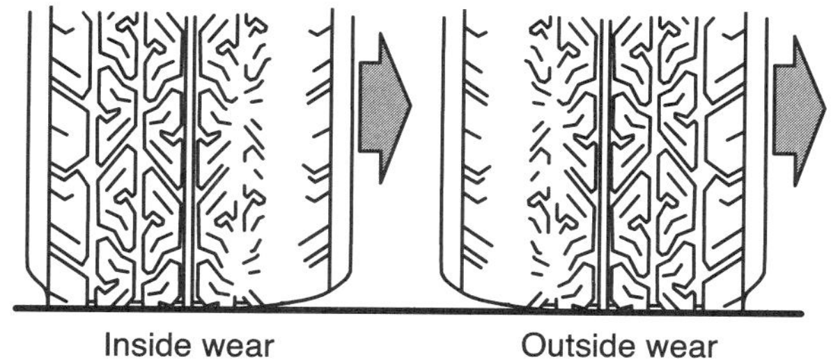


**Figure 1.** Inside and outside wear is caused by incorrect camber, which can be caused by many different suspension problems.

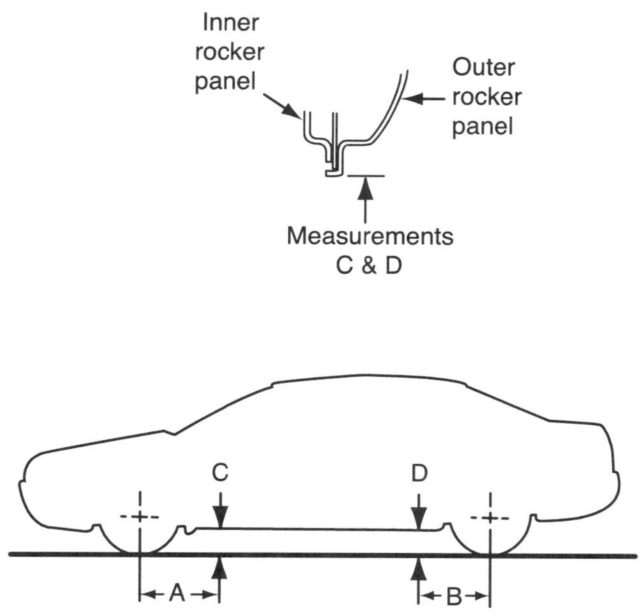


**Figure 2.** Typical front and rear curb riding height measurement locations.

Reduced curb riding height on the front suspension may cause decreased directional stability. If the curb riding height is reduced on one side of the front suspension, the vehicle may pull to one side. Reduced rear suspension height increases steering effort and causes rapid steering wheel return after turning. Harsh riding can be caused by a reduced curb riding height.

If the curb riding height is less than specified, the control arms and bushings should be inspected and replaced as necessary. When the control arms and bushings are in normal condition, the reduced curb riding height may be caused by sagged springs that require replacement. On a torsion bar front suspension, the torsion bars may be adjusted to correct improper curb riding height.

## Shock Absorber and Strut Checks

When performing a shock absorber or strut bounce test, the vehicle's bumper is pushed downward with considerable weight applied on each corner. The bumper is released after this action, and one free upward bounce should stop the vertical chassis movement if the shock absorber or strut provides proper spring control. Some bad shock absorbers or struts will make an internal clunking, clicking, and squawking noise, or display binding. When these noises or conditions are experienced, the shock absorber or strut must be replaced.

# DETAILED INSPECTION

Raise the vehicle on a frame hoist or raise it with a floor jack and support it with jack stands. Grasp the upper and lower surfaces of the tire and shake each wheel to check for worn or loose wheel bearings. On front-wheel-drive cars, perform this check at all four wheels.

Clean all excess dirt and grease from the components of the front and rear suspensions. Visually inspect all components for visible wear or damage, including all rubber boots, bushings, and grommets. Look for cracks or signs of wear or damage.

## Shocks and Struts

Shock absorbers should be inspected for loose mounting bolts and worn mounting bushings. If these are loose or worn, rattling will be evident, and the bushings and bolts should be replaced. The bushing is permanently mounted in some shocks, and the shock must be replaced if the bushing is worn.

Shock absorbers and struts should be inspected for oil leakage. They should also be checked for physical damage such as a bent condition and severe dents or punctures **(Figure 3).** When any of these conditions is present, the shocks or struts should be replaced. The upper mounts of struts should also be carefully checked and replaced if damaged.

### Springs and Rebound Bumpers

While checking the shocks and struts on a vehicle, make sure you check the rebound bumpers for the springs, as well as the springs themselves. If the springs are cracked or broken, they must be replaced. Excessive wear on the rebound bumpers may be caused by improper curb riding height or worn-out shock absorbers.

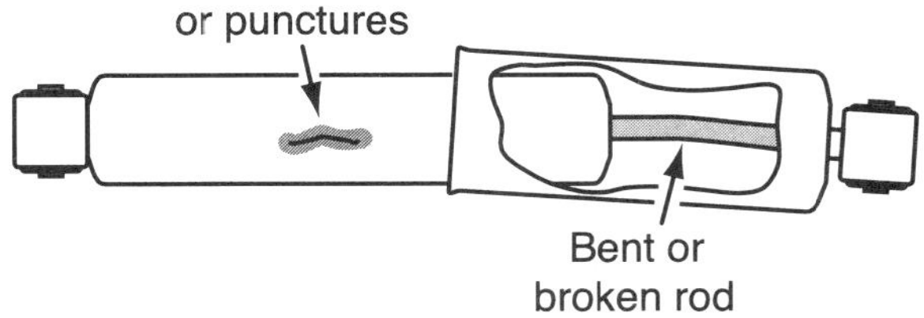


**Figure 3.** The shock absorbers and struts should be carefully inspected for damage.

## Control Arms

Upper and lower control arms should be checked for cracks, bent conditions, and worn bushings. Bent control arms or worn control arm bushings affect curb riding height and alignment angles. If the control arm bushings are worn, steering is erratic, especially on irregular road surfaces. Worn control arm bushings may cause a rattling noise. Dry or worn control arm bushings may cause a squeaking noise. Caster and camber angles are altered by worn upper and lower control arm bushings. Incorrect caster or camber angles may cause the vehicle to pull to one side. Tire wear may be excessive when the camber angle is not within specifications.

## Strut Rods

Some front suspension systems have a strut rod connected from the lower control arm to the frame **(Figure 4).** Rubber grommets isolate the rod from the chassis components. Worn strut rod grommets may allow the lower control arm to move rearward or forward. This movement changes the caster angle, which may affect steering quality and cause the vehicle to pull to one side. A bent strut rod also causes steering pull. Worn strut rod grommets or loose mounting bolts cause a rattling noise while driving on irregular road surfaces. Inspect the strut rod grommets visually for wear and deterioration. With the vehicle lifted on a hoist, grasp the strut rod firmly and apply vertical and horizontal force to check the rod and grommets for movement. Worn grommets must be replaced. If a strut rod is bent, replace the rod.

### Stabilizer Bar

Carefully check the stabilizer bar for looseness and damage. Also, make sure to inspect the condition of the

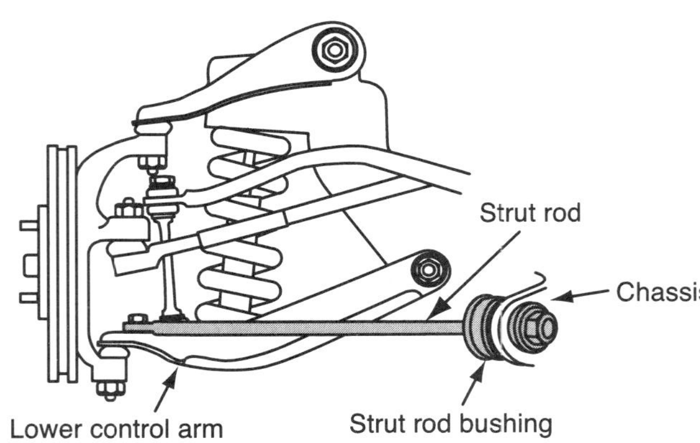


**Figure 4.** A typical strut rod.

bar's bushings. A weak stabilizer bar or worn stabilizer bar bushings can cause harsh riding, excessive body sway, and suspension noise.

## BALL JOINTS

Begin your inspection of a ball joint by looking to see if the ball joint has a wear indicator. If it does, check the placement of the grease fitting. If it is recessed, the ball joint is worn and should be replaced. On other ball joints, the grease fitting retainer extends a short distance through the ball joint surface. On this type of joint, replacement is necessary if the grease fitting shoulder is flush with or inside the ball joint cover **(Figure 5).** On some vehicles, check the grease fitting for movement in the ball joint with the vehicle's weight on the wheels. Any movement of the fitting may indicate the need to replace the ball joint.

Carefully look at the joint's boot. A damaged boot or joint seal will allow lubricant to leak out and allow dirt to enter and contaminate the lubricant. If the boot is damaged, the ball joint should be replaced.

If no boot damage is evident, gently squeeze the boot. If the boot is filled with grease, it will be somewhat firm. If the joint has a grease fitting and is not filled with grease, use a grease gun and refill the joint. Fill the joint until fresh grease is seen flowing out of the boot's vent. If too much grease is forced into the joint or it is forced in too quickly, the boot can unseat or tear.

**You Should Know** *Regular chassis lubrication at the vehicle manufacturer's recommended service interval is one of the keys to long ball joint life.*

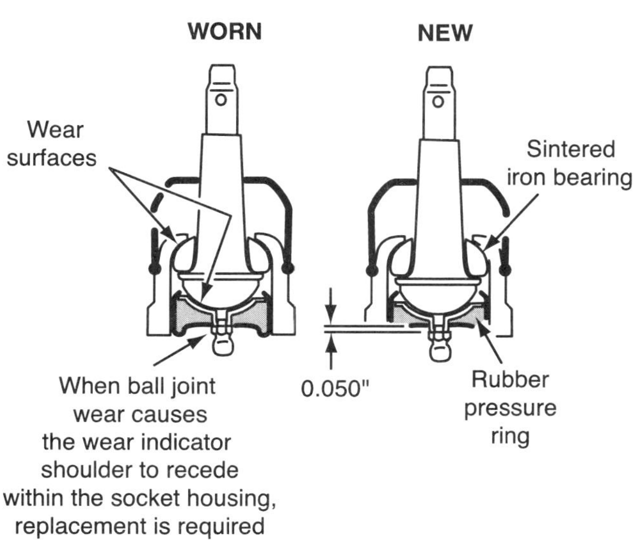


**Figure 5.** The position of the grease fitting indicates the condition of the ball joint.

Ball joints should be checked for excessive wear. Load-carrying joints will have some slope when the weight of the vehicle is taken off them. Follower joints should never have play. To check a load-carrying joint, it must be unloaded.

### Ball Joint Unloading

On many suspension systems, looseness is not apparent until the weight has been removed from the ball joint. When the coil spring is positioned between the lower control arm and chassis, place a floor jack near the outer end of the lower control arm and raise the tire off the floor **(Figure 6).** Be sure the rebound bumper does not contact the control arm or frame.

When the coil springs are positioned between the upper control arm and the chassis, place a steel block between the upper control arm and the frame. With this block in place, raise the tire off the floor with a floor jack under the front crossmember **(Figure 7).**

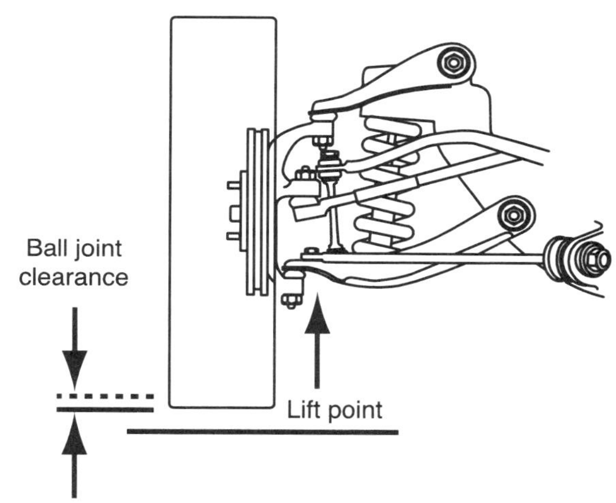


**Figure 6.** The position of the floor jack when checking ball joint wear when the vehicle's spring is between the lower control arm and the chassis.

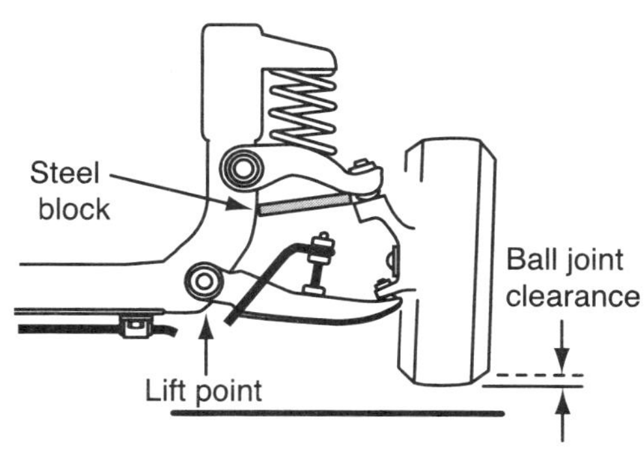


**Figure 7.** The position of the floor jack when checking ball joint wear when the vehicle's spring is between the upper control arm and the chassis.

## Ball Joint Vertical Measurements

The manufacturer may provide ball joint vertical and radial (horizontal) tolerances. A dial indicator is the most accurate ball joint measuring device. Always install the dial indicator at the recommended location for ball joint measurement. When measuring ball joint vertical movement in a compression-loaded ball joint, attach the dial indicator to the lower control arm and position its stem on the lower side of the steering knuckle beside the ball joint stud **(Figure 8).** Depress the stem approximately 0.250 in. (6.35 mm), and zero the dial indicator. Place a pry bar under the tire and pry upward while observing the vertical ball joint movement on the dial indicator. If this movement is more than specified, the ball joint should be replaced.

On a tension-loaded ball joint, clean the top end of the lower ball joint stud, and position the dial indicator stem against the top end of this stud. Depress the dial indicator plunger approximately 0.250 in. (6.35 mm), and zero the dial indicator. Lift upward with a pry bar under the tire and observe the dial indicator reading. If the vertical ball joint movement exceeds the manufacturer's specifications, replace the ball joint.

## Ball Joint Radial Measurements

Make sure the front wheel bearings are adjusted properly prior to measuring radial ball joint movement. Connect the dial indicator to the lower control arm of the ball joint being checked and position the dial indicator stem against the edge of the wheel's rim **(Figure 9).** While an assistant grasps the top and bottom of the raised tire and attempts to move the tire and wheel horizontally in and out, observe the reading on the dial indicator **(Figure 10)** and compare it to specifications. Excessive readings indicate the ball joint should be replaced.

Figure 8. The position of the dial indicator to measure vertical ball joint movement on a compression-loaded ball joint.

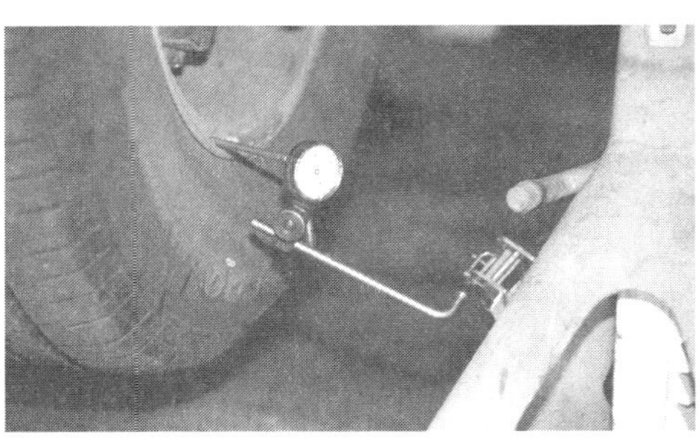

Figure 9. The position of the dial indicator to measure radial ball jooint movement.

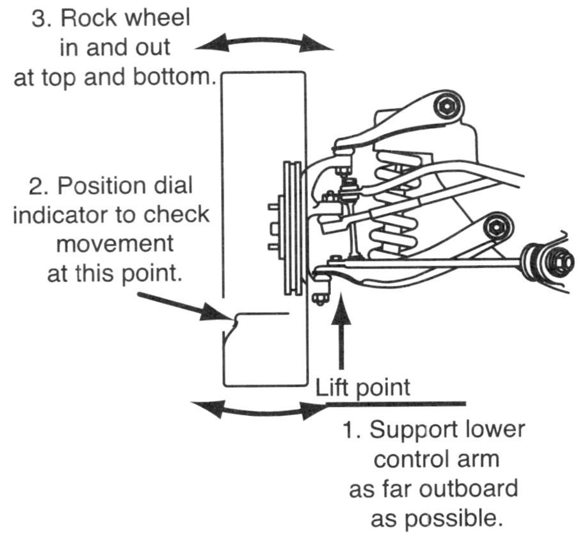


Figure 10. Measuring ball joint movement.

### MacPherson Strut Suspensions

The lower ball joint on a MacPherson strut front suspension should be checked for radial movement with a dial indicator when the tire is lifted off the floor. Grasp the front tire at the top and bottom and rock it inward and outward while a coworker visually checks for movement in the front wheel bearing. If there is movement in the front wheel bearing, adjust or replace the bearing.

Then position a dial indicator against the inner edge of the rim at the bottom **(Figure 11).** Push the bottom of the

Figure 11. To measure ball joint movement on a strut suspension, set the dial indicator so its plunger rests against the inner edge of the rim at the bottom.

**Figure 12.** With the indicator in place, push the bottom of the tire inward and set the dial indicator reading to zero.

**Figure 13.** With the indicator zeroed, grasp the bottom of the tire, pull it outward, and read the dial indicator.

tire inward **(Figure 12),** and set the dial indicator reading to zero. Now, grasp the bottom of the tire, pull outward, and read the dial indicator **(Figure 13).** If the dial indicator reading is more than specified, replace the lower ball joint.

### Twin I-Beam Axles

To check the ball joints on a twin I-beam axle **(Figure 14),** lift the front end of the vehicle with a floor jack and place safety stands near the outer end of the I-beams. While an assistant grasps the bottom of the wheel and moves it in and out, watch for movement between the lower part of the axle jaw and the spindle lower arm. If this movement exceeds 1/32 inch, replace the lower ball joint. As the assistant grasps the top of the wheel and moves it in and out, watch for movement between the upper axle jaw and the spindle upper arm. If this movement exceeds 1/32 inch, replace the upper ball joint.

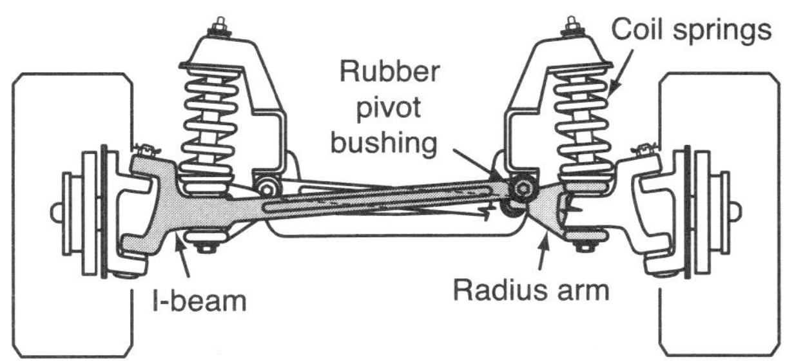


**Figure 14.** A twin I-beam suspension system.

## SUMMARY

- Begin diagnosis with gathering as much information from the customer as possible.
- Conduct a road test to confirm the customer's complaint.
- Noises can be caused by worn or broken bushings, grommets, insulators, and mounts.
- If the vehicle rolls to one side when traveling on a rough surface, check for a weakened or disconnected stabilizer bar and worn stabilizer bar bushings.
- When a vehicle is not stable when its direction is changed, check the vehicle for weak or broken springs or worn bushings and anchors.
- Begin a visual inspection by inspecting the tires.
- The curb riding height is determined mainly by spring condition.
- Shock absorbers should be inspected for loose mounting bolts and worn bushings.
- Inspect the control arms for damage, and check the control arm bushings for wear.
- Check all suspension bushings for looseness, splits, cracks, misplacement, and noises.
- The placement of the grease fitting on some ball joints may indicate whether the joint is worn or not.
- A damaged ball joint boot normally means the ball joint should be replaced.
- Load-carrying joints will have some free movement when the weight of the vehicle is taken off them. Follower joints should never have play. To check a load-carrying joint, it must be unloaded.
- Specifications may give ball joint vertical and radial (horizontal) tolerances. Both are measured with a dial indicator.

## REVIEW QUESTIONS

1. The shou der on the ball joint grease fitting is inside the ball joint cover. Technician A says the ball joint should be replaced. Technician B says a longer grease fitting should be installed. Who is correct?
   A. Technician A
   B. Technician B
   C. Both Technician A and Technician B
   D. Neither Technician A nor Technician B
2. While discussing ball joint unloading on a short/long arm suspension system with the coil springs between the lower control arm and the chassis, Technician A says a steel spacer should be installed under the upper control arm. Technician B says a floor jack should be placed under the lower control arm to unload the ball joints. Who is correct?
   A. Technician A
   B. Technician B
   C. Both Technician A and Technician B
   D. Neither Technician A nor Technician B
3. While diagnosing a customer's complaint of steering instability on a MacPherson strut front suspension with wear indicator-type ball joints, an inspection of the ball joints shows the grease fittings to be solid. Technician A says no movement of the fitting means the ball joint is good. Technician B says ball joint wear may not be apparent with the vehicle's weight on the joint. Who is correct?
   A. Technician A
   B. Technician B
   C. Both Technician A and Technician B
   D. Neither Technician A nor Technician B
4. A customer says her rear-wheel-drive car has excessive body sway while cornering. In diagnosing this problem, which of the following components should you check first?
   A. Control arm bushings
   B. Strut rod bushings
   C. Stabilizer bar bushings
   D. Shock absorber bushings
5. Technician A says a bent strut rod can cause a car to pull to one side. Technician B says a deteriorated strut rod bushing can cause steering and braking problems. Who is correct?
   A. Technician A
   B. Technician B
   C. Both Technician A and Technician B
   D. Neither Technician A nor Technician B

# Chapter 31 Front Suspension Service

## INTRODUCTION

The replacement of springs, shock absorbers, and struts is discussed in other chapters. This chapter covers the replacement and service of the other parts of a typical front suspension system. There are many variations in the configuration of a front suspension, therefore always refer to the appropriate service manual before replacing parts in the front suspension.

**You Should Know** *You and the owner of your shop are liable for any accident that results from service to the suspension and steering system. Perform all work with the customer's safety in mind.*

When working on a strut-type suspension, make sure you check the strut pivot bearing **(Figure 1).** Move the bearing with your hand. If the bearing is hard to move or seems to bind, it must be replaced. When replacing the bearing, make sure the correct side is up. Manufacturers normally mark the up side with paint or some other marking. Also, make sure you check all rubber insulators for deterioration and other damage and replace them if necessary.

Also, make sure you mark the eccentric camber bolts before loosening them. Returning the bolts in the same location will help maintain the correct camber angle after reassembly.

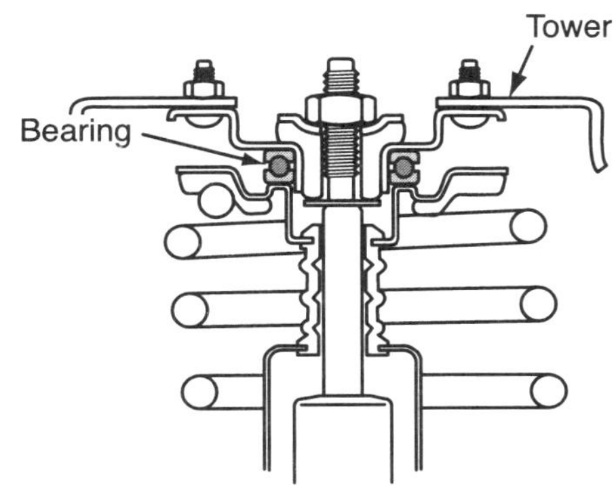


**Figure 1.** Location of the pivot bearing in a strut assembly.

## STEERING KNUCKLES

Although a steering knuckle is related to both the suspension and steering system, the knuckle is a key component of the suspension system. If it is damaged or worn, it must be replaced. Make sure to check the ball joint and tie-rod end openings in the knuckle for wear and out-of-round. Replace the knuckle if these conditions are present.

### SLA Front Suspension Systems

To replace a steering knuckle on an SLA front suspension system with the coil spring positioned between the lower control arm and the frame, remove the outer tie-rod cotter pin and nut and the outer tie-rod end from the steering arm. Then, remove the cotter pins from the upper and lower ball joint nuts and loosen (but do not remove) the nuts.

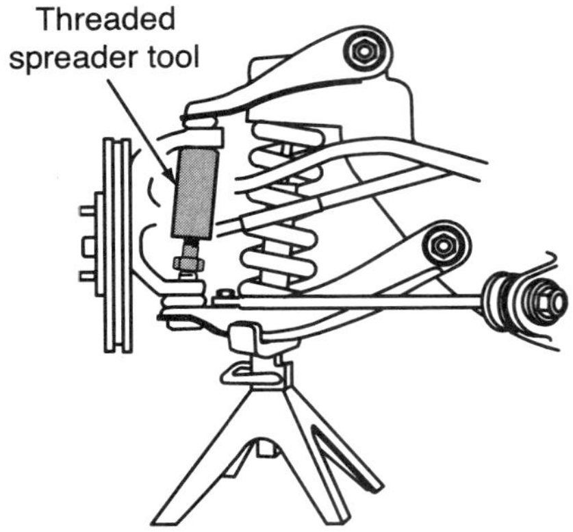


**Figure 2.** A threaded spreader tool may be required to push the upper and lower ball joints from the knuckle.

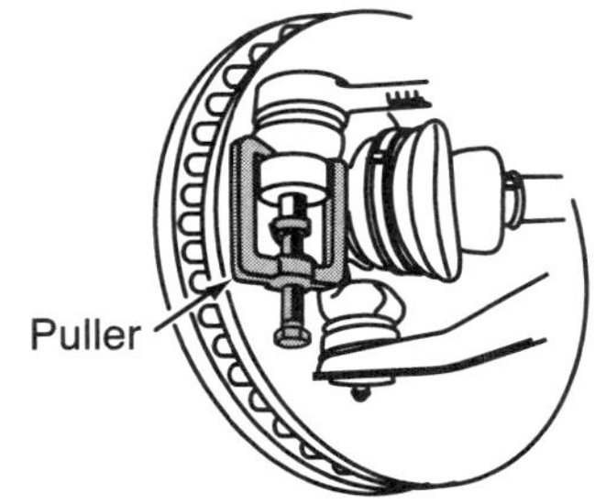


**Figure 3.** A special tool (puller) is used to disconnect the tie-rod end from the steering knuckle.

> **You Should Know** *During knuckle replacement, the floor jack should support the spring's tension when the ball joints are disconnected from the knuckle. Chain the coil spring to the lower control arm as a safety precaution.*

Support the lower control arm with a floor jack and raise the floor jack until the spring tension is supported by the jack. Remove both ball joint nuts and remove the ball joint from the steering knuckle. A threaded spreader tool may be used to push the upper and lower ball joints from the knuckle **(Figure 2).**

## Strut Front Suspension Systems

Prepare the vehicle by loosening the wheels' lug nuts. On some vehicles, it may be necessary to remove the drive axle, therefore the axle nut needs to be loosened before lifting some vehicles. Then raise the vehicle and remove the front wheels, brake calipers, and brake rotors. Remove the outer end of the drive axle from the steering knuckle and hub. On some models, a puller is required for this operation.

If necessary, remove the inner end of the drive axle from the transaxle with a pulling or prying action. On some transaxles, the differential cover must be removed and axle circlips compressed prior to drive axle removal from the transaxle. Always check the service manual before disconnecting the drive axle.

Make sure the vehicle's weight is supported on the hoist with the front suspension dropped downward. Remove the mounting bolts in the lower control arm. Then, remove the cotter pin from the tie-rod nut and the outer tie-rod nut. Use a puller to disconnect the tie-rod end from the steering knuckle **(Figure 3).** If an eccentric cam is used on one of the strut-to-knuckle bolts, mark the cam and bolt's position to the strut and remove the strut-to-knuckle bolts.

Remove the knuckle from the strut and lift the knuckle out. Pry the dust deflector from the knuckle with a large flat-bladed screwdriver. Use a puller to remove the ball joint from the steering knuckle.

Install the ball joint into the new steering knuckle. Use the proper driving tool to install the dust deflector in the steering knuckle. Then, fit the knuckle onto the strut. If an eccentric cam is used on one of the struts-to-knuckle bolts, align the eccentric and bolt to the marks made previously, then install and tighten the bolts.

Install the outer tie-rod end into the steering knuckle and tighten the nut. Install a new cotter pin in the tie-rod nut. Now install the mounting bolts in the lower control arm. Install the outer end of the drive axle into the steering knuckle and hub and the inner end into the transaxle.

Attach the brake rotor and caliper assembly. Install the front wheel and drive axle nut, then lower the vehicle and tighten the lug nuts and drive axle nut to specifications.

## BALL JOINTS

Ball joints may be threaded, pressed, bolted, or riveted into the control arms **(Figure 4).** The ball joint replacement procedure varies depending on the type of suspension and the method of ball joint attachment. Always follow the manufacturer's recommended procedure for replacing ball joints.

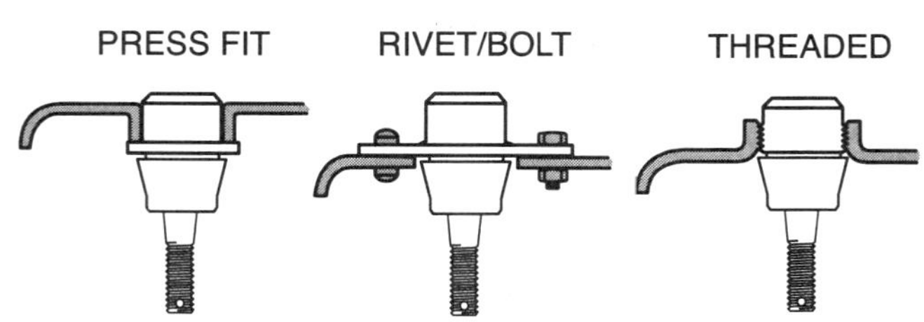


**Figure 4.** Ball joints may be threaded, pressed, bolted, or riveted into the control arms.

## Replacement Procedure

Remove the wheel and place a floor jack under the outer end of the lower control arm. With the floor jack, raise the lower control arm until the ball joints are unloaded. Remove other components such as the brake caliper, rotor, and drive axle as required to gain access to the ball joints. Then remove the cotter pin in the ball joint or joints that require replacement, and loosen, but do not remove, the ball joint stud nuts.

> You Should Know *Some ball joints have a tapered stud, and a nut is threaded onto the top of this stud to retain the ball joint in the steering knuckle. Other ball joints have a straight stud, and a pinch bolt extends through the steering knuckle and a notch in the side of the ball joint stud to hold the ball joint in the steering knuckle.*

Loosen the ball joint stud tapers in the steering knuckle. A threaded expansion tool is available for this purpose. Remove the ball joint nut and lift the knuckle off the ball joint stud. Block or tie up the knuckle and hub assembly to access the ball joint.

If the ball joint is riveted to the control arm, drill and punch out the rivets, and bolt the new ball joint to the control arm. If the ball joint is pressed into the lower control arm, remove the ball joint dust boot and use a pressing tool to remove and replace the ball joint **(Figure 5).** If the ball joint housing is threaded into the control arm, use a proper size socket to remove and install the ball joint. The replacement ball joint must be torqued to specifications. If a minimum torque cannot be obtained, the control arm threads are damaged and control arm replacement is necessary. If the ball joint is bolted to the lower control arm, install the new ball joint and tighten the bolt and nuts to the specified torque.

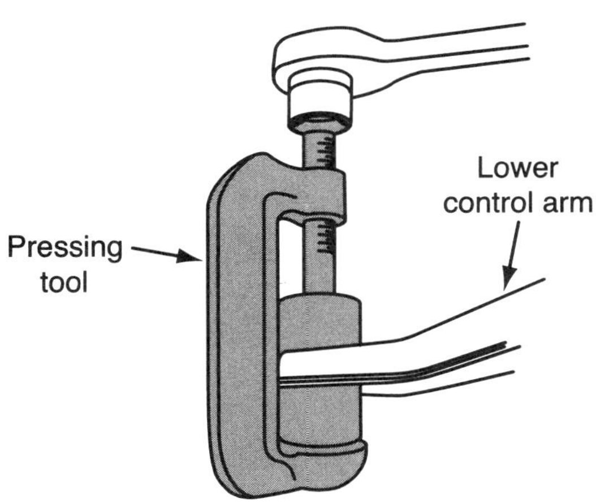


**Figure 5.** Removing a ball joint that is pressed into the control arm.

Clean and inspect the ball joint stud tapered opening in the steering knuckle. If this opening is out-of-round or damaged, the knuckle must be replaced. Check the fit of the ball joint stud in the steering knuckle opening. This stud should fit snugly in the opening, and only the threads on the stud should extend through the knuckle. If the ball joint stud fits loosely in the knuckle tapered opening, either this opening is worn or the wrong ball joint has been supplied.

Install the ball joint stud in the steering knuckle opening, making sure the stud is straight and centered. Install the stud nut and tighten this nut to the specified torque. Install a new cotter pin through the stud and nut. Do not loosen the nut to align the nut and stud openings.

Reassemble the components, making sure the wheel nuts are tightened to the specified torque. After ball joint replacement, the front suspension alignment should be checked.

> You Should Know *Never back off a ball joint stud to align the cotter pin openings in the nut and stud. This may cause the ball joint stud to loosen, resulting in a suspension failure. Always tighten the nut to the next hole to install the cotter pin.*

# CONTROL ARMS

Replacing an upper or lower control arm may be necessary if there was damage to or excessive wear on the original one. Often wear and damage are obvious during an inspection, other times it is evident when the ball joints or steering knuckle are being replaced. Always follow the manufacturer's recommended procedure for replacing control arms.

## Rebound Bumpers

> You Should Know *Excessively worn rebound bumpers indicate worn-out shock absorbers, reduced curb riding height, or driving continually on irregular road surfaces.*

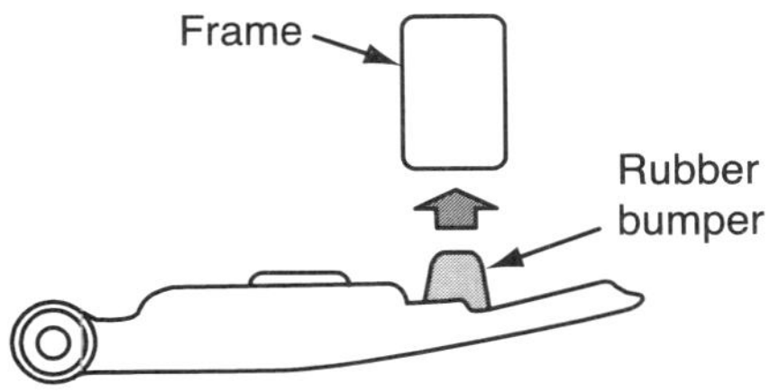


**Figure 6.** During an inspection of a front suspension, the rebound bumpers should be carefully checked.

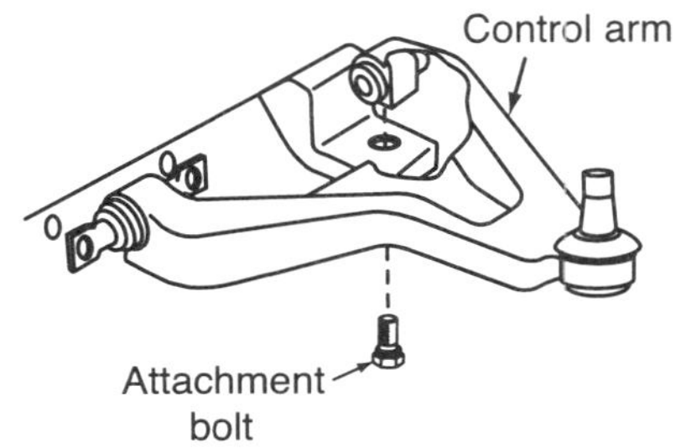


**Figure 8.** Remove the bolt and nut from the rear side of the lower control arm.

If, during the inspection of the suspension, the rebound bumpers **(Figure 6)** are found to be worn, they should be replaced. These can be replaced without removing or replacing the control arms. Rebound bumpers are usually bolted to the lower control arm or to the chassis. Inspect the rebound bumpers for cracks, wear, and flattened conditions. If the rebound bumpers must be replaced, remove the mounting bolts and the bumper. Install the new bumper and tighten the mounting bolts to the specified torque.

## Control Arm Replacement on Strut Suspensions

A strut-type suspension system has one control arm, a lower one, on each side. Typically to replace this arm, lift the front of the vehicle with a floor jack, and place jack stands under the chassis. Lower the vehicle onto the stands and allow the front suspension to drop downward. Remove the front wheel and the front fender apron seal, if necessary.

Remove the cotter pin and nut from the outer tie-rod end, and use a puller to remove the tie-rod end from the steering arm **(Figure 7).** Now remove the stabilizer brackets from the lower control arm. Remove the lower ball joint nuts and bolt from the lower control arm.

Remove the two attaching bolts at the front side of the lower control arm. Now remove the bolt and nut on the rear side of the lower control arm **(Figure 8),** and remove the lower arm bushing stopper from the lower arm shaft.

**Figure 7.** Removing the outer tie-rod end.

To install a new control arm, reverse the procedure for removing it. Make sure to tighten all bolts and nuts to the specified torque, and install new cotter pins as required.

## Lower Control Arm Replacement on SLA Suspensions

**You Should Know** *During control arm and spring replacement, the coil spring tension must be supported by a compressing tool or floor jack. Serious personal injury or property damage may occur if the spring tension is released suddenly.*

To replace a lower control arm on a vehicle with an SLA suspension and coil with the coil springs positioned between the lower control arm and the frame, lift the vehicle until the tires are a short distance off the floor, and allow the front suspension to drop downward. If the vehicle is raised with a floor jack, make sure the vehicle is securely supported by safety stands.

Disconnect the lower end of the shock absorber. On some applications, the shock absorber must be removed. Then disconnect the stabilizer bar from the lower control arm.

Install a spring compressor **(Figure 9),** and tighten it until the spring is compressed. Make sure all of the spring tension is supported by the compressing tool.

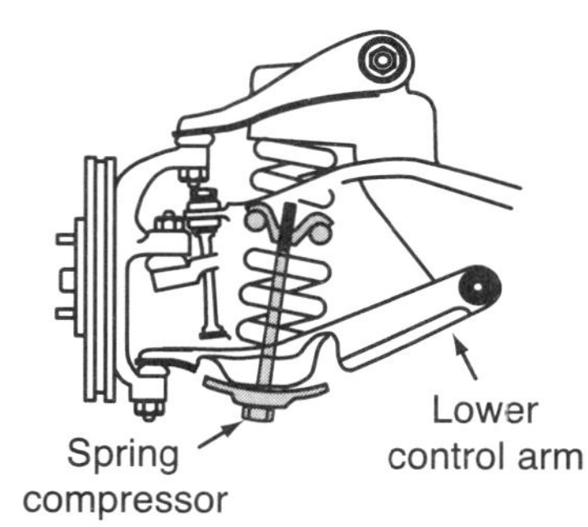


**Figure 9.** A spring compressor installed on an SLA suspension system.

Place a floor jack under the lower control arm, and raise the control arm. Then, remove the lower ball joint cotter pin and nut, and use a threaded expansion tool to loosen the lower ball joint stud. Now slowly lower the floor jack to allow the control arm and coil spring to drop slowly. Once the vehicle's weight is off the control arm, remove the lower control arm inner mounting bolts, and remove the lower control arm.

Rotate the spring compressing tool bolt to release the spring's tension, and remove the spring from the control arm. Inspect all control arm bushings. Loose or worn bushings must be replaced. Check the upper and lower spring insulators for cracks and wear, and inspect the spring seat areas in the chassis and lower control arm. Worn or cracked insulators must be replaced.

To reassemble the suspension, reverse the procedure for disassembly. Make sure the coil spring and insulators are properly seated in the lower control arm and in the upper spring seat, and that all bolts are tightened to specifications. After the suspension is reassembled, check and correct the alignment angles of the front suspension.

## Upper Control Arm Replacement on SLA Suspensions

To replace an upper control arm on a vehicle with an SLA suspension and coil with the coil springs positioned between the lower control arm and the frame, lift the vehicle until the tires are a short distance off the floor, and allow the front suspension to drop downward. If the vehicle is raised with a floor jack, make sure the vehicle is securely supported by safety stands.

Remove the front wheel and tire. Then, place a floor jack under the lower control arm it. Remove the cotter pin and loosen the upper ball joint nut, but do not remove it. Use a threaded expansion tool to loosen the ball joint stud in the control arm. Make sure the floor jack is supporting the spring tension.

**You Should Know** ***Since the floor jack is supporting the spring tension, do not lower the floor jack until the ball joint is reconnected and the ball joint nut is tightened to the specified torque.***

Remove the ball joint nut and the inner control arm mounting bolts. If shims are located on the inner control arm mounting bolts, note the positions of the shims **(Figure 10).** Then, remove the control arm from the chassis. Inspect the control arm shaft and bushings for wear, and replace all worn parts.

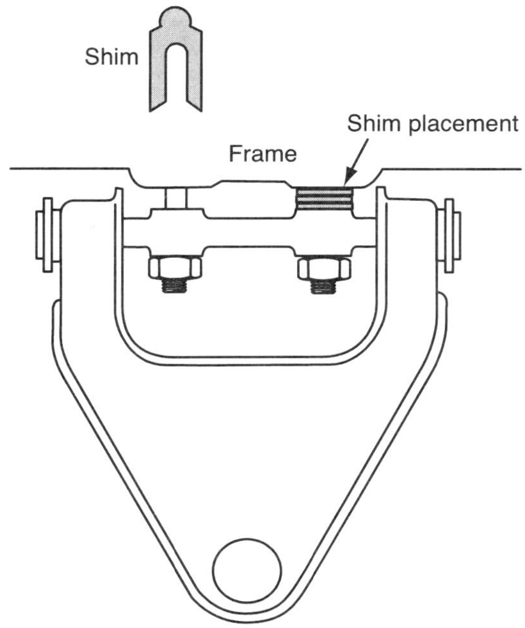


**Figure 10.** If shims are located on the inner control arm mounting bolts, note their positions so they can be installed in the same location during reassembly

Install the control arm and place the original number of shims on the inner mounting bolts. Tighten these bolts to specifications. Install the ball joint stud into the steering knuckle and tighten the nut to the specified torque. Next, install the cotter pin into the upper ball joint. Install the front wheel and tighten the lug nuts or bolts to the specified torque. After the suspension is reassembled, check and correct the alignment angles of the front suspension.

## CONTROL ARM BUSHINGS

All suspension bushings should be checked for wear, looseness, and deterioration. Bushings are important for providing quiet suspension operation and preventing the transmission of vibrations to the chassis and passenger compartment **(Figure 11).** Bushings also are critical to

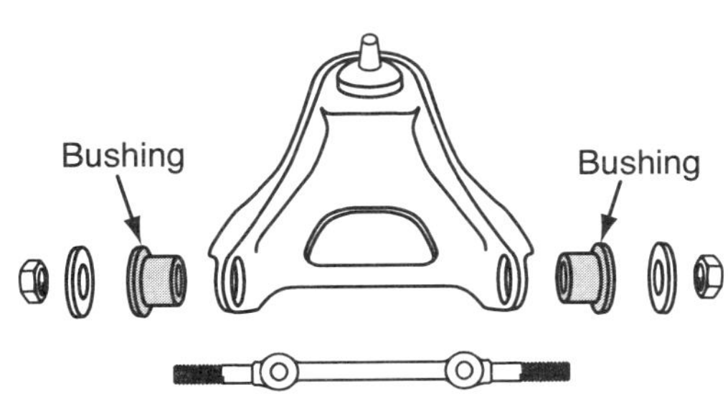


**Figure 11.** A control arm bushing and shaft assembly.

maintaining proper wheel alignment. Control arm bushing removal and replacement varies with the type of suspension system. Always follow the manufacturer's recommended procedure.

## Bushing Replacement

Bushings are pressed in and out of their bores. Special tools are required to this. The removal tool is installed over the bushing after the correct size tool has been selected. Tightening the tool pushes the bushing out of the control arm. A spacer is installed between the bushing support lugs to prevent distorting these lugs during the bushing removal process. Some bushings are threaded and are simply turned in and out of their bores **(Figure 12).**

The same process is used to press a new bushing into the control arm. As the installation tool is tightened, the bushing moves into its bore. When installing new bushings, make sure they stay straight while they are being pressed in. The same spacer that is used during the bushing removal is installed between the bushing lugs during bushing installation.

When a bushing is contained in a steel sleeve, press only on the outer sleeve during the removal and replacement procedure.

After the front control arm bushing is installed, a bushing flaring tool is used to flare the bushing **(Figure 13).** Flaring retains the bushing securely **(Figure 14).** Notice that the spacer is also installed between the bushing lugs during the flaring process.

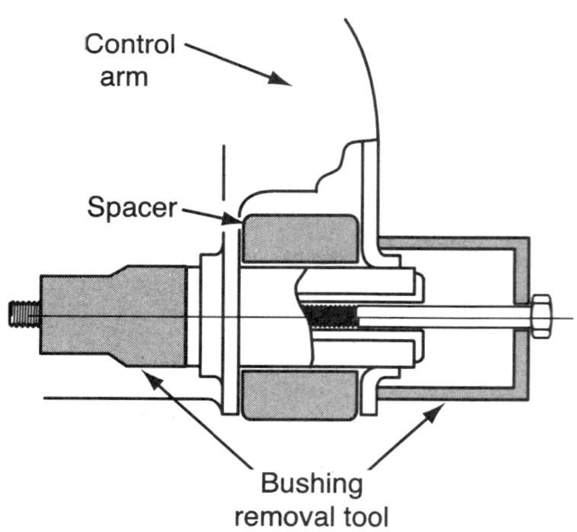


**Figure 12.** Removing a control arm bushing.

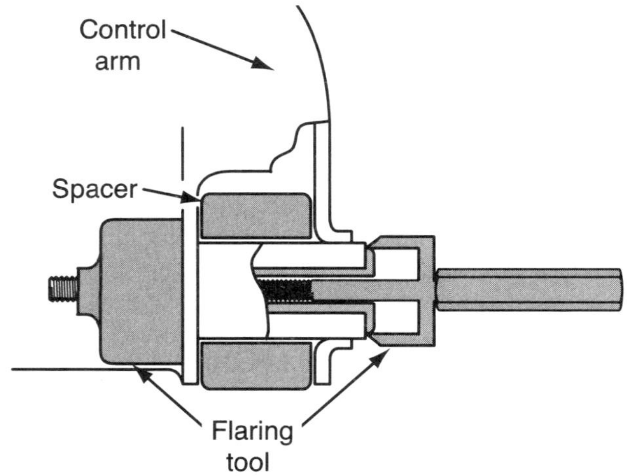


**Figure 13.** A control arm bushing flare tool.

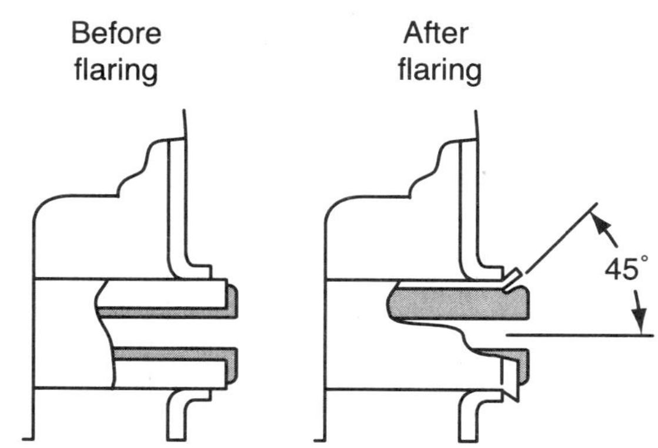


**Figure 14.** The effects of flaring a control arm bushing.

## STRUT RODS

To replace a strut rod, lift the vehicle on a hoist. Then, remove the strut rod nut from the front end of the rod and the strut rod bolts from the lower control arm. Now pull the strut rod rearward to remove the rod. Remove the bushings from the opening in the chassis.

Visually inspect the strut rod, bushings, washers, and retaining bolts. Replace all worn parts. Reverse the removal procedure to reinstall the strut rod. Tighten the strut rod nut and bolts to the specified torque. Since strut rod and bushing conditions affect front caster, check the alignment of the front suspension after strut rod service.

## SUMMARY

- If the openings in a steering knuckle are worn or out-of-round, it should be replaced.
- When removing a steering knuckle on an SLA suspension system, a threaded spreader tool may be used to push the upper and lower ball joints from the knuckle.
- Ball joints may be threaded, pressed, bolted, or riveted into the control arms.
- Some ball joints have a tapered stud, and a nut is threaded onto the top of this stud to retain the ball joint in the steering knuckle. Other ball joints have a straight stud, and a pinch bolt extends through the steering knuckle and a notch in the side of the ball joint stud to hold the ball joint in the steering knuckle.
- Replacing an upper or lower control arm may be necessary if there was damage to or excessive wear on the original one. Often, wear and damage are obvious during an inspection, other times it is evident when the ball joints or steering knuckle are being replaced.
- Excessively worn rebound bumpers indicate worn-out shock absorbers, reduced curb riding height, or driving continually on irregular road surfaces.
- A strut-type suspension system has one control arm, a lower one, on each side.
- To replace a lower control arm on a vehicle with an SLA suspension and coil spring, the control and spring must be removed.
- To replace an upper control arm on a vehicle with an SLA suspension and coil with the coil springs positioned between the lower control arm and the frame, the control arm must be removed while a floor jack is supporting the tension of the spring.
- All suspension bushings should be checked for wear, looseness, and deterioration. Bushings are important for maintaining wheel alignment, providing quiet suspension operation, and preventing the transmission of suspension vibration to the chassis and passenger compartment.
- Bushings are pressed in and out of their bores with special tools.

## REVIEW QUESTIONS

1. *True or False:* The spring must be compressed when replacing a rebound bumper on a lower control for a SLA suspension system.
2. *True or False:* The condition of the strut rod and its bushings can affect front wheel alignment.
3. While discussing ball joint installation, Technician A says the ball joint nut may be backed off to install the cotter pin. Technician B says only the ball joint threads should extend above the opening in the steering knuckle. Who is correct?
   A. Technician A
   B. Technician B
   C. Both Technician A and Technician B
   D. Neither Technician A nor Technician B
4. When installing a new threaded ball joint in a lower control arm, the technician can only torque the ball joint to 90 ft-lbs. The necessary repair for this problem is to:
   A. weld the ball joint into the control arm.
   B. place Loctite on the ball joint threads.
   C. replace the lower control arm.
   D. install a larger diameter ball joint.
5. When removing and replacing control arm bushings,
   A. if the bushing is contained in a steel sleeve, press on the rubber bushing.
   B. a spacer should be installed between the control arm support lugs during bushing removal and installation.
   C. the same tool is used for bushing removal and replacement.
   D. bushing flaring is necessary to remove the bushing.

# Chapter 32 Rear Suspension Diagnosis

## INTRODUCTION

The rear suspension has an important role in maintaining vehicle safety, ride quality, and tire life. Proper diagnosis results from a solid understanding of the problem, logical thinking, a complete road test, and inspection of the key components.

## DIAGNOSIS

Diagnosis is looking at systems that are not working properly and finding out why. Through an understanding of the purpose and operation of the system, you can accurately diagnose problems.

There are diagnostic aids in service manuals. These are either symptom based or flow charts. Flow charts guide you through a step-by-step process. As you answer the questions given at each step, you are told what your next step should be. Symptom-based diagnostic charts focus on an explanation of the problem and offer the possible causes of the problem. Sometimes the diagnostic aids are a combination of the two—a flow chart based on clearly defined symptoms.

When these diagnostic aids are not available or prove to be ineffective, most good technicians conduct a visual inspection and then take a logical approach to finding the cause of the problem. This approach relies on critical thinking skills as well as system knowledge. Logical diagnosis follows these steps:

1. Gather information about the problem. Find out when and where the problem happens and what exactly happens.
2. Verify that the problem exists. Take the vehicle for a road test and try to duplicate the problem if possible.
3. Thoroughly define what the problem is and when it occurs. Pay strict attention to the conditions present when the problem happens. Also pay attention to the entire vehicle; another problem may be evident to you that was not evident to the customer.
4. Research all available information and knowledge to determine the possible causes of the problem. Try to match the exact problem with a symptoms chart or think about what is happening and match a system or some components to the problem.
5. Isolate the problem by testing. Narrow down the probable causes of the problem by checking the obvious or easy-to-check items.
6. Continue testing to pinpoint the cause of the problem. Once you know where the problem should be, test until you find it!
7. Locate and repair the problem, then verify the repair. Make sure the problem is history before returning it to the customer.

### Vibration Diagnosis

Vibrations are diagnosed with the same basic diagnostic procedure that is used for diagnosing defects in other parts of the vehicle. To aid in diagnostics, try to gather the answers to the following questions when interviewing the customer **(Figure 1):**

1. At what vehicle speed is the vibration most noticeable?
2. Where the vibration is felt?
3. Is there a sound associated with the vibration?

**Figure 1.** Thoroughly interview the customer to understand his or her concern.

4. Does engine load affect the vibration, such as acceleration?
5. Does vehicle load (number of passengers, weight in the trunk) affect the vibration?
6. Does the vibration occur in all transmission drive ranges?
7. When was the vibration first noticed?
8. Did the vibration occur right after another service was performed on the vehicle?

## Road Test

Conduct a road test to verify the customer's concern and to observe any and all operating conditions that exist when the problem is evident. Make sure you drive safely and obey all laws.

## Diagnosis of Harsh Ride

A harsh ride can be caused by reduced curb riding height **(Figure 2)** which may be caused by sagged springs, worn control arm bushings, or bent control arms. A harsh ride may also be caused by worn-out struts or shock absorbers that cause excessive chassis oscillations.

## Noise Diagnosis

A squeaking noise in the rear suspension may be caused by a worn or dry suspension bushing **(Figure 3)** or a defective strut or shock absorber. Worn spring silencers or pads can also be the cause of the squeaking noise.

A rattling noise from the rear suspension may be caused by worn or missing suspension bushings such as control arm, tie-rod, strut or shock absorber, track bar, stabilizer bar, trailing arm, or strut rod bushings. Rattles can also be caused by broken springs or spring insulators, or worn shackles and/or brackets.

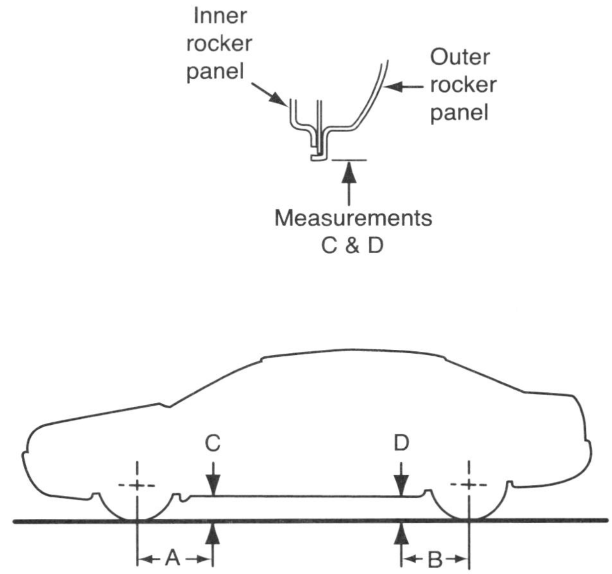


**Figure 2.** Typical points for checking curb riding height.

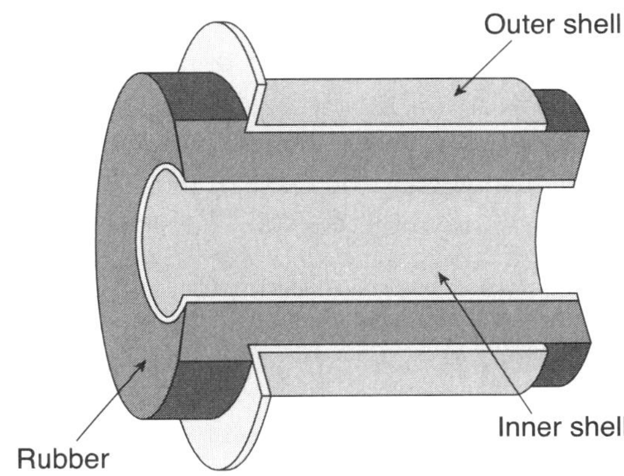


**Figure 3.** The part of a typical suspension bushing.

## Sway and Lateral Movement Diagnosis

Excessive body sway or roll when one wheel strikes a road irregularity may be caused by a weak stabilizer bar or loose stabilizer bar bushings. Body sway can also be caused by worn spring bushings, shackles, and/or brackets.

If lateral (sideways) movement is experienced on the rear of the chassis, the track bar or track bar bushings may be defective.

# LOWER CONTROL ARM AND BALL JOINT DIAGNOSIS

Worn bushings on the lower control arms may cause incorrect rear wheel camber or toe, which results in rear tire

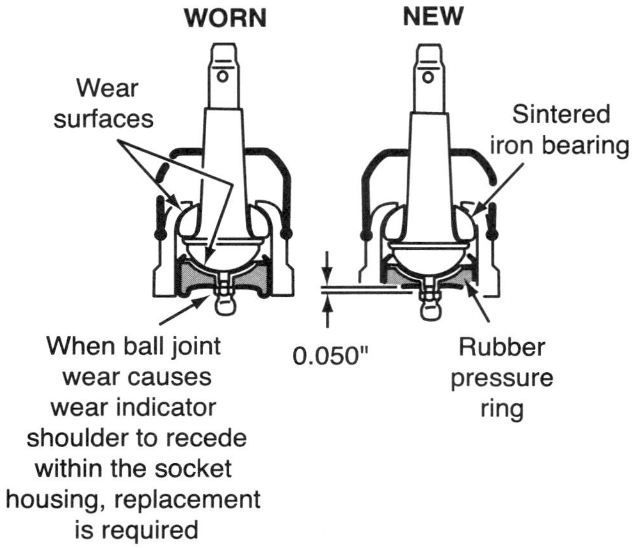


**Figure 4.** When ball joints with wear indicators are in normal condition, there is 0.050 in. (1.27 mm) between the grease nipple shoulder and the cover.

wear and steering pull. Bent lower control arms must be replaced. When ball joints with wear indicators are in normal condition, there is 0.050 in. (1.27 mm) between the grease nipple shoulder and the cover **(Figure 4).** If the ball joint is worn, the grease nipple shoulder is flush with or inside the cover. A worn ball joint causes improper rear wheel toe and/or camber, which may result in tire tread wear or steering pull.

## TRACK BAR DIAGNOSIS

Some rear suspension systems have a track bar to control lateral chassis movement. Rubber mounting bushings insulate the track bar from the chassis components. Worn track bar mounts and bushings may cause rattling and excessive lateral chassis movement.

To check a track bar, lift the vehicle on a hoist or floor jack with the rear suspension at its normal riding height. If a floor jack is used, place safety stands under the rear axle to support the vehicle's weight. Grasp the track bar firmly and apply vertical and horizontal force. If there is movement in the track bar mountings, track bar, or bushing, replacement is essential.

Another track bar checking method is to leave the vehicle on the shop floor and observe the track bar mounts as an assistant applies side force to the chassis or rear bumper. If there is lateral movement in the track bar bushings or brackets, replace the bushings and check the bracket bolts. Bent track bars must be replaced.

### Stabilizer Bar Diagnosis

Worn stabilizer bar mounting bushings, grommets, or mounting bolts cause a rattling noise as the vehicle is driven on irregular road surfaces **(Figure 5).** A weak stabilizer bar or worn bushings and grommets cause harsh riding and excessive body sway while driving on irregular road surfaces. Worn or very dry stabilizer bar bushings may cause a squeaking noise on irregular road surfaces. All stabilizer bar components should be visually inspected for wear.

## TIE-ROD INSPECTION

Rear tie rods, sometimes referred to as adjustment links, should be inspected for worn grommets, loose mountings, and bent conditions **(Figure 6).** Loose tie-rod bushings or a bent tie rod will change the rear wheel tracking and result in reduced directional stability. Worn tie-rod bushings also cause a rattling noise on road irregularities.

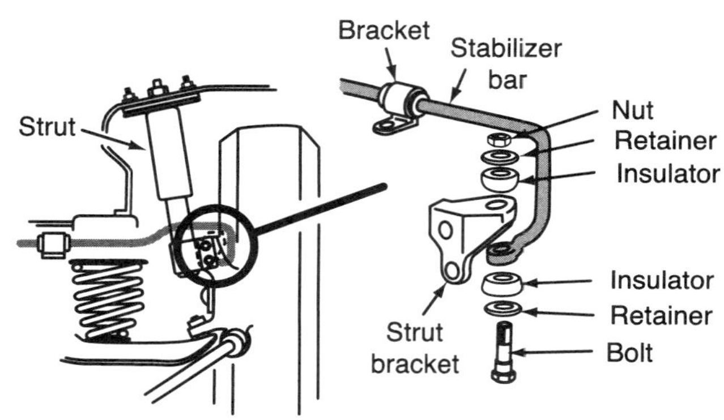


**Figure 5.** The rear stabilizer bar assembly should be carefully inspected.

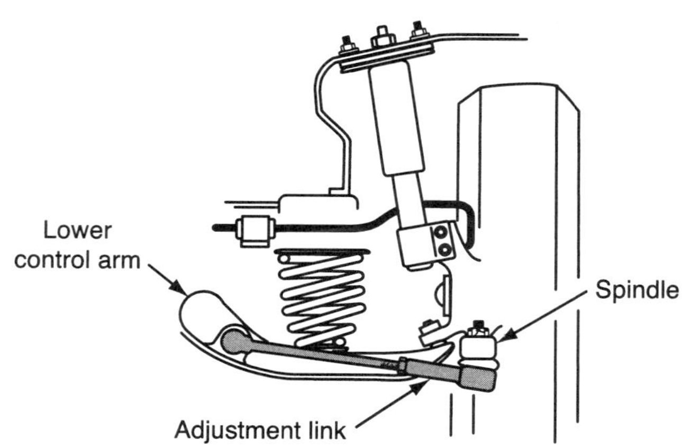


**Figure 6.** An adjustment link (tie rod) for a rear suspension.

## SUMMARY

- With an understanding of the purpose and operation of the system, you can accurately diagnose problems.
- Good technicians conduct a visual inspection and take a logical approach to find the cause of a problem.
- Diagnosis should begin with gathering as much information as possible from the customer.
- A harsh ride can be caused by reduced curb riding height and worn-out struts or shock absorbers.
- A squeaking noise in the rear suspension may be caused by a worn or dry suspension bushing, a defective strut or shock absorber, or worn spring silencers or pads.
- A rattling noise may be caused by worn or missing suspension bushings, broken springs or spring insulators, or worn shackles and/or brackets.
- Excessive body sway may be caused by a weak stabilizer bar or loose stabilizer bar bushings, worn spring bushings, shackles, and/or brackets.
- Lateral movement can be caused by a worn or damaged track bar or track bar bushings.

## REVIEW QUESTIONS

1. While discussing lateral movement of the rear chassis, Technician A says this problem may be caused by loose stabilizer bar bushings. Technician B says this problem may be caused by loose track bar bushings. Who is correct?
   A. Technician A
   B. Technician B
   C. Both Technician A and Technician B
   D. Neither Technician A nor Technician B
2. While discussing excessive rear body sway or roll when driving on road irregularities, Technician A says this problem may be caused by worn spring insulators. Technician B says this problem may be the result of sagged rear springs. Who is correct?
   A. Technician A
   B. Technician B
   C. Both Technician A and Technician B
   D. Neither Technician A nor Technician B
3. Name five problems that could cause a rattle from the rear suspension.
4. The most likely cause of excessive rear chassis lateral movement is/are:
   A. a weak stabilizer bar.
   B. worn track bar bushings.
   C. worn out struts.
   D. sagged rear coil springs.
5. A squeak in the rear suspension could be caused by all of the following *except:*
   A. the suspension bushing.
   B. weak spring leaves.
   C. worn spring anti-friction pads.
   D. a defective shock absorber.

# Chapter 33 Rear Suspension Service

## INTRODUCTION

Many rear suspension components can create a safety hazard if they are not serviced properly. Always follow the appropriate service instructions for the vehicle being serviced. After servicing the rear suspension, rear wheel toe and camber must be adjusted to specifications to provide normal tire tread life.

> **You Should Know** *You and the owner of your shop are liable for any accident that results from service to the suspension and steering system. Always follow the recommended procedures.*

Some independent rear suspension systems have ball joints that perform a similar function as the front ball joints, and they should be inspected in the same way. Also, some independent suspension systems have components that would normally be seen only on vehicles with four-wheel-steering **(Figure 1).** Components such as tie-rod ends may appear and serve the same purpose as if they were on the front suspension. They are used to adjust the angle of the wheels for stable straight-ahead performance.

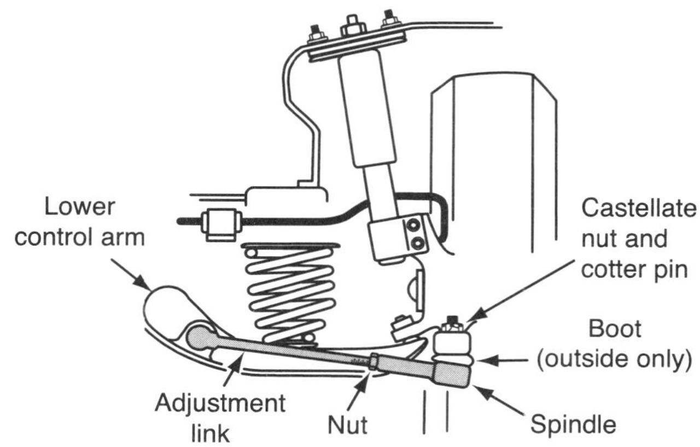

**Figure 1.** An IRS with control arms, ball joints, and adjustment links (tie rods).

## LOWER CONTROL ARM AND BALL JOINTS

The procedure for removal and installation of the lower control arm varies with the type of vehicle and suspension. Always follow the recommended procedure for the particular vehicle.

### Removal

The following is a typical lower control arm removal procedure:

1. Lift the vehicle on a hoist with the chassis supported on the hoist and the control arms dropped downward. The vehicle may be lifted with a floor jack, and the chassis supported on safety stands.
2. Remove the wheel and tire assembly.
3. Remove the stabilizer bar from the knuckle bracket.
4. Remove the parking brake cable retaining clip from the lower control arm.

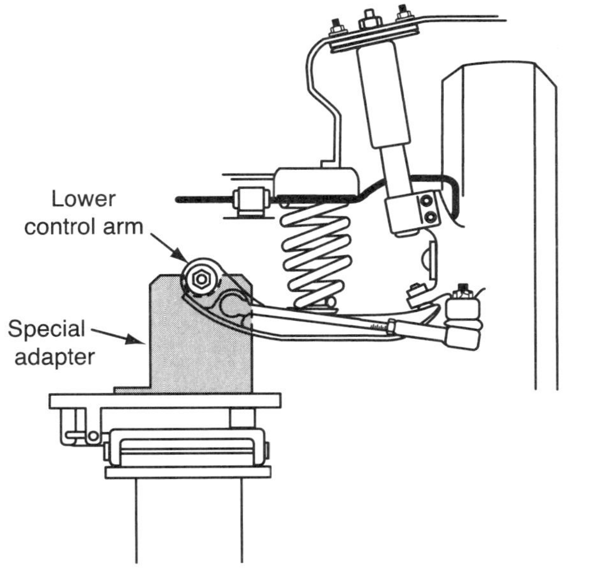


**Figure 2.** A special adapter should be used to support the lower control arm.

5. If the car has ELC, disconnect the height sensor link from the control arm.
6. Install a special tool to support the lower control arm **(Figure 2).**
7. Place a transmission jack under the special tool and raise the jack enough to remove the tension from the control arm bushing retaining bolts. If the car was lifted with a floor jack and supported on safety stands, place a floor jack under the special tool.
8. Place a safety chain through the coil spring and around the lower control arm.
9. Remove the bolt from the rear control arm bushing.
10. Be sure the jack is raised enough to relieve the tension on the front bolt in the lower control arm and remove this bolt.
11. Slowly lower the jack and allow the control arm to pivot downward. When the tension is released from the spring, remove the safety chain, coil spring, and insulators.
12. Check the spring for distortion and proper free length. If the spring has a vinyl coating, check it for scratches and nicks. Check the spring insulators for cracks and wear.
13. Proceed to remove the nut on the inner end of the suspension adjustment link, and disconnect this link from the lower control arm **(Figure 3).**
14. Remove the cotter pin from the ball joint nut, and loosen, but do not remove, the nut from the ball joint stud.
15. Use a special ball joirt removal tool to loosen the ball joint in the knuckle.

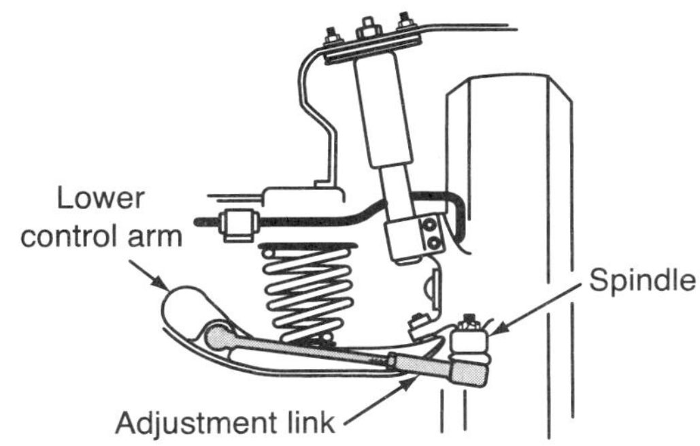


**Figure 3.** The adjustment link is connected between the spindle and the inner end of the control arm.

16. Remove the ball joint nut and the lower control arm.
17. Check the lower control arm for bends, distortion, and worn bushings.
18. A special ball joint pressing tool is used to press the ball joint from the lower control arm **(Figure 4),** and the same pressing tool with different adapters is used to press the new ball joint into the control arm.

**You Should Know** *When disconnecting a linkage joint, such as the ends of the suspension adjustment link, do not use a wedge-type tool (also known as a pickle fork) between the joint and the attached part. This method of joint removal may damage the joint seal.*

**Figure 4.** Removing a ball joint from the lower control arm.

19. Remove the nut and cotter pin on the outer end of the suspension adjustment link, and use a special puller to remove this link from the knuckle.
20. Remove the suspension adjustment link and inspect the joints. If the joints are loose or the seals are damaged, joint replacement is necessary. The joint studs must fit snugly in the knuckle and lower control arm openings. When the joint studs are worn, joint replacement is necessary, and worn stud openings in the knuckle or lower control arm require component replacement.

## Installation

The following is a typical lower control arm installation procedure:

1. Install the ball joint stud in the knuckle and install the nut on the ball joint stud. Tighten the ball joint nut to the specified torque, and then tighten the nut an additional two thirds turn. If necessary, tighten the nut slightly to align the nut castellations with the cotter pin hole in the ball joint stud, and install the cotter pin **(Figure 5).**
2. Snap the upper insulator on the coil spring. Install the lower spring insulator and the spring in the lower control arm **(Figure 6).**
3. Be sure the top of the coil spring is properly positioned.
4. Install the special tool on the inner ends of the control arm, and place a transmission or floor jack under the special tool.
5. Slowly raise the transmission jack until the control arm bushing openings are aligned with the openings in the chassis.

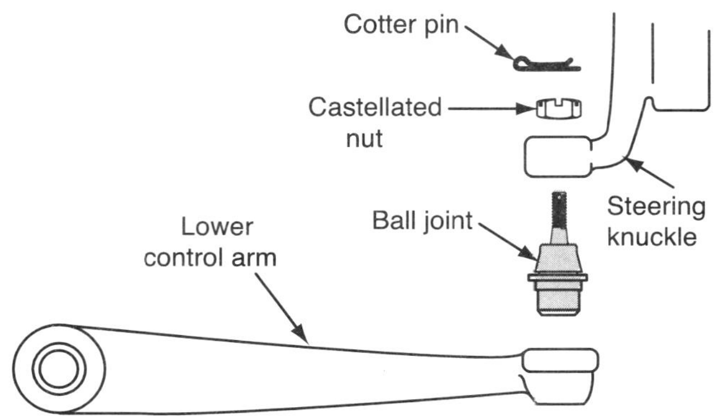


**Figure 5.** Install the ball joint and secure it in place, after torquing, with a new cotter pin.

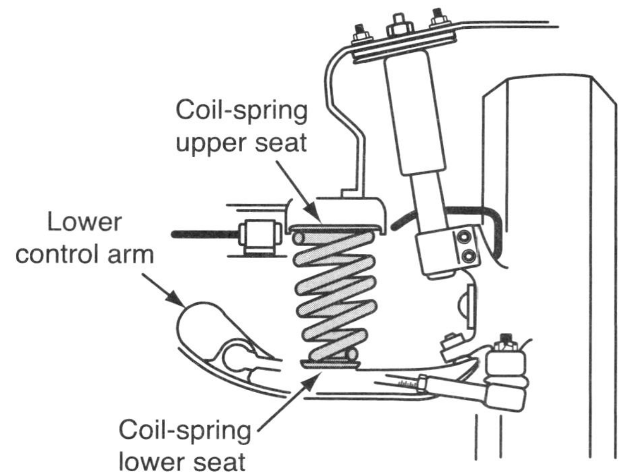


**Figure 6.** Make sure the spring is properly positioned with its insulators.

**You Should Know** *The pivot bolts and nuts at the inner ends of the lower control arm must be tightened to specifications with the vehicle's weight on its wheels and the suspension at normal curb height. Failure to do this may affect ride quality and steering characteristics.*

6. Install the bolts and nuts in the inner ends of the control arm. Do not torque them at this time.
7. Install the stabilizer bar-to-knuckle bracket and tighten the fasteners to the specified torque.
8. Install the parking brake retaining clip.
9. If the vehicle has ELC, install the height sensor link, and tighten the fastener to the specified torque.
10. Install the suspension adjustment link and tighten the retaining nuts to the specified torque. Install cotter pins as required.
11. Remove the transmission jack and install the wheel-and-tire assembly.
12. Lower the vehicle onto the floor and tighten the wheel hub nuts and lower control arm bolts and nuts to the specified torque.

**You Should Know** *When the rear coil springs are mounted on the struts, never loosen the upper strut nut until all the spring tension is removed from the upper support with a spring compressor. If this nut is removed with spring tension on the upper support, the spring tension turns the spring and the upper support into a dangerous projectile, which may cause serious personal injury and/or property damage.*

## TRACK BAR REPLACEMENT

When the track bar is replaced, remove the mounting bolts, bushings, grommets, and track bar. Inspect the mounting bolt holes in the chassis for wear. After the track bar is installed with the proper grommets and bushings, tighten the mounting bolts to the specified torque.

**You Should Know** ***On rear suspension systems with an inverted U-channel, the stabilizer bar inside the U-channel sometimes breaks away where it is welded to the end plate in the U-channel. This results in a rattling, scraping noise when the car is driven over road irregularities.***

## TIE ROD REPLACEMENT

Rear tie rods should be inspected for worn grommets, loose mountings, and bent conditions. Loose tie-rod bushings or a bent tie rod will change the rear wheel tracking and result in reduced directional stability. Worn tie-rod bushings also cause a rattling noise on road irregularities. When the rear tie rod is replaced, remove the front and rear rod mounting nuts. The lower control arm or rear axle may have to be pried rearward to remove the tie rod. Check the tie-rod grommets and mountings for wear, and replace parts as required. When the tie rod is reinstalled, tighten the mounting bolts to specifications, and check the rear wheel toe.

# SUMMARY

- A safety hazard can result from incorrectly servicing suspension parts.
- After servicing the rear suspension, rear wheel toe and camber must be adjusted to specifications to provide normal tire tread life.
- To replace or service a control arm and ball joint, the spring must be removed first.
- The pivot bolts and nuts at the inner ends of the lower control arm must be tightened with the vehicle's weight on its wheels and the suspension at normal curb height.

# REVIEW QUESTIONS

1. Why should you never loosen a ball joint nut when installing a cotter pin?
2. What is the purpose of rear tie rods on a vehicle that does not have four-wheel steering?
3. What problem can result from a bent tie rod?
4. All these statements about rear suspension systems with lower control arms and ball joints are true *except:*
   A. Worn lower control arm bushings do not affect tire tread wear.
   B. Some rear suspension ball joints have wear indicators.
   C. On some rear suspensions, the ball joint is pressed into the lower control arm.
   D. On some vehicles, the vehicle's weight must be resting on the tires when the inner, lower control arm bolts are tightened.
5. Rear suspension adjustment links:
   A. may be lengthened or shortened to adjust rear wheel camber.
   B. are connected from the lower control arm to the knuckle.
   C. may be loosened on each end with a wedge-type tool.
   D. prevent fore-and-aft control arm and wheel movement.

# Chapter 34 Electronic Suspension Service

## INTRODUCTION

Most electronic suspension servicing requires special procedures for individual component replacement that are covered the vehicle's service manual. Serviceable components include the air compressor, mounting brackets, sensors, air springs, air lines and connections, struts, strut mounts, control arm components, shock absorbers, and stabilizer bars.

Failure to follow procedures might result in a sudden failure of the air spring or suspension system. Suspension fasteners are important attaching parts. New fasteners must be used in the place of old ones whenever they are loosened or removed. They must be replaced with fasteners of the same part number or an equivalent.

## PRELIMINARY INSPECTION

Prior to diagnosing a computer-controlled suspension system, a preliminary inspection should be performed. This inspection may locate a defect that is the cause of the problem. If an inspection is not done, much time may be wasted. If the inspection does not locate any problems, further diagnosis is required. Follow these steps to conduct a preliminary diagnosis:

1. Talk to the customer and find out the exact complaint regarding vehicle operation.
2. If necessary, road test the vehicle to identify the exact symptoms.
3. Inspect the vehicle for any work that was completed recently, including body work. Wires or air lines may have been damaged by a collision.
4. Check all of the electrical connections in the suspension system **(Figure 1).**
5. If the vehicle has an air suspension system, listen for air leaks.

**Figure 1.** Check all electrical connections and wiring in the computer-controlled suspension system.

6. Check all the air lines for cracks, breaks, damage, and kinks.
7. Measure the vehicle curb (trim) height.

### Basic Diagnosis

If the visual inspection reveals no problems, diagnosis by symptom may lead to the cause of a problem. Make sure the regular (non-controlled) suspension system is in good condition before diagnosing the electronic system.

As with all computerized systems, the computer stores diagnostic trouble codes (DTCs) that can be retrieved with a scan tool. If a DTC indicates a faulty component or circuit, that should be tested with a lab scope or digital multimeter (DMM). Always follow the specified procedure for retrieving the codes, as well as for testing and replacing individual parts.

## AIR SYSTEMS

In an air suspension system, the air springs are inflated or deflated to provide a constant vehicle trim height. Variable rate air springs at each corner of the vehicle are integral with the struts. Rotary height sensors send inputs to the air suspension control module.

## Diagnosis

> You Should Know
> *The system control switch, typically in the trunk must be turned off prior to hoisting, jacking, or towing the vehicle. If the front of the chassis is lifted with a bumper jack, the rear suspension moves downward. The electronic air suspension system will attempt to restore the rear trim height to normal and may cause the front of the chassis to fall off the bumper jack resulting in personal injury or vehicle damage.*

If the air suspension warning lamp is illuminated with the engine running, the control module has detected a defect in the system. Each manufacturer has a unique procedure for setting the system into a self-test, always follow that procedure.

If defects are found during the test sequence, specific electrical tests should be performed to locate the problem.

> You Should Know
> *When fault codes are provided on a priority basis, always recheck the system for codes after a fault is corrected. Some computers are only capable of storing a specific number of codes in order of priority. If only one code is displayed, there could be another fault of lesser priority in the system.*

### Spring Fill Diagnostics

The spring fill diagnostic routine allows a technician to force the control module to fill and vent each air spring. This procedure can identify inoperative components such as air spring solenoids or the module.

## Air Spring Removal and Installation

> You Should Know
> *When air spring valves are being removed, always rotate the valve to the first stage until all the air escapes from the air spring. Never turn the valve to the second (release) stage until all air is released from the spring.*

Many components such as control arms, shock absorbers, and stabilizer bars are diagnosed and serviced in the same way as a conventional suspension system. However, air spring service procedures are different. Follow these steps for air spring removal:

1. Turn the electronic air suspension switch off.
2. Hoist the vehicle and allow the suspension to drop, or lift the vehicle with a floor jack and place safety stands under the chassis. With the vehicle on safety stands, allow the suspension to drop.
3. Disconnect the air line from the spring solenoid valve and rotate the valve to the first stage to allow the air to escape from the spring.
4. Disconnect the lower spring retainer and remove the spring from the chassis.
5. Before installing an air spring, it must be properly folded over the piston at the bottom of the membrane **(Figure 2).**
6. Install the spring and connect the lower spring retainer. Make sure the top of the spring is properly seated in its seal.

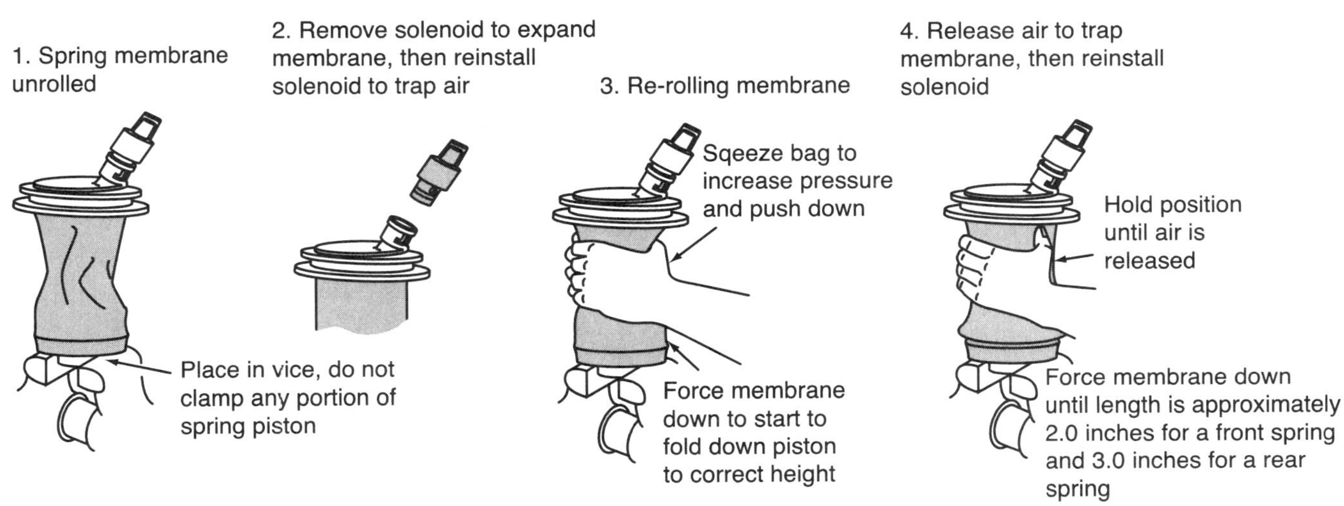


**Figure 2.** The procedure for folding an air spring prior to installation.

7. Properly position the spring and make sure there are no creases in the membrane.
8. Reconnect the air line to the valve, and rotate the valve to the second stage.

## Air Spring Inflation

> **You Should Know** *Do not allow the suspension to compress the spring until it is inflated.*

To inflate the air spring on a typical system, put the vehicle on a hoist and lower just enough so its weight is on the suspension. Turn on the air suspension system switch. Then move the ignition switch from OFF to RUN for 5 seconds with the driver's door open and the other doors shut. Then turn the switch OFF. Ground the diagnostic lead.

Apply the brake pedal and turn the ignition to the run position. The warning lamp will flash every 2 seconds to indicate the fill mode. To fill a rear spring or springs, close and open the driver's door once. After a 6-second delay, the rear spring will fill for 60 seconds. To fill a front spring or springs, close and open the driver's door twice. After a 6-second delay, the front spring will fill for 60 seconds. When the springs require filling, fill the rear springs first. The spring fill mode is terminated when the diagnostic lead is disconnected from ground.

## Trim Height Adjustment

Once the air springs are filled, check the vehicle's trim height. If it is incorrect, the front and rear height sensors may be adjusted to obtain the proper trim height.

## Line Service

The air lines have quick disconnect fittings and should be released by pushing downward and holding the plastic release ring, and then by pulling outward on the nylon line **(Figure 3).** Simply push the nylon line into the fitting until it seats to reconnect an air line.

If a leak occurs in a nylon line, a sharp knife may be used to cut the defective area out of the line. A service fitting is available for line repairs. After the defective area is cut out from the line, push the two ends of the line into the service fitting **(Figure 4).**

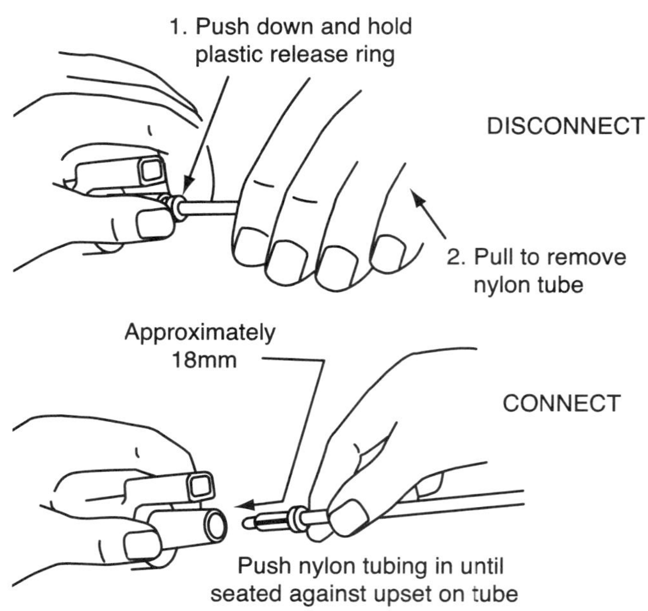


**Figure 3.** Removing an air line from an air spring valve or compressor outlet.

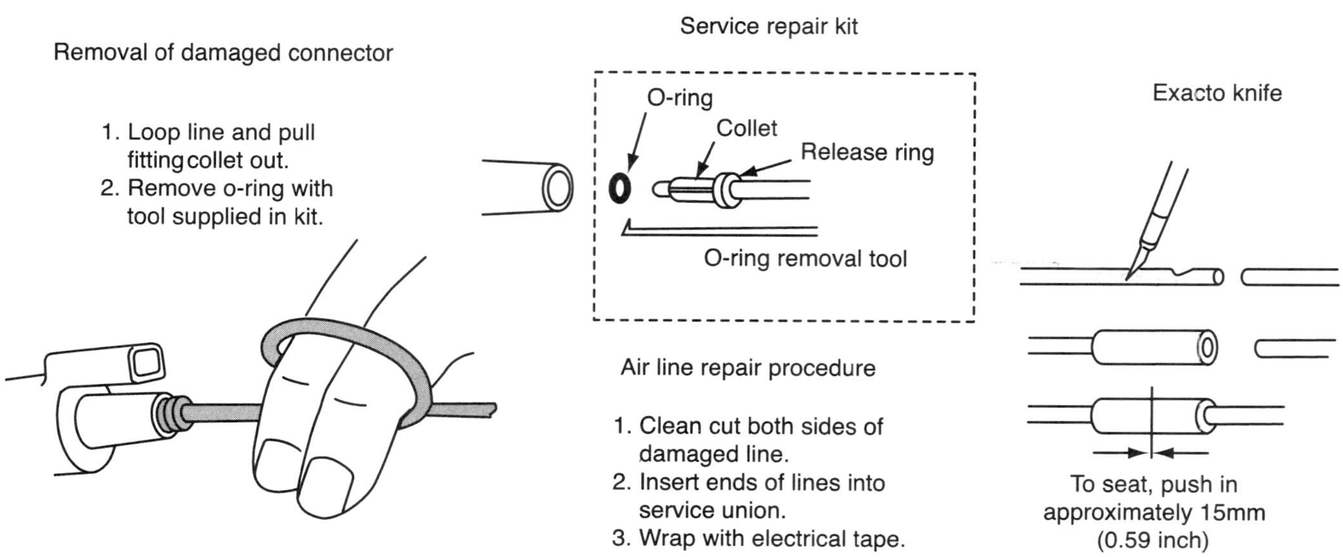


**Figure 4.** Air line repair.

## Air Suspension System with Speed-Leveling Capabilities

Before performing an electronic diagnosis, inspect the system for mechanical or electronic defects. Turn the ignition switch off and leave the air suspension switch on. Connect a scan tool to the air suspension data link connector (DLC). Record any DTCs displayed.

Follow these steps to complete the self diagnostics:

1. Connect a battery charger to the vehicle's battery. Set the charger to a low charging rate.
2. Open the trunk.
3. Connect the scan tool to the DLC.
4. Close all of the vehicle's doors and leave the driver's window down so the ignition switch is accessible.
5. Turn the air suspension switch OFF and then turn it ON.
6. Turn the ignition switch from the OFF to the RUN position.
7. Select Air Suspension Diagnosis and Auto Test Mode on the scan tool.
8. While the test is running, if the air suspension module detects any faults, the test will be interrupted and DTCs representing the fault will be displayed on the scan tool.
9. When the test is complete, open the driver's door, turn the steering wheel one quarter turn in each direction, and close the driver's door. After doing this, open and close the passenger's door. Press the appropriate button on the scan tool to continue the auto test.
10. Record the DTCs displayed on the scan tool.
11. Turn the ignition switch off and disconnect the scan tool.

## Automatic Ride Control (ARC), 4WD Vehicles

This air suspension system operates only when additional weight is added to the vehicle or if the driver selects four-wheel-drive. The height sensors send normal trim height, below-normal trim height, or above-normal trim height voltage signals to the ARC module.

**You Should Know** *During this test procedure, compressor operation may lower the battery's voltage and cause inaccurate results. The use of a battery charger prevents this problem.*

Inspect the system for mechanical or electronic defects. Repair anything found during the inspection before proceeding.

The scan tool is connected to the DLC. Be sure the ignition switch is off when connecting or disconnecting the scan tool. All doors, including the rear tailgate, must be closed, and the transfer case selector switch must be in two-wheel drive. With the scan tool connected, turn the ignition switch on, and obtain and record the continuous DTCs **(Figure 5)**. Continu-

| DTC | CAUSED BY | DESCRIPTION | ACTION |
|---|---|---|---|
| B1318 | Battery Low Voltage | ARC Module | Go to Pinpoint Test U. |
| B1342 | ECU is Defective | ARC Module | Go to Pinpoint Test V. |
| B1485 | Brake Pedal Position Switch Input Short to Battery | ARC Module | Go to Pinpoint Test W. |
| B1565 | Door Ajar Input Short of Power | ARC Module | Go to Pinpoint Test X. |
| C1439 | Acceleration Input Signal Circuit Failure | ARC Module | Go to Pinpoint Test F. |
| C1724 | Height Sensor Power Circuit Failure | ARC Module | Go to Pinpoint Test G. |
| C1725 | Front Pneumatic Failure | ARC Module | Go to Pinpoint Test H. |
| C1726 | Rear Pneumatic Failure | ARC Module | Go to Pinpoint Test J. |
| C1756 | Front Height Sensor Power Circuit Failure | ARC Module | Go to Pinpoint Test K. |
| C1760 | Rear Height Sensor Power Circuit Failure | ARC Module | Go to Pinpoint Test K. |
| C1770 | Vent Solenoid Circuit Failure | ARC Module | Go to Pinpoint Test L. |
| C1830 | Air Compressor Relay Circuit Failure | ARC Module | Go to Pinpoint Test M. |
| C1845 | Front Fill Solenoid Circuit Failure | ARC Module | Go to Pinpoint Test N. |
| C1865 | Rear Fill Solenoid Circuit Failure | ARC Module | Go to Pinpoint Test N. |
| C1869 | Rear Gate Solenoid | ARC Module | Go to Pinpoint Test P. |
| C1901 | Right Rear Shock Absorber Circuit Failure | ARC Module | Go to Pinpoint Test Q. |
| C1905 | Left Rear Shock Absorber Circuit Failure | ARC Module | Go to Pinpoint Test R. |
| C1909 | Right Front Shock Absorber Circuit Failure | ARC Module | Go to Pinpoint Test S. |
| C1913 | Left Front Shock Absorber Circuit Failure | ARC Module | Go to Pinpoint Test T. |
| P1807 | 4X4 High Range Output Short to Ground | ARC Module | Go to Pinpoint Test D. |
| P1808 | 4X4 Low Range Output Circuit Failure (4.0L), Ride Control Switch Input Circuit Failure (5.0L) | ARC Module | Go to Pinpoint Test E. |

**Figure 5.** Examples of available DTCs from an ARC system.

ous DTCs represent faults in the ARC system that are present at the time of testing.

During the test, active commands may be used to operate specific components to determine if the system responds properly. If there are any defects, DTCs will be set.

## Rear Electronic Level Control (ELC)

This system maintains rear suspension trim height regardless of rear suspension load. The rear height sensor may be adjusted to correct the trim height **(Figure 6).** Trim height is measured between the top of the rear axle housing and the frame. Position the vehicle on a level surface over an open service pit or equivalent. Place the transmission in Park and apply the parking brake. Make sure the air suspension switch is on, then open the driver's door and turn the ignition switch to the RUN position. Then, add weight to the rear bumper and wait for the compressor to run and restore the rear suspension to the trim height. After the compressor stops and the suspension is at trim height, remove the weight from the rear bumper and close the driver's door. Allow the rear suspension drop to trim height.

Rock the rear suspension laterally and allow the suspension to stabilize. Then check the height and compare it to specifications.

If a trim height adjustment is required, the upper height sensor mounting bolts may be loosened and the bracket moved downward to increase trim height or upward to decrease trim height **(Figure 7).**

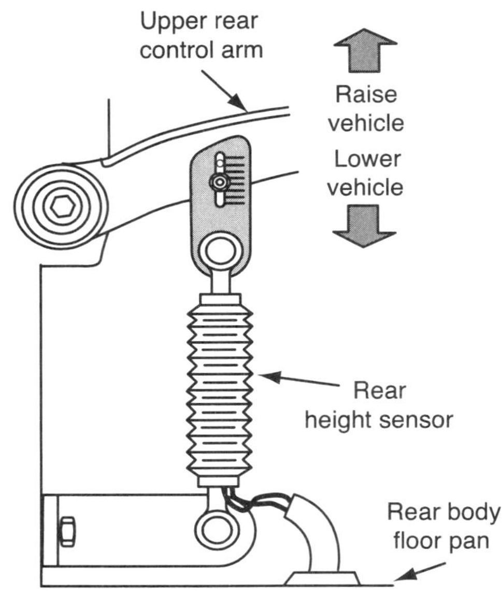


**Figure 6.** Rear suspension trim height may be adjusted by loosening the attaching bolt on the top height sensor bracket.

## ELECTRONICALLY CONTROLLED SHOCK ABSORBERS AND STRUTS

Many vehicles are equipped with an active suspension system. In these systems, an actuator controls the action of the shock or strut.

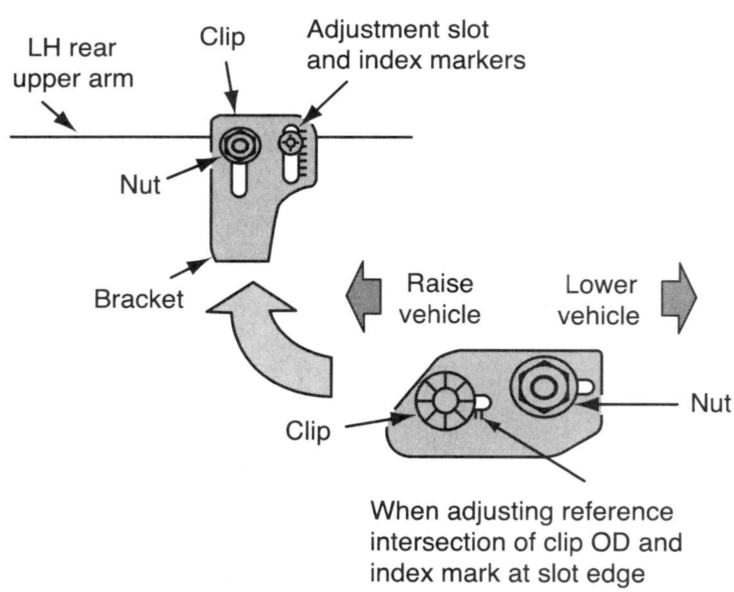


**Figure 7.** The procedure for adjusting rear trim height.

## Programmed Ride Control (PRC) System

In the programmed ride control system, the damping action of the front and rear struts and shocks is automatically controlled to provide improved ride and handling characteristics under various driving conditions.

The control module monitors the system while the vehicle is driven. If a defect occurs, the mode indicator light starts flashing. When the mode selector switch is moved to the opposite position and then returned, erroneous codes are cleared from the memory. If the mode indicator light continues flashing, a self-test should be performed.

The self-test connector has two terminals that must be connected together to conduct the self-test. Turn the ignition off, and then position the mode select switch to the Auto position. Jump across the two terminals and start the engine. After the engine has been running for 20 seconds or more, remove the jumper. Now, count the indicator light flashes to obtain the DTCs.

## Computer Command Ride (CCR) System

If a defect occurs in the CCR system, the service ride control lamp is illuminated. Should the defect correct itself, the lamp goes out, but a DTC is retained in the control module's nonvolatile electronically erasable programmable read-only memory (EEPROM). When a DTC is set in memory, the control module performs a 1-second retest at 3-minute intervals or when the driver select switch position is changed.

A jumper wire is connected between terminals A and C in the DLC to obtain the DTCs **(Figure 8)**. After the wire is installed and the ignition switch turned on, the service lamp will flash the DTCs. Each DTC flashes three times and the codes are displayed in numerical order. This sequence repeats until the ignition switch is turned off or the jumper wire is removed. Voltmeter and ohmmeter tests may be necessary to locate the exact cause of the problem.

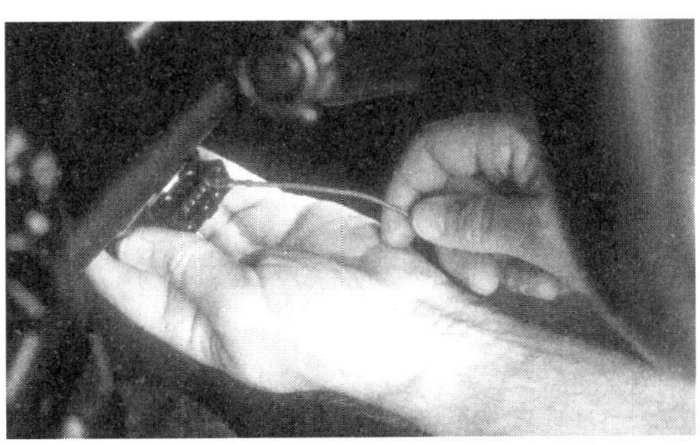

**Figure 8.** A jumper wire is connected between terminals A and C in the DLC to obtain the DTCs from the system.

After a defect is corrected, DTCs may be cleared by alternately connecting and disconnecting the jumper wire between DLC terminals A and C three times with the ignition switch on. Each connection must be held for 1 second, and there should be a 1-second pause between them. When the DTC erasing procedure is completed, turn the ignition switch off and check for DTCs. If the defect is corrected and the DTCs are erased, only code 12 flashes.

## Road Sensing Suspension (RSS) System

The road sensing suspension system controls damping forces of the front and rear struts and shocks in response to various road and driving conditions. To enter the diagnostic mode of this system, turn the ignition switch on and simultaneously press the off and warmer buttons on the climate control center. Hold these buttons until a display shows on the instrument panel cluster. Do not proceed with the diagnostics unless all of the segments are illuminated.

> You Should Know *During computer system diagnosis, use only the recommended test equipment to prevent damage to computer system components.*

Once diagnostics is entered, the HI and LO buttons in the climate control center are used to select or reject test displays. Pressing the HI button may be compared to a "yes" input, the LO button may be considered a "no" input. The fault codes are a three-digit number with a one-letter prefix and a one-letter suffix.

Any fault codes in the real time damping (RTD) module are prefixed by the letter S. When a code has a C suffix, it is a current code. An H suffix indicates a history code and represents an intermittent fault. The fault codes are displayed in numerical order.

If there are no fault codes, NO S CODES appears in the IPC display. If NO S DATA is displayed, the RTD module cannot communicate with the IPC display. When the AUTO button is pressed any time, the system changes from the diagnostic mode to normal operation.

> You Should Know *Do not supply voltage to or ground any circuit or component unless instructed to do so by the manufacturer. This action may damage system components.*

After the fault codes are displayed, the technician may select RTD data, inputs, outputs, or clear codes. If RTD data is selected, the data from a number of inputs will be displayed in numerical order beginning with number SD01 **(Figure 9).** This means the vehicle must be driven to obtain input from this sensor. This sensor input appears as a numerical range from –100 to 100 percent. As the vehicle is accelerated and decelerated, sensor input should change within the specified range. Always consult the service manual for detailed diagnosis and specifications. The next test is selected by pressing the HI button, and the display goes back to the previous test when the LO button is pressed.

**You Should Know** *When removing, replacing, or servicing an electronic component on a vehicle, always disconnect the negative battery cable before starting the service.*

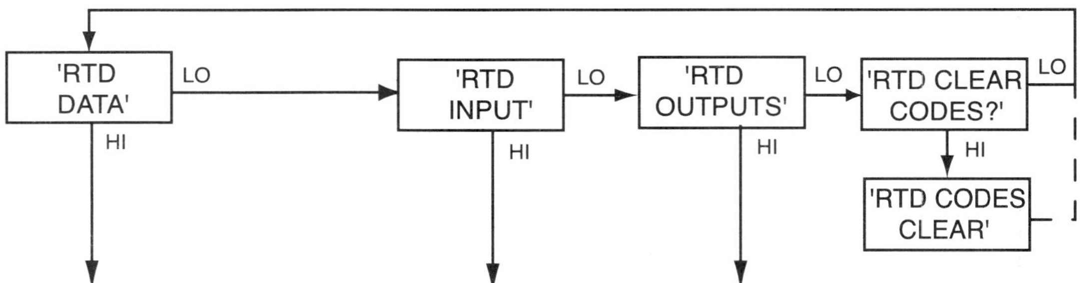


| RTD DATA | | | |
|---|---|---|---|
| Parameter Number | Parameter | Display Range | Units |
| SD01 | RF Accelerometer | -100 to 100% | Travel |
| SD02 | LF Accelerometer | -100 to 100% | Travel |
| SD03 | RR Accelerometer | -100 to 100% | Travel |
| SD04 | LR Accelerometer | -100 to 100% | Travel |
| SD05 | SSS Solenoid PWM | 0 to 100 | % |
| SD06 | SSS Solenoid Feedback | 0 to 100 | % |
| SD11 | RF Position Sensor | -100 to 100% | Travel |
| SD12 | LF Position Sensor | -100 to 100% | Travel |
| SD13 | RR Position Sensor | -100 to 100% | Travel |
| SD14 | LR Position Sensor | -100 to 100% | Travel |
| SD20 | Vehicle Speed | 0 to 159 | MPH |
| SD50 | Battery Volts | 0 to 16.3 | Volts |
| SD96 | HPC PROM ID | | Code |
| SD97 | COP PROM ID | | Code |
| SD98 | TMS320 PROM ID | | Code |
| SD99 | EEPROM ID | | Code |

| RTD INPUT | |
|---|---|
| Input # | Input |
| S101 | Lift/Drive Discrete |

| RTD OUTPUTS | |
|---|---|
| Output # | Output |
| SO00 | No Outputs |
| SO01 | RF Damper |
| SO02 | LF Damper |
| SO03 | RR Damper |
| SO04 | LR Damper |
| SO05 | SSS Solenoid |
| SO06 | ELC Compressor |
| SO07 | ELC Exhaust Valve |
| SO08 | Cycle All |

| KEY |
|---|
| To Select Another Test Press:<br>Hi - To Increment<br>Lo - To Decrement<br>* PROM ID Code Number Identifies an Individual Calibration and Is Periodically Updated; Refer to Latest Service Publication for Correct ID Number |

**Figure 9.** Optional test modes including RTD data, inputs, outputs, and clear codes.

## SUMMARY

- Prior to diagnosing a computer-controlled suspension system, conduct an inspection of the system.
- The systems' computer stores diagnostic trouble codes (DTCs) that can be retrieved.
- In an air suspension system, the air springs are inflated or deflated to provide a constant vehicle trim height.
- The air suspension system control switch must be turned off prior to hoisting, jacking, or towing the vehicle.
- If the air suspension warning lamp is illuminated with the engine running, the control module has detected a defect.
- Height sensors may be adjusted to obtain proper trim height.
- The programmed ride control system automatically changes the damping action of the front and rear struts and shocks.
- When removing, replacing, or servicing an electronic component on a vehicle, always disconnect the negative battery cable before starting the service procedure.

## REVIEW QUESTIONS

1. Technician A says terminals A and B should be connected in the DLC to obtain fault codes on a CCR system. Technician B says a flashing mode indicator light indicates a system defect in a PRC system. Who is correct?
   A. Technician A
   B. Technician B
   C. Both Technician A and Technician B
   D. Neither Technician A nor Technician B
2. The owner of a car with an automatic air suspension says the Check Air Suspension light is illuminated. A preliminary inspection does not reveal any air leaks and the measured trim eight is to specifications. Technician A says the next step is to connect a scan tool. Technician B says the next step is to complete the drive cycle. Who is correct?
   A. Technician A
   B. Technician B
   C. Both Technician A and Technician B
   D. Neither Technician A nor Technician B
3. On a vehicle with an electronic air suspension system, a bumper jack is used to lift one corner of the vehicle to change a tire. Technician A says the electronic air suspension switch should be in the ON position. Technician B says the air spring should be deflated on the corner of the vehicle being lifted. Who is correct?
   A. Technician A
   B. Technician B
   C. Both Technician A and Technician B
   D. Neither Technician A nor Technician B
4. Technician A says the suspension switch must be turned off before raising any corner of a car with an electronic air suspension. Technician B says the ignition switch must not be turned on while any corner of a car with electronic air suspension is raised. Who is correct?
   A. Technician A
   B. Technician B
   C. Both Technician A and Technician B
   D. Neither Technician A nor Technician B
5. Technician A says when diagnosing the ARC system, the ignition switch must be on and the driver's door open. Technician B says when diagnosing the ARC system, the transfer case must be in 4WD. Who is correct?
   A. Technician A
   B. Technician B
   C. Both Technician A and Technician B
   D. Neither Technician A nor Technician B

# Section 7

## Steering Systems



**Interesting Fact**

*For many years, rear-wheel drive vehicles were equipped with parallelogram steering linkages. The smaller, lighter front-wheel drive cars introduced in large numbers in the late 1970s and 1980s required a more compact, lighter steering gear. The rack and pinion steering gear is used almost exclusively on front-wheel drive cars because of its reduced weight and space requirements.*

## SECTION OBJECTIVES

After you have read, studied, and practiced the contents of this section, you should be able to:

- Describe the similarities and differences between parallelogram, worm and roller, and rack-and-pinion steering linkage systems.
- Describe the common four-wheel steering systems.
- Explain how the steering column provides driver safety during a frontal collision.
- Explain the purposes of the pitman arm and the idler arm.
- Define the purpose of a steering gear.
- List two possible mounting positions for the rack-and-pinion steering gear.
- Describe the operation of an electronic variable orifice (EVO) steering system.
- Disable and enable a supplemental restraint system (SRS).
- Remove and replace a steering wheel; center/time a SRS coil (clock spring).
- Diagnose steering column noises, looseness, and binding concerns (including tilt mechanisms); determine necessary action.
- Inspect steering shaft universal joint(s), flexible coupling(s), collapsible column, lock cylinder mechanism, and steering wheel; perform necessary action.
- Inspect and replace a pitman arm, relay (center link/intermediate) rod, idler arm and mountings, and steering linkage damper.
- Inspect, replace, and adjust tie-rod ends (sockets), tie-rod sleeves, and clamps.
- Adjust manual or power non-rack-and-pinion worm bearing preload and sector lash.
- Remove and replace manual or power rack-and-pinion steering gear; inspect mounting bushings and brackets.
- Inspect and replace manual or power rack-and-pinion steering gear inner tie-rod ends (sockets) and bellows boots.
- Diagnose electronic power steering systems.
- Perform a preliminary inspection on a four-wheel steering (4WS) system.
- Perform a trouble code diagnosis on a 4WS system.

# Chapter 35 Steering Columns and Linkages

## INTRODUCTION

The steering wheel and column produces the necessary force to turn the steering gear. The exact type of steering wheel and column depends upon the year and the car manufacturer. The steering column, also called a steering shaft, relays the movement of the steering wheel to the steering gear. The steering linkage transfers the motion of the steering gear to the wheels.

## STEERING WHEELS AND COLUMNS

The steering wheel provides the turning effort. Universal joints **(Figure 1)** are used as couplings to transfer the rotation of the steering wheel to the steering gear located at an angle to the column. Support brackets are used to hold the steering column in place.

The parts of the steering column must minimize freeplay and provide steering control. Steering wheel freeplay is the amount of movement before the front wheels start to move to the right or left.

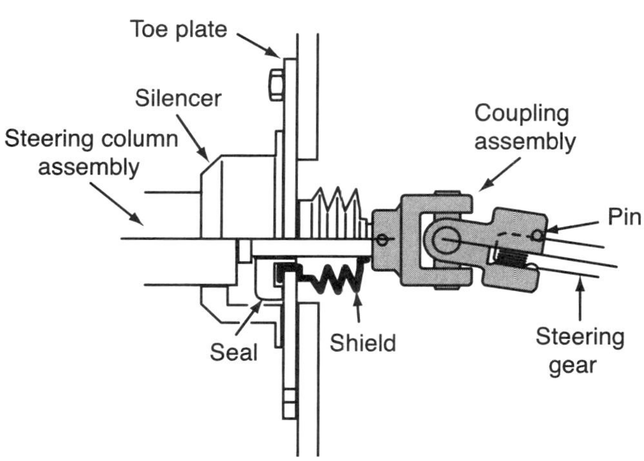


**Figure 1.** Universal joints are used as couplings to transfer the rotation of the steering wheel to the steering gear located at an angle to the column.

### Safety Features

Since 1968, all steering columns have some method of column collapse during a collision. Some vehicles have plastic pins that shear off the column jacket, gearshift tube, and steering shaft if the driver hits the steering wheel. This allows the column to collapse away from the driver **(Figure 2).** In other vehicles, the column-to-instrument-panel mounting is designed to allow column movement if the driver hits the steering wheel. There are also

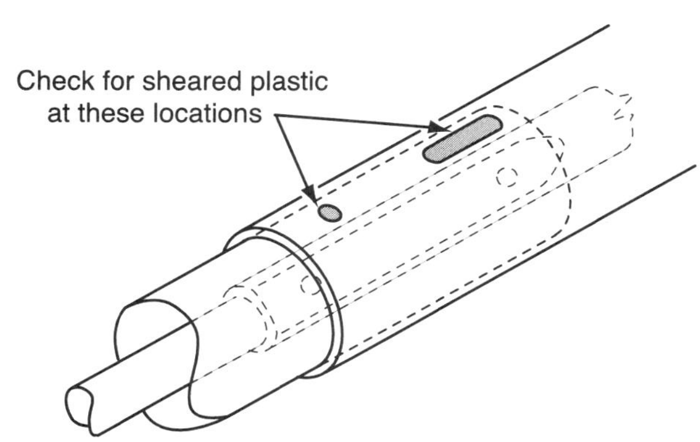


**Figure 2.** Injection plastic in a collapsible outer steering column jacket.

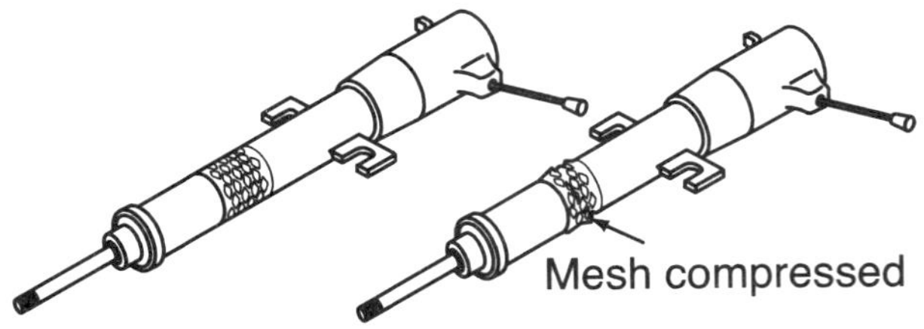


**Figure 3.** Collapsible steel mesh may be used in a steering column. The mesh compresses during frontal impact to allow the steering column to move away from the driver.

collapsible steel mesh **(Figure 3)** or accordion-pleated devices that give way under pressure.

After a vehicle has been in an accident, the steering column should be checked for evidence of collapse. Although the car can be steered with a collapsed column that has been pulled back, the collapsed portion must be replaced.

## Designs

Steering columns may be classified as non-tilting, tilting, and tilting/telescoping. Both fixed and tilt columns may house an emergency warning flasher control, a turn signal switch, ignition key, lights (high/low beams), horn, windshield wipers and washers, an anti-theft device that locks the steering system, and an air bag clock spring **(Figure 4).** On automatic transmission-equipped vehicles, the transmission linkage is also locked with the ignition switch.

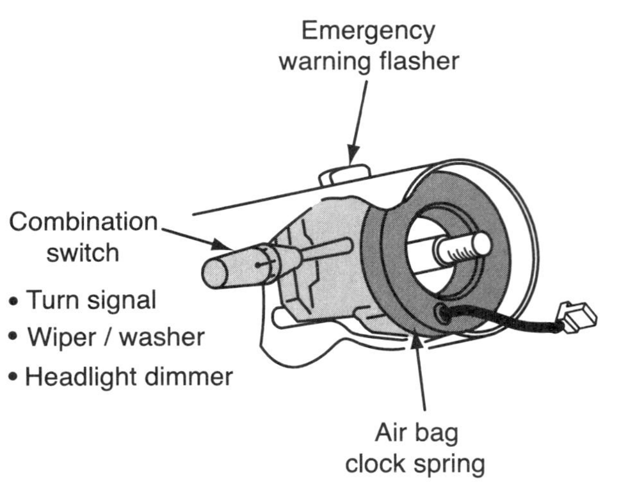


**Figure 4.** A turn signal switch, hazard warning switch, dimmer switch, and wipe/wash switch mounted in a steering column.

The ignition switch cylinder is usually mounted in the upper right side of the column housing, and the ignition switch is bolted on the lower side of the housing. An operating rod connects the ignition switch cylinder to the ignition switch. Ignition switches are integral with the lock cylinder in some steering columns.

The turn signal and hazard warning switch are mounted on top of the steering column. Lugs on the bottom of the steering wheel are used to cancel the signal lights after a turn is completed. On many vehicles, the signal lever also operates the wipe/wash and dimmer switches.

If the gearshift is mounted in the steering column, a tube extends from the gearshift housing to the shift lever at the lower end of the steering column. This shift lever is connected through a linkage to the transaxle or transmission shift lever. A lock plate is attached to the upper steering shaft, and a lever engages the slots in this plate to lock the steering wheel and gearshift when the gearshift is in Park and the ignition switch is in the Lock position.

Tilt steering columns facilitate driver entry to and exit from the front seat and allow the driver to position the steering wheel to suit individual comfort requirements **(Figure 5).** A tilting/telescoping steering column allows the driver to tilt and extend or retract the steering wheel, which provides the driver with more steering wheel position choices.

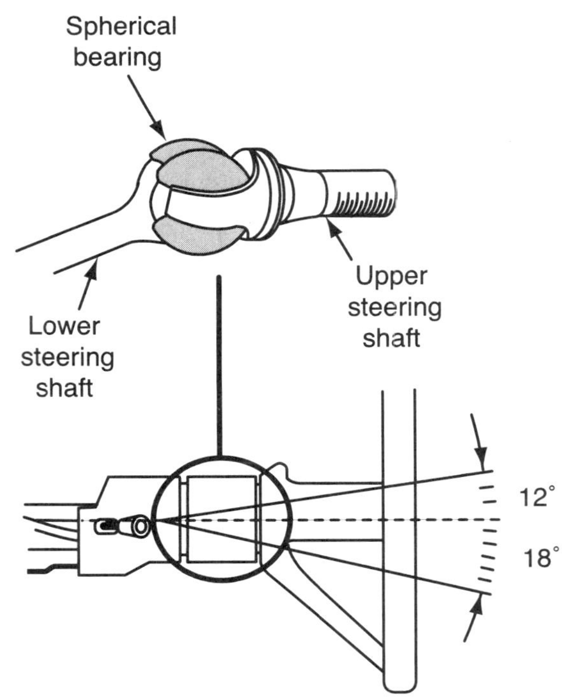


**Figure 5.** A tilt steering column.

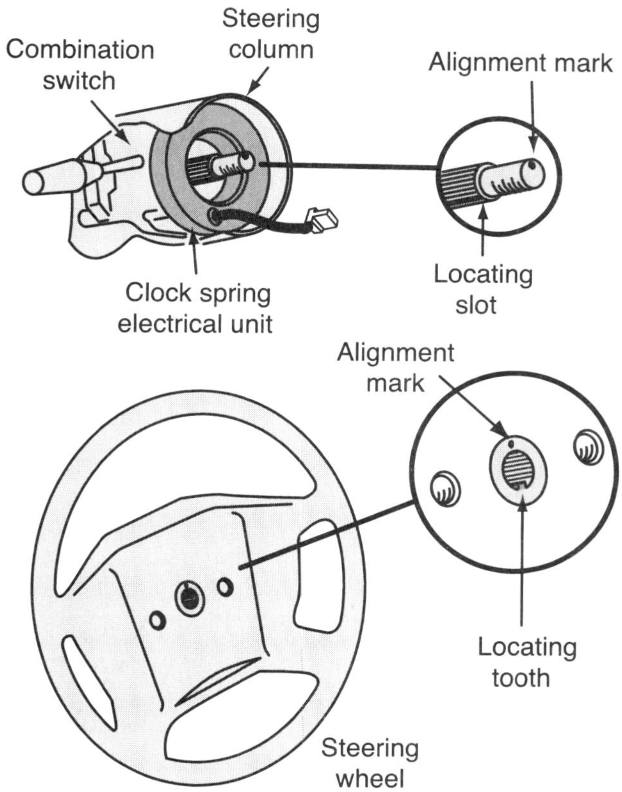


**Figure 6.** The steering wheel is splined to the steering column and is retained with a nut.

## Steering Wheels

The steering wheel is usually attached to the steering column by a bolt or nut. Most steering wheels and shafts have alignment marks that must be aligned when the steering wheel is installed. The steering wheel splines fit on matching splines on the upper steering shaft, and a nut retains the wheel to the shaft **(Figure 6).** When the steering wheel spokes are in their normal position, centered, the front wheels should be in the straight-ahead position.

## Air Bags

Most cars and minivans are equipped with air bags. All air bag-equipped vehicles have a driver's side air bag located in the upper side of the steering wheel, and some vehicles also have a passenger's side air bag mounted in the instrument panel. When the vehicle is involved in a frontal collision, the air bag deploys in a few milliseconds. As the driver is thrust forward during the collision, the inflated air bag prevents the driver from striking the steering wheel, windshield, or instrument panel. Since air bag components are mounted in the steering column, the column is a very important part of the vehicle's safety equipment.

A steering wheel pad containing the air bag deployment module is mounted in the top of the steering wheel. The module must be disarmed and removed before the steering wheel can be removed. This module contains the air bag and the inflator device. A clock spring electrical connector, or spiral cable, is mounted under the steering wheel. This component contains a ribbon-type conductor that maintains constant electrical contact between the air bag module and the air bag electrical system during steering wheel rotation.

## Conventional Non-tilt Steering Columns

Many steering columns contain a two-piece steering shaft connected by two universal joints. A jacket and shroud surround the steering shaft. The upper shaft is supported by two bearings in the jacket. A toe plate, seal, and silencer surround the lower steering shaft and cover the opening where the shaft extends through the floor. The lower steering shaft is surrounded by a shield underneath the toe plate.

The lower universal joint couples the lower shaft to the steering gear. In some steering columns, a flexible coupling is used in place of the lower universal joint **(Figure 7).**

## Tilt Steering Columns

Tilt steering columns have a short upper steering shaft connected to the steering wheel. This upper steering shaft is connected through a universal joint to the lower steering mainshaft. An upper column tube surrounds the upper steering shaft, and the lower steering mainshaft is supported on bushings in the lower column tube. In some steering columns, a spherical bearing acts as a pivot

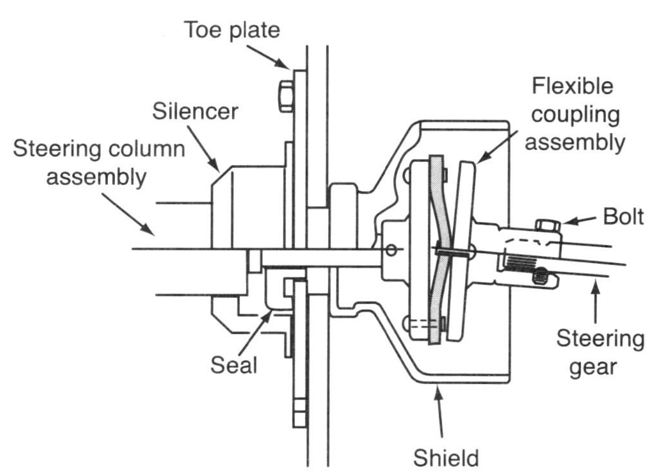


**Figure 7.** The construction of a steering column with a flexible coupling.

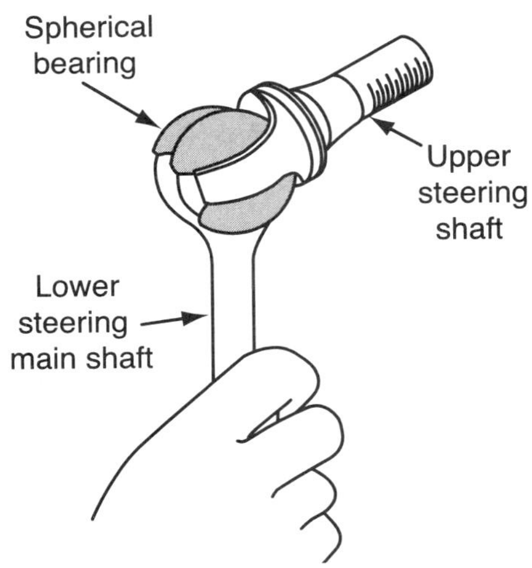


**Figure 8.** A spherical bearing acts as a pivot between the upper steering shaft and the lower steering mainshaft in some tilt steering columns.

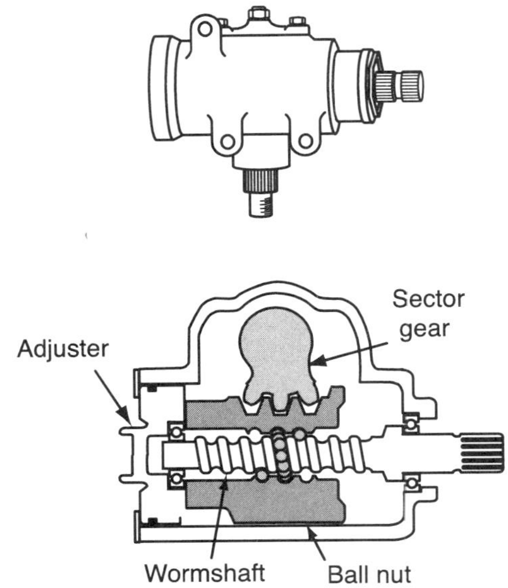


**Figure 9.** A cutaway of a manual recirculating ball steering gear.

between the upper steering shaft and the lower steering mainshaft **(Figure 8).**

When the steering wheel is tilted, the wheel and upper steering column tube pivot on two bolts connected between the upper and lower column tubes. During this tilting motion, the upper steering shaft pivots on the universal joint connected between the upper steering shaft and lower steering mainshaft. A release lever on the side of the steering column allows the wheel and upper column to tilt. This allows the driver to position the steering wheel for greater comfort and easier movement in and out of the driver's seat.

## STEERING LINKAGES

The term **steering linkage** is applied to the system of pivots and connecting parts that is placed between the steering gear and the steering arms attached to the front or rear wheels, controlling the direction of vehicle travel. The steering linkage transfers the motion of the steering gear output shaft to the steering arms, turning the wheels to maneuver the vehicle.

The type of front-wheel suspension (independent wheel suspension as compared with a solid front axle) greatly influences steering geometry. All currently used steering systems can be classified as having a parallelogram or rack and pinion steering linkage.

### Parallelogram Steering Linkage

The parallelogram steering linkage arrangement was the most commonly type used on passenger cars. Now it is found mostly on larger cars, pickups, and larger SUVs. It is used with a recirculating ball steering gear **(Figure 9)** and can be one of two distinct configurations: behind the front suspension **(Figure 10A)** or placed ahead of the suspension **(Figure 10 B).**

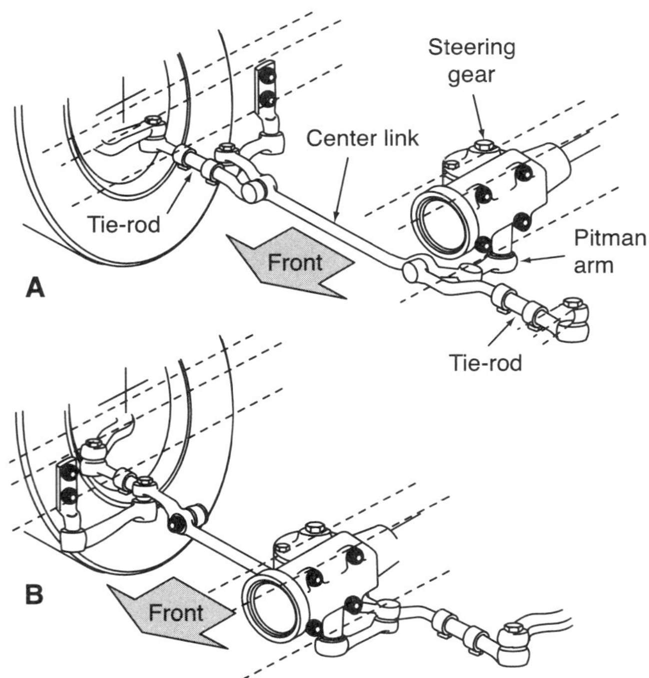


**Figure 10.** A parallelogram steering system mounted (A) behind the suspension and (B) ahead of the suspension.

Regardless of steering linkage mounting position, the steering linkage has the same components. The main components are the pitman arm, idler arm, links, and tie rods.

A parallelogram steering linkage has tie rods mounted parallel to the lower control arms. The tie-rods have ball socket assemblies at each end. One end is attached to the wheel's steering arm and the other end to the center link.

### Pitman Arm

The **pitman arm** connects the steering gear to the center link. Motion from the steering wheel is transmitted to the pitman arm. This arm transfers that movement to the steering linkage. This forces the steering linkage to move to the right or left, and the linkage moves the front wheels in the desired direction. The pitman arm also positions the center link at the proper height to maintain the parallel relationship between the tie-rods and the lower control arms. This reduces the chances for unsteady toe settings or **bump steer.** Toe, a critical alignment factor, is a term that defines how well the tires point to the direction of the vehicle.

There are two basic types of pitman arms: wear and nonwear. Wear-type pitman arms have ball sockets and studs at the outer end, and the stud fits into the center link opening **(Figure 11).** The ball stud and socket are subject to wear, and the pitman arm should be replaced if the ball stud is loose. Nonwear arms have tapered holes at the center link and normally need to be replaced only if they have been damaged or have been mounted with excessive tolerance. The opening in the inner end of both types of pitman arms has serrations that fit over matching serrations on the steering gear shaft. The shaft may need to be indexed before it is removed. A nut and lock washer retain the pitman arm to the steering gear shaft.

### Idler Arm

The idler arm assembly is normally attached on the opposite side as the center link and is connected from

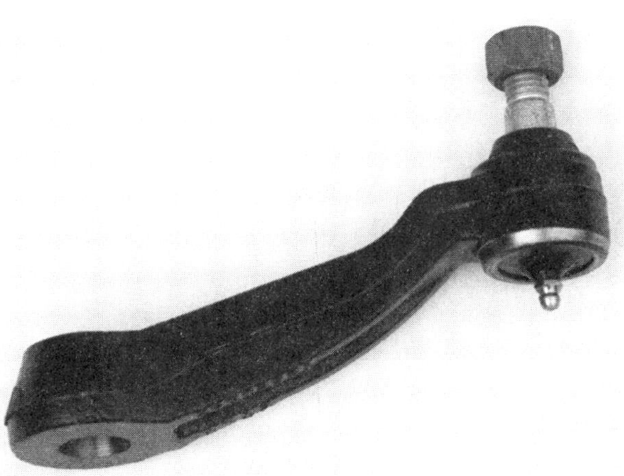

**Figure 11.** A pitman arm.

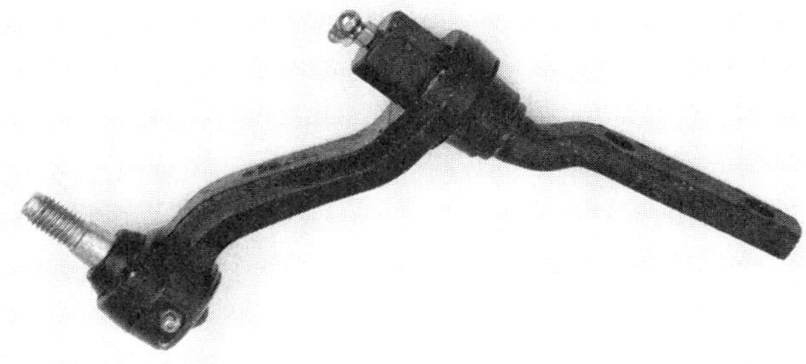

**Figure 12.** An idler arm.

the pitman arm and to the frame. It helps maintain the parallel relationship between the tie rods and lower control arms. An idler arm support is bolted to the frame or chassis, and the idler arm is connected from the bracket to the center link. In some idler arms, a ball stud on the outer end of the arm fits into the center link. In other idler arms, a ball stud in the center link fits into the idler arm **(Figure 12).**

A pivot built into the arm permits sideway movement of the linkage. Idler arms normally wear more than pitman arms because of this pivot function. Worn bushings or stud assemblies on idler arms permit excessive vertical movement in the idler arms.

### Center Links

A **center link** may be referred to as a **drag link,** steering link, or intermediate link. It controls sideway linkage movement, which changes the wheel's direction. The center link, along with the pitman arm and idler arm, provides the proper height for the tie rods. If center links are not mounted at the correct height, toe is unstable and a condition known as the toe change or bump steer will occur.

Center links, like pitman arms, can be characterized as wear or nonwear. Some center links have tapered openings in each end, and the studs of the pitman arm and idler arm fit into the openings. This type of center link may be called a taper end or nonwear link. They normally need to be replaced only if they have been damaged in an accident or through excessive wear. Many wear-type center links have ball sockets in each end with tapered studs extending from the sockets. These tapered studs fit into the pitman arm and idler arm and are retained with castellated nuts and cotter pins.

**You Should Know** *If a center link is nonwear, the pitman arm is normally a wear arm. If the link is a wear type, the arm is usually nonwear.*

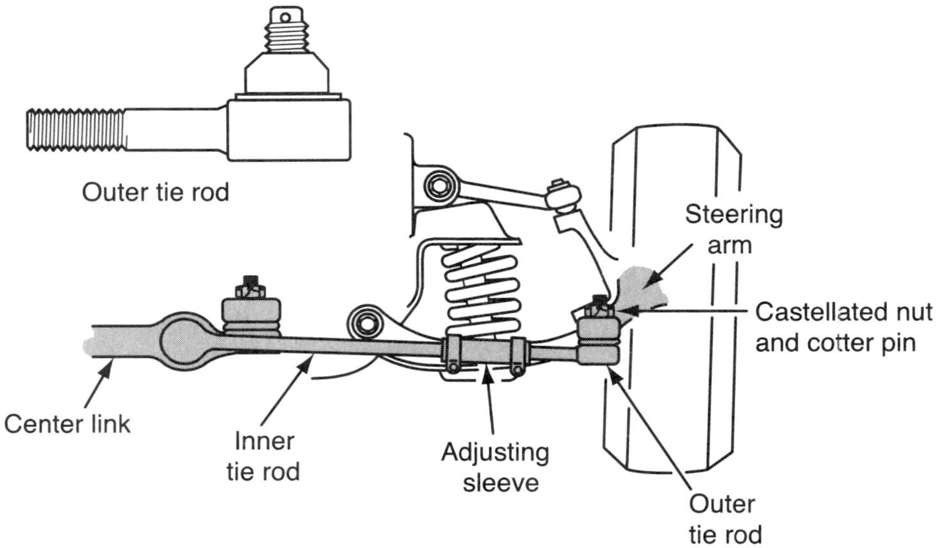


**Figure 13.** Location and parts of a tie-rod assembly.

### Tie Rods

Tie rods are actually assemblies that make the final connections between the steering linkage and steering knuckles. They consist of inner tie-rod ends, which are connected to the opposite sides of the center link, outer tie-rod ends, which connect to the steering knuckles, and adjusting sleeves or bolts, which join the inner and outer tie-rod ends permitting the tie-rod length to be adjusted for correct toe settings **(Figure 13).**

A ball socket is at the inner end to each tie rod and is mounted in the center link. A castellated nut and cotter pin retain the tie rods to the center link. Some outer tie-rod ends have a ball stud that is surrounded by an upper hardened steel bearing and a high-strength polymer lower bearing seat **(Figure 14).** The upper bearing provides strength and durability, and the lower bearing seat provides smooth rotation of the ball stud in the tie-rod end. An internal spring at lower bearing seat supplies self-adjusting action and constant tension on this seat.

A seal in the upper part of the tie-rod end seals the ball stud to prevent contaminants from entering it. Contaminants such as dirt and moisture can enter and cause rapid part failure. A special bonded ball stud, in which no boot is used, is available for use on certain light-duty two-wheel-drive and four-wheel-drive trucks. Some vehicles have rubber encapsulated outer tie-rod ends in which a rubber bushing surrounds the lower end of the ball stud.

A threaded sleeve is mounted on the outer end of each tie rod **(Figure 15).** Each tie-rod sleeve has a left-hand and right-hand thread where it is threaded onto the tie-rod end and the tie rod. Therefore, sleeve rotation changes the tie-rod length and provides toe adjustment. Clamps are used to tighten the tie-rod sleeves.

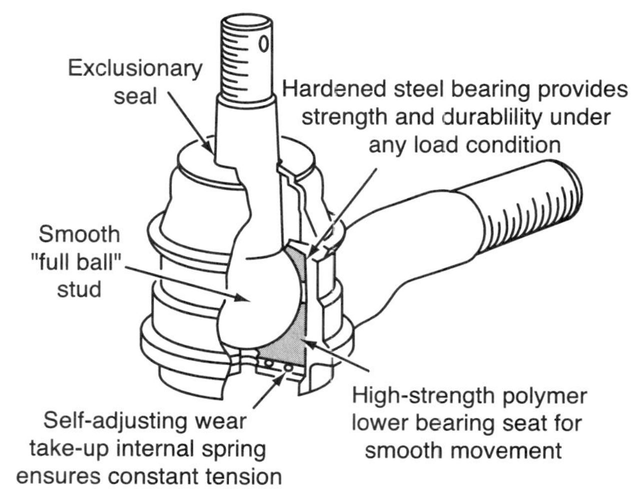


**Figure 14.** An outer tie-rod end with a hardened steel upper bearing and high-strength polymer lower bearing.

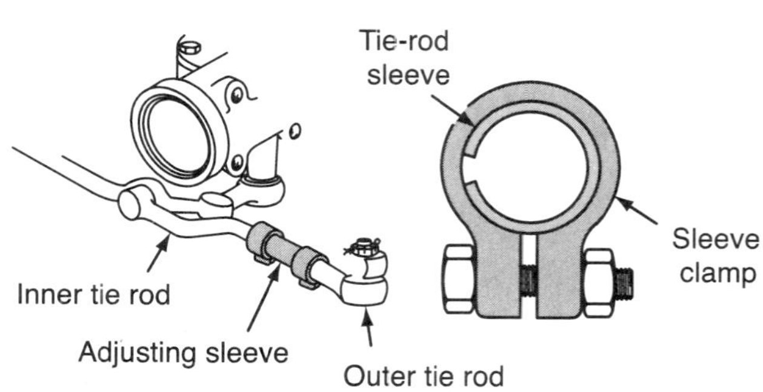


**Figure 15.** An adjusting sleeve and sleeve clamp for a tie rod.

## Rack-and-Pinion Steering Linkage

Most FWD vehicles have rack-and-pinion steering **(Figure 16).** Since the rack is linked directly to the steering arms, this type of steering linkage provides good road feel. The rack is a rod with teeth on one side, and it slides horizontally on bushings inside the gear housing. The rack teeth are meshed with teeth on a pinion gear, and this pinion gear is connected to the steering column. When the steering wheel is turned, the pinion rotation moves the rack sideways. Tie rods are connected directly from the ends of the rack to the steering arms.

Since the rack is connected directly to the tie rods, the rack replaces the center link in a parallelogram steering linkage. An inner tie-rod end connects each tie rod to the rack, and bellows boots are clamped to the gear housing. The boots keep dirt out of the joints. The inner tie-rod end contains a spring-loaded ball socket. Some inner tie-rod ends contain a bolt and bushing **(Figure 17).** Others have a mirror-finished ball and a high-strength polymer bearing to ensure low torque, minimal friction, and extended life.

The outer tie-rod ends are threaded onto the rack. A hardened alloy steel rod extends from the ball to the outer tie-rod end and provides maximum strength and durability.

Figure 16. A rack-and-pinion steering system.

Figure 17. An inner tie rod and outer tie-rod end for a rack-and-pinion steering system.

The rack and pinion steering gear may be mounted on the front crossmember or attached to the cowl behind the engine. Rubber insulating bushings surround the steering gear, and these bushings are clamped to the crossmember or cowl. The rack and pinion steering gear is mounted at the proper height to position the tie rods and lower control arms parallel to each other.

## Steering Damper

A **steering damper** may be found on some parallelogram steering linkages. The steering damper is similar to a shock absorber and is connected from one of the steering links to the chassis or frame **(Figure 18).** When a front wheel strikes a bump, a shock is transferred from a wheel to the steering linkage, steering gear, and steering wheel. The steering damper helps absorb this road shock and prevents it from reaching the steering wheel. Heavy-duty steering dampers are available for severe road conditions such as those encountered by 4WD vehicles.

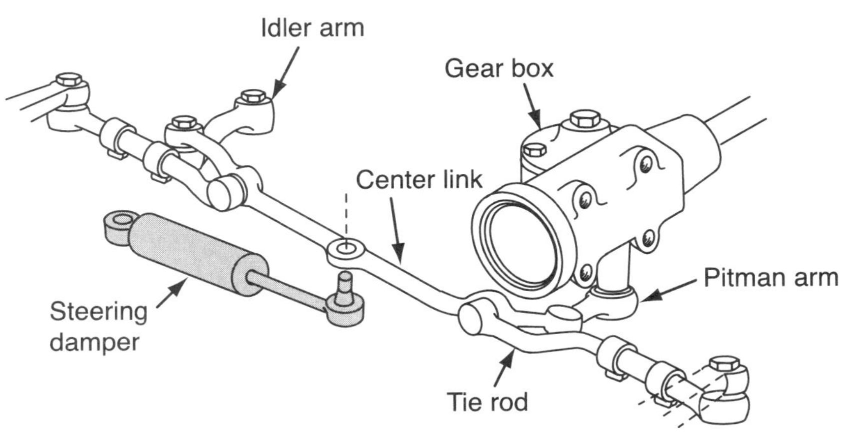


Figure 18. A steering damper and linkage.

# SUMMARY

- The components of a manual steering system include the steering linkage, steering gear, and steering column and wheel.
- The steering wheel and column produces the necessary force to turn the steering gear.
- Steering columns help provide steering control, driver convenience, and driver safety.
- Steering columns provide some method of energy absorption to protect the driver during a frontal collision.
- Many steering wheels and columns have an air bag deployment.
- Tilt steering columns provide increased driver convenience while driving, as well as getting in or out of the driver's seat.
- A clock spring electrical connector supplies positive electrical contact between the air bag module in the steering wheel and the air bag electrical system.
- The ignition switch, dimmer switch, signal light switch, hazard switch, and wipe/wash switch may be mounted in the steering column.
- Steering linkage is the system of pivots and connecting parts placed between the steering gear and steering arms attached to the front wheels. The steering linkage transfers the motion of the steering gear output shaft to the steering arms.
- In a parallelogram steering linkage, the tie rods are parallel to the lower control arms.
- The parallelogram steering linkage minimizes toe change as the control arms move.
- A rack-and-pinion steering linkage has few friction points and is light and compact.
- A steering damper helps absorb road shock.

# REVIEW QUESTIONS

1. The main job of the idler arm is to:
   A. support the left side of the center link.
   B. support the right side of the center link.
   C. support the pitman arm.
   D. keep both ends of the steering system level.
2. What connects the steering column to the wheels?
   A. Steering shaft
   B. Steering linkage
   C. Gearbox
   D. Pitman arm
3. Explain how a collapsible steering column protects the driver in a frontal collision.
4. List the five main components in a parallelogram steering linkage, and explain the purpose of each component.
5. In a parallelogram steering linkage, the tie rods are parallel to the ________________ ________________ ________________.
6. In a rack and pinion steering system, the rack is connected directly to the ________________________ ________________.
7. While discussing idler arms,
   A. some idler arms contain a tapered roller bearing.
   B. the idler arm bracket is bolted to the upper control arm.
   C. a worn idler arm has no effect on front wheel toe.
   D. a partially seized idler arm bearing increases steering effort.
8. In a parallelogram steering linkage, the tie rods are parallel to the lower control arms to:
   A. improve ride quality.
   B. provide longer steering linkage life.
   C. extend shock absorber and spring life.
   D. reduce toe change during upward and downward front wheel movement.
9. A rack and pinion steering gear:
   A. has tie rods that connect the rack directly to the steering arms.
   B. may be bolted to the vehicle frame.
   C. has inner tie-rod ends that are pressed onto the rack.
   D. has more friction points compared to a parallelogram steering linkage.
10. While discussing steering linkages and dampers,
    A. in a rack-and-pinion steering gear, the rack positions the tie rods parallel to the lower control arms.
    B. a rack-and-pinion steering gear is used on most rear-wheel drive cars.
    C. a steering damper is used on many front-wheel drive vehicles.
    D. a defective steering damper may cause excessive steering effort.

# Chapter 36 Manual Steering Gears

## INTRODUCTION

The steering gear changes the rotational motion of the steering wheel to a reciprocating motion to move the steering linkage **(Figure 1).** The steering gear and wheel also provide a mechanical advantage that allows the driver to turn the wheels with little effort.

Three types of steering gears have been used: the recirculating ball, worm and roller, and rack and pinion, which incorporate the linkage and steering gear as a single unit.

In the early 1900s, steering gears were a worm and gear or worm and **sector** (part of a gear) design. In 1923, the Ross cam and lever steering gear was introduced. The cam in this gear was a spiral groove machined into the end of the steering shaft. A pin on the pitman shaft was mounted in the groove. When the steering wheel and shaft were turned, the pin was forced to move, and this action rotated the pitman shaft. This design prevented serious **kickback** at the steering wheel. (Steering kickback is the movement of the steering wheel caused by a front wheel striking a road irregularity.) However, this design created a considerable amount of friction and required higher steering effort.

In the mid-1920s, the Saginaw Steering Division of General Motors Corporation developed the **worm and**

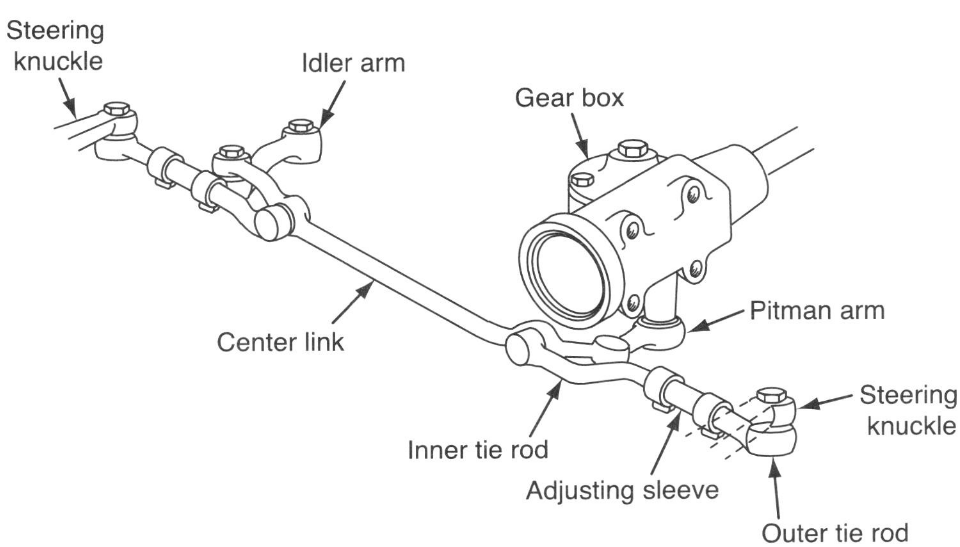


**Figure 1.** The steering gear changes the rotational motion of the steering wheel to a reciprocating motion to move the steering linkage and thereby turn the wheels.

**roller** steering gear. In this steering gear, the sector became a roller, which greatly reduced friction and steering effort. The worm and roller steering gear was the forerunner of the recirculating ball steering gear that has been widely used on rear-wheel drive cars for many years.

During the late 1970s and 1980s, the domestic automotive industry converted much of its production from larger RWD cars to smaller, lightweight, and more fuel-efficient FWD cars. Rack and pinion steering gears were fitted in these compact cars because they are lighter and more compact than the recirculating ball steering gears and parallelogram steering linkages.

## RECIRCULATING BALL STEERING

In a recirculating ball steering gear, the steering wheel and column are connected to the worm shaft, which is supported by bearings in the steering gear housing. A seal above the upper **worm shaft** bearing prevents oil leaks, and there is an adjusting plug on the upper worm shaft bearing to adjust worm shaft **bearing preload.** Proper preloading of the worm shaft bearing eliminates worm shaft **endplay** and prevents steering gear **freeplay** and vehicle wander.

A ball nut is mounted over the worm shaft, and internal threads or grooves on the ball nut match the grooves on the worm shaft. Ball bearings run in the ball nut and worm shaft grooves **(Figure 2).** These bearings recirculate through two outside loops, referred to as ball return guide tubes. As the steering wheel is rotated, the worm shaft rotates, causing the ball nut to move up or down the worm shaft. Since the gear teeth on the ball nut are meshed with the gear teeth on the sector shaft, the movement of the nut causes the sector (pitman) shaft, via the ball bearings, to rotate and swing the pitman arm.

The two separate circuits of ball bearings result in a near friction-free movement of the ball nut and worm shaft

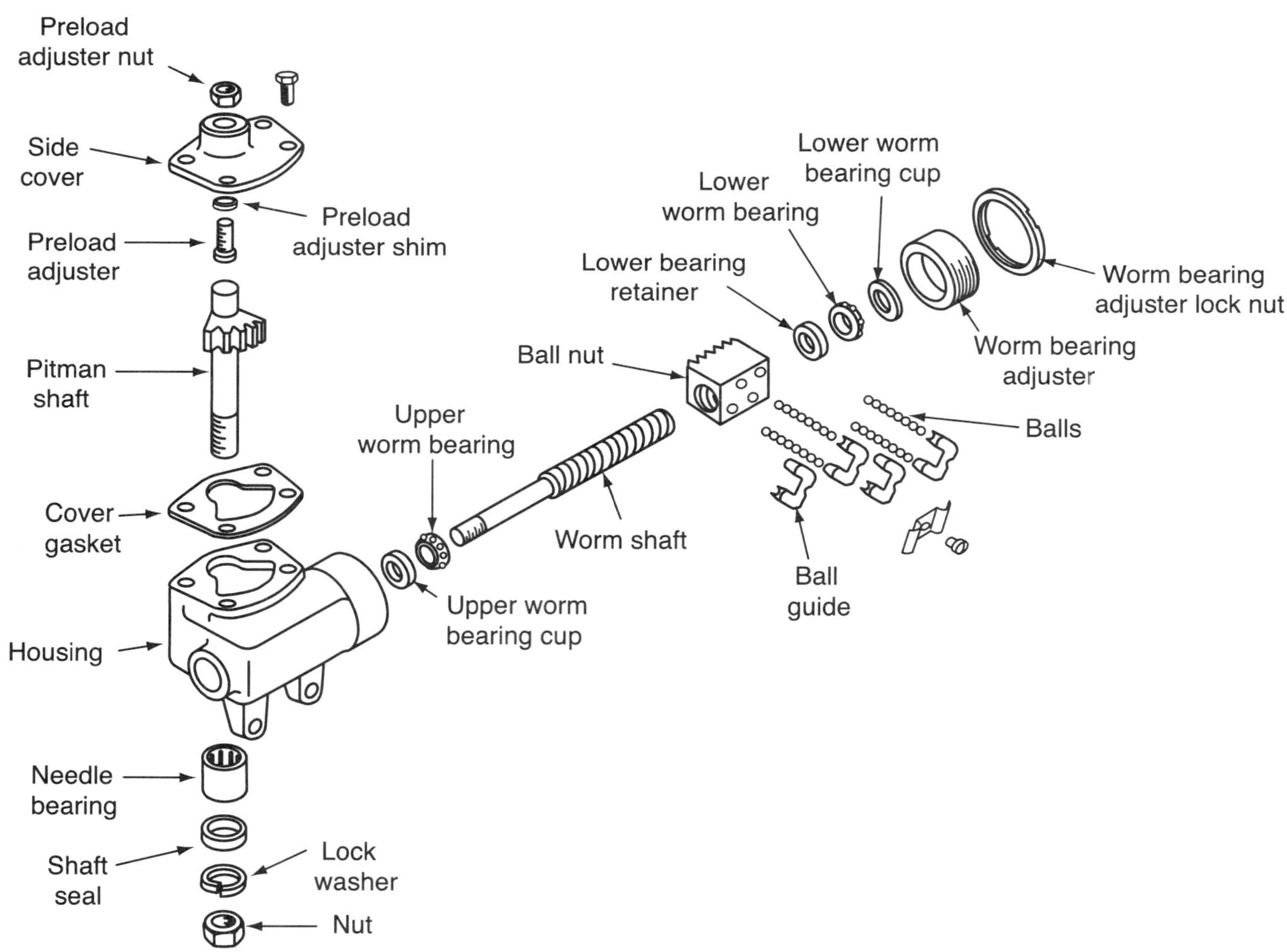


**Figure 2.** A manual recirculating ball steering gear assembly.

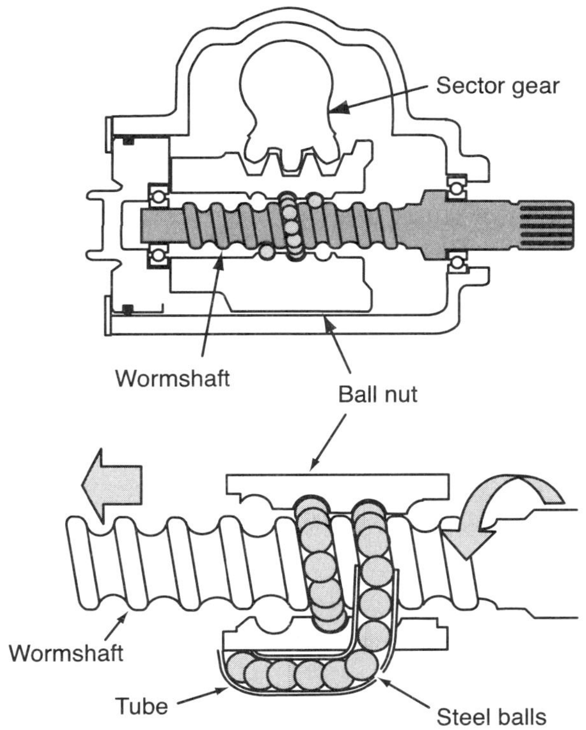


**Figure 3.** A cutaway of a recirculating ball steering gear.

**(Figure 3).** When the steering wheel is turned, the ball bearings roll in the thread grooves of the worm shaft and ball nut. When the ball bearings reach the end of their respective circuit, they enter the guide tubes and are returned to the other end of the circuit.

## Operation

When the worm shaft is rotated by the steering wheel, the ball nut is moved up or down on the worm shaft. The gear teeth on the ball nut are meshed with matching teeth on the pitman shaft sector. Therefore, ball nut movement causes the pitman shaft sector to rotate. The front wheels are turned by the pitman shaft sector through the pitman arm and steering linkage. The lower end of the pitman shaft sector is usually supported by a bushing or a needle bearing in the steering gear housing. A bushing in the sector cover supports the upper end of this shaft.

When the front wheels are straight ahead, an **interference fit** exists between the sector shaft and ball nut teeth. This interference fit eliminates gear tooth **backlash** when the front wheels are straight ahead and provides the driver with a positive feel of the road. Proper axial adjustment of the sector shaft is necessary to obtain the necessary interference fit between the sector shaft and worm shaft teeth. A sector shaft adjuster screw is threaded into the sector shaft cover to provide axial sector shaft adjustment **(Figure 4).** When two gears have an interference fit, the gear teeth are

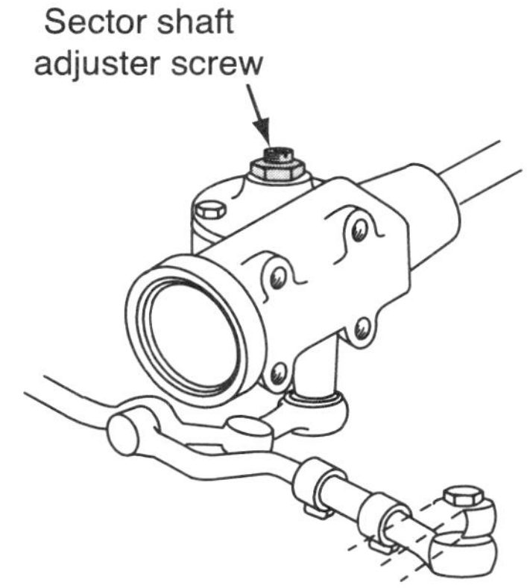


**Figure 4.** Typical steering gear adjustment screw locations.

meshed together so there is a slight tension, or preload, on the teeth.

> You Should Know
>
> *Proper adjustment between the worm gear and the pitman gear is important for adequate steering response.*

## Steering Ratios

Manual recirculating ball steering gears have sector gear teeth designed to provide a constant ratio, whereas power recirculating ball steering gears usually have sector gear teeth with a variable ratio **(Figure 5).** A constant ratio steering gear has a sector gear with identical gear teeth. In a variable ratio steering gear, one tooth is longer than the two on either side of it. This arrangement changes the amount of mechanical advantage according to the position of the wheel.

The **steering gear ratio** in a constant ratio manual steering gear is usually 15:1 or 16:1, whereas the average variable ratio steering gear ratio may be 13:1. Gear ratio refers to the relationship between the rotation of the drive and driven gears. If 13 turns of the drive gear are necessary

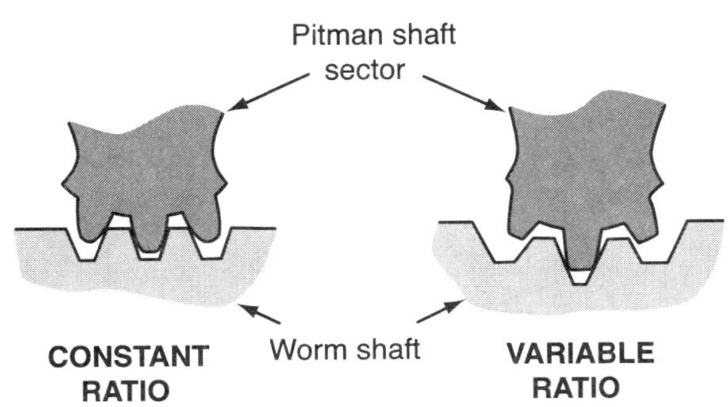


**Figure 5.** Constant and variable ratio steering gears.

to obtain 1 turn of the driven gear, the gear ratio is 13:1. When the same types of steering gears are compared, a higher numerical ratio provides reduced steering effort and increased steering wheel movement in relation to the amount of front wheel movement.

*A steering gear with a lower numerical ratio may be called a faster steering gear compared to a steering gear with a higher numerical gear ratio.*

## Mounting

Many recirculating ball steering gears are bolted to the frame with hard steel bolts **(Figure 6).** These bolts must be tightened to the specified torque.

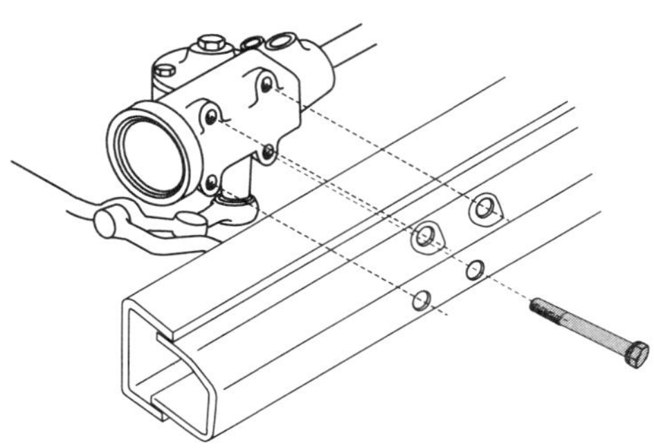

**Figure 6.** A steering gear mounting to the frame of a vehicle.

*Hard steel bolts may be used for steering gear mounting. Bolt hardness is indicated by ribs or numerical markings on the bolt head. Never substitute softer bolts for the original harder bolts. They may break, allowing the steering box to detach from the frame. This will result in the loss of steering control.*

## WORM AND ROLLER GEARS

The **worm and roller** gearbox is similar to the recirculating ball except a single roller replaces the balls and ball nut. This reduces internal friction. The steering linkage typically includes a pitman arm, center link, idler arm, and two tie-rod assemblies. The function of these components is the same as the parallelogram steering linkage.

In operation, the steering shaft rotates the worm gear. The worm gear, in turn, engages the roller, causing the roller shaft to turn. The shaft, in turn, moves the pitman arm.

## RACK-AND-PINION STEERING GEARS

In a rack-and-pinion steering system, steering input is received from a pinion gear attached to the steering column. This gear moves a toothed rack that is attached to the tie rods.

The **rack** is a toothed bar that slides back and forth in the housing. The rack replaces the idler and pitman arms of a parallelogram steering system. Bushings support the rack in the steering gear housing. Sideways movement of the rack pulls or pushes the tie rods and steers the wheels **(Figure 7).**

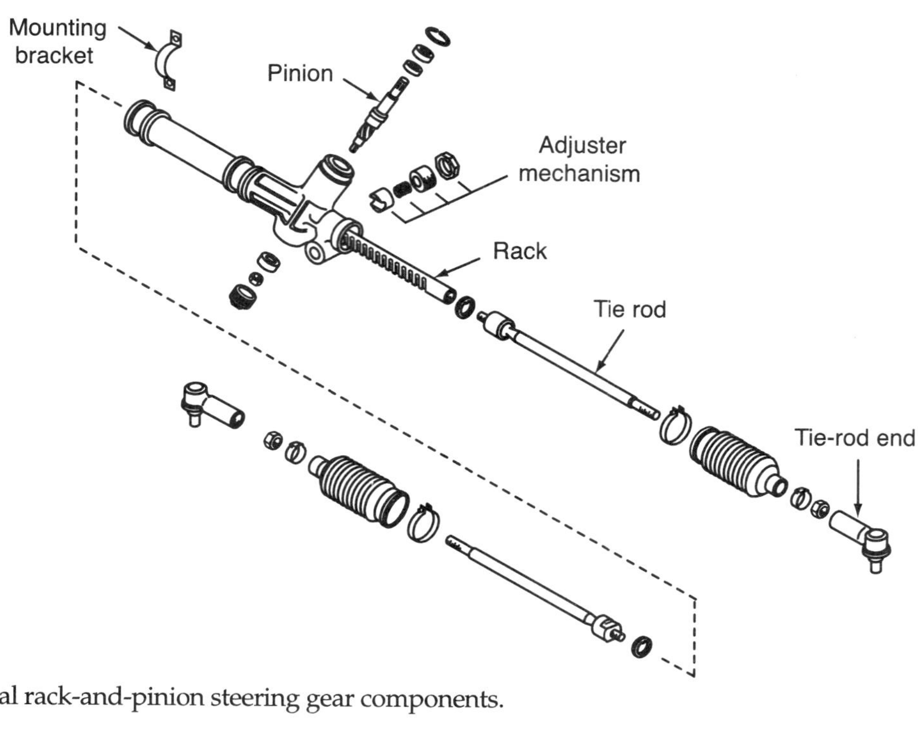


**Figure 7.** Manual rack-and-pinion steering gear components.

The **pinion** is a toothed shaft mounted in the housing, and its teeth are meshed with the rack's teeth. The upper end of the pinion shaft is connected to the steering shaft from the steering column. Therefore, steering wheel rotation moves the rack sideways to steer the front wheels. The pinion is supported on a ball bearing in the steering gear housing.

## Tie Rods

The tie rods are similar to those used on parallelogram steering linkages. A spring-loaded ball socket on the inner end of the tie rod is threaded onto the rack. When these ball sockets are torqued to the vehicle manufacturer's specification, a preload is placed on the ball socket. A bellows boot is clamped to the housing and tie rod on each side of the steering gear, and these boots keep contaminants out of the ball socket and rack **(Figure 8).**

A tie-rod end is threaded onto the outer end of each tie-rod. A jamb nut locks the outer tie-rod end to the tie rod.

## Rack Adjustment

A rack bearing is positioned against the smooth side of the rack. A spring is located between the rack bearing and the rack adjuster plug that is threaded into the housing. This adjuster plug is retained with a lock nut. The rack bearing adjustment sets the preload between the rack and pinion teeth, which affects steering harshness, noise, and kickback.

## Steering Gear Ratio

> **You Should Know** *Rotation of the steering wheel from extreme left to extreme right is called lock-to-lock, or stop-to-stop.*

When the steering wheel is rotated from lock-to-lock, the front wheels turn about 30 degrees in each direction from the straight-ahead position. Therefore, the total front wheel movement from left to right is approximately 60 degrees. With a steering ratio of 1:1, 1 degree of steering wheel rotation would turn the front wheels 1 degree, and 30 degrees of steering wheel rotation in either direction would result in lock-to-lock front wheel movement. With this steering ratio, the slightest steering wheel movement would cause the vehicle to swerve. The steering gear must have a ratio that allows more steering wheel rotation in relation to front wheel movement.

A steering ratio of 15:1 is acceptable; this ratio provides 1 degree of front wheel movement for every 15 degrees of steering wheel rotation. To calculate the steering ratio, divide the **lock-to-lock** steering wheel rotation in degrees by the total front wheel movement in degrees. For example, if the lock-to-lock steering wheel rotation is 3.5 turns, or 1,260 degrees, and the total front wheel movement is 60 degrees, the steering ratio is 1,260 ÷ 60 = 21:1. As a general rule, large, heavy cars have higher numerical steering ratios than small, lightweight cars.

## Gear Mounting

Large rubber bushings are positioned between the steering gear and the mounting brackets. These bushings help prevent the transfer of road noise and vibration from the steering gear to the chassis and passenger compartment. The rack-and-pinion housing may be attached to the front crossmember or to the cowl. Proper mounting is important to maintain the parallel relationship between the tie rods and the lower control arms.

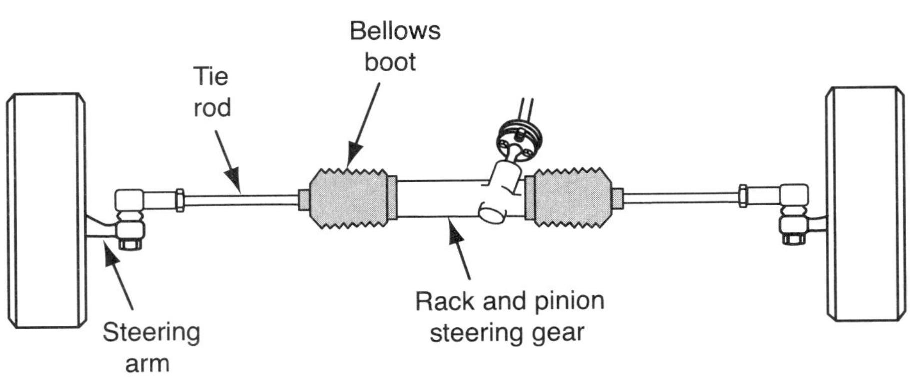


**Figure 8.** Rubber bellows boots are clamped to the housing and tie rod on each side of the steering gear to keep contaminants out of the ball socket and rack

## SUMMARY

- In a manual recirculating ball steering gear, a worm shaft adjusting plug provides a worm shaft preload adjustment.
- In a recirculating ball steering gear, rotation of the steering wheel causes the ball nut to move up and down on the worm shaft.
- With the front wheels straight ahead in a recirculating ball steering gear, an interference fit exists between the ball nut teeth and the sector shaft teeth.
- Axial sector shaft adjustment provides the proper interference fit between the ball nut teeth and the sector shaft teeth.
- Sector shaft teeth may have a constant ratio design or a variable ratio design.
- Steering ratio is the relationship between steering wheel movement and front wheel movement to the right or left.
- Rack-and-pinion steering systems are lighter and more compact than recirculating ball steering gears and parallelogram steering linkages.
- The rack replaces the idler arm and pitman arm of a parallelogram steering linkage.
- The inner tie rods are connected to the rack through spring-loaded inner ball sockets.
- The rack bearing and adjuster plug maintains proper preload between the rack-and-pinion teeth.
- A rack-and-pinion steering gear may be mounted on the firewall or front crossmember.
- A rack-and-pinion steering system has fewer friction points than a recirculating ball steering gear and parallelogram steering linkage.
- Reducing the number of friction points in a steering system provides more road feel and reduces the steering system's ability to isolate road vibration and noise.

## REVIEW QUESTIONS

1. Explain why a rack-and-pinion steering gear provides improved feel of the road compared to a recirculating ball steering gear and parallelogram steering linkage.
2. The preload on the rack-and-pinion teeth affects steering harshness, noise, and ____________________.
3. Define the term gearbox ratio.
4. Describe how a rack-and-pinion steering, parallelogram steering, and worm and roller system operate.
5. In a manual recirculating ball steering gear, the interference fit between the ball nut and sector shaft teeth is obtained by proper ____________________ sector shaft adjustment.
6. A variable ratio steering gear varies the amount of ____________________ ____________________ in relation to steering wheel position.
7. While discussing the purpose of worm shaft bearing adjustment, Technician A says the worm shaft bearing preload is adjusted with a screw in the sector shaft cover. Technician B says proper worm shaft bearing preload prevents steering kickback. Who is correct?
   A. Technician A
   B. Technician B
   C. Both Technician A and Technician B
   D. Neither Technician A nor Technician B
8. In a manual recirculating ball steering gear,
   A. the interference fit between the sector shaft teeth and ball nut teeth is present when the front wheels are straight ahead.
   B. the proper interference fit between the sector shaft teeth and ball nut teeth is obtained by radial sector shaft movement.
   C. the interference fit between the sector shaft teeth and ball nut teeth becomes tighter when the front wheels are turned to the right or left.
   D. tightening the worm shaft adjusting plug tightens the interference fit between the sector shaft teeth and ball nut teeth.

# Chapter 37 Steering System Diagnosis

## INTRODUCTION

It is important to realize that many steering concerns are caused by systems other than the steering system. Although customers may describe their concerns in different ways, the most common problems and their typical causes are discussed next.

### Excessive Steering Wheel Play

Excessive steering wheel play is apparent when there is too much steering wheel movement before the wheels begin to turn. A small amount of play is normal. This can be caused by the following:

- Loose, worn, or damaged steering linkages or tie-rod ends
- Loose, worn, or damaged steering column U-joints
- Loose, worn, or damaged steering column bearings
- Damaged or worn steering gear
- Worn idler arm or loose mounting
- Loose worm bearing preload adjustment
- Loose sector lash adjustment
- Worn steering gears
- Loose steering gear mounting bolts

### Feedback

When the driver feels the surface of the road through the steering wheel, this is called **feedback.** This problem can be caused by the following:

- Loose, worn, or damaged steering linkages or tie-rod ends
- Loose, worn, or damaged steering column U-joints
- Loose or damaged steering gear mounting bolts
- Damaged or worn steering column bearings
- Loose suspension bushings, fasteners, or ball joints

### Hard Steering

When extra effort is needed to turn the steering wheel, the cause may be simply an absence of power assist. Hard steering can also be caused by the following:

- A faulty power steering pump
- Column misaligned
- Damaged or faulty steering column bearings
- Seized steering column U-joints
- Internal problems in the power rack assembly
- Faulty suspension components
- Inadequately inflated tires
- Binding idler arm
- Tight worm bearing preload adjustment
- Tight sector lash adjustment
- Low steering gear lubricant
- Worn steering gears or bearings

### Uneven Steering Effort

When steering effort seems to become greater at some spots as the steering wheel is turned, the problem may be caused by:

- Worn steering gears or bearings
- Worn rack or pinion teeth
- Damaged rack assembly due to collision

### Pulling or Drifting

A pull is a tugging sensation felt at the steering wheel. The driver must push the steering wheel to keep the vehicle going straight. Drifting is a condition where the vehicle slowly moves to one side of the road when the driver's hands are taken off the steering wheel. This can be caused by the following:

- Improper frame alignment
- Brake system problem
- Worn or binding suspension components, especially springs
- Poor wheel alignment
- Unevenly loaded or overloaded vehicle
- Loose, worn, or damaged steering linkages or tie-rod ends
- Out-of-balance steering gear valve
- Torque steer
- Tire problems
- Binding strut bearing

## Shimmy

When the wheels shimmy, large, consistent rotational oscillations are felt at the steering wheel. These motions are caused by the lateral movement of the tires, which can result from:

- Loose, worn, or damaged steering linkages or tie-rod ends
- Loose, worn, or damaged suspension parts
- Out-of-balance tires
- Excessive wheel runout
- An unevenly worn or damaged tire
- Loose wheel bearings

## Nibble

Nibble is similar to a shimmy and results from the interaction of the tires to the road's surface. A customer may describe this as slight rotational oscillations of the steering wheel. This can be caused by loose, worn, or damaged steering linkages, tie-rod ends, or suspension parts.

## Sticking Steering or Poor Return

Poor returnability and sticky steering describes the steering wheel's resistance to return to center after a turn. This problem can be caused by the following:

- Binding steering column U-joints
- Loose, worn, or damaged steering linkages or tie-rod ends
- Binding steering gear
- Loose, damaged or worn suspension parts
- Poor wheel alignment
- Binding steering column bearings

## Wandering

When a vehicle wanders, the driver must turn the steering wheel to the left and right to keep the vehicle going straight on a level road. This can be caused by the following:

- Loose or worn suspension components
- Poor wheel alignment
- Unevenly loaded or overloaded vehicle
- Loose or damaged steering gear mounting bolts
- Loose steering column U-joint bolts
- Loose, worn, or binding steering linkages or tie-rod ends
- Improper steering gear preload adjustment
- Worn rack or pinion teeth

## Noises

There may be abnormal noises that accompany the turning of the steering wheel. The cause of these noises is best identified by paying close attention to where the noise is coming from. Some noises may be caused by tires or interference between the steering wheel and the steering column covers. Others can result from a faulty power steering pump or system.

- A hissing noise during low speed turning, such as while parking, is normal.
- A chirping or squealing normally means the drive belt is loose or slipping.
- Improperly routed hoses or loose mounting bolts on the power steering pump or the steering gear can cause a rattle.
- A groaning or howling can indicate a restriction in the hydraulic system.
- A sticking control valve will cause a buzzing sound.
- Rattling on rough road surfaces can be caused by loose or worn steering column couplings or U-joints, loose or worn idler arms and tie-rod ends, and a loose steering gear.
- A clunking sound while moving a tilt steering wheel can be caused by a worn universal joint, spherical bearing, tilt bumper, or damaged air bag clock spring.

## Diagnosis

Diagnosis should begin with trying to duplicate the customer's concern. This is done on a road test; make sure you drive carefully and cautiously, especially since the vehicle has a control problem. Drive the vehicle under conditions similar to the owner's normal driving. While it may be somewhat inconvenient to seek out a particular type of road surface, this may be the only way to verify the condition. Before going on the road test, conduct a thorough safety inspection of the vehicle that includes the tires.

Once the road test has been completed and it has been determined that there is an abnormal condition, conduct a thorough inspection.

# VISUAL INSPECTION

Begin the inspection of the steering system by inspecting the tires. Check for correct pressure, construction, size, wear, and damage, and for defects that include ply separations, sidewall knots, concentricity problems, and force problems. Tire wear patterns are good indicators of steering and suspension problems.

Also, check the tire and wheel assemblies for radial and lateral runout, static, and dynamic imbalance. Check the adjustment of the wheel bearings.

Continue by checking the drive belt for the power steering pump. If the belt shows evidence of cracking or glazing, it should be replaced. Adjust belt tension to specifications.

Also, check the fluid level and condition. The fluid is checked at the pump's reservoir with a dipstick attached to the reservoir cap. Check the fluid level after the engine has been run at idle for 2 or 3 minutes and the steering wheel has been cycled from lock to lock several times. This warms the fluid and gives a more accurate reading.

Examine the condition of the fluid. Check for evidence of contamination, such as solid particles or water. If the fluid is contaminated or has a burnt odor, the system should be flushed.

With the ignition OFF, wipe off the outside of the power steering pump, pressure hose, return hose, and the steering gear **(Figure 1).** Start the engine and turn the steering wheel several times from stop to stop. Check for leaks. Fluid leakage will not only cause abnormal noises, but may result in unequal and abnormal steering efforts. Other sources of fluid leaks are a worn pitman or worm shaft seal, defective cylinder or pinion end rack seals, or a bad control valve housing gasket.

On all systems, carefully check all of the mechanical parts of the steering and suspension system. Worn ball joints can cause erratic steering and premature tire wear. Bad suspension bushings will allow excessive wheel movement that can affect braking, handling, and wheel alignment. If any part is found to be defective, it should be replaced.

## Steering Column Checks

Steering columns can be the source of noises and vibrations if their mounting bolts and brackets are not securely tightened. If the steering wheel is not in its proper position, check the column to see if there is collision damage. The column jacket, gearshift tube, and steering shaft should be checked for sheared injected plastic. In addition, measure the distance from the end of the bearing assembly to the lower edge of the upper column jacket.

## Dry Park Check

An excellent overall check for worn or loose steering components is the **dry park check.** With the full weight of the vehicle on the wheels, have an assistant rock the steering wheel back and forth. Look for looseness in tie-rod, center link, idler arm, or pitman arm sockets **(Figure 2)** at both sides of the vehicle.

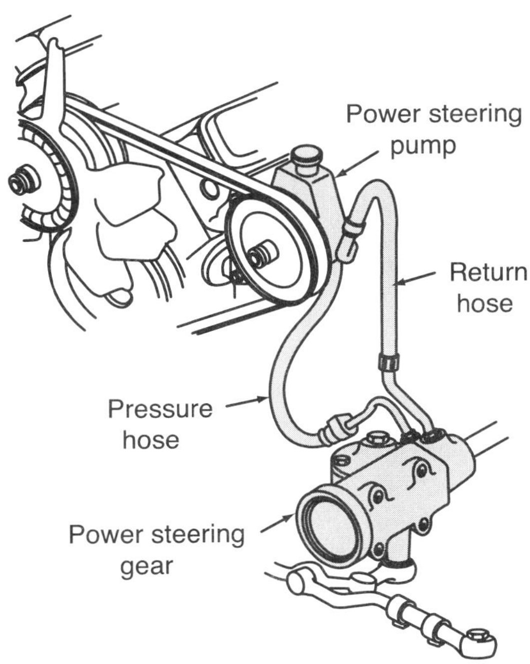


**Figure 1.** Visually inspect the power steering drive belt, pump, pressure hose, return hose, and steering gear.

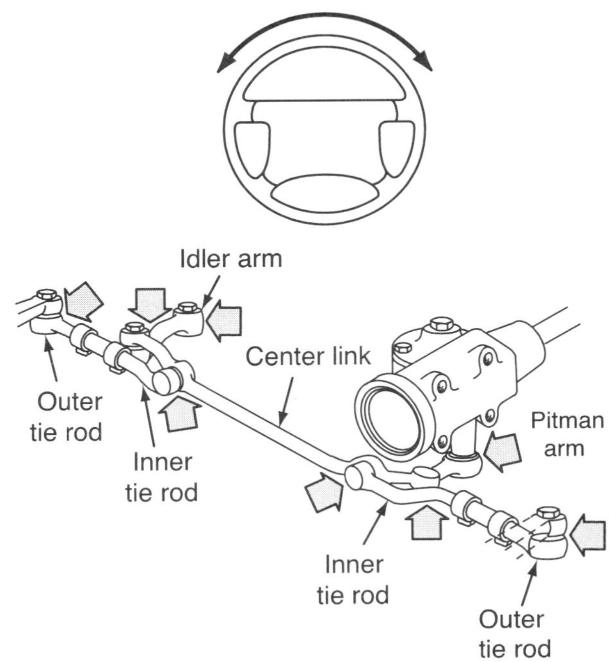


**Figure 2.** Circles areas indicate where a dry park check of the steering linkage should be made.

## Turning Effort

If excessive turning effort is required, a pull scale should be used to read the actual force required to turn the wheel **(Figure 3).** Compare the results to specifications. If the effort is excessive, check the entire steering system before performing a power steering pressure test.

### Tie-Rod Articulation Effort

The effort required to move the tie-rod or its inner ball socket should be checked with a pull scale if excessive steering effort or looseness is noted during the road test. If the effort is not within the specified limits, the tie-rod must be replaced.

## Steering Wheel Freeplay

With the engine stopped and the front wheels in the straight-ahead position, move the steering wheel in each direction with light finger pressure. Measure the amount of steering wheel movement before the front wheels begin to turn **(Figure 4).** This movement is referred to as **steering wheel freeplay.** Excessive steering wheel freeplay is caused by worn steering shaft universal joints or flexible coupling. Other causes are worn steering linkage mechanisms and a worn or out-of-adjustment steering gear.

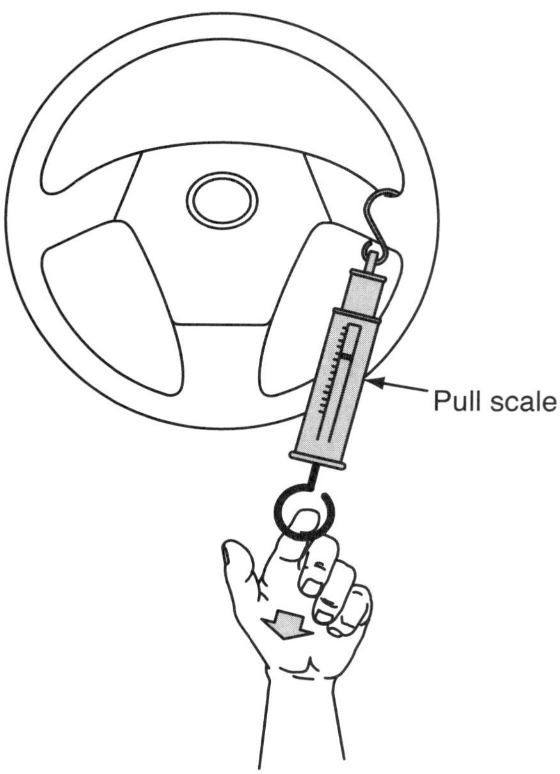


**Figure 3.** A pull scale is used to measure steering effort.

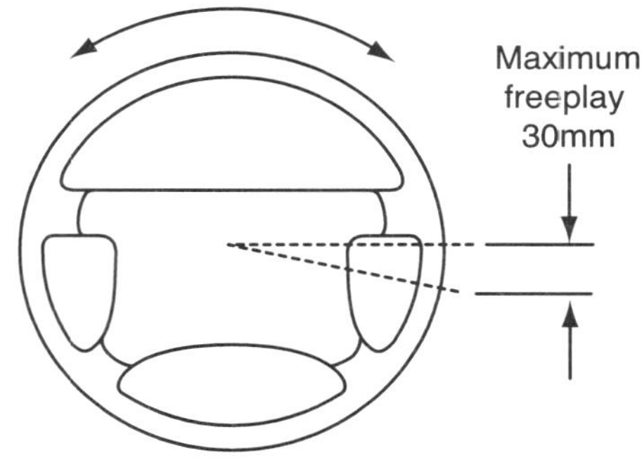


**Figure 4.** Steering wheel freeplay is the amount the steering wheel moves before the front wheels begin to turn.

A worn universal joint or flexible coupling in the steering column may also cause rattling noises. With the vehicle's weight resting on the front suspension, watch the flexible coupling or universal joint as an assistant turns the steering wheel one-half turn in each direction. The flexible coupling or universal joint must be replaced if it has excessive freeplay.

## RACK AND PINION STEERING

When inspecting a rack-and-pinion steering gear, check all steering system components, including the tires, and follow these steps:

1. With the front wheels straight ahead and the engine stopped, measure the maximum steering wheel freeplay and compare your findings to specifications. Excessive steering wheel freeplay indicates worn steering components.
2. With the vehicle sitting on the shop floor and the front wheels straight ahead, have an assistant turn the steering wheel about one-quarter turn in both directions. Watch for looseness in the steering shaft's flexible coupling or universal joints. If looseness is observed, replace the coupling or universal joint.
3. Grasp the pinion gear shaft at the flexible steering coupling and try to move it in and out of the gear. If it can be moved, a pinion bearing preload adjustment may be required **(Figure 5).** If the steering gear does not have provisions for pinion bearing preload, replace the necessary steering gear components.
4. Carefully inspect the rack housing. Have an assistant turn the steering wheel about one-half turn in both directions and watch for movement of the housing in its mounting bushings. If there is any movement, replace the bushings. Worn or damaged bushings can cause loose and erratic steering.
5. Road test the vehicle and check for excessive steering effort. A bent steering rack, tight rack bearing adjust-

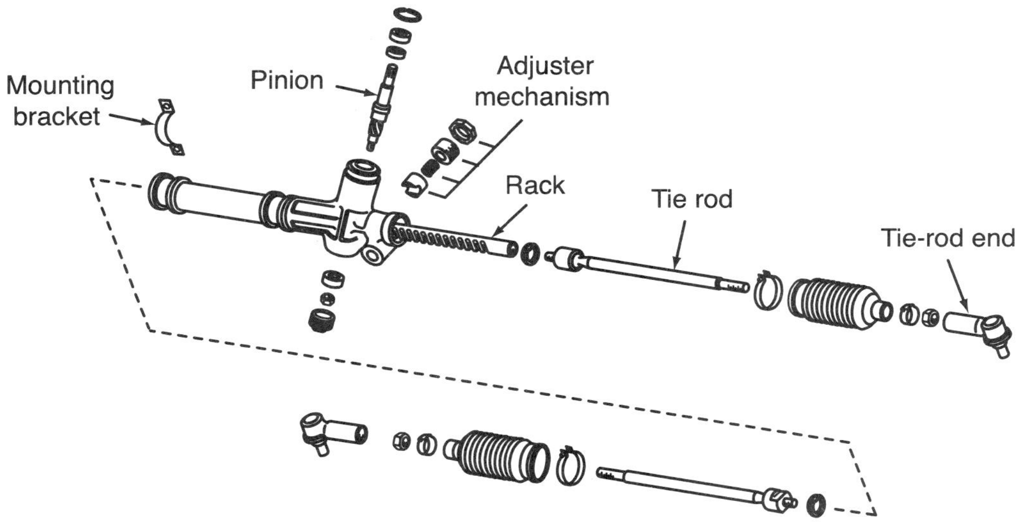


**Figure 5.** An exploded view of a rack and pinion steering gear with a pinion bearing preload adjuster.

ment, or damaged front drive axle joints in FWD cars may cause excessive steering effort. Also, check for abnormal under- or over-steer. These can result from a misalignment of the rack and pinion assembly due to collision damage.

6. Inspect the bellows boots for cracks, splits, leaks, and proper clamp installation. Replace any boot that is damaged. If the boot clamps are loose or improperly installed, tighten or replace the clamps. Boots should be inspected each time undercar service, such as oil and filter change or chassis lubrication, is performed. Any sign of fluid leaking from the boots indicates a leak in the hydraulic system of a power rack assembly.
7. Check the inner tie-rod socket assemblies inside the bellows **(Figure 6).** To do this, loosen the inner boot clamp and pull the boot back. The inner tie-rod socket

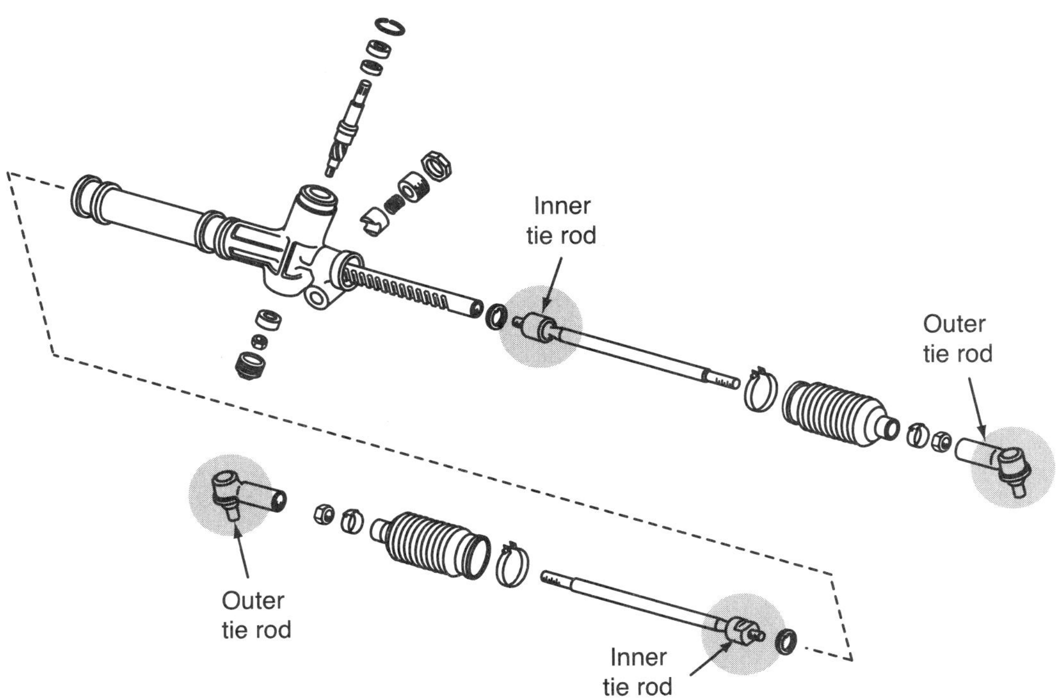


**Figure 6.** The wear points on the inner and outer tie rods in a rack and pinion steering gear.

can also be checked by squeezing the boot until the inner socket can be felt. Then, push and pull on the tire. If looseness is felt, the tie rod should be replaced. On some vehicles, the boot is made of hard plastic.

> **You Should Know** *Bent steering components must be replaced. Never straighten steering components because this action may weaken the metal and result in sudden component failure, serious personal injury, and vehicle damage.*

For this type of boot, lock the steering wheel and push and pull on the tire. Observe the tie rod; if it moves, replace it.

8. Grasp each outer tie-rod end and check for vertical movement **(Figure 7).** While an assistant turns the steering wheel one-quarter turn in each direction, watch for looseness in the outer tie-rod ends. If any looseness or vertical movement is present, replace the tie-rod end. During the dry park check, grab each tie rod end and feel for any roughness that would indicate internal rusting. Check the outer tie-rod end seals for cracks and proper installation of the nuts and cotter pins. Cracked seals must be replaced. Inspect the tie rods for a bent condition.

## RECIPROCATING BALL STEERING GEARS

Reciprocating ball steering gears are part of a parallelogram steering linkage system and have far more wear points than a rack and pinion steering system. The dry park check is still recommended to see if there are worn or loose steering components.

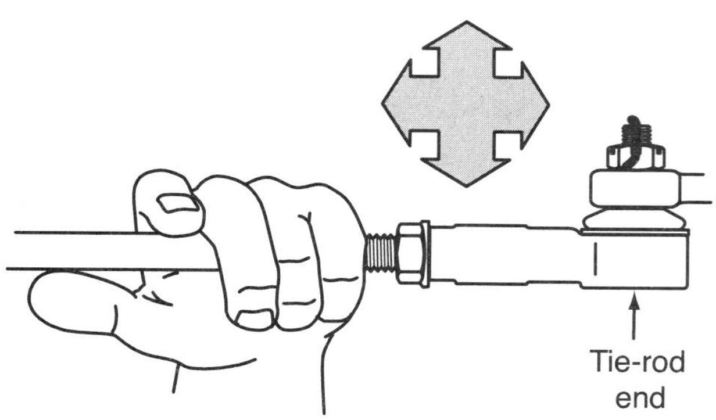


**Figure 7.** Check the outer tie-rod ends for looseness.

### Tie Rod Assembly

Worn tie-rod ends result in incorrect toe-in settings, scalloped and scuffed tires, wheel shimmy, understeering, or front-end noise and tire squeal on turns.

Grasp the tie-rod end firmly. Push vertically with the stud and inspect for movement at the joint with the steering knuckle. Any movement over 1/8 inch or observation of damaged or missing parts, such as seals, is sufficient evidence that replacement is necessary.

Adjusting sleeves resemble a piece of internally threaded pipe. They have a slot or separation that runs either their entire length or just part way. Adjusting sleeves also have two crimping or squeezing clamps located at each end to lock the toe adjustment. Badly rusted, worn, or damaged adjusting sleeves should be replaced.

> **You Should Know** *After adjusting the tie-rod sleeves, make sure they do not interfere with anything as they go through their full movement as the steering wheel is turned.*

Check the seals on each tie-rod end for damage and cracks. Cracked seals allow dirt to enter the pivoted joints, which result in rapid wear. If looseness or damaged seals are found on the tie rods, the components must be replaced.

An additional check of the tie-rods can be made by rotating each tie-rod end to feel for roughness or binding, which could indicate that the socket has probably rusted internally. A special puller is often required to separate a tie-rod end from the steering knuckle.

### Center Link

Worn or bent center links can cause front-end shimmy, vehicle pull to one side, or change in the toe setting, causing excessive tire wear. When inspecting the center link **(Figure 8)** look closely to see if it is bent or damaged. Grasp

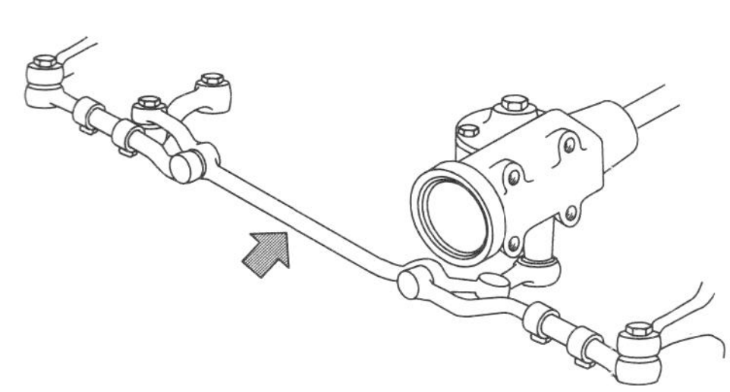

**Figure 8.** Carefully check the center link for looseness.

the center link firmly and try moving it in all directions. If the ball sockets are loose on either end of the rod, the center link assembly should be replaced. Tapered openings seldom wear but should be checked for enlargement caused by a loose connection. A bent or damaged center link must be replaced.

Check the seals at each pivot on the center link for damage and cracks. Cracked seals allow dirt to enter the pivoted joints, which result in rapid wear. If looseness or damaged seals are found on any pivoted joint on the center link, the link should be replaced.

## Pitman Arm

Because of its function, the pitman arm **(Figure 9)** is the most heavily stressed point in the system. To inspect the pitman arm, grasp it and vigorously shake it to detect any looseness. Check the socket to reveal any damage or looseness. Either condition must be corrected by replacing the worn part. Their removal normally requires the use of a special puller.

## Idler Arm

A worn or damaged idler arm **(Figure 10)** can cause steering instability, uneven tire wear, front-end shimmy, hard steering, excessive play in steering, or poor returnability. The idler arm is the weakest link in the steering linkage and wears quicker than the rest of the system.

To check an idler arm, place the vehicle with its suspension normally loaded on the ground or on an alignment rack. Check the idler arm ends for worn sockets or deteriorated bushings. Grasp the center link firmly with your hand at the idler arm end. Push up with approximately a 25-pound load. Then, pull down with the same load. The allowable movement of the idler arm and support assembly in one direction is –1/8 inch, for a total acceptable movement of 1/4-inch.

You Should Know *The test forces should not exceed 25 pounds, as even a new idler arm might be forced to show movement due to steel flexing when excessive pressure is applied.*

The load and movement can be accurately measured by using a dial indicator or pull spring scale located as near the center link end of the idler arm as possible. Attach the magnetic base of a dial indicator to the frame near the idler arm and position the dial indicator stem against the upper side on the outer end of the idler arm. Preload the dial indicator stem and zero the dial. Use a pull scale to apply 25 lbs. (11.34 kg) of force downward and upward on the idler arm **(Figure 11).** Check the total vertical movement on the dial

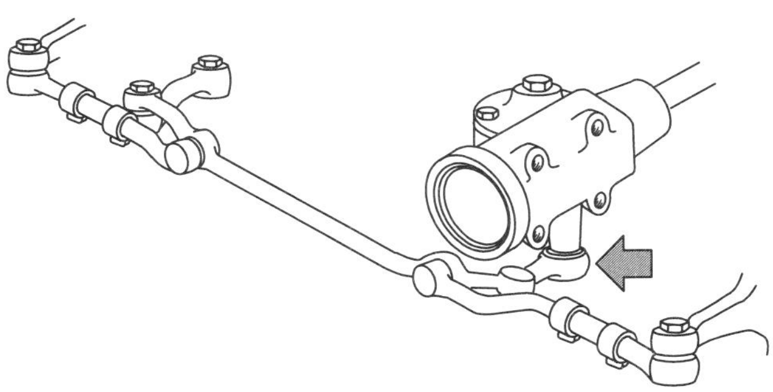

**Figure 9.** A pitman arm.

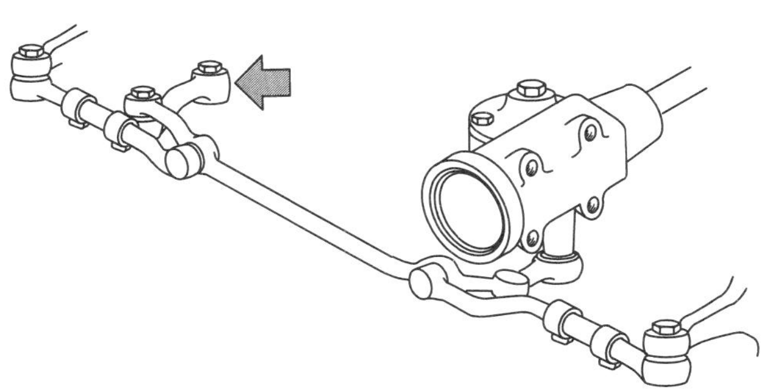

**Figure 10.** An idler arm.

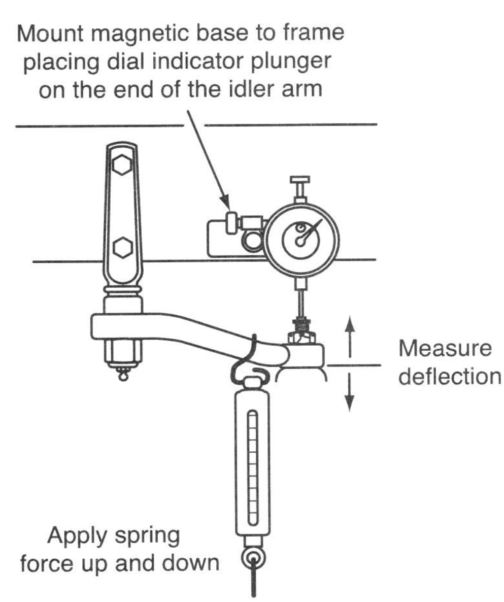


**Figure 11.** Using a pull scale and dial indicator to check idler arm wear.

indicator. If this vertical movement exceeds the manufacturer's specifications, replace the idler arm. Typical maximum vertical idler arm movement from the downward to the upward position is 0.250 in. (63.5 mm). The idler arm should always be replaced if it fails this test.

Binding idler arm bushings may also result in hard steering, a squawking noise when the front wheels are turned, and poor steering wheel returnability..

> **You Should Know** *A binding idler arm may suddenly break off and cause complete loss of steering control, vehicle damage, and personal injury.*

## Steering Damper

The steering dampers are generally nonadjustable, non-refillable, and are not repairable. Inspect the mountings and assembly for damage (such as being bent) and fluid leaks. A light film of fluid is permissible on the body of the damper near the shaft seal. A dripping damper should be replaced. A bad steering damper may cause wheel shimmy.

## Steering Arm Diagnosis

If the front rims have been damaged, the steering arms should be checked for a bent condition. Measure the distance from the center of the tie-rod stud to the edge of the rim on each side. Unequal readings may indicate a bent steering arm. Bent steering arms must be replaced.

## Worm and Roller Steering

Since the worm and roller steering linkage components are almost identical to those of a parallelogram linkage, the same methods of inspection are used.

# SUMMARY

- On a road test, the customer's concern should be verified if possible.
- Begin a visual inspection with an inspection of the tires.
- Check the power steering belt and fluid level before checking other steering components.
- Check the steering column for looseness or damage due to a collision.
- Conduct a dry park check to check for loose components.
- Check steering wheel freeplay to identify worn steering shaft universal joints or flexible couplings, steering linkage parts, and a worn or out-of-adjustment steering gear.
- On rack-and-pinion steering gears, check the tie-rod ends and the rack's mounting brackets. Also, inspect the boots for cracks, splits, leaks, and proper clamp installation.
- On parallelogram steering linkages, check the tie rods, center link, pitman arm, idler arm, steering arm, and steering damper.

# REVIEW QUESTIONS

1. Technician A says that the manual steering gear is the probable cause of a shimmy. Technician B says it could be loose steering linkage. Who is correct?
   A. Technician A
   B. Technician B
   C. Both Technician A and Technician B
   D. Neither Technician A nor Technician B
2. Technician A says adjusting the steering gear too tightly can cause hard steering. Technician B says adjusting the steering gear too tightly can cause poor returnability. Who is correct?
   A. Technician A
   B. Technician B
   C. Both Technician A and Technician B
   D. Neither Technician A nor Technician B
3. Technician A says a worn or damaged idler arm can cause front-end shimmy. Technician B says a worn or damaged idler arm can cause hard steering. Who is correct?
   A. Technician A
   B. Technician B
   C. Both Technician A and Technician B
   D. Neither Technician A nor Technician B
4. Worn tie-rod ends can result in ___________.
   A. scalloped and scuffed tires
   B. the vehicle pulls to one side
   C. poor returnability
   D. all of the above

5. An owner's concern indicates excessive turning effort. Technician A checks the alignment of the vehicle's frame. Technician B uses a pull scale. Who is correct?
   A. Technician A
   B. Technician B
   C. Both Technician A and Technician B
   D. Neither Technician A nor Technician B
6. Which one of the following is *not* a probable cause of wheel shimmy?
   A. Binding steering column U-joints
   B. Loose wheel bearings
   C. Out-of-balance tires
   D. Loose, worn, or damaged suspension parts
7. A preliminary steering inspection reveals that the steering wheel has nearly 1/2 inch of movement in either direction before the pitman arm begins to move. All of the following could cause this problem *except:*
   A. coupling joints.
   B. U joints.
   C. a tight or frozen shaft bearing.
   D. a worn or out-of-adjustment steering gear.
8. While discussing manual recirculating ball steering gear diagnosis, Technician A says excessive steering wheel freeplay may be caused by a worm shaft bearing preload adjustment that is tighter than normal. Technician B says excessive steering wheel freeplay may be caused by loose steering gear mounting bolts. Who is correct?
   A. Technician A
   B. Technician B
   C. Both Technician A and Technician B
   D. Neither Technician A nor Technician B
9. While discussing steering problems, Technician A says a "jerky" steering wheel and a "clunking" noise could indicate worn steering column U-joints. Technician B says lack of assist and a "growling" noise in a fluid-filled steering pump could indicate a hose or pump internal restriction. Who is correct?
   A. Technician A
   B. Technician B
   C. Both Technician A and Technician B
   D. Neither Technician A nor Technician B
10. While discussing steering effort, Technician A says a tight rack bearing adjustment may cause excessive steering effort. Technician B says a loose inner tie-rod end may cause excessive steering effort. Who is correct?
    A. Technician A
    B. Technician B
    C. Both Technician A and Technician B
    D. Neither Technician A nor Technician B

# Chapter 38 Air Bag Service

## INTRODUCTION

An air bag is much like a nylon balloon that quickly inflates to stop the forward movement of the occupant's upper body. An air bag's job during a crash is normally over in less than 1 second after it began.

The systems and parts used to deploy an air bag vary with the year and manufacturer of the vehicle, as well as the location of the air bag. An air bag is inflated or deployed by rapid expansion (an explosion) of a gas. The gas is fired by igniter, commonly called a **squib.**

Different manufacturers also call their air bag systems different names, such as **supplemental inflatable restraint (SIR)** and **supplemental restraint system (SRS).** All late-model vehicles have a driver-side and a passenger-side air bag, some have more **(Figure 1).**

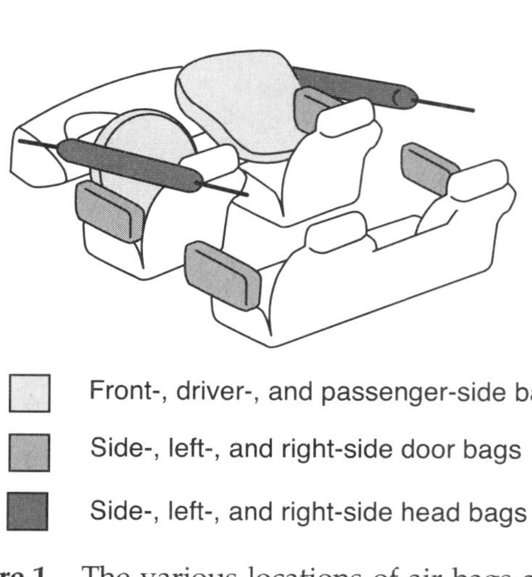


**Figure 1.** The various locations of air bags available today.

Passenger-side air bag modules are located in the vehicle's dash. These air bags are very similar in design and operation as the driver's side. However, many manufacturers use a different set of sensors. The actual capacity of gas required to inflate the passenger side air bag is much greater because the bag must span the extra distance between the occupant and the dashboard. The steering wheel and column make up this difference on the driver's side.

## ELECTRICAL SYSTEM COMPONENTS

The electrical circuit of an air bag system includes impact sensors and an electronic control module **(Figure 2).** The

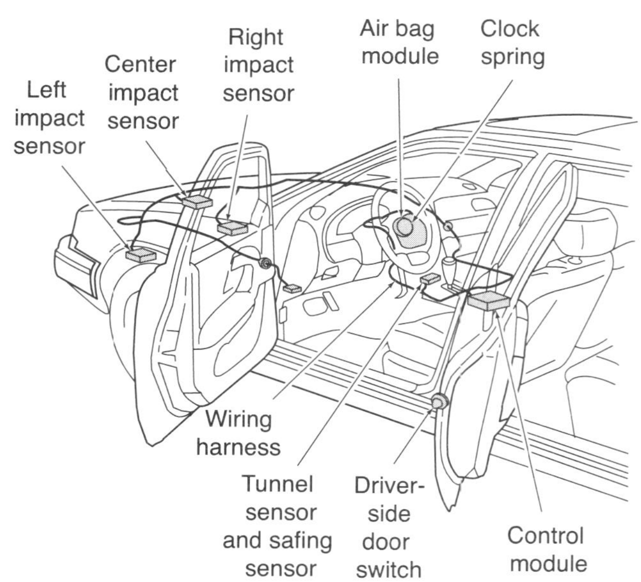


**Figure 2.** The typical layout of an air bag system.

electrical system conducts a system self-check to let the driver know that it is functioning properly, detects an impact, and sends a signal that inflates the air bag.

## Sensors

To prevent accidental deployment of the air bag, most systems require that at least two sensor switches be closed to deploy the air bag **(Figure 3).** The number of sensors used in a system depends on the design of the system. Some systems use a single sensor and others use up to five. The sensors are normally located in the engine and passenger compartments.

Typically, ignition of the air bag only occurs when an outside (impact or crash) sensor and an inside (safing or arming) sensor is closed. Once the sensors are closed, the electrical circuit to the igniter is complete. The igniter starts the chemical chain reaction that produces heat. The heat causes the generant to produce nitrogen gas, which fills the air bag.

### Roller-Type Sensors

This sensor has a roller located on a ramp **(Figure 4).** One terminal of the sensor is connected to the ramp. The second terminal is connected to a spring contact extending through an opening in the ramp but not touching the ramp. The roller is held against a stop by springs. If the vehicle is involved in a collision, the roller moves up the ramp and strikes the spring contact. This completes the circuit and causes the air bag to deploy.

### Mass-Type Sensor

This sensor contains a normally open set of contacts, a ball, and a sensing mass that is held in place by a magnet **(Figure 5).** When there is sufficient force, the ball breaks loose of the magnet and makes contact with the electrical contacts to complete the circuit.

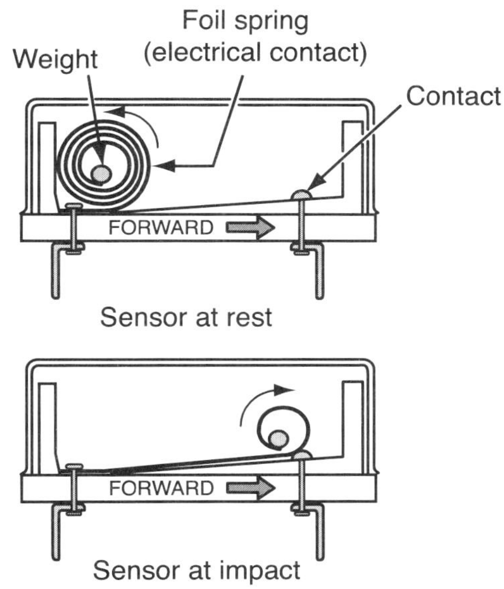


**Figure 4.** A roller-type air bag sensor.

### Accelerometer

This sensor contains a piezoelectric element that is distorted during a collision. This element generates an analog voltage in relation to deceleration forces. The sensor also senses the direction of impact force. Typically, this sensor is located in the air bag's computer.

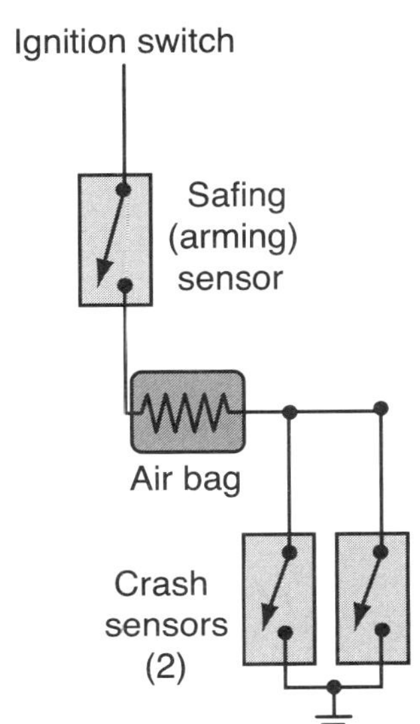


**Figure 3.** A simple air bag circuit with sensors.

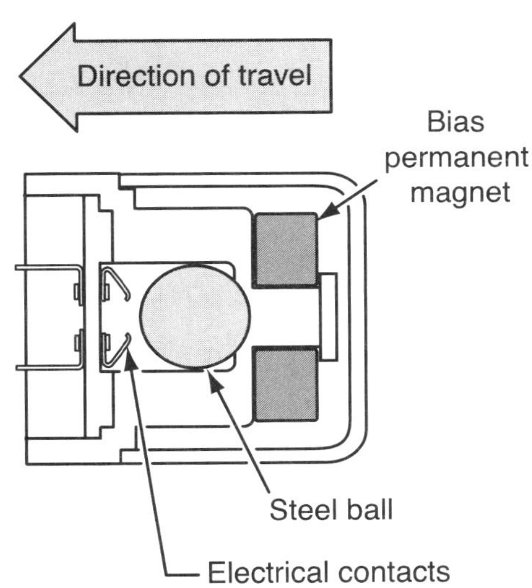


**Figure 5.** A typical ball and magnet sensor for an air bag system.

## Diagnostic Monitor Assembly

The **air bag diagnostic monitor (ASDM)** constantly monitors the readiness of the SIR electrical system. If the module determines there is a fault, it illuminates the warning lamp.

The diagnostic module also supplies backup power to the air bag module in the event that the battery or cables are damaged during the accident. The stored charge can last up to 30 minutes after the battery has been disconnected.

> You Should Know *The backup power supply must be depleted before any air bag service is performed. To deplete this backup power, disconnect the positive battery cable and wait at least 30 minutes. Refer to the service manual to determine exactly how long you should wait.*

## Wiring Harness

For identification and safety purposes, the electrical harnesses of the SIR system has yellow connectors. Single-stage air bags have one inflator and a pair of wires that connect to the air bag module. Two-stage air bags have two inflators and two pairs of wires connected to the module.

## Clock Spring

The **clock spring,** sometimes called a spiral cable, allows electrical contact to the air bag module at all times. The clock spring is designed to provide voltage to the module regardless of steering wheel position. It is located between the steering wheel and the column **(Figure 6).**

The wires of the air bag's electrical system are connected from the clock spring to a conductive ribbon. The other end of the ribbon is connected to the air bag module. When the steering wheel is turned, the ribbon coils and uncoils without breaking the electrical connection.

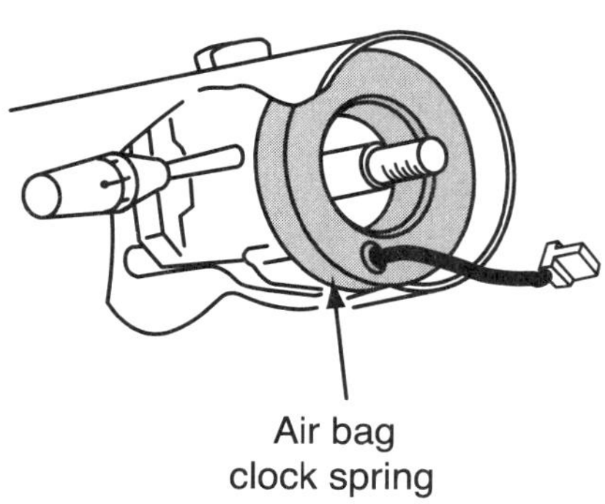


**Figure 6.** The electrical parts in a steering column, including the clock spring (spiral cable).

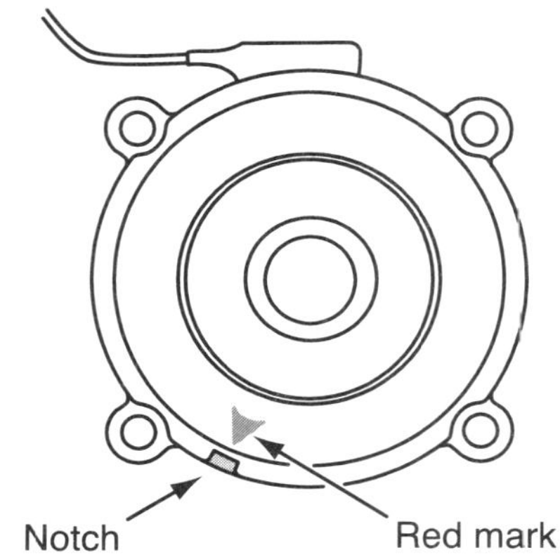


**Figure 7.** Centering the clock spring.

> You Should Know *Failure to center a clock spring prior to installation may cause a broken conductive tape in the clock spring.*

When installing the clock spring, make sure to turn its electrical connector clockwise three turns, and align the red mark on the center part of the spring's face with the notch in the cable circumference **(Figure 7).** This centers the clock spring electrical connector.

## SIR or Air Bag Readiness Light

The readiness light lets the driver know the air bag system is ready to do its job. The warning lamp is operated by the diagnostic module. The readiness lamp illuminates briefly when the ignition key is turned from off to run. The lamp should go out once the engine is running. A malfunction in the system causes the light to stay on or to flash. Some systems have a tone generator that sounds if there is a problem in the system or if the readiness light is not functioning.

On other vehicles, the air bag warning light flashes seven to nine times when the ignition switch is turned on and after the engine is started. Warning light operation varies with the make and model year of the vehicle. Always check the service manual for expected air bag warning light operation. When the warning light does not work as specified, the air bag system is defective, and the air bag or bags will probably not deploy if the vehicle is involved in a collision.

## AIR BAG MODULE

The air bag module is the air bag and inflator assembly packaged into a single unit. The module is located in the steering wheel or in the dash panel for the passenger side. The various types of side protection air bags have the module located at the point where the bag is deployed.

Inflation of an air bag is typically accomplished through a release of nitrogen gas. The igniter **(Figure 8)** is part of the inflator assembly. It starts a chemical reaction to inflate the air bag. At the center of the igniter assembly is the squib, which contains **zeronic potassium perchlorate (ZPP).** When voltage is supplied through the squib, an electrical arc forms between two pins. This ignites a canister of gas and causes a rapid expansion of the gas to deploy or inflate the air bag.

The inflation assembly is composed of a gas generator (called a generant) containing sodium azide and copper oxide or potassium nitrate propellant. During ignition, large quantities of hot, expanding nitrogen gas are produced very quickly. This expanding nitrogen gas quickly inflates the air bag.

> **You Should Know** *Wear gloves and eye protection when handling a deployed air bag module. Sodium hydroxide residue may remain on the bag and can cause a skin irritation.*

Not all air bags use nitrogen gas; some use a solid propellant and compressed argon gas. Argon has a stable structure, cools quickly, and is inert as well as nontoxic.

A mounting plate and retainer ring attach the air bag assembly to the inflator to keep the entire air bag module connected to the steering wheel.

The bag itself is made of a thin nylon fabric folded into the steering wheel, dash, seat, or door. The powdery substance released from the air bag when it is deployed is regular cornstarch or talcum powder. These powders are used to keep the air bag lubricated and pliable while they are in storage. The module must be serviced as a unit.

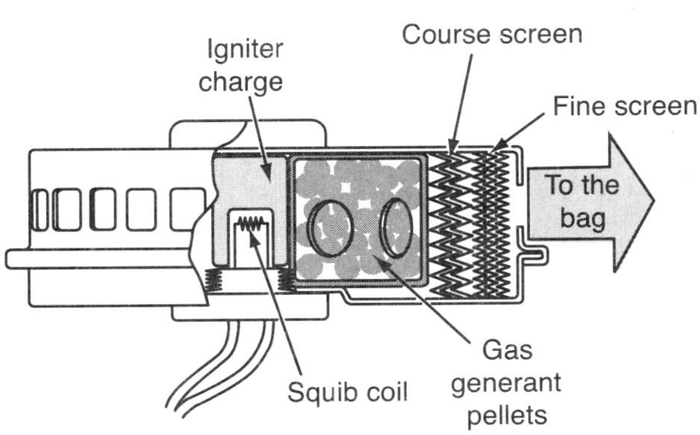


**Figure 8.** The operation of a typical air bag inflator.

## DIAGNOSIS

Perform a system check by observing the warning light. Typically, a problem is indicated by any of the following warning lamp conditions:

- If the lamp does not come on when the ignition is initially turned on
- If the lamp does not steadily light while the engine is cranking
- If the lamp remains on but does flash when the ignition is turned on
- If the lamp flashes seven to nine times and then remains on when the ignition is turned on
- If the lamp comes on while the engine is running

If any of these conditions is present, the air bag system needs to be checked. A thorough inspection of the sensors is the best place to start. Damage from a collision or from mishandling during a non-related repair can set up a fault that will disarm the air bag system.

## Retrieving Trouble Codes

The control module of most air bag systems will store trouble codes that can be retrieved by either a scan tool or flash codes. Since 1996, most systems require the use of a scan tool.

### Scan Tool DTC Retrieval

Normally an air bag system will store two types of faults. Active DTCs will turn the air bag warning lamp on, whereas stored codes are intermittent problems and probably will not turn on the warning lamp.

To retrieve codes, connect the scan tool to the DLC and turn the ignition on. Record all stored and active codes. Diagnose the cause of the codes in order from the lowest number to the highest.

### Flash Codes

On vehicles that display codes with the warning light or on the digital instrument panel, follow the specified procedure to retrieve the codes. Typically, the cause of an intermittent problem must be found through normal diagnostics and reasoning.

## SERVICING THE AIR BAG SYSTEM

Whenever working on or around air bag systems, it is important to follow all safety warnings. Examples of these warnings follow:

1. Wear safety glasses when servicing the air bag system.
2. Wear safety glasses when handling an air bag module.
3. Wait at least 30 minutes after disconnecting the battery before beginning any service on or around the air bag system. The reserve energy module is capable of

storing enough power to deploy the air bag for up to 30 minutes after battery voltage is lost.

4. Handle all air bag sensors with care.
5. When carrying a live air bag module, face the trim and bag away from your body.
6. Do not carry the module by its wires or connector.
7. When placing a live module on a bench, face the trim and air bag up.
8. Deployed air bags may have a powdery residue on them. Sodium hydroxide is produced by the deployment reaction and is converted to sodium carbonate when it comes into contact with atmospheric moisture. It is unlikely that sodium hydroxide will still be present. However, wear safety glasses and gloves when handling a deployed air bag. Wash your hands immediately after handling the bag.
9. A live air bag module must be deployed before disposal. Because the deployment happens by an explosive process, improper disposal may result in injury and fines. A deployed air bag should be disposed of according to EPA and manufacturer procedures.
10. Do not use a battery- or AC-powered voltmeter, ohmmeter, or any other type of test equipment not specified in the service manual.

**You Should Know** *A two-stage air bag may appear to be fully deployed when only its first stage has deployed. Make sure two-stage air bags have been fully deployed before handling them. Always assume that any deployed two-stage air bag has an active stage two. Improper handling or servicing can activate the inflator module and cause personal injury.*

## Service Guidelines

Any type of wiring concerns call for careful removal of the air bag module and installation of a simulator in its place. The simulator eliminates any chance of air bag deployment during diagnosis. If a condition is present in the system that would cause deployment of the air bag, the simulator will indicate it and will remain in that mode until the problem is fixed.

An air bag module is serviced as a complete assembly. A crash sensor that has been damaged should be replaced. It is a good idea to replace the entire set if a failure or degradation of any single sensor is found.

Before removing an air bag module or steering wheel, disconnect the negative battery cable and tape the terminal to prevent accidental connection with the battery **(Figure 9).** A piece of rubber hose can be used instead of tape. Then, remove the SIR fuse from the fuse panel. Wait at least 30 minutes to allow reserve energy to dissipate.

After the wait, remove the connector position assurance (CPA) from the yellow electrical connector at the base of the steering column. Then, disconnect the yellow electrical connector **(Figure 10).** Remove the bolts that secure the module, then rotate the horn lead one-quarter turn and disconnect it. Disconnect the electrical connectors at the module, and then remove the module **(Figure 11).** Replacement is the reverse procedure.

Keep in mind, the steering column's clock spring should be in its correct position at all times. Failure to do so can cause damage to the enclosure, wiring, or module. The clock spring should be replaced any time it has been removed.

Before returning the vehicle to the customer, make sure the sensors are firmly fastened to their mounting fixtures with their arrows facing forward. Also, make sure all fuses are correctly rated and replaced.

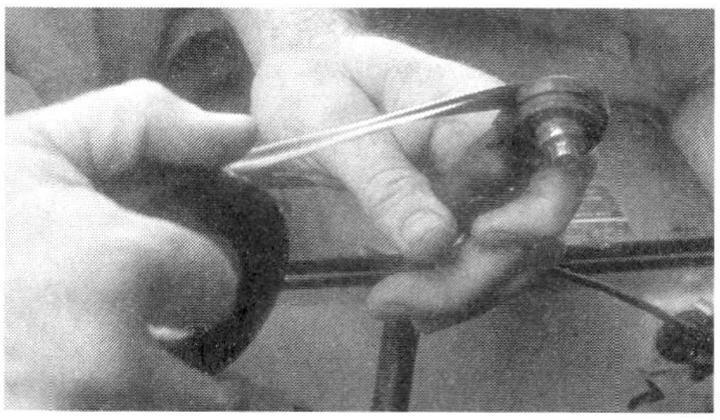

**Figure 9.** After the negative cable is disconnected, tape the cable's terminal to prevent accidental circuit completion.

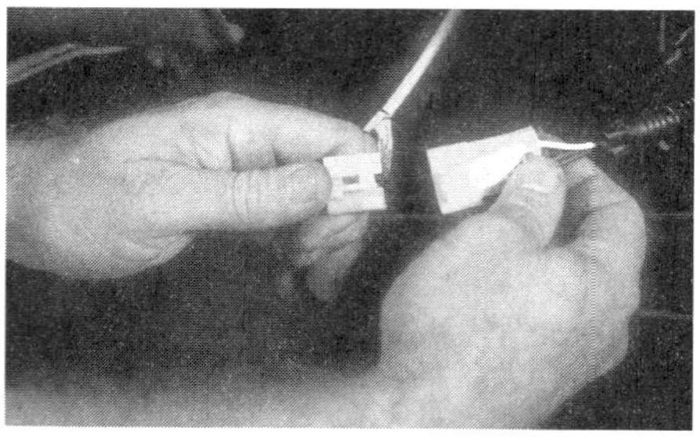

**Figure 10.** Disconnect the yellow two-wire electrical connector.

**Figure 11.** Lift the air bag module from the steering wheel.

## SUMMARY

- An air bag is inflated or deployed by rapid expansion of a gas that is fired by an igniter or squib.
- The electrical circuit of an air bag system includes impact sensors and a control module.
- The air bag module is the air bag and inflator assembly packaged into a single unit.
- A system check of an air bag system consists of observing the air bag warning light.
- The trouble codes stored in a control module can be retrieved by a scan tool.
- Before doing any work on an air bag system, disconnect the battery.
- Many air bag systems have a backup power supply circuit.
- Care must be taken when removing a live (not deployed) air bag. Be sure the bag and trim cover are pointed away from you.

## REVIEW QUESTIONS

1. If the clock spring's electrical connector is not properly centered before it is installed,
   A. the steering may pull to one side.
   B. the clock spring electrical connector may be broken.
   C. the air bag may deploy without the vehicle being in an accident.
   D. the steering may have a binding condition.
2. What air bag system device contains the ZPP?
3. *True or False:* The air bag diagnostic monitor supplies backup power to the air bag module in the event that the battery or cables are damaged during the accident.
4. *True or False:* The powdery substance released from the air bag when it is deployed is sodium hydroxide, which can cause skin irritation.
5. Technician A says that the module assembly will disarm the air bag system if certain faults occur. Technician B says that at least two sensors must signal the air bag in order to trigger it. Who is correct?
   A. Technician A
   B. Technician B
   C. Both Technician A and Technician B
   D. Neither Technician A nor Technician B
6. Which of the following is *false*?
   A. The air bag igniter assembly is a spark-plug-type device with two pins that current must jump across.
   B. The air bag igniter assembly creates a spark that ignites a canister of gas generating zerconic potassium perchlorate.
   C. The air bag igniter assembly is housed in the module assembly.
   D. None of the above.
7. Technician A checks the clock spring electrical connections for signs of damage when replacing a deployed air bag module. Technician B back probes an air bag system with a multimeter to determine if the system is in good working order. Who is correct?
   A. Technician A
   B. Technician B
   C. Both Technician A and Technician B
   D. Neither Technician A nor Technician B
8. Technician A says ignition of the air bag only occurs when an outside sensor and an inside sensor is closed. Technician B says the safing sensor determines if the collision is severe enough to inflate the air bag. Who is correct?
   A. Technician A
   B. Technician B
   C. Both Technician A and Technician B
   D. Neither Technician A nor Technician B
9. Technician A says the steering column clock spring should be maintained in its correct index position at all times. Technician B says failure to keep the clock spring straight ahead or in the neutral position relative to the steering wheel can cause damage to the enclosure, wiring, or module. Who is correct?
   A. Technician A
   B. Technician B
   C. Both Technician A and Technician B
   D. Neither Technician A nor Technician B
10. Which of these statements about air bag sensors is *not* true?
    A. Roller-type sensors rely on a roller held in place with a magnet. The circuit is closed when the roller moves and strikes the spring contact.
    B. Mass-type sensors rely on a ball and a magnet. The circuit is complete when the ball makes contact with the electrical contacts in the sensor.
    C. An accelerometer contains a piezoelectric element that generates an analog voltage in relation to the severity of the deceleration forces.
    D. An accelerometer also senses the direction of an impact force.

Chapter 39

# Steering Column and Linkage Service

## INTRODUCTION

The steering wheel and column seldom require service; however, they can cause problems. Steering linkages are serviced frequently because they are subjected to road conditions.

### STEERING WHEEL

A steering wheel can be damaged by abuse or as a result of a collision. It must be removed to be replaced and when repairs are made to the ignition switch, lock mechanisms, multi-function switches, and steering columns. The procedure for removing and installing a steering wheel varies with the model of vehicle. Always follow the recommended procedure. Also, be sure to follow all precautions for working with and around air bags.

#### Steering Wheel Removal

A typical way to remove a steering wheel begins with making sure the front wheels are straight ahead and the ignition switch in its LOCK position. Then disconnect the negative battery cable and wait the specified length of time to ensure the air bag cannot accidentally deploy.

After the wait, loosen and remove the air bag module retaining screws. Then, pull the module from the steering wheel and disconnect the module's electrical connector **(Figure 1).** Do not pull on the wires in the steering column. Place the module face up on the workbench. Now, disconnect the air bag wiring retainer in the steering wheel if necessary.

With a socket and ratchet, remove the steering wheel retaining nut **(Figure 2).** Observe the matching alignment

**Figure 1.** Lift the module away from the steering wheel so its electrical connector can be disconnected.

**Figure 2.** Loosen and remove the steering wheel's retaining bolt.

**Figure 3.** Check for alignment marks on the steering wheel and steering shaft.

marks on the steering wheel and steering shaft. If alignment marks are not obvious, make new alignment marks with a center punch and a hammer **(Figure 3).**

> **You Should Know** *Do not pull on the steering wheel in an attempt to remove it from the steering shaft. The steering wheel may suddenly come off, resulting in personal injury.*

Install a steering wheel puller with its bolts threaded into the bores in the steering wheel (do not use a jaw-type puller). Tighten the puller's nut to remove the steering wheel **(Figure 4).** Check the condition of the steering wheel. If the steering wheel is bent or cracked, replace it.

> **You Should Know** *If the steering wheel puller bolts are too long, they will extend through the steering wheel and damage the clock spring.*

**Figure 4.** A puller attached to the steering wheel.

Align the marks on the steering wheel and steering shaft, and install the steering wheel onto the shaft. Install the steering wheel retaining nut and tighten it to the specified torque.

Install the air bag wiring retainer in the steering wheel. Then, hold the air bag deployment module near the top of the steering wheel and connect the air bag module's electrical connector. Now install the module in the top of the steering wheel and tighten the retaining bolts or screws. The negative battery cable should now be reconnected.

## STEERING COLUMNS

To determine if the energy-absorbing steering column components are functioning as designed or if repairs are required, a close inspection should be made. An inspection is called for in all cases where damage is evident or whenever the vehicle is being repaired due to a front-end collision. If damage is evident, the affected parts must be replaced.

To perform services on the steering column's upper-end components, it is not necessary to remove the column. The steering wheel, horn components, directional signal switch, ignition switch, and lock cylinder can be removed with the column remaining in the vehicle.

### Steering Column Removal and Replacement

Steering column removal and replacement procedures vary with the vehicle make, type of steering column, and gearshift lever position. Always follow the recommended procedure in the service manual. The following is a typical procedure.

1. Set the parking brake.
2. Disconnect the battery negative cable. If the vehicle is equipped with an air bag, wait the prescribed amount of time before proceeding.
3. Place the front wheels in the straight-ahead position and remove the ignition key.
4. Remove the cover under the steering column and the lower finish panel if necessary.
5. Disconnect all wiring connectors from the steering column.
6. If the vehicle has a column-mounted gearshift lever, disconnect its linkage at the lower end of the column. If there is a floor-mount gearshift, disconnect the shift interlock.
7. Remove the retaining bolt or bolts in the lower universal joint or flexible coupling.
8. Remove the steering-column-to-instrument-panel mounting bolts.
9. Carefully remove the steering column. Be careful not to damage upholstery or paint.

10. Install the steering column and insert the steering shaft into the lower universal joint.
11. Install the steering-column-to-instrument-panel mounting bolts. Make sure the steering column is properly positioned, and tighten these bolts to the specified torque.
12. Install the retaining bolt or bolts in the lower universal joint or flexible coupling, and tighten the bolts to the specified torque.
13. Connect the gearshift linkage and/or shift interlock.
14. Connect all the wiring harness connectors to the steering column connectors.
15. Install the steering column cover and the lower finish panel.
16. Reconnect the negative battery cable.
17. Road test the vehicle and check for proper steering column operation.

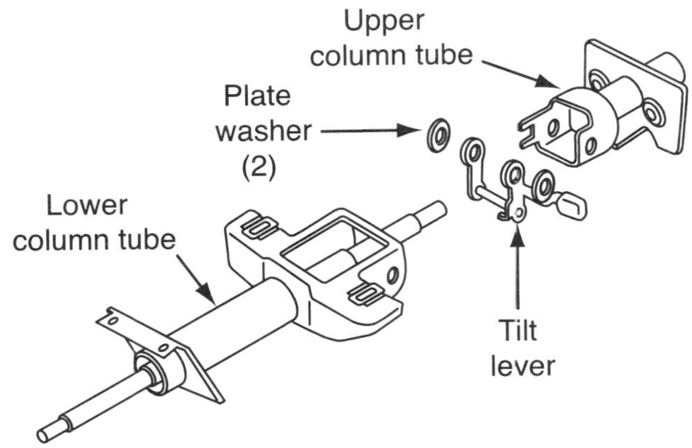


**Figure 5.** The basic assembly of a typical tilt-type steering column.

## Steering Column Flexible Coupling and Universal Joint Service

A worn universal joint or flexible coupling in the steering column can cause rattling noises. These noises may occur while driving straight on irregular road surfaces. With the vehicle's weight resting on the front suspension, watch the flexible coupling or universal joint as an assistant turns the steering wheel one-half turn in each direction. If the vehicle has power steering, the engine should be running with the gear selector in park. The flexible coupling or universal joint must be replaced if it has some freeplay.

If the flexible coupling must be replaced, loosen the coupling-to-steering-gear-stub-shaft bolt. Disconnect the steering column and move it rearward until the flexible coupling can be removed from the steering column shaft. Remove the coupling-to-steering-shaft bolts and disconnect the coupling from the shaft. When the new coupling and steering column are installed on some vehicles, the clearance between the coupling clamp and steering gear adjusting plug should be about 1/16 in. (1.5 mm). This specification may vary.

## Tilt Steering Column Removal and Replacement

The procedure for removing and installing a tilt-type steering column will vary with the model of vehicle, so always refer to the appropriate service manual before beginning. Make sure the negative cable of the battery is disconnected.

The covers and panel around the column must be removed, then the column can be unbolted from the instrument panel and the lower universal or flexible joint. The steering column has additional joints and a control lever that allows it to tilt **(Figure 5).** These must be disconnected. Also, remember to tighten all bolts and nuts to specifications.

# STEERING LINKAGES

The steering linkage should be carefully checked since it can be the source of many steering and handling problems. Any defective part must be replaced. Before assembling any part of the linkage, check all tapered holes for out-of-roundness and wear. Clean all bores that the studs mount into. Firmly install the tapered stud into its tapered hole. The stud must seat firmly without rocking. If the parts do not fit properly, a mating part is worn and must be replaced, or the correct parts are not being used.

## Pitman Arm Replacement

Some pitman arms have a ball socket joint on the outer end. The threaded extension on this ball socket fits into the center link. On other steering linkages, the ball socket joint is in the center link, and the threaded extension fits into the pitman arm. If the pitman arm is bent, the tie rod is not parallel to the lower control arm, and the pitman arm should be replaced.

To replace a typical pitman arm, make sure the front wheels are straight ahead. Remove the cotter pin and nut from the ball socket joint on the outer end of the pitman arm. Then, remove the ball socket extension from the pitman arm or center link with a tie-rod end puller.

Loosen the pitman-arm-to-pitman-shaft nut. Index the shaft and use a puller to pull the arm loose from the shaft

**Figure 6.** A puller being used to remove a pitman arm from the shaft of the steering gear assembly.

**(Figure 6).** Then remove the nut, lock washer, and pitman arm. Check the shaft's splines. If they are damaged, the shaft or steering gear box must be replaced.

Reverse the steps for removal to install a pitman arm. The pitman arm-to-shaft nut and ball socket extension nut must be tightened to the specifications. Make sure the pitman arm is in the correct position on the shaft splines. Install a new cotter pin in the ball socket extension.

## Center Link Replacement

A bent center link must be replaced. If the ball socket joints are loose on either end of the rod, the center link should be replaced. The center link should also be replaced if its ball stud openings are worn.

To replace a typical center link, remove the cotter pins from the tie-rod-to-center-link nuts and idler arm and pitman-arm-to-center-link nuts. Then remove the inner tie-rod inner, idler-arm-to-center-link, and pitman-arm-to-center-link ball socket extensions.

Use a tie-rod end puller to pull the inner tie rods, idler arm, and pitman arm from the center link, then remove the center link.

Reverse this procedure to install a center link. Tighten all the ball socket nuts to the specified torque, and install new cotter pins in all the nuts. If the ball sockets have grease fittings, lubricate them with a grease gun and chassis lubricant.

## Idler Arm Replacement

When an idler arm has excessive vertical movement, excessive toe change, front tire tread wear, excessive steering wheel freeplay, reduced steering control, and front end shimmy results. An idler arm with excessive vertical movement must be replaced.

A binding idler arm may suddenly break off and cause complete loss of steering control, therefore a binding idler arm should be replaced. Binding idler arm bushings can result in hard steering, a squawking noise, and poor steering wheel returnability.

The following is a typical idler arm removal and replacement procedure.

1. Remove the idler-arm-to-center-link cotter pin and nut.
2. Remove the center link from the idler arm.
3. Remove the idler arm bracket mounting bolts and remove the idler arm. If the idler arm's mount has slotted holes, make a reference mark so the new idler arm can be placed in the same position. A misalignment of the idler arm will affect bump steer.
4. Some idler arms contain steel bushings, others have rubber bushings. If the idler arm has a steel bushing, thread the bracket into the bushing until the specified clearance is obtained between the center of the lower bracket bolthole and the surface of the upper idler arm **(Figure 7).** If this clearance is incorrect, front wheel toe is affected.
5. Install the idler arm bracket to frame bolts and tighten the bolts to the specified torque. Be sure that lock washers are installed on the bolts.
6. Install the center link into the idler arm, and tighten the mounting nut to the specified torque. Install the cotter pin in the nut.
7. If the idler arm contains a grease fitting, lubricate as required.
8. After idler arm replacement, front wheel toe should be checked.

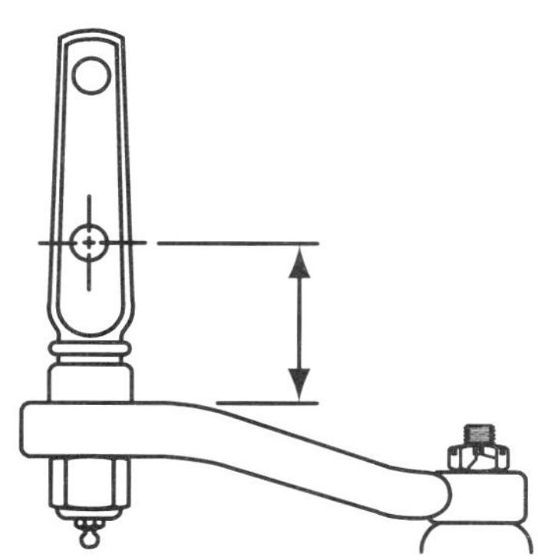

**Figure 7.** On some idler arms, the clearance between the center of the lower bracket bolt and the upper surface of the idler arm must meet specifications.

## Tie-Rod End Replacement

Worn tie-rod ends can cause excessive steering wheel freeplay, incorrect front wheel toe settings, tire squeal on turns, tread wear on front tires, front wheel shimmy, and a rattling noise on road irregularities.

To replace a tie-rod end, raise the vehicle with its tires supported by the lift. If the vehicle has rubber-encapsulated tie-rod ends, make sure the front wheels are pointed straight ahead. Measure from the top center of the tie-rod end ball stud to the outer end of the tie-rod sleeve. Record the measurement. Then loosen the nut on the tie-rod sleeve bolt.

The cotter pin and nut must be removed prior to tie-rod end replacement. A puller is used to remove the tie-rod end from the steering arm **(Figure 8).** On tie-rod ends with rubber-encapsulated ball studs, check for looseness of the ball stud and rubber capsule in the outer housing. If there is any looseness, replace the tie-rod end.

After the tie-rod end is removed from the steering arm, install two nuts on the stud's threads and tighten them against each other. Use a socket and a torque wrench to rotate the ball stud through a 40-degree arc **(Figure 9).** If the ball stud turning torque is less that 20 ft.-lb (27 N.m), replace the tie-rod end.

The tie-rod clamp must be loosened before the tie rod can be removed. Count the number of turns required to remove the tie-rod end from the sleeve and install the new tie rod with the same number of turns. Toe *must* be checked after the steering linkage components are replaced.

Before a new tie-rod end is installed, center the stud in the tie-rod end. Then, push the tie-rod end into the steering arm opening. With the tie-rod installed in the steering arm opening, only the threads should be above the steering arm surface. If the machined surface of the stud is visible above the surface or if the stud fits loosely in the steering arm, the opening is worn or the tie-rod end is not correct for that application. The tie-rod nut must be tightened to

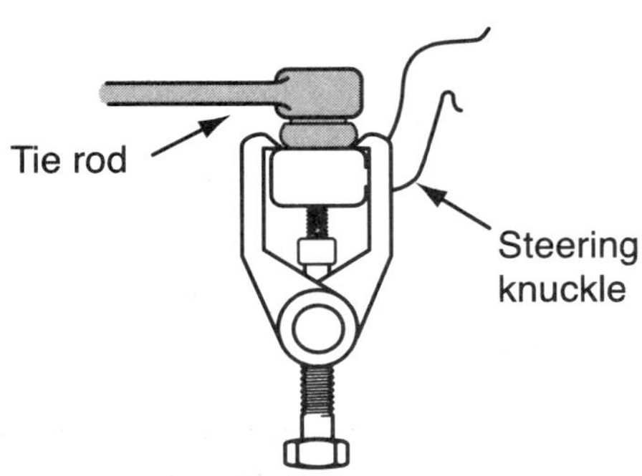


**Figure 8.** A puller is used to separate the tie-rod end from the arm of the steering knuckle.

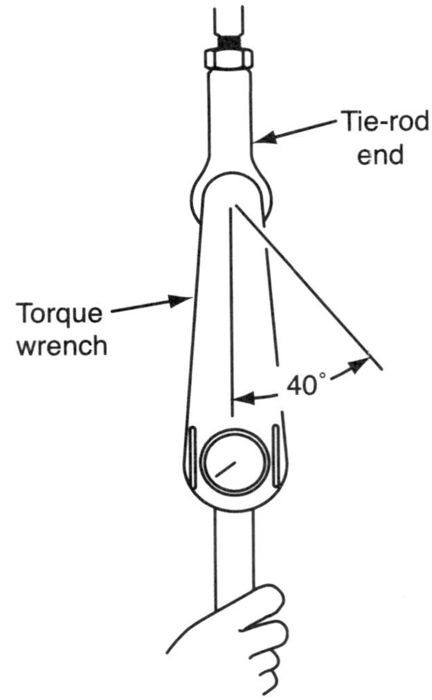


**Figure 9.** Measuring the turning torque on a rubber-encapsulated tie-rod end.

specifications, and a new cotter pin installed through the tie-rod end and nut openings **(Figure 10).** The tie-rod nut must never be loosened from the specified torque to install the cotter pin.

Another method of positioning a replacement tie-rod end is to measure the distance from the center of the tie-rod stud to the end of the sleeve prior to removal. When a new tie rod is installed, be sure this measurement is the

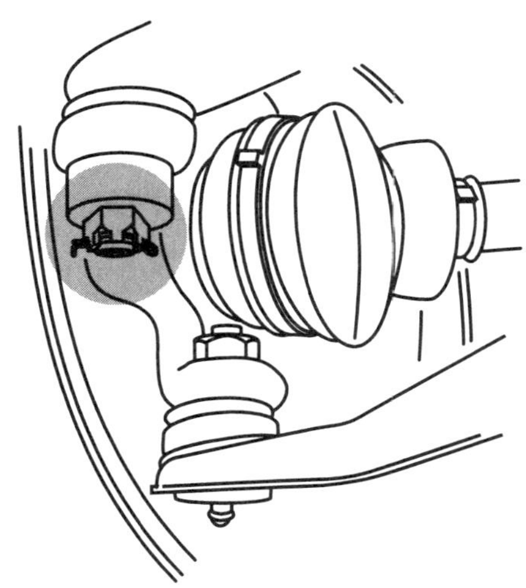

**Figure 10.** The cotter pin and nut for a tie-rod end.

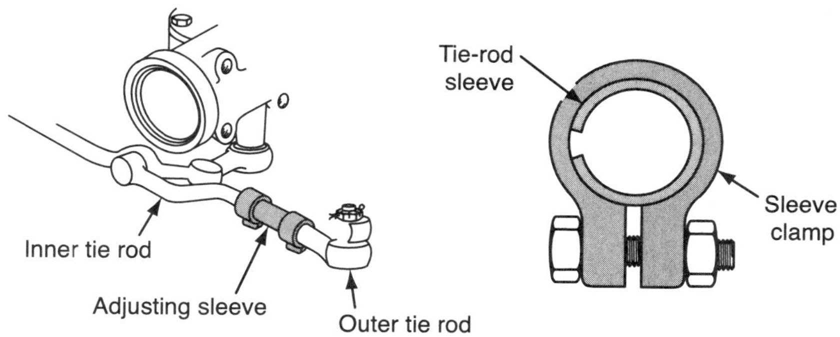


**Figure 11.** The proper slot and clamp position for tie-rod sleeves.

same. The slots in the tie-rod sleeve must be positioned away from the opening in the sleeve clamps **(Figure 11).** Leave the sleeve clamps loose until front wheel toe is checked, and then tighten the sleeve clamp bolts to the specified torque. A special tool is available to rotate the tie-rod sleeves and set the front wheel toe **(Figure 12)**.

**Figure 12.** A tie-rod sleeve adjusting tool.

## SUMMARY

- A steering wheel is removed to replace it or to service the steering column, ignition switch, lock mechanisms, and multi-function switches.
- The steering wheel is indexed to the steering column.
- Steering columns are removed by removing their covers, disconnecting all electrical connectors and linkages, and unbolting them from the lower universal joint or flexible coupling and the instrument panel.
- The procedure for removing and installing a tilt-type steering column is similar to that of a non-tilt column with the exception of the various joints used in the tilt mechanisms.
- Some idler arms contain steel bushings, and others are equipped with rubber bushings.
- Front toe should be checked after any part has been replaced in the steering linkage.

## REVIEW QUESTIONS

1. While discussing steering wheel removal, Technician A says a steering wheel should be marked in relation to the steering shaft. Technician B says the steering wheel should be grasped firmly with both hands and pulled from the shaft. Who is correct?
   A. Technician A
   B. Technician B
   C. Both Technician A and Technician B
   D. Neither Technician A nor Technician B
2. When diagnosing and servicing tie-rod ends,
   A. rubber encapsulated tie-rod ends should be tightened with the wheels turned fully to the right or left.
   B. the machined part of the tie-rod stud should be visible above the surface of the steering arm.
   C. the nut on the tie-rod stud may be loosened to install the cotter pin.
   D. if the turning torque on a rubber-encapsulated tie-rod end is less that specified, the tie-rod end must be replaced.
3. *True or False:* An adjustment must be made on an idler arm with a steel bushing to ensure the specified clearance is obtained between the center of the lower bracket bolthole and the upper idler arm surface.
4. *True or False:* When assembling the steering linkage, if a tapered ball stud fits tightly into its bore, the part with the bore should be replaced.
5. Which of the following parts do *not* require a special tool to separate them from their mating part?
   A. Center link
   B. Steering damper
   C. Idler arm
   D. Pitman arm

# Chapter 40 Recirculating Ball Steering Gear Service

## INTRODUCTION

Proper adjustment of a recirculating ball steering gear is critical to its operation. Adjustments are made whenever the steering gear has been serviced and when symptoms dictate the need. Loose steering gear adjustments cause excessive steering wheel freeplay and steering wander. Tight steering gear adjustments cause excessive steering effort.

Defective worm shaft bearings cause uneven turning effort and steering gear noise. Excessive steering effort may be caused by worn steering gears. A rattling noise from the steering gear may be caused by loose mounting bolts or worn steering shaft U-joints. In addition, excessive steering wheel freeplay may be caused by loose worm shaft bearing preload or sector shaft lash adjustments, worn steering gears, loose steering gear mounting bolts, and/or a worn steering shaft U-joint or flexible coupling.

### Lubricant

Proper lubrication is critical to the operation of a non-power steering gear. To check the level of the lubricant in the steering gear, remove the oil-level plug from the steering gear. The lubricant should be level with the bottom of the plug opening. If the oil level is low, fill the steering gear with the specified steering gear lubricant. A steering gear will cause binding and hard, uneven steering effort if the oil level is low. If the oil level is low, check the sector shaft and worm shaft seals or leaks. Leaking seals must be replaced.

## REMOVAL AND REPLACEMENT

Prior to removing the steering gear, disconnect the battery's ground cable. Wait the specified time to avoid accidental deployment of the air bag, then raise the vehicle with the front wheels in the straight-ahead position.

> **You Should Know** *Recirculating ball steering gears are often mounted near the exhaust manifold, which may be extremely hot. Wear protective gloves, and use caution when inspecting, adjusting, and servicing the steering gear.*

Remove the pitman arm nut and washer. Mark the pitman arm position in relation to the pitman shaft with a center punch, and use a puller to remove the pitman arm. Loosen the sector shaft backlash adjuster lock nut, and back off the adjuster screw one-quarter turn. Gently turn the steering wheel in one direction until it stops, and then turn the steering wheel back one-half turn toward the center position.

> **You Should Know** *When the steering linkage is disconnected from the gear, do not turn the steering wheel hard against the stops. Doing this may damage the ball guides in the steering gear.*

To remove the steering gear, disconnect the flexible coupling at the worm shaft. Then loosen and remove the steering-gear-to-frame mounting bolts. The steering gear may be cleaned externally with solvent or in a parts washer.

To reinstall the steering gear, put in position and install the mounting bolts. Then, connect the flexible coupling to the worm shaft. Make sure the steering gear is filled with the specified steering gear lubricant. Connect the pitman arm according to the alignment marks made during removal. Install and secure all parts that were removed, and then connect the battery. Make sure all bolts are tightened to the specified torque.

## DISASSEMBLY AND REASSEMBLY

Place the gear in a vise, and follow these steps to disassemble the steering gear:

1. Rotate the worm shaft until it is in the centered position, and use a center punch to mark the worm shaft in relation to the housing.
2. Remove the sector lash adjusting screw lock nut and cover attaching screws. Turn the lash adjusting screw clockwise through the cover to remove the cover and gasket.
3. Remove the lash adjusting screw and **selective shim** from the end of the pitman shaft and remove it from the housing. Place all components on clean shop towels.
4. Use a brass punch and hammer to loosen the worm shaft bearing adjuster lock nut and remove it with the bearing. Remove the bearing retainer with a slotted screwdriver.
5. Remove the worm shaft, ball nut assembly, and ball nut return guide. Turn the ball nut over and rotate the worm shaft to remove the steel balls. Remove the ball nut from the worm shaft.
6. Clean all parts in solvent, and inspect them for wear, scoring, or pitting.
7. Pry the pitman shaft seal from the housing with a slotted screwdriver.
8. Check for movement between the pitman shaft and the bearing in the housing. If movement is present, use the proper bearing driver to remove the bearing **(Figure 1)** and install a new one **(Figure 2).** If the pitman shaft is worn or scored in the bearing contact area, shaft replacement is necessary.
9. Remove the steering shaft seal with a slotted screwdriver and use the proper seal driver to tap the new seal into the housing.
10. If the upper and lower worm shaft bearing cups in the housing or adjuster nut are scored, pitted, or worn, install new bearing cups with the proper driver.

### Reassembly

Clean all parts and secure replacement parts for all damaged or worn components. Follow these steps to reassemble the steering gear:

1. Lubricate all seals, bushings, and bearings with the recommended lubricant.

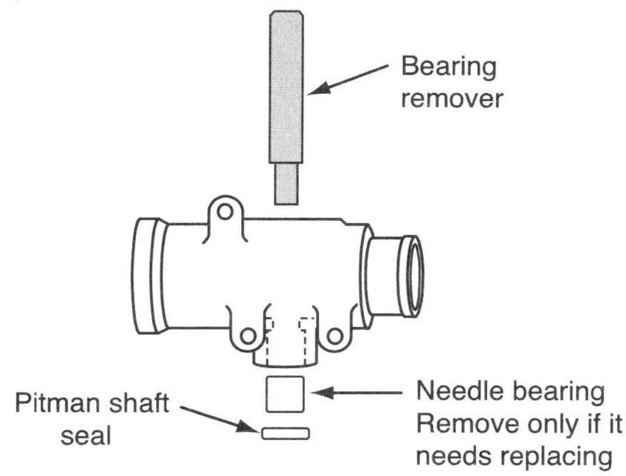


**Figure 1.** Removing the pitman shaft bearing with a driving tool.

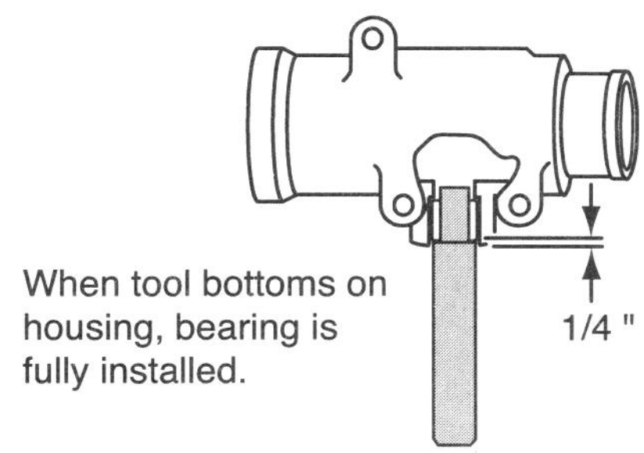


**Figure 2.** Installing a pitman shaft.

2. Position the ball nut on the worm shaft and install the balls in the return guides and ball nut. Place an equal number of balls in each ball nut circuit. Install the return guide clamp and tighten the screws.
3. Install the upper bearing over the worm shaft and center the ball nut on the worm gear. Slide the worm shaft assembly into the gear housing. Be careful not to damage the upper worm shaft seal.
4. Place the lower bearing on the worm shaft adjuster plug with its retainer, and install the adjuster and lock nut in the housing. Tighten the adjuster enough to hold the bearing in place.
5. Install the pitman shaft adjusting screw and shim on the pitman shaft. Measure the endplay between the screw and the slot. If this measurement exceeds specifications, replace the shim with one of the correct thickness.
6. Install the pitman shaft and adjusting screw in the housing.
7. Install the sector cover and gasket on the adjusting screw and turn the adjusting screw counterclockwise. Continue turning the adjusting screw until the cover and gasket contact the housing and the screw extends

5/8 to 3/4 inch above the cover. Install—but do not tighten—the adjusting screw lock nut. Install and tighten the cover screws.

8. Tighten the sector lash adjusting screw until the sector teeth and the ball nut teeth are meshed. There should be some movement between the teeth. Final adjustment is made later.
9. Wrap masking tape around the pitman shaft splines to protect the seal, and install the pitman shaft seal with the proper driver.
10. Fill the steering gear housing with the recommended lubricant, and turn the worm shaft fully in each direction several times to check for binding.

# ADJUSTMENTS

After the steering gear is assembled or to correct problems, the worm shaft bearing preload and sector lash must be adjusted **(Figure 3).**

## Worm Shaft Bearing Preload Adjustment

If the worm shaft bearing preload adjustment is loose, steering wheel freeplay is excessive. This results in steering wander. Steering effort increases if worm shaft bearing preload is too tight. Use the following procedure to adjust the worm shaft bearing preload.

1. Loosen the worm shaft (sector lash) adjuster plug lock nut with a brass punch and hammer **(Figure 4)** or appropriate tool **(Figure 5),** and tighten the plug until all worm shaft end play is removed. Then loosen the plug one-quarter turn.
2. Turn the worm shaft fully to the right with an inch-pound torque wrench, and then turn the worm shaft one-half turn toward the center position.
3. Tighten the adjuster plug until the specified bearing preload is indicated on the torque wrench **(Figure 6).**

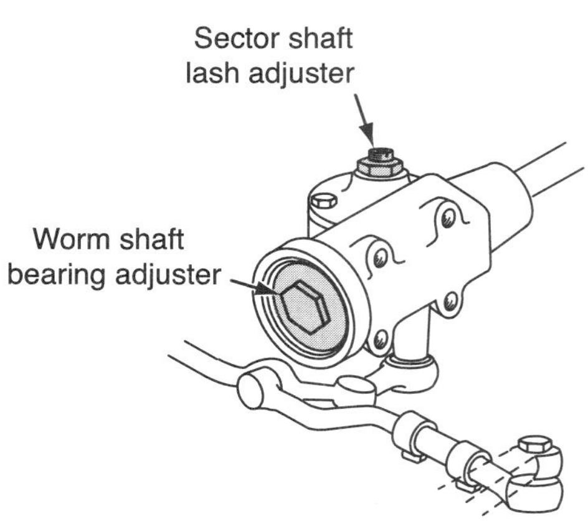


**Figure 3.** Points on the steering gear to make adjustments.

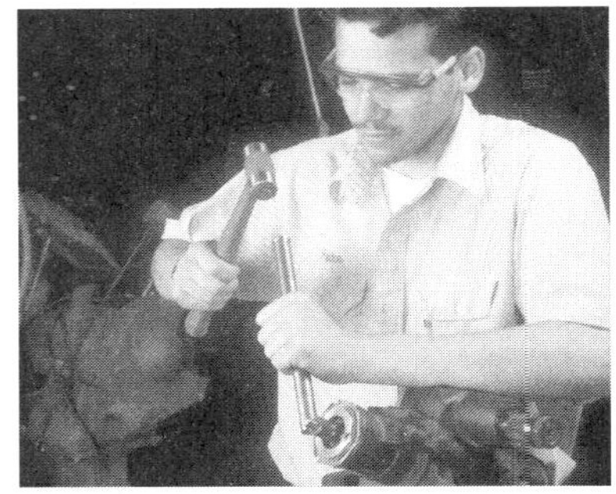

**Figure 4.** Remove the worm shaft bearing locknut with a hammer and brass punch.

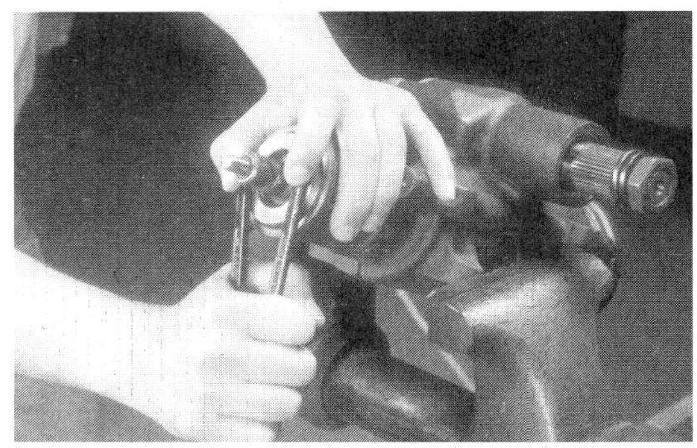

**Figure 5.** A special tool can be used to turn the worm shaft adjuster plug.

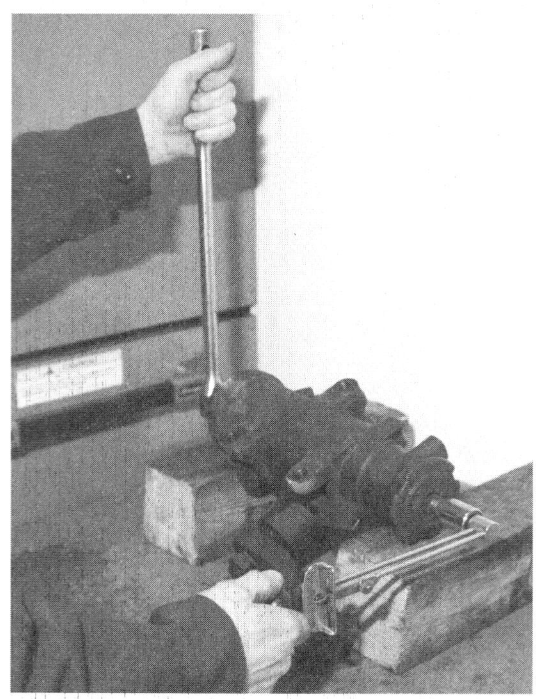

**Figure 6.** Adjusting worm shaft bearing preload.

The specification on some steering gears is 5 to 8 in.-lb (0.56 to 0.896 N.m). Always use the specified preload.

4. Tighten the adjuster plug lock nut to specifications.

## Pitman Sector Shaft Lash Adjustment

When the sector shaft lash adjustment is too loose, steering wheel freeplay is excessive and the vehicle wanders. If the sector lash adjustment is too tight, steering effort is increased, especially with the front wheels in the straight-ahead position. The following procedure may be used when the pitman shaft sector lash is adjusted.

1. Turn the backlash adjuster screw out until it stops, and then turn it in one turn.
2. Rotate the worm shaft from one stop to the other, and carefully count the number of shaft rotations.
3. Starting at either stop, turn the stub shaft back two-thirds of the total number of turns. In this position, the flat on the stub shaft should be facing upward **(Figure 7),** and the master spline on the pitman shaft should be aligned with the pitman shaft backlash adjuster screw **(Figure 8).**
4. Connect an inch-pound torque wrench to the worm shaft. Turn the stub shaft through a 45-degree arc on each side of the position in step 3. Read the torque as the stub shaft turns through the center position.
5. Turn the pitman backlash adjuster screw until the torque wrench reading is 6 to 10 in. lb (0.44 to 1.12 N.m) more than the worm shaft bearing preload torque in step 4 **(Figure 9).**
6. Hold the pitman shaft adjuster screw in this position and tighten the lock nut to the specified torque.

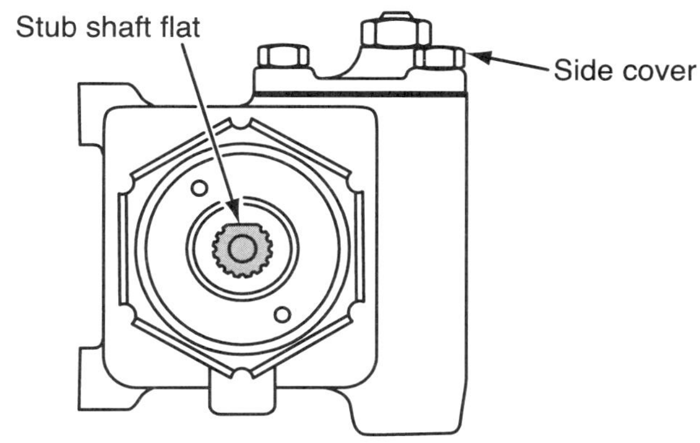


**Figure 7.** A stub shaft flat facing upward and parallel with the side cover.

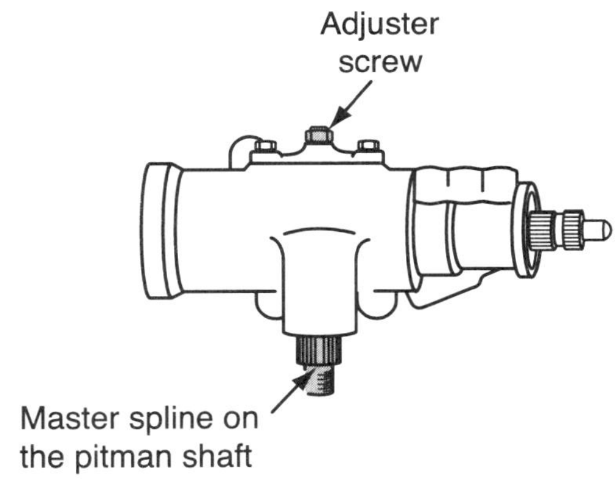


**Figure 8.** The pitman shaft master spline aligned with the pitman backlash adjuster screw.

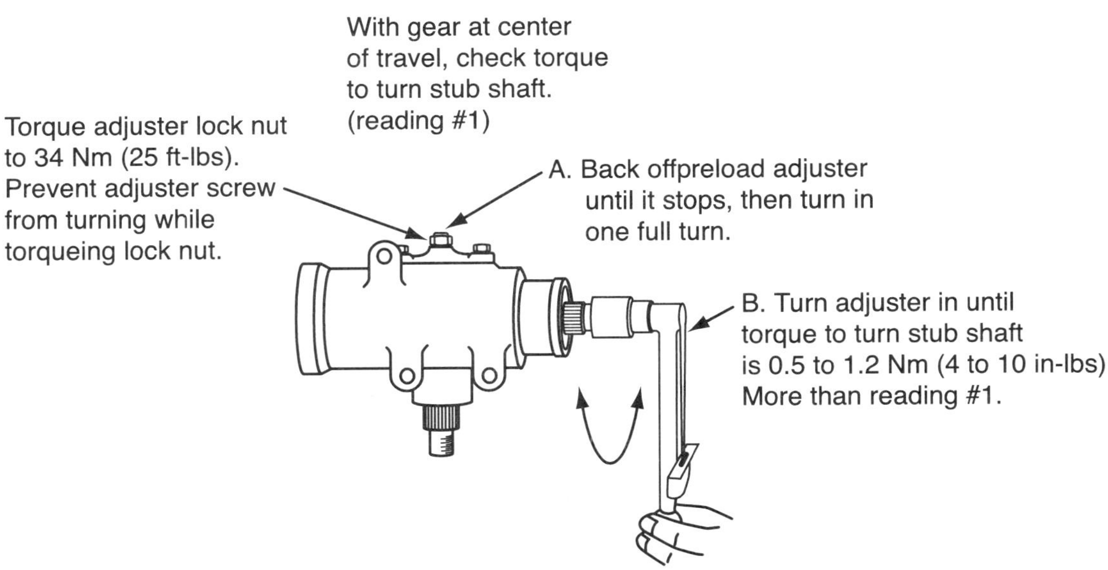


**Figure 9.** Adjusting the pitman backlash adjuster screw.

## SUMMARY

- Loose steering gear adjustments result in excessive steering wheel freeplay and wander.
- Tight steering gear adjustments cause excessive steering effort.
- Low lubricant levels can cause binding and hard, uneven steering effort.
- The adjustments of a steering gear should be checked before and after the steering gear has been removed from the vehicle.
- The worm shaft bearing preload adjuster plug is rotated until the correct worm shaft turning torque is obtained with the worm shaft positioned one-half turn from the full-right or full-left position.
- The sector lash adjusting screw is tightened until the correct worm shaft turning torque is obtained as the worm shaft is rotated through the center position.

## REVIEW QUESTIONS

1. While diagnosing a manual recirculating ball steering gear, Technician A says the worm shaft bearing preload adjustment is performed with the sector lash adjuster screw backed off one-quarter turn. Technician B says the worm shaft bearing preload adjustment is performed with the worm shaft turned fully to the right or left. Who is correct?
   A. Technician A
   B. Technician B
   C. Both Technician A and Technician B
   D. Neither Technician A nor Technician B
2. While discussing manual recirculating ball steering gear service, Technician A says if the end play between the sector lash screw and the slot is excessive, selective shims are available to correct the end play. Technician B says excessive steering effort may be caused by a wheel alignment problem. Who is correct?
   A. Technician A
   B. Technician B
   C. Both Technician A and Technician B
   D. Neither Technician A nor Technician B
3. *True or False:* If the upper and lower worm shaft bearing cups in the steering gear housing or adjuster nut are scored, pitted, or worn, use a slide-hammer-type puller to remove these cups. Install the new bearing cups with a hammer.
4. While servicing a manual recirculating ball steering gear, Technician A says the worm shaft bearing preload adjuster plug is rotated until the correct worm shaft turning torque is obtained with the worm shaft positioned at the full-right or full-left position. Technician B says the sector lash adjusting screw is tightened until the correct worm shaft turning torque is obtained as the worm shaft is rotated through the center position. Who is correct?
   A. Technician A
   B. Technician B
   C. Both Technician A and Technician B
   D. Neither Technician A nor Technician B
5. Outline the procedure for removing a recirculating ball steering gear.

# Chapter 41 Rack-and-Pinion Steering Service

## INTRODUCTION

There are four wear points in a rack and pinion steering gear: the inner and outer tie-rod ends on both sides of the assembly **(Figure 1).** The first step in manual or power rack-and-pinion steering gear diagnosis is a thorough inspection of the entire system. All steering system components, such as the inner and outer tie-rod ends, boots, mounting bushings, couplings or universal joints, ball joints, tires, and steering wheel freeplay, must be checked.

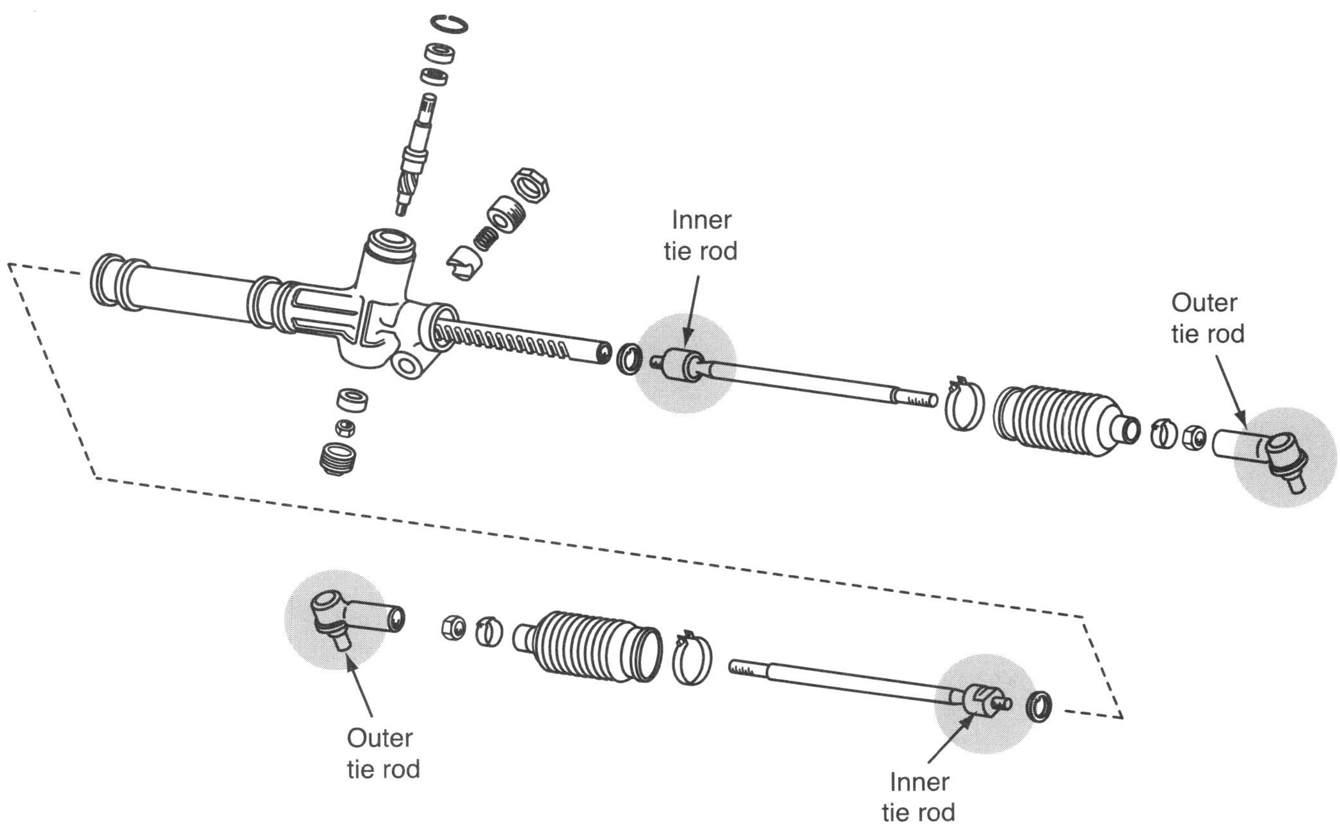


**Figure 1.** The wear points of a rack and pinion steering gear.

Road test the vehicle and check for excessive steering effort. A bent steering rack, tight rack bearing adjustment, or damaged front drive axle joints in a front-wheel drive car may cause excessive steering effort. It is sometimes difficult to separate the diagnosis of manual steering gears, steering columns, and suspension systems because some noises and symptoms may be caused by problems in any of these components or systems.

## STEERING GEAR REMOVAL AND REPLACEMENT

The replacement procedure is similar for manual and power rack-and-pinion steering gears. On some vehicles, the front cross member or engine support cradle must be lowered to remove the steering gear. Always follow the procedure given for the vehicle being serviced. The following is a typical rack-and-pinion steering gear removal and replacement procedure.

1. Place the front wheels in the straight-ahead position and remove the ignition key from the ignition switch to lock the steering wheel.
2. Place the driver's seat belt through the steering wheel to prevent the steering wheel from moving **(Figure 2)** or use a steering wheel lock. Locking the steering wheel will keep the clock spring centered on air bag-equipped vehicles.
3. Lift the front end with a floor jack, and place safety stands under the vehicle's chassis. Lower the vehicle onto the safety stands.
4. Place punch marks on the lower universal joint and the steering gear pinion shaft so they may be reassembled in the same position **(Figure 3).**
5. Remove the universal joint bolts and disconnect the joint.
6. Remove the cotter pins from the outer tie-rod ends.
7. Loosen, but do not remove, the tie-rod end nuts.
8. Use a tie-rod end puller to loosen the outer tie-rod ends in the steering arms.
9. Remove the tie-rod end nuts and remove the tie-rod ends from the arms.
10. Remove the four stabilizer bar mounting bolts.
11. Disconnect and cap the power steering lines. Then remove the steering gear mounting bolts **(Figure 4).**
12. Remove the steering gear assembly through the side of the vehicle.

**Figure 2.** Wrap the driver's seat belt around the steering wheel to prevent it from moving.

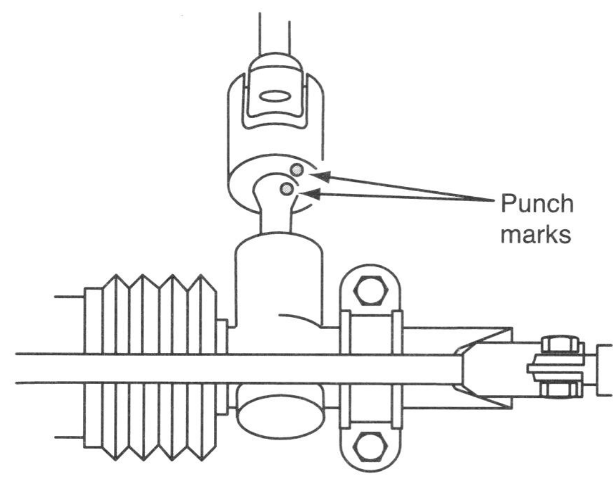


**Figure 3.** Make sure there are alignment marks on the universal joint and pinion shaft before disconnecting the two.

**Figure 4.** Removing the mounting bolts for the steering gear.

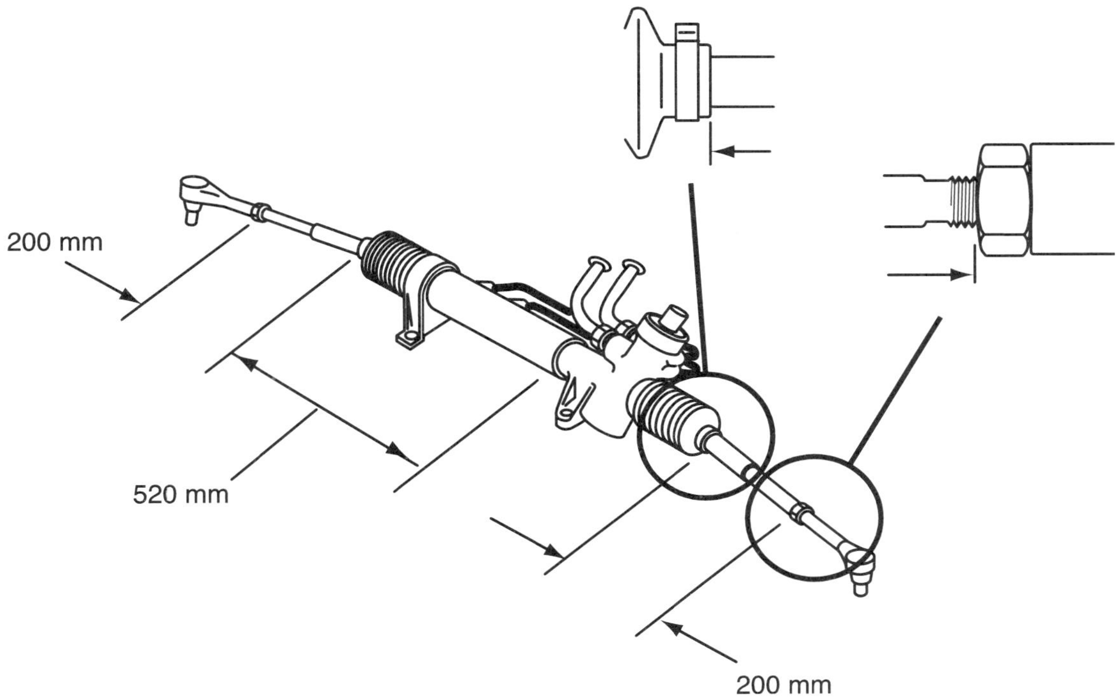


**Figure 5.** Measure and record the distance between the tie rods and housing before servicing the tie rods.

13. To reinstall the gear, position the right and left tie rods at the specified distance from the steering gear housing **(Figure 5).** Measure this distance with a tape measure.
14. Move the steering gear into position on the vehicle.
15. Install the pinion shaft in the universal joint with the punch marks aligned.
16. Tighten the upper and lower universal joint bolts to the specified torque.
17. Install the steering gear mounting bolts, and tighten these bolts to the specified torque.
18. Install the stabilizer bar mounting bolts, and torque these bolts to specifications.
19. Install the outer tie-rod ends into the steering knuckles.
20. Tighten the nuts to the specified torque. Install new cotter pins into the nuts.
21. Check the front wheel toe and adjust as necessary.
22. Tighten the outer tie-rod end jam nuts to specifications and tighten the outer boot clamps.
23. Road test the vehicle and check for proper steering gear operation and steering control.

## MANUAL RACK-AND-PINION STEERING GEAR DISASSEMBLY

With the steering gear removed, clean the outside of the assembly. Then clamp the center of the steering gear housing in a soft-jaw vise. Do not apply excessive force to the vise. The assembly can now be disassembled. Pay attention to the condition of all parts as they are removed, and replace any part that is damaged. The following is the procedure for disassembling a typical manual rack and pinion steering gear:

1. Insert a short piece of wire through each outer tie-rod end cotter pin hole and connect a pull scale to this wire. Pull upward on the scale to check the tie-rod articulation effort. If this effort is not within specifications, replace the inner tie-rod end.
2. Mark the position of the outer tie rod and jam nut with masking tape wrapped around the tie-rod threads next to the jam nut **(Figure 6).**

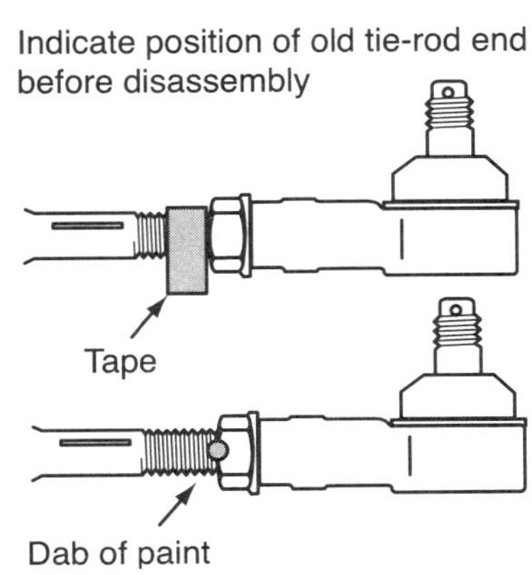


**Figure 6.** Mark the position of the outer tie-rod end and the jam nut.

3. Loosen the jam nuts and remove the outer tie-rod ends.
4. Remove the inner and outer boot clamps and pull the boots from the tie rods.
5. Straighten the lock plate where it is bent over the inner tie-rod end **(Figure 7).**
6. Hold the rack and use a wrench to remove the inner tie rod from the rack **(Figure 8).**
7. A jam nut is used in place of the lock ring on some inner tie-rod ends, and a roll pin retains some inner tie-rod ends to the rack. Various inner tie-rod socket removal procedures are required depending on the socket design.
8. Rotate the pinion shaft until the specified distance is obtained between the end of the rack and the steering gear housing. Mark pinion shaft position in relation to the housing with a center punch. If the specified distance from the end of the rack to the housing is not available, rotate the pinion shaft fully clockwise. Then rotate it fully counterclockwise and count the number of turns. Turn the pinion shaft clockwise one-half that number of turns. Measure the distance from each end of the rack to the housing and record this measurement.
9. Loosen the adjuster plug lock nut and remove the adjuster plug and spring.

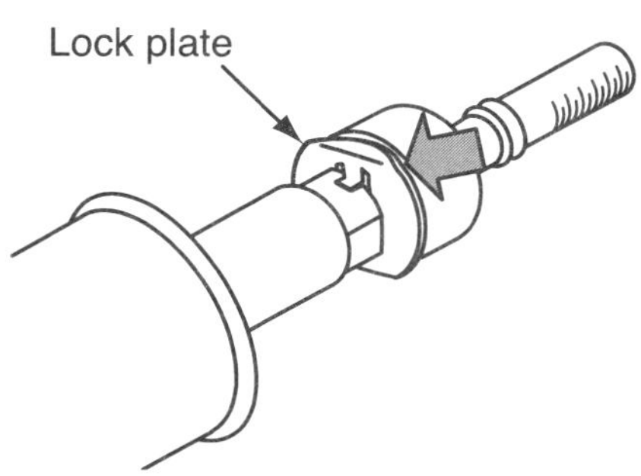


**Figure 7.** Straighten the lock plate where it is bent over the inner tie-rod end.

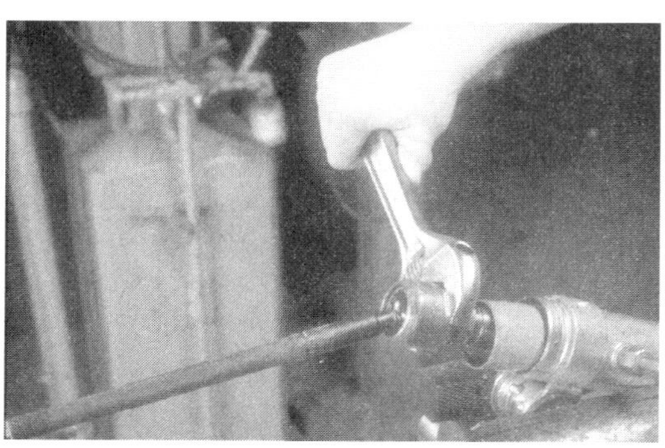

**Figure 8.** Hold the rack and loosen the inner tie-rod end.

10. Remove the rack bearing through the adjuster plug opening.
11. Clean the surface around the pinion shaft seal and pierce the pinion shaft seal at one of the two round areas on the seal's surface.
12. Pry the seal from the housing.
13. Remove the pinion shaft retaining ring with snap ring pliers.
14. Place the end of the pinion shaft in a soft-jaw vise, and tap the gear housing with a soft hammer to remove the pinion shaft and bearing from the housing.
15. Remove the rack from the housing.
16. Use an approved solvent to clean all steering gear components except the inner tie-rod assemblies, and blow-dry the parts with compressed air.

## Inspection of Manual Rack and Steering Gear Components

After the steering gear components are removed and cleaned, they must be inspected.

### Inner Tie-Rod Sockets

If the inner tie-rod ends were loose during the on-car inspection or the inner tie-rod end articulation effort did not meet specifications, replace these ends.

### Pinion and Bearing Assembly

If the pinion bearing is loose on the pinion shaft, replace the pinion and bearing assembly. A pinion shaft with worn or chipped teeth must be replaced. Inspect the pilot bearing contact area on the pinion shaft. Wear, pitting, or scoring in this area indicates that a new pinion shaft is required. If the pinion bearing is rough, loose, or noisy, replace the pinion shaft and bearing assembly.

### Pilot Bearing

Check the pilot bearing in the steering gear housing. If this bearing is worn or scored, replace the pilot bearing and the pinion shaft and bearing assembly. Use an appropriate bearing driver to remove and replace the pilot bearing in the steering gear housing.

### Rack Bushing

If the rack bushing is worn, bushing replacement is necessary. Remove the bushing retaining ring prior to bushing removal. Position the puller fingers behind the bushing and operate the slide hammer on the puller to remove the bushing. If a puller is not available, an appropriate bushing driver or socket may be used to drive the bushing out of the housing.

### Mounting Bushings

If the mounting bushings are loose, replace the bushings. Always replace the bushings in pairs. If the bushings are in satisfactory condition, do not disturb them.

## MANUAL RACK-AND-PINION STEERING GEAR ASSEMBLY

Once all required replacement parts are obtained, make sure you have the recommended lubricants available. Also, refer to the procedure for assembly given in the service manual. What follows is a typical procedure for reassembling a manual rack-and-pinion steering gear:

1. Install the rack in the housing with the teeth parallel to the pinion shaft.
2. Move the rack until the specified distance is obtained between the end of the rack and the housing. If this specification is not available, position the rack ends at the distance from the housing recorded during the disassembly procedure.
3. With anhydrous calcium grease, lubricate the pilot bearing and the pilot bearing contact area on the pinion shaft.
4. Install the pinion shaft with the punch marks aligned on the shaft and housing.
5. Install the pinion shaft retaining ring with the beveled edge facing upward.
6. Apply a coating of anhydrous calcium grease on the top of the pinion shaft bearing.
7. Use the proper seal driver to install the pinion shaft seal.
8. Apply a coating of lithium-based grease to the rack bearing and install the bearing.
9. Apply lithium-based grease to the ends of the preload spring and the adjuster plug threads, and install the preload spring and adjuster plug.
10. Turn the adjuster plug until it bottoms, then rotate it 45 to 60 degrees counterclockwise.
11. Rotate the pinion shaft with a torque wrench and observe the turning torque. If necessary, rotate the adjuster plug to obtain the specified turning torque.
12. Tighten the adjuster plug lock nut to the specified torque.
13. Coat both ends of the rack and fill the rack's teeth with lithium-based grease.
14. Rotate the pinion shaft fully in each direction several times. Apply more grease to the rack teeth after each complete pinion shaft rotation.
15. Rotate the inner tie-rod ends onto the rack until they bottom.
16. Install the inner tie-rod pins or stake these ends **(Figure 9)** as required. Always use a wooden block to support the opposite side of the rack and inner tie-rod end while staking.
17. If jam nuts are used on the inner tie-rod ends, be sure they are properly tightened.
18. Place a new large boot clamp over each end of the gear housing.
19. Install the boots. Make sure they are seated in the housing and the tie-rod undercuts.
20. Install and tighten the large inner boot clamps **(Figure 10).**

**Figure 9.** Support the inner tie rod and stake both sides of the inner tie-rod joint with a hammer and punch.

**Figure 10.** Install the bellows boot with new clamps.

**Figure 11.** Make sure the index marks are aligned and then install the jam nut and outer tie-rod end.

21. Install the outer boot clamps on the tie rods, but do not install them on the boots until the steering gear is installed and the toe is adjusted.
22. Install the jam nuts and the outer tie-rod ends **(Figure 11).** Align the marks placed on these components during the disassembly procedure. Leave the jam nuts loose until the steering gear is installed and the toe is adjusted.
23. Install the steering gear in the vehicle and check the front wheel toe.
24. Then tighten the outer bellows boot clamps and the outer tie-rod end jam nuts.

## SUMMARY

- The inner and outer tie-rod ends on both sides of a rack and pinion steering gear are the wear points.
- The replacement procedure is similar for manual and power rack and pinion steering gears.
- After the steering gear components are removed and cleaned, they must be inspected.
- Once all required replacement parts are obtained, prepare for the reassembly of the steering gear assembly by making sure all required lubricants are available.

## REVIEW QUESTIONS

1. While discussing steering effort, Technician A says a tight rack bearing adjustment may cause excessive steering effort. Technician B says a loose inner tie-rod end may cause excessive steering effort. Who is correct?
   A. Technician A
   B. Technician B
   C. Both Technician A and Technician B
   D. Neither Technician A nor Technician B
2. While discussing steering gear removal and replacement on air bag-equipped vehicles, Technician A says the steering column should be locked in the centered position to prevent air bag deployment. Technician B says the rack should be centered in the housing prior to steering gear installation. Who is correct?
   A. Technician A
   B. Technician B
   C. Both Technician A and Technician B
   D. Neither Technician A nor Technician B
3. While discussing manual rack and pinion steering gear lubrication, Technician A says pilot bearing should be lubricated with anhydrous calcium grease. Technician B says the rack and rack teeth should be coated with lithium-based grease. Who is correct?
   A. Technician A
   B. Technician B
   C. Both Technician A and Technician B
   D. Neither Technician A nor Technician B
4. Technician A says the pinion and bearing assembly should be replaced if the pinion bearing is loose on the pinion shaft. Technician B says a pinion shaft with worn or chipped teeth must be replaced. Who is correct?
   A. Technician A
   B. Technician B
   C. Both Technician A and Technician B
   D. Neither Technician A nor Technician B
5. Where are the typical wear areas for a rack-and-pinion steering gear?

# Chapter 42 Four-Wheel Steering Systems

## INTRODUCTION

A few manufacturers have offered **four-wheel steering (4WS)** systems, where the rear wheels help to turn the car by electrical, hydraulic, or mechanical means.

Most cars tend to understeer, although some will oversteer. If a car could automatically compensate for an understeer/oversteer problem, the driver would enjoy nearly neutral steering under varying conditions. By design, 4WS is intended to provide near-neutral steering with the following advantages.

- The vehicle's cornering behavior becomes more stable and controllable at high speeds and on wet or slippery road surfaces **(Figure 1)**.

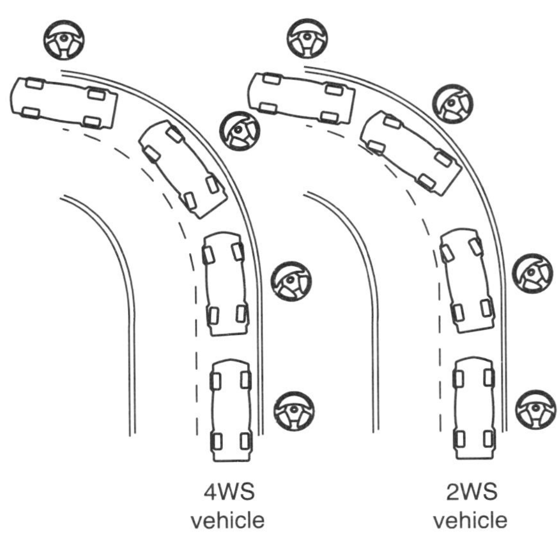


**Figure 1.** A comparison of 2WS and 4WS during cornering.

**Interesting Fact**

*The four-wheel steering concept has been researched by automotive engineers for several decades. In 1962, an engineer representing one of the Japanese manufacturers stated at the Japanese Automotive Engineering Association's technical meeting, "A major improvement in control and stability may be anticipated by means of an automatic rear-wheel steering system." In the late 1970s, Honda and Mazda were actively engaged in 4WS development. However, it was not until the late 1980s that 4WS was introduced on some Honda and Mazda cars.*

- The vehicle's steering response is quicker and more precise throughout the vehicle's speed range.
- The vehicle's straight-line stability at high speeds is improved. Negative effects of road irregularities and crosswinds on the vehicle's stability are minimized.
- Stability in lane changing at high speeds is improved. High-speed slalom-type operations become easier. The vehicle becomes less likely to go into a spin even in situations when the driver must make a sudden and relatively large change of direction.
- The vehicle's turning circle is greatly reduced. Therefore, maneuvering on narrow roads and during parking becomes easier.

## FOUR-WHEEL STEERING THEORY

When turning, the driver is putting into motion a complex series of forces. These forces compete with each other during steering. The tires are subject to grip, momentum, and steering input when making a movement other than straight-ahead driving. Grip holds the tires to the road, and momentum moves the car straight ahead. Steering input causes the front wheels to turn.

With a front-steered vehicle, the rear end is always trying to catch up to the directional changes of the front wheels. The car momentarily resists the turning motion, causing a tire slip angle to form. Once the vehicle begins to respond to the steering input, cornering forces are generated. The vehicle sways as the rear wheels attempt to keep up with the cornering forces already generated by the front tires. This is referred to as rear-end lag because there is a time delay between steering input and vehicle reaction. When the front wheels are turned back to a straight-ahead position, the vehicle must again try to adjust by reversing the forces developed by the turn. As the steering is turned, the body sways as the rear wheels again try to keep up with the cornering forces generated by the front wheels. As a normal part of operating a vehicle, the driver learns to adjust to these forces without thinking about them.

The idea behind four-wheel steering is that a vehicle requires less driver input for any steering maneuver if all four wheels are steering the vehicle. With 4WS, all four wheels react to the steering input, causing slip angles to form at all four wheels. The entire vehicle moves in one direction rather than the rear half attempting to catch up to the front. The vehicle responds more quickly to steering input because rear-wheel lag is eliminated. There is also less sway when the wheels are turned back to a straight-ahead position.

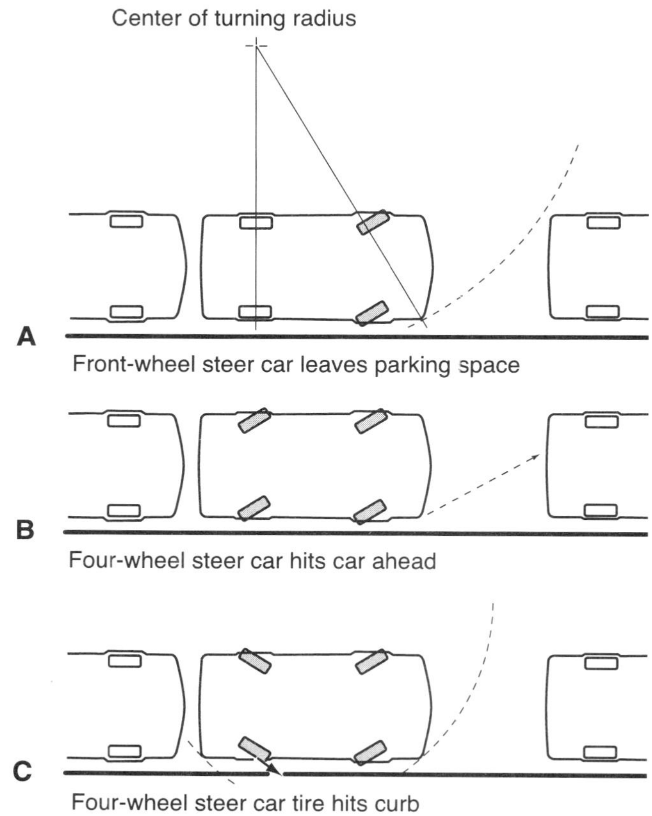


**Figure 2.** Parallel parking with conventional front-wheel steering and four-wheel steering.

## Operation

Rear-wheel steering may be controlled in relation to vehicle speed or the amount of steering wheel rotation. At low vehicle speeds or with a considerable amount of steering wheel rotation, the rear wheels are steered in the opposite direction to the front wheels. When the vehicle is operating at higher speeds or with a small amount of steering wheel rotation, the rear wheels are steered in the same direction as the front wheels.

When a vehicle is cornering at higher speeds, centrifugal force tends to move the rear of the vehicle sideways. This action causes the rear tires to slip sideways on the road surface. This process may be called sideslip. The vehicle speed and the severity of the turn determine the amount of sideslip. If sideslip is excessive, the car may spin around, causing the driver to lose control. When the 4WS system steers the rear wheels in the same direction as the front wheels at higher speeds, sideslip is reduced and vehicle stability is improved.

If a car with conventional front-wheel steering is parallel parked at a curb between two vehicles, it may be driven from the parking space without hitting the car in front if the front wheels are turned all the way to the left **(Figure 2A).** When the same car is equipped with 4WS and the rear wheels steer in the same direction as the front wheels at low speed, the car will not steer out of the parking space without striking the vehicle in front **(Figure 2B).**

When the car in the same parking space has a 4WS system that steers the rear wheels in the opposite direction to the front wheels at low speed, the car steers out of the parking space with plenty of distance between the vehicle parked in front **(Figure 2C).** When the rear wheels steer in the opposite direction to the front wheels, the rear wheels steer toward the curb. This action causes the right rear tire to strike the curb if the car is parked close to the curb. Therefore, the maximum rear steering angle must be considerably less than the maximum front-wheel steering angle to help prevent this problem. A car with 4WS has a smaller turning circle compared to a vehicle with front wheel-steering. This improves maneuverability while parking.

## MECHANICAL 4WS

In a mechanical 4WS system, two steering gears are used—one for the front, and other for the rear. A shaft con-

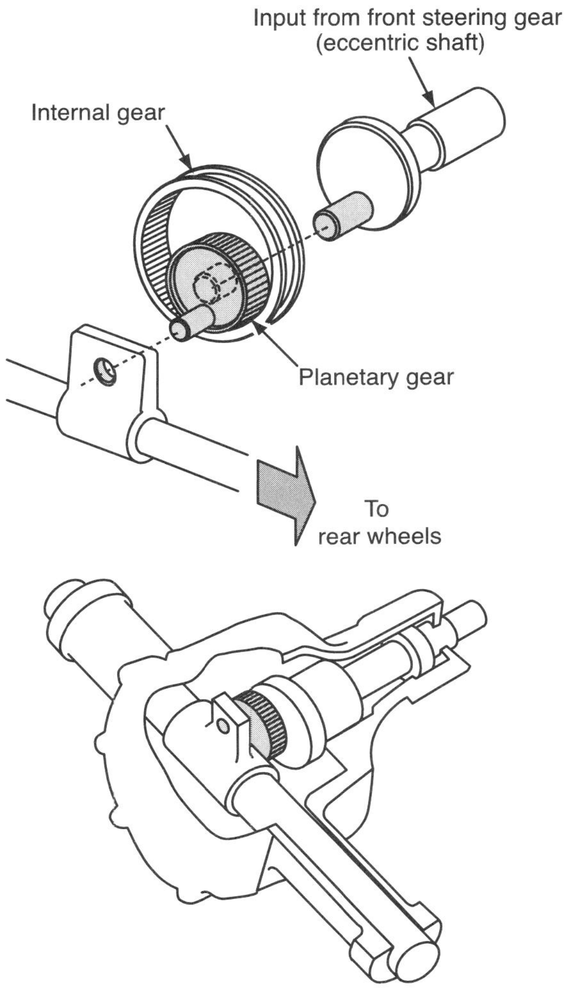


**Figure 3.** Inside the rear-steering gearbox is a simple planetary gear setup.

nects the two steering gearboxes and terminates at an eccentric shaft. This shaft is fitted with an offset pin **(Figure 3)** that engages a second offset pin that fits into a planetary gear. Mechanical 4WS is steering-angle sensitive and is not sensitive to vehicle road speed.

## HYDRAULIC 4WS

The hydraulically operated 4WS system, shown in **Figure 4,** is a simple design. The rear wheels turn only in the same direction as the front wheels. The system only activates at speeds above 30 mph and does not operate when the vehicle moves in reverse.

A two-way hydraulic cylinder turns the wheels. Fluid is supplied by a rear steering pump that is driven by the differential. A tank in the engine compartment supplies the rear steering pump with fluid.

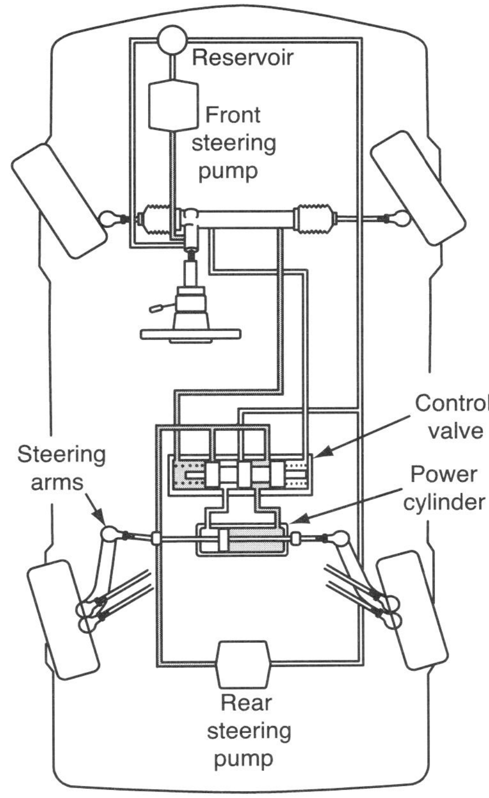


**Figure 4.** A simple hydraulic 4WS system.

When the steering wheel is turned, the front steering pump sends fluid under pressure to the rotary valve in the front rack and pinion unit. This forces fluid into the front power cylinder, and the front wheels turn in the direction steered. The fluid pressure varies with the turning of the steering wheel. The faster and farther the steering wheel is turned, the greater the fluid pressure.

The fluid is also fed under the same pressure to the control valve where it opens a spool valve in the control valve housing. As the spool valve moves, it allows fluid from the rear steering pump to move through and operate the rear power cylinder. The higher the pressure on the spool, the farther it moves. The farther it moves, the more fluid it allows through to move the rear wheels.

## ELECTRO/HYDRAULIC 4WS

Several 4WS systems combine electronic controls with hydraulics to make the system sensitive to both steering angle and road speeds. In this design, a speed sensor and steering wheel angle sensor sends information to the electronic control unit (ECU). Based on this information, the ECU commands the hydraulic system to steer the rear wheels.

At moderate road speeds, the rear wheels are steered momentarily counterphase, through neutral, then in phase with the front wheels. At high road speeds, the rear wheels turn only in phase with the front wheels.

Another electro/hydraulic 4WS system is shown in **Figure 5.** The basic working elements of the design are the control unit, a stepper motor, a swing arm, a set of beveled gears, a control rod, and a control valve with an output rod.

The stepper motor moves the control yoke. A swing arm is attached to the control yoke. Stepper motor action eventually causes a push-or-pull movement of its output shaft to steer the rear wheels. The position of the control yoke varies with vehicle speed.

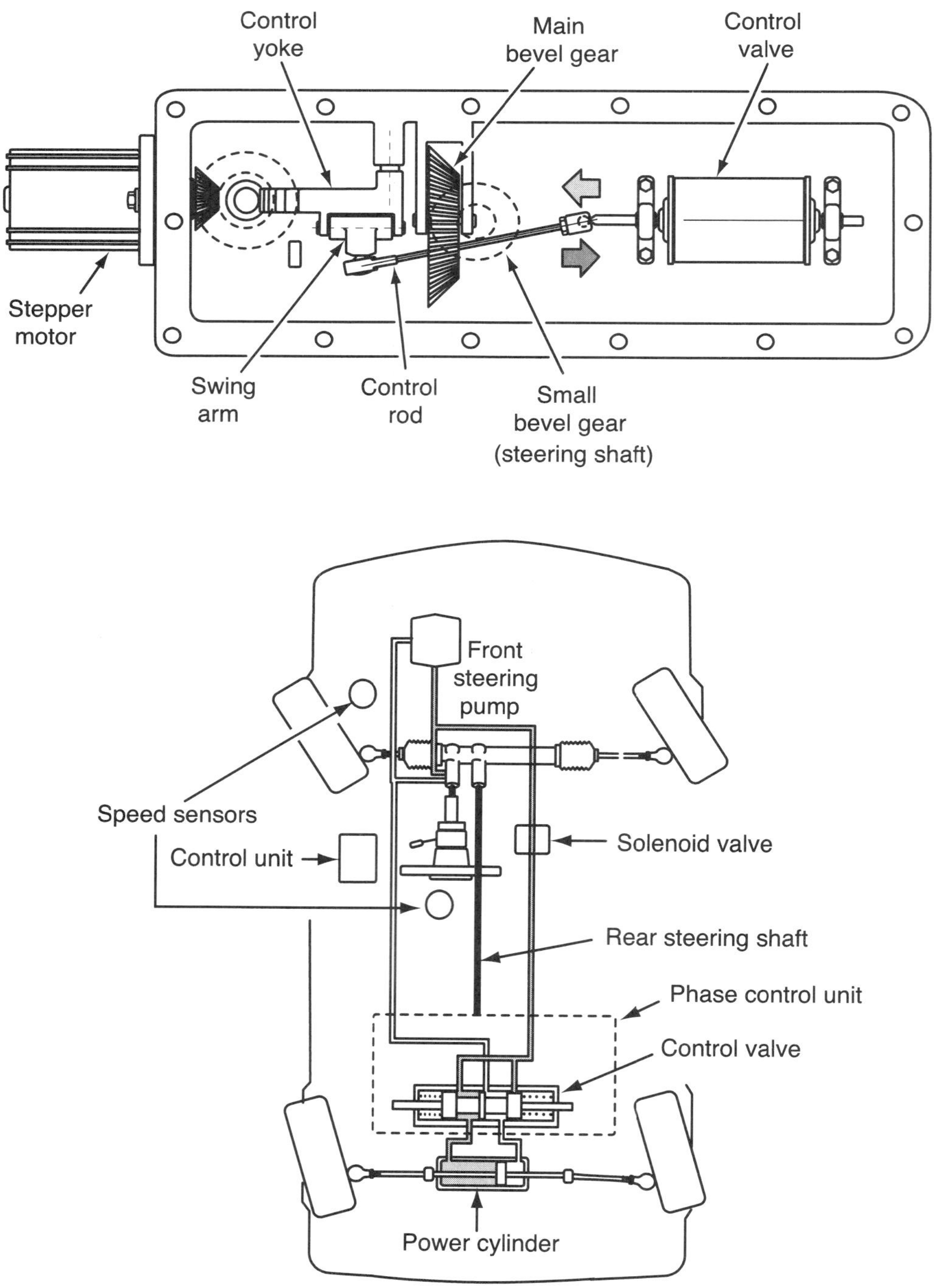


**Figure 5.** An electronically and hydraulically controlled 4WS system with a stepper motor and a control yoke.

## ELECTRONIC/ELECTRICAL 4WS

In an electronic 4WS system, there is no mechanical connection between the front steering gear and the rear steering actuator. This rear steering actuator is controlled by a 4WS control unit **(Figure 6).** The control unit uses steering wheel rotational speed, vehicle speed, and front steering angle information to calculate and control the rear steering angle.

### Rear Steering Actuator

The rear steering actuator may be compared to an electric steering gear. It contains an electric motor that drives a steering rack through a ball screw mechanism **(Figure 7).** Tie rods are connected from the steering actuator to the rear steering arms and spindles. A return spring inside the actuator moves the rear wheels to the straight-ahead position when the ignition switch is turned off or when a defect occurs in the 4WS system. A main rear wheel angle sensor and a sub-rear wheel angle sensor are mounted on top of the rear steering actuator.

### Operation

When the engine is running, the 4WS control unit continually receives information from all the input sensors. If the steering wheel is turned, the 4WS control unit analyzes information from the vehicle speed sensor, main steering wheel angle sensor, sub-front wheel angle sensor, main rear wheel angle sensor, sub-rear wheel angle sensor, and the rear wheel speed sensors. The 4WS control unit calculates

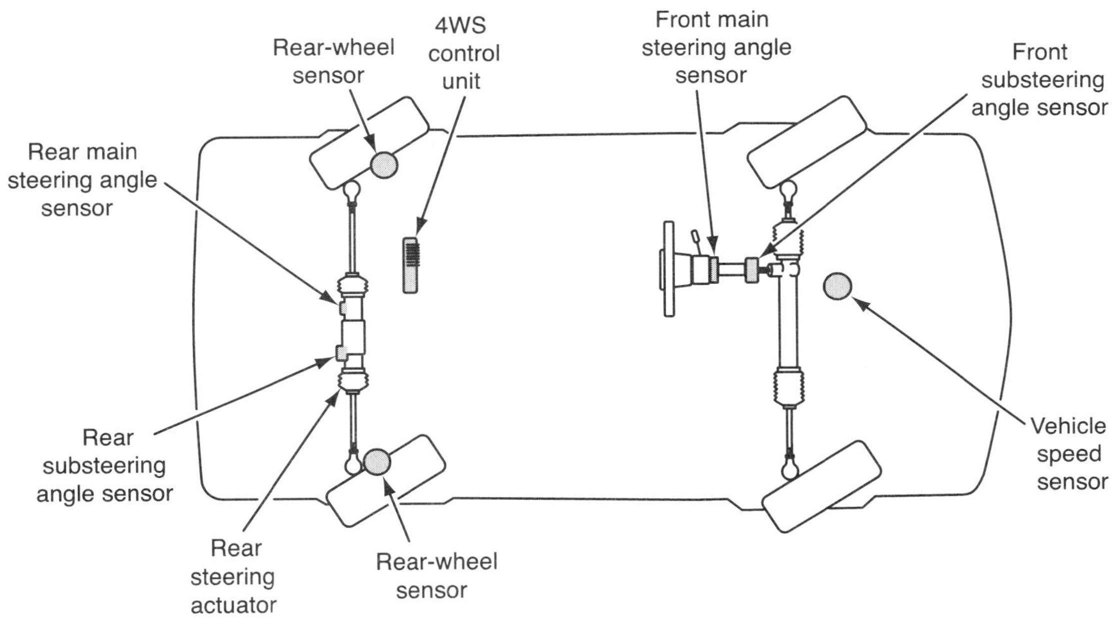


**Figure 6.** An electronically controlled 4WS with the control unit mounted in the trunk.

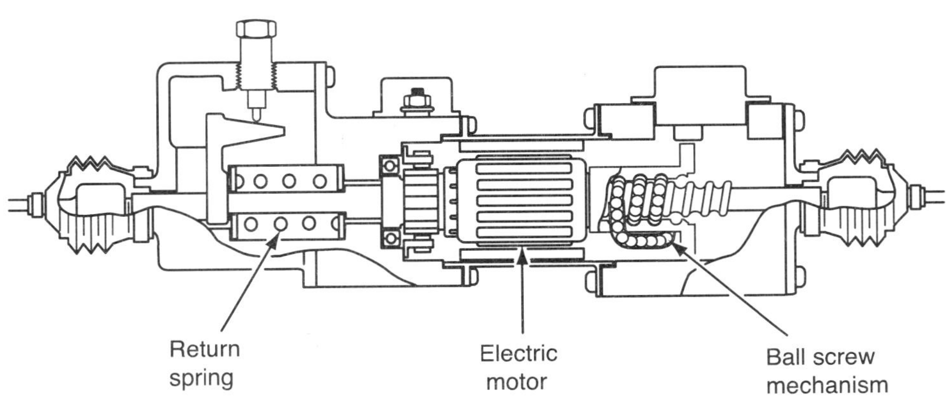


**Figure 7.** A rear-steering actuator, electronically controlled 4WS system.

Figure 8. A quadrasteer rear steering setup.

the proper rear wheel steering angle and then sends battery voltage to the rear steering actuator motor to provide this rear steering angle.

Battery voltage is sent to the rear steering actuator motor through two heavy-duty output transistors. One of these transistors conducts current during a right turn, whereas the other transistor is activated during a left turn. The main rear wheel angle sensor and the sub-rear wheel angle sensor send feedback signals to the 4WS control unit indicating the proper rear steering angle has been supplied.

## Quadrasteer

**Quadrasteer** is a four-wheel steering system that improves low-speed maneuverability (decreasing the turning radius), high-speed stability, and trailering capabilities for full-size pickups, vans, and SUVs. The system combines normal front-wheel steering with an electrically powered and electronically controlled rear-wheel steering system **(Figure 8).** The system uses wheel position and vehicle speed sensors and a central control module.

At low speeds, the rear wheels turn in the opposite direction of the front wheels. At moderate speeds, the rear wheels remain straight. While at high speeds, the rear wheels turn in the same direction as the front wheels. If the system were to fail, the truck will default to normal two-wheel steering.

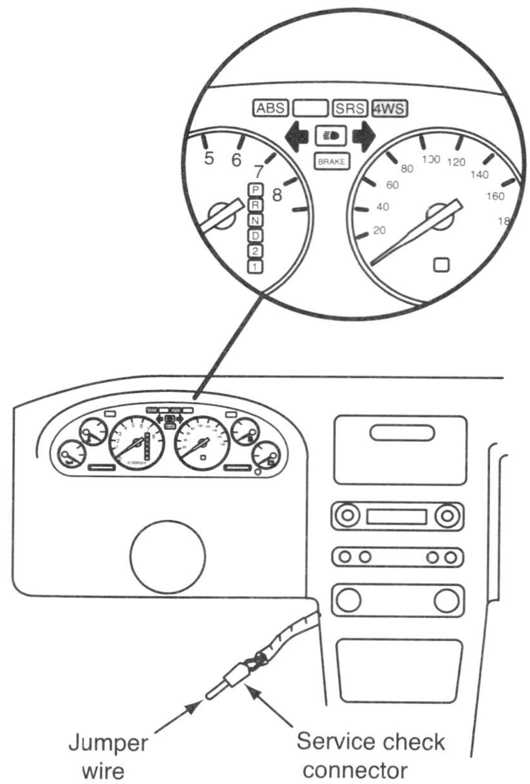


Figure 9. The warning light for a 4WS system.

## FAIL-SAFE PROGRAMS

All 4WS systems have fail-safe measures. For example, with the electro/hydraulic setup, the system automatically counteracts possible causes of failure (both electronic and hydraulic) and converts the entire steering system to a conventional two-wheel steering type. Specifically, if a hydraulic defect reduces the pressure level (by a movement malfunction or a broken driving belt), the rear-wheel steering mechanism is automatically locked in a neutral position, activating a low-level warning light.

In the event of an electrical failure, it would be detected by a self-diagnostic circuit integrated in the four-wheel-steering control unit. The control unit neutralizes the hydraulic pressure, thereby switching the system to two-wheel steering. The failure would be indicated by the system's warning light in the main instrument display **(Figure 9).**

## FOUR-WHEEL STEERING SERVICE

On any 4WS system, there must be near-perfect compliance between the position of the steering wheel, front wheels, and rear wheels. It is usually recommended that the car be driven about 20 feet in a straight line, then the position of the front/rear wheels is checked with respect to steering wheel position. The base reference point is a strip of masking tape on the steering wheel hub and the steering column. When the wheel is positioned dead center, draw a line down the tape. Run the car a short distance straight to see if the reference line holds. If not, corrections are needed, such as repositioning the steering wheel.

Even severe imbalance of a rear wheel on a speed-sensitive 4WS system can cause problems and make basic troubleshooting a bit frustrating.

If the 4WS indicator lamp is lit and there is no rear-wheel steering, suspect an electrical problem in the control system.

Prior to diagnosis of an electronic 4WS system, a preliminary system inspection should be performed.

## Damper Control

When the 4WS control unit enters the fail-safe mode, a quick return of the rear wheels to the straight-ahead position would adversely affect steering under certain conditions. To prevent this action, the 4WS control unit energizes the damper relay when it enters the fail-safe mode. The rear steering actuator motor is spun by the steering shaft as this shaft is moved to the centered position by centering spring force. This causes the armature of the motor to act as a voltage generator. The voltage generated by the armature is fed back through the damper relay to the motor's armature. When this happens, the motor rotation is slowed and the return spring slowly moves the rear steering shaft to the straight-ahead position.

# ROAD TEST

The 4WS control unit stores a fault code and illuminates the 4WS indicator light if a defect occurs in the system, even if the defect is temporary. If the 4WS light is not illuminated during the road test, the system is satisfactory electronically and does not require further electronic diagnosis. The troubleshooting procedures in the service manual assume that the problem is present at the time of diagnosis.

Inspect all electronic and mechanical parts of the system. Make sure the wires, connectors, sensors, and actuators are not physically damaged **(Figure 10).**

**Figure 10.** Raise the vehicle on a lift, and check all electrical connectors on the front steering gear and rear-steering actuator.

# DTC RETRIEVAL

On some vehicles, a jumper wire must be connected across two connectors to display the DTCs. Others can display the trouble codes on the scan tool. If the vehicle is designed to flash codes on the instrument panel, the 4WS indicator light should be observed. The codes are normally given in numerical order. Always follow the exact 4WS service and diagnostic procedures in the service manual.

> **You Should Know** *When diagnosing a computer system, never connect or ground any terminals unless instructed to do so in the vehicle manufacturer's service manual.*

## Flash Codes

The Honda Prelude is an example of a vehicle with 4WS, although it is no longer available in the U.S. However, it is discussed here because it used the indicator lamp to display DTCs. The 4WS control unit actually had two processing units that are referred to as the main and subprocessing units. Each processing unit can store a maximum of ten trouble codes. If the trouble code diagnosis is performed with the engine running, the code display indicates whether the codes are stored in the main or subprocessor. When the service connector terminals are connected with a jumper wire **(Figure 11)** and the engine is started, the 4WS indicator light **(Figure 12)** will

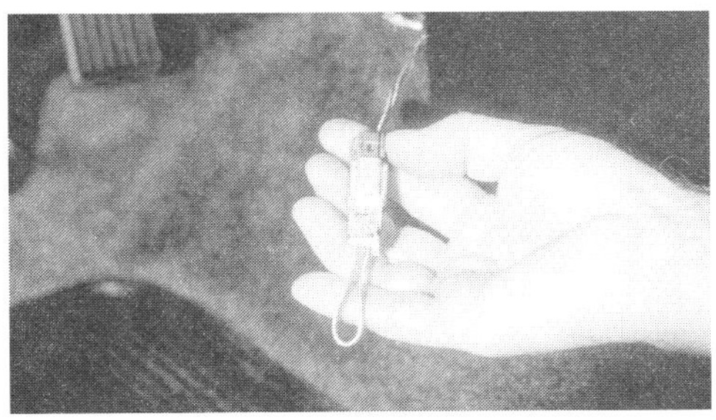

**Figure 11.** To retrieve the DTCs, connect a jumper wire between the terminals in the service check connector.

**Figure 12.** Observe the flashes of the 4WS indicator light to retrieve the DTCs.

follow a distinct sequence. This sequence is used to test the system.

### Scan Tool Diagnostics

Vehicles equipped with OBD II and 4WS (such as GM trucks with Quadrasteer) will display data and DTCs related to the four-wheel steering system. The information given in the appropriate service manual must be carefully compared to the displayed codes and serial data on the scan tool. Remember, the inputs from some of the sensors used in the 4WS system are shared with other control systems. If these are faulty, DTCs may be set in other systems.

### SENSORS

Depending on the system, most sensors used in a 4WS system are used in other systems. Some are unique to a 4WS system and are critical. These sensors have the responsibility of tracking the angle of the front and rear wheels. If the input from these sensors is wrong, the vehicle may handle and steer poorly. Typically, the front main steering, front substeering, and rear substeering sensors are adjustable, whereas the rear main steering sensor is not adjustable.

## SUMMARY

- Four-wheel steering (4WS) advantages include cornering capability, steering response, straight-line stability, lane changing, and low-speed maneuverability.
- The rear-wheel steering system may be electronically or electronically and hydraulically controlled.
- Steering the rear wheels in the opposite direction to the front wheels at low speed provides a shorter turning circle, or radius, for easier maneuvering.
- Steering the rear wheels in the same direction as the front wheels at higher speeds reduces rear sideslip and improves vehicle stability while cornering or changing lanes.
- In an electronically controlled 4WS system, there is no mechanical connection between the front steering gear and the rear steering gear.
- In an electronically controlled 4WS system, the rear steering actuator has an electric motor controlled by the control unit in response to various input sensor signals.
- Prior to diagnosis of a 4WS system, a preliminary inspection should be performed.
- If the 4WS control unit senses a system problem, it will switch to a fail-safe mode.

## REVIEW QUESTIONS

1. What are the basic features of all four-wheel steering systems?
2. By design, 4WS is intended to provide __________ __________.
   A. over steering
   B. near-neutral steering
   C. variable-power steering
   D. under steering
3. While discussing 4WS principles, Technician A says a 4WS system reduces vehicle sideslip. Technician B says a 4WS system provides a smaller turning circle. Who is correct?
   A. Technician A
   B. Technician B
   C. Both Technician A and Technician B
   D. Neither Technician A nor Technician B
4. While discussing the fail-safe function, Technician A says the 4WS indicator light is illuminated during the fail-safe function. Technician B says the rear wheels steer normally when the 4WS control unit enters the fail-safe mode. Who is correct?
   A. Technician A
   B. Technician B
   C. Both Technician A and Technician B
   D. Neither Technician A nor Technician B
5. While discussing the fail-safe function and damper control, Technician A says the rear wheels move instantly to the centered position when the 4WS control unit enters the fail-safe mode. Technician B says the return spring moves the rear wheels away from the centered position. Who is correct?
   A. Technician A
   B. Technician B
   C. Both Technician A and Technician 3
   D. Neither Technician A nor Technician B

# Section 8

## Power Steering



**Interesting Fact**

*Steering gear design progressed from the crude, high-friction worm and gear of the early 1900s to the Ross cam and lever gear and Saginaw worm and roller gear of the 1920s. The Saginaw worm and roller steering gear was the forerunner of the modern, low-friction recirculating ball steering gear. Today, many vehicles have rack-and-pinion steering boxes, and nearly all vehicles sold have power-assisted steering.*

## SECTION OBJECTIVES

After you have read, studied, and practiced the contents of this section, you should be able to:

- Describe the main components in a power-assisted rack-and-pinion steering system, and explain the steering gear mounting position.
- Describe the difference between a Hydroboost power steering system and an integral power steering system.
- Explain the operation of the spool valve and rotary valve.
- List three different types of power steering pump rotor designs.
- Describe the power steering pump operation while the vehicle is turning a corner.
- Explain the advantages of an EVO steering system.
- List the input sensors in an EVO steering system.
- Inspect power steering fluid levels and condition.
- Flush, fill, and bleed a power steering system.
- Remove, inspect, replace, and adjust a power steering pump belt.
- Inspect and replace power steering hoses and fittings.
- Perform a power steering pump pressure test.
- Remove and reinstall a power steering pump.
- Remove and reinstall a power steering pump pulley and check pulley and belt alignment.
- Remove and replace power steering pump integral reservoirs.
- Remove, replace, and check power steering pump rotating components.
- Remove and replace power steering pump seals and O-rings.
- Diagnose power recirculating ball steering gear problems.
- Remove and replace power recirculating ball steering gears.
- Diagnose and repair oil leaks in power recirculating ball steering gears.
- Disassemble, repair, and reassemble power recirculating ball steering gears.
- Diagnose power rack-and-pinion steering gear problems.
- Perform a manual or power rack-and-pinion steering gear inspection.
- Diagnose and adjust components of electronically controlled steering systems using a scan tool.

# Chapter 43 Power Steering Designs

## *INTRODUCTION*

Power steering systems are designed to reduce the amount of effort required to turn the steering wheel. They also reduce driver fatigue and make it easier to steer the vehicle at slow speeds, particularly during parking. Most power steering systems use fluid pressure to assist the driver in turning the wheels; however, electrical-assist systems are becoming more common.

## TYPES OF POWER-ASSISTED STEERING SYSTEMS

There are several types power steering systems in use today. The most common are the external-piston, integral-piston, and power-assisted rack and pinion system **(Figure 1)**.

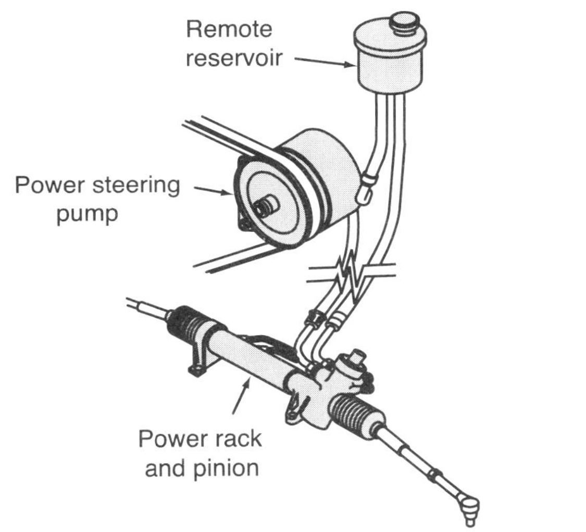


**Figure 1.** A power-assisted rack-and-pinion steering system.

### External-Piston Linkage System

The **external-piston linkage** (non-integral) system is generally found on older cars. It has a power-steering pump and reservoir, control valve, power cylinder, and four hoses. The control valve and pressure cylinder are separate components not incorporated in the steering gear. The pressure hose connects the pump to the control valve and directs pressurized fluid into the booster hoses to the power cylinder mounted on the steering linkage **(Figure 2)**.

The control valve directs fluid under pressure to the appropriate side of the power cylinder piston, which applies force to the steering linkage to provide steering assist. Since this linkage-type power steering is not widely used, it will not be discussed in detail.

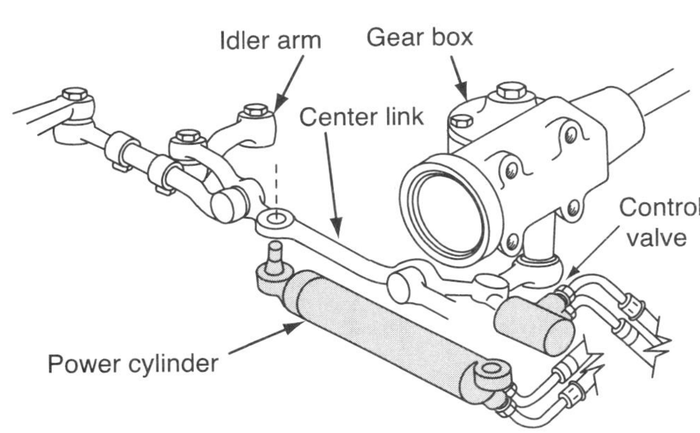


**Figure 2.** A linkage-type power-assisted steering system.

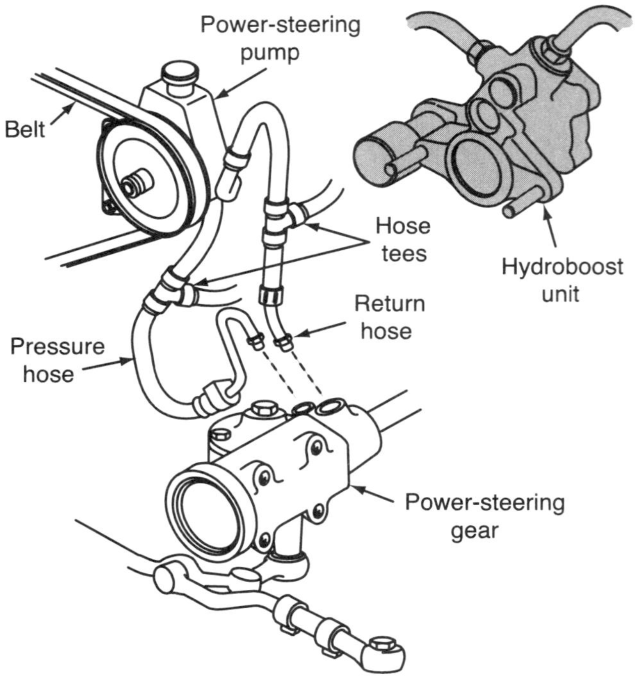


**Figure 3.** An integral power-assisted steering system. In the upper right is a typical Hydroboost power-assisted steering and brake unit.

## Integral Piston System

The **integral piston** system is one of the most common power-steering systems in use today. It has a power-steering pump and reservoir, power-steering pressure and return hoses, and steering gear. The power cylinder and control valve are in the steering gear housing. A high-pressure hose and a return hose are connected from the pump to the steering gear **(Figure 3).**

### Hydroboost Power-Assisted Steering System

On some recent cars and light trucks, the hydraulic fluid from the power-steering pump is also used for the power brakes. This system is called the Hydroboost system. These systems are similar to integral units but have hydraulic lines routed from the pump to the steering gear, and another set of lines is connected from the pump to the brake master cylinder.

## Power-Assisted Rack-and-Pinion System

Most FWD vehicles have a power rack and pinion power steering system. The power cylinder and control valve are in the rack housing. Turning the steering wheel moves the control valve directing pressure to either end of the power piston. The system uses a pressure hose from the pump to the control valve and a return line to the pump reservoir.

In many systems, the reservoir is part of the pump. Some systems have a remote reservoir. This is commonly found on small cars with limited underhood space. The reservoir is placed in a convenient position, and a return hose is connected from the steering gear to the reservoir. An additional return hose connects the reservoir to the pump **(Figure 4).**

## Electrical/Electronic Systems

**Electronically controlled power-steering (EPS)** systems use electronics to control the amount of hydraulic boost in relationship to road speed. The input from sensors is used by a control module to change the opening in an **electronic variable orifice (EVO)** actuator. This controls the fluid flow to the steering gear **(Figure 5).**

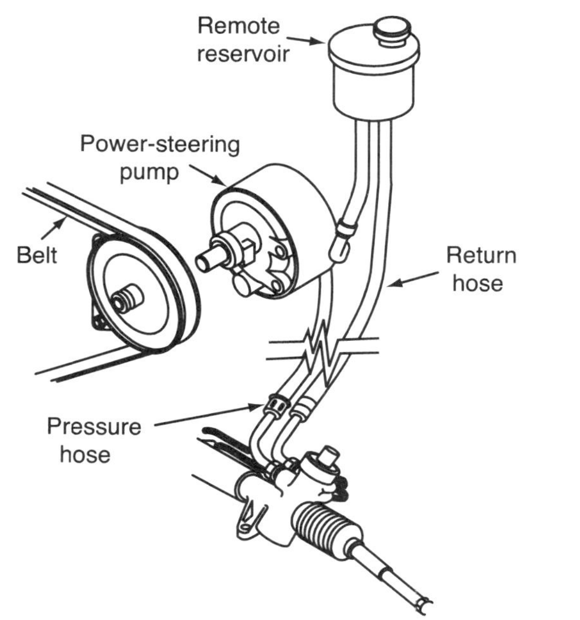


**Figure 4.** A remote reservoir on a power-assisted rack-and-pinion steering system.

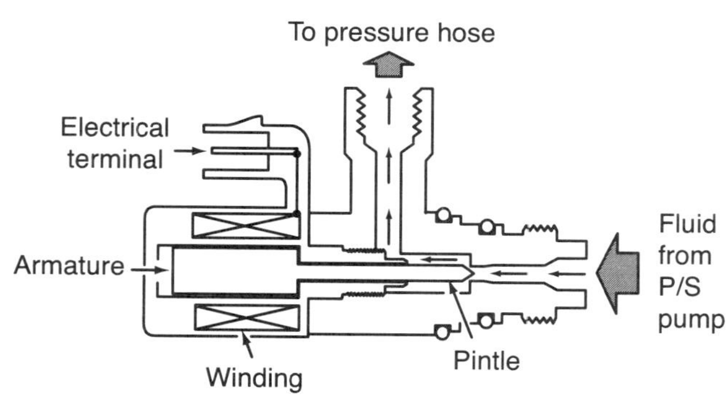


**Figure 5.** The construction and action of an EVO.

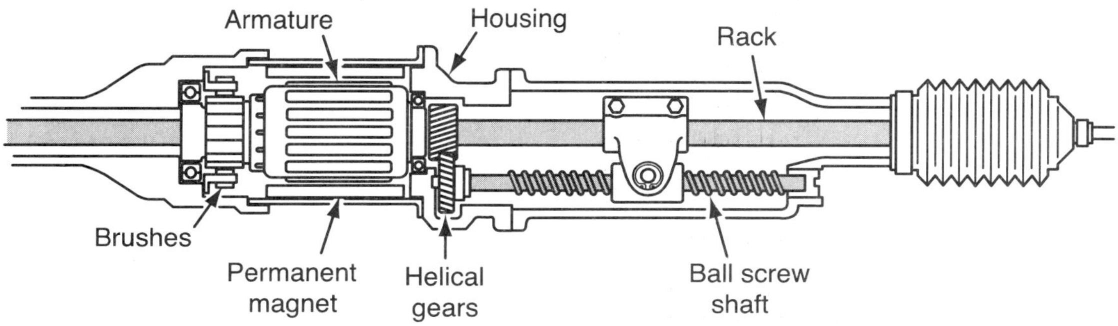


**Figure 6.** A rack-and-pinion steering gear with an electric assist motor.

The electric/electronic rack and pinion unit does not have the hydraulic pump, hoses, and fluid associated with conventional power-steering systems. They rely on electronic controls and an electric motor located in the rack **(Figure 6).** The DC motor's armature has a hollow shaft that allows passage of the rack through it. The armature is mechanically connected to the ball nut through an internal/external spline arrangement.

## MAIN COMPONENTS

Several parts of a manual steering system are used in power steering systems, such as the steering linkage. Parts are normally added to that system to provide for power assist. These are the power-steering pump, flow control and pressure relief valves, reservoir, spool valves and power pistons, hydraulic hose lines, and gearbox or assist assembly on the linkage.

### Power-Steering Pump

The power steering pump develops hydraulic flow to provide the force needed to operate the steering gear. The pump is belt driven from the engine's crankshaft. It is usually mounted near the front of the engine. The pump assembly includes a reservoir and an internal flow control valve. A dipstick is mounted in the reservoir to check fluid level **(Figure 7).** The drive pulley is normally pressed onto the pump's shaft.

There are four basic types of power steering pumps: roller, vane, slipper, and gear. All of these operate in the same way. Hydraulic fluid is stored in a reservoir. Fluid is routed to and from the pump by hoses and lines. Excessive pressure is controlled by a relief valve.

### Flow Control and Pressure Relief Valves

The pressure relief valve controls the output of the pump. This valve is needed due to the changes in engine rpm and the need for consistent steering ability in all engine speed ranges. It is located in a chamber exposed to pump outlet pressure at one end and supply hose pressure at the other. A spring is used at the supply pressure end to help maintain a balance.

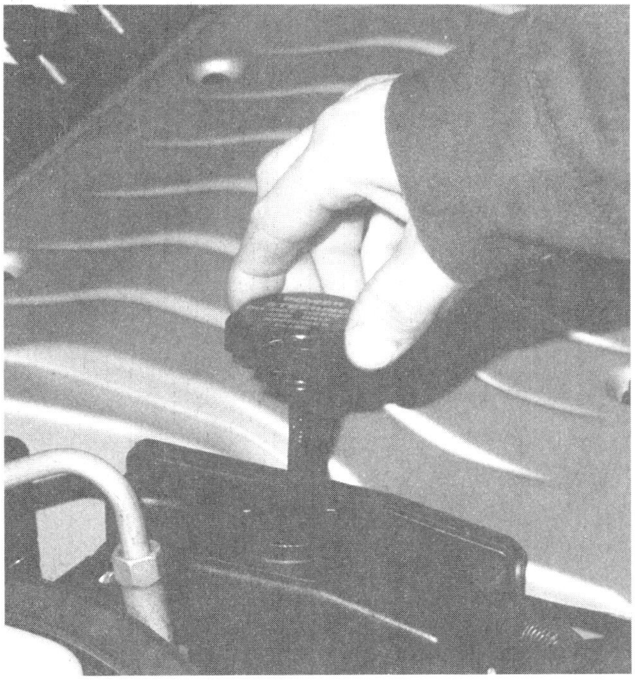

**Figure 7.** The dipstick for a power steering fluid reservoir.

As the fluid leaves the pump, it is forced through an orifice that causes a slight drop in pressure. This reduced pressure, aided by the springs, holds the flow control valve in the closed position. All pump flow is sent to the steering gear.

When engine speed increases, the pump can deliver more flow than is required. Since the outlet orifice restricts the amount of fluid leaving the pump, the difference in pressure at the two ends of the valve becomes greater until pump outlet pressure overcomes the combined force of supply line pressure and spring force. The valve is pushed down against the spring, opening a passage that returns the excess flow back to the inlet side of the pump.

A spring and ball inside the flow control valve are used to relieve pump pressure. This prevents damage due to excessive pressure when the steering wheel is held against the stops.

When outlet pressure reaches a preset level, the pressure relief ball is forced off its seat. This opens the control valve wider and permits more pressure to flow back to the pump inlet and pressure is held at a safe level.

## Power-Steering Hoses

Power steering hoses transmit power (fluid under pressure) from the pump to the steering gearbox and back to the pump reservoir **(Figure 8).** Hoses also, through material and construction, serve as additional reservoirs and as sound and vibration dampers.

Hoses are generally made with a reinforced synthetic rubber (neoprene) material coupled to metal tubing. The pressure side must be able to handle pressures up to 1,500 psi. For that reason, wherever there is metal tubing to a rubber connection, the connection is crimped. Pressure hoses are also subject to surges in pressure and pulsations from the pump. The reinforced construction permits the hose to expand slightly and absorb changes in pressure.

Two internal diameters of hose **(Figure 9)** may be used on the pressure side; the larger diameter is at the pump end. It acts as a reservoir and an accumulator absorbing pulsations. The smaller diameter or return hose reduces the effects of kickback from the gear itself. By restricting fluid flow, it also maintains constant backpressure on the pump, which reduces pump noise. If the hose has one diameter, the gearbox performs the damping functions internally.

Because of working fluid temperature and adjacent engine temperatures, these hoses must be able to withstand temperatures up to 300°F. Due to various weather conditions, they must also tolerate sub-zero temperatures as well. Hose material is specially formulated to resist breakdown or deterioration due to oil or temperature conditions.

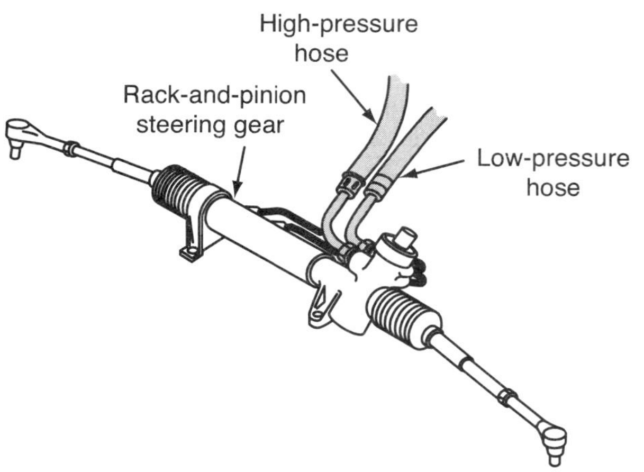


**Figure 8.** A power steering pump to gear hoses.

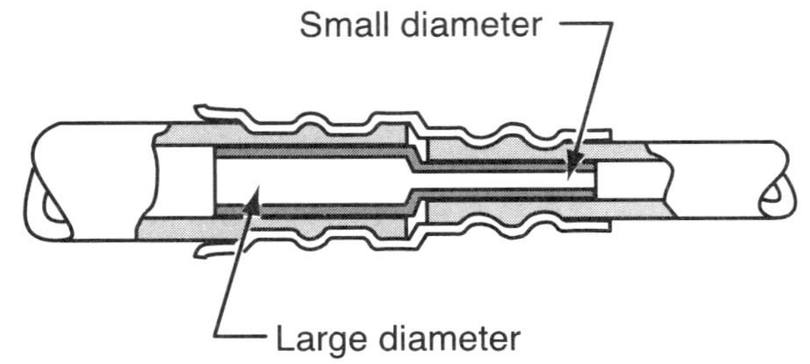


**Figure 9.** Power steering hoses may have two different internal diameters.

> **You Should Know** *Hoses must be routed away from engine manifolds. Power steering fluid is very flammable and if it contacts hot engine parts, it could start an underhood fire.*

## Power Steering Gearbox

An integral power steering system has a spool valve and power piston integrated with the gearbox. The **spool valve** directs the oil pressure to the left or right power chamber to steer the vehicle. The spool valve is actuated by a lever or a small torsion bar.

In external linkage systems, the control valve is connected directly to the steering center link and pitman arm. Any movement of the steering wheel and pitman arm compresses the centering spring and moves the valve spool. This opens and closes a series of ports directing pressurized fluid from the pump to one side of the power cylinder piston.

The power cylinder is attached to the center link, and the piston shaft is attached to the frame. As the fluid is directed to one side of the piston by the control valve, power assist is provided to turn the wheels in that direction. Two lines connect the cylinder to the control valve. Each functions as the return line and supply line, depending on the direction of the turn.

## POWER RECIRCULATING BALL STEERING GEARS

In a power recirculating ball steering gear, a torsion bar is connected between the steering shaft and worm shaft. Since the wheels are resting on the road's surface, they resist turning, and the parts attached to the worm shaft also

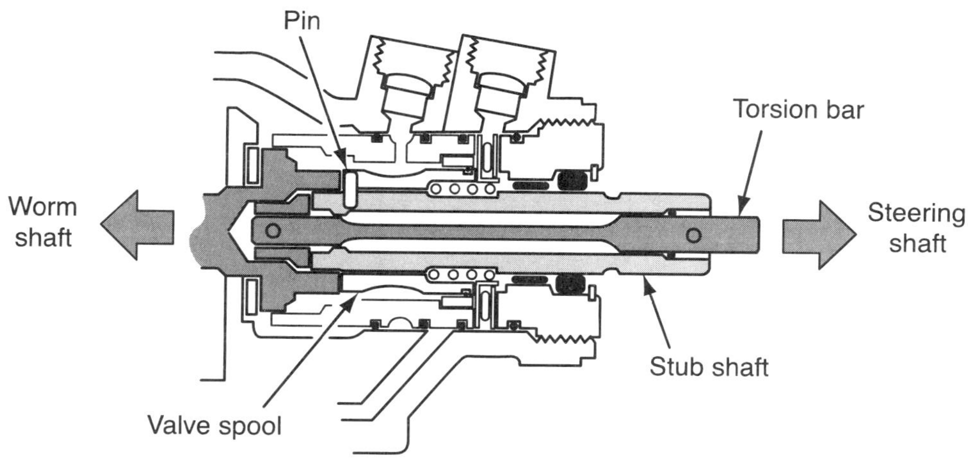


**Figure 10.** A torsion bar and stub shaft.

resist turning. This resistance causes torsion bar deflection when the wheels are turned. The worm shaft is connected to the rotary valve body, and the torsion bar pin connects the torsion bar to the worm shaft. The upper end of the torsion bar is attached to the steering shaft and wheel. A stub shaft is mounted inside the rotary valve, and a pin connects the outer end of the shaft to the torsion bar. The pin on the inner end of the stub shaft is connected to the spool valve in the center of the rotary valve **(Figure 10).**

## Operation

When the car is driven with the front wheels straight, oil flows from the power steering pump through the spool valve, **rotary valve,** and low-pressure return line to the pump inlet. In this position, oil pressure is equal on both sides of the recirculating ball's piston, and the oil acts as a cushion to prevent road shocks from reaching the steering wheel.

If the driver makes a left turn, torsion bar deflection moves the valve spool to allow oil flow to be directed through the rotary valve to the left-turn holes in the spool valve. Since the fluid is directed from the left-turn holes to the upper side of the recirculating ball piston **(Figure 11),** hydraulic pressure assists the driver in turning the wheels to the left. When the driver makes a right turn, torsion bar deflection moves the valve spool so oil flows through the spool valve, rotary valve, and a passage in the housing to

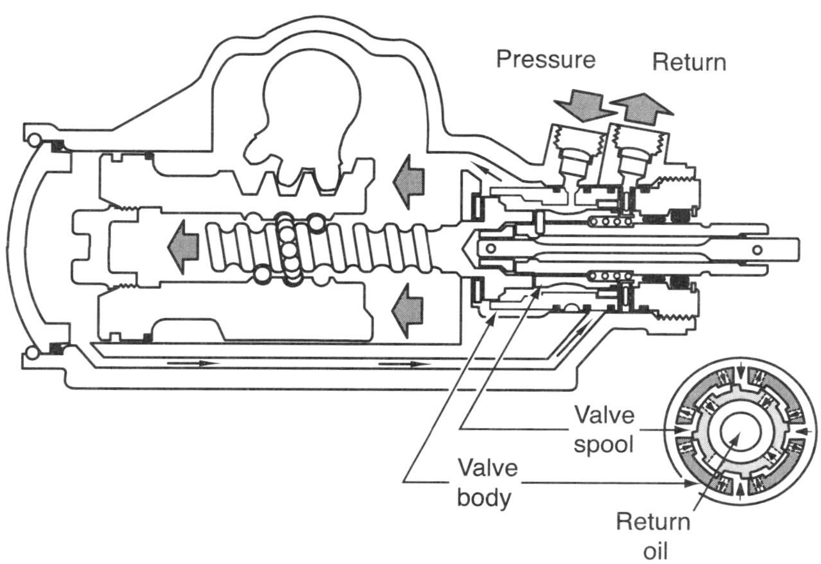


**Figure 11.** Power steering gear fluid flow during a left turn.

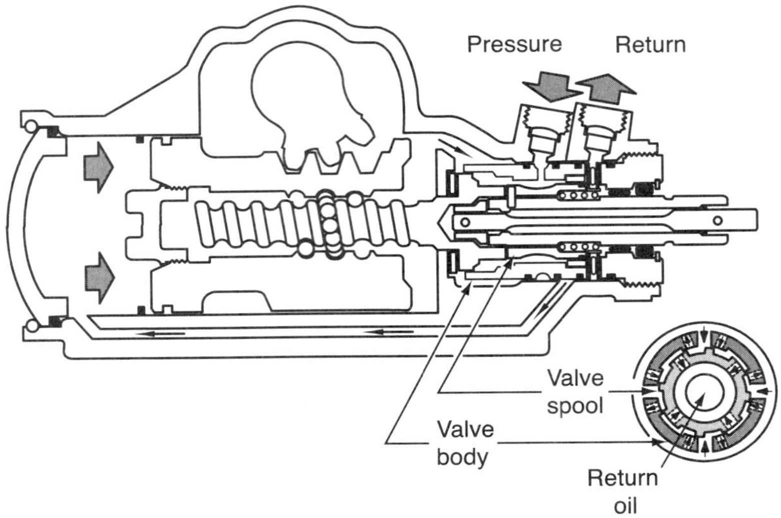


**Figure 12.** Power steering gear fluid flow during a right turn.

the pressure chamber at the lower end of the ball nut piston **(Figure 12).** Hydraulic pressure is applied to the lower end of the recirculating ball piston to help the driver turn the wheels.

During a turn, if a wheel strikes a bump and pushes the wheel in the opposite direction that is steered into, the recirculating ball piston tends to move against hydraulic pressure and forces oil back out the pressure inlet port. This creates kickback, but a **poppet valve** in the pressure inlet fitting closes and prevents this kickback action.

## POWER RACK-AND-PINION STEERING GEARS

The housing is sealed on either side of the rack piston **(Figure 13)** to form two separate hydraulic chambers for the left and right turn circuits. Hydraulic lines are connected to each end of the housing. When the steering wheel is turned, the rotary valve creates a pressure differential on either side of the rack piston. This causes the rack to move toward the lower pressure and reduces the total effort required to turn the wheels.

### Operation

On most vehicles, the steering gear is mounted behind the front wheels. Therefore, rack movement to the left is necessary for a right turn, and movement to the right causes a left turn. When turning left, fluid is pumped into the left side of the fluid chamber and is exhausted from the right chamber area. The pressure on the left side of the piston helps the pinion move the rack to the right **(Figure 14A).** During a right turn, fluid is pumped into the right side of the chamber, and fluid leaves the left side. Thus, hydraulic pressure is

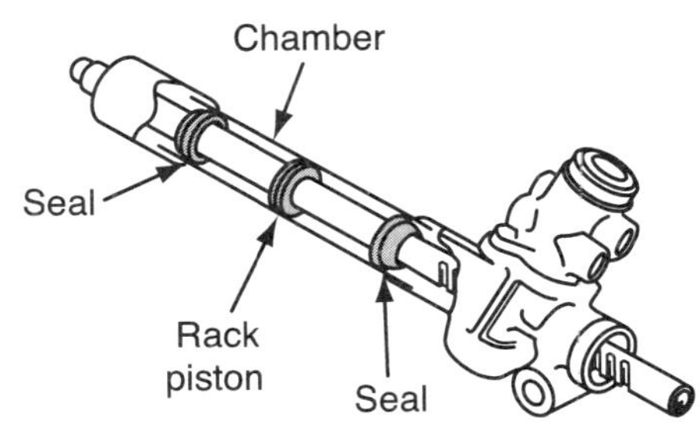


**Figure 13.** A hydraulic chamber in a power rack-and-pinion steering gear.

exerted on the right side of the piston to assist the pinion gear in moving the rack to the left **(Figure 14B).**

Fluid direction is controlled by a rotary valve attached to the pinion assembly. A stub shaft on the pinion assembly is connected to the steering shaft and wheel. The pinion is connected to the stub shaft through a torsion bar that

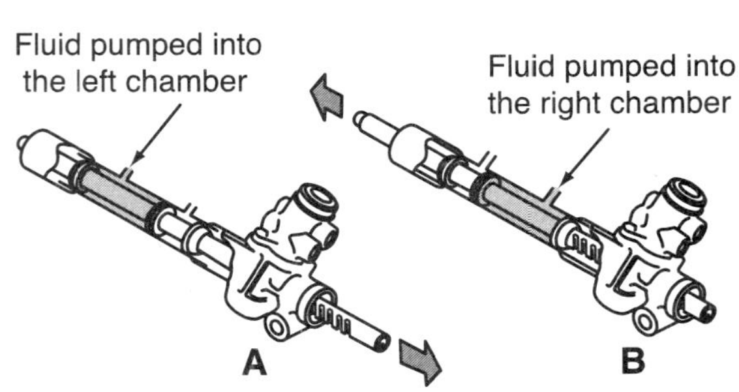


**Figure 14.** Rack movement during right and left turns.

twists when the steering wheel is rotated, and it springs back to the center position when the wheel is released. The rotary valve body contains an inner spool valve that is mounted over the torsion bar.

When the front wheels are in the straight-ahead position, fluid flows from the pump through the high-pressure hose to the center rotary valve body passage. Fluid is then routed through the valve body to the low-pressure return hose and the pump reservoir. Teflon rings or O-rings seal the rotary valve to the steering gear housing.

If hydraulic pressure is not available from the pump, the steering system operates like a manual system, but steering effort is higher. When the torsion bar is twisted to a designed limit, tangs on the stub shaft engage with drive tabs on the pinion. This action mechanically transfers motion from the steering wheel to the rack.

## SUMMARY

- Power steering units are designed to reduce the amount of effort required to turn the steering wheel.
- The most common power steering systems are the integral, linkage, Hydroboost, and power-assisted rack and pinion systems.
- The major components of a power steering system are the steering linkage, power steering pump, flow control and pressure relief valves, reservoir, spool valves and power pistons, hydraulic hose lines, and gearbox or assist assembly on the linkage.
- In a power recirculating ball steering gear, the spool valve movement inside the rotary valve is controlled by torsion bar deflection.
- During a turn with a power rack and pinion steering gear, fluid pressure is directed to the opposite side of the piston and fluid is released from the other side.
- The fluid pressure in a power rack and pinion steering gear is directed to the appropriate side of the rack piston by the spool valve and rotary valve position.
- The position of the spool valve is controlled by the torsion bar during turns.
- If there is no hydraulic pressure in a power rack and pinion steering gear, stop tangs on the stub shaft and pinion provide steering action, but steering effort is much higher.

## REVIEW QUESTIONS

1. What is an integral power steering system?
2. Describe the flow of power steering fluid with the engine running and the front wheels straight ahead in a power recirculating ball steering gear.
3. Describe the fluid movement in the rack piston chamber during a right turn.
4. What is the primary purpose of power steering hoses?
   A. To lubricate the steering gear box
   B. To relieve pressure
   C. To transmit power through fluid under pressure
   D. None of the above
5. While discussing power rack-and-pinion steering gear operation, Technician A says in a power rack and pinion steering gear torsion bar, twisting moves the pinion assembly vertically. Technician B says torsion bar twisting moves the spool valve vertically. Who is correct?
   A. Technician A
   B. Technician B
   C. Both Technician A and Technician B
   D. Neither Technician A nor Technician B

# Chapter 44 Power Steering Pumps

## INTRODUCTION

Central to all power steering systems, except for pure electrical systems, is the hydraulic pump. The pump supplies fluid under pressure for the system. The pump's hydraulic system is a closed system. This means the fluid that goes out and comes back without waste. The power steering pump supplies the necessary pressure to assist steering.

## POWER STEERING PUMP DRIVE BELTS

The power steering pump drive belt is a simple, but very important, part of the power steering system. A power steering pump in perfect condition will not produce the required pressure for steering assist if the drive belt is slipping.

Many power steering pumps are driven by a V-belt that surrounds the crankshaft pulley and power steering pump **(Figure 1)**. The V-belt may also drive other components, such as the water pump. The sides of a V-belt are the friction surfaces that drive the pump. If the sides of the belt are worn and the lower edge of the belt contacts the bottom of the pulley, the belt will slip. The power steering pump pulley, crankshaft pulley, and any other pulleys driven by the V-belt must be properly aligned. If misaligned, excessive belt wear occurs.

A ribbed V-belt, often called a serpentine belt, is used on many vehicles. This belt may be used to drive all the belt-driven components on an engine. The ribbed V-belt is much wider than a conventional V-belt, and the underside of the belt has a number of small ribbed grooves. The smooth backside of the ribbed V-belt may also be used to drive components. Most ribbed V-belts have spring-loaded automatic belt tensioners **(Figure 2)** that eliminate peri-

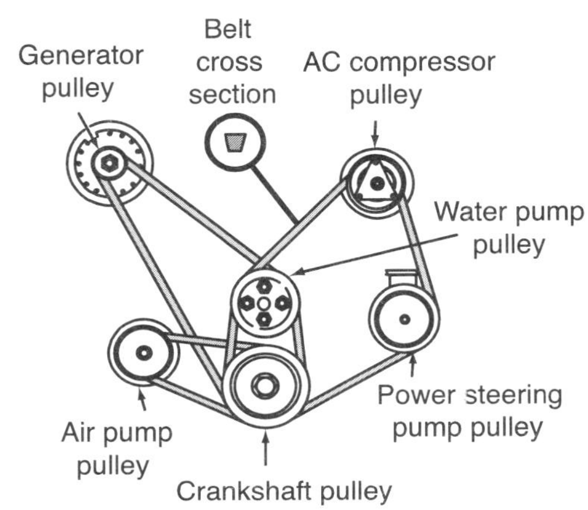


**Figure 1.** V-belts are often used to drive the power steering pump, as well as other accessories. An engine can have three or more V-belts.

**Figure 2.** A belt tensioner for a ribbed V-belt.

odic belt tension adjustments. Regardless of the type of belt, the belt tension is critical.

**You Should Know** *Always keep hands, tools, and equipment, such as shop light cords, away from rotating belts and pulleys. If these become entangled in the belts, personal injury and equipment damage may result. Double-check these areas before starting the engine during and after service.*

## POWER STEERING PUMP DESIGNS

A power steering pump may have a vane, roller, or slipper-type rotor assembly, but all pumps operate on the same basic principle. Many vane-type power steering pumps have flat vanes that seal the pump rotor to the elliptical pump cam ring **(Figure 3).** Other vane-type power steering pumps have rollers to seal the rotor to the cam ring. In some pumps, inverted, U-shaped slippers are used for this purpose. The major differences in these pumps are in the rotor design and the method used to seal the pump rotor in the elliptical pump ring.

**You Should Know** *Power steering reservoirs are typically made of very thin sheet metal or plastic, therefore, avoid prying on any reservoir. This can damage or puncture the reservoir.*

A balanced pulley is pressed on the steering pump's drive shaft. This pulley and shaft are belt-driven by the engine. A large O-ring seals the front of the reservoir to the pump housing **(Figure 4).** Smaller O-rings seal the bolt fittings on the back of the reservoir. The combination cap and

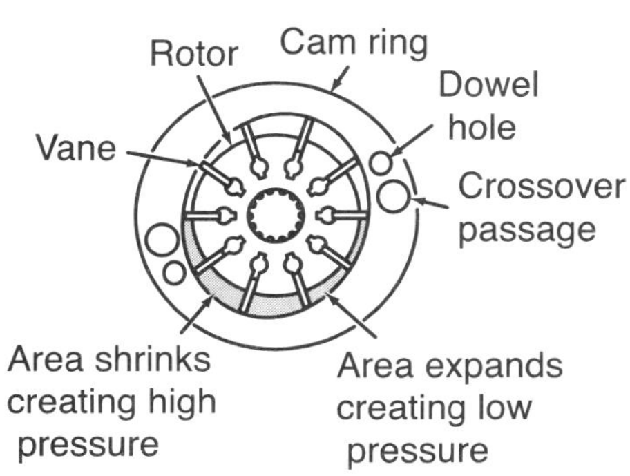


**Figure 3.** The cam ring, rotor, and vanes of a vane-type power steeering pump.

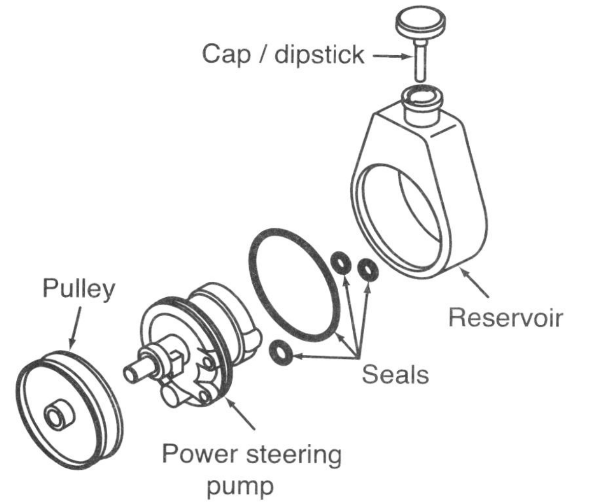


**Figure 4.** Typical power steering pump housing and reservoir seals.

dipstick keeps the fluid in the pump and vents the reservoir to the atmosphere.

The rotating components inside the pump housing include the shaft and rotor with the vanes mounted in the rotor slots. A seal between the output shaft and the housing prevents oil leaks around the shaft. As the pulley drives the pump shaft, the vanes rotate inside an elliptically shaped opening in the cam ring. The cam ring remains in a fixed position inside the pump housing. A pressure plate is installed in the housing behind the cam ring **(Figure 5).**

A spring is positioned between the pressure plate and the end cover, and a retaining ring holds the end cover in the pump housing. The flow control valve is mounted in the

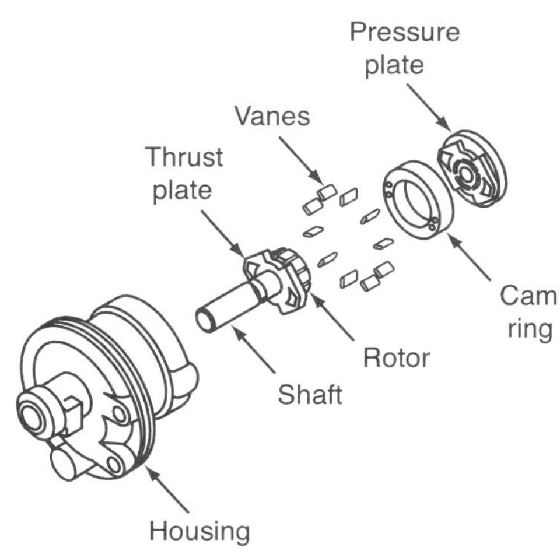


**Figure 5.** A power steering pump housing with shaft and rotor, vanes, cam ring, and pressure plate.

pump housing, and a magnet is positioned on the pump housing to pick up metal filings rather than allowing them to circulate through the power steering system.

The flow control valve is controlled by spring pressure and fluid pressure. Any dirt or roughness on the valve results in erratic pump pressure. The flow control valve contains a pressure relief ball. High-pressure fluid is forced past the control valve to the outlet fitting. A high-pressure hose connects the pump to the inlet fitting on the steering gear. A low-pressure hose returns the fluid from the steering gear to the inlet fitting in the pump's reservoir.

## POWER STEERING PUMP OPERATION

As the belt rotates the rotor and vanes, centrifugal force causes the vanes to slide out of the slots in the rotor. The vanes follow the elliptical surface of the cam. When the area between the vanes expands, a low-pressure area is present between the vanes, and fluid flows from the reservoir into the space between the vanes. As the vanes approach the higher portion of the cam, the area between the vanes shrinks and the fluid is pressurized.

High-pressure fluid is forced into two passages on the thrust plate. These passages reverse the direction of the fluid. Fluid is discharged through the cam crossover passages and pressure plate openings to the high-pressure cavity and flow control valve. The fluid is discharged from the flow control valve through the outlet fitting **(Figure 6).** Though most of the fluid is discharged from the pump, some fluid returns to the bottom of the vanes past the flow control valve and through the bypass passage.

A machined **venturi** orifice is located in the outlet fitting. A venturi is a narrow area in a pipe through which a liquid or gas can flow. When liquid or gas flow increases in the venturi, the pressure drops proportionally. If liquid or gas flow decreases, the pressure in the venturi increases.

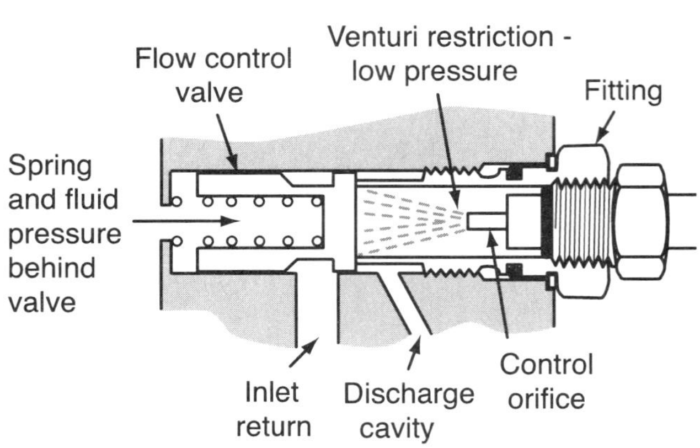


**Figure 6.** Power steering pump fluid flow to the flow control valve and outlet fitting.

### Operation at Idle Speed

At idle speed with the wheels straight ahead, the flow of fluid from the pump to the steering gear is relatively low. Under this condition, the steering gear valve is positioned so the fluid discharged from the pump is directed through the valve and the low-pressure hose to the pump's inlet. With low fluid flow, the pressure is higher in the venturi and control orifice. This higher pressure is applied to the spring side of the flow control valve. This helps keep the flow control valve closed **(Figure 7).** As a result, pump pressure remains low, and all of the fluid is discharged from the pump through the steering gear valve and back to the inlet of the pump.

### Operation at Higher Engine Speeds

When engine speed is increased, the pump delivers more fluid than is needed. This increase in fluid flow creates a pressure decrease in the venturi. This pressure reduction allows the pump discharge pressure to force the flow control valve partially open. The excess fluid from the pump is now routed past the flow control valve to the pump inlet. If the steering wheel is turned, the discharged fluid rushes into the pressure chamber inside the steering gear.

### Operation During Pressure Relief

If the steering wheel is turned (in either direction) until the steering linkage makes contact with the steering stops, the rack piston is stopped. When this occurs, pump pressure could become extremely high and damage hoses or

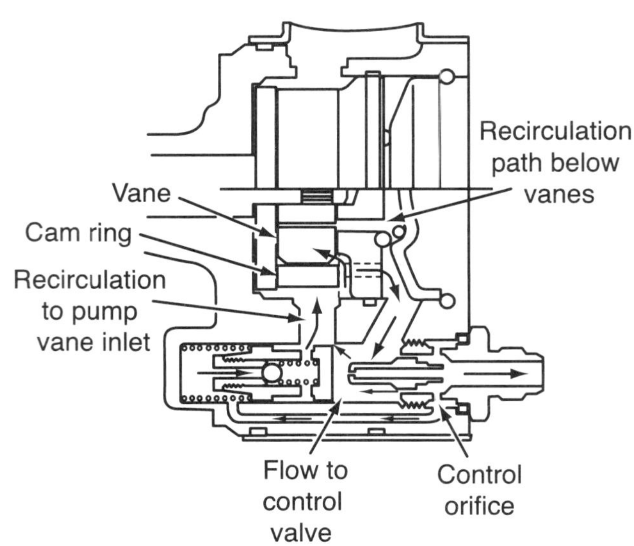


**Figure 7.** The flow control valve operation under various conditions.

other components. When the rack piston stops, the flow from the pump decreases, but the high pressure from the pump is still directed through the steering gear's valve to the piston. Since the flow of fluid through the venturi and control orifice is reduced, a higher pressure is present. This high pressure is also supplied to the spring end of the flow control valve. At a predetermined pressure, the pressure relief ball in the center of the flow control valve is unseated and the flow control valve moves to the wide-open position. This allows some of the high pressure to return to the inlet of the pump, which limits pump pressure to a maximum safe value.

## SUMMARY

- A ribbed V-belt or a V-belt may be used to drive the power steering pump.
- The friction surfaces are on the sides of a V-belt.
- A ribbed V-belt may be used to drive all the belt-driven components. This allows these components to be on the same vertical plane.
- Proper belt tension is important for power steering pump operation.
- A power steering pump may have a vane, roller, or slipper-type rotor assembly, but all three types of pumps operate on the same basic principle.
- The flow control valve in a power steering pump is moved toward the closed position by spring pressure and fluid pressure from the venturi in the pump outlet fitting.

## REVIEW QUESTIONS

1. Explain how the fluid pressure is produced in a vane-type power steering pump.
2. The lower edge of a V-belt should never contact the ______________________.
3. When fluid movement through the venturi increases, the pressure in the venturi ______________.
4. While discussing power steering pump drive belts, Technician A says the smooth backside of a ribbed V-belt is never used as a drive surface. Technician B says a power steering pump will not develop full pressure if the belt is slipping. Who is correct?
   A. Technician A
   B. Technician B
   C. Both Technician A and Technician B
   D. Neither Technician A nor Technician B
5. Which of the following statements about power steering pump operation is *correct?*
   A. If the engine is idling and the front wheels are straight ahead, the flow control valve is near the closed position.
   B. When the engine is running at 2,000 rpm and the steering wheel is turned fully to the right, the pressure relief ball is closed.
   C. When the pressure relief ball is open, most of the pump flow is directed through the return hose to the pump inlet.
   D. If the pressure relief ball sticks closed, pump pressure becomes very low when the front wheels are turned fully to the right.

# Chapter 45 Electronic Steering Systems

## INTRODUCTION

Power steering makes steering easier at low speeds, especially while parking. However, higher steering efforts are desirable at higher speeds in order to provide improved steering feel. **Electronically controlled power steering (EPS)** systems **(Figure 1)** provide both of these benefits. The assist provided by these systems is tapered off by electronic control as road speed increases or when other conditions are present. Thus, these systems require well under 1 pound of steering effort at low road speeds and 3-plus pounds of effort at higher road speeds.

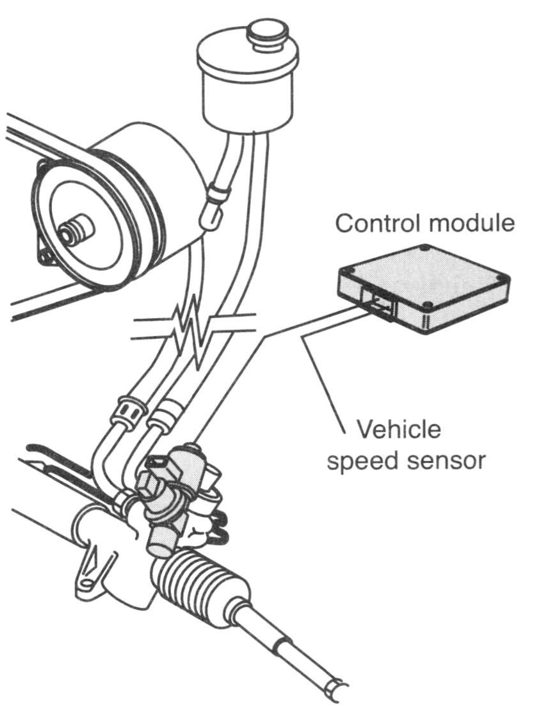


**Figure 1.** An electronically controlled power steering system.

Electrically assisted steering systems offer the same benefits as electronically controlled systems. However, these systems are rack and pinion steering gears with an electric motor for steering assist. The system is light and compact compared to a rack and pinion steering gear with hydraulic steering assist because a pump and hoses are not required. Since the system does not rely on the engine for power, engine power loss is minimized and fuel consumption is reduced.

### ELECTRONICALLY CONTROLLED POWER STEERING SYSTEMS

A typical electronically controlled system has a power steering gearbox, pump, pressure hose, return hose, control module, and control valve. The amount of hydraulic pressure used to boost steering is controlled by a solenoid valve that is identified as its pressure control valve (PCV). The PCV is controlled by the control module.

> **You Should Know** *The pressure control valve (PCV) is not to be confused with the positive crankcase ventilator (PCV) which is an emission control device.*

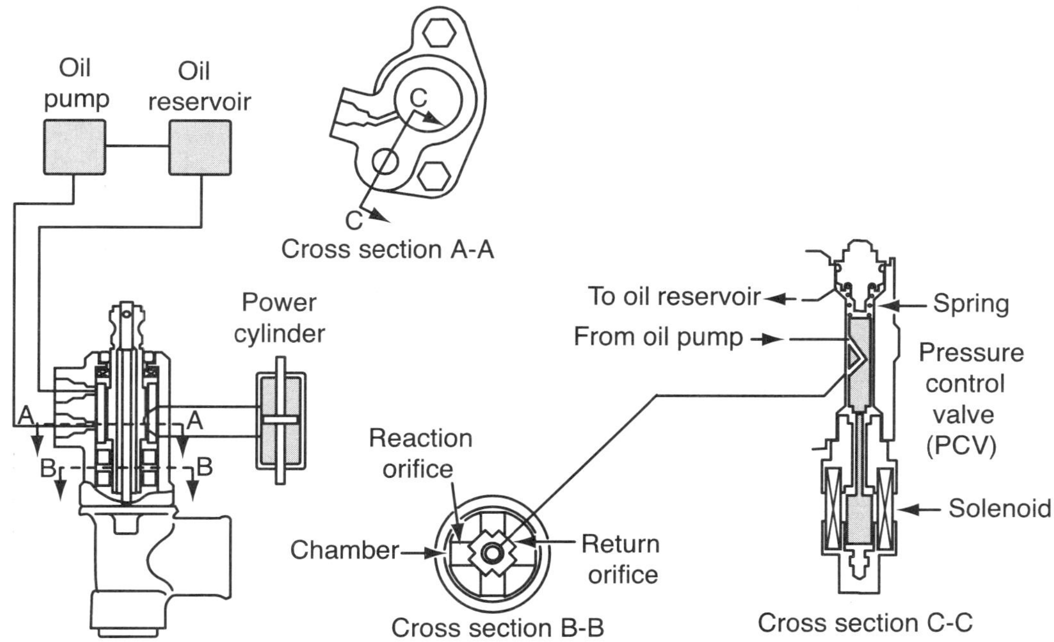


**Figure 2.** The working components in an electronically controlled power steering system.

The PCV **(Figure 2)** is exposed to spring tension on the top and plunger force on the bottom. The plunger slips inside an electromagnet. By varying the electrical current to the electromagnet, the upward force exerted by the plunger can be varied as it works against the opposing spring. Current flow to the electromagnet varies with road speed.

Some systems control the amount of assist by varying the on time of the solenoid. This is referred to as **pulse width modulation (PWM).** When the solenoid is off, the pump supplies full power assist. This reduces steering effort during low-speed maneuvers and parking.

The control module supplies a pulse width modulated voltage to the solenoid in the pump. The control module also provides a ground connection for the solenoid.

When the vehicle is operating at low speeds, the control module supplies a signal to cycle the solenoid faster to allow high pump pressure. This provides maximum power assist during cornering and parking. As the vehicle's speed increases, the solenoid cycles less and the pump provides a lower amount of assist. This gives the driver better road feel during high speeds.

## Magnasteer

Cadillac has a system, called Magnasteer, that adjusts steering effort according to lateral acceleration and vehicle speed. A **lateral accelerometer** measures lateral G forces and sends signals to the unit's controller. When the cornering rate (lateral acceleration) increases, steering effort is automatically increased to give more road feel to the driver. A digital steering angle sensor informs the controller as to how much and how fast the steering wheel has been moved. The controller also monitors vehicle speed and adjusts the steering effort accordingly.

The Magnasteer system uses a conventional power rack-and-pinion gear assembly **(Figure 3).** A spool valve and electromagnetic device, activated by the controller, are incorporated with the pinion to control the amount of assist provided by the system.

## Active Steering

During some situations, active steering improves vehicle stability by turning the wheels more or less

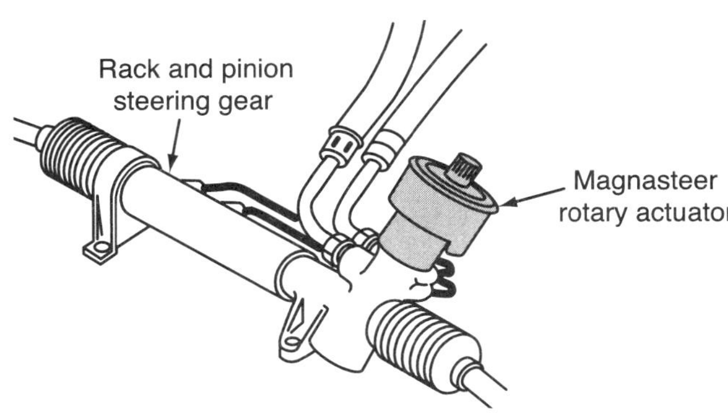


**Figure 3.** The actuator for a Magnasteer system.

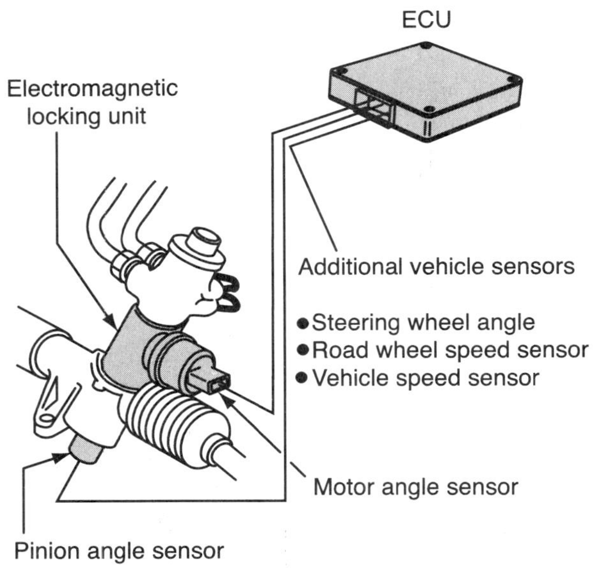


**Figure 4.** The main components of an active steering system.

sharply than commanded by the turn of the steering wheel. The system can even adjust the steering to quickly respond to the threat of skidding **(Figure 4).** Dependent on vehicle speed, the system also allows for a variable steering ratio.

Current active steering systems are not true **steer-by-wire** systems. There is still a mechanical connection between the steering wheel and vehicle's wheels. The systems have an overriding drive built into the steering column. This drive is controlled by an electric motor controlled by the system's computer. The computer determines whether the steering angle needs to be changed and by how much. If the system fails, a planetary gear unit rotates directly with the steering wheel.

## ELECTRONIC VARIABLE ORIFICE STEERING

The **electronic variable orifice (EVO)** steering system is standard on many late-model vehicles. It may also be called the **variable effort steering (VES)** system.

The EVO system provides high-power steering assistance during low-speed cornering and parking and normal power steering assistance at higher speeds. The controller supplies a pulse width modulated voltage to the actuator solenoid in the power steering pump. The controller also provides a ground connection for the solenoid **(Figure 5).**

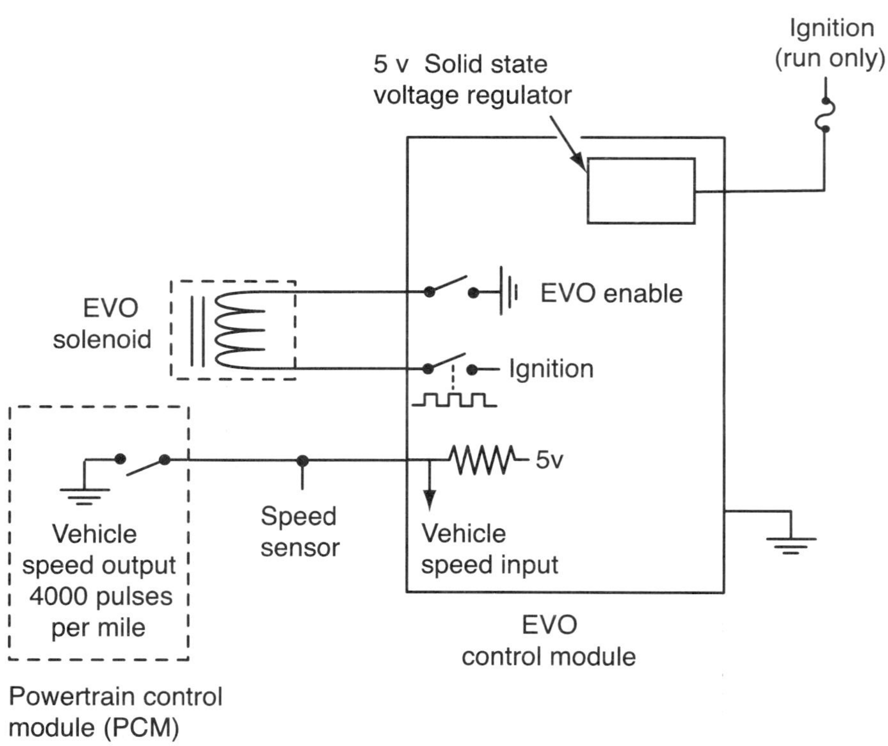


**Figure 5.** An electronic variable orifice (EVO) steering system.

## Input Sensors

> *Never short across or ground any terminals in a computer-controlled system unless instructed to do so by the service manual. Doing this can damage the components.*

The steering wheel rotation sensor is mounted on the steering column, and a shutter disc attached to the steering shaft rotates through the sensor as the steering wheel is rotated. A row of slots is positioned near the outer edge of the shutter disc **(Figure 6).** When these slots rotate through the sensor, a rotation speed signal is sent to the control module.

The **vehicle speed sensor (VSS)** is mounted in the transaxle or transmission and sends a vehicle speed signal to the control module.

## Control Module

The control module continually monitors the input signals from the sensors. Some models have an EVO steering system combined with a rear air suspension system. The inputs and outputs for these systems are connected to the same control module.

## Output Control

A varying current flow is sent from the control module through the EVO actuator **(Figure 7).** As the control module changes the current flow, the actuator supplies a variable pressure to the spool valve in the pump.

When the vehicle is operating at low speeds, the controller supplies a PWM signal to position the actuator solenoid plunger to allow high fluid pressure. At higher speeds, the controller changes the PWM signal to the actuator solenoid to lower the pressure.

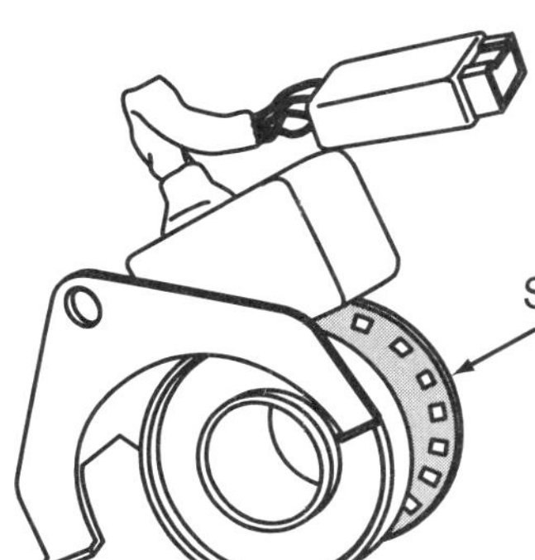


**Figure 6.** A steering wheel rotation sensor.

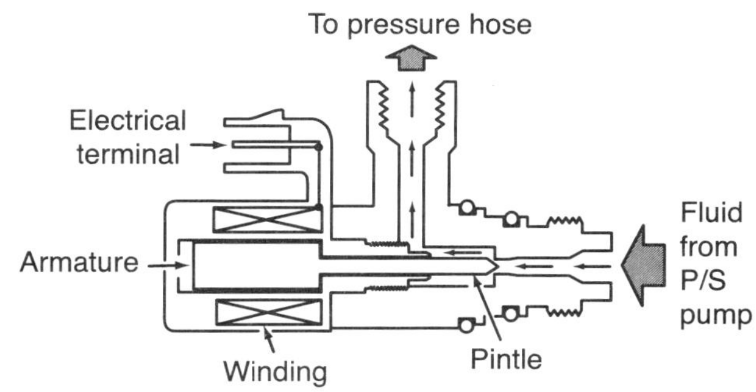


**Figure 7.** An electronic variable orifice (EVO) actuator removed from the power steering pump.

## ELECTRIC/ELECTRONIC RACK-AND-PINION SYSTEM

In **electric/electronic power steering (EPS) (Figure 8),** the outboard housing and rack are designed so that the rotary motion of the armature can be transferred to linear movement of the rack through a ball nut with thrust bearings. The armature is mechanically connected to the ball nut through an internal/external spline arrangement.

The system has the ability to change the rotational direction of the electric motor while at the same time being able to deliver the necessary amount of current to meet torque requirements. The higher the current, the greater the force exerted on the rack. The direction of the turn is controlled by changing the polarity of the motor.

Unlike conventional power steering, electric/electronic units provide power assistance even when the engine stalls since the power source is the battery rather than the engine-driven pump. It also eliminates hydraulic oil, which means no leaks.

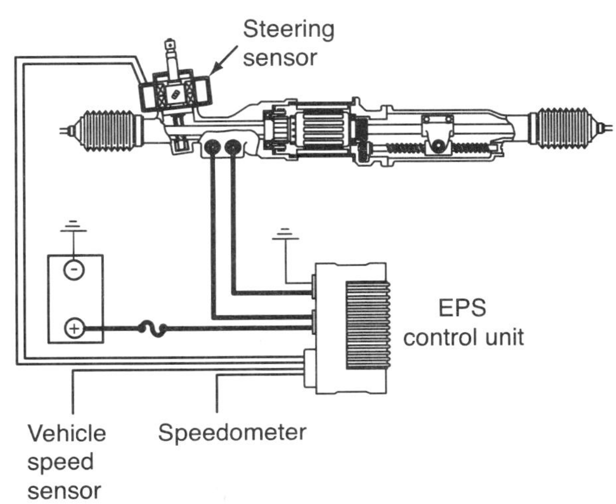


**Figure 8.** A steering gear with electric assist and related system components.

## System Components

The steering sensor is mounted in the pinion shaft. It sends voltage signals to the EPS control unit in relation to the direction, amount, and torque of the steering wheel rotation.

A vehicle speed sensor sends signals to the transmission control module (TCM). These signals are transmitted from the TCM to the EPS control unit. The speedometer also sends voltage signals to the control unit. The signals from the speedometer act as a backup in case the signals from the VSS are inaccurate or nonexistent.

## Electric Motor and Steering Gear Operation

The armature in the electric motor is hollow, and the rack extends through its center. Ball bearings are mounted between the outer diameter of the armature and steering gear housing to support the armature. The teeth on the armature gear are in constant mesh with a matching gear on the ball screw shaft **(Figure 9).** A recirculating ball screw nut is mounted on the ball screw shaft. The recirculating ball screw is bolted to the steering gear rack. A permanent magnet is mounted in the steering gear housing and surrounds the armature core. There is a small clearance between the armature core and the permanent magnet.

When the steering wheel is turned, electric current creates strong magnetic fields around the armature windings. These magnetic fields react with the magnetic field of the permanent magnet and cause armature rotation in the proper direction to supply steering assist. When the armature rotates, the gear on the armature shaft drives the gear on the ball screw shaft. Ball screw shaft rotation moves the recirculating ball screw nut on the shaft. Since the recirculating ball screw nut is bolted to the steering gear rack, movement of this ball screw nut provides steering assist.

The power module can reverse the rotation of the armature to provide steering assist in either direction by reversing the polarity of the current in the armature.

## Steering Gear Motor Current Control

The power module in the EPC control unit contains a driving circuit, current sensor, **field effect transistor (FET)** drive circuit, power relay, and fail-safe relay. When the ignition switch is turned on and the engine is cranked, the EPS control unit closes the power relay and fail-safe relay to turn on the EPS system. The EPS system remains operational if the engine stalls and the ignition switch is on. This feature reduces the possibility of a collision resulting from loss of power steering assist during an engine stall.

When the driver turns the steering wheel, the signals from the steering sensor inform the contro unit that steering assist is needed. The signals also inform the control unit the direction and amount of steering assist required.

When the control unit receives signals indicating that steering assist is required, the control unit signals the FET drive circuit in the power module. The FET drive circuit operates the four FET transistors in the FET bridge to supply the proper direction and amount of current to the motor to provide the necessary direction and amount of steering assist **(Figure 10).**

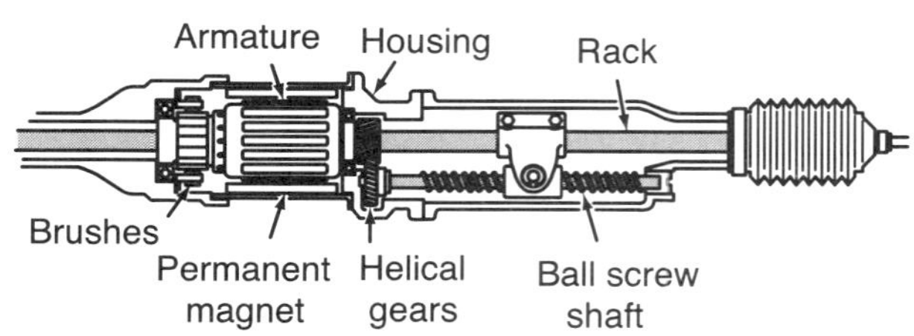


**Figure 9.** A steering gear and electric motor.

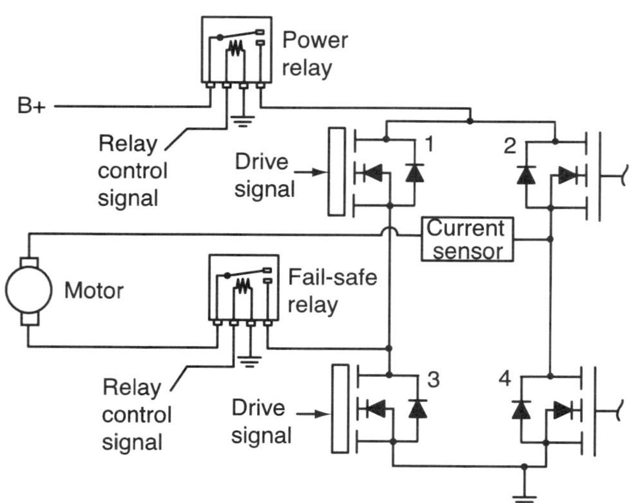


| Steering condition | FET (1) | FET (2) | FET (3) | FET (4) | Motor operation |
|---|---|---|---|---|---|
| Steering to right | PWM | OFF | OFF | ON | Operates in direction steering to the right |
| Straight ahead | OFF | OFF | OFF | OFF | Stops |
| Steering to left | OFF | PWM | ON | OFF | Operates in direction steering to the left |

**Figure 10.** A drive circuit and field effect transistor (FET) bridge in the power module.

If the motor's current exceeds a predetermined amount for the current operating condition, the control unit signals the drive circuit in the power module to reduce the current to prevent motor overheating.

If an electrical defect is present in the system, the control unit will illuminate an EPS indicator light to inform the driver that a fault is present in the system. The control unit also opens the fail-safe and power relays to make the EPS system inoperative. When this happens, manual steering without power assist is available.

## SUMMARY

- In electronically controlled power steering systems, the hydraulic boost is tapered off by electronic control as road speed increases or when other conditions are present.
- Magnasteer adjusts steering effort according to lateral acceleration and vehicle speed.
- Active steering improves vehicle stability by turning the wheels more or less sharply than commanded by the turn of the steering wheel during some situations.
- The electronic variable orifice steering system provides high steering assistance during low-speed cornering and parking and normal assistance at higher speeds.
- As vehicle speed increases, the power steering control module operates a solenoid so steering assist is gradually reduced to provide increased road feel and improved handling.
- A pulse width modulated (PWM) voltage signal has a variable on time.
- The two inputs in an EVO system are steering wheel rotation and vehicle speed.
- Electronic rack and pinion systems replace the hydraulic pump, hoses, and fluid associated with conventional power steering systems with electronic controls and an electric motor located concentric to the rack itself.

## REVIEW QUESTIONS

1. Explain the operation of each input sensor in an electronic variable orifice (EVO) steering system.
2. In an EVO steering system, the actuator solenoid is mounted in the ______________________________ ______________ ______________.
3. If an electrical defect occurs in the electronic power steering (EPS) system,
   A. the system continues to operate normally, but the EPS warning light is illuminated.
   B. the system continues to operate with a slightly reduced power steering assist.
   C. the EPS control unit shuts the system down by opening the fail-safe and power relays.
   D. the EPS control unit locks the armature in the steering gear to prevent armature and screw shaft damage.
4. In an electronic power steering (EPS) system,
   A. the EPS control unit determines the direction of steering wheel rotation from the steering sensor signals.
   B. the variable differential transformer in the steering sensor is connected to the steering rack.
   C. The slider in the steering sensor is at the lower end of its travel when driving straight ahead.
   D. the vehicle speed sensor (VSS) voltage signal is sent directly to the EPS control unit.
5. Technician A says electronically controlled power steering systems allow for power assistance even when an engine stalls. Technician B says an electronic rack and pinion system provides power assistance even when the engine stalls. Who is correct?
   A. Technician A
   B. Technician B
   C. Both Technician A and Technician B
   D. Neither Technician A nor Technician B

46

Chapter

# General Diagnosis and Service of Power Steering Systems

## INTRODUCTION

Diagnosis and service of power steering systems is similar to manual systems except for the hydraulic system and controls. For the most part, power steering systems are quite trouble free and require little regular service except for checking fluid levels and occasionally flushing the hydraulic system.

## DIAGNOSIS OF POWER STEERING SYSTEMS

Diagnosis should begin with trying to duplicate the customer's concern. This is done on a road test. Make sure you drive carefully and cautiously, especially since the vehicle has a control problem. Before taking the road test, do a thorough vehicle safety inspection.

### Visual Inspection

Begin the inspection by checking the tires. Check for correct pressure, construction, size, wear, and damage, and for defects that include ply separations, sidewall knots, concentricity problems, and force problems. Check all of the mechanical parts of the steering and suspension system. Worn ball joints can cause erratic steering and premature tire wear. Bad suspension bushings will allow excessive wheel movement that can affect braking, handling, and wheel alignment. If any part is defective, it should be replaced.

### Symptom Diagnosis

Some power steering problems may be caused by the steering gear, pump, or column. Use the following as a guideline for determining the cause of common power steering problems:

If the vehicle has excessive steering wheel free play, check for:

- Loose thrust bearing preload adjustment
- Loose sector lash adjustment
- Loose steering gear mounting bolts
- Worn steering gears or components
- Worn flexible coupling or U-joints

If the vehicle seems to require an abnormal amount of effort to turn the steering wheel, check for:

- Tight sector lash adjustment
- Tight thrust bearing adjustment
- Low fluid level
- Worn steering gears, bearings, or components
- Contaminated fluid
- Restricted hoses
- Loose belt
- Defective pump
- Low pump pressure
- Sticking, worn flow control valve

If the effort to turn the steering wheel changes with direction and/or the amount of rotation, check for:

- Low fluid level
- Worn steering gears, bearings, or components

If the vehicle is hard to steer and the fluid is low, check for:

- Leaking hoses
- Worn pitman shaft seal
- Worn end-plug O-ring seal
- Worn side-cover O-ring seal
- Worn thrust-bearing O-ring seal
- Worn worm shaft seal
- Worn cylinder end rack seal

- Worn pinion end rack seal
- Worn pinion seal, upper or lower
- Defective control valve housing gasket
- Defective upper strut bearing

## Noises

There may be abnormal noises when the steering wheel is turned. The cause is best identified by paying attention to where the noise is coming from. Some noises may be caused by tires or interference between the steering wheel and the steering column covers. Others can result from a faulty power steering pump or system.

- A hissing noise during low-speed turning, such as while parking is normal.
- A chirping or squealing normally means the drive belt is loose or slipping or the pump pulley is worn.
- Improperly routed hoses, loose steering shaft U-joints or flexible coupling, loose mounting bolts on the power steering pump or the steering gear can cause a rattle.
- A groaning or howling can indicate a restriction in the hydraulic system.
- A sticking control valve will cause a buzzing sound.
- A growling noise from the pump when the engine is running can be caused by a low fluid level, defective pump bearing or components, or air in the power steering system.

> **You Should Know** *If a belt is making an abnormal noise even though it is properly adjusted, the noise is usually the result of two factors: the belt is glazed or hard, or the pulleys are not aligned. To determine the cause of the noise, mist the belt with water. If the noise stops temporarily, the pulleys are out of alignment. If the noise continues, the belt is glazed or hard.*

## POWER STEERING PUMP BELT SERVICE

Power steering belt condition and tension are important for satisfactory power steering pump operation. Correct belt tension is essential for long belt and pump life and quiet operation. Loose belts may slip and not drive their components at their proper speed, resulting in low pump pressure and hard steering. A loose, dry, or worn belt may also cause squealing and chirping, especially during acceleration and cornering. Excessive tension can also cause noise as the driven components rotate under stress. This stress can lead to premature bearing and bushing failure in water pumps, generators, and power steering pumps.

The power steering pump belt should be checked for tension, cracks, oil soaking, worn or glazed edges, and tears **(Figure 1).** If any of these are found, the belt should be replaced.

Since the friction surfaces are the sides of a V-type belt, wear occurs there. If the edges are worn, the belt may be rubbing the bottom of the pulley and the belt should be replaced.

> **You Should Know** *If you use a shop light to examine the belt, make sure the light's cord is not or cannot become tangled or caught in the drive belts. If caught in the belts, the cord can become a serious safety hazard. In fact, shop light cords are the most common items to get caught in belts.*

### Checking Belt Tension

Belt tension can be checked by measuring belt deflection. Press on the belt at its longest span to measure belt deflection, which should be $^1/_2$ inch per foot of free span.

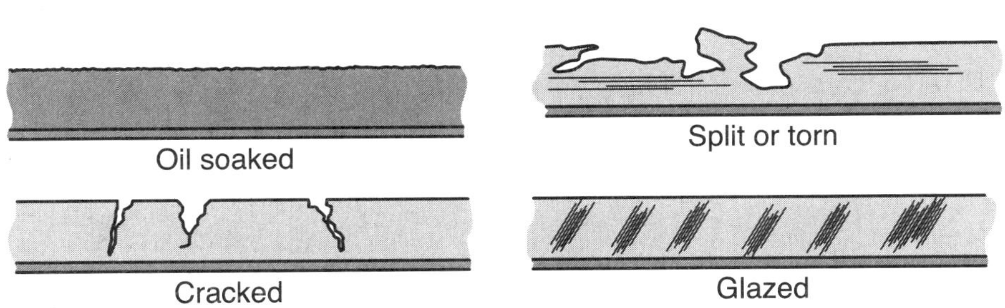


**Figure 1.** If a V-belt has any of these defects, it should be replaced.

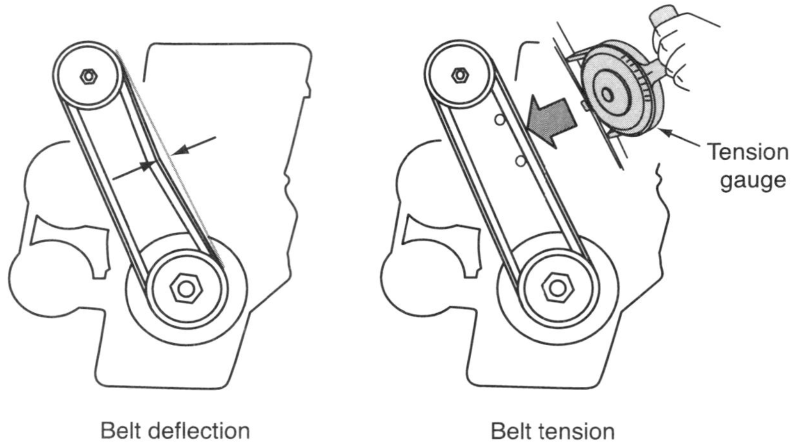


**Figure 2.** Two methods for checking belt tension.

Belt tension can also be checked with a belt tension gauge placed over the belt **(Figure 2)** in the center of the longest span. The tension on the gauge should equal specifications.

## Adjusting Power Steering Belt Tension

To adjust the tension of a V-belt, loosen the power steering pump bracket or tension adjusting bolt and pump mounting bolts. Check the bracket and pump mounting bolts for wear. If these bolts or bolt openings in the bracket or pump housing are worn, replacement is necessary.

Pry against the pump's ear and hub to tighten the belt **(Figure 3).** Some pump brackets have a 1/2-inch square opening in which a breaker bar can be installed to move the pump and tighten the belt **(Figure 4).** Hold the pump in the desired position and tighten the bracket or tension adjusting bolt. Once tightened, recheck the belt tension. If the tension is not correct, readjust it and tighten the tension adjusting bolt and the mounting bolts to the specified torque.

**You Should Know** *Do not pry on the reservoir. This can damage the reservoir.*

## Ribbed V-Belts

Some power steering pumps are driven by a ribbed V-belt. Many ribbed V-belts have an automatic tensioning pulley; therefore, a tension adjustment is not required. As the belt wears or stretches, the spring moves the tensioner

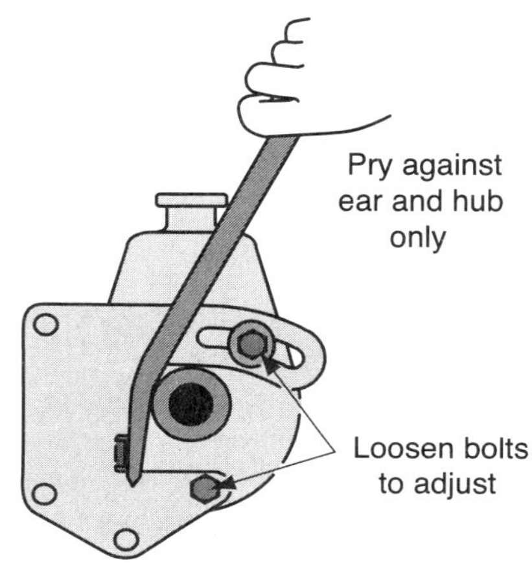


**Figure 3.** Pry against the pump ear and hub with a pry bar to tighten the belt.

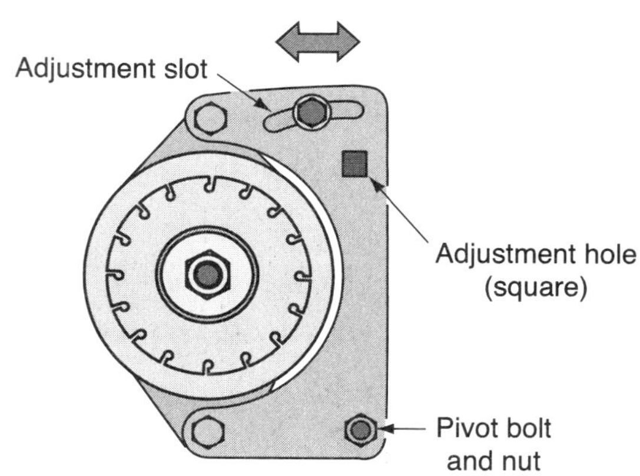


**Figure 4.** The square adjustment hole should be used to tighten the belt.

pulley to maintain belt tension. Some of these tensioners have a belt length scale that indicates new belt range and used belt range **(Figure 5)**. If the indicator on the tensioner is outside the used belt range, belt replacement is required. Many belt tensioners have a 1/2-inch drive opening in which a ratchet or flex handle may be installed to move the tensioner pulley off the belt during belt replacement **(Figure 6).**

Belt manufacturers claim the primary cause of belt wear is a worn tensioner. Therefore, they are recommending that a new belt tensioner be installed each time a new belt is installed.

A ribbed V-belt should be checked to make sure it is installed properly on each pulley and not glazed, cracked, or damaged. The tension on a ribbed V-belt may be checked with a belt tension gauge in the same way as the tension of a V-belt.

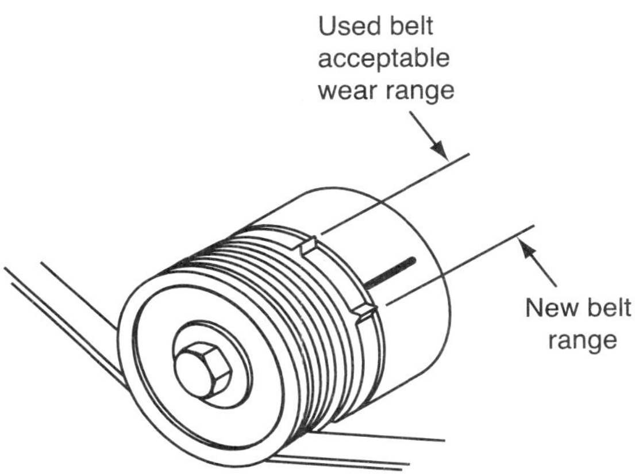


**Figure 5.** The belt tension scale on a belt tensioner.

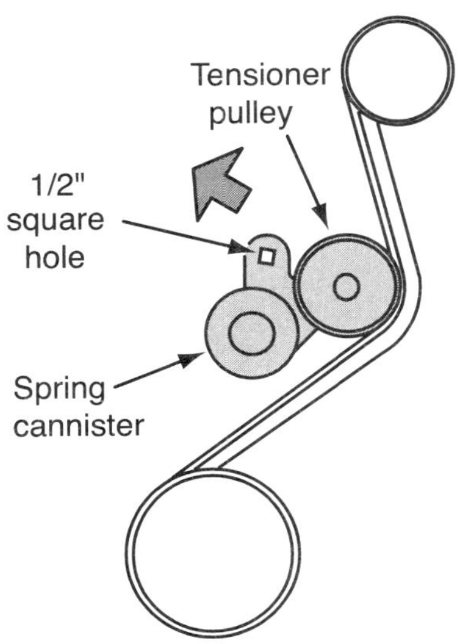


**Figure 6.** This tensioner pulley has a 1/2-inch square hole to aid in adjusting tension.

## Drive Belt Pulleys

Inspect the grooves of the drive pulleys for rust, oil, wear, and other damage. If a pulley is damaged, it should be replaced. In many cases, rust, dirt, and oil can be cleaned off the pulley and should be removed before installing a new belt.

Also, check the alignment of the pulleys. Misalignment of the V-pulleys reduces the belt's service life and can cause thrown belts and screech. The edges of the pulleys must be in line when a straightedge is placed on the pulleys **(Figure 7).** The alignment of the pulleys should be within 1/16 in. (1.59 mm) per foot of the distance across the face of the pulleys.

Repeated belt failure can be caused by misaligned pulleys or extremely high pump pressure caused by a sticking pressure relief valve or a restriction in the high pressure hose.

## Belt Replacement

V-belt replacement is a straightforward procedure, but care must be taken to install it correctly and with the correct tension. Before removing the old drive belt, disconnect the electric radiator cooling if the vehicle has one. Loosen the components that have adjusting slots for belt tension and slip the old belt off.

Always use the exact replacement belt. The size of a belt is typically given, along with the part number, on the belt container **(Figure 8).** You can also verify the size by

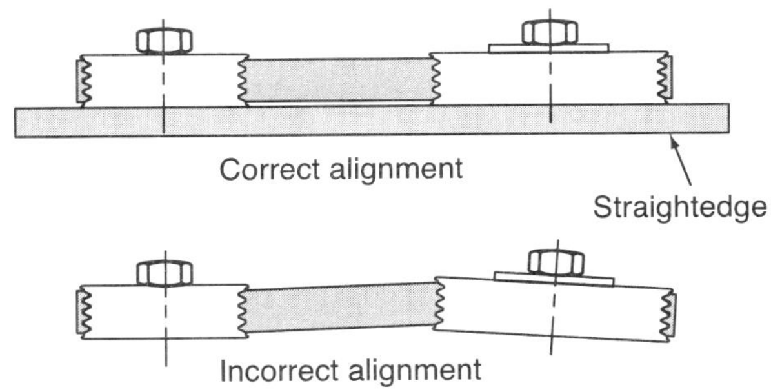


**Figure 7.** Using a straightedge to check pulley alignment.

**Figure 8.** A new drive belt.

physically comparing the two. This, however, does not account for any belt stretch that may have occurred. Therefore, use this as only a comparison. The best way to select a replacement belt is through the catalog and/or by matching the numbers on the old belt to the numbers on the new belt.

Place the new belt around the pulleys. Then, tighten the mounting bolts that were loosened during belt removal. Adjust the tension of the belt and retighten all mounting hardware. Overtightening a drive belt can result in premature accessory bearing failure, front crankshaft main bearing failure, premature belt breakage and wear, and idler bearing failure.

Before removing a serpentine belt, locate a belt routing diagram in a service manual or on the underhood decal. Compare the diagram with the routing of the old belt. If the actual routing is different from the diagram, draw the existing routing on a piece of paper. To install the new belt, wrap it according to instructions. Make sure the ribs of the belt are seated in the matching grooves in the pulleys. Once the belt is routed, put tension on the belt and adjust it to specifications. Tighten all bolts and nuts to keep the tension. Then recheck the tension.

After installing new belts, the engine should be run for 10 to 15 minutes to allow the belts to seat and reach their initial stretch condition. Recheck the tension again after 5,000 miles.

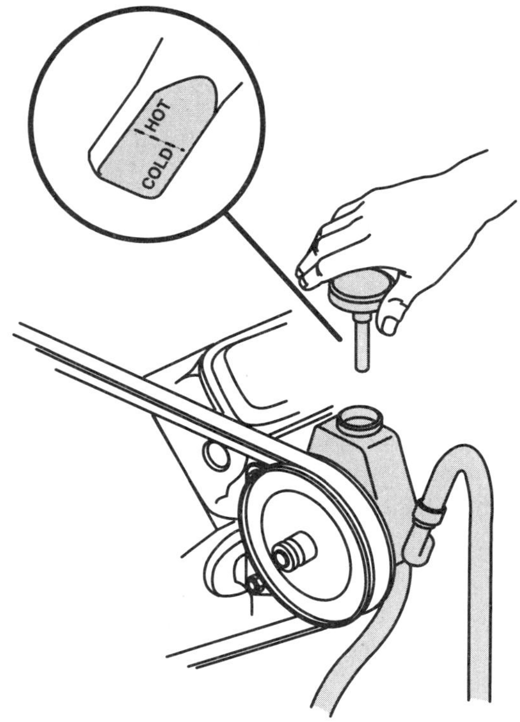


**Figure 9.** The hot and cold markings on a power steering fluid dipstick.

## POWER STEERING PUMP FLUID SERVICE

Check the power steering fluid level and condition. The fluid is checked with the dipstick attached to the reservoir cap. Most manufacturers recommend power steering fluid or automatic transmission fluid in power steering systems. Always use the type of power steering fluid recommended.

Power steering fluid can be checked either hot or cold. Some manufacturers recommend checking the fluid level with the fluid at an ambient temperature of 176°F (80°C). The dipstick is typically marked for HOT and COLD fluid levels **(Figure 9).** Some have separate markings on each side of the dipstick. Make sure to check the level on the right side of the dipstick. Fluid level will vary with temperature.

With the engine idling at 1,000 rpm or less, turn the steering wheel slowly and completely in each direction several times. Wipe all dirt from the outside of the dipstick and reservoir and then remove the dipstick

> **You Should Know** *Never mix oil types. Any mixture or any unapproved oil can lead to seal deterioration and leaks.*

Examine the fluid. Check for evidence of contamination, such as solid particles or water. If these conditions exist or if the fluid has a burnt odor, the system should be flushed.

Also, check the fluid for evidence of air trapped in the system. If the fluid looks foamy, it is likely that air is in the system. To verify this, run the engine until it reaches normal operating temperature. Then turn the steering wheel to the left and to the right several times without hitting the stops. If there is air in the system, bubbles will appear in the fluid reservoir.

With the ignition OFF, wipe off the outside of the power steering pump, pressure hose, return hose, fluid cooler, and the steering gear. Start the engine and turn the steering wheel several times from stop to stop. Check for leaks. Fluid leakage will not only cause abnormal noises, but may result in unequal and abnormal steering efforts.

### Bleeding Air from the System

When air is in the fluid, a growling noise may be heard in the pump, and steering effort may be increased or erratic. To remove the air, the system must be bled. The exact method of bleeding depends on the type of system; always follow the recommended procedures.

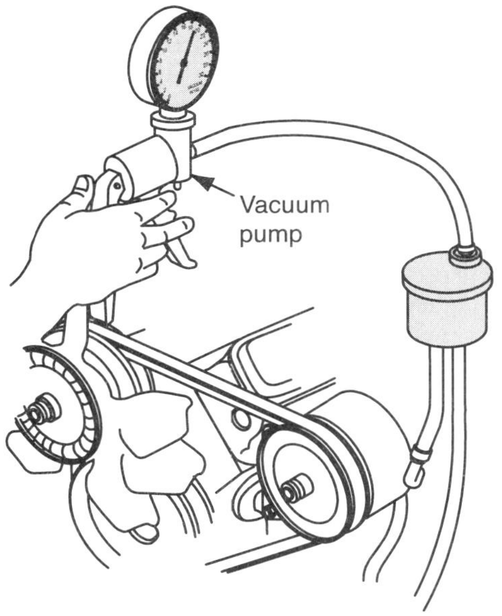


**Figure 10.** Using a hand-held vacuum pump to bleed the power-steering system.

One procedure involves connecting a vacuum pump to the cap's opening in the reservoir **(Figure 10).** While the engine is running, vacuum is applied to the reservoir and maintained for about five minutes. The vacuum is then released and the reservoir refilled with fluid. Vacuum is again applied to the reservoir and the steering wheel is cycled from stop to stop every 30 seconds for at least 5 minutes. After this period of time, the vacuum is released and the reservoir cap installed. Then the system is checked for leaks.

Another way to bleed the system involves running the engine at 1,000 rpm and turning the steering wheel fully in each direction three or four times. Each time the steering wheel is turned to the right or left, hold it there for 2 to 3 seconds before turning it in the other direction. Check for fluid foaming in the reservoir. Continue to turn the steering wheel back and forth until no more foaming is produced, then check the fluid level. Shut the engine off and make sure the fluid level does not increase more than 0.020 in. (5 mm) **(Figure 11).**

If the power steering fluid is contaminated with moisture, dirt, or metal particles, the system should be drained and flushed, and new fluid should be installed.

## Power Steering System Draining and Flushing

Hydraulic fluid is easily contaminated by moisture and dirt; flushing replaces the old fluid with its contaminants with fresh fluid. It is also wise to flush the system after you have replaced or repaired a part in the system. Before flushing the system, disable the engine's ignition or fuel injection system. Then disconnect the power steering return hose and plug the reservoir. Attach an extension hose between the power steering return hose and an empty container **(Figure 12).** Fill the reservoir with the correct fluid.

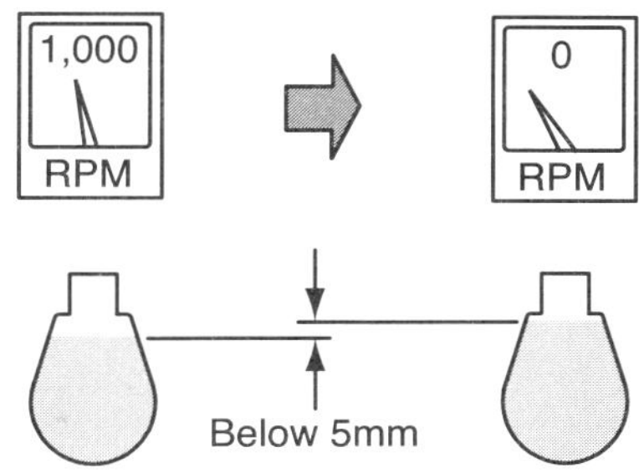


**Figure 11.** The fluid level should be up slightly after bleeding the system and the engine is shut off.

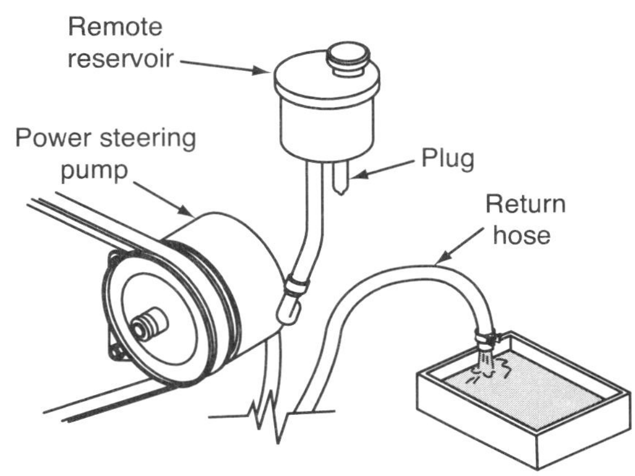


**Figure 12.** The pump end of the return hose is inserted into an empty container when the system is being drained and flushed.

Turn the steering wheel from stop to stop while cranking the engine until the fluid in the return hose is clean. Never crank the engine for more than 5 seconds at a time. Add fluid to the reservoir to make sure it does not empty. Once the fluid is clear, refill the reservoir and lower the vehicle.

## CHECKING AND SERVICING POWER STEERING LINES AND HOSES

Hoses should also be carefully inspected for leaks, dents, sharp bends, cracks, and swelling. Always replace power steering hoses with an exact replacement hose. Never attempt to patch or seal a leak in a hose or the hose's fittings. Lines and hoses must not rub against other components. This can wear a hole in the line or hose. Many high-pressure lines are made from high-pressure steel-braided hose with molded steel fittings on each end.

## Power Steering Hose Replacement

When power steering hose replacement is required, follow these steps:

1. With the engine stopped, remove the return hose at the power steering gear, and allow the fluid to drain into a drain pan.
2. Loosen and remove all hose fittings from the pump and steering gear.
3. Remove all hose-to-chassis clips.
4. Remove the hoses from the chassis, and cap the pump and steering gear fittings.
5. If O-rings are used on the hose ends, install new O-rings. Some lines have gaskets. The old gasket must be pried out of the fitting before installing the new lines. Lubricate O-rings with power steering fluid.
6. Install the new hose by reversing the steps for removal. Make sure to tighten all fittings to the specified torque. Be sure all hose-to-chassis clips are in place. Do not position hoses where they rub on other components.
7. Fill the pump reservoir to the full mark with the manufacturer's recommended fluid. Bleed air from the power steering system. Then check the fluid level in the reservoir and add fluid as required.

# SUMMARY

- A loose power steering belt causes low pump pressure, hard steering, and belt chirping, or squealing, especially on acceleration.
- Belt tension is checked with a belt tension gauge or by checking belt deflection.
- Most manufacturers recommend power steering or automatic transmission fluid in the power steering system.
- If the power steering fluid is contaminated with moisture, dirt, or metal particles, the system should be drained and flushed and new fluid installed.
- Air can be bled from the system by turning the steering wheel fully in each direction several times with the engine running. Each time the wheel is turned fully to the right or left, it should be held in that position for 2 to 3 seconds.
- Power steering pump lines should be checked for leaks, dents, sharp bends, cracks, and contact with other components.

# REVIEW QUESTIONS

1. Technician A says the tension on a power steering pump V-belt should be checked with a tension gauge. Technician B says that some ribbed V-belts have an automatic tensioning pulley that eliminates the need for tension adjustments. Who is correct?
   A. Technician A
   B. Technician B
   C. Both Technician A and Technician B
   D. Neither Technician A nor Technician B
2. While discussing power steering fluid level checking, Technician A says foaming in the remote reservoir may indicate air in the power steering system. Technician B says many car manufacturers recommend checking the power steering fluid level with the fluid warmed up. Who is correct?
   A. Technician A
   B. Technician B
   C. Both Technician A and Technician B
   D. Neither Technician A nor Technician B
3. A power steering pump has repeated belt failures. The power steering pump V-belt tension and pulley alignment are satisfactory, and the V-belt is not bottomed in the pulley. The most likely cause of this problem is:
   A. a pressure relief valve sticking open.
   B. worn pulley friction surfaces on the pulley.
   C. a worn thrust plate in the power steering pump.
   D. an intermittent restriction in the high pressure hose.
4. A vehicle has continual excessive steering effort, but there is no noise and the fluid level in the reservoir is correct. Technician A says the cause of the problem could be a stuck flow control valve. Technician B says the cause of the problem could be air in the power steering fluid. Who is correct?
   A. Technician A
   B. Technician B
   C. Both Technician A and Technician B
   D. Neither Technician A nor Technician B
5. There is air in the power steering fluid. To correct this problem, you should:
   A. with the engine running, crack the pressure line at the pump to release trapped air.
   B. with the engine running, turn the steering wheel full left and right.
   C. with the engine running, drain all fluid from the system and refill.
   D. with the engine stopped, raise the wheels and remove the return hose from the gear.

# Chapter 47 Pump Diagnosis and Service

## INTRODUCTION

If the preliminary tests indicate an internal pump or gear malfunction, a steering system analyzer can be used to look at the operation of the pump, rotary valve, and internal seals. The analyzer provides a readout of pump pressure and flow, pump relief pressure, and seal integrity. If a power steering analyzer is not available, a pressure gauge can be used to check the system.

### Power Steering Pump Oil Leaks

The possible sources of power steering pump oil leaks **(Figure 1)** are the drive shaft seal, reservoir O-ring seal, high-pressure outlet fitting, and the dipstick cap. When a leak is present at the high-pressure outlet fitting, tighten this fitting to the specified torque. If the leak still occurs, replace the O-ring seal and retighten the fitting. Correct any leaks before conducting a pressure test. After correcting the leak, conduct a pressure test to make sure all problems have been corrected. A leak in the hydraulic system will give inaccurate pressure readings.

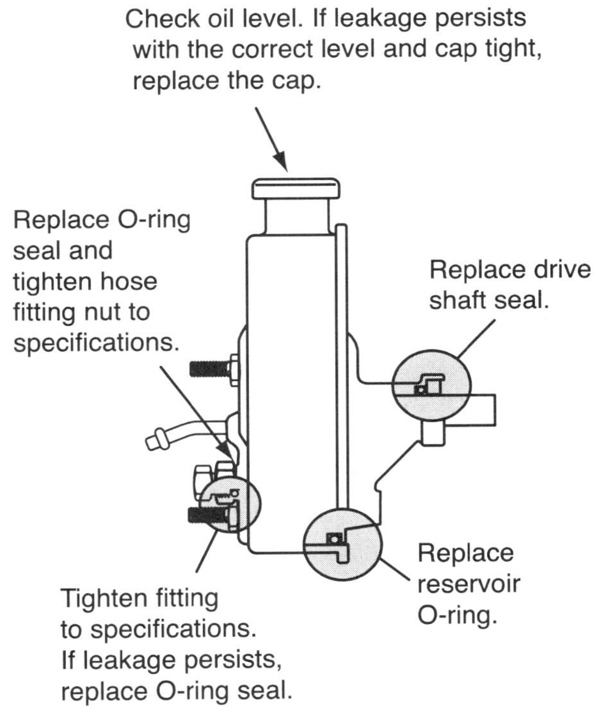


**Figure 1.** Common causes of oil leaks in a power steering pump.

## TESTING POWER STEERING PUMP PRESSURE

Since there are variations in pump pressure test procedures and specifications, the manufacturer's test procedures and specifications must be used. If pump pressure is low, steering effort is increased. Erratic power steering pump pressure causes variations in steering effort, and the steering wheel may jerk as it is turned. Since a power steering pump will never develop the specified pressure if the belt is slipping, the belt tension must be checked and adjusted, if necessary, prior to a pressure test.

With the engine turned off, disconnect the pressure line from the pump, and connect the gauge side of the pressure gauge to the outlet fitting. Then, connect the valve side of the gauge to the pressure line **(Figure 2).**

Start the engine and fully turn the steering wheel in each direction two or three times to bleed air from the system. Make sure the fluid level is correct and the fluid temperature is at least 176°F (80°C). A thermometer may be inserted in the pump's fluid reservoir to measure its temperature **(Figure 3).**

**You Should Know** *Do not allow the fluid to become too hot during the pressure test. Wear protective gloves, and always shut the engine off before disconnecting gauge fittings.*

Allow the engine to idle and observe the reading on the gauge. The reading should be between 30 to 80 psi (200 to 550 kPa). If the pressure is low, the pump may be faulty. If the pressure is too high, the problem may be restricted hoses.

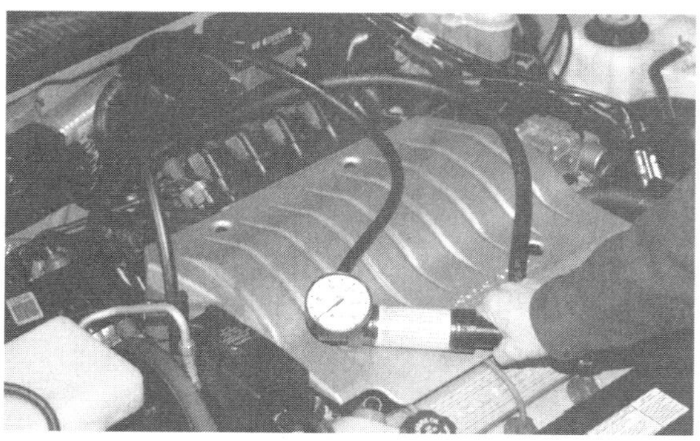

**Figure 2.** A pressure gauge connected to a power steering pump.

**Figure 3.** With a thermometer inserted into the reservoir, watch the temperature of the fluid.

While the engine is still idling, close the pressure gauge valve for no more than 10 seconds and observe the gauge reading **(Figure 4).** With the valve closed, the pressure should increase by the specified amount. If the pressure is too high, a faulty pressure relief valve is suggested. If the pressure is too low, the pump may be bad.

Turn the pressure gauge valve to the fully open position **(Figure 5).** If the pressure gauge reading does not

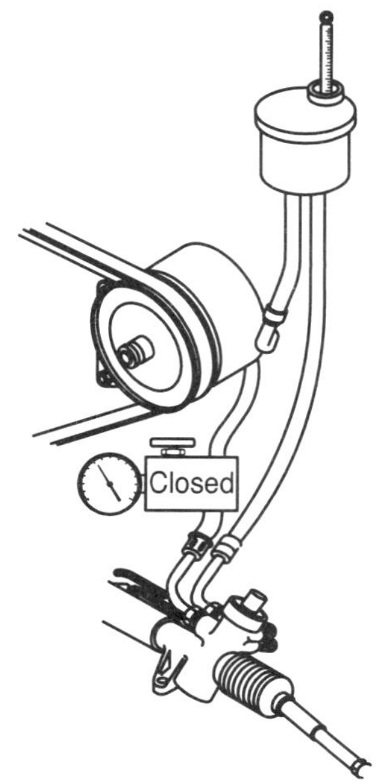


**Figure 4.** Close the pressure gauge valve for no more than 10 seconds, and observe the pressure gauge reading.

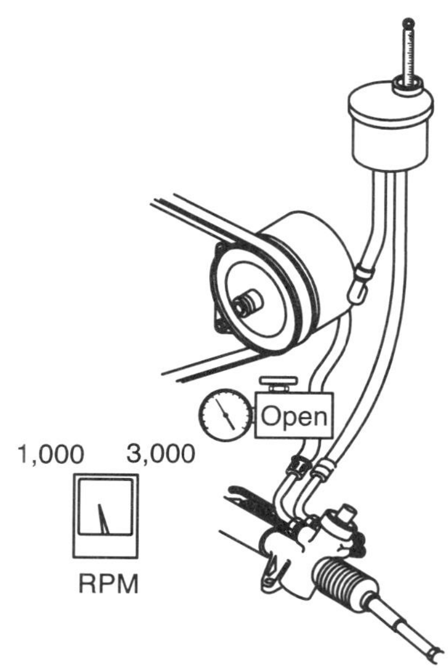


**Figure 5.** Turn the pressure gauge valve to the fully open position, and observe the pressure gauge reading.

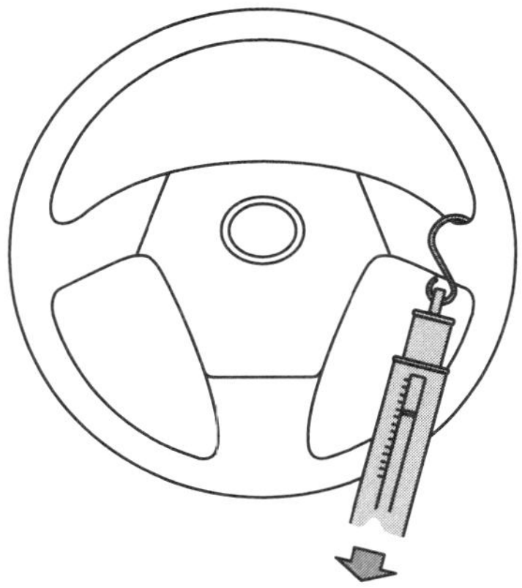

Figure 6. Measuring steering effort.

equal the specifications, repair or replace the power steering pump.

Now, check the pressure with the engine running at 1,000 rpm and 3,000 rpm, and record the pressure difference between the two readings. If the pressure difference does not equal specifications, repair or replace the flow control valve in the power steering pump.

With the engine still running, turn the steering wheel fully in one direction and observe the pressure while holding the steering wheel in this position. If the pressure is less than specified, the steering gear housing has an internal leak and should be repaired or replaced.

After conducting the pressure test, check the steering effort at the steering wheel. Connect a spring scale to the steering wheel and measure the effort in both directions **(Figure 6).** If the pump pressure is satisfactory but the steering effort is greater than specifications, the power steering gear should be repaired.

## POWER STEERING PUMP REPLACEMENT

If a growling noise comes from the power steering pump after the fluid level is checked and air has been bled from the system, the pump bearings or other components are defective and the pump should be replaced or repaired. When the pump pressure is lower than specified, the pump should be replaced or repaired. The following is a typical procedure for removing and installing a power steering pump; always refer to the appropriate service manual.

1. Disconnect the power steering return hose from the remote reservoir or pump. Allow the fluid to drain into a drain pan. Discard the used fluid according to local laws.
2. Loosen the bracket or belt tension adjusting bolt and pump mounting bolt.
3. Remove the drive belt.
4. Remove the hoses from the pump, and cap the fittings and hoses.
5. Remove the belt tension adjusting and mounting bolts, and then remove the pump.
6. Check the pump mounting bolts and boltholes for wear.
7. Reverse steps 1 through 5 to install the power steering pump. Tighten the belt and pump mounting and bracket bolts to specifications. If O-rings are used on the pressure hose, replace the O-rings. Be sure the hoses are not contacting the exhaust manifold, catalytic converter, or exhaust pipe during or after pump replacement.
8. Fill the pump reservoir, and bleed air from the power steering system.

## POWER STEERING PUMP SERVICE

Although power steering pumps are not typically rebuilt by technicians, some parts are replaceable. The actual parts that can be replaced depend on the make of pump. Common parts that can be replaced are discussed next.

### Power Steering Pump Pulley Replacement

If the pulley wobbles while it is rotating, it is undoubtedly bent and should be replaced. Worn pulley grooves and/or cracks also indicate that the pulley should be replaced. A pulley that is loose on the pump's shaft must be replaced. Never hammer on the pump's drive shaft during pulley removal or installation. This will damage the internal parts of the pump.

If the pulley is pressed onto the pump's shaft, a special puller is required to remove it **(Figure 7),** and a

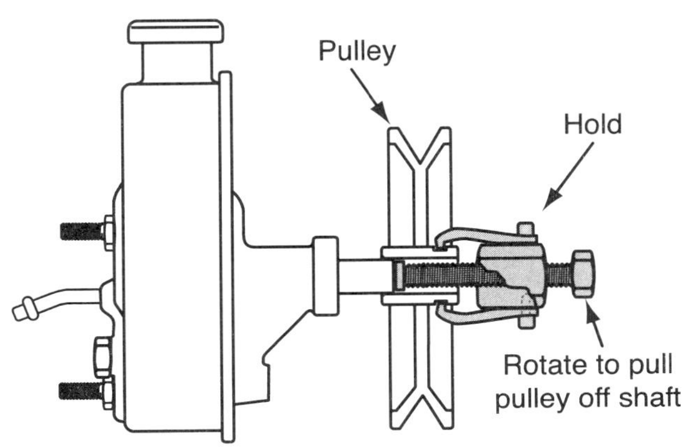


Figure 7. A puller should be used to remove a pressed-on pulley.

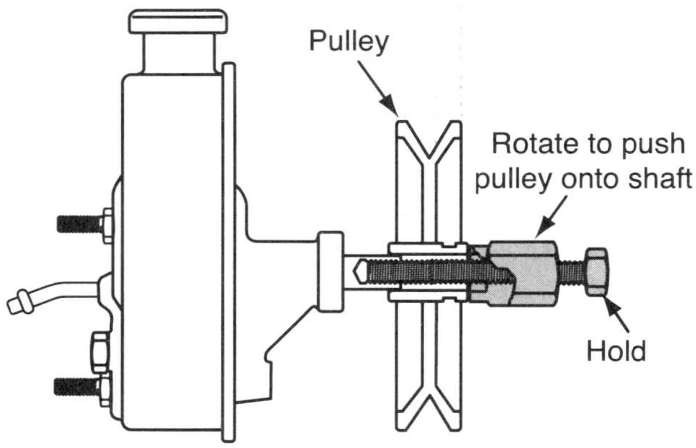


**Figure 8.** A special driver or tool should be used to install a pressed-on pulley.

pulley installation tool is used to install the pulley **(Figure 8).**

If the power steering pump pulley is retained with a nut, mount the pump in a vise. Always tighten the vise on one of the pump's mounting bolt surfaces. Do not tighten the vise with excessive force. Use a special holding tool to keep the pulley from turning, and loosen the nut with a box-end wrench **(Figure 9).** Remove the nut, pulley, and **woodruff key.** Inspect the pulley, shaft, and woodruff key for wear. Replace all worn components.

## Remove and Replace the Power Steering Pump Reservoir

In many cases, the reservoir is an actual part of the pump and cannot be replaced. However, some reservoirs can be replaced. If the reservoir is remote, replacement is quite straightforward. The following procedure is typical for replacing an integral reservoir:

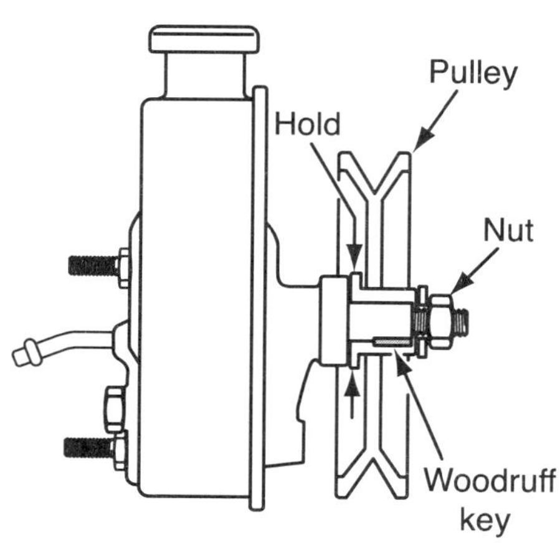


**Figure 9.** The retaining hardware for a bolt-on pulley.

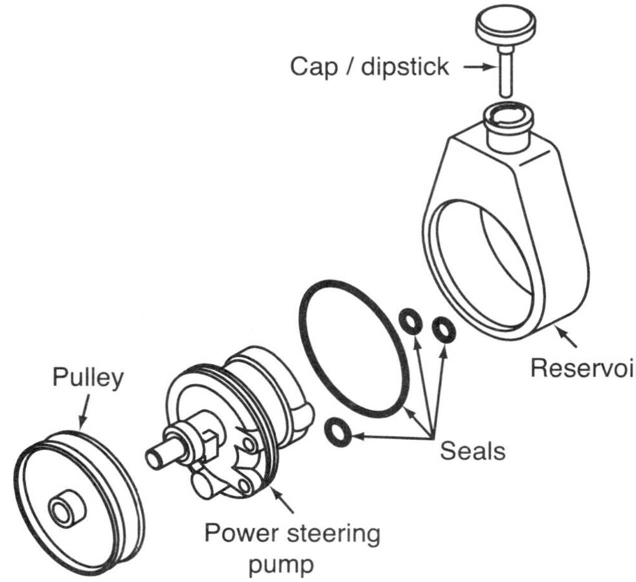


**Figure 10.** The seals and O-rings in a pump should be well lubricated before the pump is assembled.

1. Remove the filler cap and drain the oil from the reservoir.
2. Remove the bolts or studs and the fitting from the rear of the reservoir.
3. Rock the reservoir by hand or tap it gently with a soft hammer to remove it.
4. Clean all parts and discard O-rings and seals. Inspect all sealing surfaces.
5. Lubricate the new seals and O-rings with the recommended power steering fluid and install them **(Figure 10).** Make sure the O-ring is positioned correctly.

## Remove and Replace the Flow Control Valve and End Cover

To replace the flow control valve and end plate, remove the retaining ring with a slotted screwdriver and punch. Then remove the flow control valve, end cover, spring, and magnet. Check the flow control valve for burrs. Remove minor burrs with crocus cloth, and clean the flow control valve in solvent. Damaged or worn flow control valves must be replaced. Inspect the end cover sealing surface for damage. Also, check the pump's drive shaft for corrosion and damage. Remove any corrosion with crocus cloth. Clean all parts and lubricate the end cover with

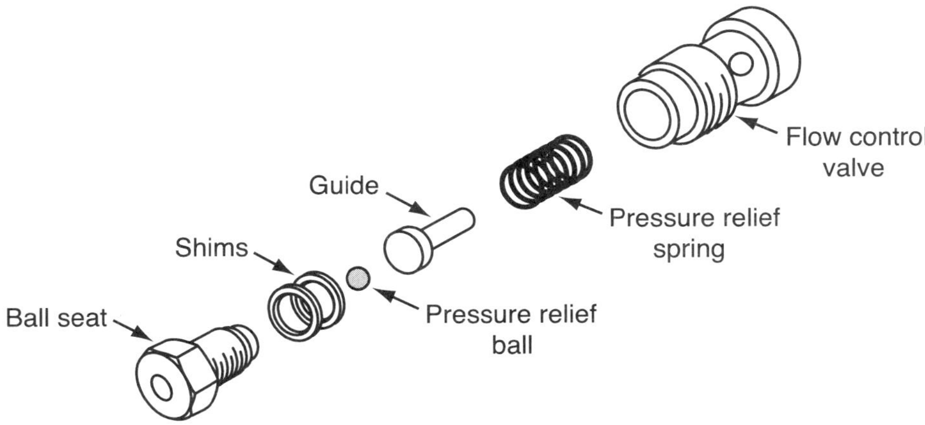


**Figure 11.** The pressure relief valve with a hex-head ball seat.

power steering fluid. Make sure to clean the magnet with a shop towel. To reassemble, install the flow control valve, end cover, retaining ring, and related components.

## Remove and Replace the Pressure Relief Valve

Follow this procedure to service the pressure relief valve:

1. Wrap a shop towel around the land end of the flow control valve and clamp this end in a soft-jawed vise. Be very careful not to damage the valve lands.
2. Remove the hex-head ball seat **(Figure 11).** Clean the components in solvent. A worn or damaged pressure relief ball, spring, guide, or seat must be replaced.
3. Reinstall the new or cleaned components, and then install the ball seat.

## Servicing Power Steering Pump Rotating Components

The rotating parts of a power steering pump vary with the type and the manufacturer of the pump. What follows is a simplified general procedure for replacing commonly found parts.

1. Remove the end cover and spring.
2. Pull the pressure plate, pump ring, rotor, and shaft from the housing **(Figure 12).**
3. Remove and discard the rotor lock ring.
4. Remove the rotor and thrust plate from the shaft.
5. Examine all the wear surfaces. Replace parts that are worn, scuffed, or scored. Check vanes and vane slots in the rotor for wear.
6. Clean all components and install the inner thrust plate, rotor, and a new lock ring on the shaft.

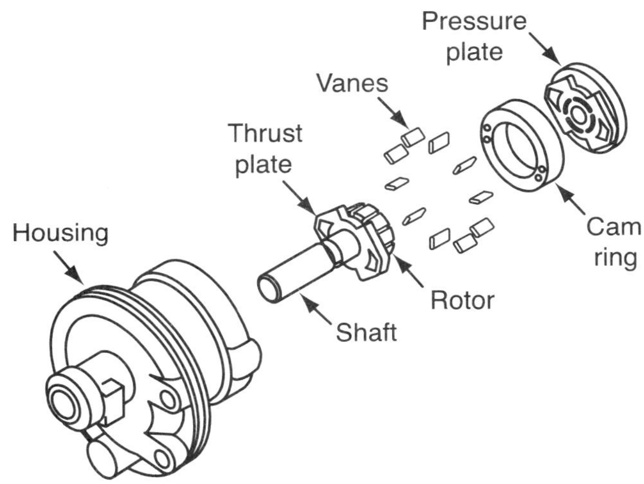


**Figure 12.** An exploded view of the pressure plate, cam ring, rotor, and shaft assembly.

7. Remove the drive shaft seal from the pump housing with a slotted screwdriver.
8. Lubricate and then install the new seal with the proper driver.
9. Lubricate the new O-rings, and position them in the housing.
10. Install the thrust plate, shaft, and pump rotor assembly in the housing, and install the pump ring.
11. Install the vanes in the rotor, and lubricate all parts with power steering fluid.
12. Install and seat the pressure plate.
13. Install the pressure plate spring, end cover, and retaining ring into the housing.

# SUMMARY

- A pressure gauge set is connected in series with the pump pressure hose to check pressure.
- The power steering pump pressure may be checked at 1,000 rpm and 3,000 rpm to check the operation of the pump flow control valve.
- After the power steering pump pressure is checked, steering gear leakage is checked by turning the steering wheel fully in one direction and observing the pressure reading. If the pressure is less than specified, the steering gear has an internal leak.
- With the engine idling, a spring scale may be attached at the outer end of the steering wheel cross bar to measure steering effort.
- The possible sources of power steering pump leaks are the drive shaft seal, reservoir O-ring seals, high-pressure outlet fitting, dipstick cap, and low-pressure inlet pipe.
- When a power steering pump is replaced, the mounting bolts and boltholes must be checked for wear.
- Power steering pump pulleys may be pressed onto the pump shaft or retained on the shaft with a woodruff key and nut.
- Never tap on the pump shaft while removing or replacing the pump pulley.
- A sticking flow control valve causes erratic power steering pump pressure.
- A defective power steering pump must be overhauled or replaced.

# REVIEW QUESTIONS

1. While discussing the steering pump pressure test, Technician A says the pressure gauge and valve should be connected in the power steering pump return hose to check pump pressure. Technician B says the pressure gauge valve should be closed for 30 seconds during the power steering pump pressure test. Who is correct?
   A. Technician A
   B. Technician B
   C. Both Technician A and Technician B
   D. Neither Technician A nor Technician B
2. A power steering system has the specified pressure with the pressure gauge valve closed and with the engine running at 1,000 and 3,000 rpm and the pressure gauge valve open. The power steering pump pressure is lower than specified with the front wheels turned fully in either direction. The cause of this problem could be:
   A. a defective steering gear.
   B. a stuck flow control valve.
   C. a stuck pressure relief valve.
   D. worn pump vanes and cam ring.
3. All of these statements about power steering pump pulley replacement are true *except:*
   A. If the pulley is retained with a nut, a woodruff key prevents pulley rotation on the shaft.
   B. If the pulley is pressed onto the shaft, a special puller must be used to remove the pulley.
   C. If the pulley is pressed onto the shaft, when reinstalling the pulley a soft hammer may by used to drive the pulley onto the shaft.
   D. If the pulley fits loosely on the pump's shaft, it should be replaced.
4. While discussing the steering pump pressure test, Technician A says connect the valve side of the gauge to the pressure line to check pump pressure. Technician B says normal specified pressure for a power steering pump pressure test is between 30 to 80 psi (200 to 550 kPa). Who is correct?
   A. Technician A
   B. Technician B
   C. Both Technician A and Technician B
   D. Neither Technician A nor Technician B
5. A power steering pump is typically replaced or repaired when all of these conditions exist, except:
   A. if a growling noise comes from the power steering pump.
   B. the pump bearings are defective.
   C. power steering pump pressure is lower than specified and the drive belt is in good condition and has the correct tension.
   D. the mounting boltholes are worn.

48

Chapter

# Recirculating Ball Power Steering Service

## INTRODUCTION

Much of the service to a power recirculating ball steering gear is the same as that for a manual unit. Because power units are assisted by hydraulics, there are some unique parts and, therefore, services. These are covered in this chapter.

## GEAR DIAGNOSIS

You Should Know *Recirculating ball steering gears are often mounted near the exhaust manifold, which may be extremely hot. Wear protective gloves and use caution when inspecting, adjusting, and servicing the steering gear.*

### Power Recirculating Ball Steering Gear Oil Leak Diagnosis

When the oil-level plug is removed from the steering gear, the lubricant should be level with the bottom of the plug opening. If the oil level is low, fill the steering gear with the specified steering gear lubricant. A steering gear will cause binding and hard, uneven steering effort if the oil level is very low. If the level is low, check the sector shaft seal and the worm shaft seal area for leaks. Leaking seals must be replaced. Common locations for oil leaks include:

1. Cover O-ring seal **(Figure 1)**
2. Adjuster plug seal

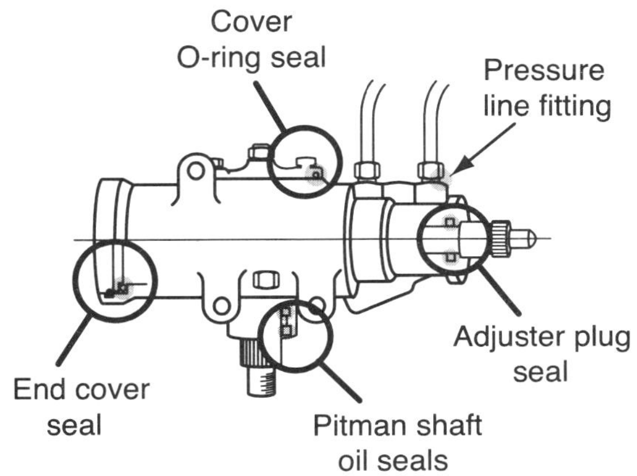


**Figure 1.** Potential leak spots in a power reciprocating ball steering gear.

3. Pressure line fitting
4. Pitman shaft oil seals
5. End cover seal

If an oil leak is present at any of these areas, complete or partial steering gear disassembly and seal or O-ring replacement is necessary.

## POWER RECIRCULATING BALL STEERING GEAR REPLACEMENT

When the power steering gear is replaced, proceed as follows:

1. Disconnect the hoses from the steering gear, and cap the lines and fittings to prevent dirt from entering the system.

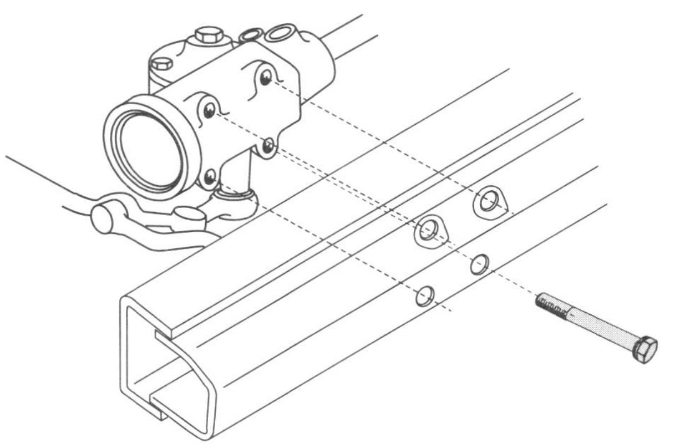

Figure 2. Remove the steering gear mounting bolts, and remove the steering gear from the chassis.

2. Remove the pitman arm nut and washer, and mark the pitman arm in relation to the shaft with a center punch. Use a puller to remove the pitman arm.
3. Disconnect the steering shaft from the worn shaft.
4. Remove the mounting bolts, and remove the steering gear **(Figure 2).**
5. Reverse steps 1 through 4 to install the steering gear. All bolts must be tightened to the specified torque. Be sure the pitman arm is installed in the original position.

## GEAR ADJUSTMENTS

A loose worm shaft thrust bearing preload adjustment or sector lash adjustment causes excessive steering wheel freeplay and steering wander. The following procedures assume the steering gear is removed from the vehicle for service. The adjustments should be made whenever diagnostics indicate there is an adjustment problem or when the gear has been disassembled for other service. Always follow the recommended procedure given in the service manual.

### Worm Shaft Thrust Bearing Preload Adjustment

When the worm shaft thrust bearing preload adjustment is performed, remove the worm shaft thrust bearing adjuster plug lock nut with a hammer and brass punch **(Figure 3).**

Turn this adjuster plug inward, or clockwise, until it bottoms, then tighten the plug to 20 ft.-lb (27 N.m). Place an index mark on the steering gear housing next to one of the holes in the adjuster plug **(Figure 4).**

Measure 0.50 in. (13 mm) counterclockwise from the index mark, and place a second index mark at this position. Then, rotate the adjuster plug counterclockwise until the hole in the adjuster plug is aligned with the second index mark placed on the housing **(Figure 5).** Install and

Figure 3. Remove the worm shaft thrust bearing adjuster plug locknut with a brass punch and hammer.

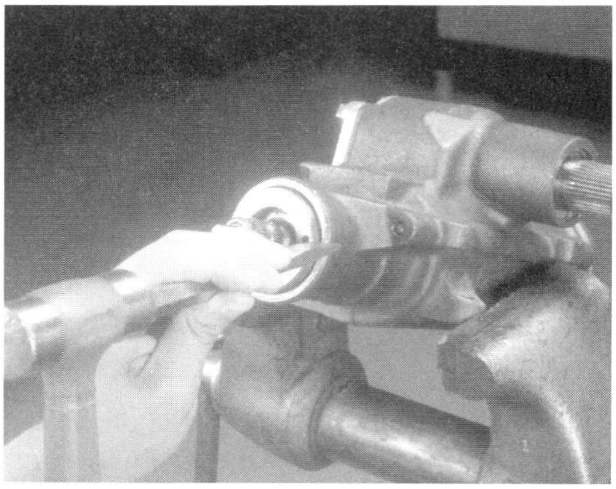

Figure 4. Place an index mark on the steering gear housing next to one of the holes in the adjuster plug.

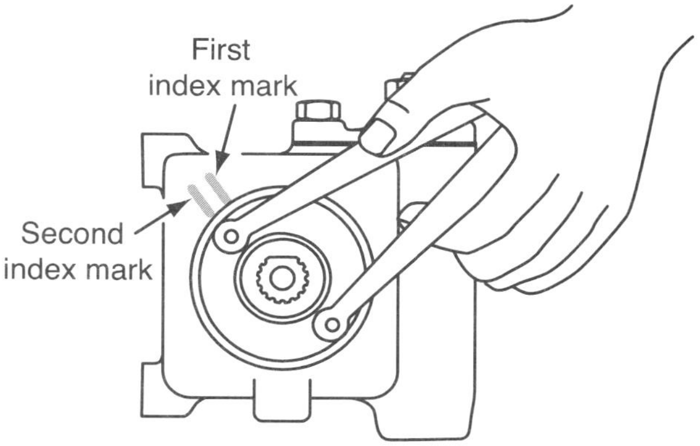


Figure 5. Align the adjuster plug hole with the second index mark on the housing.

**Figure 6.** Install and tighten the worm shaft adjuster plug locknut.

tighten the adjuster plug lock nut to the specified torque **(Figure 6).**

## Pitman Sector Lash Adjustment

To adjust the pitman sector lash on a power reciprocating ball steering gear:

1. Rotate the worm shaft from stop to stop and count the number of turns.
2. Starting at either stop, turn the worm shaft two-thirds of the total number of turns. In this position, the flat on the worm shaft should be facing up and the master spline on the pitman shaft aligned with the pitman shaft backlash adjuster screw.
3. Turn the pitman shaft backlash adjuster screw fully counterclockwise, and then turn it clockwise one turn.
4. Use an inch-pound torque wrench to turn the worm shaft through a 45 degree arc on each side of the position in step 2. Read the over-center torque as the worm shaft turns through the center position **(Figure 7).**

**Figure 7.** With an inch-pound torque wrench, measure the over-center torque as the worm shaft turns through the center position.

5. Continue to adjust the pitman shaft adjuster screw until the torque is 6 to 10 in-lbs (0.6 to 1.2 Nm) more than the torque in step 4.
6. Hold the pitman shaft adjuster screw in this position and tighten the lock nut to the specified torque.

# STEERING GEAR SERVICE

Prior to any disassembly procedure, clean the steering gear with solvent or in a parts washer. The steering gear service procedures vary depending on the make of gear. Always follow the manufacturer's recommended procedure given in the appropriate service manual.

## Cover O-Ring Replacement

The following is a typical procedure for replacing the cover O-ring:

1. Loosen the pitman backlash adjuster screw lock nut, and remove the cover bolts.
2. Rotate the pitman backlash adjuster screw clockwise to remove the cover from the screw **(Figure 8).**
3. Discard the O-ring and inspect the matching surfaces for metal burrs and scratches.
4. Lubricate a new O-ring with power steering fluid and install the O-ring.
5. Rotate the pitman backlash adjuster screw counterclockwise into the cover until it is properly positioned on the housing. Turn this adjuster screw fully counterclockwise, and then one turn clockwise.
6. Install and tighten the bolts to the specified torque.
7. Adjust the pitman sector shaft lash.

## End Plug Seal Replacement

To replace the end plug seal, insert a punch into the access hole in the housing to unseat the retaining ring,

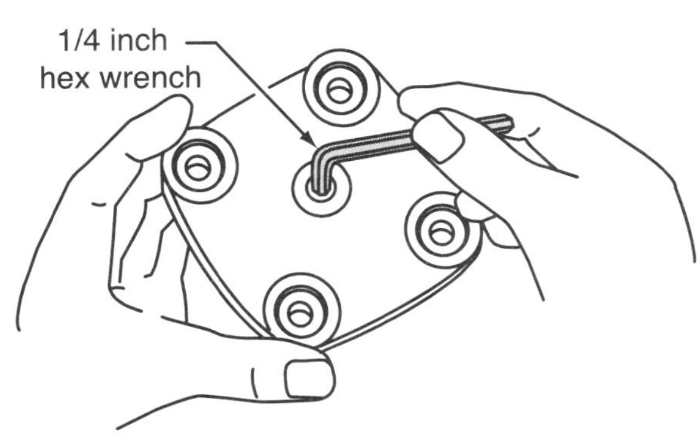


**Figure 8.** Removing the steering gear cover.

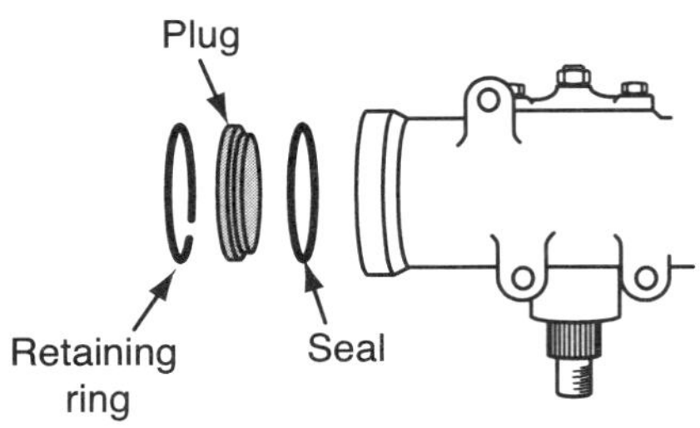


**Figure 9.** A typical steering gear end plug, seal, and retaining ring.

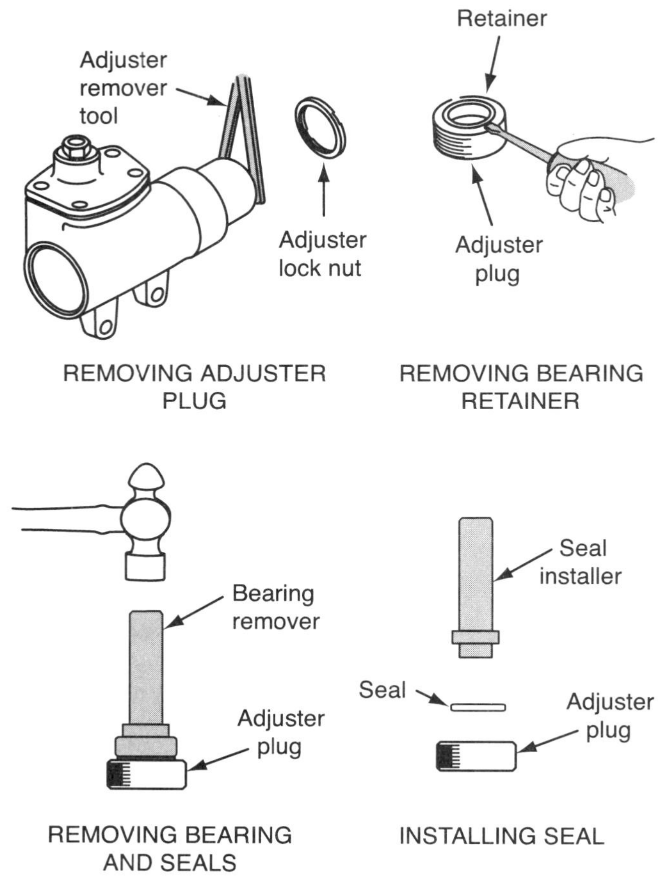


**Figure 10.** Removing and replacing the worm shaft adjuster plug, bearing, and seal.

and remove the ring **(Figure 9).** Then, remove the end plug and seal.

Clean the end plug and seal contact area in the housing with a shop towel. Lubricate the new seal with power steering fluid and install the seal. Install the end plug and retaining ring.

## Worm Shaft Bearing Adjuster Plug Seal and Bearing Replacement

To replace the worm shaft bearing adjuster plug seal and bearing, remove the adjuster plug lock nut. Using the appropriate special tool, remove the adjuster plug. With a screwdriver, pry at the raised area of the bearing retainer to remove it from the adjuster plug **(Figure 10)**. Then, place the adjuster plug face down on a suitable support, and use the proper driver to remove the needle bearing, dust seal, and lip seal.

**You Should Know** *The bearing identification number must face the driving tool to prevent bearing damage during installation.*

Place the adjuster plug outside face up on a suitable support, and use the proper driver to install the needle bearing dust seal and lip seal. Then, install the bearing retainer in the adjuster plug, and lubricate the bearing and seal with the recommended power steering fluid. Install the adjuster plug and lock nut, and adjust the worm shaft bearing preload.

## SUMMARY

- A loose worm shaft bearing preload adjustment or sector lash adjustment causes excessive steering wheel freeplay, steering wander, and reduced road feel.
- If the worm shaft bearing preload adjustment or sector lash adjustment is tighter than specified on a power recirculating ball steering gear, steering effort is higher than normal.
- Oil leaks in a power recirculating ball steering gear may occur at the end plug O-ring seal, side-cover O-ring seal, pitman shaft seal, or worm shaft thrust bearing seal.
- On some power recirculating ball steering gears, the worm shaft thrust bearing adjuster plug is tightened until the correct worm shaft turning torque is obtained with the worm shaft positioned one-half turn from the full-right or full-left position.
- On a power recirculating ball steering gear, the sector lash screw is tightened until the correct worm shaft turning torque is obtained as the worm shaft is rotated through the center position.

# REVIEW QUESTIONS

1. While diagnosing a loose feeling when steering a vehicle with power steering: Technician A says the problem could be caused by air in the system. Technician B says the problem could be caused by tight worm shaft thrust bearing preload adjustment. Who is correct?
   A. Technician A
   B. Technician B
   C. Both Technician A and Technician B
   D. Neither Technician A nor Technician B
2. *True or False:* A loose worm shaft thrust bearing preload adjustment or sector lash adjustment causes excessive steering wheel freeplay and steering wander.
3. A power recirculating ball steering gear has excessive kickback on the steering wheel. The most likely cause of this problem is:
   A. a loose sector lash adjustment.
   B. a loose worm shaft bearing preload adjustment.
   C. a defective poppet valve in the steering gear.
   D. a worn pitman shaft bearing.
4. When adjusting a power recirculating ball steering gear:
   A. the worm shaft bearing adjuster plug should be bottomed and then backed off until the specified worm shaft turning torque is obtained.
   B. the sector lash adjustment screw should be tightened one turn before the worn shaft bearing preload is adjusted.
   C. the pitman shaft backlash adjuster screw is turned fully counterclockwise and then one turn clockwise prior to the backlash adjustment.
   D. the pitman shaft is positioned one-half turn from the fully-right or fully-left position prior to the pitman shaft backlash adjustment.
5. Technician A says a hissing noise from the power steering gear is normal if the steering wheel is at the end of its travel or when the steering wheel is rotated with the vehicle standing still. Technician B says if the steering wheel jerks or surges when the steering wheel is turned with the engine running, the power steering pump belt should be checked. Who is correct?
   A. Technician A
   B. Technician B
   C. Both Technician A and Technician B
   D. Neither Technician A nor Technician B

# Chapter 49 Rack-and-Pinion Power Steering Service

## INTRODUCTION

Nearly all rack and pinion steering systems are power assisted. The procedures for removal and general inspection of the steering gear are the very similar to those for a manual rack and pinion. With the addition of the power assist system, there are some unique procedures for diagnostics and service. These are the topic of this chapter. Remember, some power steering problems may be caused by defects in the steering gear, pump, and column.

### DIAGNOSIS

The same effort should be required to turn the steering wheel in either direction. A pressure gauge connected to the high-pressure hose should also indicate the same pressure as the steering wheel is turned in each direction. Lower power assist in one direction may be caused by defective rack seals **(Figure 1).** Steering effort

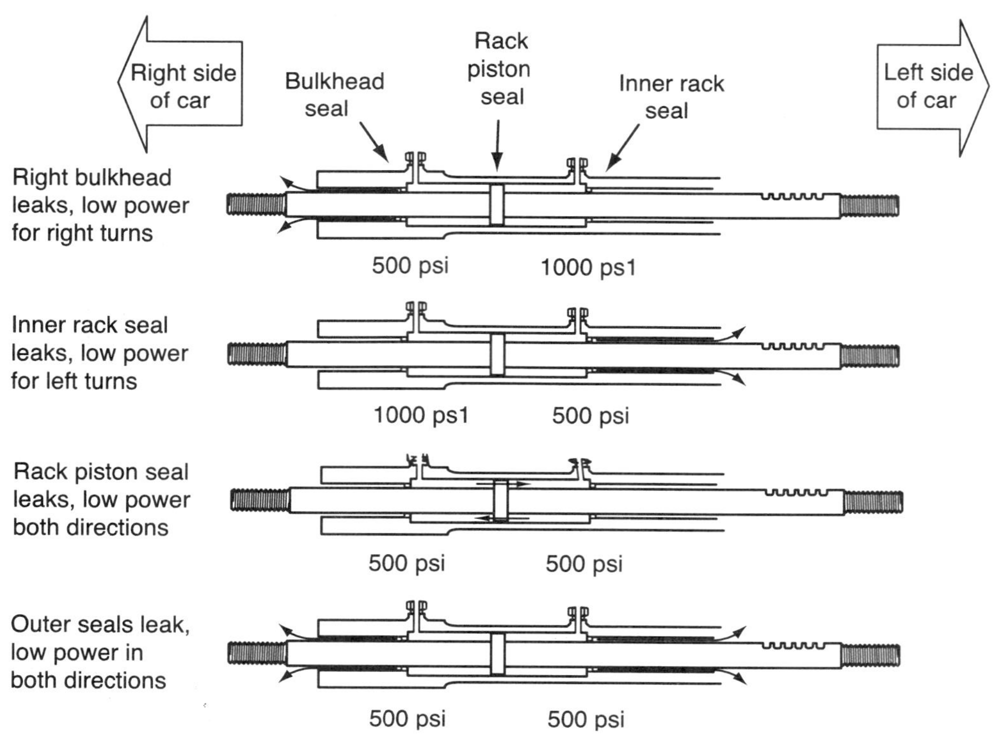


**Figure 1.** Effects of defective rack seals on steering effort imbalance and low power assist.

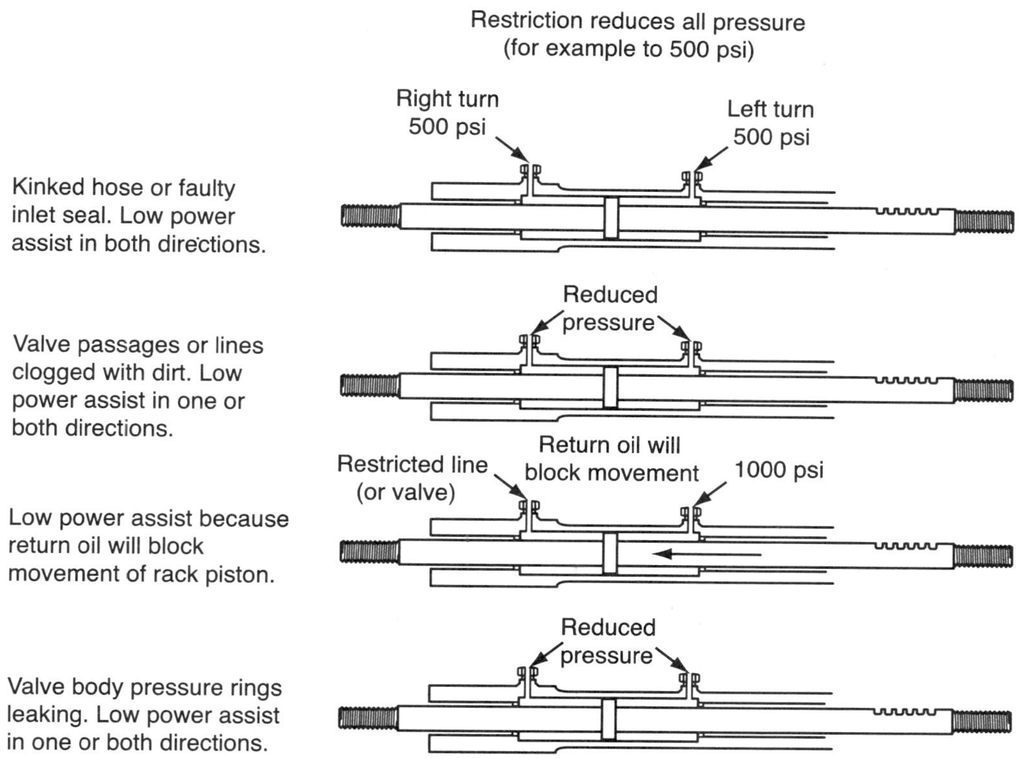


**Figure 2.** Effects of worn rotary valve rings and seals or restricted lines and hoses on steering effort.

imbalance in both directions may be caused by defective rotary valve rings and seals or restricted hoses and lines **(Figure 2).**

> **You Should Know** *To provide equal turning effort in both directions, the rack must be centered with the front wheels straight ahead.*

## Fluid Leak Diagnosis

When checking a power rack and pinion steering gear for leaks **(Figure 3),** make sure the power steering reservoir is at the specified level and the fluid is at normal operating temperature before checking the system. If necessary, rotate the steering wheel several times from lock to lock to bring the fluid to normal operating temperature.

If fluid leaks from the cylinder end of the power steering gear, the outer rack seal is leaking. The inner rack seal is

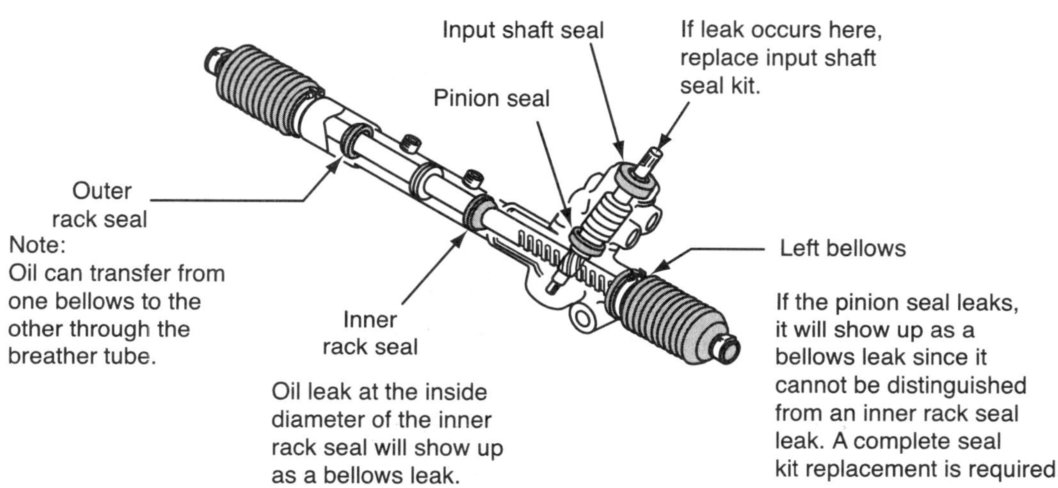


**Figure 3.** Possible leakage points on a power rack-and-pinion steering gear.

defective if fluid leaks from the pinion end of the housing when the rack reaches the left internal stop. A leak at one rack seal may result in leaks from both boots because the fluid may travel through the breather tube between the boots.

If a fluid leak occurs at the pinion end of the housing and this leak is not worse when the steering wheel is turned, the pinion seal is defective.

If the pinion coupling leaks, the input shaft seal is leaking. This seal and the pinion seal should be replaced. The pinion seal must be replaced if the pinion is removed. Also, when the rack is removed, the inner and outer rack seals and pinion seal must be replaced.

If fluid leaks from the fittings, the fittings must be torqued to specifications. If the leak is still present, the line and fitting should be replaced.

## POWER RACK-AND-PINION STEERING GEAR DISASSEMBLY

The procedure for removing a manual and power rack-and-pinion steering gear is similar, except for disconnecting the power steering lines and hoses. The correct procedure for disassembling the unit varies with the vehicle and steering gear manufacturer. Always follow the recommended procedure given in the service manual. The following is a typical power rack-and-pinion steering gear disassembly procedure:

1. Install a holding tool to the housing, and clamp the tool in a vise **(Figure 4).**
2. Use the proper line wrench to disconnect the left and right turn tubes.
3. Remove the O-rings from the ends of the turn tubes.
4. Mark the outer tie-rod ends, jam nuts, and tie rods **(Figure 5).**
5. Loosen the jam nuts and remove the tie-rod ends and jam nuts.
6. Loosen the inner and outer clamps on both boots and remove the boots **(Figure 6).**
7. Clamp the inner tie-rod socket lightly in a soft-jaw vise, and use a hammer and chisel to unstake the claw washers on the inner tie-rod sockets.

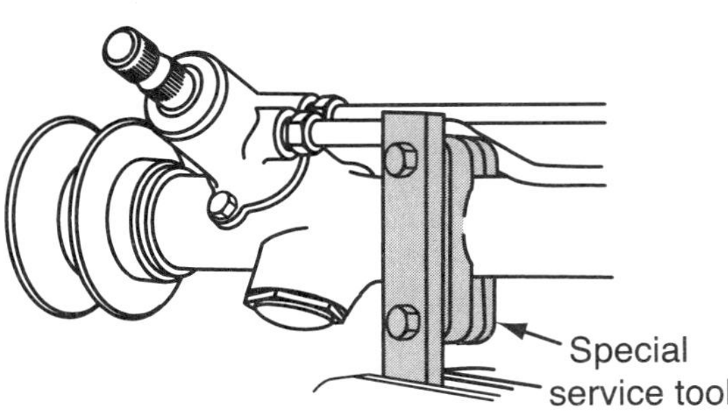


**Figure 4.** A steering gear holding tool.

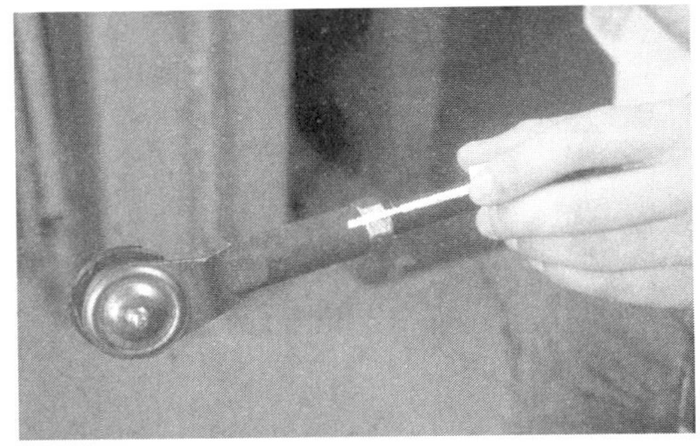

**Figure 5.** Place an index mark on the outer tie-rod end, jam nut, and tie rod before separating them.

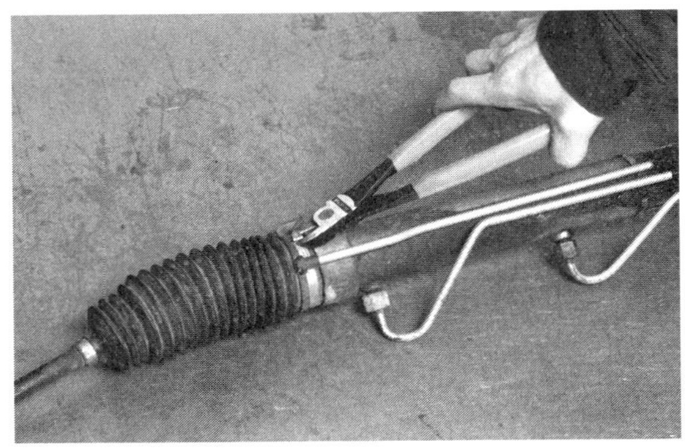

**Figure 6.** Removing the bellows boot clamps.

8. Hold the rack with an adjustable wrench and loosen the inner tie-rod ends with the proper tool.
9. Remove the inner tie-rod ends and the claw washers.

> **You Should Know** *On some power rack-and-pinion steering gears, the inner tie-rod ends are staked after they are installed on the rack. In other assemblies, a pin retains the inner tie-rod ends to the rack, and this pin must be removed before the inner tie-rod ends are removed. When the inner tie-rod ends are replaced, a new retaining pin must be installed.*

10. Rotate the pinion shaft until the specified distance is obtained between the end of the rack and the steering gear housing.
11. Mark the position of the pinion shaft in relation to the housing with a center punch. If that specification is not available, rotate the pinion shaft fully clockwise.

12. Rotate the pinion shaft fully counterclockwise and count the number of turns.
13. Turn the pinion shaft clockwise one-half this number of turns. Measure the distance from each end of the rack to the housing and record this measurement.
14. Use a center punch to mark the pinion shaft position in relation to the housing.
15. Use the proper tool to loosen the adjuster plug lock nut.
16. Remove the adjuster plug with the proper tool.
17. Remove the spring, rack guide, and the seat from the adjuster plug opening.
18. Use a socket and breaker bar to remove the pinion housing cap.
19. Hold the top of the pinion shaft and loosen the self-locking nut on the end of the pinion shaft **(Figure 7).** Remove this self-locking nut.
20. Remove the dust cover from the control valve housing and mark this housing in relation to the steering gear housing.
21. Remove the control valve housing from the steering gear housing.
22. Use a soft hammer to tap the pinion shaft from the control valve housing.
23. Place matching marks on the bracket and steering gear housing **(Figure 8).** Use a slotted screwdriver to pry the clasp open or the housing's bracket.

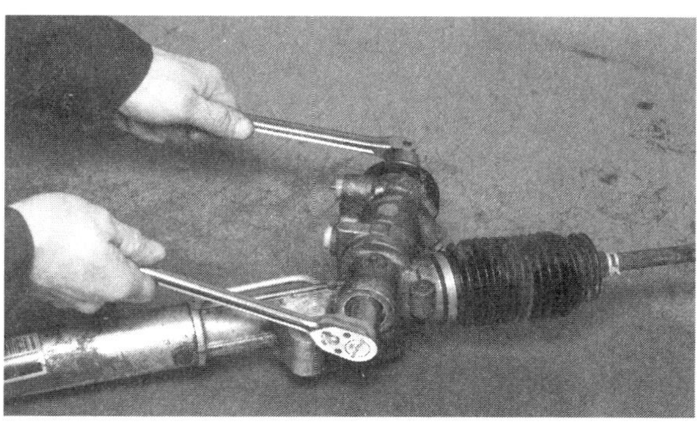

**Figure 7.** Loosening the pinion shaft self-locking nut.

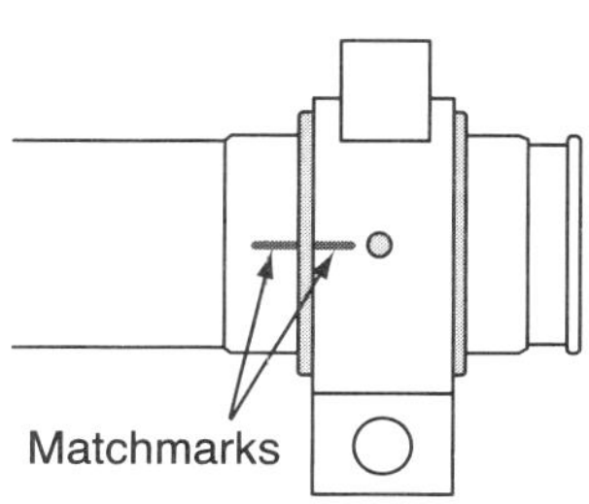


**Figure 8.** Index marks should be made on bracket and steering gear housing.

24. Remove the bushing and bracket from the housing and remove the bushing from the bracket.
25. Use an end stopper removal tool to rotate the cylinder end stopper clockwise until the wire end comes through the hole in the gear housing.
26. Rotate the end stopper counterclockwise with the same tool and remove the wire through the gear housing opening.
27. Use a brass bar and a hammer to tap the rack and rack bushing from the housing.
28. Remove the O-ring from the housing.
29. Drive the cylinder side oil seal and spacer out of the housing with the appropriate driving tool and a brass bar.

**You Should Know** *A scored cylinder surface in the steering gear housing may cause temporary excessive steering effort after the car sits overnight.*

## POWER RACK-AND-PINION STEERING GEAR PARTS INSPECTION

Clean all components except the inner tie-rod ends and any sealed bearings with solvent. Blow dry the parts with compressed air.

Inspect the rack for tooth wear or damage. The rack or complete steering gear must be replaced if the rack is damaged. After inspection, check the straightness of the rack by checking its runout. Place the rack on V-blocks and slowly rotate it with the stem of a dial indicator positioned near the center of the rack. If the runout exceeds specifications, replace the rack or complete steering gear.

Inspect the rack piston. If the outer circumference is scored or worn, replace the rack. Inspect the rack piston's contact area in the housing for scoring, pitting, and wear. If these are present, replace the housing or complete steering gear.

With the proper tool, drive the oil seal and upper bearing out of the control valve housing. Inspect the housing. If the control valve housing is scored, pitted, or worn in the ring contact area, replace the housing. A new seal should be installed in the control valve housing with the proper tool. With the new seal in place, use the proper tool to install the new upper bearing.

Check the lower and center bearings. If they are scored or worn, they should be replaced. Also, check the bearing contact areas of the pinion shaft. If the pinion shaft is worn, scored, or pitted in these areas, replace it. If the pinion shaft teeth are worn, chipped, or pitted, replace the pinion shaft. If the lower and center bearings must be replaced, drive the lower bearing out of the housing with a hammer and brass bar. Use the appropriate bearing puller to remove the center bearing. Press the new center bearing into the housing with the proper tool. Then, press the new lower bearing into the housing with the correct tool.

Pull the rack bushing oil seal from the housing with a seal puller. Coat the new seal with power steering fluid and use the correct tool to install the seal.

> **You Should Know** *Be careful not to score or mark the rack the piston while removing and replacing the piston ring or O-ring. A scored rack piston must be replaced.*

Pry the Teflon ring and O-ring from the rack piston with a small screwdriver. Coat the new Teflon ring with power steering fluid and install it on the expanding tool. Some manufacturers recommend placing the rings in boiling water for 10 minutes before installing them. Remove the Teflon ring from the expanding tool and install the ring on the rack piston. Wrap your hand around the Teflon ring and squeeze it down into its groove.

Remove the Teflon rings from the control valve and install new ones in the same way as the rack piston. When all the Teflon rings are installed on the control valve, coat them with power steering fluid and install the compressing tool over the rings **(Figure 9).**

> **You Should Know** *Do not score or mark the control valve while removing the Teflon rings. Scored control valves must be replaced.*

## POWER RACK-AND-PINION STEERING GEAR ASSEMBLY

To assemble a typical power rack-and-pinion gear assembly:

1. Coat the new cylinder side oil seal with power steering fluid.
2. Place one layer of vinyl tape on the seal driving tool extension, and install the seal and spacer on the tool. Press the cylinder side oil seal and spacer into the housing.
3. Check the rack surface and rack teeth for small metal burrs. Remove any burrs with crocus cloth, and wipe the rack with a clean shop towel.
4. Place the rack installation tool over the rack and coat the tool with power steering fluid. Slide the rack and installation tool into the housing.
5. Remove the installation tool.
6. Coat a new rack bushing O-ring with power steering fluid and install it in the bushing.
7. Push the rack bushing and cylinder end stopper into the housing until the wire installation hole appears **(Figure 10).**
8. Insert the end of a new wire into the wire installation hole. Use the proper tool and a ratchet to rotate the cylinder end stopper clockwise until the outer wire end disappears in the wire installation hole.
9. Connect a hose between the left and right turn ports on the steering gear housing.
10. Attach a vacuum pump hose to the hose and operate the pump until 15.75 in Hg (53.3 kPa) is obtained. After 30 seconds, observe the vacuum gauge. If the gauge reading decreased, the rack housing seal or rack bushing seal must be leaking and the rack should be removed to recheck these seals.
11. If the bushing and bracket were removed from the rack housing, install a new bushing in the bracket. Be sure the projection on the bushing aligns with the bracket hole.
12. Install the bracket and bushing on the housing and align the marks placed on the housing and bracket during disassembly.
13. Place the housing bracket in a vise and tighten the vise to fasten the bracket clasp.
14. Wind a layer of vinyl tape over the serrations on the upper end of the pinion shaft.

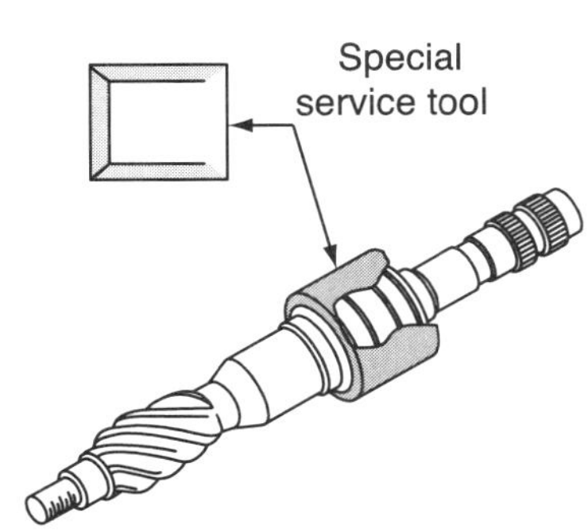


**Figure 9.** The compressing tool for Teflon rings.

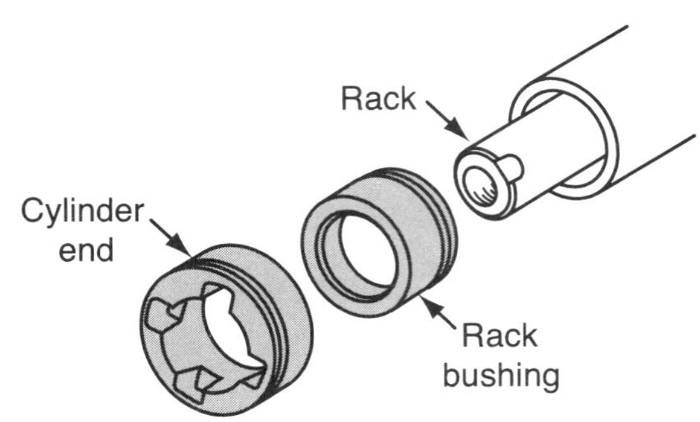


**Figure 10.** Installing the rack bushing and cylinder end stopper.

15. Coat the Teflon rings with power steering fluid, and push the control valve and pinion shaft assembly into the housing **(Figure 11).**
16. Coat a new oil seal with power steering fluid and press the seal into the lower side of the control valve housing with the appropriate driving tool.
17. Position the rack ends at the specified distance from the housing or at the distance measured during disassembly.
18. Install a new control valve housing gasket on the steering gear housing.
19. Align the marks on the control valve and housing and the pinion shaft and housing.
20. Install the control valve housing and tighten the two housing bolts to the specified torque. Install the housing dust cover.
21. Install a new self-locking nut on the lower end of the pinion shaft. Use the proper tool to hold the upper end of the pinion shaft and tighten the self-locking nut to the specified torque.
22. Coat the first two or three threads on the housing cap with sealant.
23. Install the housing cap and tighten it to the specified torque.
24. Use a center punch to stake the steering gear housing in two locations **(Figure 12).**
25. Coat the rack guide seat and rack guide with power steering fluid.
26. Install the rack guide seat, rack guide, and spring.
27. Apply sealant to the first two or three threads on the rack spring cap.
28. Install the rack spring cap and tighten it to the specified torque.
29. Use the proper tool to rotate the rack spring cap counterclockwise.
30. Turn the pinion shaft fully in each direction, and repeat this action.
31. Loosen the rack spring cap until there is no tension on the rack guide spring.
32. Place the proper turning tool and a torque wrench on top of the pinion shaft.
33. Tighten the rack spring cap while rotating the pinion shaft back and forth. Continue tightening the rack spring cap until the specified turning torque is reached.
34. Apply sealant to the first two or three threads on the lock nut.
35. Install the lock nut on the rack spring cap.
36. Use the proper tool to hold the rack spring cap, and tighten the lock nut to the specified torque.
37. Install a new claw washer on the inner tie-rod end and install the inner tie-rod end on the rack. Hold the rack with an adjustable wrench, and use a torque wrench with the correct tool to tighten the inner tie-rod end to the specified torque.
38. Clamp the inner tie-rod end in a soft-jaw vise and stake the claw washer with a brass bar and a hammer. Repeat this procedure on both inner tie-rod ends.
39. Check the vent hole in each end of the rack, and be sure they are not clogged with grease or other material.
40. Install the boots, clamps, and clips. When the inner boot clamps are tightened, there must be a minimum of 0.20 in. (0.08 mm) between the ends of the clamps.
41. Install the outer tie-rod ends and jam nuts. Align the marks placed on the tie-rod ends, jam nuts, and tie rods made during disassembly. Leave the jam nuts loose until the steering gear is installed and front wheel toe is adjusted.
42. Coat the new turn tube O-rings with power steering fluid and install the O-rings on the right and left turn tubes.
43. Tighten the left and right turn tube fittings to the specified torque.

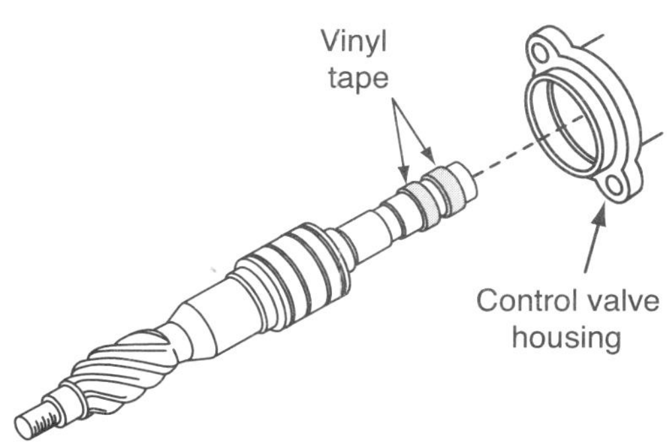


**Figure 11.** Installing the control valve and pinion shaft assembly in the control valve housing.

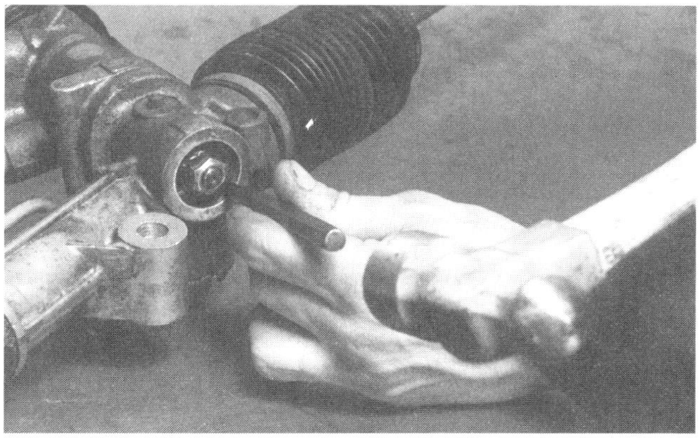

**Figure 12.** Staking the steering gear housing.

## SUMMARY

- Prior to steering gear removal, the steering wheel should be centered and the ignition key removed from the ignition switch. This locks the steering column and maintains the clock spring or spiral cable in the centered position on an air bag-equipped vehicle.
- Always hold the rack while loosening the inner tie-rod ends.
- Prior to steering gear disassembly, the ends of the rack should be positioned at the specified distance from the steering gear housing, and the pinion shaft position should be marked in relation to the housing.
- The rack adjuster plug must be adjusted until the correct turning torque is obtained on the pinion shaft.

## REVIEW QUESTIONS

1. When diagnosing and servicing power rack-and-pinion steering gears:
   A. the rack should be not held when the outer tie-rod ends are loosened on the rack.
   B. in some steering gears, claw washers lock the tie rods in position on the rack.
   C. a straightedge must be used to check the rack for a bent condition.
   D. bent tie rods may be straightened in a hydraulic press and then reinstalled.
2. While discussing power rack-and-pinion steering gear adjustment, Technician A says the self-locking nut on the pinion shaft should be tightened to the specified torque. Technician B says the rack spring cap should be rotated to obtain the specified turning torque on the pinion shaft. Who is correct?
   A. Technician A
   B. Technician B
   C. Both Technician A and Technician B
   D. Neither Technician A nor Technician B
3. A customer says his front-wheel-drive car with power rack and pinion steering is hard to steer for several minutes the first time the car is driven during the day. Technician A says the cause of the problem could be a plugged spool valve spring. Technician B says the cause of the problem could be a scored steering gear cylinder. Who is correct?
   A. Technician A
   B. Technician B
   C. Both Technician A and Technician B
   D. Neither Technician A nor Technician B
4. Technician A says an oil leak at one rack seal may result in oil leaking from both boots. Technician B says the inner rack seal is defective if oil leaks from the pinion end of the housing when the rack reaches the left internal stop. Who is correct?
   A. Technician A
   B. Technician B
   C. Both Technician A and Technician B
   D. Neither Technician A nor Technician B
5. Wandering, or poor straight-ahead tracking, may be caused by all of the following *except:*
   A. worn or loose rack mounting bushings.
   B. rack piston seal leaks.
   C. steering gear off center.
   D. loose or worn tie-rod ends.

Chapter 50

# Diagnosis and Service of Electronic Steering

## INTRODUCTION

Today, electronically controlled and electrically operated power assist steering systems comprise nearly 15 percent of the new vehicles sold. It is anticipated that 90 percent of passenger cars, light-duty trucks, and SUVs will have electronically controlled power steering. In an electronically controlled system, the amount of hydraulic fluid pressure used to boost steering is controlled by a solenoid valve that is controlled by the control module.

As with any power-steering system, the first step when servicing electric/electronic power steering systems is to look for fluid leaks, damaged components, a slipping drive belt, and so on. Only after these things have been checked and no problems have been found should the electronics be suspected as the problem. Check the appropriate service manual for the correct troubleshooting procedures of the electronics.

## COMPUTERS

Most electronically controlled or electrically operated steering systems are controlled by the **powertrain control module (PCM)** or **body control module (BCM).** Often, a separate module monitors and operates the steering system, but this module is part of or directly tied to the powertrain or body computer.

In the process of controlling the various systems, the computer continuously monitors operating conditions for possible system malfunctions. If the conditions fall outside of certain limits, the computer will detect the malfunction and a trouble code will be stored.

If a malfunction results in improper system operation, the computer may minimize the effects of the malfunction by using fail-safe action. During this mode of operation, the computer will control the system based on programmed values instead of the input signals.

### Service Bulletin Information

During diagnostics, service manuals and service bulletin information **(Figure 1)** is essential. If you do not have this information, many hours of diagnostic time can be wasted. This information is available from different suppliers of CD-ROM and paper TSBs.

### OBD II System Diagnosis and Service

All late-model vehicles have an on-board diagnostics second-generation (OBD II) system. Diagnosing and servicing OBD II systems is much the same as diagnosing and servicing any other electronic engine control system, except OBD II systems give more useful information and each make and model vehicle does not have its own basic testing and servicing procedures.

OBD II diagnosis is best done with a strategy-based procedure based on the flow charts and other information given in the service manuals.

**Figure 1.** Service manuals and technical service bulletins are a must when diagnosing computerized systems.

## Entering Diagnostics

There are as many ways to perform computer diagnostics as there are automobile manufacturers. Nearly all vehicles need a scan tool. The scan tool is plugged into the diagnostic connector for the system. Some manufacturers provide a single diagnostic connector **(Figure 2),** and the technician chooses the system through the scan tool. Other vehicles have different diagnostic connectors for the different control systems.

On systems that do not retain codes after the ignition is switched off, operate the vehicle until the problem occurs again, then retrieve the codes before switching the ignition off.

Use the recommended procedure to retrieve **diagnostic trouble codes (DTCs).** Once the DTCs are retrieved, follow the appropriate diagnostic chart to isolate the fault.

## Diagnosis with a Warning Lamp

On pre-OBD II systems, when the ignition switch is turned on, the electronic power steering (EPS) light in the instrument panel should illuminate **(Figure 3).** This proves that the bulb and related circuit are functioning normally. When the engine starts, the EPS light should turn off if there are no defects. The light should remain off while the engine is running.

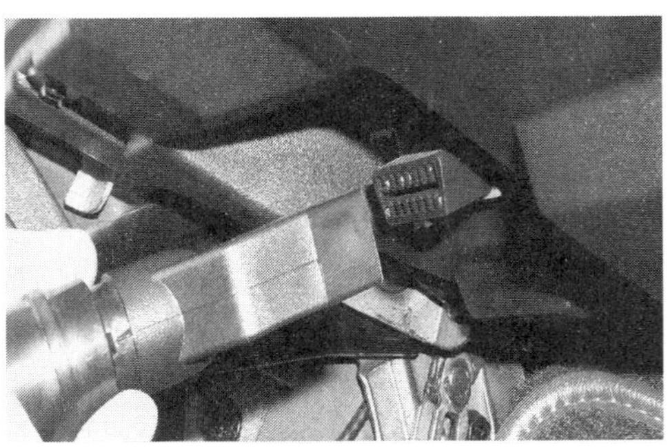

**Figure 2.** Plugging a scan tool connector into the diagnositic connector.

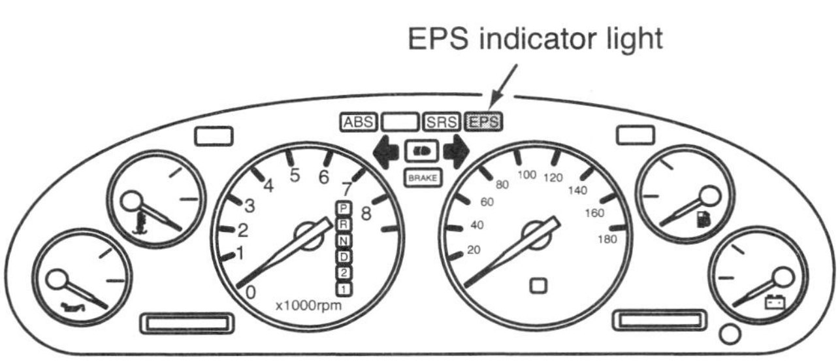


**Figure 3.** An example of an elecronic power steering (EPS) warning lamp.

If a defect occurs in the EPS system, the control unit turns on the EPS light to alert the driver that a defect exists. Under this condition, the EPS control unit shuts the system down and electric assist is no longer available, but manual steering is provided. When the control unit senses an electrical defect in the system, a DTC is stored in its memory. The control module's memory is erased if battery voltage is disconnected from the control unit.

If the EPS light is on with the engine running, turn the ignition switch off for 10 seconds, then road test the vehicle and see if the EPS light comes back on. If the EPS light comes on again, proceed with the diagnosis. Locate the EPS service check connector. With the ignition switch off, connect a jumper wire across the two designated terminals **(Figure 4)** of the service check connector. When the ignition switch is turned on, the EPS light will begin to flash DTCs. Each DTC is indicated by a long light flash or flashes followed by a short flash or flashes. If more than one DTC is stored in the memory, the DTCs are flashed in numerical order. The DTC sequence continues until the ignition switch is turned off. Be sure to remove the jumper wire after the ignition switch is turned off.

A DTC indicates a defect in a certain area. The defect may be a sensor, or the problem could be the connecting wires from the sensor to the control unit. Tests usually have to be performed with a voltmeter or ohmmeter to locate the exact cause of the defect. Detailed diagnostic charts are provided in the service manual.

## Scan Tool Diagnosis

The scan tool must fit the DLC on an OBD II system. Connect the scan tool. Enter the vehicle identification infor-

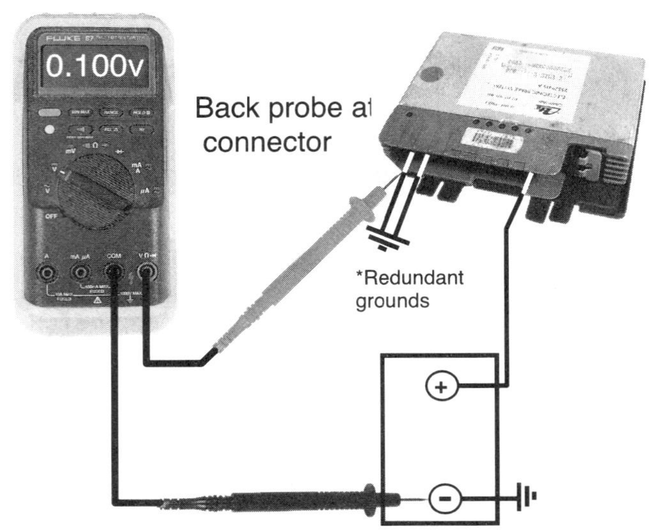


**Figure 4.** Checking the ground circuit of the computer.

mation, and then retrieve the DTCs. The scan tool can also be used to clear DTCs, perform output control tests, and read serial data.

> **You Should Know** *Do not clear DTCs unless directed by a diagnostic procedure. Clearing DTCs will also clear valuable freeze-frame and failure records data.*

The next step in diagnostics depends on the outcome of the DTC display. If DTCs were displayed, proceed by following the designated DTC table to make an effective diagnosis and repair. If no DTCs were displayed, match the symptom to the symptoms listed in the symptom tables and follow the diagnostic paths to complete the repair, or refer to the applicable component/system check. If there is not a matching symptom, analyze the complaint, then develop a plan for diagnostics. Utilize the wiring diagrams and theory of component and system operation. Call technical assistance for similar cases where repair history may be available. It is possible that the customer's concern is normal for that vehicle.

All DTCs are displayed as a five-character alphanumeric code in which each character has a specific meaning. The first character is the prefix letter that indicates the area of the fault. The second character is a number that indicates whether the DTC that follows is a standard or common code or a manufacturer-specific code. The third character indicates the subsystem of the area that the fault lies in. The fourth and fifth characters identify the specific detected fault. These characters not only indicate the component or circuit, but also give a description of the type of fault. With this sort of information, a technician knows what to check.

## Visual Inspection

Check the battery and its cables, the vehicle's wiring harnesses, connectors, ground points, and the charging system for loose or damaged connections. Also, check the connectors for signs of corrosion. Check all sensors and actuators for physical damage. Check the wiring for signs of burned or chafed spots, pinched wires, or contact with sharp edges or hot exhaust parts. Also, check all vacuum hoses for pinches, cuts, or disconnects.

## Isolating Computerized Control Problems

Electronic control problems are usually caused by defective sensors and, to a lesser extent, output devices. Check the input sensors and wiring first, then the output devices and their wiring and, finally the computer. These can be used to check individual system components:

***Visual Checks***—Look for obvious problems. Any part that is burned, broken, cracked, corroded, or has any other visible problem must be replaced before continuing the diagnosis.

***Ohmmeter Checks***—Most sensors and output devices can be checked with an ohmmeter.

***Voltmeter Checks***—Many sensors, output devices, and their wiring can be diagnosed by checking the voltage to them, and in some cases, from them.

***Lab Scope Checks***—The activity of sensors and actuators can be monitored on a lab scope. Often problems elsewhere in the system will cause a device to behave abnormally. These situations are identified by the trace on a scope and by the technician's understanding of a scope and the device being monitored **(Figure 5).**

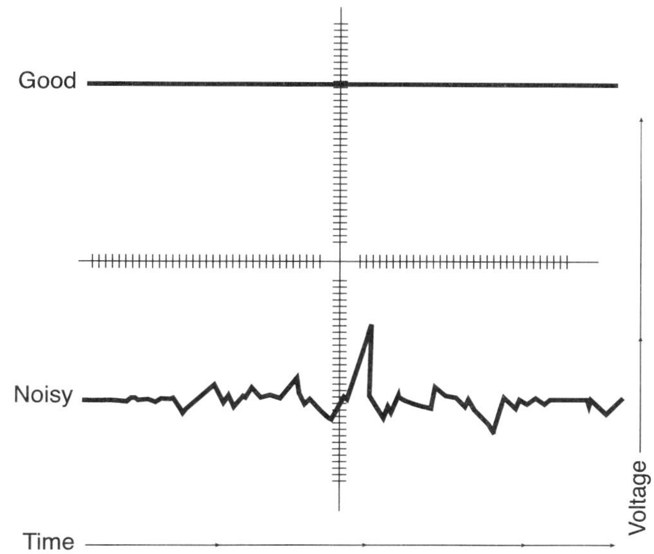


**Figure 5.** (Top) A good voltage signal displayed on a lab scope. (Bottom) A voltage signal with noise.

# DIAGNOSIS OF COMPUTER VOLTAGE SUPPLY AND GROUND WIRES

Control modules cannot operate properly unless they have good ground connections and have the correct voltage at the required terminals. The voltage at the battery terminal of a control module should be 12 volts with the ignition switch off. If 12 volts are not available, check the fuse and related circuit.

Ground wires usually extend from the computer to a ground connection on the engine or battery. With the ignition switch on, connect a voltmeter from the battery

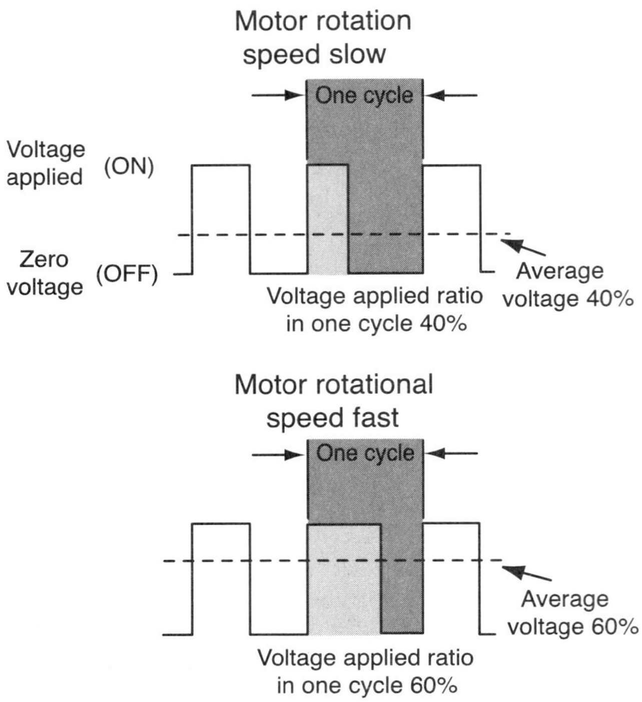


**Figure 6.** A pulse-width modulated (PWM) voltage signal.

ground to the computer ground **(Figure 6).** The voltage drop across the wire should be 30 millivolts or less. If the voltage reading is greater than that, repair the ground wires or connection.

Also check the ground and positive connection at the battery. Checking the condition of the battery and its cables should always be part of the inspection.

A voltage drop test is a quick way of checking the condition of any wire. Connect a voltmeter across the wire or device being tested. Place the positive lead on the most positive side of the circuit, then turn on the circuit. There should be a zero volt reading across any wire unless it is a resistance wire, which is designed to drop voltage. Even then, check the drop against specifications to see if it is dropping too much.

Poor grounds can allow **electromagnetic interference (EMI)** or noise to be present on the reference voltage signal. This noise causes small changes in the voltage going to the sensor. Therefore, the signal from the sensor will have these voltage changes. The best way to check for noise is to use a lab scope.

## TESTING INPUT SENSORS

Some systems control the amount of assist through pulse-width modulation. These systems rely on input signals from the vehicle speed sensor to the control module.

There are many types of sensors; their design depends upon what they are monitoring. Some sensors are simple on-off switches. Others are a form of variable resistor that changes resistance according to temperature changes. Some sensors are voltage or frequency generators, while others send signals according to the rotational speed of a device. Knowing what they are measuring and how they respond to changes are the keys to being able to test an input sensor.

### Switches

Switches are turned on and off through an action of a device or by the actions of the driver. Some of these switches are grounding switches that complete a circuit when they are closed. Others send a 5- or 12-volt signal to the PCM when they are closed.

Grounding switches always send a digital signal to the PCM; the switch's circuit is either on or off. These circuits contain a fixed resistor to limit the circuit's current and prevent voltage spikes in the circuit as the switch opens and closes. When the switch is closed, the voltage signal to the PCM is low or zero. When the switch is open, there is a high-voltage signal.

A switch can be tested with an ohmmeter. Disconnect the switch, then connect the ohmmeter across the switch's terminals. When the switch is closed, there should be zero ohms of resistance across the terminals. When the switch is open, the ohmmeter should give an infinite reading. If the switch reacts in any other way, it is bad and should be replaced.

Most grounding switches react to some mechanical action to open or close. There are some, however, that respond to changes in conditions. These respond to changes in pressure or temperature. An example of this type of switch is the power steering pressure switch. This switch informs the PCM when pressures reach a particular point. When the pressure exceeds that point, the PCM knows there is an additional load on the engine and will increase idle speed.

To test this type of switch, watch it on a DMM or lab scope. With the engine at idle speed, turn the steering wheel to its stop on one side. The voltage should drop as soon as the pressure in the power steering unit has reached a high level. If the voltage does not drop, either the power steering assembly is incapable of producing high pressures or the switch is bad.

Temperature responding switches operate in the same way. When a particular temperature is reached, the switch opens. This switch is tested by removing and submerging it in heated water. Watch the ohmmeter as the temperature increases; a good switch will open when the temperature reaches a specified amount. If the switch fails this test, it should be replaced.

### Variable Resistor-Type Sensors

Many sensors send a voltage signal to the PCM in direct response to changes in operating conditions. The voltage signal changes as the resistance in the sensor changes. Most often, these variable resistor-type sensors are thermistors and potentiometers.

These switches can be checked with an ohmmeter. The action of the switch should be smooth. If there is no change or if the change is erratic, the switch should be replaced.

The wiring to the sensor can also be checked with an ohmmeter. With the wiring connectors disconnected from the sensor and the computer, connect the ohmmeter from each sensor terminal to the correct computer terminal. The sensor wires should have less resistance than specified. If the wires have high resistance, the wires or connectors must be repaired.

## Generating Sensors

Some sensors generate a voltage or frequency signal in response to changing conditions. The VSS is a voltage-generating sensor. Common frequency generator-type sensors are some MAP sensors, some MAF sensors, and reference signals.

### Hall Effect Switches

To test a Hall effect sensor, disconnect its wiring harness. Connect a voltage source of the correct low-voltage level across the positive and negative terminals of the Hall layer, then connect a voltmeter across the negative and signal voltage terminals.

Insert a metal feeler gauge between the Hall layer and the magnet. Make sure the feeler gauge is touching the Hall element. If the sensor is operating properly, the meter will read close to battery voltage. When the feeler gauge blade is removed, the voltage should decrease. On some units, the voltage will drop to near zero. Check the service manual to see what voltage you should observe when installing and removing the feeler gauge.

When observing a Hall effect sensor on a lab scope, pay attention to the downward and upward pulses. These should be straight. If they appear at an angle, this indicates the transistor is faulty causing the voltage to rise slowly. The entire square wave from a Hall effect unit should be flat. Any change from a normal trace means the sensor should be replaced.

### Magnetic Pulse Generators

Ignition pickup coils, wheel speed sensors, and VSSs are permanent magnet voltage-producing sensors. The procedures for testing these are similar.

A defective vehicle speed sensor **(Figure 7)** may cause a variety of problems, including improper adjustments to the amount of power assist applied to the steering gear.

Prior to VSS diagnosis, the vehicle should be lifted on a hoist so the drive wheels are free to rotate. Backprobe the VSS output wire and connect the voltmeter leads from this wire to ground. Select the 20 V AC scale on the voltmeter, then start the engine.

Place the transaxle in drive and allow the drive wheels to rotate. If the VSS voltage signal is not 0.5 volts or more, replace the sensor. When the VSS provides the specified voltage signal, backprobe the VSS terminal at the PCM and repeat the voltage signal test with the drive wheels rotating. If 0.5 volts is available at this terminal, the trouble may be in the PCM.

When 0.5 volts is not available at this terminal, turn the ignition switch off and disconnect the wire from the VSS to the PCM. Connect the ohmmeter leads across the wire. The meter should read zero ohms. Repeat the test with the ohmmeter leads connected to the VSS ground terminal and the PCM ground terminal. This wire should also have zero ohms resistance. If the resistance in these wires is more than specified, repair the wires.

The condition of a speed sensor can be checked by going through the diagnostic routines for the system. If this test indicates that a sensor is faulty, the sensor should be replaced. Speed sensors can also be checked with an ohmmeter. Most manufacturers list a resistance specification. The resistance of the sensor is measured across the sensor's terminals **(Figure 8).** The typical range for a good sensor is 800 to 1400 ohms of resistance.

## TESTING ACTUATORS

Most actuators are electromechanical devices that convert the commands from the computer to mechanical action. Actuators are used to open and close switches, control vacuum flow to other components, and operate valves depending on the system.

Most actuators can be tested with a scan tool. Using a scan tool, serial data can be used to diagnose outputs. The displayed data should be compared against specifications to determine the condition of any actuator. Also, when an actuator is suspected as being faulty, make sure the inputs related to the control of that actuator are within normal range. Faulty inputs will cause an actuator to appear faulty.

Because many actuators operate with 5 to 7 volts, never connect a jumper wire from a 12-volt source unless told to do so by the service procedure. Some actuators are tested with a voltmeter by testing for input voltage to the

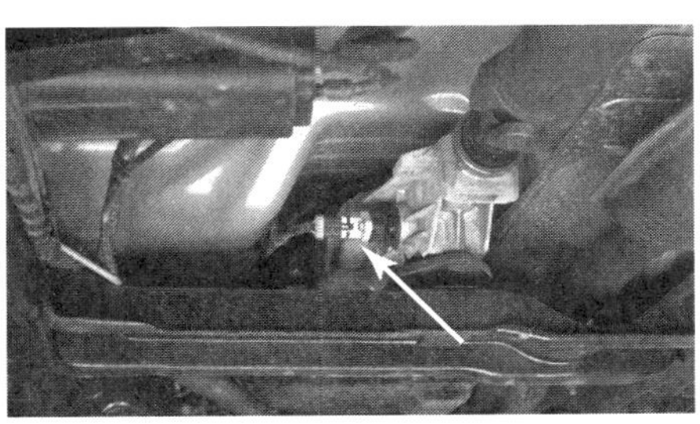

**Figure 7.** A VSS mounted in the tail shaft housing of a transmission.

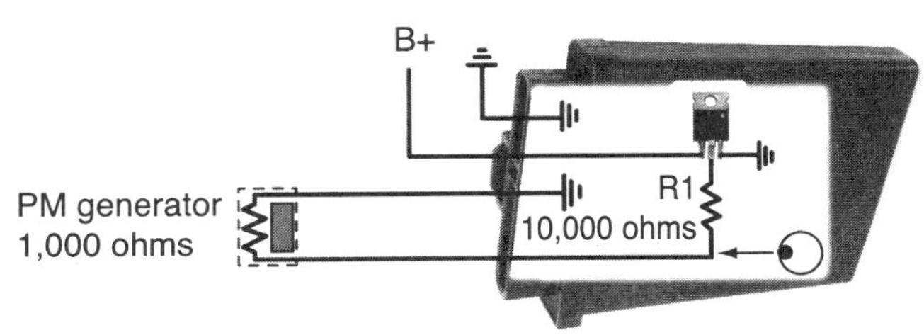


**Figure 8.** The resistance values and check points for a VSS.

actuator. If there is the correct amount of input voltage, check the condition of the ground. If both of these are good, then the actuator is faulty.

## Testing Actuators with a Lab Scope

Most computer-controlled circuits are ground-controlled circuits and the on-time pulse is a downward pulse in a scope trace. On positive-feed circuits where the computer is supplying the voltage to turn a circuit on, the on-time pulse is the upward pulse. One complete cycle is measured from one on-time pulse to beginning of the next on-time pulse.

Most actuators are solenoids. The computer controls the solenoid's action by controlling the pulse width of the control signal **(Figure 9).**

Some actuators are controlled by pulse-width modulated signals **(Figure 10)**. These devices are controlled by varying the pulse width, signal frequency, and voltage levels.

Both waveforms should be checked for amplitude, time, and shape. You should also observe changes to the pulse width as operating conditions change. A bad waveform will have noise, glitches, or rounded corners. You should be able to see evidence that the actuator immediately turns off and on according to the commands of the computer.

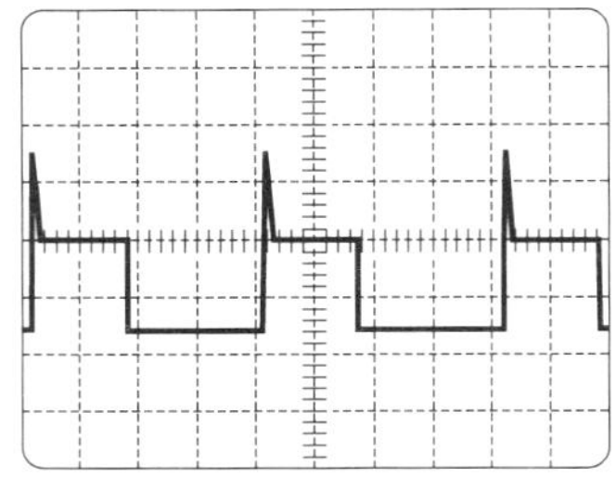

**Figure 9.** A typical control signal for an actuator.

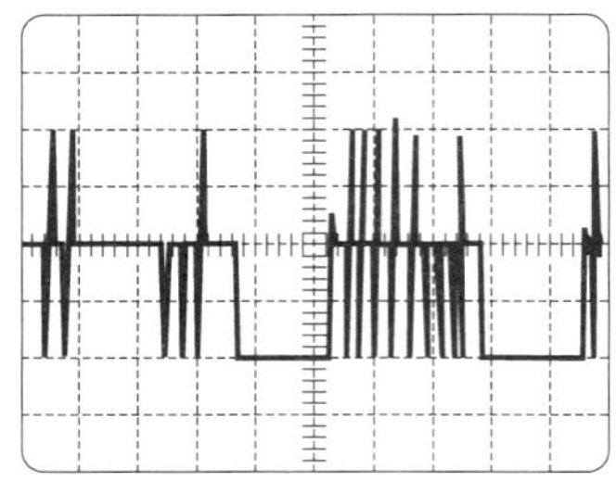

**Figure 10.** A typical pulse-width modulated control signal.

# SUMMARY

- A trouble code (DTC) is set to indicate the portion of the system that is at fault.
- To check components, perform visual, ohmmeter, voltmeter, and lab scope checks.
- When diagnosing problems, service bulletin information is absolutely essential.
- Most intermittent problems are caused by faulty electrical connections or wiring.
- PCMs (OBD II and earlier designs) cannot operate properly unless they have good ground connections and the correct voltage at the required terminals.
- A voltage drop test is a quick way of checking the condition of any wire.
- Poor grounds can also allow EMI or noise to be present on the reference voltage signal. The best way to check for noise is to use a lab scope.
- Most actuators are electromechanical devices that convert commands to mechanical action.
- Most systems allow for testing of the actuator through a scan tool.
- If an ohmmeter reads infinity, there is an open, and a low resistance reading indicates a short.
- Actuators can be accurately tested with a lab scope.

# REVIEW QUESTIONS

1. List the four ways individual components can be checked.
2. When diagnosing electronic power steering (EPS) systems:
   A. the EPS system enters the diagnostic mode when the ignition switch is cycled three times in a 5-second interval.
   B. if the EPS warning light is on with the engine shut off and the ignition switch on, there is a defect in the EPS system.
   C. the EPS warning light flashes diagnostic trouble codes (DTCs) in numerical order.
   D. the EPS control unit memory has the capability to store only one DTC.
3. When diagnosing EPS systems:
   A. the DTCs may be erased from the EPS control unit by disconnecting the stop light fuse for 10 seconds.
   B. a DTC indicates the exact cause of an electrical defect in the EPS system.

# Section 9

## Wheel Alignment



**Interesting Fact**

*Early attempts at rear-wheel alignment were slow and lacked precision. These attempts at rear-wheel alignment included the use of a track bar and even backing the rear wheels of a car onto a front-wheel aligner to align the rear wheels. To meet the need for fast, accurate front- and rear-wheel alignment, wheel alignment manufacturers designed computer wheel aligners. The technology in this equipment has greatly improved since the first models were introduced.*

## SECTION OBJECTIVES

After you have read, studied, and practiced the contents of this section, you should be able to:

- Describe the variables that affect wheel alignment.
- Explain why four-wheel alignment is essential on front-wheel drive unitized body cars.
- List the causes of rear axle offset.
- Define rear axle sideset, setback, and dogtracking.
- Diagnose front and rear tire tread wear caused by inaccurate rear wheel toe and thrust line settings.
- Define total four-wheel alignment, and explain the advantage of this type of alignment.
- Describe the relationship between camber and vehicle directional stability.
- Explain how front suspension camber and caster may be adjusted to compensate for road crown.
- Describe the effects of positive and negative caster on directional control and steering effort.
- Explain the steering axis inclination (SAI) angle and the included angle.
- Describe toe in and toe out on front suspension systems.
- Explain how the front suspension system is designed to provide toe out on turns.
- Perform a prealignment inspection.
- Diagnose wheel alignment and tire wear problems.
- Measure and adjust front-wheel camber and caster.
- Measure and adjust front- and rear-wheel setback.
- Measure and adjust steering axis inclination (SAI).
- Measure front-wheel toe and turning radius.
- Diagnose bent front struts.
- Measure and adjust rear-wheel camber and toe.
- Check and correct front engine cradle position.
- Adjust front-wheel toe.
- Center steering wheel.

Chapter 51

# Alignment Geometry

## INTRODUCTION

Suspension and steering systems are designed to provide control with acceptable driver effort and road feel. A vehicle must have predictable directional control. The wheels must be easy to turn, and tire wear should be minimized. These are achieved when front- and rear-wheel alignment angles are within specifications.

Wheel alignment provides driving safety and normal tire wear. When alignment angles are incorrect, the vehicle will swerve, skid during hard braking, and have reduced tire life.

## ROAD VARIABLES

Vehicles are subjected to many road variables that affect wheel alignment. Some of these variables are road crown (the curvature of the road surface), bumps and holes, natural crosswinds or crosswinds created by other vehicles, heavy loads or unequal weight distribution, road surface friction and conditions such as ice, snow, and water, tire traction and pressure, side forces while cornering, drive axle forces in FWD vehicles, and the relationship between suspension parts as the front wheels turn and move vertically when road bumps and holes are encountered.

Wheel alignment angles are designed to disregard road variables. The angles also control the tracking of the rear wheels in relation to the front wheels.

### Importance of Four-Wheel Alignment

Vehicles with unibodies and independent or semi-independent rear suspension systems are more likely to experience wheel alignment problems than RWD vehicles with frames and one-piece rear axle housings. Therefore, new designs make four-wheel alignment a necessity.

The driver uses the steering wheel to turn the front wheels and steer the vehicle in the desired direction. However, to a large extent, the rear wheels determine the direction of the vehicle. When the **thrust line** is positioned at the geometric centerline and the front and rear wheels are parallel to the vehicle's geometric centerline, the vehicle moves straight with minimum guidance from the steering wheel **(Figure 1).**

### Result of Improper Rear-Wheel Alignment

**Rear axle offset** occurs on RWD vehicles with a one-piece rear axle housing and FWD vehicles with a trailing arm rear suspension. If the rear axle is offset, the thrust line is no longer at the vehicle's centerline. The **thrust angle** is the angle between the geometric centerline and the thrust line.

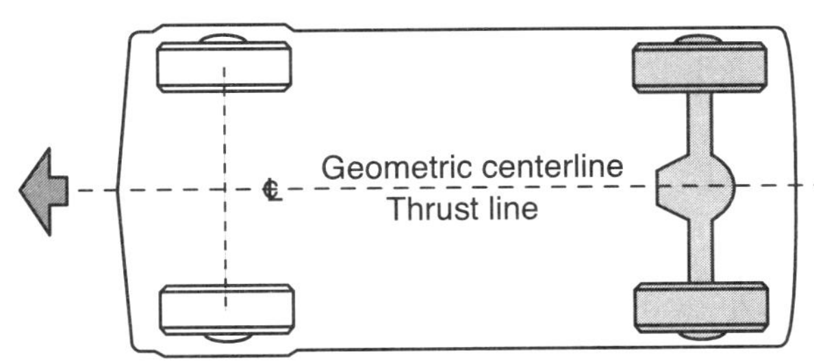


**Figure 1.** The front and rear wheels parallel to vehicle centerline.

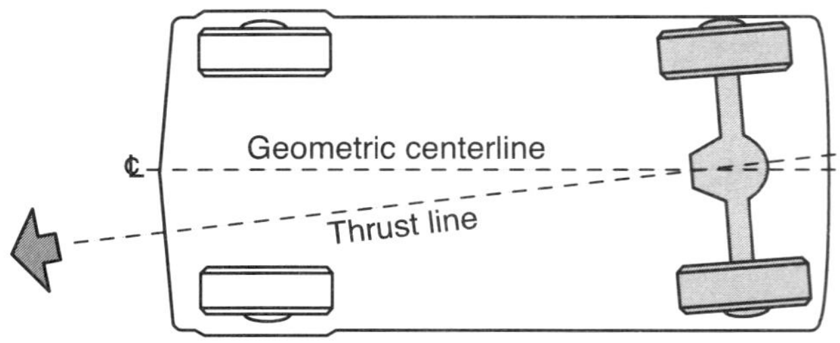


**Figure 2.** Rear axle offset and improperly positioned thrust line.

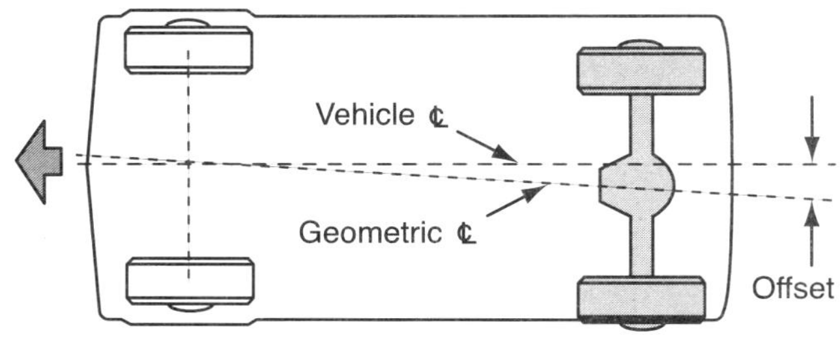


**Figure 4.** Rear axle offset occurs when the rear axle assembly has moved straight sideways and the geometric centerline is shifted away from the proper vehicle centerline position.

With the offset problem shown in **Figure 2,** the thrust line is positioned to the left of the geometric centerline. If the driver's hands are removed from the steering wheel, the steering pulls to the left.

The left front wheel toes out and the right front toes in when the vehicle is driven straight. This happens because the front wheels try to compensate for the rear suspension. The front wheels turn slightly to the left to compensate for the drift caused by the rear suspension problem. As a result, both the front and rear tires may have feathered tire wear.

The ideal correction for this problem is to reposition the rear suspension so the thrust line is at the vehicle's centerline, then set the front wheel toe in to the thrust line.

Causes of rear axle offset are a broken center bolt in rear leaf spring, worn shackles, bent frame, bent subframe or floor section, worn trailing arm bushings, or bent trailing arms.

If the left rear wheel has excessive toe out, the thrust line is moved to the left of the geometric centerline **(Figure 3).** This has the same effect as rear axle offset. Improper toe on one rear wheel occurs most often on independent rear suspensions.

Some of improper toe on one rear wheel are a bent one-piece rear axle, bent U section in trailing arm suspension, bent rear lower control arm, worn rear lower control arm bushings, bent rear spindle, or improper rear toe adjustment.

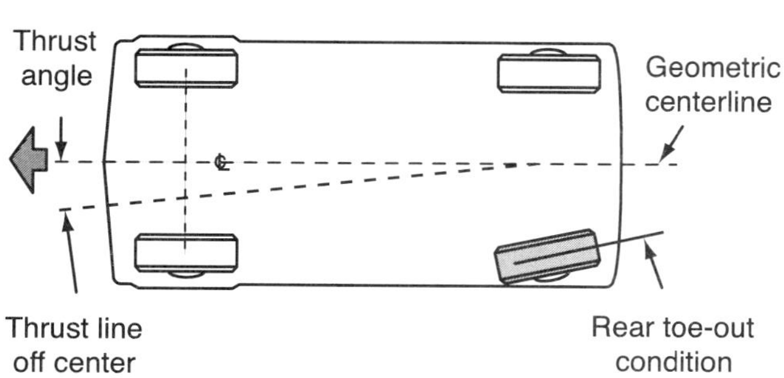


**Figure 3.** Excessive toe out on the left rear wheel moves the thrust line to the left of the geometric centerline.

If there is rear axle **sideset,** the rear wheels are parallel to each other but the axle has moved so the geometric centerline is no longer positioned at the centerline **(Figure 4).** Since the rear wheels do not follow directly behind the front wheels when the vehicle is driven straight, the vehicle dogtracks. Rear axle sideset will cause steering pull if it is severe.

**Setback** is a condition where one front or rear wheel is moved rearward in relation to the opposite wheel **(Figure 5).** This is usually caused by collision damage.

## Geometric Centerline Alignment

**Toe** refers to the angle between the plane of a front or rear wheel and a reference line. Either the geometric centerline or the thrust line may be used as a toe reference line. When the geometric centerline is used as a reference, toe on each front wheel is adjusted using the geometric centerline as a reference. This may provide a satisfactory alignment if

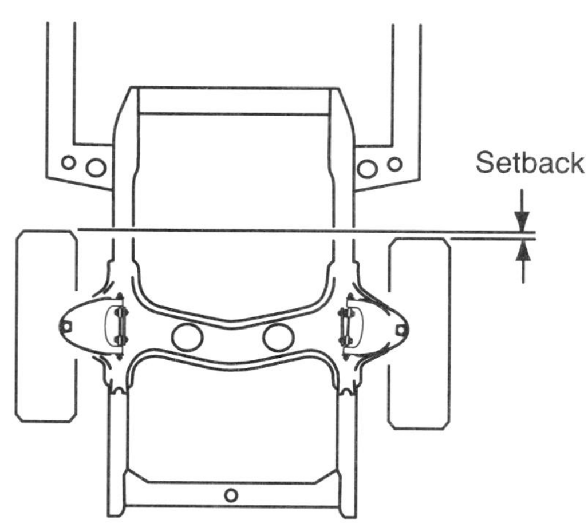


**Figure 5.** Front-wheel setback occurs when one front wheel is moved rearward in relation to the opposite rear wheel.

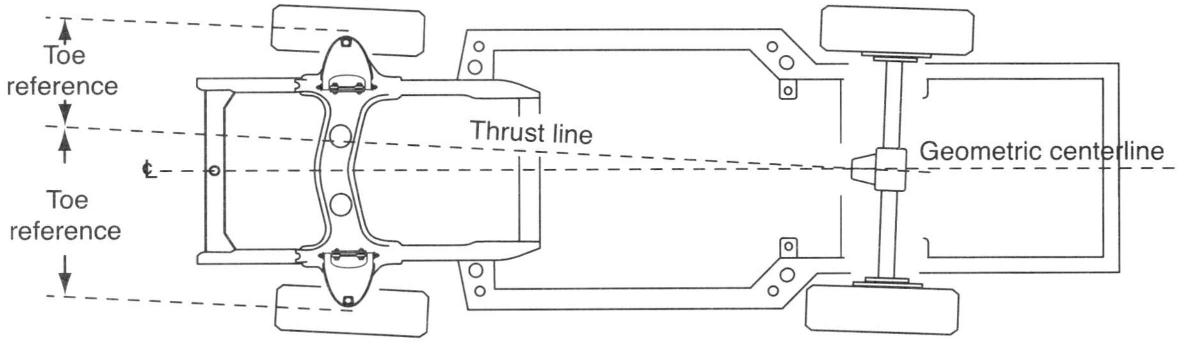


**Figure 6.** Thrust line alignment with the front-wheel toe referenced to the thrust line.

the rear wheels are properly positioned and the thrust line is at the vehicle's centerline. However, if the rear wheels are not positioned properly, steering problems will occur.

## Thrust Line Alignment

The thrust line created by the rear wheels is used as a reference for front wheel toe adjustment **(Figure 6).** When front wheel toe is adjusted to the thrust line, neither the front nor rear wheels are parallel to the geometric centerline. Therefore, none of the four wheels are facing straight when the vehicle is driven straight. This results in feathered wear on the tires' treads.

## Four-Wheel Alignment

During four-wheel alignment, the position of the thrust line is measured to the geometric centerline. All four wheels are then parallel to the geometric centerline, and the steering wheel is centered as the vehicle is driven straight.

# CAMBER FUNDAMENTALS

Positive camber exists when the top of the tire is tilted outward from the true vertical line of the wheel assembly. Negative camber occurs when the tire and wheel centerline tilts inward in relation to the wheel assembly true vertical centerline **(Figure 7).**

## Jounce and Rebound

During jounce, the top of the wheel moves outward and creates a more positive camber angle. During wheel rebound, the suspension system will have a slightly negative camber.

## Cornering Forces

During hard cornering at high speeds, centrifugal force attempts to move the vehicle to the outside of the turn. Under this condition, more vehicle weight is transferred to the side at the outside of the curve. Therefore, the outside of the front suspension is forced downward and the inside is lifted upward. When this occurs, the camber on the inside wheel becomes less positive and the tire grips the road surface to help prevent sideways skidding.

Simultaneously, the camber on the outside wheel reaches a 0-degree position or moves to a slightly negative position. This causes the outside edge of the tire to grip the road, which helps to prevent sideways skidding.

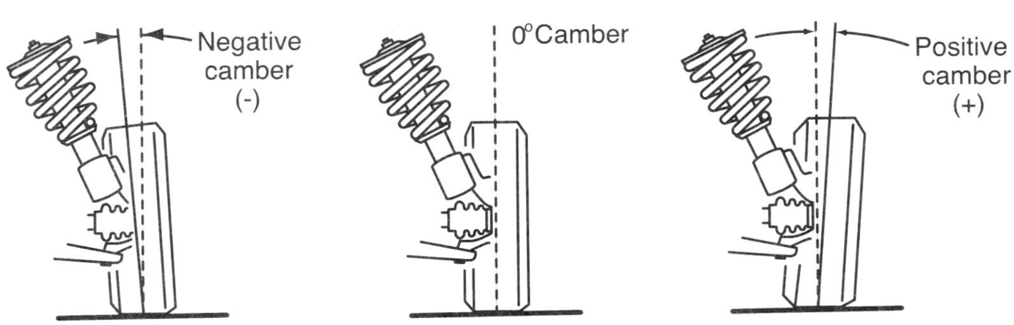


**Figure 7.** Negative and positive camber angles on a MacPherson strut front suspension system.

## Tire Tread Wear

When the wheels are set to the correct camber angle, they will remain at or very close to 0-degree camber and provide maximum tire tread life.

If a front wheel has excessive positive camber, the wheel is tilted outward and the vehicle's weight is concentrated on the outside edge of the tire. As a result, the outside edge of the tire has a smaller diameter than the inside tire edge. Therefore, the outside edge must complete more revolutions of travel and will slip and scuff on the road as it revolves.

Excessive negative camber tilts the wheel inward and puts the vehicle's weight on the inside edge of the tire. This causes wear and scuffing on the inside edge of the tire **(Figure 8).**

## Road Crown

A wheel that is tilted tends to steer in the direction it is tilted. When camber angles are equal on both front wheels, the steering forces are equal and the vehicle tends to move in a straight line. If camber on the wheels is unequal, the vehicle will drift to the side with the most positive camber.

Crowned highways prevent water buildup on the surface. When a vehicle is driven on a crowned road, it is driven on a slight slope, which causes the vehicle to pull toward the right. Some manufacturers use camber to offset this pull.

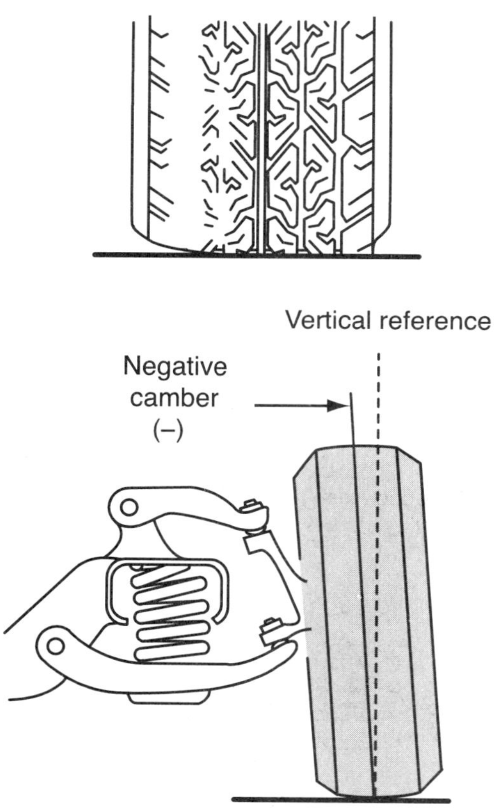


**Figure 8.** Tire tread wear caused by incorrect camber adjustment.

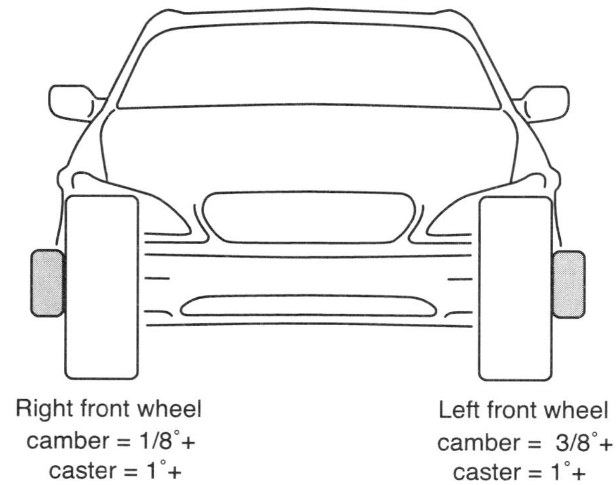


**Figure 9.** Camber setting to offset road crown.

The left front wheel may have 1/4 degree to 1/2 degree more positive camber than the right front wheel. If camber is used to offset road crown, the caster on both front wheels must be the same to maintain directional stability **(Figure 9).**

## CASTER FUNDAMENTALS

The caster line intersects the center of the upper and lower ball joints. Positive caster occurs when the caster line is tilted backward toward the rear of the vehicle in relation to the vertical centerline of the wheel and spindle viewed from the side **(Figure 10).**

Negative caster occurs when the centerline of the upper strut mount and ball joint is tilted toward the front of the vehicle in relation to the spindle and wheel vertical centerline.

### Effects of Positive Caster

Positive caster projects the vehicle's weight ahead of the wheel centerline, whereas negative caster projects the

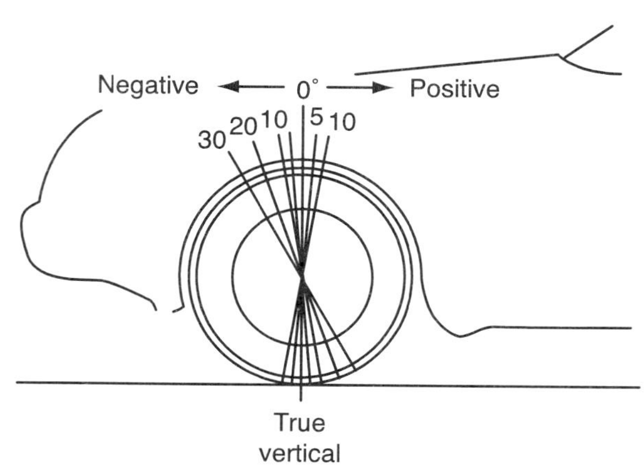


**Figure 10.** Positive and negative caster.

vehicle's weight behind the wheel centerline. Since positive caster causes a larger tire contact area behind the caster pivot point, this contact area tends to follow the pivot point. This also tends to return the wheels to a straight-ahead position after a turn and to maintain a straight-ahead position. Positive caster increases steering effort because the tendency of the tires to remain in the straight position must be overcome during a turn. However, positive caster helps maintain vehicle directional stability.

Excessive positive caster increases steering effort and creates a very rapid steering wheel return. With a caster angle of 0 degrees, the front spindles rotate horizontally in relation to the road. However, a positive caster angle causes the left spindle to tilt toward the road during a left turn.

This downward movement tends to drive the tire into the road. Since this cannot happen, the left side of the suspension and chassis is lifted. When the driver begins to return the wheel to the centered position, gravity forces the vehicle's weight to its lowest position, which helps return the steering wheel to the straight-ahead position. Excessive positive caster increases the left front spindle downward tilt during a left turn, which increases suspension and chassis lift. Therefore, excessive positive caster increases steering effort. The same action occurs at the right front spindle, but the spindle tilts downward during a right turn.

Harsh riding may be caused by excessive positive caster because the caster line is actually aimed at some road irregularities. Excessive positive caster may cause the front wheels to shimmy at low speeds.

When the caster line is aimed directly at the road, road shock is transmitted through the ball joint and upper strut mount to the suspension and chassis. A small degree of positive caster or a negative caster allows the front wheel to roll into a road depression without the caster line being aimed at the hole in the road.

Most FWD vehicles have positive caster. If one front wheel has more positive caster than the other, the steering pulls toward the side with the least positive caster. Some manufacturers recommend less positive caster on the left front wheel to offset road crown. If caster is used to compensate for road crown, the camber on both front wheels should be the same.

## Effects of Negative Caster

Negative caster moves the centerline of the upper strut mount and lower ball joint behind the vertical centerline of the wheel's contact with the road. Tire friction causes the tire to pivot at the centerline of the upper strut mount and ball joint where the tire meets the road. As a result, the wheel pulls away from its straight position and decreases directional stability.

Negative caster reduces steering effort, but improves ride quality. This is because the wheel rolls into the hole without the caster line being aimed at the hole.

## Effects of Suspension Height on Caster

When the rear springs are sagged or overloaded, the caster of the front wheels is more positive. If rear suspension height is above specifications, front wheel caster is less positive.

On many SLA front suspension systems, the inner front of the upper control arm is higher than the inner rear of the arm. This causes front wheel caster to be more positive if the front height is lowered. When the front suspension's height is above specifications, front wheel caster is less positive.

On most FWD vehicles, the top strut mount is tilted rearward. If the front suspension height is lowered, the caster becomes less positive. Conversely, if the suspension height is raised, positive caster increases. Negative caster contributes to reduced directional control, does not contribute to front wheel shimmy, and reduces road shock transmitted to the chassis.

# STEERING AXIS INCLINATION (SAI)

Improper **steering axis inclination (SAI)** angles may cause hazardous steering conditions while braking or accelerating. On FWD vehicles, SAI refers to the inward tilt of a line through the center of the top strut mount and the center of the lower ball joint in relation to the true vertical line through the center of the tire. These two lines are viewed from the front of the vehicle, and the SAI line always tilts inward in relation to the true vertical line.

Many rear-wheel drive cars have an SLA front suspension system with unequal length upper and lower control arms and a ball joint mounted in each control arm. The steering knuckle and spindle pivot on the ball joints as the wheels are turned. On this type of front suspension, the SAI line runs through the upper and lower ball joint centers. The **included angle** is the sum of the SAI angle and the camber angle **(Figure 11).**

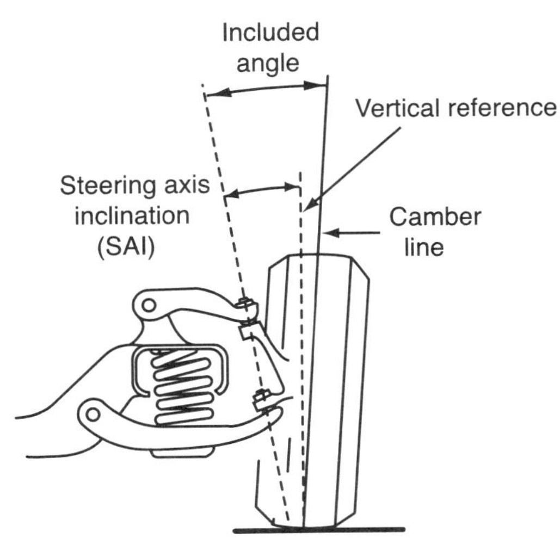


**Figure 11.** Steering axis inclination and included angle.

If the camber angle is positive, this angle is added to the SAI angle to obtain the included angle. A negative camber angle must be subtracted from the SAI angle to calculate the included angle. On some truck front suspension systems, the steering knuckle pivots on a king pin that is mounted in an I-beam-type front axle. The SAI line on this type of suspension is referred to as a **kingpin inclination (KPI)** line. This invisible line runs through the center of the king pin.

## SAI Purpose

When the SAI angle is tilted toward the center of the vehicle and the wheels are straight ahead, the height of the spindle is raised closer to the chassis. This action lowers the height of the vehicle because of gravity. When the front wheels are turned, each spindle moves through an arc that tries to force the tire into the ground. Since this reaction cannot take place, the chassis lifts when the wheels are turned. When the steering wheel is released after a turn, the vehicle weight has a tendency to settle to its lowest point. Therefore, SAI helps return the wheels to the straight-ahead position after a turn, and it also tends to maintain the wheels in the straight-ahead position. However, SAI does increase steering effort because the chassis has to lift slightly on turns.

The vehicle's weight is projected through the SAI line to the road. If a front suspension is designed with a vertical (0-degree) SAI line, the vehicle's weight is projected through the SAI line to a distance inside the true vertical tire centerline. Severe tire scuffing will occur because the wheel and tire pivot around the SAI line during a turn.

With a 0-degree SAI line, greater steering effort is required during a turn and stress on the steering system increases. This causes excessive road shock and kickback on the steering wheel during a turn because the distance between the SAI line and the tire's vertical line returns the wheels to the straight-ahead position.

When the SAI line meets the vertical centerline at the road's surface, tire wear is reduced, steering effort is decreased, and stress on steering components is reduced.

The point at which the SAI line and the tire vertical centerline intersect can be positioned closer together with excessive positive camber. However, this would result in rapid wear on the outside edge of the tire. A correct SAI line reduces the need for excessive positive camber.

MacPherson strut front suspension systems have a much greater SAI angle (12 degrees to 18 degrees) compared to SLA front suspensions (6 degrees to 8 degrees). FWD vehicles require a higher SAI angle for directional stability. Positive caster also provides directional stability. Since SAI helps maintain the wheels in the straight position, SAI reduces the need for excessive positive caster.

## SAI and Safety Factors

On MacPherson strut suspension systems, an incorrect SAI angle may indicate that the upper strut mount is out of position, the lower control arm is bent, or the center crossmember is shifted. Any of these defects may be caused by collision damage.

When the SAI angles are unequal on the left and right, serious handling problems may occur. These problems include torque steering during hard acceleration, steering pull during sudden stops, and bump steer. **Torque steer** is the tendency to pull to one side on hard acceleration because of unequal length drive axles on a front-wheel-drive vehicle. Unequal SAI angles aggravate torque steer. Bump steer refers to unequal toe and/or camber changes that jerk a vehicle to one side during jounce and rebound.

If the SAI angle is correct and the included angle and camber are less than specified, the strut or spindle may be bent. On systems that require upper strut movement to adjust camber, the SAI angle changes with this movement. When an eccentric bolt between the strut and steering knuckle is used for camber adjustment, the SAI angle will not change when camber is adjusted.

## TOE

When the distance between the rear inside tire edges is greater than the distance between the inside tire edges, the front wheels have toe in. Toe out occurs when the distance between the inside front tire edges exceeds the distance between the inside rear tire edges **(Figure 12).**

## Toe Setting for Front-Wheel-Drive and Rear-Wheel-Drive Vehicles

The front wheels are adjusted to a slight toe in or toe out so the wheels are parallel to each other when the vehicle is driven. A slight amount of lateral movement always exists in steering linkages. Forces acting on the front wheels try to compress or stretch the steering linkage as the vehicle is driven. Whether a compressing or stretching action occurs depends on whether the steering linkage is at the rear edge or front edge of the front wheels.

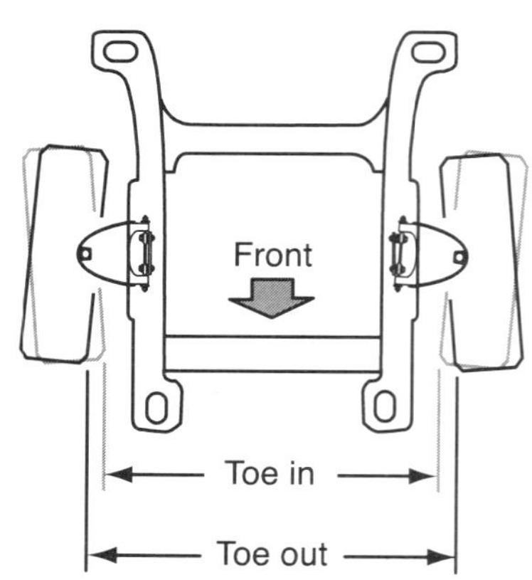


**Figure 12.** Toe in and toe out.

On a FWD vehicle, front drive axle torque forces the front wheels to an increased toe-in position. Therefore, manufacturers usually specify a slight toe-out position on the front wheels.

On RWD vehicles, friction from the road moves the wheels toward a toe-out position. Because of this, manufacturers usually specify a slight toe-in adjustment.

## Toe Adjustment and Tire Wear

When the front toe is not adjusted to specifications, front tire tread wear is excessive. If one of the tie rods is not parallel to the lower control arms, toe change is not equal during front wheel jounce and rebound and may result in bump steer. Excessive toe out causes wear on the inside and a sharp feathered edge on the outside of the tire.

Worn steering linkage parts can cause incorrect and erratic toe-in settings. If the front springs become weak, the front suspension height is lowered. When this occurs, the pitman arm and the idler arm move downward with the chassis. This moves the tie rods to a more horizontal position which tends to push outward on the steering arms and increase front wheel toe in.

# OTHER CONSIDERATIONS

During wheel alignment, other dimensions or angles must be considered. Many of these are not part of the alignment process but must be considered while doing an alignment.

## Scrub Radius

**Scrub radius** is the distance from the point where the tire's vertical line contacts the road to the location where the line through the upper strut center and the ball joint center meets the road. Positive scrub radius occurs when the line meets the road inside the tire's vertical centerline. A negative scrub radius occurs when the line meets the road outside the tire and hub's centerline **(Figure 13).** Scrub radius affects steering stability and returnability but cannot be measured on alignment equipment.

Conventional SLA front suspension systems usually have positive scrub radius. Many FWD vehicles have a negative scrub radius.

## Turning Radius

**You Should Know** *Turning radius may be referred to as cornering angle or **toe out on turns.***

When a vehicle turns a corner, the front and rear wheels must turn around a common center with respect to the **turning radius (Figure 14).** On most front suspension systems, the front wheels pivot independently at different

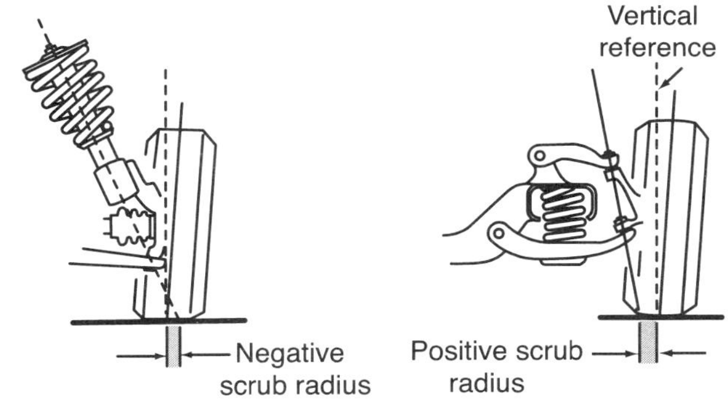


**Figure 13.** Scrub radius.

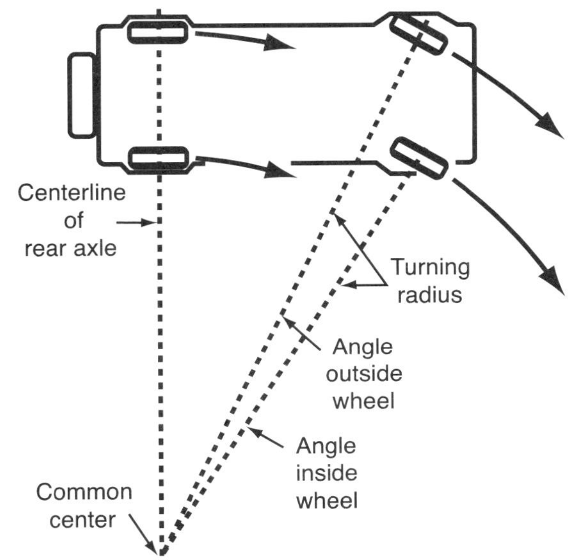


**Figure 14.** Front and rear wheels turning around a common center.

distances from the center of the turn, and the wheels turn at different angles. The inside front wheel turns at a sharper angle than the outside wheel. This is because the inside wheel is actually ahead of the outside wheel. Both front wheels remain perpendicular to their turning radius, which prevents tire scuffing.

Steering arms are connected from the tie-rod ends to the front spindles to keep the front wheels parallel to each other when the vehicle is driven straight. If the steering linkage is at the rear edge of the front wheels, the steering arms are closer. When the steering linkage is positioned at the front of the front wheels, the steering arms are closer together at their spindle pivot point than at the tie-rod connecting point.

When the front wheels are turned on a vehicle with the steering linkage at the rear edge of the front wheels, the angle formed by the inside steering arm and linkage increases, whereas the angle of the outside steering arm and linkage decreases. The inside steering arm moves through the longer arc, and the outside steering arm moves through a shorter one.

Therefore, the inside wheel turns at a sharper angle than the outside wheel. The inside front wheel always has a sharper angle, regardless of the direction it is turning. The sharper inside wheel angle causes the inside wheel to "toe out." Therefore, the term toe out on turns is used for steering action. If the front wheel turning angle is increased during a turn, the amount of toe out on the inside wheel increases proportionally.

## Slip Angle

During a turn, centrifugal force causes the tires to slip a certain amount. The amount of slip increases with speed and the sharpness of the turn. This slip action causes the actual center of the turn to be considerably ahead of the theoretical turn center. The slip angle varies with vehicle weight and the type of suspension.

# REAR-WHEEL ALIGNMENT

Incorrect rear-wheel alignment can be caused by collision damage that results in a bent frame or distorted unitized body, a leaf-spring eye that is unwrapped or spread open, leaf-spring shackles that are broken, bent, or worn, broken leaf springs or leaf-spring center bolts, worn rear upper or lower control arm bushings, worn trailing arm bushings or dislocated trailing arm brackets, and/or bent components such as radius rods, control arms, struts, and rear axles.

## Rear-Wheel Camber

Many FWD vehicles have a slightly negative rear-wheel camber to provide improved cornering stability **(Figure 15).**

## Rear-Wheel Toe

On a FWD vehicle, driving forces tend to push the rear wheel spindles back. Therefore, these wheels are designed with zero toe in or a slight toe in **(Figure 16).** Correct rear-wheel toe is important for normal tire life.

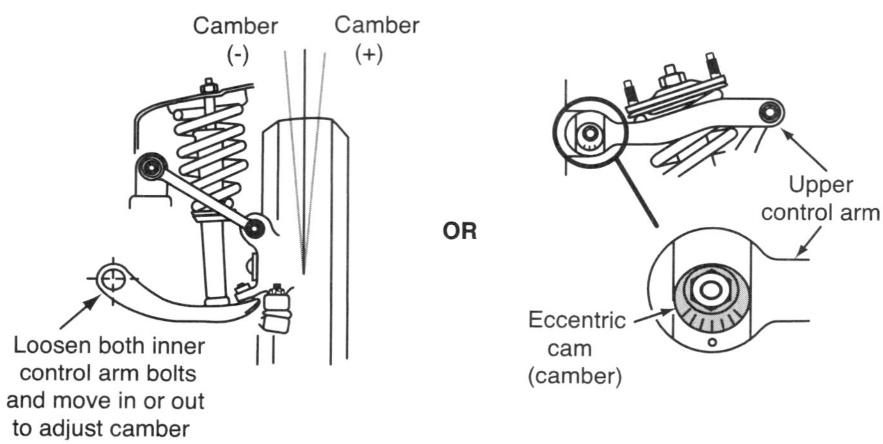


**Figure 15.** Rear-wheel camber.

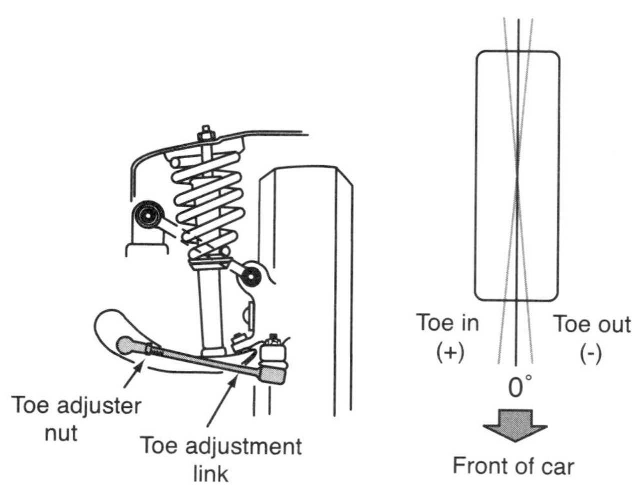


**Figure 16.** Rear-wheel toe.

## SUMMARY

- The vehicle geometric centerline is an imaginary line through the exact center of the front and rear wheels.
- The thrust line is an imaginary line at a 90-degree angle to the rear wheel centerline and projected forward.
- A rear axle offset occurs when the complete rear axle is rotated slightly, moving one rear wheel backward and the opposite rear wheel forward.
- Rear axle sideset is a condition where the rear axle assembly has moved straight sideways and the geometric centerline is not positioned at the true vehicle centerline.
- Setback occurs when one front or rear wheel is moved backward in relation to the opposite wheel.
- Directional stability is the ability of a vehicle to travel straight without being steered.
- Camber is the tilt of a line through the center of the tire and wheel in relation to the vertical centerline of the tire and wheel.
- Road crown causes a vehicle steering to drift to the right.
- Caster is the tilt of a wheel based on a line drawn through the center of the upper and lower ball joints or lower ball joint and upper strut mount.
- SAI is the angle of a line drawn through the center of the upper strut mount and lower ball joint in relation to the true vertical centerline of the tire.
- Scrub radius is the distance between the SAI and true vertical tire centerline at the road.
- Front-wheel toe is the distance between the front edges of the tires compared to the distance between the rear edges of the tires.
- Toe out on turns is the turning angle of the wheel on the inside of the turn compared to the turning angle of the wheel on the outside of the turn.
- The two rear-wheel alignment angles are camber and toe.

## REVIEW QUESTIONS

1. Describe how the caster adjustment may be used to compensate for road crown.
2. A 3-degree difference in the SAI angle on each side of the front suspension may cause increased ____________ during hard acceleration.
3. Most rear-wheel drive cars have a ____________ scrub radius.
4. Front-wheel toe on many RWD cars is adjusted to a ____________ setting.
5. During a turn, the ____________ front wheel turns at a sharper angle.
6. While diagnosing suspension and wheel alignment problems:
   A. road crown has no affect on vehicle steering or wheel alignment.
   B. vehicle loads have no effect on wheel alignment.
   C. steering angles are designed to reduce tire wear and provide directional control.
   D. wheel alignment angles do not affect riding quality.
7. While driving straight, a FWD car pulls to the right. The most likely cause is:
   A. more positive camber on the left front wheel compared to the right front wheel.
   B. sagged front springs and improper front wheel toe setting.
   C. less positive caster on the right front wheel compared to the left front wheel.
   D. the SAI on the right front wheel is $1^1/_2$ degree more than the SAI on the left.
8. While discussing front wheel caster, Technician A says excessive positive caster increases steering effort and causes harsh riding. Technician B says excessive negative caster causes front wheel shimmy. Who is correct?
   A. Technician A
   B. Technician B
   C. Both Technician A and Technician B
   D. Neither Technician A nor Technician B
9. While adjusting front-wheel camber and diagnosing camber-related problems:
   A. during front wheel jounce travel, the positive camber increases.
   B. excessive positive camber causes premature wear on the inside edge of the tire.
   C. steering will pull to the left if right camber is +1 1/2 degree and left camber is + $^1/_2$ degree.
   D. excessive positive camber on both front wheels may cause front wheel shimmy.
10. While discussing rear wheel camber and toe on FWD cars, Technician A says many have a zero toe or a slight toe in. Technician B says many have $^1/_2$ degree positive camber. Who is correct?
    A. Technician A
    B. Technician B
    C. Both Technician A and Technician B
    D. Neither Technician A nor Technician B

# Chapter 52 Pre-Alignment Diagnosis and Inspection

## INTRODUCTION

Drivers may experience a variety of symptoms related to incorrect wheel alignment or defective brakes **(Figure 1).** Not all the symptoms are related to wheel alignment, but the customer may request an alignment to correct these problems. The technician must diagnose and correct the cause of the specific complaint. Wheel alignment angles that affect tire wear are toe and camber **(Figure 2).** The most likely symptoms of incorrect toe or camber settings are pulling to one side while driving, wander, feathered tire wear, or tire wear on one side.

## ROAD TEST

In many cases, a road test is necessary to verify customer complaints. During the road test, the vehicle should be driven under the conditions cited by the customer as to when the complaint occurred. While road testing the vehicle, listen for any unusual noises related to the suspension and steering systems. Test the vehicle using a variety of

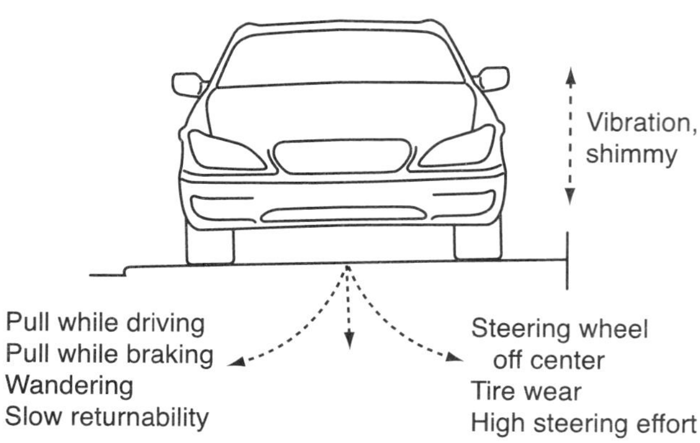


**Figure 1.** Common complaints related to wheel alignment.

| Problem | Effect |
|---|---|
| Incorrect camber setting | Tire wear<br>Ball joint/wheel bearing wear<br>Pull to side of most positive/ least negative camber |
| Too much positive caster | Hard steering<br>Excessive road shock<br>Wheel shimmy |
| Too much negative caster | Wander<br>Weave<br>Instability at high speeds |
| Unequal caster | Pull to side most negative/ least positive caster |
| Incorrect SAI | Instability<br>Poor return<br>Pull to side of lesser inclination<br>Hard steering |
| Incorrect toe setting | Tire wear |
| Incorrect turning radius | Tire wear<br>Squeal in turns |

**Figure 2.** The effect of incorrect wheel alignment.

driving conditions from slow-speed driving and cornering to normal cruising speed driving on a straight, level road surface. Make sure to check for the following abnormal steering conditions:

- Vertical chassis oscillations are the result of defective shock absorbers or struts.
- Chassis lateral waddle may be caused by excessive wheel or tire runout.
- Steering pull or steering drift is the tendency of the steering to gradually pull to the right or left when the vehicle is driven straight ahead on a level road. Steering pull or drift may be caused by improper caster or camber adjustments, severe front wheel setback condition, improper camber caused by improperly positioned cradle, setback caused by improperly positioned cradle, rear suspension overload, improper toe setting on one rear wheel, thrust line positioned to the left or right of the geometric centerline, and improper SAI on one side of front suspension.

**You Should Know** *If steering pull or drift is evident during the road test, drive the vehicle on the same road in the opposite direction to determine if a crosswind is causing the steering pull.*

- Steering pull during braking or accelerating is typically caused by improper SAI on one side of front suspension.
- Steering wander occurs when the vehicle tends to steer in either direction rather than straight ahead on a level road surface. This may be caused by improper caster adjustment.
- Erratic steering on road irregularities can be caused by improper SAI on one side of the front suspension.
- High steering effort and binding is normally caused by excessive positive caster on both front wheels, a loose power steering belt, low power steering fluid level, worn front tires, and/or a defective steering gear.
- Bump steer occurs if the tie rods are not the same height, meaning one of the tie rods is not parallel to the lower control arm. This condition may be caused by a bent or worn idler arm or pitman arm on a parallelogram steering linkage. On a rack-and-pinion steering system, worn steering gear mounting bushings may cause this unparallel condition between one of the tie rods and the lower control arm.
- Torque steer results from unequal wear on front tires and different types of front tires or tire treads. FWD vehicles with unequal drive axles produce some torque steer on hard acceleration. Some front-wheel drive vehicles, especially those with higher horsepower, have equal length front drive axles to reduce torque steer.
- Memory steer may be caused by a binding condition in the steering column or in the steering shaft universal joints. A binding upper strut mount may result in memory steer. Negative caster or reduced positive caster also causes memory steer.
- Most steering systems are designed so the steering wheel returns to within 30 degrees to 60 degrees of center after a 180-degree turn. If the steering wheel returnability is not satisfactory and memory steer is evident, the steering shaft dash seal or the steering shaft universal joints may be binding.
- Steering wheel return **(Figure 3)** problems can be caused by improper caster setting and binding in the steering column or U-joints.
- The steering wheel is not centered while driving straight ahead; this condition can be caused by improper tie-rod sleeve adjustment.
- Oversteer or understeer while cornering at high speed may be caused by improper rear wheel camber.

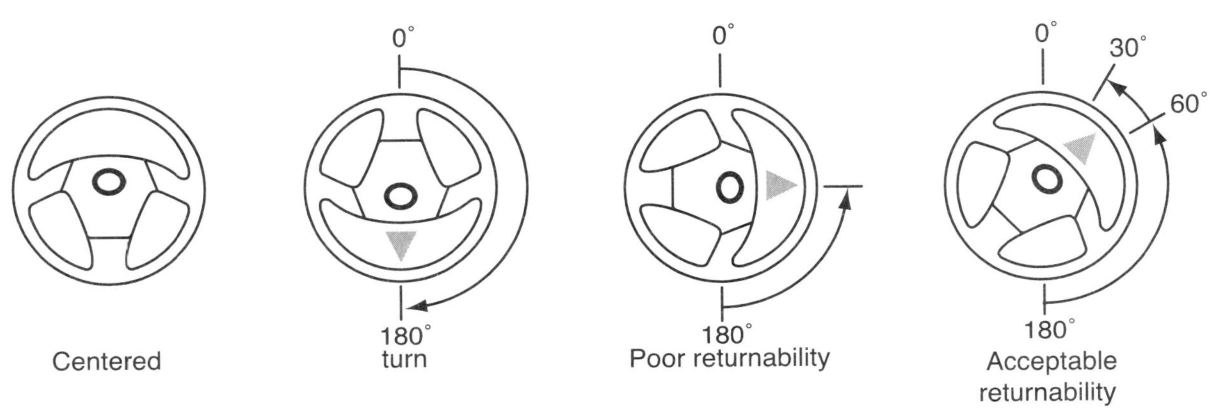


**Figure 3.** Steering return.

## PREALIGNMENT INSPECTION

You should attempt to identify the exact suspension or steering problem before performing a wheel alignment or other suspension and steering work. In addition to diagnosing the problem by the symptoms, the cause of problems may be discovered by a thorough inspection of the systems **(Figure 4).** The cause or causes of the customer's complaints must be diagnosed before aligning the wheels. These problems should also be corrected dur-

PREALIGNMENT INSPECTION CHECKLIST

Owner ______________________ Phone ______________ Date __________

Adress ______________________________ VIN ______________________

Make ________ Model ________ Year _____ Lic. number __________ Mileage __________

| 1. Road test results | Yes | No | Right | Left |
|---|---|---|---|---|
| Above 30 MPH | | | | |
| Below 30 MPH | | | | |
| Bump steer | | | | |
| When braking | | | | |
| Steering wheel movement Stopping from 2-3 MPH(Front) | | | | |
| Vehicle steers hard | | | | |
| Strg wheel returnability normal | | | | |
| Strg wheel position | | | | |
| Vibration | Yes | No | Frnt | Rear |
| | | | | |

| 2. Tire pressure | Specs Frnt ____ Rear ____ |
|---|---|
| Record pressure found | |
| RF _____ LF _______ RR _______ LR _______ | |

| 3. Chassis height | Specs Frnt ____ Rear ____ |
|---|---|
| Record height found | |
| RF _____ LF _______ RR _______ LR _______ | |

| | Yes | No |
|---|---|---|
| Springs sagged | | |
| Torsion bars adjusted | | |

| 4. Rubber bushings | OK |
|---|---|
| Upper control arm | |
| Lower control arm | |
| Sway bar / stabilizer link | |
| Strut rod | |
| Rear bushing | |

| 5. Shock absorbers/struts | Frnt | Rear |
|---|---|---|
| | | |

| 6. Steering linkage | Frnt OK | Rear OK |
|---|---|---|
| Tie rod ends | | |
| Idler arm | | |
| Center link | | |
| Sector shaft | | |
| Pitman arm | | |
| Gearbox/rack adjustment | | |
| Gearbox/rack mounting | | |

| 7. Ball joints | | OK |
|---|---|---|
| Load bearings | | |
| Specs | Readings | |
| Right _____ Left _____ | Right _____ Left _____ | |
| Follower | | |
| Upper strut bearing mount | | |
| Rear | | |

| 8. Power steering | OK |
|---|---|
| Belt tension | |
| Fluid level | |
| Leaks/hose fittings | |
| Spool valve centered | |

| 9. Tires/wheels | OK |
|---|---|
| Wheel runout | |
| Condition | |
| Equal tread depth | |
| Wheel bearing | |
| 10. Brakes operating properly | |

| 11. Alignment | Spec | | Initial reading | | Adjusted reading | |
|---|---|---|---|---|---|---|
| | R | L | R | L | R | L |
| Camber | | | | | | |
| Caster | | | | | | |
| Toe | | | | | | |

| Bump steer | Toe change right wheel | | Toe change left wheel | |
|---|---|---|---|---|
| | Amount | Direction | Amount | Direction |
| Chassis down 3" | | | | |
| Chassis up 3" | | | | |

| | Spec | | Initial reading | | Adjusted reading | |
|---|---|---|---|---|---|---|
| | R | L | R | L | R | L |
| Toe-out on turns | | | | | | |
| SAI | | | | | | |
| Rear camber | | | | | | |
| Rear total toe | | | | | | |
| Rear indiv. toe | | | | | | |
| Wheel balance | | | | | | |
| Radial tire pull | | | | | | |

**Figure 4.** A reporting sheet for a prealignment inspection.

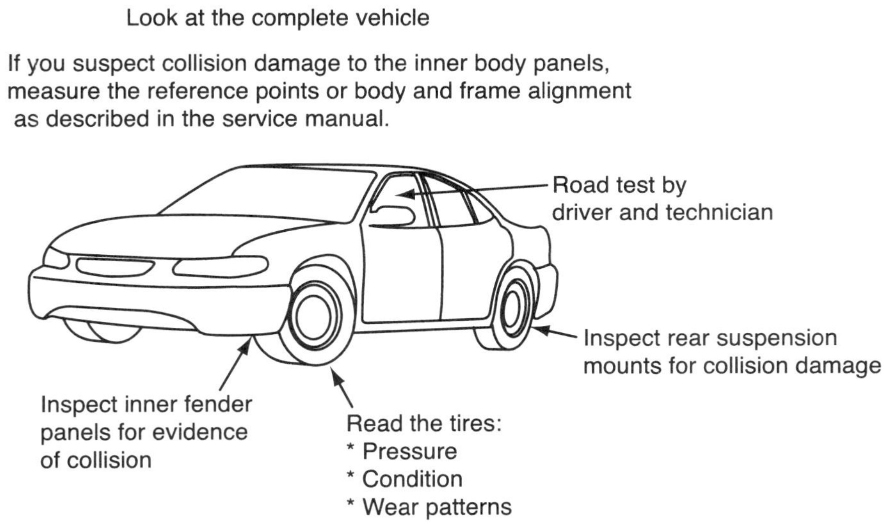


**Figure 5.** Inspecting for collision damage.

ing the wheel alignment. Worn suspension and steering components must be replaced before a wheel alignment is performed. Since collision damage may affect wheel alignment angles, an inspection for collision damage on the vehicle should be completed prior to a wheel alignment **(Figure 5).**

*The following should be checked with the vehicle sitting on the floor of the shop:*

1. Check for excessive mud adhered to the chassis. Remove heavy items from the trunk and passenger compartment that are not considered in the vehicle curb weight. If heavy items (such as tool display cases or merchandise) are normally carried in the vehicle, these items should be left in the car during a wheel alignment.
2. Inflate tires to the recommended pressure and note any abnormal tread wear or damage on each tire. Various types of tire tread wear patterns indicate specific alignment or balance defects.
   a. Feathered wear **(Figure 6)** on front tire treads—Improper front-wheel toe
   b. Feathered wear on rear tires—Improper rear-wheel toe setting
   c. Excessive front tire tread wear, scuffing—Improper turning radius
   d. Excessive wear on the edges of the tire—Excessive positive or negative camber
3. Make sure all tires are the same size.
4. Check the front tires and wheels for radial runout.
5. Check the suspension ride height. If this measurement is not within specifications, check for broken or sagged springs. On a torsion bar suspension system, check the torsion bar adjustment.
6. When the wheels are in the straight-ahead position, rotate the steering wheel back and forth to check for play in the steering column, steering gear, or linkages.
7. Check the shock absorbers or struts for loose mounting bushings and bolts. Examine each shock absorber or strut for leakage.
8. Check the condition of each shock absorber or strut with a bounce test at each corner of the vehicle.

*The following checks should be performed with the vehicle raised and the suspension supported:*

1. Check the front wheel bearings for lateral movement. On front-wheel drive cars, perform this check on all four wheel bearings. Wheel bearings must be properly

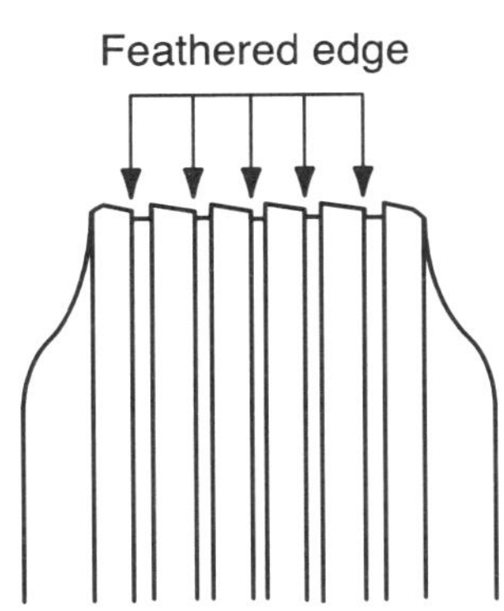


**Figure 6.** A feathered edge on a tire.

adjusted prior to a wheel alignment. Clean, repack, or adjust the wheel bearings as necessary.
2. Measure the ball joint radial and axial movement. If excessive movement exists in either direction, ball joint replacement is required. Be sure the suspension is supported correctly during ball joint diagnosis.
3. Inspect the control arms for damage and check the control arm bushings for wear.
4. Check all the steering linkages and tie-rod ends for looseness.
5. Check for worn stabilizer mounting links and bushings.
6. Check for loose steering gear mounting bolts and worn mounting brackets and bushings.

## SUMMARY

- Diagnosis by symptom is a good place to start diagnosing problems.
- A road test may be necessary to verify customer complaints prior to a wheel alignment.
- Worn suspension and steering components must be replaced before a wheel alignment is performed.
- After suspension components such as struts have been replaced, a wheel alignment should be performed.
- A prealignment inspection of all steering and suspension components must be completed before a wheel alignment procedure.
- Improper camber settings cause wear on one side of the front tire treads.
- Incorrect toe settings will cause abnormal tire wear patterns.

## REVIEW QUESTIONS

1. While performing a prealignment inspection, Technician A says improper front-wheel bearing adjustment may affect wheel alignment angles. Technician B says worn ball joints have no effect on wheel alignment angles. Who is correct?
   A. Technician A
   B. Technician B
   C. Both Technician A and Technician B
   D. Neither Technician A nor Technician B
2. When discussing unsatisfactory steering wheel returnability, Technician A says the rack-and-pinion steering gear mounts may be worn. Technician B says this problem may be caused by interference between the dash seal and the steering shaft. Who is correct?
   A. Technician A
   B. Technician B
   C. Both Technician A and Technician B
   D. Neither Technician A nor Technician B
3. While discussing steering diagnosis, Technician A says uneven half-shaft axle lengths may cause a vehicle to pull to one side when accelerating. Technician B says abnormal toe changes can cause a vehicle to pull to one side. Who is correct?
   A. Technician A
   B. Technician B
   C. Both Technician A and Technician B
   D. Neither Technician A nor Technician B
4. While performing a prealignment inspection, Technician A says a prealignment inspection should include checking the vehicle interior for heavy items. Technician B says tools and other items normally carried in the vehicle should be included during an alignment. Who is correct?
   A. Technician A
   B. Technician B
   C. Both Technician A and Technician B
   D. Neither Technician A nor Technician B
5. "Feathering" type wear of a rear tire is likely caused by:
   A. improper rear-wheel camber alignment.
   B. improper rear tire inflation.
   C. improper rear-wheel balance.
   D. improper rear-wheel toe alignment.

# Chapter 53 Four-Wheel Alignment

## *INTRODUCTION*

During four-wheel alignment, the thrust line position is measured in relation to the geometric centerline of the vehicle. After alignment, all four wheels are parallel to the geometric centerline, and the steering wheel is centered when the vehicle is driven straight.

## PRELIMINARY PROCEDURE

The vehicle should be on an alignment ramp with a turntable under each front tire and slip plates under the rear tires **(Figure 1).** The slip plates allow movement of the rear wheels during rear-wheel camber and toe adjustments. If these adjustments are made with the tires resting on the ramp, the tires cannot move, and it is impossible to accurately adjust the wheels. The center of the front wheel spindles should be in line with the zero mark on the turning plates. The locking pins for the turning plates should be left in place and the parking brake applied.

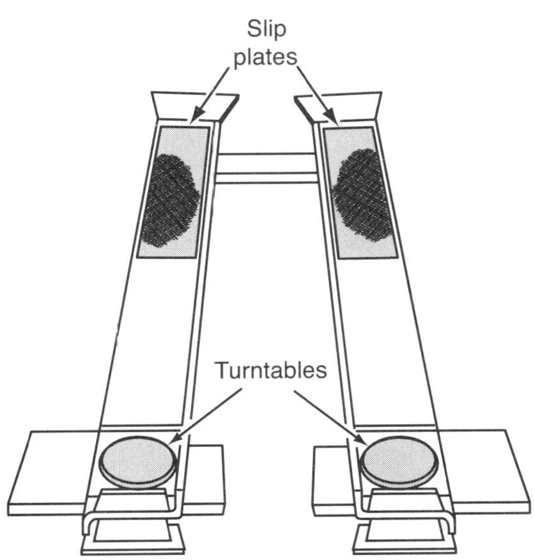


**Figure 1.** An alignment ramp with turntables and slip plates.

> **You Should Know** *When driving a vehicle onto an alignment ramp, make sure no one is standing in front of the vehicle.*

Four-wheel alignment equipment uses sensors (or **wheel units)** installed on all four wheels with self-centering rim clamps. Front wheel sensors have optical arms that project ahead of the tires to provide front toe readings **(Figure 2).** Some rear wheel sensors also have optical arms

**Figure 2.** A wheel unit clamped to a front wheel.

that project behind the rear tires to measure rear wheel set-back. Securely mount the rim clamps on each wheel. You may need to remove the hubcaps before installing the clamps.

> **You Should Know** *Since wheel units contain sensitive electronic circuits, excessive shock, such as dropping them on the floor, may damage them.*

Some wheel sensors have touch keypads to allow the entry of commands from each unit. The front wheel units have arms that project in front of the tires to transmit signals between them. When a blocked beam prompt appears on any screen, the beam between two wheel units is blocked. This can be caused by a person standing in the beam, an open car door, or a suspension jack. The blocked beam prompt must be eliminated before any tests are completed.

On some systems, the reference signals between the wheel sensors are provided by light-emitting diodes (LEDs) or electronic signals. Cables connect the wheel units to the aligner.

Some wheel sensors contain a microprocessor and a high-frequency transmitter that acquire measurements and process data and send this data to a receiver in the wheel aligner. This type of wheel sensor does not need cables. The wheel sensor signals are virtually uninterruptible, even by solid objects. When these wheel sensors are stored on the alignment machine, a "docking station" feature charges the batteries in the sensors.

Most wheel sensors provide wheel runout compensation. Since the wheel unit is clamped to the rim, a bent rim will affect alignment readings. Older wheel aligners required that the runout at each wheel be manually checked and compensated for. Newer wheel units use audible and visual prompts on the monitor screen to indicate if the wheel sensors require leveling or calibration. On those units that require manual adjustment, the adjustment knob on the rim clamp should be pointing toward the top of the wheel, and the bubble on the wheel unit should be centered. A set knob on the wheel unit allows it to be locked in this position.

Some wheel aligners have a display that indicates if any of the wheel sensors are not level **(Figure 3).** These sensors may be leveled by pressing a control on each wheel sensor.

## Monitor

The monitor of the wheel aligner can display a variety of screens that vary with the models of equipment. Typically, there is a main menu screen. This screen contains

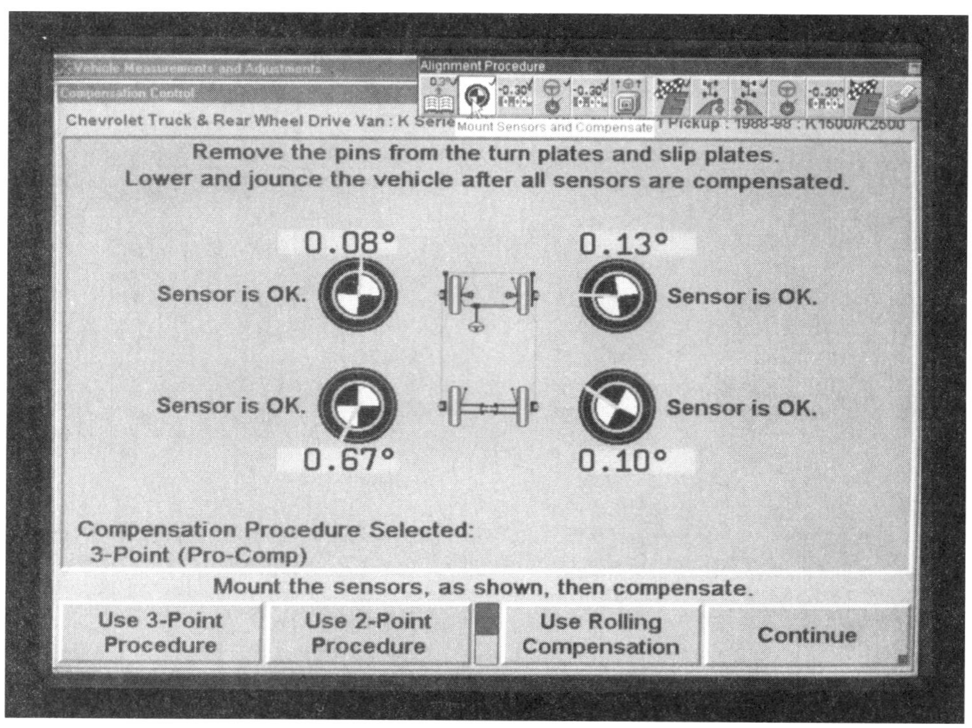


**Figure 3.** An example of a runout compensation screen.

instructions and basic setup options. Input is made by touch-screen technology or a keyboard on the aligner.

Vehicle information and access to the correct specifications are critical to proper wheel alignment. Some wheel aligners have specifications contained on CDs. With the appropriate CD installed in the aligner, specifications are gathered by the aligner after the VIN and/or other information is inputted.

Many aligners begin their sequence with the basics for completing a prealignment inspection. Some aligners have optical encoders in the wheel sensors that measure ride height when this measurement is selected and the ride height attachment on each wheel sensor is lifted until it touches the lower edge of the fender **(Figure 4).** The ride height measurement is displayed on the screen.

**You Should Know** *The ride height must be within specifications before proceeding with the wheel alignment. Improper curb riding height affects many of the other suspension angles.*

## WHEEL ALIGNMENT SCREENS

**You Should Know** *Many wheel aligners can print any single screen. It may be helpful to print out the wheel alignment angle screen before and after adjustments are made. These printouts can be shown to the customer.*

On some aligners, the monitor will direct the technician to position the front wheels straight ahead, lock the steering wheel, apply the brake pedal depressor, and level and lock the wheel units.

The brake pedal depressor is an adjustable rod installed between the front edge of the front seat and the brake pedal. If the vehicle has power brakes, the engine should be running when the depressor is used to apply the brakes. Some steering wheel holders are installed between the steering wheel and the top of the front seat.

After wheel runout has been compensated for, some wheel aligners automatically display camber, toe, setback, and thrust line measurements. After this, the monitor directs the technician to swing the front wheels outward a

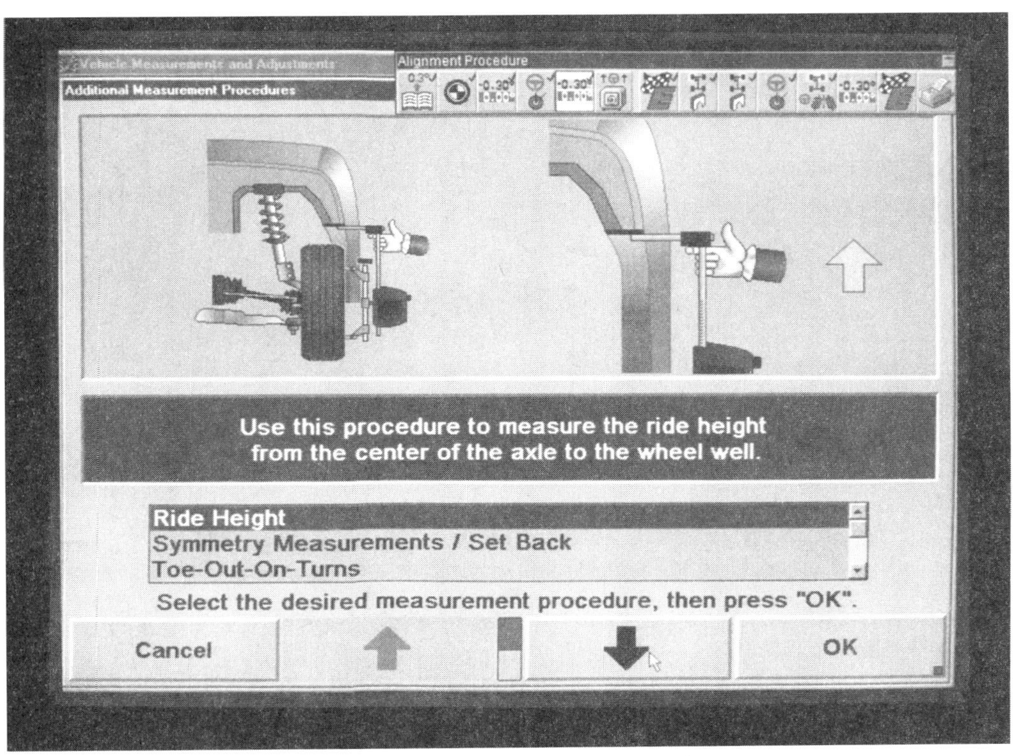


**Figure 4.** On-monitor instructions for measuring ride height.

specific amount. The steering wheel holder must be released and the lock pins removed from the front turntables before the wheels can move. This front wheel swing may be referred to as a caster/SAI swing. The amount of front wheel swing varies with the make of aligner. Older wheel aligners required the technician to read the degrees of wheel swing on the turntable's degree scales. On some newer wheel aligners, the required amount of wheel swing is illustrated on the screen **(Figure 5).**

After the front wheel swing procedure, all the front- and rear-wheel alignment angles are displayed, including caster, steering axis inclination (SAI), and included angle **(Figure 6).** Some wheel aligners highlight any wheel angles that are not within specifications. If cross camber

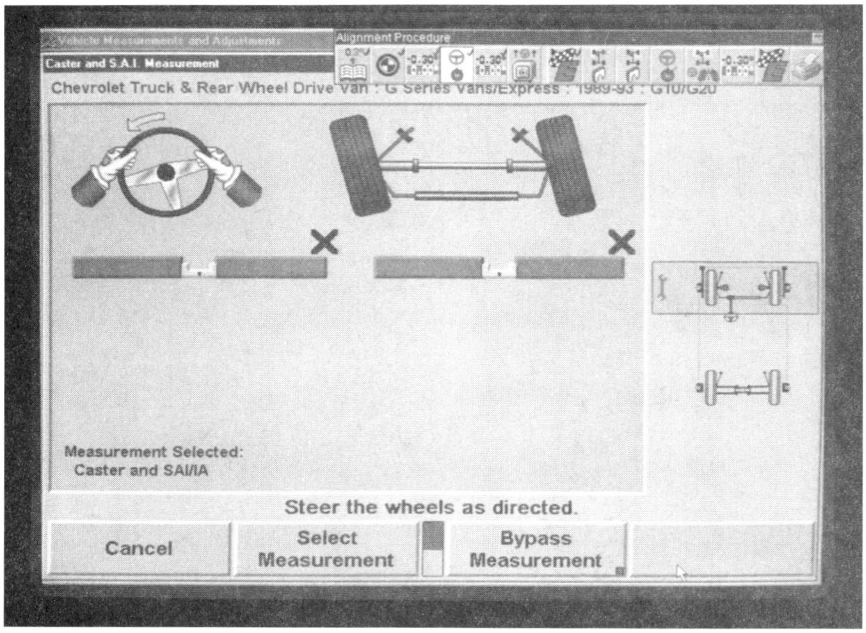


**Figure 5.** Wheel swing instructions shown on a monitor.

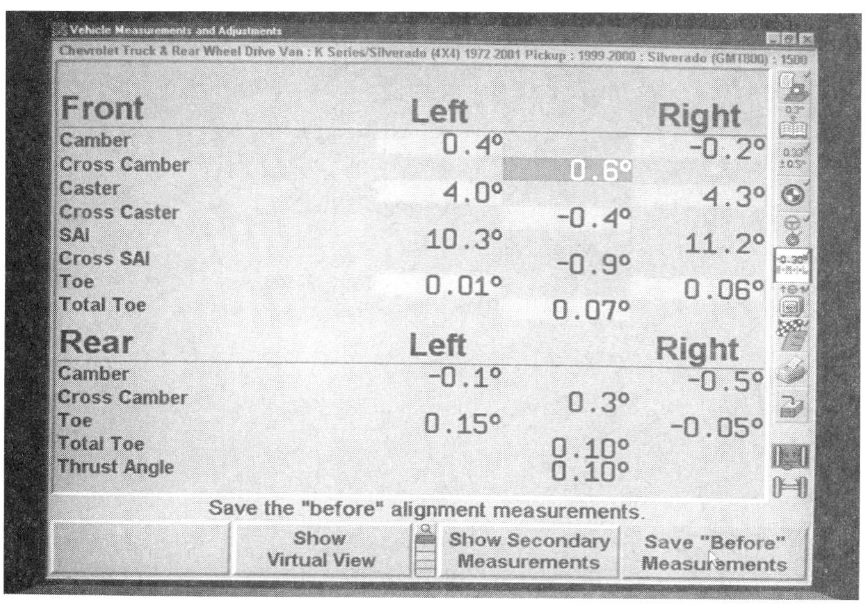


**Figure 6.** The angles measured by the aligner show up on one screen.

and cross caster are displayed on the screen, these readings indicate the maximum difference allowed between the right and left side readings.

## Adjustment Screens

The technician may select adjustment screens that provide bar graph readings of camber, caster, and toe. An arrow on the bar graph shows the amount and direction that the actual measurement is from the preferred specification. As the alignment angle is adjusted, the arrow moves. If an alignment angle moves from the out-of-specification range to within the specification range, the bar graph color changes from red to green **(Figure 7).** When the arrow on the bar graph is in the zero position, the alignment angle is at the preferred specification. Some aligners have a jack and hold feature that allows the suspension to be lifted to make an adjustment while maintaining accurate displays on the adjustment screen. Other wheel aligners have a remote display that may be connected to the aligner and taken under the car for close-up viewing while performing suspension adjustments **(Figure 8).** The remote display duplicates the bar graphs shown on the monitor screen.

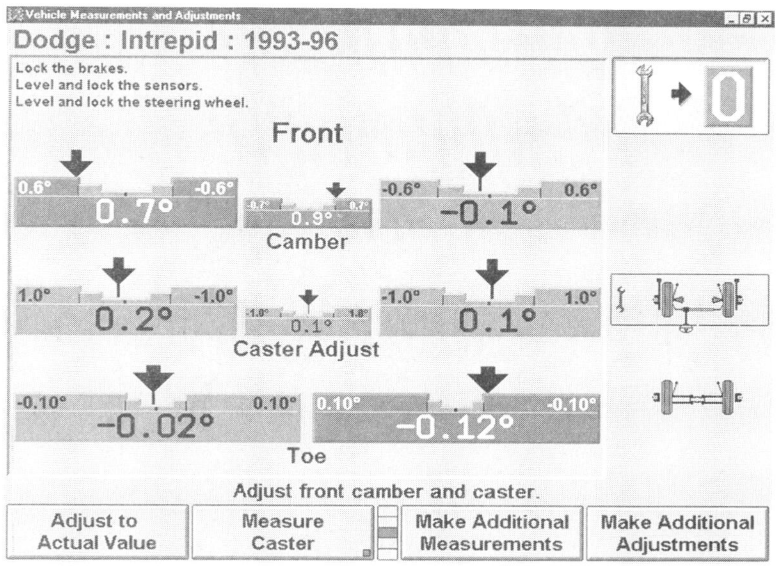


**Figure 7.** An example of an alignment angle adjustment screen.

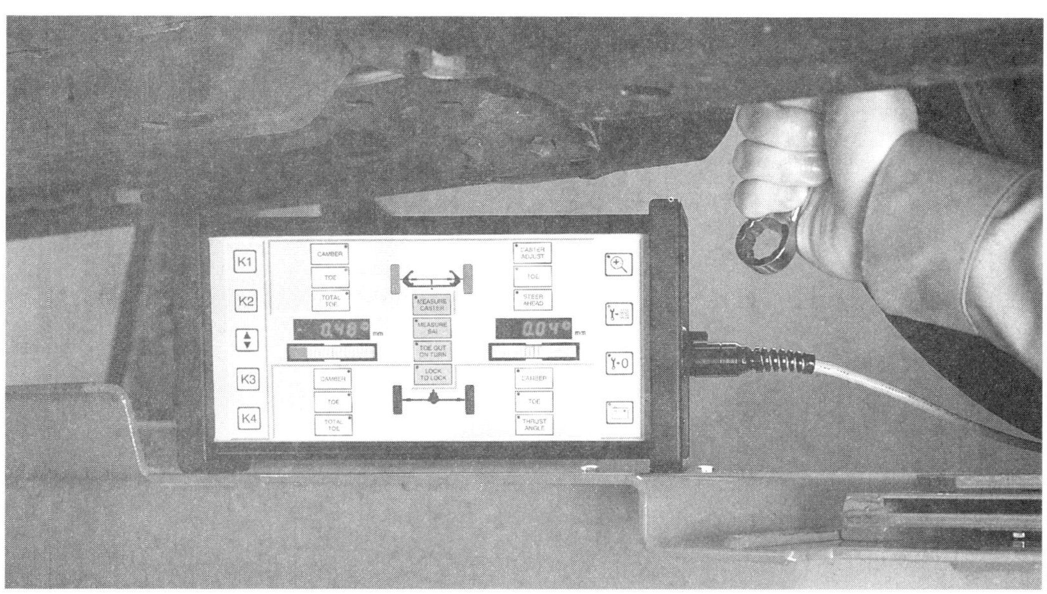


**Figure 8.** A remote display for a wheel aligner.

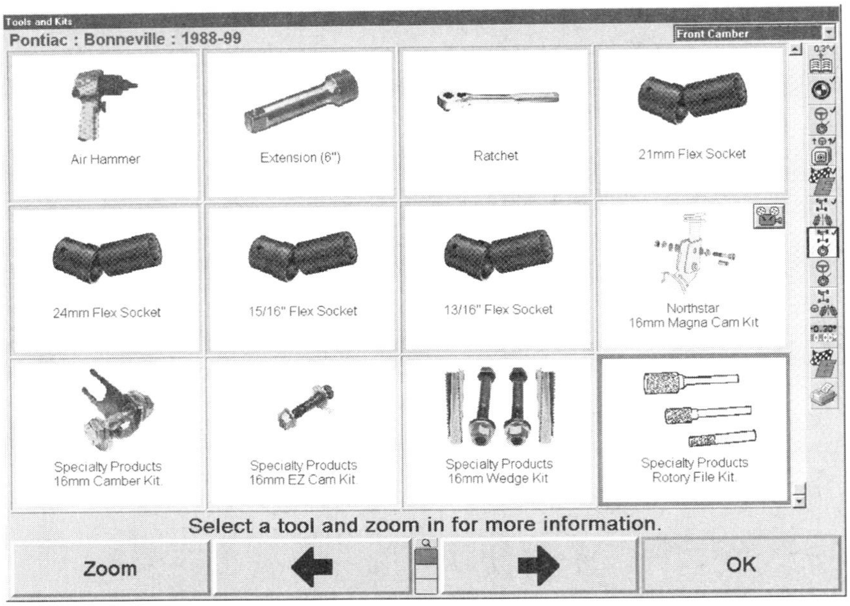


**Figure 9.** Wheel alignment tools and kits display on a monitor.

Some wheel aligners provide symmetry angle measurements that help determine if out-of-specification readings may have been caused by collision or frame damage. These symmetry angle measurements display axle offset, right or left lateral axle sideset and track width difference, front- and rear-wheel setback, and wheelbase difference.

## Diagnostic Screens

The aligner may have a capability to display the necessary wheel alignment adjustment tools for the vehicle **(Figure 9).** The kits displayed will include special components such as adjustment shims that are available for alignment adjustments on that particular vehicle.

The technician may select digital photos that indicate how to perform wheel alignment adjustments. Live-action videos may also be available. A part finder database is available in some aligners. This database allows access to part numbers and prices from many manufacturers.

On many FWD cars, the rear wheel camber and toe are adjusted with shims. Some aligners have shim display screens that indicate the required thickness of shim and the proper position of the shim. On some aligners, a light pen can be used to change the orientation angle of the shim on the monitor while observing the resulting change in camber and toe.

# FOUR-WHEEL ALIGNMENT

The proper procedure for doing an alignment is important since adjusting one wheel angle may change another angle. For example, if the positive caster is increased, the top of the spindle is moved rearward. If the steering linkage is at the rear edge of the front wheels, this rotates the rear of the steering arm downward, which moves the wheel toward a toe-out position. Therefore, other alignment angles such as camber, caster, and SAI must be adjusted within specifications before toe is checked. The wheel alignment procedure is especially critical on four-wheel independent suspension systems.

**You Should Know** *Since rear-wheel alignment plays a significant role in aiming or steering the vehicle, rear-wheel alignment should be corrected before adjusting front-suspension angles.*

## Camber

Various methods are used to adjust camber.

### Shims

Some manufacturers use a shim-type camber adjustment between the upper control arm mounting and the inside of the frame. Increasing the shim thickness moves the camber setting to a more negative position, whereas decreasing shim thickness changes the camber toward a more positive position. Shims of equal thickness should be added or removed on both upper control arm mounting bolts to change the camber setting without changing caster.

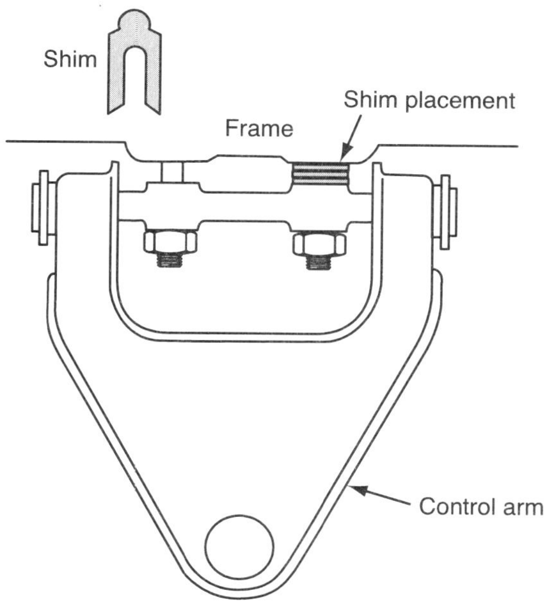


**Figure 10.** Adjuster camber with shims located between the upper control arm mounting and the outside of the frame.

On other vehicles, shims are placed between the upper control arm mounting and the outside of the frame **(Figure 10).** Increasing the shim thickness increases positive camber.

### Eccentric Cams

Some vehicles have a cam-type camber adjustment on the inner ends of the upper control arm. Other vehicles have a cam adjustment on the inner ends of the lower control arms. If the cam adjustment on the control arms does not provide enough camber adjustment, aftermarket upper control arm shaft kits are available that provide an extra 1 1/2 degree of camber adjustment. Some MacPherson strut front-suspension systems have a cam on one of the steering knuckle-to-strut bolts to adjust camber **(Figure 11).** Aftermarket camber adjustment kits that provide 2 1/2 degree of camber adjustment are also available for many MacPherson strut suspension systems. On some double wishbone front-suspension systems, a graduated cam on the inner end of the lower control arm provides a camber adjustment **(Figure 12).**

### Slotted Strut Mounts and Frames

On some strut front suspension systems, the upper strut mounts may be loosened and moved inward or outward to adjust camber. Other strut front-suspension systems do not provide a camber adjustment. For these, aftermarket manufacturers sell slotted upper strut bearing plates **(Figure 13).** If front wheel camber is not correct, check for worn components such as ball joints or upper strut mounts. Also, check the engine cradle mounts. Since the inner ends of the lower control arms are attached to the engine cradle on many FWD vehicles, a cradle that is out of position may cause improper camber readings.

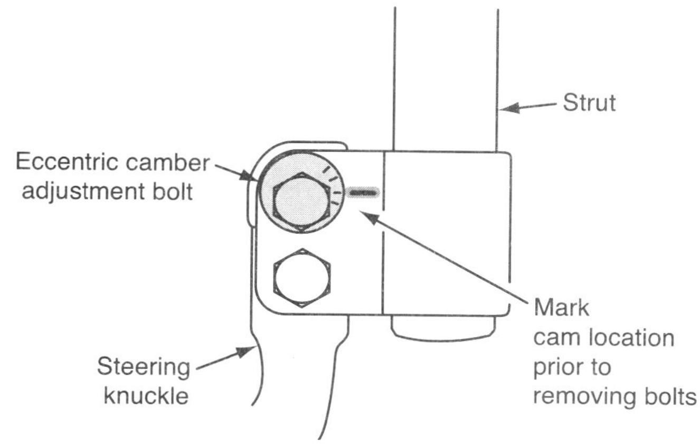


**Figure 11.** Eccentric steering knuckle-to-strut camber adjusting bolt.

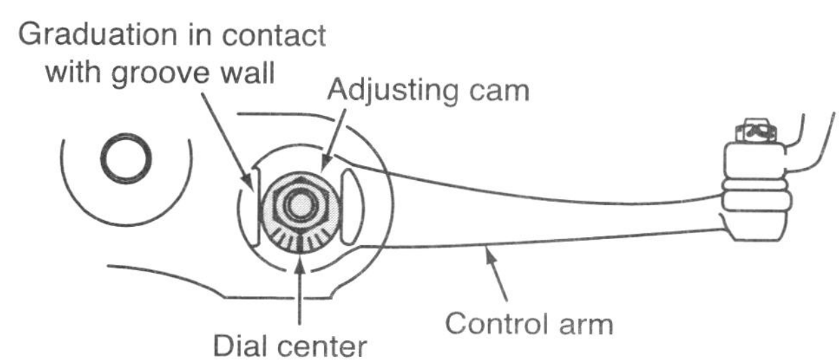


**Figure 12.** Camber adjustment for a double wishbone suspension.

**Figure 13.** A slotted upper strut bearing allows for camber adjustments.

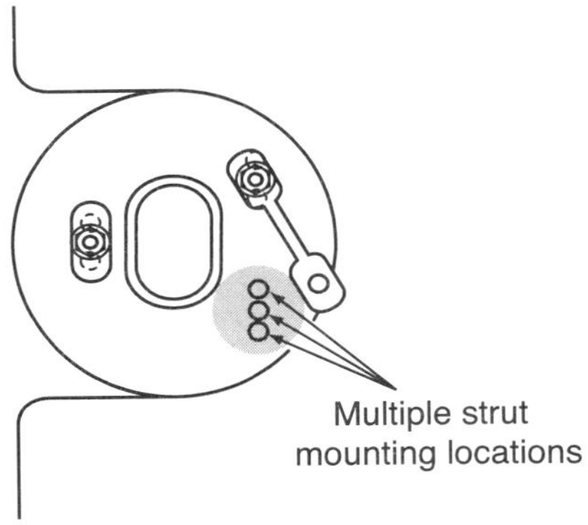


**Figure 14.** The camber plate is locked in place with a pop rivet on some upper strut mounts. The pop rivet may be removed to release the camber plate and make adjustments to camber and caster.

**Figure 15.** A camber adjusting sleeve for a twin I-beam front suspension

A camber plate is locked in place with a pop rivet on some upper strut mounts. The pop rivet is removed to release the camber plate **(Figure 14).** After the upper strut mount bolts are loosened, the mount is moved to adjust the camber. When the camber is set to specifications, the strut mount bolts should be tightened to the correct torque and a new pop rivet installed.

On some SLA front-suspension systems, the boltholes in the frame are elongated for the inner shaft on the upper control arm. When these bolts are loosened, the upper control arm can be moved inward or outward to adjust camber.

### Eccentric Bushing or Ball Joint Eccentric

In some SLA front-suspension systems, an eccentric bushing in the upper end of the steering knuckle is used to adjust camber. On other SLA systems, a ball joint eccentric is be rotated to adjust camber.

### Twin I-Beam Front-Suspension Systems

On twin I-beam front-suspension systems, a special clamp-type bending tool with a hydraulic jack is used to bend the I-beams and change the camber. The clamp and hydraulic jack must be positioned properly to adjust the camber. On some twin I-beam front-suspension systems, camber sleeves are available to adjust camber. These sleeves are located in the top hole in the outer end of the I-beams, and the upper ball joint studs extend through the openings in the camber sleeves **(Figure 15).**

> **You Should Know** *Improper use of bending equipment may result in personal injury. Always follow the equipment and vehicle manufacturer's recommended procedures.*

## Caster

Many of the same techniques used to adjust camber are also used to adjust caster. Keep in mind that without careful planning, you can change your camber adjustments while correcting caster, and vice-versa.

### Shims

On a SLA front-suspension system, the same shims are used for both camber and caster. If the shims are located between the upper control arm shaft and the inside of the frame, subtract shims from the front mounting bolt and add shims to the rear mounting bolt to increase positive caster. After caster is adjusted, the camber should be rechecked.

### Strut Rod Length Adjustment

On some suspension systems, the nuts on the forward end of the strut rod are adjusted to lengthen or shorten the strut rod to change the caster **(Figure 16).** Shorten the strut rod to increase the positive caster.

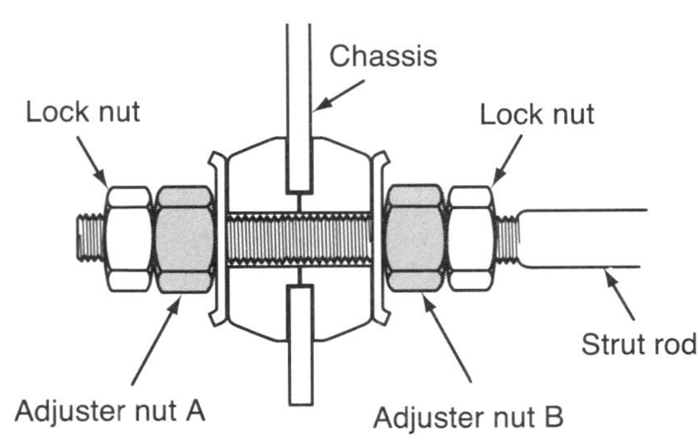


**Figure 16.** Caster may be adjusted by changing the length of the strut rod.

### Eccentric Cams

The eccentric cams on the inner ends of the upper or lower control arms may be used to adjust camber or caster. If an eccentric bushing in the outer end of the upper control arm is rotated to adjust camber, this same eccentric also adjusts caster. On some double wishbone front-suspension systems, the pivot adjuster mounting bolt nuts must be loosened under the compliance pivot, and the graduated cam can then be rotated to adjust caster.

### Slotted Strut Mounts and Frames

If a caster adjustment is required and the suspension does not have provisions for an adjustment, check for worn components such as upper strut mounts and ball joints. In addition, check the engine cradle mounts. Since the inner ends of the lower control arms are attached to the engine cradle on many front-wheel-drive cars, a cradle that is out of position may cause improper caster readings.

The slots in the frame at the upper control arm shaft mounting bolts provide camber and caster adjustment. If the upper strut mount is adjustable on a strut suspension, the mount retaining bolts are loosened and the mount moved forward or rearward to adjust caster. On some suspension systems, one of the upper strut mount boltholes must be elongated with the proper size drill bit and a round file before caster can be adjusted.

**You Should Know** *After any alignment angle adjustment, always be sure the adjustment bolts are tightened to the specified torque. Loose suspension adjustment bolts will result in improper alignment angles, steering pull, and tire wear. Loose suspension bolts may also cause reduced directional stability, which can put the driver in a dangerous position.*

### Twin I-Beam Front-Suspension Systems

The same bending tool and hydraulic jack is used to adjust camber and caster on twin I-beam front-suspension systems. The tool is installed on the radius rods connected from the I-beams to the chassis to adjust caster. If the hydraulic jack is placed near the rear end of the radius rod and the clamp arm is positioned over the center area of this rod, the rod is bent downward in the center when the jack is operated. This tilts the I-beam and caster line rearward to increase positive caster.

If the hydraulic jack is placed in the center area of the radius arm and the clamps are positioned at the ends of the arm, the arm is bent upward when the jack is operated. This tilts the I-beam and caster line forward to decrease positive caster.

On some twin I-beam front-suspension systems, sleeves can be used for camber and caster adjustments. These sleeves are located in the top hole in the outer end of the I-beams. The upper ball joint studs extend through the openings in the camber sleeves.

## Steering Axis Inclination (SAI)

Steering axis inclination (SAI) is not considered adjustable. If the SAI is not within specifications, the upper strut tower may be out of position, the lower control arm may be bent or the center crossmember or cradle, could be shifted. In most cases, these defects are caused by collision damage. When the SAI angle is correct but the camber and included angle are less than specified, the strut or spindle may be bent. The difference in the included angles on the front wheels must not exceed 1 1/2 degree.

## Setback

Computerized wheel aligners can measure the setback of the front wheels. If a FWD vehicle has experienced collision damage on the left front, the left side of the cradle or subframe may be driven rearward. This type of damage also moves the bottom of the left front strut rearward to a setback condition and shifts the caster to a negative position.

With this problem, it may be possible to adjust the caster back to the specified setting, but that does not correct the setback problem. If the caster is adjusted and the setback problem is ignored, the steering may pull to one side while driving straight, especially if the setback is severe. The cradle must be straightened and a unitized body must be pulled back to the original measurements to correct this problem. A severely bent, distorted cradle should be replaced.

Collision damage may push a front cradle sideways. This may increase the positive camber on one side of the car and move the camber toward a negative position on the opposite side. When this condition exists, inspect and correct the cradle position.

## Toe

When adjusting toe, apply penetrating oil to the tie-rod adjusting sleeves and sleeve clamp bolts. Loosen the tie-rod adjusting sleeve clamp bolts enough to allow the clamps to partially rotate. One end of each tie-rod sleeve contains a right-hand thread; the opposite end has a left-hand thread. When the tie-rod sleeve is rotated, the complete tie rod, sleeve, and tie-rod end assembly is lengthened or shortened. Use a tie-rod sleeve rotating tool to turn the sleeves until the toe in on each

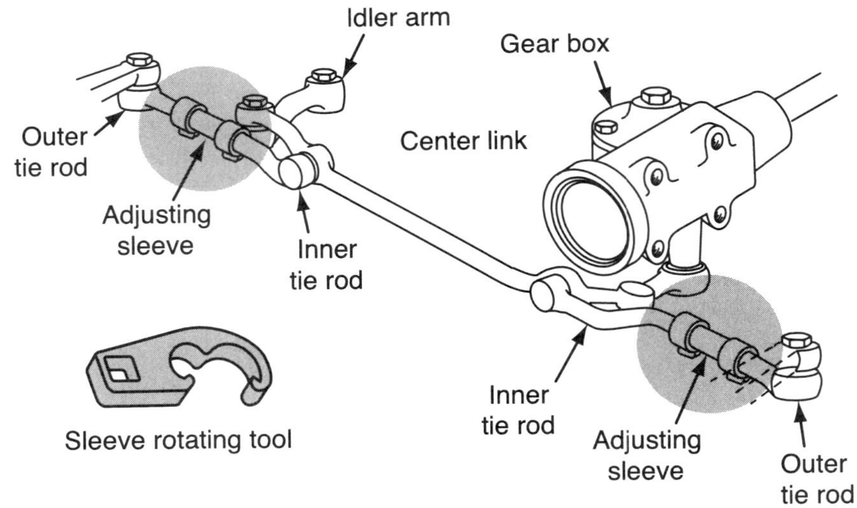


**Figure 17.** Rotate the tie-rod sleeves to adjust toe.

front wheel is equal to one-half the total toe specification **(Figure 17).**

> **You Should Know** *Do not heat tie-rod sleeves to loosen them. This may weaken the sleeves, resulting in sudden sleeve failure, loss of steering control, personal injury, and vehicle damage.*

After adjusting the toe, the adjusting sleeve clamps must be turned so the openings in the clamps are positioned away from the slots in the adjusting sleeves **(Figure 18).** Replace clamp bolts that are rusted, corroded, or damaged, and tighten these bolts to the specified torque.

On some rack-and-pinion steering gears, the tie-rod lock nut is loosened and the tie rod is rotated to adjust toe on each front wheel. Prior to rotating the tie rod, the small outer boot clamp should be removed to prevent it from twisting during tie-rod rotation. After the toe-in adjustment, the tie-rod nut must be torqued to specifications and the small boot clamp reinstalled.

On other rack-and-pinion steering gears, the tie-rod and outer tie-rod end have internal threads and a threaded adjuster is installed in these threads. One end of the adjuster has a right-hand thread, and the opposite end has a left-hand thread. Clamps on the outer tie-rod end and tie rod secure these components to the adjuster. A hex-shaped nut is designed into the center of the adjuster **(Figure 19).** After the clamps are loosened, an open-end wrench is placed on this hex nut to rotate the adjuster and change the toe setting.

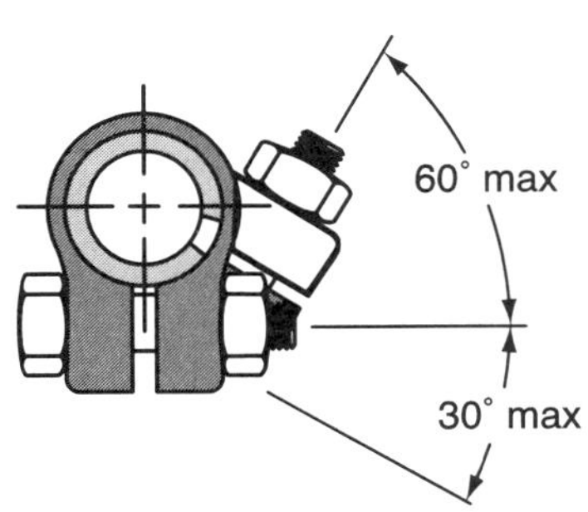


**Figure 18.** Proper tie-rod adjusting sleeve clamp position.

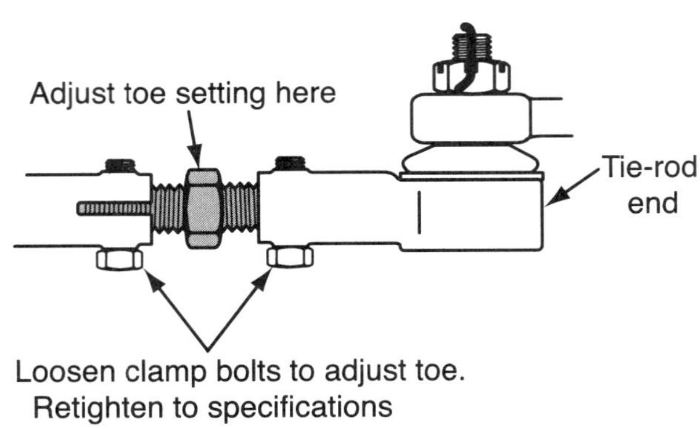


**Figure 19.** A rack-and-pinion steering gear with an external thread to adjuster and internal threads on the tie rod and outer tie-rod end.

After toe is adjusted, the toe must be within specifications with the steering wheel in its centered position. If the steering wheel is not centered, a centering adjustment is required.

> **You Should Know** *The most accurate check of a properly centered steering wheel is done while driving the vehicle straight during a road test.*

## Turning Radius

If the turning radius is not within specifications, excessive front tire tread wear occurs while cornering. Tire squeal may also occur while cornering. Make sure the other alignment angles are set to specifications before checking the turning radius or toe out in turns.

With the front wheels on the center of the turntables, apply the brakes with a brake pedal jack. Then, remove the turntable locking pins and make sure the turning radius gauges are in the zero position with the wheels straight ahead. Turn the right front wheel inward toward the center of the vehicle until the turning radius gauge indicates 20 degress. Read the turning radius on the left front wheel and record the reading. On many steering systems, the turning radius on the wheel on the inside of the turn is 2 degrees more than the turning radius on the wheel on the outside of the turn.

Turn the left front wheel inward toward the center of the vehicle until the turning radius gauge reads 20 degrees. Read the turning radius on the right front turntable radius gauge and record the readings. Compare the turning radius readings on each front wheel to specifications. Turning radius or toe out on turns is not adjustable. If the toe out on turns is not within specifications, the steering arms are bent or components, such as tie-rod ends, are worn.

## Steering Linkage Height

If the steering linkage on each side is not at the same height, bump steer may occur. When bump steer occurs, abnormal toe changes occur during wheel jounce and rebound. To check for abnormal toe changes, make sure the toe is adjusted to specifications and the steering wheel is locked. Pull the chassis downward approximately 3 in. (7.6 cm) and observe the toe change. It is acceptable if the toe setting on each front wheel remains at the original reading or if each front wheel toes in or toes out an equal small amount.

During wheel jounce, if one front wheel toes inward or outward while the opposite wheel remains at the original setting or if one front wheel toes in and the opposite front wheel toes out, the steering linkage height is not equal and must be corrected. Push the chassis upward 3 in. (7.6 cm) and check the toe change. If the steering linkage height is equal, the toe setting on each front wheel remains the same or moves the same amount to a toe-in or toe-out position.

On some vehicles, steering linkage height can be adjusted with shims between the rack-and-pinion steering gear and the chassis. Also on some vehicles, the idler arm can be moved upward or downward to adjust the steering linkage height. On other vehicles, if the steering linkage height is unequal, there is no adjustment and bent or worn steering components are indicated.

## Steering Wheel Centering Procedure

When all the front-suspension alignment angles are adjusted to specifications, road test the vehicle. When the vehicle is driven on a relatively smooth, straight road, there should be directional stability with no drift to the right or left, and the steering wheel must be centered. If steering wheel centering is necessary, follow this procedure:

1. Lift the front of the vehicle.
2. Lower the vehicle onto safety stands and place the front wheels in the straight-ahead position.
3. Use a piece of chalk and mark each tie-rod sleeve in relation to the tie rod. Then loosen the sleeve clamps.
4. Position the steering wheel in the position it was in while driving straight during the road test.
5. Then turn the steering wheel to its centered position and note the direction of the front wheels.
6. If the steering wheel points to the left side while driving straight, use a tie-rod sleeve to shorten the left tie rod and lengthen the right. One-quarter turn on a tie-rod sleeve moves the steering wheel position approximately 1 inch. Turn the tie-rod sleeves enough to bring the steering wheel to the centered position. For example, if the steering wheel spoke is 2 inches off center, turn each tie-rod sleeve one-half turn.
7. If the steering wheel favors the right side while driving straight, lengthen the left tie rod and shorten the right one.
8. Mark each tie-rod sleeve in its new position in relationship to the tie rod.
9. Make sure the sleeve clamp openings are positioned properly, and tighten the clamp bolts to the specified torque.
10. Lift the vehicle with a floor jack and remove the safety stands.
11. Lower the vehicle and check the steering wheel position during a road test.

# REAR SUSPENSION ADJUSTMENTS

Incorrect rear-wheel alignment can be caused by collision damage that results in a bent frame or distorted unitized body, a leaf-spring eye that is unwrapped or spread open, leaf-spring shackles that are broken, bent, or worn, broken leaf springs or leaf-spring center bolts, worn rear upper or lower control arm and trailing arm bushings, dislocated trailing arm brackets, and bent components such as radius rods, control arms, struts, and rear axles.

## Rear-Wheel Camber Adjustment

Improper rear-wheel camber causes excessive wear on the edges of the rear tires. Cornering stability may be affected by improper rear-wheel camber, especially while cornering at high speeds. Improper rear-wheel camber on FWD vehicles may change the understeer or oversteer characteristics of the vehicle. Rear-wheel camber may cause steering pull. For example, excessive positive camber on the right rear wheel causes the rear suspension to drift to the right and steering to pull to the left.

On some semi-independent rear-suspension systems, camber and toe are adjusted by inserting different sizes of shims between the rear spindle and the spindle mounting surface. These shims are retained by the spindle mounting bolts. The shim thickness is changed between the top or bottom of the spindle to adjust camber **(Figure 20).**

Many rear camber shims are circular and are available in a wide variety of configurations to fit various rear wheels. Some wheel aligners indicate the thickness of shim required and the proper shim position. These same shims also adjust rear-wheel toe **(Figure 21).**

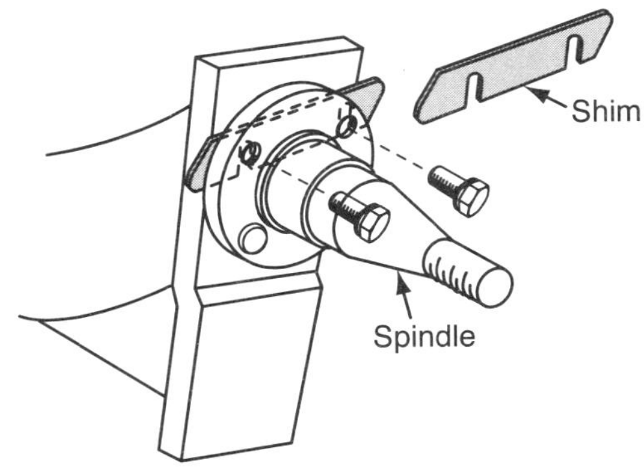


**Figure 20.** An adjustment shim for rear-wheel camber.

On some independent rear-suspension systems with a lower control arm and ball joint, the strut-to-knuckle bolts are loosened and a special camber tool installed on the lower strut-to-knuckle bolt and the rear side of the strut. Rotate the adjusting bolt in the back of the tool to adjust the camber, and then tighten the strut-to-knuckle nuts to the specified torque.

On other rear-suspension systems, an adjustment wedge can be installed between the top of the knuckle

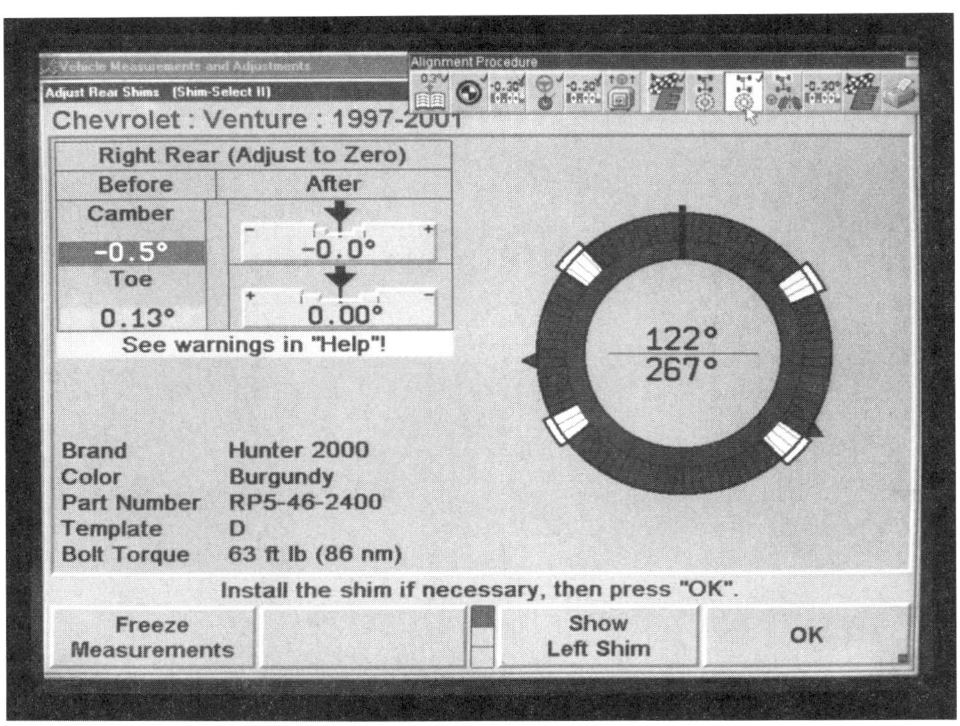


**Figure 21.** Some wheel aligners indicate the thickness of shim required and the proper shim position.

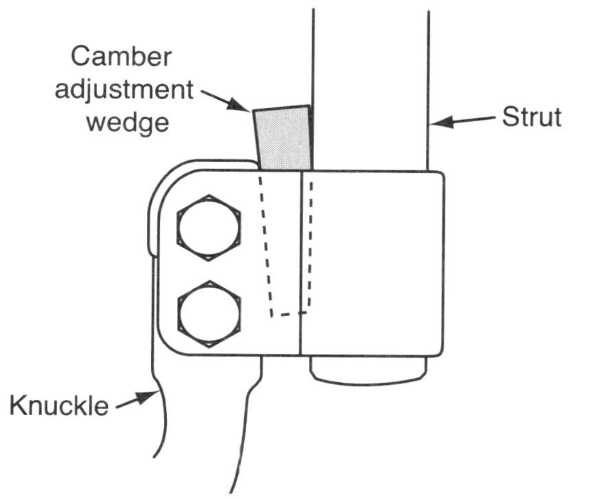


**Figure 22.** A camber adjusting wedge installed between the top of the rear knuckle and the strut.

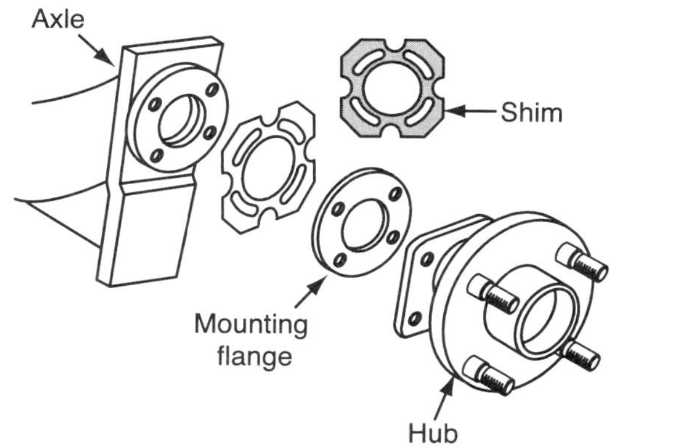


**Figure 23.** Rear-wheel toe on some vehicles is adjusted by changing the thickness of a shim between the rear spindle and the mounting flange.

and the strut to adjust camber **(Figure 22).** On some double wishbone rear-suspension systems, graduated cams on the inner ends of the lower control arms provide camber adjustment.

## Rear-Wheel Toe Adjustment

Improper rear-wheel toe moves the thrust line away from the geometric centerline, which causes steering pull and feathered tire wear. If the left rear wheel has excessive toe out, the thrust line is moved to the left of the geometric centerline. Excessive toe in on the left rear wheel moves the thrust line to the right of the geometric centerline. When the thrust line is moved to the left of the geometric centerline, the steering pulls to the right, whereas a thrust line positioned to the right of the geometric centerline results in steering pull to the left.

Improper rear-wheel toe adjustment on a non-driving rear axle causes feathered tire wear. The toe may be adjusted by changing the thickness of the tapered shim between the front and rear edges of the rear spindle mounting flange **(Figure 23).** These shims are also used to adjust rear-wheel camber.

On some independent rear suspension systems, an eccentric star wheel on the rear lower control rod is used to adjust the rear-wheel toe. On other systems with a lower control arm and ball joint, the tie rod is rotated to adjust the rear-wheel toe **(Figure 24).**

On some FWD vehicles, rear-wheel toe is adjusted by loosening the toe link bolt and moving the wheel in or out to adjust the toe. On other FWD vehicles, rear wheel toe is adjusted by rotating the inner adjustment link cam bolt **(Figure 25).**

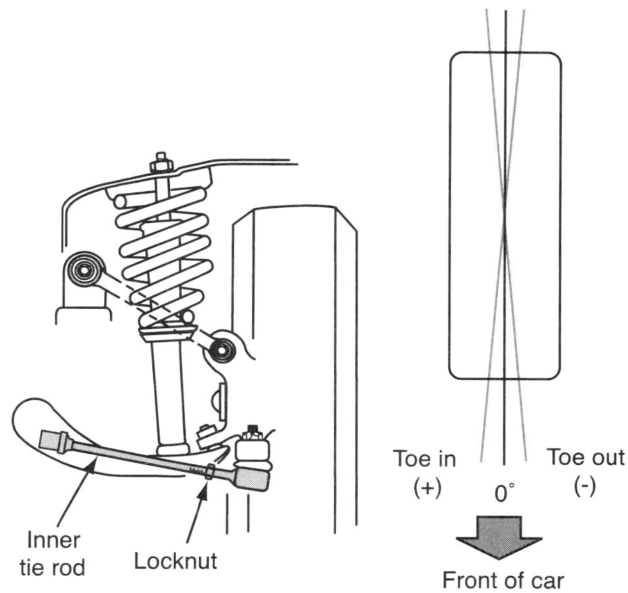


**Figure 24.** Rotating the tie rod to change tie-rod length and to adjust rear-wheel toe.

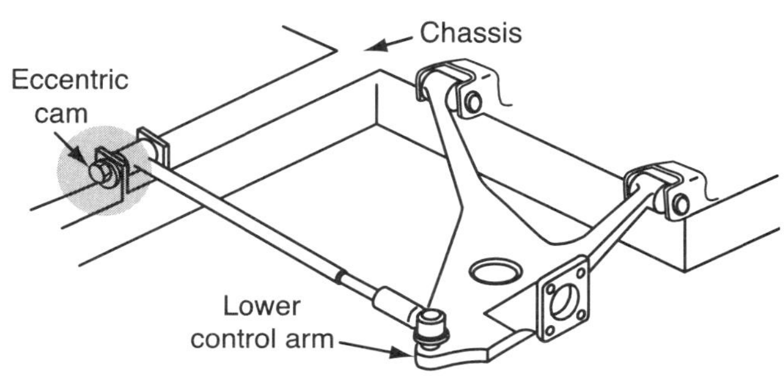


**Figure 25.** The cam bolt on an inner adjustment link is used to adjust rear toe.

## SUMMARY

- Four-wheel alignment should always begin at the rear wheels.
- An automatic or manual wheel runout compensation procedure must be performed at each wheel because wheels with excessive runout affect the position of the rim clamps and the alignment readings.
- Shim select screens show the proper location and thickness of shims required to bring some alignment angles within specifications.
- The camber and caster must be within specifications before checking toe.
- Different methods for adjusting camber include shims, eccentric cams, slotted strut mounts and frames, eccentric bushings or ball joints, and upper strut mount camber plates.
- A special tool is available to bend the I-beams and adjust some twin I-beam front suspensions.
- Caster adjustments on various suspension systems include shims, strut rod length, slotted upper strut mounts and frames, and eccentric cams.
- If the SAI is not within specifications, the upper strut tower may be out of position, the lower control arm may be bent, or the front cradle may be out of position.
- While checking front-wheel toe, the wheels must be straight ahead with the steering wheel locked and the brakes applied.
- Steering wheel centering may be necessary after a toe adjustment.
- On many rear-suspension systems of FWD vehicles, shims are positioned between the spindle and the spindle mounting surface to adjust rear-wheel camber and toe.
- On some rear-suspension systems with a lower control arm and ball joint, a special tool is installed on the strut and lower strut bolt to adjust camber.
- A camber wedge is installed between the strut and knuckle to adjust camber on some rear-suspension systems.
- A star wheel is located on the inner end of the lower control rod to adjust the toe on some rear-suspension systems.
- On other rear-suspension systems, the tie-rod nut may be loosened and the tie rod rotated to adjust rear-wheel toe.
- When all four wheels are parallel to the geometric centerline and the thrust line is positioned on the geometric centerline, the rear wheels track directly behind the front wheels.

## REVIEW QUESTIONS

1. *True or False:* When preparing to do a four-wheel alignment, the vehicle should be positioned on an alignment ramp with a slip plate under each tire.
2. Technician A says rear-wheel toe may be adjusted by changing the thickness of the tape red shim between the front and rear edges of the rear spindle mounting flange. Technician B says on systems with a lower control arm and ball joint, the nut on the tie rod is loosened and the tie rod is rotated to adjust the rear-wheel toe. Who is correct?
   A. Technician A
   B. Technician B
   C. Both Technician A and Technician B
   D. Neither Technician A nor Technician B
3. While discussing turning radius measurement, Technician A says a bent steering arm will cause the turning radius to be out of specification. Technician B says if the turning radius is not within specifications, tire tread wear is excessive while cornering. Who is correct?
   A. Technician A
   B. Technician B
   C. Both Technician A and Technician B
   D. Neither Technician A nor Technician B
4. All these statements about front-wheel toe change during wheel jounce and rebound are true *except:*
   A. If one front wheel toes in and the opposite front wheel toes out during front-wheel jounce and rebound, the tie-rod height is unequal.
   B. If both front wheels toe in or toe out a small, equal amount during front-wheel jounce and rebound, the tie rods are parallel to the lower control arms.
   C. The improper toe change during front-wheel jounce and rebound may cause bump steer.
   D. Improper toe change during front-wheel jounce and rebound may be caused by a worn upper strut mount.
5. While discussing turning radius, Technician A says incorrect turning radius may often be noted by tire squeal while cornering. Technician B says to adjust

turning radius toe out on turns with the inner wheel turned to the stop. Who is correct?

A. Technician A
B. Technician B
C. Both Technician A and Technician B
D. Neither Technician A nor Technician B

6. A customer returns to your shop and says the car continues to drift after you have completed a front end alignment. To correct this problem, you should:
   A. check the manual steering gear for possible miscentering of the sector gear.
   B. check rear-wheel alignment.
   C. ask the customer to fill the fuel tank.
   D. check the steering column flex coupling.
7. While discussing camber adjustments on a MacPherson strut front-suspension system, Technician A says an eccentric cam on the strut-to-steering-knuckle bolt may be used to adjust camber. Technician B says an eccentric on the lower ball joint stud may be used to adjust camber. Who is correct?
   A. Technician A
   B. Technician B
   C. Both Technician A and Technician B
   D. Neither Technician A nor Technician B
8. When diagnosing and adjusting front-wheel caster:
   A. on some suspensions, the radius rod should be shortened to decrease positive caster.
   B. on some twin I-beam suspensions, the I-beams may be bent to adjust caster.
   C. on some double wishbone suspensions, a graduated cam on the pivot adjuster is rotated to adjust caster.
   D. the steering pulls to the side of the suspension that has the most positive caster.
9. Unequal SAI angles on the left and right sides of the front suspension may cause:
   A. tread wear on the front tires.
   B. brake pull during sudden stops.
   C. bump steer on a rough road.
   D. steering wander while driving straight ahead.
10. All these statements about adjusting front-wheel toe are true *except:*
    A. If the positive caster is increased on the right front wheel, this wheel moves toward a toe-out position.
    B. The tie-rod sleeves may be heated with an acetylene torch to loosen them.
    C. The opening in the sleeve clamp must be positioned away from the slots in the tie-rod sleeve.
    D. The front-wheel toe should be measured with the front wheels straight ahead.

# Appendix A

## ASE PRACTICE EXAM FOR SUSPENSION AND STEERING SYSTEMS

1. After new tires and new alloy rims are installed, the owner complains about steering wander and steering pull in either direction while braking. Technician A says there may be brake fluid on the front brake linings. Technician B says the replacement rims may have a different offset than the original rims. Who is correct?
   A. Technician A
   B. Technician B
   C. Both Technician A and Technician B
   D. Neither Technician A nor Technician B
2. The outside edge of the left front tire on a rear-wheel drive car is badly scalloped. Technician A says the cause could be worn ball joints. Technician B says the cause could be incorrect tire pressure. Who is correct?
   A. Technician A
   B. Technician B
   C. Both Technician A and Technician B
   D. Neither Technician A nor Technician B
3. The owner of a large rear-wheel drive sedan says the front tires squeal loudly during low-speed turns. The most probable cause of this condition is:
   A. excessive positive camber.
   B. negative caster adjustment.
   C. improper steering axis inclination (SAI).
   D. improper turning angle.
4. A mini pickup has a severe shudder when the vehicle is started from a stop with a load in the bed. Technician A says the problem may be worn spring eyes. Technician B says the problem may be axle torque wrap-up. Who is correct?
   A. Technician A
   B. Technician B
   C. Both Technician A and Technician B
   D. Neither Technician A nor Technician B
5. A cyclic noise (moaning, whining, or howling) that changes pitch with road speed and is present whenever the vehicle is in motion may be caused by any of the following *except:*
   A. worn differential gears.
   B. rear axle bearings.
   C. incorrect driveshaft runout.
   D. off-road tire tread pattern.
6. Technician A says hard steering may be caused by low hydraulic pressure due to a stuck flow control valve in the pump. Technician B says this may be caused by low hydraulic pressure due to a worn steering gear piston ring or housing bore. Who is correct?
   A. Technician A
   B. Technician B
   C. Both Technician A and Technician B
   D. Neither Technician A nor Technician B
7. While discussing tire tread wear, Technician A says a scalloped pattern of tire wear indicates an out-of-round wheel or tire. Technician B says uneven wear on one side of a tire may indicate radial force variation. Who is correct?
   A. Technician A
   B. Technician B
   C. Both Technician A and Technician B
   D. Neither Technician A nor Technician B
8. A front-wheel drive car makes chattering noises only during a hard right turn. Technician A says the problem is probably caused by an upper spring seat binding. Technician B says the problem is probably caused by a defective strut bearing. Who is correct?
   A. Technician A
   B. Technician B
   C. Both Technician A and Technician B
   D. Neither Technician A nor Technician B

9. A customer says he equipped the rear of his 1998 minivan with inflatable air shocks to handle heavier trailer hitch tongue weights, but the van's rear end still drags when he hooks up his ski boat trailer. Technician A says the problem is the air shocks are not heavy duty. Technician B says the problem is heavy-duty springs are also needed with the air shocks. Who is correct?
   A. Technician A
   B. Technician B
   C. Both Technician A and Technician B
   D. Neither Technician A nor Technician B
10. When a vehicle pulls to one side, any of the following problems may be the cause *except:*
    A. worn ball joints.
    B. reduced curb height.
    C. bent strut rod.
    D. improper turning angle.
11. A customer says when he applied the brakes hard on his front-wheel drive car, "the whole car shook." Technician A says the problem could be worn ball joints. Technician B says the problem could be worn or loose strut rod bushings. Who is correct?
    A. Technician A
    B. Technician B
    C. Both Technician A and Technician B
    D. Neither Technician A nor Technician B
12. A customer with a sport utility vehicle says after an off-road outing, his vehicle pulls to the left on acceleration. Which of the following could cause this problem?
    A. Broken leaf-spring center bolt
    B. Loose steering unit
    C. Stuck brake pad
    D. Stuck in 4WD
13. While discussing electronic air suspension, Technician A says the compressor must be running when a corner of the vehicle is lifted from the ground. Technician B says on a car with electronic air suspension, the switch should be in the "on" position. Who is correct?
    A. Technician A
    B. Technician B
    C. Both Technician A and Technician B
    D. Neither Technician A nor Technician B
14. Front wheel "shimmy"—a side-to-side movement of the front wheels that is felt in the steering wheel—may be caused by any of the following *except:*
    A. worn tie-rod ends.
    B. wheel/tire imbalance.
    C. rack bushings and rack alignment.
    D. tight sector shaft adjustment.
15. A customer says her rear-wheel drive car "nearly shakes the wheel out of my hands" on rough roads. All of the following could cause or contribute to this problem *except:*
    A. worn ball joints.
    B. a damaged axle housing.
    C. a worn steering damper.
    D. worn or loose gear box mounts.
16. Technician A says raising the suspension height in the rear of a vehicle will affect the front suspension geometry. Technician B says raising the suspension height in the rear of a vehicle could lead to premature rear-spring failure. Who is correct?
    A. Technician A
    B. Technician B
    C. Both Technician A and Technician B
    D. Neither Technician A nor Technician B
17. A vehicle with recirculating ball steering has excessive steering wheel freeplay. Technician A says a loose worm bearing preload adjustment may cause the problem. Technician B says loose column U-joints may cause the problem. Who is correct?
    A. Technician A
    B. Technician B
    C. Both Technician A and Technician B
    D. Neither Technician A nor Technician B
18. The engine has recently been replaced in a front-wheel drive car with power rack-and-pinion steering. The customer now complains about excessive steering effort. A preliminary check of the steering revealed the fluid level was okay and the belts were not slipping. Technician A says perhaps the rack was knocked out of alignment when the engine was installed. Technician B says perhaps the pressure line was bent or pinched when the engine was installed. Who is correct?
    A. Technician A
    B. Technician B
    C. Both Technician A and Technician B
    D. Neither Technician A nor Technician B
19. If a power steering system has proper pump pressure but steering effort is more than specified, all of the following could cause or contribute to the problem *except:*
    A. worn front tires.
    B. cold fluid.
    C. hose restrictions.
    D. a steering gear malfunction.

20. The customer says his vehicle has a rapid "thumping" noise and a vibration in the steering wheel that is most noticeable when the vehicle is at a steady speed in a long curve. Technician A says the problem could be caused by a tire defect. Technician B says the problem could be improper dynamic wheel balance. Who is correct?
    A. Technician A
    B. Technician B
    C. Both Technician A and Technician B
    D. Neither Technician A nor Technician B
21. A customer returns to your shop and says she just had steering work done on the car the day before and it has been pulling to the left since then. A check of the records shows that lower control arm ball joints and both rubber encapsulated tie-rod ends were replaced. Technician A says the tie-rod end may have been installed with the wheels turned slightly away from center. Technician B says the alignment should have been checked after installing the ball joints. Who is correct?
    A. Technician A
    B. Technician B
    C. Both Technician A and Technician B
    D. Neither Technician A nor Technician B
22. While discussing excessive vehicle noise and vibration, Technician A says suspension and driveline vibrations may be amplified by the resonance of the body or steering wheel. Technician B says noises that phase in and out of hearing range may be two separate noises that alternately amplify then cancel each other. Who is correct?
    A. Technician A
    B. Technician B
    C. Both Technician A and Technician B
    D. Neither Technician A nor Technician B
23. Power steering fluid is leaking from the high-pressure outlet fitting on a nonsubmerged power steering pump. To correct this problem, you should:
    A. replace the cover seal.
    B. replace the O-ring.
    C. replace the fitting.
    D. replace the check valve.
24. A customer says the steering wheel of her front-wheel drive car does not return to a center turn. Technician A says the problem is memory steer and could be caused by binding in the steering shaft universal joints. Technician B says memory steer in a front-wheel drive car may also be caused by binding upper strut mounts. Who is correct?
    A. Technician A
    B. Technician B
    C. Both Technician A and Technician B
    D. Neither Technician A nor Technician B
25. A sport utility vehicle with a power recirculating ball steering gear makes a rattling noise when the steering wheel is turned. This noise is noticeable whether the car is in motion or standing still. Technician A says the cause of the problem could be a loose pitman arm or shaft. Technician B says the cause of the problem could be the high-pressure power steering hose touching some part of the vehicle. Who is correct?
    A. Technician A
    B. Technician B
    C. Both Technician A and Technician B
    D. Neither Technician A nor Technician B
26.. What can cause a rear-wheel drive car to have a steering problem?
    A. Tire tread wear
    B. Loose outer tie-rod ends
    C. Worn idler arm
    D. All of the above
27. Many vehicles have too much play in the steering wheel, usually caused by loose steering components. A check of these components should be:
    A. done with vehicle weight on the tires.
    B. done with a computerized alignment system.
    C. completed after a front wheel alignment.
    D. performed with the vehicle raised and the wheels unsupported.
28. While diagnosing shock absorbers, Technician A says the "bounce test" is the sure way to quickly pinpoint bad shocks and struts. Technician B says oil film on the lower chamber of a shock or strut indicates leakage and requires replacement. Who is correct?
    A. Technician A
    B. Technician B
    C. Both Technician A and Technician B
    D. Neither Technician A nor Technician B
29. While discussing front suspension angles, Technician A says turning angle is fixed by the steering and suspension system design. Technician B says toe out on turns may be adjusted with the steering stopper bolts. Who is correct?
    A. Technician A
    B. Technician B
    C. Both Technician A and Technician B
    D. Neither Technician A nor Technician B
30. Technician A says worn stabilizer bushings may cause squeaks as the vehicle moves down a bumpy road. Technician B says worn or dry control arm bushings may squeak as the vehicle moves down a bumpy road. Who is correct?
    A. Technician A
    B. Technician B
    C. Both Technician A and Technician B
    D. Neither Technician A nor Technician B

31. While diagnosing tire thump, Technician A says this problem may be caused by cupped tire treads. Technician B says a heavy spot in the tire may cause the problem. Who is correct?
    A. Technician A
    B. Technician B
    C. Both Technician A and Technician B
    D. Neither Technician A nor Technician B
32. Technician A says underinflation causes excessive wear on a tire's outer edge. Technician B says underinflation causes a harsher-than-normal ride. Who is correct?
    A. Technician A
    B. Technician B
    C. Both Technician A and Technician B
    D. Neither Technician A nor Technician B
33. Technician A says improper static balance will cause wheel shimmy. Technician B says excessive wheel lateral runout will cause wheel shimmy. Who is correct?
    A. Technician A
    B. Technician B
    C. Both Technician A and Technician B
    D. Neither Technician A nor Technician B
34. Technician A says excessive play in the steering wheel is apparent when there is too much steering wheel movement before the wheels begin to turn. Technician B says a small amount of play in the steering wheel is normal. Who is correct?
    A. Technician A
    B. Technician B
    C. Both Technician A and Technician B
    D. Neither Technician A nor Technician B
35. Technician A says abnormal steering feedback can be caused by a faulty power steering pump. Technician B says feedback can be caused by damaged or worn steering column bearings. Who is correct?
    A. Technician A
    B. Technician B
    C. Both Technician A and Technician B
    D. Neither Technician A nor Technician B
36. Which of the following is the *least* likely to cause hard steering?
    A. A faulty power steering pump
    B. Damaged or faulty steering column bearings
    C. Seized steering column U-joints
    D. Overinflated tires
37. Technician A says steering pull is a tugging sensation felt at the steering wheel. The driver must push the steering wheel in the opposite direction to keep the vehicle going straight. Technician B says drifting is a condition where the vehicle quickly moves to one side of the road when the driver's hands are taken off the steering wheel. Who is correct?
    A. Technician A
    B. Technician B
    C. Both Technician A and Technician B
    D. Neither Technician A nor Technician B
38. Which of the following would be the least likely cause for a steering wheel's resistance to return to center after a turn?
    A. Binding steering column U-joints
    B. Binding steering gear
    C. Low power steering pump pressure
    D. Loose, damaged, or worn suspension parts
39. Constant velocity universal joints are packed with:
    A. as much grease as can be forced into the boot.
    B. a specified type and amount of grease.
    C. 90-weight gear oil.
    D. none of the above.
40. If a front-wheel drive car pulls to one side, the trouble could be:
    A. a bent axle shaft.
    B. a defective CV joint.
    C. a dragging wheel bearing.
    D. all of the above.

# Appendix B

## NATEF CORRELATIONS

### A. General Suspension and Steering Systems Diagnosis

A.1 Identify and interpret suspension and steering concern; determine necessary action. See Chapter 9.

A.2 Research applicable vehicle and service information, such as suspension and steering system operation, vehicle service history, service precautions, and technical service bulletins. See Chapter 3.

A.3 Locate and interpret vehicle and major component identification numbers (VIN, vehicle certification labels, calibration decals). See Chapter 4.

### B. Steering Systems Diagnosis and Repair

B.1 Disable and enable supplemental restraint system (SRS). See Chapter 38.

B.2 Remove and replace steering wheel; center/time supplemental restraint system (SRS) coil (clock spring). See Chapter 39.

B.3 Diagnose steering column noises, looseness, and binding concerns (including tilt mechanisms); determine necessary action. See Chapter 39.

B.4 Diagnose power steering gear (non-rack-and-pinion) binding, uneven turning effort, looseness, hard steering, and fluid leakage concerns; determine necessary action. See Chapter 46.

B.5 Diagnose power steering gear (rack-and-pinion) binding, uneven turning effort, looseness, hard steering, and fluid leakage concerns; determine necessary action. See Chapter 46.

B.6 Inspect steering shaft universal joint(s), flexible coupling(s), collapsible column, lock cylinder mechanism, and steering wheel; perform necessary action. See Chapter 39.

B.7 Adjust manual or power non-rack-and-pinion worm bearing preload and sector lash. See Chapter 40.

B.8 Remove and replace manual or power rack-and-pinion steering gear; inspect mounting bushings and brackets. See Chapter 41.

B.9. Inspect and replace manual or power rack-and-pinion steering gear inner tie-rod ends (sockets) and bellows boots. See Chapter 41.

B.10 Inspect power steering fluid levels and condition. See Chapter 46.

B.11 Flush, fill, and bleed power steering system. See Chapter 46.

B.12 Diagnose power steering fluid leakage; determine necessary action. See Chapter 46.

B.13 Remove, inspect, replace, and adjust power steering pump belt. See Chapter 46.

B.14 Remove and reinstall power steering pump. See Chapter 47.

B.15 Remove and reinstall power steering pump pulley; check pulley and belt alignment. See Chapter 47.

B.16 Inspect and replace power steering hoses and fittings. See Chapter 46.

B.17 Inspect and replace pitman arm, relay (center link/intermediate) rod, idler arm and mountings, and steering linkage damper. See Chapter 39.

B.18 Inspect, replace, and adjust tie-rod ends (sockets), tie-rod sleeves, and clamps. See Chapter 39.

B.19 Diagnose and adjust components of electronically controlled steering systems using a scan tool; determine necessary action. See Chapter 50.

## C. Suspension Systems Diagnosis and Repair

*1. Front Suspension*

C.1.1 Diagnose SLA suspension system noises, body sway, and uneven riding height concerns; determine necessary action. See Chapter 30.

C.1.2 Diagnose strut suspension system noises, body sway, and uneven riding height concerns; determine necessary action. See Chapter 30.

C.1.3 Remove, inspect, and install upper and lower control arms, bushings, shafts, and rebound bumpers. See Chapter 31.

C.1.4 Remove, inspect, install, and adjust strut rods (compression/tension) and bushings. See Chapter 31.

C.1.5 Remove, inspect, and install upper and lower ball joints. See Chapter 31.

C.1.6 Remove, inspect, and install steering knuckle assemblies. See Chapter 31.

C.1.7 Remove, inspect, and install SLA suspension system coil springs and spring insulators. See Chapter 24.

C.1.8 Remove, inspect, install, and adjust suspension system torsion bars; inspect mounts. See Chapter 24.

C.1.9 Remove, inspect, and install stabilizer bar bushings, brackets, and links. See Chapter 24.

C.1.10 Remove, inspect, and install strut cartridge or assembly, strut coil spring, insulators (silencers), and upper strut bearing mount. See Chapter 25.

C.1.11 Lubricate suspension and steering systems. See Chapter 9.

*2. Rear Suspension*

C.2.1 Remove, inspect, and install coil springs and spring insulators. See Chapter 24.

C.2.2 Remove, inspect, and install transverse links, control arms, bushings, and mounts. See Chapter 33.

C.2.3 Remove, inspect, and install leaf springs, leaf-spring insulators (silencers), shackles, brackets, bushings, and mounts. See Chapter 24.

C.2.4 Remove, inspect, and install strut cartridge or assembly, strut coil spring, and insulators (silencers). See Chapter 25.

*3. Miscellaneous Service*

C.3.1 Inspect, remove, and replace shock absorbers. See Chapter 24.

C.3.2 Remove, inspect, and service or replace front- and rear-wheel bearings. See Chapter 19.

C.3.2 Remove, inspect, and service or replace front- and rear-wheel bearings. See Chapter 20.

C.3.3 Test and diagnose components of electronically controlled suspension systems using a scan tool; determine necessary action. See Chapter 34.

## D. Wheel Alignment Diagnosis, Adjustment, and Repair

D.1 Differentiate between steering and suspension concerns using principles of steering geometry (caster, camber, toe, etc.). See Chapter 52.

D.2 Diagnose vehicle wander, drift, pull, hard steering, bump steer, memory steer, torque steer, and steering return concerns; determine necessary action. See Chapter 9.

D.3 Perform prealignment inspection; perform necessary action. See Chapter 52.

D.4 Measure vehicle riding height; determine necessary action. See Chapter 9.

D.5 Check and adjust front- and rear-wheel camber; perform necessary action. See Chapter 53.

D.6 Check and adjust caster; perform necessary action. See Chapter 53.

D.7 Check and adjust front-wheel toe; adjust as needed. See Chapter 53.

D.8 Center steering wheel. See Chapter 53.

D.9 Check toe out on turns (turning radius); determine necessary action. See Chapter 53.

D.10 Check SAI (steering axis inclination) and included angle; determine necessary action. See Chapter 53.

D.11 Check and adjust rear-wheel toe. See Chapter 53.

D.12 Check rear-wheel thrust angle; determine necessary action. See Chapter 53.

D.13 Check for front-wheel setback; determine necessary action. See Chapter 53.

D.14 Check front cradle (subframe) alignment; determine necessary action. See Chapter 53.

## E. Wheel and Tire Diagnosis and Repair

E.1 Diagnose tire wear patterns; determine necessary action. See Chapter 12.

E.2 Inspect tires; check and adjust air pressure. See Chapter 12.

E.3 Diagnose wheel/tire vibration, shimmy, and noise; determine necessary action. See Chapter 12.

E.4 Rotate tires according to manufacturer's recommendations. See Chapter 13.

E.5 Measure wheel, tire, axle, and hub runout; determine necessary action. See Chapter 13.

E.6 Diagnose tire pull (lead) problem; determine necessary action. See Chapter 12.

E.7 Balance wheel and tire assembly (static and dynamic). See Chapter 14.

E.8 Dismount, inspect, repair, and remount tire on wheel. See Chapter 13.

E.9 Reinstall wheel; torque lug nuts. See Chapter 13.

E.10 Inspect and repair tire. See Chapter 13.

# Appendix C

## USCS AND METRIC CONVERSIONS

Linear Measurements
- 1 meter (m) = 39.37 inches (in.)
- 1 centimeter (cm) = 0.3937 inch
- 1 millimeter (mm) = 0.03937 inch
- 1 inch = 2.54 centimeters
- 1 inch = 25.4 millimeters
- 1 mile = 1.6093 kilometers

Area (Square) Measurements
- 1 square inch = 6.452 square centimeters
- 1 square centimeter = 0.155 square inches

Volume Measurements
- 1 cubic inch = 16.387 cubic centimeters
- 1000 cubic centimeters = 1 liter (l)
- 1 liter = 61.02 cubic inches
- 1 gallon (gal.) = 3.7854 liters

Weight Measurements
- 1 ounce (oz.) = 28.3495 grams (g)
- 1 pound (lb.) = 453.59 grams
- 1000 grams = 1 kilogram
- 1 kilogram = 2.2046 pounds

Temperature Measurements
- 1 degree Fahrenheit (°F) = $\frac{9}{5}$ (°C + 32 degrees)
- 1 degree Celsius (°C) = $\frac{5}{9}$ (°F – 32 degrees)

Pressure Measurements
- 1 pound per square inch (psi) = 0.07031 kilograms (kg) per square centimeter
- 1 kilogram per square centimeter = 14.22334 pounds per square inch
- 1 bar = 14.504 pounds per square inch
- 1 pound per square inch = 0.06895 bars

Torque Measurements
- 10 pounds per foot = 13.558 newton meters (N•m)
- 1 N•m = 0.7375 ft.-lb
- 1 ft.-lb = 0.138 kg m
- 1 kg cm = 7.233 ft.-lb
- 10 kg cm = 0.98 N•m

# Appendix D

## FRACTIONS OF INCHES TO DECIMAL AND METRIC EQUIVALENTS

| Fraction | Decimal (in) | Metric (mm) |
|---|---|---|
| 1/64 | 0.015625 | 00.39688 |
| 1/32 | 0.03125 | 00.79375 |
| 3/64 | 0.046875 | 01.19062 |
| 1/16 | 0.0625 | 01.58750 |
| 5/64 | 0.078125 | 01.98437 |
| 3/32 | 0.09375 | 02.38125 |
| 7/64 | 0.109375 | 02.77812 |
| **1/8** | **0.125** | **03.1750** |
| 9/64 | 0.140625 | 03.57187 |
| 5/32 | 0.15625 | 03.96875 |
| 11/64 | 0.171875 | 04.36562 |
| 3/16 | 0.1875 | 04.76250 |
| 13/64 | 0.203125 | 05.15937 |
| 7/32 | 0.21875 | 05.55625 |
| 15/64 | 0.234375 | 05.95312 |
| **1/4** | **0.250** | **06.35000** |
| 17/64 | 0.265625 | 06.74687 |
| 9/32 | 0.28125 | 07.14375 |
| 19/64 | 0.296875 | 07.54062 |
| 5/16 | 0.3125 | 07.93750 |
| 21/64 | 0.328125 | 08.33437 |
| 11/32 | 0.34375 | 08.73125 |
| 23/64 | 0.359375 | 09.12812 |
| **3/8** | **0.375** | **09.52500** |
| 25/64 | 0.390625 | 09.92187 |
| 13/32 | 0.40625 | 10.31875 |
| 27/64 | 0.421875 | 10.71562 |
| 7/16 | 0.4375 | 11.11250 |
| 29/64 | 0.453125 | 11.50937 |
| 15/32 | 0.46875 | 11.90625 |
| 31/64 | 0.484375 | 12.30312 |
| **1/2** | **0.500** | **12.7000** |
| 33/64 | 0.515625 | 13.09687 |
| 17/32 | 0.53125 | 13.49375 |
| 35/64 | 0.546875 | 13.89062 |
| 17/16 | 0.5625 | 14.28750 |
| 37/64 | 0.578125 | 14.68437 |
| 19/32 | 0.59375 | 15.08125 |
| 39/64 | 0.609375 | 15.47812 |
| **5/8** | **0.625** | **15.8750** |
| 41/64 | 0.640625 | 16.27187 |
| 21/32 | 0.65625 | 16.66875 |
| 43/64 | 0.671875 | 17.06562 |
| 11/16 | 0.6875 | 17.46250 |
| 45/64 | 0.703125 | 17.85937 |
| 23/32 | 0.71875 | 18.25625 |
| 47/64 | 0.734375 | 18.65312 |
| **3/4** | **0.750** | **19.05000** |
| 49/64 | 0.765625 | 19.44687 |
| 25/32 | 0.78125 | 19.84375 |
| 51/64 | 0.796875 | 20.24062 |
| 13/16 | 0.8125 | 20.63750 |
| 53/64 | 0.828125 | 21.03437 |
| 27/32 | 0.84375 | 21.43125 |
| 55/64 | 0.859375 | 21.82812 |
| **7/8** | **0.875** | **22.22500** |
| 57/64 | 0.890625 | 22.62187 |
| 29/32 | 0.90625 | 23.01875 |
| 59/64 | 0.921875 | 23.41562 |
| 15/16 | 0.9375 | 23.81250 |
| 61/64 | 0.953125 | 24.20937 |
| 31/32 | 0.96875 | 24.60625 |
| 63/64 | 0.984375 | 25.00312 |
| **2/2=1** | **1.000** | **25.40000** |

# Appendix E

## GENERAL TORQUE SPECIFICATIONS

Note: The values in this chart should only be used when manufacturer's specifications are *not* available. Also, the values are valid only when SAE 10 oil is used to lubricate the threads of the bolt.

| Bolt Diameter in Inches | Torque: ft.-lb SAE 2 | SAE 5 | SAE 8 |
|---|---|---|---|
| 1/4 | 7 | 10 | 14 |
| 5/16 | 14 | 21 | 30 |
| 3/8 | 24 | 37 | 52 |
| 7/16 | 39 | 60 | 84 |
| 1/2 | 59 | 90 | 128 |
| 9/16 | 85 | 130 | 184 |
| 5/8 | 117 | 180 | 255 |
| 3/4 | 205 | 320 | 450 |
| 7/8 | 200 | 515 | 730 |
| 1 | 300 | 775 | 1090 |

| Bolt Diameter in Millimeters | Torque: kg cm* kg m Property Class: | | | | | | | | | |
|---|---|---|---|---|---|---|---|---|---|---|
| | **4.6** | **4.8** | **5.6** | **5.8** | **6.6** | **6.8** | **6.9** | **8.8** | **10.9** | **12.9** |
| 6 | 49* | 63* | 61* | 79* | 74* | 95* | 103* | 126* | 172* | 206* |
| 8 | 119* | 153* | 148* | 178* | 178* | 230* | 250* | 306* | 417* | 500* |
| 10 | 235* | 303* | 294* | 379* | 353* | 455* | 495* | 606* | 8.2 | 10 |
| 12 | 411* | 529* | 427* | 662* | 616* | 7.9 | 8.6 | 10.5 | 14 | 17 |
| 14 | 654* | 8.4 | 8.2 | 10.5 | 10 | 12 | 13 | 17 | 23 | 27 |
| 16 | 10 | 13 | 12 | 16 | 15 | 20 | 21 | 26 | 36 | 43 |
| 18 | 14 | 18 | 17 | 23 | 21 | 27 | 30 | 36 | 49 | 59 |
| 22 | 27 | 35 | 34 | 44 | 41 | 52 | 57 | 70 | 95 | 114 |

# Bilingual Glossary

**Abrasion** Wearing or rubbing away of part.
***Abracion*** Gastar o raspar en alguna parte.

**Acceleration** An increase in velocity or speed.
***Acelerar*** Un incremento en velocidad o carrera.

**Accelerometer** An input sensor that senses vehicle acceleration in a computer-controlled suspension system.
***Aceleradometro*** El sensorador que siente la aceleracion del vehiculo en un sistema de suspencion controlado por computadora.

**Active suspension system CA** A computer controlled suspension system with double-acting solenoids at each wheel.
***Sistema de suspencion activo*** Sistema de suspencion activo controlado por computadora con solenoides en cada rueda.

**Actuator** A control device that delivers mechanical action in response to an electrical signal.
***Actuador*** Control devisador que envia accion mecanica en respuesta a una senal electrica.

**Adaptive Learning** The ability of a computer to monitor the drivers' habits and the operating conditions of its system and make adjustments to its program to correct for them.
***Aprendizaje Adaptivo*** La habilidad de una computadora de detectar los habitos del chofer y las condiciones de operacion del sistema y hacer ajustes al programa para corregir.

**Additive** In automotive oils, a material added to the oil to give if certain properties; for example, a material added to engine oil to lessen its tendency to congeal to thicken at low temperatures.
***Aditivo*** En aceites para carro, es un material agregado al aceite para dar ciertas propiedades, por ejemplo: un material agregado al aceite del motor para disminuir la tendencia a cuajar o engruesar a bajas temperatures.

**Adhesives** Chemicals used to hold gaskets in place during the assembly of an engine. They also aid the gasket in maintaining a tight seal by filling in the small irregularities on the surfaces and by preventing the gasket from shifting due to engine vibration.
***Adhesivos*** Quimicos usados para mantener las partes en su lugar durante el asamble del motor. Tambien ayudan a las partes a mantenerse fuertemente selladas sintiendo las irregularidades mas pequenas en las superficies, y previenen que las partes se muevan con la vibracion del motor.

**Adjustable strut** strut with a manually operated adjustment for strut firmness.
***Circuito adjustable*** Un circuito operado manualmente para firmesa del mismo circuito.

**Aeration** The process of mixing air into a liquid.
***Airacion*** El proseso de dissolver aire con liquido.

**Air bag deployment module** The air bag and deployment canister assembly that is mounted in the steering wheel for the drivers side air bag or in the dash panel for the passenger's side air bag.
***Bolsa de aire Modula de despliege*** El asamble de la lata de despliege de la bolsa de aire que esta montada en la manejadera, en el lado del chofer o en el panel del lado del pasajero para la bolsa de aire.

**Air bag system** A system that uses impact sensors, vehicles on-board computer, an inflation module, and a nylon bag in the steering column and dash to protect the driver and passenger during a head on collision.
***Sistema de Bolsa de Aire*** Sistema que usa sensores de impacto de computadora en el vehiculo, modula de inflacion y una bolsa de nylon en la columna de la manejadera, pera proteger al chofer y al pasajero durante un impacto.

**Air spring** An air filled membrane that replaces the conventional coil springs in an air suspension system.
***Resorte de Aire*** Una membrana llena de aire que remplaza el resorte convencional en un sistema de suspencion de aire.

**Air spring solenoid valve** An electrically operated solenoid that allows air to flow in and out of an air spring.
***Resorte de vavluula de solenoide de Aire*** Solenoide operado electricamente, que permite que el aire entre y salga del resorte de aire.

**Alignment** An adjustment to a line or made to bring into a line.
***Alineamiento*** Ajuste de alineamiento hecho para ponerlo en una linea.

**All season tires** Tires with special tread designed to improve traction on snow or ice while providing acceptable noise levels on smooth road surfaces.
***Llantas para toda ocacion*** Lantas disenadas con una clase de cordon especial para mejorar la traccion en la nieve o hielo, proporcionando niveles de ruido acceptable en superficie de caminos planos.

**All-wheel drive (AWD)** System of driving up to four wheels on the vehicle based on traction conditions changes direction.
***Traccion de cuatro llantas (AWD)*** Sistema de manejar de cuatro llantas en el vehiculo basado en condicion de traccion.

**Alternating current (AC)** Electrical current that changes direction between positive and negative.
***Corriente de alternador (CA)*** Corriente electrica que cambia de positivo a negativo.

**Alternator** An alternating current generator that produces electrical current and forces it into the battery to recharge it.
***Alternador*** Un AC (Alternador de Corriente) generador que produce corriente electrica y la lleva a la bateria para cargarla.

**Ammeter** The instrument used to measure electrical current flow in a circuit.
***Medidor*** El intrumento usado para medir la corriente elctrica que sale de un cirquito.

**Ampere** The unit for measuring electrical current:usually called an amp.
***Amperor*** La unidad que se usa para medir corriente electrica usuamente llamada amp.

**Amplitude** A measurement of a vibration's intensity.The height of a waveform signal strength, or the maximum measured value of a signal.
***Amplitud*** Medimiento de la intensidad de vibracion. El peso de una senal de fuerza o la medida maxima del valor de una senal.

**Analog** A nondigital measuring method that uses a needle to indicate readings. A typical dashboard guage with a moving needle is an analog instrument.
***Analo*** Un metodo de medida no digital que usa una aguja para senalar medidas con aguja que se mueve, es un instrumento analo.

**Analog signal** A voltage signal that varies within a given range (from high to low,including all points in between).
***Senal de analo*** Una senal de voltage que varia en su rango (De arriba a abajo incluyendo los puntos de enmedio).

**Antifriction bearing** A bearing desaigned to reduce friction.This type of bearing normally uses ball or roller inserts to reduce the friction.
***Soporte de ant friccion*** Un soporte disenado para el rose de partes. Este tipo de rose es usualmente como bola o rollon insertado que reduce el rose o friccion.

**Antilock brake control module** The computer that controls the ABS operation. May be used on some systems for traction control.
***Modula de cotroldel freno*** La computdora que controla la operacion ABS. Puede ser usado para sistemas de control de rose o friction.

**Antilock brake system (ABS)** Abrake system that modulates hydraulic pressure to one or more systems for traction control.
***Sistema Antilock del freno (ABS)*** Un sistema de freno que modula presion hidraulica a uno o mas ejes, para prevenir que esos ejes se atranquen cuando frenan.

**Antiseize compound** A type of lubrication that prevents dissimilar metals from reacting with one another and seizing.
***Antiasir compuesto*** Un tipo de lubricacion que previene que metales diferentes reaccionen adheriendose unos con otros.

**Anti-sway bar** A suspension geometry that will resist a vehicles tendency to drop or squat on the rear springs when accelerating.
***Barra de antitorcer*** Una suspencion geometrica que resiste a la tendencia de un vehiculo de doblarse en un resorte cuando se acelera.

**Aqueous** Solution that is mostly water.
***Aguoso*** Una solucion compuesta mayormente de agua.

**Aspect ratio** The height of a tire, from bead to tread expressed as a percentage of the tire's section width.
***Aspecto de rango*** La altura de una llanta de arriba hacia abajo, expresando como porcentaje la anchura de la llanta.

**Atmospheric pressure** The weight of the air at sea level (about 14.7 pounds per square inch or less at higher altitudes.
***Presion atmosferica*** El peso del aire al nivel del mar (cerca de 14.7 libras por pulgada cuadrada o menos a altos niveles).

**Atom** The smallest particle of an element in which all the chemical characteristics of the element are present.
***Atomo*** La particula mas pequena en la que todas las caracteristicas quimicas de un elemento estan presentes.

**Axial** Having the same direction or being parallel to the axis or rotation.
***Eje*** Llevar la misma direccion o estar paralelo al eje de rotacion.

**Axial load** A type of load placed on a bearing that is parallel to the axis of the rotating shaft.
***Eje de peso*** Tipo de peso localizado en el soporte que esta paralelo al eje de rotacion.

**Axial play** Movement that is parallel to the axis or rotation.
***Eje de movimiento*** Movimiento que es paralelo al eje de rotacion.

**Axle** The shaft or shafts of a machine upon which the wheels are mounted.
***Eje*** La linea central de una parte rotando, una parte simetrica o figura circular.

**Axle housing** Designed in the removable carrier or integral carrier types to house the drive pinion, ring gear, differential and axle shat assemblies.
***Caja del eje*** Disenado en el portador removable o portador integral, encajar el pinon, anillo de cambio diferencial, y eje de los asambles de cambio.

**Axel shaft** Shaft on which the road wheels are mounted.
***Eje astil*** Astil en el cual las llantas estan montadas.

**Axle-shaft end thrust** A force exerted on the end of an axle shaft that is most pronounced when the vehicle turns corners or curves.
***Eje astil fuerte*** Un force fuerte al final del astil del eje, que se pronuncia mas cuando el vehiculo da vueltas en esquinas y curves.

**Axle shaft tubes** These tubes are attached to the axle housing center section to surround the axle shaft.
***Tubos-del eje*** Estos tubos estan unidos a la caja del eje, seccin central que rodea el astil del eje y los soportes.

**Backlash** The clearance or play between two parts, such as meshed gears.
***Hueco Negro*** El espacio o hueco enmedio de dos partes, tales como engranes y la marcha.

**Balance** Having equal weight distribution. The term is usually used to describe the weight distribution around the circumference and between the front and backsides of a wheel.
***Balance*** Tener peso distribuido igualmente. El termino se usa para desribir la distribucion del peso alrededov de la circunferencia y enmedio del frente y los lados de atras de la llanta.

**Ball bearing** An anti-friction bearing consisting of a hardened inner and outer race with hardened steel balls that roll between the two races and support the load or shaft.
***Bola de soporte*** Un soporte de anti-rose que consiste en una bola hecha de fierro fuertemente duro que soporta el peso enmedio de las partes.

**Ball joint** Pivot point for turning a front wheel to the right or left. Ball joints can be considered either non-loaded or loaded when carrying the car's weight.
***Bola de coyontura*** Un pivote con punto para voltear la manejadera para la derecha o para la izquierda. La bola coyontura puede ser considerada cargada o no cargada cuando llevan el peso de un carro.

**Ball joint wear indicator** A visual indicator of ball joint internal wear.
***Bola coyontura o indicador*** Un indicador visual de la bola coyontura de uso internal.

**Base** A solution that has an excess of OH ions, also called an alkali. Also, the center layer of a bipolar transistor.
***Base*** Una solucion que tiene un exeso de OH jonico, tambien conocido como alkali, y tambien la capa central de un transistor.

**Battery** A device for storing energy in chemical form so it can be released as electricity.
***Bateria*** Un deviso para guardar la energia en forma quimica para que pueda ser transmitida en forma de electricidad.

**Bead** The portion of the tire that helps keep it contact with the rim of the wheel. It also provides the air seal on the tubeless tires.
***Caja*** La porcion de la llanta que ayuda a conservarla en contacto con el girador. Tambien sella salidas de aire en llantas sin camara.

**Bead filler** A piece of rubber placed above the bead that reinforces the sidewall and acts as a rim extender.
***Pieza proveedora*** Una pieza de plastico localizada sobre la parte que refuerza la pared de lado y actua como borde extendido.

**Bead wire** A group of circular wire strands molded into the inner circumference of the tire that anchors the tire from the rim.
***Caja de cables*** Un grupo de cables acordonados y modulados en la circunferencia interior de la llanta que ancla la misma.

**Bearing** The supporting part that reduces friction between a stationary and rotating part, or between two moving parts.
***Soporte*** La parte del soporte que reduce rose enmedio de una parte inmovil y una que gira, o entre dos partes que se mueven.

**Bearing cage** A spacer tat keeps the balls or rollers in a bearing in proper position between the inner and outer races.
***Cja de soporte*** Un espacio que conserva las bolas o rolers en un soporte en posicion apropiada entre las races de adentro y de afuera.

**Bearing caps** In the differential, caps held in place by bolts or nuts which, in turn, hold bearings in place.
***Tapas de soporte*** En el diferencial las tapas se quedan en su lugar por tornillos o nudos los cuales en turno, retienen los soportes en su lugar.

**Bearing clearance** The amount of space left between a shaft and the bearing surface for lubricating oil to enter.
***Despejo del soporte*** La cantidad de espacio que hay entre un soporte y una parte para la entrada de la lubricacion de aceite.

**Bearing cone** The inner race, rollers, and cage assembly of a tapered roller bearing. Cone s and cups must always be replaced in matched sets.
***Soporte de cono*** Una parte de adentro, como los rolers, espacio hecho para un soporte roler. Conos y copas deben ser remplazados con la misma clase de partes.

**Bearing cup** The outer race of a tapered roller bearing or ball bearing.
***Soporte de copa*** La parte de afuera de un soporte de roler o bola de soporte.

**Bearing end play** The designed looseness in a bearing assembly.
***Espacio del soporte*** La soltura disenada en el asamble de un soporte.

**Bearing hub unit** A complete wheel bearing and hub unit to which the wheel rim is bolted.
***Unidad de un soporte*** Un completo soporte de la llanta y unidad del eje, en el que la llanta esta atornillada.

**Bearing race** The surface upon which the rollers or balls of a bearing rotate. The outer race is the same as the cup, and the inner race is the same as the one closest to the axle shaft.
***Soporte de la marcha*** La superficie en la cual los roles y las bolas de un soporte giran. La marcha de afuera es la misma que la de adentro, y la de adentro es la mas cercana al astil o eje.

**Bearing seals** Circular metal rings with a sealing lip on the inner edge. Bearing seals are usually mounted on both sides of the bearing, to prevent dirt and moisture from entering the bearing while sealing the lubricant in the bearing.
***Sellos de soporte*** Anillos de metal con substancia de sellar en la orilla de adentro.Los soportes sellados usualmente estan montados el los lados de del soporte, para prevenir de que el polvo y la humedad entren al soporte mientras que se sella el lubricante en el soporte.

**Bellows** A movable cover or seal, usually rubber like material, that is pleated or folded like an accordion to allow for expansion and contraction.
***Tapas*** Una tapadera movable o sello, usualmente como material de plastico, que esta doblada como accordion para permitir expancion y contraccion.

**Belted bias ply** A type of tire construction with the cord plies, wound by an angle to the center of the tire and fiberglass, or steel belts mounted under the tread area of the tire.
***Bandas aplicadas*** Es una forma de construir las llantas con cordon bien atado, en angulo al centro de la llanta, de fibra de vidrio o bandas de fierro montadas debajo del cordon de la llanta.

**Bias** A diagonal line of direction. In relationship to tires, bias means that belts and plies and laid diagonally or crisscrossing each other.
***Bandas*** Una linea de direccion diagonal en relacion con las llantas, bandas quiere decir que los cordones estan puestos diagonalmente o cruzandose unas con otras.

**Bias ply tire** Tire construction in which the cords in the body plies of the carcass, run from the bead to bead, at an angle of 90 degrees as in a radial-ply tire.
***Llantas aplicadas*** La construccion de la llanta en la cual los cordones del cuerpo de la llanta van de ladao a lado, en un angulo de 90 grados en una llanta radial.

**Binder** An adhesive, or glue, used in brake linings to bond all the other materials together.
***Trapa*** Un adhesivo, o pegamento,usado para pegar o juntar otros materials.

**Bleeding** The process of removing air from a closed hydraulic system.
***Vaciar*** El proceso de sacar aire de un sistema hidraulico.

**Bolt diameter** The measurement across the major diameter of a bolts threaded area, or across the bolt shank.
***Diametro del tornillo*** La medida mayor cruzando el diametro de un tornillo de un lado a otro.

**Bolt head** The part of a bolt that the socket or wrench fits over in order to torque or tighten the bolt. Of the carcass run from the bead to bead, at an angle of 90 degrees as in a radial-ply tire.
***Cabeza del tornillo*** La parte del tornillo en la que se coloca el socket, o llave para girar y apretar el tornillo.

**Bolt shank** The smooth area of a bolt from the bottom surface of the head to the start of the threads.
***Cuerpo del tornillo*** La parte suave de un tornillo de la punta hacia arriba, a la cabeza donde esta la rosca.

**Bolt torque** The turning effort required to offset resistance, as the bolt is being tightened.
***Rosca del tornillo*** La parte que permite que el tornillo se enrosque y apreiete para resistencia.

**Brinneling** Rough lines worn across a bearing race, or shaft due to impact loading due to impact loading, vibration, or inadequate lubrication.
***Raspando*** Tremendas lineas o rasgunos en algunas partes, a causa del vehiculo con mucha carga o por vibracion, o por lubriacion inadecuada.

**Bump steer** The tendency of the steering to veer suddenly in one direction, when one or both front wheels strike a bump.
***Topes de vuelta*** La tendencia del volante de voltearse para una direccion ,cuando una o las dos llantas de adelante topan en un bordo en el camino.

**Burnishing** The process of applying frictional materials to each other, to create a desired wear pattern.
***Puliendo*** El proceso de aplicar materials friccionales en cada parte, para crear un molde deseado.

**Burr** A feather edge of metal left on a part, being cut with a file or other cutting tool.
***Filo*** Un filo ligero que queda en alguna de las partes, al cortarla con una lima u otra herramienta de cortar.

**Bushing** Cylindrical lining used as a bearing assembly made of steel, brass, bronze, or plastic.
***Forrando*** Una ligera capa de metal en una parte hecha de fierro, brazz , bronze, nylon o plastico.

**Cage** A spacer used to keep the balls or rollers in proper relation to one another. In a constant-velocity joint, the cage is an open metal framework that surrounds the balls to hold them in position.
***Caja*** Un espacio usado para conservar las bolas y rollers en posicion apropiada uno con otro.En constante velocidad: la caja en un marco que dispone las bolas a permanecer en una sola posicion.

**Chamber** A suspension alignment term used to define the amount that the centerline of a wheel is tilted, inward or outward from the true vertical plane of the wheel.
***Cambero*** Alineamiento de suspencion; es un termino usado para definir la cantidad de centramiento en un eje ya, sea hacia afuera o hacia adentro desde el plano vertical de la llanta.

**Carcass** The steel beads around the rim and layers of cords, or plies that are bonded together to give a tire its shape and strength.
***Carcas*** Cordones de fierro en la llanta y capas de cordon, que estan atadas juntas para dar a la llanta forma y fuerza.

**Carcass plies** in a tire, carcass plies surround both beads and extend around the inner surface of the tire, to provide load-carrying capabilities.
***Carcas tenacillas*** En una llanta, carcass tenacillas son las que rodean las dos roscas de la llanta para sostenerlas y dar capacidad de carga.

**Case harden** To harden the surface of steel, the carburizing method used on low-carbon steel or other allows to make the case or outer layer of the metal harder than its core.
***Caja dura*** Para endurecer una superficie de fierro, el metodo carburizador de low-carbon u otro, para endurecer mas el fierro.

**Casing** Layers of side wall and undertread rubber added to a tire carcass.
***Cubrir*** Capas de paredes de plastico no tratadas, agregadas a las carcass de la llanta.

**Castellate** Formed to resemble a castle battlement, as in a castellated nut.
***Encasillar*** Formada para reforzar algun nudo o tornillo.

**Castellated nut** A nut with six raised portions or notches through which a cotter pin, can be inserted to secure the nut.
***Nudo encasillado*** Nudo con seis porciones de pequenas barritas que pasan enmedio, que son insertadas para asegurar el nudo.

**Caster** A measurement expressed as an angle of the forward and rearward, tilt of the top of the wheel spindle.
***Ruedecilla*** Una medida expresada como angulo para atras o para delante, arriba de la broca de la llanta.

**Caustic** Something that has the ability to eat away at something through chemical action.
***Corrosivo*** Una substancia que tiene la habilidad de calcomer substancias con accion quimica.

**C-clip** A clip used to retain the drive axles, in some rear-axle assemblies.
***C –broche*** Un broche en forma de clip, que se usa para retener algunos de los ejes traseros asamblados.

**Center link** A steering linkage component connected to between the pitman and idler arm.
***Centro del anillo*** Un anillo que da vuelta enmedio de una barra inmovil.

**Center take-off power rack-and-pinion steering gear** A type of steering linkage, in which both tie rods are connected to the center rack in a rack-and-pinion steering gear.
***Centro quitado rack y engrane del pinon rotador*** Un tipo de rotador en el que dos varillas estan conectadas en el centro del rack y el engrane del pinon rotador.

**Central processing unit (CPU)** The calculating part of any computer that makes logical decisions, by comparing conditioned input with data in memory.
***Unidad de proceso central (CPU)*** La parte calculadora de cada computadora que hace decisions logicas, comparando condiciones puestas en la memoria.

**Centrifugal force** The force that acts on a rotating body, that tends to move outward and away from the center of the rotation. The force increases as speed increases.
***Fuerza centrifugal*** La fuerza actuando en un cuerpo girando, que tiende a moverse hacia afuera y aparte del centro de rotacion. La fuerza aumenta conforme la rotacion acelera.

**Chamfer face** A beveled surface on a shaft or part that allows for easier assembly. The ends of FWD drive shaft, are often chambered to make installation of the CV joints easier.
***Cara del chamfer*** Superficie biselada en un astil o parte que permite un asamble mas facil. Los astiles de doble traccion, frecuentemente son biselados para que se asamblen en las coyonturas facilmente.

**Chase** To straighten up or repair damaged threads.
***Carrera*** Para reforzar o reparar roscas maltratadas.

**Chassis** The vehicle frame, suspension, and running gear. On FWD cars, it includes the control, struts, springs, trailing arms, sway bars, shocks, steering knuckles, and frame.
***El chasis*** El marco del vehiculo, suspencion, y engrane corredor. En carros de doble traccion; incluye el control de brazos, riostra, resortes, brazos jaladores, barras de balanceo, circuitos, tornillos giradores y marco.

**Circlip** A split-steel snapring that fits into a groove to hold various parts in place. Circlips are often used on the ends on FWD drive shafts, to retain retian the constant-velocity joints.
***Circlipe*** Una barilla de fierro que entra en asamble, para sostener varias partes en su lugar.Circlipes se usan seguido para las orillas de astiles de doble traccion, para retener las coyonturas en constante velocidad.

**Circuit breaker** A resettable circuit protection device, that automatically opens in response to high current.
***Cortador de circuito*** Un circuito asentable devisador de proteccion, que se abre en respuesta a altas corrientes.

**Clearance** The space allowed between two parts, such as between a journal and a bearing.
***Espacio*** El espacio entre dos partes, como enmedio del jornal y un soporte.

**Clock spring electrical connecter** A conductive ribbon in a plastic case mounted on top of the steering column that maintains electrical contact between the air bag inflator module and the air bag electrical system.
***Resorte electrico del reloj conectador*** Un cordon conductivo en una caja de plastico, montado arriba de la columna giradora que mantiene contacto electrico entre la modula del girador de la bolsa de aire y el sistema electrico de la bolsa de aire.

**Clutch** A devise for connecting and disconnecting the engine from the transmission, or for a similar purpose in other units.
***Cloche*** Un devisador para conectar y desconectar el motor de la transmision, o para un proposito similar en otras unidades.

**Coil spring** A heavy wire like steel coil used to support the vehicle weight while allowing for suspension motions. On FWD cars the front coil springs are mounted around the MacPherson struts. On the rear suspension, they may be mounted to the rear axle, to trailing arms, or around rear struts.
***Resorte de ciol*** Un pesado coil como cable de fierro usado para soportar el peso del vehiculo permitiendo movimientos de suspencion. En carros de doble traccion, los resortes del coil de enfrente estan montados alrededor de MacPherson movimiento. En la suspencion trasera, pueden ser montados en el eje trasero, en los brazos o alrededor del pavoneo trasero.

**Cold inflation pressure** The tire inflation pressure after a tire has been standing for three hours or driven less than 1 mile after standing for three hours.
***Presion de inflacion fria*** La presion de la inflacion de la llanta que ha estado parada por tres horas o corrida por menos de 1 milla despues de parada por tres horas.

**Collapsible steering column** A steering column that is designed to collapsible to help reduce driver injury when impacted by the driver in a collision.
***Columna giradora movible*** Una columna giradora que es disenada para pararse y ayudar a reducir lastimaduras al chofer cuando este se envuelve en un impacto.

**Collet** A collar or band into which is pushed or forced.
***Collar*** Banda en la cual algo es ampujado o forzado.

**Compact spare tire** A spare tire and a wheel that is much smaller that the other three tires on the designed for short distance driving at low speed.
***Llanta extra compacta*** Llanta y ring que son mucho mas pequenas que las otras llantas en el vehiculo y es disenada para distancias cortas y bajas velocidades.

**Complete box frame construction** Type of vehicle frame shaped like a rectangular steel box.
***Caja completa de la construccion del marco*** Un tipo de marco de vehiculo en forma de caja rectangular de fierro.

**Complete circuit** An electrical circuit that includes a path that connects the positive and negative terminals of the electrical power source.
***Circuito completo*** Un circuito electrico que incluye un medio que conecta las terminals negativa y positiva en la fuente de electrecidad.

**Compression-loaded ball joint** A suspension ball joint mounted so the vehicle is forcing the ball into the joint.
***Coyontura de compresion cargada*** Una suspension de coyontura montada para que el peso del vehiculo force la bola en la coyontura.

**Computer** An electric device that receives information, stores information, processes information, and communicates information.
***Computadora*** Devisador electronico que recive informacion, guarda informacion, procesa informacion y comunica informacion.

**Computer command ride (CCR) system** A computer controlled suspension system that controls strut absorber firmness in relation to driving and road conditions.
***Sistema de carrera comandada por computadora (CCR)*** Una suspencion controlada por computadora,sistema que controla y afirma, y a la vez absorbe el movimiento del circuito,con relacion al manejar en las condiciones del camino.

**Constant velocity joint (also called the CV joint)** A flexible coupling between two shafts that permits each shaft to maintain the same driving or driven speed regardless of operating angle, allowing for a smooth transfer of power. The constant-velocity joint consists of an inner and outer housing with balls in between, or a tripod and yoke assembly.
***Coyontura de constante velocidad (tambien llamada C V coyontura)*** Una coneccion flexible entre dos astiles, que permite a cada uno de los astiles mantener la misma velocidad al manejar, sin importer el angulo operador, permitiendo ademas transmitir una fuerza suave.La coyontura de constante velocidad consiste en un espacio afuera y adentro con las bolas enmedio, como un yoke asamble.

**Control arm** A suspension component that links the vehicle frame to the steering knuckle or axle housing and acts as a hinge to allow up-and-down wheel motions.
***Brazo de control*** Una suspencion componente que enlaza o conecta el marco del vehiculo al nudillo rotador o eje y actua como una visagra para permitir movimiento hacia arriba y abajo a las llantas.

**Controllers** Devices such as switches or relays that direct the flow of electrons.
***Controladores*** Devisadores tales como swiches o conductos, que corre la function de los electrons.

**Convection** The transfer of heat by the movement of a heated object.
***Conveccion*** El paso de calor por el movimiento de otro objeto caliente.

**Cords** The inner materials running through the plies that produce strength in the tire. Common cord materials are fiberglass and steel.
***Lazos*** El material interior que al corer rapido produce fuerza a las llantas.Cordones communes hechos de fibra de vidrio y fierro.

**Cornering angle** The turning angle of the front wheel on the inside of a turn in relation to the turning angle of the front wheel on the outside of the turn. This angle may be called turning radius.
***Angulo coronario*** El angulo volteador enfrente del volante en la parte de afuera.Este angulo puede tambien ser llamado radio girador.

**Corrosion** Chemical action, usually caused by an acid, that eats away (decomposes) metal.
***Corrosivo*** Accion quimica,usualmente causada por un acido que calcome (descompone) un metal.

**Cotter pin** A type of a fastener made from soft steel in the form of a split pin that can be inserted in a drilled hole. The split ends are spread to lock the pin in position
***Llave de pin*** Una clase de sujetador echo de fierro suave en forma de separador que puede ser insertado en un agujero taladrado. Las orillas del separador son destendidas para atrancar el seguro en su pocision.

**Counter-clockwise rotation** Rotating the opposite direction of the hands of a clock.
***Rotacion al contrario del reloj*** Rotacion que va en direccion opuesta a las manecillas del reloj.

**Crocus cloth** A very fine polishing paper. It is designed to remove very little metal, and it is safe to use on critical surfaces.
***Tela de azafran*** Un papel muy fino de pulir. Esta disenado para remover pequenos reciduos de metal y es perfecto para superficies criticas.

**Cross-groove CV joint** A CV joint that uses six balls and inner and outer bearing races. The grooves in the races are cut on an angle.
***Ranura cruzada CV coyontura*** Una CV coyontura que usa seis bolas en soportes internos o externos. Estas ranuras estan cortadas en angulo.

**Current** Relates to the number of electrons flowing past a given point in a given amount of time
***Curso*** Relacionado con el numero de electrones que se pasan a un punto dado de tiempo.

**Cycle** One set of changes in a signal that repeats itself several times.
***Turno*** Un juego de cambios en una senal que se reoite varias veces.

**Dampen** To slow or reduce oscillations or movement.
***Humedecedor*** Para disminuir o reducer la vibracion o movimiento.

**Damper** A device used to reduce or eliminate vibrations.
***Humedecedor*** Devisador que se usa para reducer o aun eliminar vibraciones.

**Dead axle** An axle that does not rotate but merely forms a base on which to attach the wheels.
***Eje muerto*** Un eje que no da vuelta pero meramente forma una base en la que se soportan las llantas.

**Deceleration** The rate of decrease in speed.
***Decelerar*** El echo de disminuir velocidad.

**Default mode** A mode of operation that allows for a limited use of the transmission in the case of electronic or serious mechanical failure.
***Modulo de falla*** Un modulo de operacion que funciona para usar la transmission por tiempo limitado en caso de alguna falla electronica o mecanica.

**Deflection** Bending or movement away from normal due to loading.
***Defleccion*** Doblarse o moverse de lo normal a causa de la carga.

**Degree** A unit of measure equal to 1/360th of a circle.
***Degrado*** Una unidad de medida igual a 1/360 de un circulo.

**Deployed** The condition of an air bag when it has been released and expanded.
***Desplazado*** La condicion de una bolsa de aire cuando ha sido activada y expandida.

**Diagnostic trouble code (DTC)** A numerical code generated by an electric control system to indicate a problem in a circuit or subsystem or to indicate a general condition that is out of its limits.
***Codigo de diagnostico de problemas (DTC)*** Un codigo numerico generado por un sistema de control electrici para detector algun problema en un circuito o subsistema que indica una condicion general que esta fuera de los limites.

**Dial caliper** A versatile measuring instrument capable of taking inside, outside, depth, and step measurements.
***Barometro calibrador*** Un instrumento de medida que es capaz de tomar medidas hacia afuera o hacia adentro, hacia adentro y frofundo y medidas inclinadas.

**Dial indicator** A measuring instrument with the readings indicated on a dial rather than on a thimble as on a micrometer.
***Barometro indicador*** Un instrumento de medidas con lecturas indicadas en un disco mejor que en un dedal como en un micrometro.

**Diamond-shaped frame** A frame that is diamond shaped from collision damage.
***Marco en forma de diamante*** Un marco que esta en forma de diamante a causa de collision.

**Die** A tool used to restore external threads on a fastener.
***Torquel*** Llave usada para restaurar roscas externas en un sujetador o cierre.

**Diffusion** The random movement of gas particles; also ensures that any two gases sharing the same container will totally mix.
***Difusion*** La posibilidad de movimiento de particulas de gas.tambien asegura que cualquier dos clases de gases se revuelvan totalmente.

**Digital signal** A voltage signal that has only two valves—on or off.
***Senal digital*** Senal de voltage que solo tiene dos valores—prendido o apagado.

**Digitized** The process of converting an analog voltage signal to a digital equivalent that the computer can understand.
***Digitalizado*** El proceso de convertir una senal de voltage analo en una senal digital equivalente que la computadora puede entender.

**Diode** Semicnductor that allows current to flow through it in one direction only.
***Convertidor*** Semiconductor que permite a la corriente pasar en una sola direccion.

**Direct current** A type of electrical power used in moble applications. An unidirectional current of substantially constant value.
***Corriente directa (DC)*** Un tipo de fuerza electrica usada en aplicaciones moviles.Una corriente direccional de constante valor substancial.

**Directional control** The ability to steer the automobile while stopping.
***Control direccional*** La habilidad de rotar el automovil mientras se detiene.

**Directional stability** The ability to maintain straight line stopping action.
***Estabilidad direccional*** La habilidad de mantener una direccion derecha en accion de parar.

**Distortion** Warpage or change in form from the original shape.
***Distorsion*** Un cambio en una parte,cambia de la forma original.

**Data Link Connecter (DLC)** This is the connecter used to connect into a vehicle's computer system for the purpose of diagnostics. Prior to J1930, this was a commonly referred to as the ALDL.
***Base del conector (DLC)*** Este es el conectrador que se usa para conectar en el sistema de computadora de un vehiculo, para el proposito de diagnosticos.Antes de 1930, esto era comunmente referido como el ALDL.

**Digital Multimeter (DMM)** A tool that combines the voltmeter, ohmmeter, and ammeter together in one diagnostic instrument and provides a digital reading.
***Multimetro digital (DMM)*** Lava que combina el volimetro ohmmeter y ammeter juntos en un instrumento diagnostico que prove una lectura digital.

**Dogtracking** A condition where the rear wheels are not positioned directly behind the front wheels.
***Detector de trote*** Una condicion donde las llantas de atras no estan puestas directamente detras de las llantas de adelante.

**Double offset joint** Another name for a plunging ball-type joint.
***Coyontura de doble juego*** Otro nombre para un tipo de bola de coyontura.

**Double wishbone suspension** A multi-link suspension system in which the links may be shaped like a wishbone.
***Doble suspencion de wishbone*** Sistema de suspencion multi-coneccion en el cual los enlaces pueden tener forma de wishbone.

**Dowel** A metal pin attached to one object which, when inserted into a hole in another object, ensures proper alignment.
***Boton*** Un seguro de metal en un objeto en el cual, cuando se inserta dentro de un agujero en otro objeto, asegura alineamiento apropiado.

**Dowel pin** A pin inserted in matching holes in two parts to maintain those parts in fixed relation to each other.
***Boton de seguro*** Un seguro insertado en agujeros iguales para mantener algunas partes en relacion una con otra.

**Drag link** A connecting rod or link between the steering gear, pitman arm, and the steering linkage.
***Coneccion de remolque*** Una coneccion en medio del engrane rotador, brazo pitman, y la caja del rotador.

**Drive line** The universal joints, drive shaft and other parts connecting the transmission with the driving axles.
***Linea de manejo*** Las coyonturas universals, astil de manejo.

**Drive shaft** An assembly of one or more universal joints connected by a shaft or tube; used to transmit power from the transmission to the differential. Also called propeller shaft.
***Astil de manejo*** Un asamble de una o mas coyonturas conectadas o comunicadas por un astil o tubo; usado para transmitir fuerza de la transmission al diferencial.Tambien llamado el astil propeller.

**Drive train** The assembly that carries the power from the engine to the drive wheels, includes the transmission (transaction), driveline, differential and final gears, and axle shafts. The term used to refer to the operating condition in which the vehicle is accelerated with a definite application of the throttle to demand engine torque.
***Tren de manejo*** El asamble que lleva la fuerza de el motor a las llantas , include la transmission (transeje) linea de manejo, diferencial y engranes finales de manejo, y astil del eje. El termino usado para referirse a la condicion de operacion en el vehiculo acelerado con una aplicacion definida de la valvula para demander torque del motor.

**Driven gear** The gear meshed directly with the driving gear to provide torque multiplication, torque reduction, or change of direction.
***Engrane de manejo*** El engrane conectado directamente con el engrane de manejo que prove multiplicacion del torque o un cambio de direccion.

**DTC** An acronym for the diagnostic trouble codes available from the memory of the computer control systems.
***DTC*** El acronomo para el codigo diagnosticador de problemas disponible de la memoria de sistema de control de la computadora.

**Duty Cycle** The percentage of time a solenoid is energized during one complete on-off cycle during pulse width modulation.
***Duracion del cycle (apagado o prendido)*** El porcentaje de tiempo que un solenoide recive energia durante un turno de orendido o apagado,durante pulso con modulacion.

**Duty solenoid** A solenoid that has its duty cycle controlled by being togged on and off by the control system.
***Funcion del solenoide*** Un solenoide que tiene su turno de function controlada por ser prendido y apagado por el sistema de control.

**Digital volt ohm meter (DVOM)** A tool that combines the voltmeter, ohmmeter, and ammeter together in one diagnostic instrument and provides a digital reading.
***Voltio digital / ohmmeter (DVOM)*** Una llave que combina el volmetro,ohmmeter,y ammeter juntos en un instrumento de diagnostico y prove lectura digital.

**Dynamic** In motion.
***Dinamico*** En movimiento.

**Dynamic balance** The motion an object when it is in motion, for example, the dynamic balance of a rotating driveshaft.
***Balance Dinamico*** El balance de un objeto cuando esta en movimiento, por ejemplo, el balance dinamico de un astil girador.

**Dynamic pressure** Pressure that changes and/or causes something to move as pressurized fluid is present.
***Presion dinamica*** Presion que cambia y o causa que algo se mueva mientras que la salida del liquido es presionado.

**Dynamic seal** A seal used between two parts that move in relation to each other.
***Sello dinamico*** Sello usado enmedio de dos partes que se mueven en relacion una con otra.

**Eccentric** One circle within another wherein both circles do not have the same center or a circle mounted off center.
***Exentric*** Un circulo con otro circulo donde los dos circulos no tienen el mismo centro, o un circulo no montado en el centro.

**Eccentric washer** A normal-looking washer with its hole not in the center the hole is offset from the center.
***Limpiador exentrico*** Un limpiador normal con un agujero pero no en el centro. El agujero esta no centrado en el medio.

**Elastomer** Any rubber-like plastic or synthetic material used to make bellows bushings or seals.
***Elastometrico*** Cualquier material de plastico o sintetico usado para hacer fuelles, bushings y sellos.

**Electricity** The type of energy caused by the flow of electrons from one atom to another. It is the release of energy as one electron leaves the orbit of one atom and jumps into the orbit of another.
***Electricidad*** El tipo de energia causada por la salida de electrons de un atomo a otro. Es la salida de energia mientras que un electron brinca a la orbita de otro.

**Electromagnetic Interference (EMI)** A magnetic force field that influences a signal being sent to the microprocessor.
***Influenca electromagnetica (EMI)*** Un campo de fuerza magnetica que influencia una senal siendo enviada al micropresor.

**Electronic** Pertaining to the control of systems or devices by the use of small electrical signals and various semiconductor devices and circuits.
***Electrinicos*** Pertenece al control de sistemas o devisadores con el uso de pequenas senales electricas y varios semiconductores, devisadores y circuitos.

**Electronic level control (ELC)** A type of computer-controlled rear suspension leveling system.
***Control de level electronico (ELC)*** Un tipo de control computarizado del sistema del nivel de suspencion trasero.

**Electronic rotary height sensor** An input sensor that sends a signal to the computer in relation to curb riding height in a computer-controlled suspension system.
***Sensor electronico de rotacion de altura*** Un sensor que manda una senal a la computadora en relacion con las curves al manejar,en una computadora-controlada de sistema de suspencion.

**Electronic variable office (EVO)** Provides high power steering assistance during low speed cornering and parking and normal power steering assistance at higher speeds for proper road feel.
***Orificio variable electronico (EVO)*** Provee asistencia para alta fuerza de rotacion durante baja celocidad en las esquinas y al parar,tambien prove alta fuerza a altas velocidades y apropiado sentido en el camino.

**Electronically erasable programmable read only memory (EEPROM)** A chip in a computer that may be erased easily with special equipment.
***Memoria electronica borrable o programmable (EEPROM)*** Una pequena chip en una computadora que puede ser facilmente borrada con equipo especial.

**Element** A substance with only one type of atom.
***Elemento*** Una substancia con una sola clase de atomo.

**Ellipse** A compressed form of a circle.
***Eclipse*** Una forma compresa.

**Electromotive Force (EMF)** Another name for voltage; the force or pressure that exists beatween a positive and a negative point within an electrical circuit.This force is measured in units called volts.
***Fuerza electromotive (Emf)*** Otro nombre para voltage; la fuerza o presion que existe entre un punto positive y negative con un circuito electrici.Esta fuerza es medida en unidades llamadas voltios.

**Emulsification** Proces that forces water droplets to mix with grase, resulting in a thicker solution with less lubricating qualities than normal grease.
***Emulsificacion*** Un proceso que fuerza a las gotas de agua a revolverse con grasa, resultando en una espesa solucion con menos cualidades que la grasa normal.

**End clearance** distance between a set of gears and their cover, commonly measured on oil pumps.
***Final de espacio*** Una distancia entre un juego de engranes y la tapa, comunmente medida en bombas de aceite.

**End play** The amount of axial or end-to-end movement in a shaft due to clearance in the bearings.
***Movimiento final*** La cantidad de movimiento del eje en un astil por el espacio de los soportes.

**End take-off power rack-and-pinion steering gear** A rack-and-pinion steering gear in which the tie rods are connected to the ends of the rack.
***El apagador del engrane gorador en el pinon*** El engrane rotador del pinon en el cual lazos fuertes estan conectados al final del rack.

**Energy** The ability to do ability to do work.
***Energia*** La habilidad de hacer trabajo.

**Energy-absorbing lower bracket** A steering column bracket that protects the driver by allowing the column to move away from the driver when impacted by the driver in a collision.
***Energia absorvente del soporte*** Una columna de soporte rotador que protegera al chofer permitiendo que la columna se mueva lejos del chofer cuando es impactado hacia este.

**Energy absorbing steering column** A steering column that protects the driver by moving away from the driver when impacted by a collision.
***Columna rotadora absorbente de energia*** Una columna rotadora que protegera al chofer moviendose cuando el chofer se impacta con este.

**Engine Torque** A turning or twisting action developed by the engine measured in foot-pounds or kilogram-meters.
***Vuelta del motor*** Accion de voltear o torcer desenvuelta por el motor, medida en pies-libras o kilogram-metros.

**Engine vacuum** The low pressure formed by an engines pistons.
***Aspiradora del motor*** La baja presion formada por los pistons del motor.

**EPA** Environmental Protection Agency.
***E P A*** Agencia de Proteccion Ambiental.

**Equilibrium** Exists when the applied forces on an object are balanced and there is no overall resultant force
***Equilibrio*** Existe cuando la fuerza aplicada en un objeto es balanceada y no hay otra fuerza resultante.

**Erasable programmable read-only memory (EPROM)** Computer memory program circuits that can be erased and reprogrammed. Erasure is done by exposing the IC chip to ultraviolet light.
***Memoria de lectura borrable o programmable (EPROM)*** Circuitos del programa de memoria de computadora, que pueden ser borrados o reprogramados. Se borran cuando el micro chip es expuesta a la luz ultravioleta

**Etching** A discoloration or removal of some material caused by corrosion or some other chemical reaction.
***Raspando*** Una decoloracion o acabado en un material causado por corrosive o alguna otra reaccion quimica.

**Evaporate** Atoms or molecules break free from the body of the liquid to become gas particles.
***Evaporado*** Atomos o moleculas que salen del cuerpo del liquido que se convierten en particulas de gas.

**Face** The front surface of an object.
***Cara*** La superficie de enfrente de un objeto.

**Fatigue** The build up of natural stress forces in a metal part that eventually causes it to break. Stress results form bending and loading the material.
***Fatiga*** El acomulamiento de fuerzas de presion natural, en una parte de metal que eventualmente la quiebra, y esto resulta por doblar o cargar el material.

**Feel of the road** A feeling experienced by a driver during a turn when the driver has a positive feeling that the front wheels are turning in the intended direction.
***Sentir el camino*** Un sentir experimentado por el chofer durante una vuelta, cuando el chofer tiene un sentir positive que las llantas voltean en la direccion deseada.

**Feeler gauge** A feeler gauge is a thin strip of metal or plastic of known and closely controlled thickness used to measure clearances and gaps.
***Sensor*** Una caja, un sensor en un Delgado cordon de metal o plastico y espesor controlado para medir espacios y aberturas.

**Field of flux** A field that exists around around a magnet. These lines emerge from the North Pole and enter the South Pole, returning to the North Pole through the magnet itself.
***Campo de cambio*** Un campo que esta alrededor de un iman.Estas lineas senalan del polo norte al polo sur, retornando al polo norte por medio del iman.

**Fillet** The smooth curve where the shank flows into the bolt head.
***Filo*** La curva suave del cuerpo del tornillo que va a la cabeza.

**Fit** The contact between two machine surfaces.
***Medida*** El contacto entre dos superficies de maquina.

**Fixed displacement pump** Fluid pumps that maintain a particular amount of fluid flow.
***Bomba de desplazamiento*** Bombas de liquido que mantienen una cantidad particular de liquido.

**Fixed joint** A CV joint that cannot telescope or plunge to compensate for suspension travel.
***Coyontura compuesta*** Una coyontura CV que no se puede telescope o conectar para compensar suspencion travel.

**Fixed rotor** A rotor that has the hub and the rotor cast as a single part.
***Girador compuesto*** Un girador que tiene un espacio y el rotador como parte singular.

**Fixed seal** A seal for the caliper piston that is installed in a groove in the caliper bore and does not move with the piston.
***Sello compuesto*** Sello para un piston caliber que es instalado en una ranura en el agujero del caliber, y no se mueve en el piston.

**Fixed-type constant-velocity joint** A joint that cannot telescope or plunge to compensate for suspension travel. Fixed joints are always found on the outer ends of the drive shafts of FWD cars. A fixed joint may either be Rzeppa or tripod type.
***Coyontura constante de velocidad compuesta*** Una coyontura que no se puede telescope o conectar para compensar suspencion de viaje. Coyonturas compuestas son siempre encontradas en las orillas de afuera de los astiles de los carros de doble traccion. Una coyontura compuesta puede ser de tipo Rzeppa o Tripod.

**Flange** A projecting rim or collar on an object used to keep it or another object in place.
***Borde*** Un borde o collar que protégé el rim y es usado para mantener algun otro objeto en su lugar.

**Flare** An expanded shaped end on a metal tube or pipe.
***Flare*** La orilla destendida de un tuvo o pipa.

**Flat rate** A pay system in which technicians are paid for the amount of work they do. Each job has a flat-rate time.
***Pago plano*** Un sistema de pago en el cual los tecnicos son pagados de acuerdo al trabajo que ellos hacen. Cada trabajo tiene un pago plano.

**Flat-rate manuals** Literature containing figures dealing with the length of time specific repairs are supposed to require. Flat-rate manuals often contain a parts list with prices as well.
***Pago plano de manuals*** Literatura que contiene figuras , tratando con la duracion de tiempo que algunas reparaciones especiales deben de tomar. Pago plano de manuals muy seguido contienen una lista de partes que incluyen los precios.

**Fluid** Something that does not have a definite shape; therefore liquids and gases are fluids.
***Liquidos*** Algo que no tiene una definida figura, por lo tanto son liquidos, gases o fluidos.

**Force** Any push or pull exerted on an object; measured in pounds or ounces, or newtons (N) in the metric system.
***Force*** Cualquier objeto que empuja o jala, medido en libras y onzas, o en newtons, (N) en el sistema metrico.

**Four wheel drive** On a vehicle, driving axles at both the front and rear, so that all four wheels can be driven.
***Doble traccion*** En un vehiculo, los dos ejes, el de atras y el de adelante, son preparados para que las cuatro llantas sean manejadas.

**Frame** The main understructure of the of the vehicle to which everything else is attached. Most FWD cars have only a subframe for the front suspension and drive train. The body serves as the frame for the rear suspension.
***Marco*** La destructura principal de un vehiculo en el cual todo lo demas esta atado, la mayoria de los carros de doble traccion tienen solo un submarco para la suspencion de enfrente y el tren de manejo, el cuerpo sirve como marco para la suspencion trasera.

**Frame buckle** A frame that is accordion-shaped to some extent from collision damage.
***Banda de marco*** Un marco que tiene una forma de accordion, para que se extienda en caso de un impacto.

**Frame diamond condition** A condition that occurs when one side of the frame is pushed rearward or forward in relation to the opposite side.
***Condicion de marco de diamante*** Condicion que ocurre cuando un lado del marco es empujado ya sea para atras o para adelante, en relacion con el lado opuesto.

**Frame sag** A frame that is bent downward in the center from collision damage.
***Marco zag*** Marco que es doblado hacia abajo en el centro por haber sido danado.

**Frame side away** A frame that is bent sideways in the center from the collision.
***Marco de balabceo*** Un marco que se dobla de los dos lados en caso de ser danado en un ompacto.

**Frame twist** A frame that is bent from a collision so one corner is higher in relation to the other corner.
***Marco de twiste*** Un marco que es doblado y danado en un impacto y una esquina esta mas alta que la otra, con relacion al lado opuesto.

**Free diameter** Refers to tire diameter when the tire is not supporting any weight.
***Diametro libre*** Se refiere al diametro de las llantas, cuando las llantas no estan soportando ningun peso.

**Free play** Looseness in a linkage between the start of application and the actual movement of the device, such as the clearance between the clutch release yoke fingers and the release bearing housing.
***Juego libre*** Se refiere a una union enmedio del principio de aplicacion y el movimiento actual del devisador, como un espacio enmedio de la palanca de soltura del coche y la caja de soporte.

**Free wheel** To turn freely and not transmit power.
***Bolante libre*** Para voltear libremente y no transmitir fuerza.

**Frequency** The number of times, or speed, at which an action occurs within a specified time interval. In electronics, frequency indicates the number on times that a signal occurs, or repeats, in cycles per second. Cycles per second are indicated by the symbol "hertz" (Hz).
***Frecuencia*** El numero de veces, o velocidad, en los cuales una accion ocurre a un intervalo de tiempo especifico.En electronica, frecuencia indica el numero de veces que una senal ocurre, o se repite, en vueltas de sgundos.

**Friction** The force that resists motion between the surfaces of two objects or forms of matter.
***Friccion*** Una fuerza que resiste movimiento entre las superficies de dos objetos en forma de materia.

**Friction bearing** A bearing in which there is sliding contact between the moving surfaces. Sleeve bearings, such as those used in connecting rods, are friction bearings.
***Soporte de friccion*** Un soporte en el cual hay un ligero contacto entre dos partes que se mueven.

**Front-wheel drive (FWD)** The vehicle has all drive train components located at the front.
***Traccion de enfrente*** El vehiculo tiene components de cuatro traccion en el frente.

**Full-wire open-end springs** Coil spring ends that are cut straight off and sometimes flattened, squared, or ground to a D-shape.
***Resortes de cables con orillas abiertas*** Un resorte de cables que tiene las orillas cortadas y a vaces planos, o formados como una D.

**Fuse** An electrical device used to protect a circuit against accidental overload or unit malfunction.
***Fusible*** Un devisador electrico que es usado para proteger un circuito contra recargas accidentals o unidades mal funcionando.

**Fusible link** A type of fuse made of a special wire that melts to open a circuit when current draw is excessive.
***Coneccion de fusible*** Un tipo de fusible echo de un cable especial que se derrite para abrir un circuito cuando el retiro de energia es demaciado.

**FWD** Abbreviation for front wheel drive.
***FWD*** Abreviacion para traccion delantera.

**Galling** Wear caused by metal-to-metal contact in the absence of adequate lubrication. Metal is transferred from one surface to the other, leaving behind a pitted or scaled appearance.
***Raspadura*** Raspadura causada de metal a metal por la ausencia de lubricacion adecuada. El metal es transferido de una superficie a otra,dejando una apariencia deraspadura.

**Galvanic corrosion** A type of corrosion that occurs when two similar metals, such as magnesium and steel, are in contact with each other.
***Corrosion galvanica*** Untipo de corrosion que ocurre cuando dos partes similares, como magnesio y fierro, estan en contacto uno con otro.

**Gas-filled shock absorbers** Contain a gas charge and hydraulic fluid.
***Circuito absorbedor de la bomba de gas*** Cargador de gas y liquido hidraulico.

**Gasket** A layer of material, usually made of cork, paper, plastic, composition, metal, or a combination of these, placed between two parts to make a tight seal.
***Empaque*** Una capa de material, usualmente hecha de corcho, papel, plastico, compuesto, metal, o una combinacion de estos, localozados enmedio de dos partes para hacer un fuerte sello.

**Gasket cement** A liquid adhesive material, or sealer, used to install gaskets.
***Empaque de cemento*** Un material adhesive liquido, o sellador, usado para instalar empaques.

**Gear** A wheel with external or internal teeth that serves to transmit or change motion.
***Engrane*** Una ruedecilla con dientes externos o internos, que sirven un transmisor o cambian movimiento.

**Gear pumps** Positive displacements pumps that use two meshing external gears, one drive and one driven.
***Bombas de engrane*** Bombas de desplazamiento positive, que usa dos engranes atrancadores externals, y uno es manejado.

**Geometric centerline** A static dimension represented by a line through the center of the vehicle from front to rear.
***Linea central geometrica*** Una dimension estatica representada por una linea por el centro del vehiculo, viniendo del frente.

**Geometric centerline alignment** A wheel alignment procedure in which the front-wheel toe is adjusted to specifications using the geometric centerline as a reference.
***Linea central geometrica de alineamiento*** El procedimiento de una ruedecilla de alineamiento, en la cual las llantas de adelante son ajustadas para usos especificos, usando la linea central geometrica como referencia.

**Glitches** Abnormal, slight movements of a waveform on a lab scope. These can be circuit problems or noise in the circuit.
***Rechinidos*** Ligeros movimientos anormales, de una parte en forma ovalada. Estoe pueden ser problemas de circuito.

**Grade markings** Marks on fasteners that indicate strength.
***Una clase de marcas*** Marcas en las bandas que indicant fuerza.

**Graphite** Very fine carbon dust with a slippery texture used as a lubricant.
***Grafito*** Un ligero polvo de carbon con textura resbalosa que se usa como lubricante.

**Gravity bleeding** The process of letting old brake fluid out and air drain from the brake hydraulic system through a wheel bleeder screw.
***Vaciado de gravedad*** El proceso de dejar que el liquido Viejo salga del sistema de los frenos hidraulicos,por medio de un tornillo vaciador.

**Grease** A combination of oil and a thickened agent.
***Grasa*** Una combinacion de aceite y alguna substancia espesa.

**Gross vehicle weight rating** Total weight of a vehicle plus its maximum rated payload.
***Rango y costo del vehiculo*** El peso total de un vehiculo y el valor.

**Half-shaft** Either of the two drive shafts that connect the transaxle to the wheel hubs in FWD cars.
***Medio-astil*** Cualquiera de los dos astiles de manejo, que conectan el transeje a los hubes en los carros de doble traccion.

**Hall effect switch** A device that produces a voltage pulse dependent on the presence of a magnetic field.
***Efecto del switch del hall*** Un devisador que produce un pulso de voltage dependiente en la presencia de un campo magnetico.

**Hall element** An electronic element used in distributor or crankshaft pickups that generates a digital voltage as a metal blade rotates past the element.
***Elemento hall*** Un elemento electronico usado en el distribuidor del cranque de trocas que genera un voltage digital, como una navaja de metal que pasa por el elemento.

**Hardening** A process that increases the hardness of a metal, deliberately or accidentally, by hammering, rolling, carburizing, heat treating, tempering, or other physical processes.
***Endurecedor*** El proceso de incrementar el endurecimiento de un metal, deliverada o accidentalmente por martillasos, rodando, carburizando, tratando con calor, temperature,o algun otro proceso fisico.

**Hazard Communication Standard** A document published by the Occupational Safety and Health Administration (OSHA) that formed the basis of the Right-to-Know laws.
***Comunicacion standar de peligro*** Un documento publicado por La Administracion Ocupacional de Salud y Bienestar (OSHA)que forma los derechos basicos de conocer la ley.

**Hazardous waste** Waste is considered hazardous if it has one or more of the following characteristics, ignitability, corrosivity, reactivity, and EP toxicity.
***Desechos peligrosos*** Los desechos son considerados peligrosos si contienen alguna de los siguientes caracteristicas; ingitabilidad, corrosividad, reactividad, y EP toxico.

**Heat** A form of energy caused by the movement of atoms and molecules.
***Calor*** Una forma de energia causada por el movimiento de atomos y moleculas.

**Height sensor** Sends a signal to the suspension computer in relation to chassis height.
***Sensor de altura*** Manda una senal a la suspencion de la computadora en relacion a la altura del chasis.

**Height-sensing proportional valve** A proportioning valve in which hydraulic pressure is adjusted automatically according to the vertical movement of the chassis in relation to the rear axle during braking.
***Valvula proporcional del sensor de altura*** Una valvula proporcional en la que la presion hidraulica es ajustada automaticamente, de acuerdo con el movimiento vertical del chasis en relacion con el eje trasero durante frenar.

**Herringbone gear** A gear cut in a V-shaped pattern.
***Engrane de Herring*** Un engrane cortado en forma de V.

**Hertz (Hz)** The unit of measure in which frequency is most often expressed and which is equal to one cycle per second.
***Hertz (Hz)*** La unidad de medida de en cual frecuencia es muy seguido expresada y es como igual a una vuelta por Segundo.

**Horsepower** A measure of mechanical power, or the rate at which work is done. One horsepower equals 33,000 ft-lbs per minute.
***Caballo de fuerza*** Una medida de fuerza mecanica, o el rango al que un trabajo es echo.Un caballo de fuerza es igual a 33,000 pies-lbs por minuto.

**Hub** The center part of a wheel to which the wheel is attached.
***Centro de eje*** La parte central del eje, al cual la llanta esta conectada.

**Hunting gear set** A gear set in which one drive pinion gear tooth contacts every ring gear tooth after several rotations.
***Juego de engrane cazador*** Un juego de engranes en el cual, los dientes del engrane de un pinon tiene contacto con cada anillo de engrane, despues de cada rotacion.

**Hybrid valve** A spool valve that relies on spring tension and hydraulic force for movement.
***Valvula de Hybrid*** La bobina de una balvula que se apoya en un resorte de tencion y fuerza hidraulica por movimiento.

**Hydraulic press** A piece of shop equipment that develops a heavy force by use of a hydraulic piston and jack assembly.
***Compresa hidraulica*** Una pieza de equipo de taller que desenvuelve una fuerza pesada, por uso de un piston hidraulico y asamble del gato.

**Hydraulic pressure** Pressure exerted through a medium of a liquid.
***Presion hidraulica*** Presion exertada por medio de un liquido.

**Hydroboost** A hydraulic power brake system that uses that power steering hydraulic system to provide boost for the brake system.
***Hydrobusto*** Una fuerza del sistema de frenos hidraulico, que usa esa fuerza de sistema de rotacion hidraulica, para proveer estimulo para el sistema de frenos.

**Hydroboost power-assisted steering-system** Uses fluid pressure from the power steering pump to supply brake power assistance.
***Sistema de rotacion reforzado Hydroboost*** Usa presion de liquidos de la bomba de fuerza de rotacion, para suplir asistencia para el sistema de frenos.

**Hydroplane** The action of a tire rolling on a layer of water on the road surface instead of staying in contact with the pavement.
***Hydroplano*** La accion de una llanta rodando en una capa de agua, en la superficie del camino, en lugar de quedarse en contacto con el camino.

**Hygroscopic** The property of something that means it has a tendency to absorb moisture from the atmosphere.
***Hygroscopico*** La propiedad de algo que significa que tiene tendencia a absorber la humedad de la atmosfera.

**ID Inside diameter**
***Diametro de adentro***

**IDF** The commonly used acronym for independent front suspension system.
***IDF*** La acronomia comunmente usada para la independencia del sistema de suspencion de enfrente.

**Idler arm** A short, pivoted steering arm bolted to the vehicle frame and connected to the steering center link.
***Brazo Idler*** Un pivote corto, brazo rotador, unido al brazo del vehiculo y conectado a la coneccion central del rotador.

**Idler pulley** A pulley used to tension or torque the belt(s).
***Polea Idler*** Una polea usada para tencion o tuerse de la banda.

**Ignitability** The characteristic of a material that enables it to spontaneously ignite.
***Encender*** La caracteristica de un material,que puede encenderce espontaneamente.

**Impedance** Refers to the operating resistance of a component or piece of equipment. The higher the impedance, the lower the amperage.
***Inminencia*** Se refiere a la operacion de resistencia de un componente o pieza de equipo.Entre mas alta la inminencia , mas baja es el amperio.

**Impeller** The pump or driving member in a torque converter.
***Impulsador*** La bomba o miembro de manejar que es un tornillo convertidor.

**Impermeable** Materials that absorb fluid.
***Impermeable*** Material que absorve liquido.

**Inactive coils** Inactive coils located at the top and bottom ends of a coil spring induce force into the spring when a wheel strikes the road irregularly.
***Coils inactivos*** Coils inactivos localizados en la parte de arriva y de abajo del coil resorte, inducen fuerza al resorte cuando las llantas golpean un camino irregular.

**Inboard joint** The CV-joint that connects the drive axle and the transaxle of differential assembly.
***Coyontura interior*** La CV-coyontura que conecta el astil y el transeje del asamble del diferencial.

**Included angle** The sum of the angle of chamber and steering axis inclination. Also a term of measurement of the shape of a helical gear.
***Angulo incluido*** La suma del angulo del cambero y el axis rotador inclinado.Tambien un termino de medida de la forma de un engrane helical.

**Increments** Series of regular additions from small to large.
***Incrementos*** Series de adiciones regulares de pequeno a grande.

**Independent rear suspension** A rear suspension system in which one rear wheel moves upward or downward without effecting the opposite rear wheel.
***Suspencion trasera independiente*** Un sistema de suspencion trasera en el cual una llanta de atras se mueve hacia arriva o hacia abajo, afectando la otra llanta de atras.

**Index** To orenant two parts by marking them. During reassembly, the parts are arranged so that the index marks are next to each other. Used to preserve the orientation between the balanced parts.
***Index*** Organizar dos partes al hacerlas.Al asamblarlas, las partes son acomodadas para que el index marque que estan una junto a la otra. Usado para preserver la orientacion entre las dos partes balanceadas.

**Inertia** The tendency of an object in motion to keep moving and the tendency at rest to remain at rest.
***Inercia*** La tendencia de un objeto en movimiento, de seguir moviendose, y ,que soporta descanzar, mantenerse descanzando.

**Inflation pressure** Refers to the amount of air pressure required to inflate the tire. This pressure is measured in pounds per square inch (psi) or kilopascals.
***Presion de inflacion*** Se refiere a la cantidad de la presion del aire requerido para inflar una llanta.Esta presion es medida en libras por pulgada cuadrada, (psi) o kilopascals (kPa).

**Inner race** The race in the center of the ball or tapered roller bearing that supports the balls or rollers.
***Carrera interior*** La carrera en el centro de la bola o el soporte de ausia roler que soporta las bolas o rolers.

**Insulators** Materials that have more than 4 electrons in their outer ring. With these materials the force holding them in orbit is strong, and high voltages are needed to move them.
***Insoladores*** Materiales que tienen mas de 4 electrones en el anillo de afuera. Con esos materials, la fuerza que los detiene en orbita se hace fuerte, y se nesecitan altos voltages para moverlos.

**Integral** Built into, as part of a whole.
***Integral*** Integrado a, parte de algo completo.

**Integral power steering system** A power steering system in which the power-assist components are integral with the steering gear.
***Sistema del poder rotador integral*** Un poder del sistema rotador en el cual la fuerza asiste components que son integrals con el engrane rotador.

**Interference fit** A precision fit between components that provides a specific amount of friction between the components.
***Medida de interferencia*** Una precisa medida entre componentes que proveen una especifica cantidad de friccion, entre los componentes.

**Independent Rear Suspension (IRS)** A type of suspension system that allows the rear wheels of the vehicle to respond to road changes independently or without regard to the other side.
***Suspencion independiente trasera*** Un tipo de sistema de suspencion que permite que las llantas traseras del vehiculo, respondan a los cambios del camino, independientemente, sin que lo haga el otro lado.

**Jam nut** A second nut tightened against a primary nut to prevent it from working loose. Used on inner or outer tie-rod adjustment nuts and on many pinion-bearing adjustment nuts.
***Jam nut*** Un Segundo nudo apretado contra un primer nudo, para prevenir que el primero se afloge.Usado en nudos de ajuste externos o internos, y tambien en muchos soportes de pinon para ajustar nudos.

**Jounce** The up-and-down movement of the car in response to road surfaces.
***Pavoneo*** El movimiento de arriba a abajo del carro, en respuesta a la superficie del camino.

**Jumper wires** Wires used in electrical trouble-shooting.These wires typicly have connecters or alligator clips on the end to allow for quick insertion into a circuit.
***Cables cargadores*** Cables que se usan cuando hay problemas electricos. Esos cables tipicamente tienen conectadores o pinzas de lagarto en las puntas, que permiten una rapida insercion a un circuito.

**Key** A small block inserted between the shaft and hub to prevent circumferential movement.
***Llave*** Un pequeno bloque insertado enmedio del circuito y bobina, para prevenir movimiento circunferencial.

**Keyway** A groove or slot cut to permit the insertion of a key.
***Espacio de llave*** Una pequena ranura que permite la insercion de la llave.

**Kickback** A force supplied to the steering wheel when one of the front wheels strikes a road irregularly.
***Kickback*** Una fuerza suplida que va al volante, cuando una de las llantas de adelante golpea en un camino irregular.

**Kinetic energy** The energy of a mechanical word or motion.
***Energia kinetica*** La energia de una palabra de mecanica en movimiento.

**Kinetic friction** Friction between the moving objects or between one moving object and a stationary surface.
***Friccion kinetica*** Friccion entre dos objetos en movimiento, o entre uno que se mueve y uno que esta quieto.

**King pins** A metal rod or pin on which the steering knuckles turn.
***Broche de rey*** Un metal rodo o broche,en el que los nudos rotadores dan vuelta.

**Knock** A heavy metallic sound usually caused by a loose or worn bearing.
***Nock*** Un pesado sonido de metal, causado por un soporte suelto o acabado.

**Lateral accelerometer** A sensor used to measure vehicle cornering forces or lateral S forces.
***Acelerometro lateral*** Unsensor usado para medir la fuerza del vehiculo en las esquinas o fuerza S lateral.

**Lateral runout** A side-to-side variation, or wobble, as the tire and wheel are rotated.
***Corrida lateral*** Una variacion de lado a lado, mientras que la llanta va rodando.

**Lathe-cut seal** Another name for a square cut seal designed to withstand axial movement.
***Sello corte de torno*** Otro nombre para un sello o empaque que esta en forme de cuadro, disenado para el pedestal del eje.

**Lever** A device made up of a bar turning about a fixed pivot point, called the fulcrum, that uses a force applied at one point to move a mass on the other end of the bar.
***Lever*** Un devisador echo de una barra rotadora con un pivote arreglado con punto, llamado el fulcro, que usa una fuerza aplicada a un punto para mover una masa al otro lado de la barra.

**Leverage** The use of a lever and fulcrum to create a mechanical advantage, usually to increase force applied to an object. The brake pedal is the first point of leverage in a vehicle brake system.
***Sistema de palancas*** El uso de una palanca y fulcro para crear un avanze mecanico, usualmente para incrementar la fuerza aplicada a un objeto. El pedal de los frenos es el primer punto de palancas en el sistema de frenos de un vehiculo.

**Lift kit** An assortment of parts used to raise the axle housings or the body of the vehicle in order to increase the ground clearance of a vehicle.
***Equipo de levantamiento*** Un asorte de partes usadas usadas para levanter la caja del eje, o el cuerpo del vehiculo, para poder incrementar el espacio entre el suelo y el vehiculo.

**Linear rate springs** A coil spring with equal spacing between the coils, one basic shape, and constant wire diameter, having a constant deflection rate regardless of load.
***Resortes de rango de linear*** Un circuito de resorte que tiene el mismo espacio enmedio de los circuitos, una forma basica, y constante diametro de cable, teniendo una constante rango de deflección a pesar de la carga.

**Linear stiffness** Linear stiffness means the shaft will resist deflection regardless of length.
***Alineacion testible*** Alineacion testible quere decir que el astil resistira la flexion a pesar de la distancia.

**Linearity** The expression of sensor accuracy throughout its dynamic range.
***Alineacion*** La veracidad de la exprecion de un sensor por su rango dinamico.

**Lip seals** A type of seal used to seal parts that have axial movement.
***Sello lip*** Un tipo de sellador para sellar partes que tienen movimiento ejial.

**Lip-type seal** An assembly consisting of a metal or plastic casing, a sealing element made of rubber, and a garter spring to help hold the seal against a turning shaft.
***Tipo-lip sellador*** Un asamble que consiste de un metal o casco de plastico, un elemento sellador echo de plastico, y un resorte garter para ayudar a retener el sello contra el astil rodando.

**Live axle** An axle assembly that supports and drives the vehicle. Dead axles support a load and the vehicles wheels but do not drive the vehicle.
***Eje vivo*** Un asamble de astil que soporta y maneja el vehiculo.El eje muerto soporta la carga y las llantas del vehiculo, pero no maneja el vehiculo.

**Loads** Devices that use electricity to perform work, such as lights and motors.
***Cargas*** Devisadores que usan electrecidad para hacer el trabajo, como las luces y motores.

**Lock pin** Used in some ball sockets (inner tie-rod end) to keep the connecting nuts from working loose. Also used on some lower ball joints to hold the tapered stud in the steering knuckle.
***Broche de seguro*** Usado en algunos sockets de bola (fin interno) para prevenir que los nudos de coneccion se aflogen.Tambien usados para coyonturas de bola bajas que detienen el tabique en el nudo rotador.

**Lock ring** A type of sealing ring that has ends that meet or lock together during installation. There is no gap between the ends when the ring is installed.
***Anillo de seguro*** Un tipo de anillo de sello, que tiene orillas que se encuentran y se atoran juntas durante instalacion. No hay ranuras enmedio de las orillas cuando el anillo es instalado.

**Locking** A bearing fault caused by large particles of dirt that become trapped between a bearing and its race.
***Atorado*** Falla de un soporte causada por particulas de polvo que se quedan atrapadas enmedio de un soporte y su race.

**Lock nut** A second nut turned down on a holding nut to prevent loosening.
***Nudo asegurado*** Un Segundo nudo atado en otro nudo para prevenir que el primero se afloge.

**Lockplates** Metal tabs bent around nuts or bolt heads.
***Platos de seguro*** Pedazos de metal doblados alrededor del nudo, o cabeza del nudo.

**Lockwasher** A type of washer which, when placed under the head of a bolt or nut, prevents the bolt or nut from working loose.
***Wacha de seguro*** Un tipo de wacha que, cuando es pueste debajo de la cabeza del tornillo o del tornillo, previene que el tornillo o nudo se afloge.

**Lubricant** Any material, usually a petroleum product such as grease or oil, that is placed between two moving parts to reduce friction.
***Lubricante*** Cualquier material, usualmente un producto de petroleo como grasa o aceite, que es puesta enmedio de dos partes para reducer friccion.

**Lubrication** The process of reducing friction between the moving parts of an engine.
***Lubricacion*** El proceso de reducer friccion enmedio de partes que se mueven en un motor.

**MacPherson strut suspension** A suspension system in which the strut is connected from the steering knuckle to an upper strut mount, and the strut replaces the shock absorber.
***Suspencion de rostra de MacPherson*** Un sistema de suspencion en el cual,la rostra es conectada de el nuckle rotador a otra rostra montada mas arriba, y la rostra remplaza el circuito absorbedor.

**Magnetic pulse generator** An engine position sensor used to monitor the position of the crankshaft and control the flow of current to the center base terminal of the switching transistor.
***Pulso generador magnetico*** Un sensor de pocision de motor usado en el monitor de pocision.

**Magnetism** A force between two poles of opposite potential caused by the alignment of electrons.
***Magnetismo*** Una fuerza enmedio de dos polos de potencial opuesto, causado por el alineamiento de electrons.

**Mandrel** A tapered bar that acts as a template for the shaping or bending of metal.
***Mandrel*** Una barra que actua como template para una forma o dobles de metal.

**Mass** The measure of inertia of an object or form of matter or its resistance to acceleration. Also the molecular density of an object.
***Masa*** La medida de inercia de un objeto o forma de material o su resistencia a acaleracion. Tambien la densidad molecular de un objeto.

**Material Safety Data Sheets (MSDS)** Information sheets containing chemical composition and precautionary information for all products that can present a healh or safety hazard.
***Material de informacion contra el peligro*** Hojas de informacion que contienen detalles acerca de composicion e informacion acerca de los productos que representan peligro o amenaza de salud.

**Matter** Anything that occupies space.
***Materia*** Cualquier cosa que ocupa espacio.

**Maxi-fuses** A heavy-duty circuit protection device that has the appearance of a large blade-type fuse.
***Maxi-fusibles*** Un circuito de proteccion de alta potencia, devisador que tiene la apariencia de un tipo-navaja fusible.

**Mechanical advantage** Based on the principles of levers, force can be increased by increasing the length of the lever or by moving the pivot or fulcrum.
***Avanse mecanico*** Basado en la principal de las palancas, la fuerza puede ser incrementada al incrementar lo largo del la palanca o moviendo el pivote o fulcro.

**Memory steer** When memory steer occurs; the vehicle does not want to steer straight ahead after a turn because the steering does not return to the straight-ahead position.
***Memoria rotadora*** Cuando memoria rotadora ocurre, el vehiculo no vuelve a su linea derecha despues de hacer una vuelta, porque el volante no vuelve a la pocision derecha.

**Mesh** To fit together, as gear teeth.
***Mechear*** Que quedan juntas, como dientes de engrane.

**Meshing** The mating, or engaging, of the teeth of two gears.
***Mecheando*** El mache, o enlaze, de los dientes de dos engranes.

**Micrometer** A precision measuring device used to measure small bores, diameters, and thicknesses. Also called a mike.
***Micrometro*** Un devisador de medida presisa, usada para medir pequenas barras, diametros y grosuras.Tambien llamado un mike.

**Microprocessor** A digital computer, or processor, built on a single integrated circuit ship, performs functions of arithmetic logic and control logic.
***Microprocesor*** Una computadora digital, o procesador, hecho en un solo circuito integrado, proporciona funciones de aritmetica logica y control logico.

**MIL** The malfunction indicator lamp for a computer control system, prior to J1930, commonly called a Check Engine or Service Engine Soon lamp.
***Mil*** La lampara indicadora de mal function para el sistema de control de una computadora,antes de J1930, comunmente llamado Servicio de recvision de la lampara de la maquina.

**Mineral spirits** An alcohol-based liquid that leaves no surface residue after it dries.
***Espiritus minerals*** Basado en alchol liquido que no deja superficie o residuo despues que se seca.

**Misalignment** When bearings are not on the same center line.
***Desalineamiento*** Cuando los soportes no estan en la misma linea central.

**Molecule** The smallest particle of an element or compound that can exist in the free state and still retain the characteristics of the element or compound.
***Molecula*** La mas pequena particular de un elemento o compuesto que puede existir en el estado libre y todavia retener las caracteristicas de un elemento o compuesto.

**Momentum** A type of mechanical energy that is the product of an object's weight times its speed.
***Momenum*** Un tipo de energia mecanica que es el producto de el peso de un objeto.

**Mounts** Train part, such as engine or transmission mounts.
***Monte*** Parte tren, como el monte de el motor o transmission.

**Multimeter** A tool that combines the voltmeter, ohmmeter, and ammeter together in one diagnostic instrument.
***Multimetro*** Una herramienta que combina el volmetro, ohmmeter, y ammeter juntos en un instrumento de diagnostico.

**Multiple disc** A clutch with a number of driving and driven discs as compared to a single plate clutch.
***Disco multiple*** Un cloche con un numero de discos de manejar y manejados, comparados a un plato singular de cloche.

**Multiplexing** An electrical system in which voltage signals or information can be shared with two or more computers.
***Multiplex*** Un sistema electrico en el cual las senales de voltage pueden ser compartidas con dos o mas computadoras.

**Needle bearing** An anti-friction bearing using a great number of long, small-diameter rollers. Also known as quill bearing.
***Agujas de soporte*** Un soporte de antifriccion, usando un gran numero de largos, pequeno diametro rolers. Tambien conocidos como soportes quill.

**Negative scrub radius** Negative scrub radius is when the SAI line meets the road surface outside the true vertical centerline of the tire at the road surface.
***Radios negativos scrub*** Radios negativos scrub es cuando el SAI encuentra la superficie del camino fuera de la linea central vertical de la llanta en la superficie del camino.

**Negative wheel slip** Wheel speed is less than vehicle speed during breaking.
***Resbalo negetivo de la llanta*** La velocidad de la llamta es menos que la del vehiculo, al tiempo de frenar.

**Network** The channel through which several computers share.
***Coneccion*** El canal por el cual varias computadoras comparten informacion.

**Neutral** A solution that has ph of 7. Also, the condition where work is not being transferred.
***Neutral*** Una solucion que tiene ph de 7.Tambien, la condicion cuando el trabajo no es tranferido.

**Neutral safety switch** A switch wired into the ignition switch to prevent engine cranking unless the transmission shift lever is in neutral.
***Encendedor salvo neutral*** Un encendedor de cables que va al encendedor, para prevenir el cranque del motor, a menos que la palanca de velocodades este en neutral.

**Newton-Meter (N·m)** metric measurement of torque or twisting force.
***Metro neuto*** Medida metrica de tuerce o fuerza de torcer.

**Noise** An unwanted or annoying sound.
***Ruido*** Un ruido no deseado y fastidioso.

**Normal shim** A shim with designated thickness.
***Riel normal*** Un riel con una grosura disenada.

**Non-hardening** A gasket sealer that never hardens.
***No endurecedor*** Un empaque sellador que no se endurece.

**Non-integral** This is also called an add-on ABS unit and is separate from the master cylinder and power booster.
***No integral*** Esto es llamado tambien un add-on ABS unidad y esta seoarado del cilindro principal y disparador de fuerza.

**Occupational Safety and Health Administration (OSHA)** The government agency charged with ensuring safe work environments for all workers.
***Administracion de cuidado y salud ocupacional (OSHA)*** Una agencia de govierno encargada de asegurar trabajo cuidadoso y buen ambiente para los trabajadores.

**OD** Outside diameter.
***OD*** Diametro de afuera.

**Offset** Placed off center. Also, the measurement between the center of the rim and the point where a wheel's center is mounted.
***Fuera de medida*** No colocado en el centro. Tambien, la medida enmedio del centro del ring y el punto donde el centro de la llanta esta montado.

**Ohm** A unit of measured electrical resistance.
***Ohm*** Una unidad de medida electrica resistente.

**Ohmmeter** The meter used to measure electrical resistance.
***Ohmmetro*** El metro usado para medir resistencia electrica.

**Ohm's law** A basic law of electricity expressing the relationship between current, resistance, and voltage in any electrical circuit.
***Ley de Ohm*** Una ley basica de electricidad expresando la relacion entre resistencia actual y voltage en un circuito electrico.

**Oil seal** A seal placed around around a rotating shaft or other moving part to prevent leakage of oil.
***Sello de aceite*** Un sello puesto alrededor de una parte rotadora o otra parte moviendose para prevenir goteras de aceite.

**Open** An electrical circuit that has a break in the wire or is not complete.
***Abierto*** Un circuito electrico que tiene una ranura en un cable o no esta complete.

**Open-end-seals** A seal whose ends do not meet when it is installed.
***Sello de orillas abiertas*** Un sello, el cual las orillas no se encuentran cuando es instalado.

**Orfice** A small opening designed to restrict flow.
***Orificio*** Una pequena abertura disenada para restrictir liquido.

**O-ring** A type of sealing ring, usually made of rubber of rubber-like material. In use, the O-ring is compressed into a groove to provide sealing action.
***Anillo–O*** Un tipo de anillo sellador, usualmente echo de plastico o material como plastico. En el uso el anillo de O apretado en un grove para proveer accion de sellar.

**Oscillation** Any single swing of an object back and forth between the extremes of its travel.
***Osilacion*** Cualquier columpio singular de un objeto que va para atras y para adelante enmedio de los extremos.

**OSHA** The Occupational Safety and Health Administration, a government agency charged with ensuring safe work environments for all workers.
***OSHA*** La administracion ocupacional de cuidado y salud, una agencia de govierno encargada de asegurar y asegurar el bienestar de los trabajadores.

**Outboard joint** The CV joint that connects the drive axle to the wheel spindle.
***Coyontura de fuera borda*** La coyontura CV que conecta el eje de manejo a la llanta espindola.

**Out-of-round** An inside or outside diameter, designed to be perfectly round and having varying diameters when measured at different points across its diameter.
***Fuera de rueda*** Un diametro de adentro o afuera, disenado para ser perfectamente Redondo y tener varios diametros cuando medida de diferentes puntos cruzando este diametro.

**Output driver** Located in the processor and operated by the digital commands of the computer, an output driver is an electronic on/off switch used to control the ground circuit of a specific actuator.
***Manejador puesto afuera*** Localizada en el processor y operada por los commandos digitales de la computadora, un manejador puesto afuera es un encendedor electronico usado para controlar el circuito de la tierra de un actuador especifico.

**Overall ratio** The product of the transmission gear ratio multiplied by the final drive or rear axle ratio.
***De todo ratio*** El producto de el engrane de la transmision multiplicada por el final o astil trasero.

**Oxidation** Burning or combustion; the combining of a material with oxygen. Rusting is slow oxidation, and combustion is rapid oxidation.
***Oxidacion*** Quemadura o combustion; la combinacion de un material con oxigeno. Quemadura es oxidacion lenta, combustion es rapida oxodacion.

**Ozone** An unstable molecule of oxygen with three atoms instead of the normal two. Ozone oxides other elements and compounds by giving up its extra oxygen atom.
***Ozona*** Una molecula de oxigeno instable con trs atomos en vez de los dos normales.Ozona oxida otros elementos y compuestos dejando ir su oxigeno extra de atomos.

**Parallel circuit** An electrical circuit that has more than one path for the current to follow.
***Circuito paralelo*** Un circuito electrico que tiene mas de un molde para seguir lo actual.

**PCM** The power train control module of a computer control system, prior to J1930, commonly called an ECA, ECM or one of many acronyms used by the various manufacturers.
***PCM*** La modula de control del tren de fuerza del systems de control de una copmutadora, previo a J1930, comunmente llamada ECA, ECM, o uno de muchos acronomos usados por varias fabricas.

**Permanent magnet (PM) generator** A reluctance sensor. A sensor that generates a voltage signal by moving a conductor through a permanent magnet field.
***Iman permanente (PM) generador*** Un sensor reluctante. Un sensor que genera una senal de voltage moviendo un conductor por un campo magnetico permanente.

**Permeable** Materials that absorb fluids.
***Permeable*** Materiales que absorben liquido.

**pH scale** Scale used to measure how acidic or basic a solution is.
***Escala pH*** Escala usada para medir que tan acidica o basica una solucion es.

**Pinch bolt** A bolt used to squeeze a slotted bore tightly around a shaft or stud.
***Tornillo pinchador*** Un tornillo usado para apretar una cabeza plana alrededor de un astil o stud.

**Pitch** The number of threads per inch on any threaded part.
***Piche*** El numero de pasadas por pulgada o cualquier otra parte con pasada.

**Pitman arm** A steering linkage component that connects the steering gear to the linkage at the left end of the center link.
***Brazo de Pitman*** Una coneccion componente rotadora que se conecta al engrane rotador aa la coneccion del lado izquierdo del enlaze central.

**Pivot** A pin or shaft upon which another part rests or turns.
***Pivote*** Un broche o astil an el cual otra parte descanza o rota.

**Play** Movement between two parts.
***Movimiento*** Movimiento entre dos partes.

**Plunging joints** The type of CV joint that allows the half-shaft to effectively change length in response to suspension movements by having the capability of in-and-out movements.
***Coyonturas conectables*** El tipo de coyontura CV que permite que medio astil cambia efectivamente la longitude en respuesta a los movimientos de la suspencion teniendo la capacidad de movimientos de afuera y adentro.

**P-metric system** The most common modern system to specify passenger car tire sizes.
***Sistema primetrico*** El sistema mas moderno y comun para especificar las tallas de las llantas de carros de pasageros.

**Polyglycol** Polyalkylene-glycol-ether brake fluids that meet specifications for DOT 3 DOT 4 brake fluids.
***Polyglycol*** Polialkalina-glycol-eter liquido para los frenos que supple especificaciones para DOT 3 DOT 4 liquido de frenos.

**Porosity** A statement of how porous or permeable to liquids a material is.
***Porosidad*** Una pruebe de que poroso o permeable un material puede ser a los liquidos.

**Positive scrub radius** Occurs when the SAI line meets the road surface inside the true vertical centerline of the tire at the road surface.
***Rango positive scrub*** Ocurre cuando la linea de SAI se encuentra con la superficie del camino dentro de la linea vertical en la linea centro de la llanta en la superficie del camino.

**Positive wheel slip** Wheel speed is greater than the vehicle speed during acceleration.
***Slip llanta positive*** La velocodad de la llanta es mas alta que la del vehiculo durante la aceleracion.

**Potential energy** Stored energy.
***Energia potencial*** Energia guardada.

**Potentiometers** A type of variable resistor. Potentiometers provide a varying voltage signal from its tap and have three connections, one at each end of the resistance and one connected to a sliding contact with the resistor. Turning the control moves the sliding contact away from one end of the resistance, but toward the other end.
***Potenciometros*** Un tipo de resistidor variable.Potenciometros proveen una variedad de senales de voltage de su tapa y tiene tres conecciones, uno en cada orilla de la resistencia y uno conectado ligeramente junto con el resistidor.

**Power** A measure of work being done.
***Poder*** Una medida de trabajo siendo echa.

**Power train** The mechanisms that carry the power from the engine crankshaft to the drive wheels; these include the clutch, transmission, driveline, differential, and axles.
***Tren de fuerza*** El mecanismo que lleva la fuerza de la maquina del cranque a las llantas; estos incluyen el cloche, transmision, linea de manejo, diferencial, y ejes.

**Powermaster** A self-contained hydraulic power brake system with its own hydraulic reservoir and independent electrical pump; developed by General Motors for cars and trucks that could not economically use vacuum-assisted power brakes.
***Maestro de fuerza*** Una fuerza hidraulica del sistema de frenos que se contiene a si misma con su propio reservador electrico y bomba electrica independiente; desenvuelta por General Motors para carros y trocas que pudieran no usar frenos asistidos por una aspiradora.

**Powertrain Control Module (PCM)** The computer responsible for the monitoring and controlling of the engine and its related systems.
***Modula de control del tren de fuerza*** La computadora responsible de observar y controlar el motor y sus sistemas relacionados.

**Preload** A load applied to a part during assembly so as to maintain critical tolerances when the operating load is applied later.
***Precargado*** Una carga aplicada a una parte durante el asamble para mantener tolerancias criticas cuando la operacion de la carga es aplicada despues.

**Press fit** Forcing a part into an operating that is slightly smaller than the part itself to make a solid fit.
***Presionar*** Forzar una parte con una que esta operando que es ligeramente mas pequena que la otra parte que quedan bien.

**Pressure** Force exerted on a given unit of a surface area. Pressure equals force divided by area and is measured in pounds per square inch (psi) or kilopascals (kPa).
***Presion*** Fuerza exertada en una unidad dada o en una area. Precion igual a fuerza dividida por area y es medida en libras por pulgada cuadrada (psi) o kilopascals (kPa).

**Pressure bleeding** The process of using a tank filled with brake fluid and pressurized with compressed air to expel air and brake fluid from the system. Pressure bleeding is a one-person operation.
***Presion de liquido*** El proceso de usar un tanque lleno de liquido de frenos y precionada con pompresora de aire para expulsar aire y liquido de frenos del sistema. Presion de liquido es una operacion de una sola persona.

**Pressure differential** The difference between two pressures on two surfaces or in two separate areas. The pressures can be either pneumatic (air) or hydraulic.
***Presion del diferencial*** La diferencia entre dos presiones en dos lugares o en dos areas separadas. Las presiones pueden ser neumaticas (aire) o hidraulica.

**Primary shoe** The leading shoe in a duo-servo brake. The primary shoe is self-energized and applies servo action to the secondary shoe to increase its application force. Primary shoes have shorter linings than secondary shoes.
***Zapato principal*** El zapato principal entre dos-servo freno.El zapato principal es a si mismo energetizado y aplica servo accion al zapato secundario para incrementar la aplicacion de fuerza. Zapato principal tienen lineas mas cortas que los zapatos secundarios.

**Processor** The term used to describe that actual metal box that houses the computer and its related components.
***Procesor*** El termino usadi para describir la actual caja de metal que conserva la computadora y sus componentes relacionados.

**Program** The job instructions for a computer.
***Programa*** Las instrucciones de trabajo para una computadora.

**PROM** Programmable Read Only Memory. A computer IC chip that contains all the information about the vehicle and its systems so that the control system can make decisions that are based on that vehicle.
***PROM*** Memoria programable only para leer. Un micro chip de una computadora que contiene toda la informacion de un vehiculo y su sistema, para que el sistema de control pueda hacer decisions que estan basadas en el vehiculo.

**psi** Abbreviation for pounds per square inch, a measurement of pressure.
***psi*** Abreviacion para libras por pulgada cuadrada, una medida de presion.

**Pulley** A wheel with a grooved rim in which a rope, belt, or chain runs to raise something by pulling on the other end of the rope, belt, or chain.
***Jalador*** Una rueda con un ring en el cual una cuerda, banda, o cadena corre para levanter algo jalandolo en el otro lado de a cuerda, banda, o cadena.

**Pulsation** The rapid up-and-down pumping movement of the clutch pedal as the clutch engages or disengages.
***Pulsacion*** El movimiento rapido de un bombeo hacia arriba y hacia abajo, mientras el cloche se engancha o se desengancha.

**Pulse width** The amount of time that an electromachanical device, such as a solenoid, is energized. Pulse width usually is measured in milliseconds.
***Pulso de anchura*** La cantidad de tiempo que un devisador electromecanico, como un soleniode es energizado. Pulso de anchura usualmente es medida en milisegundos.

**Pulse-width modulation (PWM)** The characteristics of a continuous on-and-off cycling of a solenoid for a fixed number of times per second. While the frequency of the cycles remains constant, the ratio of on time to total cycle time varies or is modulated.
***Modulacion de pulso de achura*** Las caracteristicas de una continua ronda de encendido o apagado, de un solenoide, para un numero acomodado de veces por Segundo. Mientras que la frecuencia de la ronda permanence constante, la cantidad de veces de la ronda total, varia o puede ser modulada.

**Quadrant** A quarter of something.
***Cuadrante*** La cuarta parte de algo.

**Race** A channel in the inner or outer ring of an anti-friction bearing in which the balls or rollers roll.
***Carrera*** Un cana en el ring de adentro o de afuera de un soporte de antifriccion en el que las bolas o los roles ruedan.

**Rack-and-pinion steering** The end of the steering shaft has a pinion gear that meshes with a rack gear.
***Racke y pinon rotador*** La orilla del astil rotador tiene un engrane que mechea con los dientes del engrane.

**Radial** The direction moving straight out from the center of the circle perpendicular to the shaft or bearing bore.
***Radial*** La direccion moviendose derecho fuera del centro de el circulo perpendicular de el astil o soporte del bore.

**Radial clearance (radial displacement)** Clearance within the bearing and between the balls and races perpendicular to the shaft.
***Espacio radial ( desplazamiento radial )*** Espacio entre el soporte y enmedio de las bolas y las carreras perpendiculars del astil.

**Radial load** A load that acts at 90 degrees to the axle's center of axis rotation. Radial loading is always present whether the vehicle is moving or not.
***Carga radial*** Una carga que actua a 90 grados al axis de rotacion del eje. Carga radial es siempre presente ya sea que el carro este corriendo o no.

**Radial ply tire** Tire construction in which the cords in the body plies of the carcass run at an angle of 90 degrees to the steel beads in the inner rim of the carcass. Each cord is parallel to the radius of the tire circle.
***Llanta radial fuerte*** Construccion de la llanta en la cual los cordones de las capas del cuerpo de el carcass corren a un angulo de 90 grados, en las bolas de fierro en el interior del ring de el carcass. Cada cordon es paralelo al radio del circulo de la llanta.

**Radial runout** An outf-round condition in which the radius of the wheel or tire is not consistent from the wheel center to any point on the rim or the tread.
***Radial fuera de ronda*** La condicion de un radial fuera de ronda en el cual el radio de la llanta no es consistente de el centro de la llanta a ningun punto en el ring o la cuerda.

**Radial tire runout** Variations in tire diameter.
***Llanta radial fuera de ronda*** Variacion en el diametro de la llanta.

**Ratio frequency interference** A narrow band of frequencies within the electromagnetic interference spectrum.
***Frecuencia de la interferencia del ratio*** Una banda angosta de frecuencias en el spectrum magnetico de interferencias.

**Radius arm** A longitudinal suspension arm used to position an axle.
***Brazo del radio*** Un brazo de suspencion de longitude, usado para posisionar el eje.

**Ratio** The relation or proportion that one number bears to another.
***Radio*** La relacion o proporcion que un numero difiere del otro.

**Reactivity** The characteristic of a material that enables it to react violently with water and other materials.
***Reactividad*** La caracteristica de un material, que hace que este reaccione violentamente con agua y otros materials.

**Read only memory (ROM)** A type of memory used in automotive microcomputers to store information permanently.
***Memoria de lectura (ROM)*** Un tipo de memoria usada en autumotriz.

**Ream** A process that cleans out or slightly enlarges a hole when turned inside it.
***Rim*** Un proceso que limpia o ligeramente limpia un agujero entrando.

**Rear axle offset** A rear axle offset occurs when the complete rear is rotated slightly, moving one rear wheel backward and the opposite rear wheel forward.
***Eje trasero desnivelado*** Un eje trasero desnivelado ocurre cuando el eje trasero esta lijeramente movido y la llanta de un lado esta un poco mas adelante que la otra.

**Rear axle sideset** Rear axle sideset is a condition where the rear axle has shifted straight sideways so the geometric centerline is not at the true vehicle centerline position.
***Eje trasero ladeado*** Eje trasero ladeado es cuando el eje trasero se mueve hacia unlado y la linea geometrica no esta derecha conforme a la posicion de linea central.

**Rear wheel drive** A term associated with a vehicle where the engine is mounted at the front and the driving axle and driving wheels are mounted at the rear of the vehicle.
***Doble traccion trasera*** Un termino asociado con un vehiculo el cual el motor esta montado enfrente, y el eje de manejo y llantas de manejo estan montadas en la parte de atras del vehiculo.

**Rebound** The movement of the suspension system as it attempts to bring the car back to normal heights after jounce.
***Rebote*** El movimiento del sistema de suspencion que hace que la parte de atras del carro vuelva a la altura normal despues de botar.

**Recirclating ball steering gear** A popular low-friction steering gear box. A sector gear meshes with a ball nut that rides on ball bearings on the worm shaft to provide a smooth steering feel.
***Reciclando la bola del engrane rotador*** Una popular caja de engrane rotador de baja friccion. Un sector de engrane que mechea con el nudo de bola que corre con soportes de bola en el astil, para proveer un suave sentir rotador.

**Reference voltage** A fixed voltage supplied to the sensor by a voltage regulator inside the computer or control module. As the sensor changes the return voltage is altered and sent back to the computer for use. Most control systems operate with a 5-volt reference voltage.
***Referencia de voltage*** Un voltage acomodado y suplido para el sensor por el regulador de voltage dentro de la computadora o el modulo de control. Mientras el sensor cambia el voltage devuelto, es alterado y enviado de Nuevo a la computadora para ser usado. La mayoria de los sistemas de control operan con 5-volt referencia de voltage.

**Reference voltage (Vref) sensors** Sensors that provide input to the computer by modifying or controlling a constant, predetermined reference voltage signal.
***Referencia de voltage*** Sensores que proveen entrada a la computadora modificando o controlando un constante predeterminada referencia de senal de voltage.

**Relay** An electromagnetic device used to control a high-current circuit with a low-current control circuit.
***Relevo*** Un devisador electromagnetico usado para controlar circuito de corriente alta con un circuito de control de baja corriente.

**Resistance** An electrical term for something that opposes current flow.
***Resistencia*** Un temino electrico para algo que es opuesto a corriente.

**Retaining ring** A removable fastener used as a shoulder to retain and position a round bearing in a hole.
***Anillo retenedor*** Un cierre removible usado como reforse para retener en posicion un soporte metido en un agujero.

**Return spring** A strong spring that retracts a drum brake shoe when hydraulic pressure is released.
***Resorte retractor*** Un resorte fuerte que retracra un zapato de freno al tambor, cuando la presion hidraulica es relevada.

**RFI** Radio frequency interference, which is an unwanted voltage signal that rides on a signal.
***RFI*** Interferencia de frecuencia de radio, el cual es una senal de voltage no deseada que corre en una senal.

**Rheostat** A type of variable resistor that provides for varying amounts of voltage and current from its tap.
***Resistidor*** Un tipo de resistidor variable que prove para varias cantidades de voltage y corriente de su tapa.

**Right-to-Know Laws** Laws requiring employers to provide employees with a safe workplace as it relates to hazardous materials.
***Derecho-de-saber la ley*** Las leyes requieren que los patrones proveean a los empleados con un salvo lugar de trabajo y lo que se relaciona a material peligroso.

**Road crown** The slant of a road that allows water to drain off. Road crown causes a vehicle's steering to drift to the right.
***Corona del camino*** La inclinacion del camino que permite que el agua se drene. Corona del camino provoca que el rotor del vehiculo se resbale hacia la derecha.

**Rolled finish** The result of a manufacturing process in which a hardened metal is repeatedly rolled over with great pressure on the surface of the porous metal until the pores in the metal close.
***Acabado de roll*** El Resultado de un proceso de fabrica en el cual el metal es repetidamente enrollado con una gran presion en la superficie de los poros del metal hasta que los poros desaparecen.

**Roller bearing** An inner and outer race upon which hardened steel rollers operate.
***Soporte de roll*** Un soporte de carrea de entrada y salida, en el cual, rolles de fierro endurecido operan.

**Roller clutch** A type of overrunning clutch that uses rollers to engage or disengage.
***Cloche de roller*** Un tipo de corredor de cloche que usa rolles para engancharse o desengancharse.

**Rollers** Round steel bearings that can be used as the locking element in an overrunning clutch or as a rolling element in an antifriction bearing.
***Rolles*** Soportes de fierro redondos que pueden ser usados para atrancar un elemento en un corredoe de cloche o un elemento rodante en un soporte de antifriccion.

**Rotor** The rotating part of a disc brake that is mounted on the wheel hub and contracted by the pads to develop friction to stop the car. Also called a disc.
***Rotador*** La parte rotadora de un disco de freno, que esta montado en el cubo de la llanta, y es contractado por los zapatos de freno, para desenvolver friccion para detener el carro. Es tambien llamado, disco.

**Rotor-type pump** A fluid pump that uses an inner and outer rotor inside a housing to move the fluid.
***Rotador-tipo bomba*** Una bomba de fluido que usa un rotor que entra y sale en la caja para mover el fluido.

**Runnout** Deviation of the specified normal travel of an object. The amount of deviation or wobble a shaft or wheel has as it rotates. Runnout is measured with a dial indicator.
***Correfuera*** Desviacion del viaje normal delviaje especifico de un objeto. La magnitude de desviacion de un engrane o llanta al rodar. Desviacion es medida con un dial indicador.

**RWD** Abbreviation for rear-wheel drive.
***RWD*** Abreviacion para coble traccion trasera.

**Rzeppa joint** The most commonly used ball-type CV joint.
***Coyontura rezipa*** La mas comunmente usada coyontura de tipo-bola.

**Scan tool** A microprocessor designed to communicate with the vehicle on-board computer in order to perform diagnosis and troubleshooting.
***Llave Scan*** Un microprocessor disenado para comunicarse con la computadora que esta montada en el vehiculo, para diagnosticar los problemas de encendido.

**Scarf-cut rings** A name for Teflon-locking-end seals.
***Anillos corte-scarf*** Un nombre para los anillos que sellan sellos de Teflon.

**Score** A scratch, ridge or groove marring a finished surface.
***Talladura*** Un rasguno, rspon o ranura estropeando una superficie pulida.

**Screw pitch gauge** A tool used to check the threads per square inch (pitch) of a fastener.
***Llave de caliber de balanceo*** Una llave usada para chequear las roscas por pulgada cuadrada ( pitch ) de un asegurador.

**Scrub radius** The distance from the tire contact patch centerline to the point where the steering axis intersects the road.
***Espacio radio*** La distancia desde el contacto de la linea central del parche de la llanta al punto donde el eje rotador intersecta el camino.

**Scuffing** Scraping and heavy wear from the piston to the cylinder walls.
***Asperosidad*** Raspaduras y asperesa de el piston a las paredes del cilindro.

**Seal** A material shaped around a shaft used to close off the operating compartment of the shaft preventing oil leakage.
***Sello*** Un material en forma redonda una flecha usada para apagar el compartimiento de operacion de el astil previniendo goteras de aceite.

**Sealer** A thick, tacky compound usually spread with a brush.
***Sellador*** Un espeso, grueso compuesto usualmente untado con una brocha.

**Section width** The width of a tire across the widest point of its cross section, usually measured in millimeters.
***Seccion de anchura*** La anchura de una llanta cruzando el punto ancho de su seccion cruzando, usualmente medida en milimetros.

**Semiconductor** A material that is not a good conductor or a good insulator but can function as either when certain conditions exist.
***Semiconductor*** Un material que no es un buen conductor o buen insolador, pero puede funcionar cuando ciertas condiciones existen.

**Sensor** Any device that sends an input signal to a computer.
***Sensor*** Cualquier devisador que envia una senal a una computadora.

**Serial data** The communications to and from the computer.
***Serie de datos*** Las comunicaciones entre computadoras.

**Series circuit** An electrical circuit that has only one path for current flow.
***Series de circuito*** Un circuito electrico que tiene solo un molde para fluido corriente.

**Serpentine belts** Multiple-ribbed belts used to drive water pumps, power steering pumps, air-conditioning compressors, alternators, and emission control pumps.
***Bandas de serpentina*** Bandas con multiples costillas usadas para manejar las bombas de agua, bombas del fuerza rotadora, compresora del aire acondicionado, alternadores, y bombas de control de emision.

**Setback** A condition where one front or rear wheel is moved reward in relation to the opposite wheel.
***Contratiempo*** Una condicion donde las llantas de atras o de adelante son movidas,o estan desaniveladas una con la otra.

**Shank** The part of a shoe that protects that ball of the foot.
***Gancho*** La parte del zapato del freno que cubre la bola.

**Shim** Thin sheets, usually metal, that are used as spacers, between two parts, such as the two halves of a journal bearing.
***Riel*** Hojas delgadas, usualmente metal, que son usados como separadores, enmedio de dos partes, tales como las dos mitades de un soporte de jornal.

**Shimmy** Tire wobbles felt through the steering system.
***Rechinidos*** Rechinidos que se sienten en las llantas en el volante.

**Shock absorber** A device that dampens spring oscillations by converting the energy from spring movement into heat energy.
***Absorbedor de chokes*** Un devisador que golpea en un resorte y convierte la energia del golpe en movimiento en energia de calor.

**Short** An intentional or unintentional grounding of an electrical unit.
***Corto*** Un intencional o anintencional descarga de una unidad electrica.

**Short/long arm suspension** A suspension system using an upper and lower control arm. The upper arm is shorter than the lower arm.
***Corto-largo brazo de suspencion*** Un sistema de suspencion usando un alto y bajo brazo de control.El brazo de arriba es mas alto que el de abajo.

**Shudder** Momentary shake or quiver; can sometimes be severe.
***Temblor*** Temblor momentaneo o estremecimiento, algunas veces severo.

**Sideset** A condition where the rear axle assembly has moved straight sideways and the geometric centerline is not positioned at the true vehicle centerline.
***Ladeado*** Una condicion cuando el asamble del eje trasero se ha movido derecho hacia un lado y la geometria de la linea central no esta localizada en la verdadera linea central del vehiculo.

**Sidewalls** The sides of the tire's body.
***Paredes de lados*** Los lados del cuerpo de la llanta.

**Slip** A condition caused when a driving part rotates faster than a driven part.
***Desliz*** Una condicion causada cuando al manejar una parte rueda mas rapido que otra.

**Slip angle** The actual angle of the front wheels during a turn compared to the turning angle of the front wheels with the vehicle centerline.
***Angulo de desliz*** El angulo actual de las llantas de adelante durante una vuelta comparado a el angulo de voltear de las llantas de adelante con la linea central del vehiculo.

**Slip joint** A variable-length connection that permits the drive shaft to change its effective length.
***Coyontura de deslize*** Una longitude variable de coneccion que permite que el astil de manejo cambie a su efectiva longitude.

**Slip yoke** The transmission end of a shaft that fits into the extension housing and onto the splines of the output shaft.
***Yugo de deslize*** La orilla de la transmision de un astil que queda en una caja de extencion u en los splines de un astil.

**Snapring** Split spring-type ring located in an internal or external groove to retain a part.
***Anillobroche*** Un broche como tipo-resorte localizado en un broche interno o externo, para retener una parte.

**Solenoid valve** A mechanical valve operated by a solenoid to control the flow of liquid or gas.
***Valvula de solenoide*** Una valvula mecanica operada por un solenoide que controla la salida de liquido o gas.

**Solvent** The liquid in a solution.
***Solvente*** El liquido en una solucion.

**Spalling** A condition of a bearing caused by overloading the bearing as evidenced by pits on the bearings or their races.
***Astillado*** L acondicion de un soporte causada por la sobrecarga en un soporte que se averea durante la carreras.

**Speed** The distance an object travels in a set amount of time.
***Velocidad*** La distancia que un objeto viaja en una cierta cantidad de tiempo.

**Spindle** The shaft on which the wheels and wheel bearings mount.
***Espindola*** El astil en el que el soporte de las llantas y el volante estan montados.

**Spline** Slot or groove cut in a shaft or bore; a splined shaft onto which a hub, wheel, gear, etc., with matching splines in its bore is assembled so that the two must turn together.
***Espindol*** Espacio cortedo en un astil o tornillo; en el cual una bobina, llanta, engrane, etc.. con splines que mechean el tornillo es asamblado para que los dos rueden.

**Split lip seal** Typically, a rope seal sometimes used to denote any two-part oil seal.

***Split pin*** A round split spring steel tubular pin used for locking purposes; for example, for locking a gear to a shaft.
Seguro del split Un seguro Redondo de espiral de fierro usado para proposito de atrancar;por ejemplo, para atrancar engranes o astiles.

**Spring** A device that changes shape when it is stretched or compressed but returns to its original shape when the force is removed.
***Resorte*** Un devisador que cambia la figura cuando es estirado o compresado, pero vuelve a su figura original cuando el forcr es removido.

**Spring pads** A mounting point for a spring. Normally a spring pad is made of insulating material such as rubber.
***Cojinete de resorte*** Un punto de posicion para un resorte. Normalmente el cojinete de un resorte esta echo de un material de insolacion tal como plastico.

**Square-cut seal** A fixed seal for a caliper piston that has a square cross section.
***Empaque cuadrado de empaque*** Un sello arreglado para el calibre del piston que tiene una seccion cuadrada.

**Squeak** A high-pitched noise of short duration.
***Rechinido*** Un tremendo rechinido de corta duracion.

**Squeal** A continuous high-pitched noise.
***Chillido*** Un continuo tremendo rechinido.

**Squib** The igniter for an air bag.
***Soplido*** El soplido para una bolsa de aire.

**Squirm** To wriggle or twist about a body. When applied to tires, squirm is the wriggle or movement of the tires against the road surface. Squirm increases tire wear.
***Tuerce*** Tambalearse o torcerse en un cuerpo.Cuando se aplica a las llantas,tuerse es el movimiento o tambaleo de las llantas contra la superficie del camino. Tuerce hace que las llantas se acaben.

**Static** A form of electricity caused by friction.
***Estatica*** Una forma de electrecidad causada por friccion.

**Static balance** Balance at rest, or still balance. It is the equal distribution or weight of the wheel and tire around the axis of rotation such that the wheel assembly has no tendency to rotate by itself regardless of its position.
***Estatica de balance*** Balance en descanzo, o balance quieto. Es la igual distribucion de peso o la rueda y llanta alrededor del eje de rotacion, tal como el asamble de la rueda que no tiene tendencia a rodar por si misma a pesar de su posicion.

**Static electricity** A charge of electricity that is not moving but will quickly move when given a path. Static electricity is normally caused by friction.
***Estatica de electrecidad*** Una carga de electrcidad que no se mueve pero que se movera mpronto cuando se le de un molde. Estatica de electrecidad es normalmente causada por friccion.

**Static friction** Friction between two stationary objects or surfaces.
***Estatica de friccion*** Friccion entre dos objetos estacionarios o superficies.

**Steer-by-wire systems** The steering drive is controlled by an electronic motor, which is controlled by the systems computer. The computer determines whether the steering angle needs to be changed and by how much.
***Sistema de manea de cable*** La manea de manejar es controlada por un motor electronico, el cual es controlado por el sistema de computadora. La computadora determina si el angulo rotador necesita ser cambiado y por cuanto.

**Steering damper** A device added to some steering linkages to help stabilize the movement of the wheels as a vehicle is making a turn.
***Limpiador de rotador*** Un devisador agregado a algunas conecciones rotadoras para ayudar a estabilizar el movimiento de las llantas cuando el vehiculo hace una vuelta.

**Steering knuckle** The outboard part of the front suspension that pivots on the ball joints and lets the wheels turn for steering control.
***Nudillos de rotacion*** La parte de afuera de la suspencion de enfrente que esta en la coyontura de bola y deja que las llantas giren por control rotador.

**Stress** The force to which a material, mechanism, or component is subjected.
***Estirar*** La fuerza a la que un material, mecanismo, o componente es sometido.

**Strut** Components connected from the top of the steering knuckle to the upper strut mount that maintain the knuckle position and act as shock absorbers to control spring action in a vehicle's suspension system.
***Pavoneo*** Componentes conectados de arriba del nudillo de rotacion a la parte de arriba del monte que mantiene la posicion del nudillo y actua como absorbedor de choque para controlar el resorte de accion en el sistema de suspencion de un vehiculo.

**Sway bar** A stabilizer bar used to prevent the vehicle's body from diving into turns.
***Barra de control*** Una barra estabilizadora usada para prevenir el cuerpo del vehiculo de desviarse en las vueltas.

**Tap** A tap can cut and restore the threads in a bore.
***Punctura*** Una puncture puede cortar o restaurar las roscas en un tornillo.

**Tapered roller bearing** A specific kind of bearing that is based on tapered steel rollers held together in a cage.
***Soporte roller punctuado*** Una clase especifica de soporte que esta basado en rolles de fierro puestos juntos en una caja.

**Temper** To change the physical characteristics of a metal by applying heat.
***Templar*** Cambiar las caracteristicas fisicas de un metal aplicando calor.

**Temperature** An indication of an object's kinetic energy.
***Temperatura*** Una indicacion de la energia kinetica de un objeto.

**Tempering** The heat-treating of metal alloys, particularly steel, to result in specific properties.
***Temperando*** Tratando algunos metales con calor, particularmente fierro, para resultar con especificas propiedades.

**Tensile force** The moving force that slides or pulls an object over a surface.
***Fuerza tensil*** La fuerza de movimiento que resbala o jala un objeto en una superficie.

**Tension** Effort that elongates or "stretches" a material.
***Tension*** Fuerza que enlarga o estira un material.

**Test light** A test instrument that illuminates its lamp when there is the presence to voltage at one of its leads and connected to ground at the other lead.
***Prueba de luz*** Un instrumento de prueba que ilumina su lampara cuando hay la presencia de voltage en uno de sus plomos y conectado a la tierra con otro plomo.

**Thermistor** A type of variable resistor. Thermistors are designed to change in value as its temperature changes. Thermistors are used to provide compensating voltage in components or to determine temperature.
***Termisor*** Un tipo de resistor variable. Termisores estan disenados para cambiar conforme a los cambios de temperature. Termisores son usados para proveer voltage compensativo en componentes o para determinar temperature.

**Thread chaser** A device, similar to a die, that is used to clean threads.
***Rosca cinceladora*** Un devisador similar a un dado, que es usado para limpiar las roscas.

**Thread pitch** The number of threads in one inch of threaded bolt length. In the metric system, thread pitch.
***Grado de rosca*** El numero de roscas en una pulgada de roscas en un tornillo largo. En el sistema metrico, Thread Pitch.

**Threaded insert** A threaded coil that is used to restore the original thread size to a hole with damaged thread.
***Coneccion de rosca*** Uncoil de roscas que es usado para restaurar talla de la rosca original, a un agujero que esta danado.

**Thrust angle** The angle between the geometric centerline and the thrust line.
***Angulo de entrada*** El angulo entre la linea central geometrica y la linea de entrada.

**Thrust bearing** A bearing designed to resistor contain side or end motion as well as reduce friction.
***Soporte de entrada*** Un soporte disenado para resistir contener movimiento,tambien como reducer friccion.

**Thrust line** A line that divides the total toe angle of the rear wheels.
***Linea de lado*** Una linea que divide el total angulo de las sllantas de atras.

**Thrust line alignment** The thrust line should be positioned at the geometric centerline of the vehicle. Both the front and rear wheels should be parallel to the thrust line.
***Linea de alineamiento de lado*** El alineamiento de lado debe ser colocado a la linea central geometrica del vehiculo. Las llantas de adelante y de atras deben ser paralelas a la linea de lado.

**Thrust load** A load that acts on the axis bearing parallel with the center of axis of rotation. It is present on the driving wheels, axle shafts, and axle bearings when the vehicle turns corners or curves.
***Carga de un lado*** Una carga que actua en el soporte astil paralelo con el centro del eje de rotacion.

**Thrust washer** A washer designed to take up end thrust and prevent excessive end play.
***Rosca de lado*** Una rosca disenada para tomar y prevenir movimiento exesivo.

**Tie rod** A rod connected from the steering arm to the rack or center link, depending on the type of steering linkage.
***Varilla fuerte*** Una varilla conectada del brazo rotador de la rueda o enganche, dependiendo en el tipo de enganche rotador.

**Tire conicity** The condition where the plies and or belts are not level across the tread.
***Conicidad de la llanta*** La condicion donde las bandas o cinturones no estan a nivel cruzando la llanta.

**Toe** A suspension dimension that reflects the difference in the distance between the extreme front and extreme rear of the tires. It also refers to the inside diameter of the ring gear.
***Dedo*** Una dimension de suspencion que refleja la diferencia en la distancia entre el extreme frontal y el extreme trasero de las llantas. Tambien se refiere al diametro de adentro del anillo del engrane.

**Toe out in turns** Toe out in turns allows the front wheels to be at different angles when the vehicle is turning a corner.
***Dedo fuera-dentro vueltas*** Dedo fuera-dentro en vuelta permite a las llantas delanteras estar en diferente angulo cuando el vehiculo da vuelta en una esquina.

**Tolerance** A permissible variation between the two extremes of a specification or dimension.
***Tolerancia*** Una variacion permissible enmedio de dos extremos de una especificacion o dimencion.

**Torque** A twisting force applied to a shaft or bolt, usually measured in lb-ft. (pounds foot) or Newton per meters (Nr.M).
***Torque*** Fuerza de torcer aplicada a un astil o tornillo, usualmente medodi en libras o pies o neuton por metros.(Nr. M).

**Torque steer** A self-induced steering condition in which the axles twist unevenly under engine torque.
***Rotor torcido*** Un rotor que funciona por si mismo cuando los ejes se tuercen deigualmente, debajo del torque del motor.

**Torque stick** An extension for an impact wrench that includes the correct size socket for a wheel nut and acts as a torsion bar to limit the torque applied by the impact wrench when installing a wheel.
***Extencion torque*** Una extencion para una llave impacto que incluye la talla correcta del socket para el nudillo de una rueda y actua como una barra de torcion, para limitar el torque aplicado por el impacto de la llave al instalar una rueda.

**Torque wrenches** Torque wrenches measure how tight a nut or bolt is.
***Llaves de torque*** Las llaves de torque miden que tan apretados estan los tornillos.

**Torsion bar** A steel bar connected from the chassis to the lower control arm. As the vehicle weight pushes the chassis downward, the torsion bar twists to support this weight. Torsion bars are used in place of coil springs.
***Barra de torcion*** Una barra de fierro conectada del chasis al mas bajo brazo de control. Mientras que el peso del vehiculo empuja el chasis hacia abajo, la barra de torcion se tuerse para soportar este peso. Barras de torcion son usadas en lugar de resortes de coil.

**Toxicity** The term used to describe how poisonous a substance is.
***Toxidad*** El termino usado para describir que tan venenosa puede ser una substancia.

**Tracking** The parallel relationship between the front and rear wheels when a vehicle is moving forward.
***Direccion*** La relacion paralela entre el frente y las llantas traseras cuando el vehiculo se mueve hacia delante.

**Traction** A tires's ability to hold or grip the road surface.
***Traccion*** La habilidad de las llantas funcionar en el camino.

**Tramp** Wheel hop caused by static balance.
***Trampa*** Brincos de las llantas causada por la estatica del balance.

**Transaxle** Type of construction in which the transmission and differential are combined in one unit.
***Eje transal*** Tipo de construccion en la cual la transmision y el diferencial estan combinados en una unidad.

**Transfer case** An additional component added to a four-wheel drive vehicle in which the power from the engine is divided equally between both the front and rear wheels.
***Caja de transfer*** Un componente adicional agregado a un vehiculo de doble traccion, en el cual la fuerza del motor es dividida igualmente entre las llantas de adelante y las de atras.

**Transmission** The device in the power train that provides different gear ratios between the engine and drive wheels as well as reverse.
***Transmision*** El devisador en la fuerza del tren que prove diferentes rangos de engrane entre el motor y las ruedas y viceversa.

**Transverse** Perpendicular, or at right angles, to a front-to-back centerline. Power train layout in a front-wheel drive automobile extending from side to side.
***Transversidad*** Perpendicular, o a angulos rectos, de frente hacia atras linea central. El tren de fuerza descanza en un automovil de traccion delantera extendiendose de lado a lado.

**Tread** The layer of rubber on a tire that contacts the road and contains a distinctive pattern to provide traction.
***Cordon*** La capa de plastico en una llanta que contacta con el camino y contiene una distinctive forma para proveer traction.

**Tread contact patch** The area of the tire tread that contacts the road; determined by tire section width, diameter, and inflation pressure.
***Contacto del parche de cordon*** El area de cordon de la llanta que contacta el camino: determinada por la anchura de la llanta, diametro, y presion de inflacion.

**Tread wear indicator** A continuous bar that appears across a tire tread when the tread wears down to the last 2/32 (1/16) inch.
***Cordon indicador de acabado*** Una barra que aparece continuamente cruzando la llanta cuando los hilos se acaban en las ultimasn 2/32 ( 1-16 ) pulgadas.

**Tripod CV joint** A CV joint that uses a central hub or tripod with three trunnions placed inside housing.
***Tripode CV coyontura*** Una coyontura que usa una caja central donde tres tonniones estan en la caja mayor.

**Turning radius** Turning radius or cornering angle is the amount of toe out on turns.
***Radio volteador*** Radio volteador o angulo de coronilla es la cantidad de dedos en vueltas.

**U-joint** A four-point cross-connected to two U-shaped yokes that serve as a flexible coupling between shafts.
***Coyontura U*** Una coneccion -cruzada de cuatro -puntos en forma de –U enganches que sirven como una coneccion flexible entre astiles.

**Unibody** A stressed hull body structure that eliminates the need for a separate frame.
***Uncuerpo*** Una estructura que elimina la necesidad de un marco separado.

**United States Customary System (USCS)** A system of weights and measures used in places that do not use the metric system, such as the United States. The USCS system is also known as the English or inch system.
***Sistema Costumario de los Estados Unidos (USCS)*** Un sistema de pesos y medidas usado en lugares donde no usan el sistema metrico, como en los Estados Unidos. El sistema de USCS es tambirn conocido como el sistema Ingles o sistema de pulgada.

**Universal joint** A joint that allows the driveshaft to transmit torque at different angles as the suspension moves up and down.
***Coyontura universal*** Una coyontura que permite que el eje de manejo transmita tuerce a diferentes angulos de suspencion se muevan de arriba a abajo.

**V-belt** A rubber-like continuous loop placed between the engine crankshaft pulley and accessories to transfer rotary motion of the crankshaft to the accessories.
***Banda –V*** Una banda de plastico continua localizada enmedio de los cranques del motor, polea y accesorios para transferir movimiento de rotacion a los cranques y los accesorios.

**Vacuum** In automotive service, vacuum is generally considered to be air pressure lower than atmospheric pressure. A true vacuum, however, is a complete absence of air.
***Aspiradora*** En servicio de automovil, la aspiradora es generalmente considerada para ser aire de presion mas bajo que la presion atmosferica. Una verdadera aspiradora, comoquiera, es una completa ausencia de aire.

**Vacuum bleeding** The process of using a vacuum pump filled with brake fluid and attached to a wheel bleeder screw to draw old brake fluid and air from the system. Vacuum bleeding is a one-person operation.
***Bombeo de aspiradora*** El proceso de aspirar una bomba llena de liquido de frenos y puesta en una rueda con tornillos, y saca del sistema el liquido de frenos que esta viejo. Bombeo de aspiradora es un trabajo para una sola persona.

**Vane-type pump** A type of fluid pump that uses several sliding vanes that seal against a slide in the housing to push fluid out.
***Bomba tipo-vane*** Un tipo de bomba de fluido que usa varios vanes que se sellan contra una hcja en la caja mayor, para empujar el liquido hacia afuera.

**Variable capacity pumps** A classification of a fluid pump whose output is controllable.
***Bombas de capacidad variable*** Una clasificacion de bomba de liquido y su salida es controllable.

**Variable-rate coil spring** Rather than having a standard spring deflection rate, these springs have an average spring rate based on a load at a predetermined deflection.
***Rango variable de coil de resorte*** Mejor que tener un resorte estandar de resorte defleccion rango, estos resortes tienen un rango average de resorte basado en una carga a una predeminada defleccion.

**Vehicle identification number (VIN)** The number assigned to each vehicle by its manufacturer, primarily for registration and identification purposes.
***Numero de identificacion del vehiculo (VIN)*** El numero asignado a cada vehiculo por su fabricador, primeramente para propositos de registracion e identificacion.

**Vehicle speed sensor (VSS)** A sensor used to track the current speed and total miles traveled by a vehicle. This input is used by many computers systems in the vehicle.
***Sensor de velocidad del vehiculo (VVS)*** Un sensor usado para seguir la velocidad actual y total de millas que el vehiculo viaja. Esta entrada es usada por muchos sistemas de computadora del vehiculo.

**Velocity** The speed of an object in a particular direction.
***Velocidad*** La velocidad de un objeto, en una direccion particular.

**Vibration** A quivering, trembling motion felt in the vehicle at different speed ranges.
***Vibracion*** Un temblor o movimiento de estremecimiento que se siente en el vehiculo a diferentes rangos de velocidad.

**Viscosity** The resistance to flow exhibited by a liquid. Thick oil has greater has greater viscosity than than thin oil.
***Viscosidad*** La resistencia exibida de un liquido. Aceite espeso tiene mas grande viscocidad que el aceite Delgado.

**Volatile** The characteristic of a material that allows it to vaporize quickly.
***Volarse*** La caracteristica de un material de eveporarse rapidamente.

**Volatility** The tendency for a fluid to evaporate rapidly or pass off in the form of vapor. For example gasoline is more volatile than kerosene because it evaporates at a lower temperature.
***Volatolidad*** La tendencia de un liquido de eveporarse rapidamente o desaparecer en forma de vapor. Por ejemplo la gasoline es mas volante que kerosene, porque se evapora a bajas temperatures.

**Volt** A unit of measurement of electromotive force. One volt of electromotive force applied steadily to a conductor of 1 ohm of resistance produces a current of 1 ampere.
***Voltio*** Una unidad de medimiento de fuerza electromotive. Un voltio de fuerza electromotive aplicada quietamente a un conductor de 1 ohm de resistencia, produce una corriente de 1 ampere.

**Voltage** Electrical pressure formed by the attraction of electrons to protons.
***Voltage*** Presion electrica formada por la atraccion de electrons a protones.

**Voltage drop** The amount of voltage or energy consumed when pushing current through a resistance.
***Caida de voltage*** La cantidad de voltage o energia consumida cuando pasa corriente por una resistencia.

**Voltage generating sensors** Sensors that are capable of producing their own input voltage signal. This varying voltage signal enables the computer to monitor and adjust for changes in the computerized control system.
***Voltage generando sensores*** Sensores que son capaces de producir su propia entrada de senal de voltage. Esta senal de voltage variable hace que la computadora capte y ajuste los cambios en el sistema de control computarizado.

**Voltmeter** A tool used to measure the voltage available at any point in an electrical system.
***Volmeter*** Una llave usada para medir el voltage disponible a cualquier punto en un sistema electrico.

**V-ribbed belt** A belt that has multiple ribs on one side and is flat on the other side.
***Banda de costillas–V*** Una banda que tiene multiples costillas en un lado y esta plana en el otro lado.

**Vulcanized** A process of heating rubber under pressure to mold it into a special shape.
***Vulcanizado*** Un proceso de plastico calentado bajo presion para moldearlo en una forma especial.

**Weight** The measure of Earth's gravitational force, or pull on an object.
***Peso*** La medida de la fuerza gravitacional del universo, o jalar un objeto.

**Wheel** A circular frame or hub of an axle to which the tire is attached.
***Rueda*** Un marco circular o caja de un eje al cual la llanta esta unida.

**Wheel alignment** Wheel alignment may be defined as an adjustment and refitting of suspension parts to original specifications that ensures design performance.
***Alineamiento de las llantas*** Alineamiento de las llantas puede ser definido como un adjustable y remedida de suspencion de partes de originales especificaciones, que aseguran el exito del diseno.

**Wheelbase** The distance between the center of a front wheel and the center of a rear wheel.
***Base de rueda*** La distancia entre el centro de una llanta de enfrente y el centro de una llanta de atras.

**Wheel cylinder** The hydraulic slave cylinder mounted on the backing plate of a drum brake assembly. The wheel cylinders convert hydraulic pressure from the master cylinder to mechanical force that applies the brake shoes.
***Cilindro de la rueda*** El cilindro montado en el asamble del plato del freno del tambor. El cilindro de la llanta convierte presion hidraulica de el cilindro principal a fuerza mecanica que se aplica a los zapatos de los frenos.

**Wheel speed sensors** Speed sensing units placed at the wheels of a vehicle to monitor the relational speed of the individual wheels.
***Sensores de velocidad de las llantas*** Unidades que sienten la velocidad puesta en las llantas de un vehiculo a un monitor de velocidad racional en las ruedas individuales.

**Wheel spindle** The short shaft on the front wheel upon which the wheel bearings ride and to which the wheel is attached.
***Spindola de la rueda*** El astil corto de las llantas de enfrente en el cual soportes de rueda corren y al cual la llanta esta unida.

**Work** The amount of force applied to something and the distance over which it is applied.
***Trabajo*** La cantidad de fuerza aplicada a algo y la distancia en la que es aplicada.

**Yaw** Yaw is defined as swinging motion to the left or right of the vertical centerline or rotation around the vertical centerline.
***Yaw*** Yaw es definida como movimiento de columpio a la derecha o a la izquierda del centro de linea vertical o rotacion alrededor del centro de linea vertical.

**Yoke** A "Y" shaped assembly fitted to ends of a shaft and into which two of the four arms of a universal joint are fitted, typically used with the front U-joint of a drive shaft.
***Enganche*** Un asamble de forma de Y queda a las orillas del astil y en el cual dos de los cuatro brazos de una coyontura universal quedan, tipicamente usadas con la coyontura-U de enfrente de un eje de manejo.

**Zerk Fitting** Small check valves that allow grease to be injected into a part without letting grease out and dirt in.
***Medida Zerk*** Pequenas valvulas que permiten a la grasa ser inyectada en una parte, sin permitir a la grasa ensuciar esa parte.

**Zeronic potassium perchlorate** A chemical that reacts to ignition and causes that air bag to inflate.
***Potasio Zeronico de percolorate*** Un quimico que reacciona al encender y causa aire para inflar.

# Index

NOTE: Page numbers followed by f refer to figures; page numbers followed by t refer to tables.

## Z